An introduction to
GLOBAL ENVIRONMENTAL ISSUES

'The writing is clear, the information content is high, and the topics are timely, but what is particularly impressive about this book is the way in which attention and interest are maintained. The authors provide a strong armoury of data and information and a coherent rationale for action. Each chapter treats the facts of a topic, the issues arising from them, and management implications in a consistent and convincing way. It is a book that should empower students by bringing them up to date with important research on critical issues and by helping them understand environmental issues in their appropriate social context.'

Thom Meredith, *Associate Professor, McGill University, Canada*

'An impressive piece of work, covering a broad range of contemporary issues from an earth science perspective. The text is clear and accessible, and is written in a lively yet informative style. A timely update of a rapidly evolving field.'

Doug Benn, *Senior Lecturer in Geography, University of Aberdeen*

'This book brings the major global environmental issues into focus. The approach taken in the book is innovative and stimulating. It presents an authoritative, comprehensive and contemporary coverage of environmental issues and offers an opportunity for the reader to be brought up to date rapidly on the environmental questions of the 1990s. A particularly pleasing aspect of this book is the marriage of hard scientific data with non-judgmental and balanced discussion on interpretations of the data and the development of attitudes towards environmental management issues such as sustainable development. This fulfils the essential requirement for a complete text on environmental issues.'

Robert Bourman, *Associate Professor (in Geomorphology), University of South Australia*

Kevin T. Pickering is Reader in Sedimentology and Stratigraphy at University College London. **Lewis A. Owen** is Assistant Professor in the Department of Earth Sciences at the University of California, Riverside.

Comments on the first edition

At last a text on issues packed with scientific data, that is backed by a comprehensive bibliography and is bang up to date. Bold type and key points are the icing on an already reader-friendly text.

D. J. L. Harding, *School of Applied Sciences, Wolverhampton University*

The aim, pitch and content of *An Introduction to Global Environmental Issues* is not matched in terms of content and breadth of subject matter by any other introductory text in this field. The writing level is spot on for introductory students.

L. Dumayne, *University of Birmingham*

An extremely well illustrated and well written text book, that covers all of the fundamental aspects of environmental issues that are required in my courses.

Mike Whatley, *Leicester University*

Clear, easy to read, with great diagrams and photographs.

Lois Mansfield, *School of Environmental Management, Newton Ridge College, Cumbria*

Good coverage of subject, well structured with splendid illustrations.

David Rice, *School of Town and Regional Planning, University of Dundee*

It is well written, clearly illustrated and presents environmental principles in a simple but up-to-date manner.

Professor G. Kelling, *Department of Earth Sciences, University of Keele*

Excellent coverage and very accessible.

John Soussan, *Environment Centre, Leeds University*

Clearly set out. Covers issues comprehensively yet in a manner that is easy to understand. Excellent photographic and diagrammatic examples. Well organised and attractively presented.

Dr Fiona Tweed, *Lecturer in Physical Geography, Staffordshire University*

Each subject is introduced in a clear and readable way that assumes little prior knowledge on the part of the student. The text is well supported by excellent illustrations.

Dr Paul Elliot, *Lecturer in Science, Education and Biology, Warwick University*

This is a timely up-to-date scientific introduction to a range of important global environmental issues.

Dr Ada Pringle, *Lancaster University*

An introduction to
GLOBAL
ENVIRONMENTAL
ISSUES

Second Edition

Kevin T. Pickering
and
Lewis A. Owen

LONDON AND NEW YORK

First published 1994
Second edition published 1997
by Routledge
11 New Fetter Lane, London EC4P 4EE

Simultaneously published in the USA and Canada
by Routledge
29 West 35th Street, New York, NY 10001

© *1994, 1997 Kevin T. Pickering and Lewis A. Owen*

Typeset by Florencetype Ltd, Stoodleigh, Devon

Printed and bound in Great Britain by Butler & Tanner Ltd,
Frome and London

British Library Cataloguing in Publication Data
A catalogue record for this book is available from the British Library

Library of Congress Cataloguing in Publication Data
Pickering, K. T. (Kevin T.)
An introduction to global environmental issues/Kevin T.
Pickering and Lewis A. Owen. — 2nd ed.
p. cm.
Includes bibliographical references and index.
1. Environmental issues.   I. Owen, Lewis A.
II. Title.
GE105.P53   1997
363.7—DC20   96-42364

ISBN 0-415-14098-6
ISBN 0-415-14099-4 (pbk)

Whoever you are! you are he or she for whom the
earth is solid and liquid,
You are he or she for whom the sun
and moon hang in
the sky.

'A Song of the Rolling Earth'
from Walt Whitman, *Leaves of Grass*

Contents

Colour plates

Black and white plates

Figures

Tables

Boxes

Preface

Are **acidic deposition** (including **acid rain**) and an anthropogenically enhanced greenhouse effect the grave risk to ecosystems that some scientists and environmentalists claim, or are they an insignificant part of natural processes? Can scientists and policy makers ameliorate their effects? What were past climates like, how rapid and abrupt can climate change be, and how does such knowledge help predict future climate changes? Are human activities permanently damaging vulnerable **ecosystems** beyond recovery? Is society wasting energy resources? Are there economically viable alternative energy resources to the traditional **fossil fuels**? Does society want nuclear power? Are nuclear weapons acceptable and necessary in a civilised world? Is it possible to predict natural hazards and so to mitigate their often devastating effects? How does human activity affect the landscape? Can the world's growing population be adequately fed? Is sustainable development a myth or reality? These are issues considered in this book. The final chapter examines ways in which the Earth is managed, including a look at such diverse topics as population growth, the destruction of the rainforests and agriculture, and it is there that we suggest that there are things which can be done to make the planet more habitable – to increase the chances of human beings and other vulnerable species surviving longer. The reader may well disagree with our shopping list of action. If so, then one of the main aims of this book will have been achieved – to provide a critical and provocative look at global environmental issues.

This book is as much about scientific developments that involve global environmental issues as it is about the attitudes and implications raised. Perhaps its single most important outcome will be to stimulate discourse over the relationships between the natural world and the ways in which human activities are forcing change. When the environmental damage is done, maybe the only plea in mitigation by humankind will have to be that of Socrates in ancient Greece, who, when threatened with the death penalty, permitted himself no other superiority than that he did not presume to know what he did not.

The global issues addressed in this book should cause us to ask how we can make our planet more habitable. There are no easy solutions to these weighty questions. We encourage you to consider where your priorities lie in helping to shape the key issues for the rest of the 1990s and into the twenty-first century. And we would hope that, having formed opinions on these issues, you will act in whatever capacity you see fit, however insignificant it may seem in the global scheme of things. As fellow travellers on Spaceship Earth, we cannot duck the issues for long without forfeiting our right to criticise the words and actions of industrialists and politicians.

Having no opinions about global issues is tantamount to sticking one's head in the proverbial sands of time. And, as surely as our present existence, the sands of time will run out on us unless there is a more prudent management of this planet.

To manage the Earth more efficiently, and husband the natural resources with less waste, there is a need to understand the processes that shape the Earth. This book is concerned with presenting many of the inextricable links between the living and the inanimate world, about the way in which the forces of nature influence human activity, and also the converse. In this book we suggest actions that can be taken for humanity to become more in harmony with the pulse of the Earth. Whether or not you agree with our opinions, the arguments set out in this book are presented to stimulate debate, and to emphasise the links between the purely scientific aspects and the social sciences.

As Earth scientists, we authors wear these labels in our professional careers as university lecturers. As human beings concerned with environmental issues, we have used our scientific training and expertise to express personal opinions from a perspective that combines scientific explanations with our emotional involvement with the world in which we live. The information presented in this book does not lead to only one conclusion and a unique course of action. This book is not a cosy cornucopia of facts to be digested and regurgitated in examinations, although this certainly is the least that we hope for from the book, but it is aimed at bringing the major global environmental issues into focus in a broad context of science and society.

Using this book

To discuss and attempt to address the questions posed above, as well as many more that are pertinent to global management, this textbook is divided into ten chapters, each examining a set of major themes. Environmental issues can be studied from a variety of perspectives and they are studied within many different disciplines. These include the social sciences; politics; economics; the biological sciences; geography; geology; meteorology and climatology; oceanography; and ecology. The study of environmental issues, therefore, requires an understanding and appreciation of all of these disciplines because of the complex interrelation between both the physical and biological world, and human activities. Only when the full range of interrelated factors have been considered is it possible to have a really good understanding of environmental issues, and have the ability for effective management to be implemented.

This book is designed to be read at different levels, depending upon the reader's particular interests and in order to suit a broad range of syllabi. The book can be scanned in less than fifteen pages using the 'key points' at the end of each chapter, which allow the reader to assess the contents as well as providing useful summaries: also, the key points provide a useful *aide mémoire* for examination revision. At a more detailed level, the reader can select parts of the main text that are relevant to specific topics under the various sub-headings and in the boxes. To help make the text more accessible to the reader, the key terms have been highlighted in bold and appear in the glossary at the end of the book. For general courses in environmental science we hope that the entire text will be read. Finally, for full semester or longer courses in the environmental sciences the book includes lists of further reading at the end of each chapter, a comprehensive bibliography, questions for essays or group discussion, and an Instructor's Manual to accompany this book.

The first chapter, 'Introducing Earth', is primarily for readers who require an introduction to the basics of global systems as a background to the study of environmental issues.

Chapter 2, 'Climate change and past climates', examines the nature of climate change, and the rates and magnitudes of global climate change, and gives a résumé of the Quaternary Period. This chapter is aimed primarily at students taking a degree in Earth Science, as it contains the most challenging technical material. Chapter 2 also considers the various theories that have been developed to help explain why climates and environmental conditions have changed throughout geological time. This is also particularly important for environmental managers and policy-makers, because they must be able to distinguish between natural processes and anthropogenic effects on global climate change.

Chapter 3, 'Global atmospheric change', follows on logically from Chapter 2 by examining the effects of human activity on the present atmosphere, focusing on the anthropogenic emission of greenhouse gases and stratospheric ozone depletion. International action concerning global atmospheric change is summarised at the end of this chapter.

Chapter 4, 'Acid deposition', continues the theme of human activities and atmospheric pollution by examining the effects of the acidification of rain, surface and ground waters, and the resultant degradation of the hydrosphere and soils, along with the effects on the biosphere. As with Chapter 3, this chapter ends with a review of international action to mitigate the effects of acidic deposition.

Chapter 5, 'Water resources and pollution', emphasises the importance of water as a resource and considers the various ways in which water quality is affected by human activities. Hydro-politics are dealt with at the end of this chapter.

Chapter 6, 'Nuclear issues', continues the pollution theme while addressing the broader issues associated with nuclear power, for example through the proliferation and growth of nuclear weapons. The main ways in which nuclear energy is harnessed are considered, and there is a brief look at some of the world's worst accidents at nuclear power stations.

Chapter 7, 'Energy resources', considers the production and consumption of traditional fossil-fuel and alternative (including renewable) energy resources, together with some associated issues such as a carbon-energy tax.

Chapter 8, 'Natural hazards', examines the threats to humankind from natural processes and assesses the various ways in which their effects can be mitigated.

Chapter 9, 'Human impact on the Earth's surface', explores land degradation by considering several important topics such as biosphere degradation, soil erosion, quarrying and mining, channelisation schemes, and over-fishing. This chapter also considers the various ways of reducing land degradation and includes an examination of strategies such as environmental impact assessments and environmental audits.

The final chapter, 'Managing the Earth', attempts to provide an integrated perspective of various global issues that are inextricably linked with those outlined in other chapters. In order to provoke discussion this chapter concludes by presenting a personal manifesto that we believe could go a long way towards the chimera of sustainable development and the sensitive and prudent management of the Earth's natural environment.

Acknowledgements

The authors and publishers would like to thank the following for permission to reproduce copyrighted material:

Literary extracts: Viking Penguin, a division of Penguin Books USA Inc. and Lawrence Pollinger Limited and the Estate of Frieda Lawrence Ravagli for the extract 'In the Cities' from *The Complete Poems of D.H. Lawrence* by D.H. Lawrence, edited by V. de Sola Pinto and F.W. Roberts. Copyright 1964, 1971 by Angelo Ravagli and C.M. Weekley, executors of the Estate of Frieda Lawrence Ravagli; Little Brown for 'On the Pulse of Morning' by Maya Angelou; David Higham Associates for 'Prayer before Birth' by Louis MacNeice from *Collected Poems of Louis MacNeice*, published by Faber & Faber; Faber & Faber for the extract from 'Little Gidding' by T.S. Eliot from *Four Quartets*; Sony Music Publishing and Special Rider Music for 'A Hard Rain's a Gonna Fall' by Bob Dylan.

Plates: Ian Oswald-Jacobs Aerial Agricultural Photography; Magnum Photos; J. Jacyno; Comstock Photo Library; Rex Features; the National Gallery; Greenpeace Communications Ltd; Panos Pictures; The Environmental Picture Library; Vlaso Milankovitch; Dr M. Collinson; the British Institutions Reflection Profiling Syndicate; NASA/Lunar and Planetary Institute; Jeremy P. Richards; Rhodri Jones, Oxfam; the US Geological Survey; M. Eden; R. Robinson; Gary Nichols; R. Potter; Professor Windley; SABA Katz Pictures; Geotechnical Control Office, Hong Kong.

Finally, the authors thank the many individuals who have in some way contributed to this book, either in conversation with us or through reviewing parts of, or the entire, earlier drafts. In particular we owe a large measure of gratitude to Judith Bates, Jim Best, Dougie Brown, Bill Chaloner, Sarah Davies, Alastair Dawson, David Evans, Cathy Hayward, Catrin Jones, David Kemp, Vicky Myers, Louise Pickering, Val Saunders, Dorrik Stow and Steve Temperley for reviewing this book and making many helpful comments; to Jill Keegan for help with the quotes, and to Justin Jacyno for drafting a considerable part of the artwork. Kevin Pickering acknowledges the help and advice which was given by the UK Parliamentary Office of Science and Technology staff at Westminster, London, where during tenure of a COPUS Westminster Fellowship in 1993, the first edition was completed and upon which parts of this second edition are based. We would like to thank Dennis Hodgson for carrying out the exhausting task of copy-editing and, at Routledge, Sarah Lloyd for commissioning the second edition, Moira Taylor for steering this edition through its planning stages to deal with the various drafts, and Tristan Palmer for commissioning the first edition. Anne Owen is thanked for her unstinting desk-editing of the second edition. Last, but by no means the least, we thank our respective wives, Louise Pickering and Regina Robinson-Owen, for being so patient whilst weekends and evenings disappeared under 'pressure of work' to complete this new edition, all self-inflicted of course!

A Rock, A River, A Tree
Hosts to species long since departed,
Marked the mastodon,
The dinosaur, who left dried tokens
Of their sojourn here
On our planet floor,
Any broad alarm of their hastening doom
Is lost in the gloom of dust and ages.

But today, the Rock cries out to us, clearly,
forcefully,
Come, you may stand upon my
Back and face your distant destiny,
But seek no haven in my shadow,
I will give you no hiding place down here.

You, created only a little lower than
The angels, have crouched too long in
The bruising darkness
Have lain too long
Facedown in ignorance,
Your mouths spilling words

Armed for slaughter.
The Rock cries out to us today,
You may stand upon me;
But do not hide your face.

Maya Angelou, 'On the Pulse of Morning'
(Read by the poet at the inauguration of
William Jefferson Clinton, 20 January 1993)

Civilisation is now so advanced that it is possible to study in considerable detail the Earth and the Universe, an exciting and stimulating endeavour. Scientists can examine the Earth at all scales, from the subatomic using high-energy particle physics to cosmic scales using the most sophisticated telescopes and spacecraft. Images of Earth from Space are now familiar to all (Plate 1). Sophisticated global climate modelling and predictions about future climate change are becoming commonplace. With such technological advances and the wealth of opportunities for monitoring the natural world there is little excuse for any profligate use of raw materials and environmental degradation.

Humans, unlike other animals, have the ability, which may not be matched by the foresight, to appreciate the responsibility for the wise and prudent management of the Earth. Also, humans have the capacity to control and monitor the anthropogenic impact on the environment. Humans can observe the Earth from Space, communicate rapidly around the world and even from Space to Earth, prevent and remedy many diseases, manufacture many items that make life more comfortable and enjoyable, and construct complex urban settlements. Humans can inhabit nearly every environment on Earth.

Environmental issues concern the interaction of the natural world with human activities, the scales and rates of change in the **ecosphere** caused by natural variability and those precipitated by human activities. Environmental issues are about what has happened, the changes that have been brought about, and future predictions or prophecy about any changes in the environment that may occur as a consequence of human activities.

Broadly, there are four main components of the ecosphere that may be significantly affected on a long-term basis (decades to millennia) by human activities. First is the climate system, where human activities are causing the destruction of the **ozone layer** over large parts of the world, the production of **acidic deposition** and the emission of **greenhouse gases** and other harmful trace gases and aerosols into the atmosphere such as hydrocarbons and exhaust particulates. Second, there is the interaction between the organic and inorganic components of the ecosphere, that is the global circulation of nutrients – the **nutrient cycles**. These nutrient cycles include the mobilisation and redistribution of chemical elements, amongst the most important being those for carbon, nitrogen and phosphorus, resulting in some parts of the cycle becoming enriched while other parts are depleted. Third, humans have a profound effect on the **hydrological cycle**, for example by the withdrawal and pollution of water, anthropogenically induced droughts and floods, and activities which contribute to processes of erosion and deposition of sediment to silt up rivers and estuaries. Fourth, there is the direct or indirect human influence on the natural environment, which can lead to the extinction of endangered species, and the commensurate reduction of biological diversity – **biodiversity** – and the changes in the vegetational character of various regions of the world. Here, the main threat posed to other species is a consequence of the rapid growth of the human population in ever expanding urban developments, deforestation, the marginalisation of natural habitats in the countryside by over-intensive farming methods, and land use that is insensitive to sustainable vegetation, e.g. leading to **salinisation**, etc.

The survival and evolution of life on Earth is, in essence, about being adaptable to changing circumstances. The alternative is extinction. This appears to be a truism both for species and individuals. The conditions leading to **mass mortality** in a species may be different to those which cause the elimination of an entire species of organism. Mass mortality generally does not cause the extinction of an entire species, but rather represents a catastrophe that leaves enough of a population for recovery to some equilibrium level, perhaps similar to the pre-catastrophe value. Major earthquakes, volcanic eruptions and the impact of relatively small meteorites are examples of natural disasters that have the potential to wipe out geographically restricted populations, but which permit recovery of a species. While these arguments are true for the natural way in which life has evolved on Earth, most rational people are unlikely to countenance a nuclear holocaust or any other anthropogenically precipitated disaster as an acceptable *modus operandi*. At least, as civilised, compassionate and caring people, the life of other fellow humans demands that others are treated much as we might wish to be considered.

Scientists could take a dispassionate, seemingly objective and long-term perspective, say on a geological time scale, and say that the human species is bound to become extinct sooner or later like so many species before. It is inevitable, so why worry. The Earth will survive; the human species will not. The same philosophy could be applied equally to all the Earth's fauna and flora. With or without human intervention, various species have reached near-extinction levels. You might think that nobody could be quite so *laissez faire* about the environmental impact of human activities. But this is exactly how many human activities and attitudes could be construed. The scant regard often shown for the environment is symptomatic of the prevalent attitude that somebody else can clean up after us. Certainly the selfish side to human nature is part of our genetic make-up but humans have the ability, and many wasted opportunities, to suppress this basic instinct in favour of a more thoughtful attitude to the environment. Actually, such an approach could be rationalised as an ultimately selfish regard for the survival of the species rather than the short-term benefit of the individual at the expense of further environmental degradation. Humankind can no longer bequeath such a legacy to future generations. Concern and broad interest across all sections of society over the environmental impact of human activities is urgent, simply because the consequences of human actions which affect the natural environment appear to have increasingly serious knock-on effects. The will to translate that concern and interest into preventive or remedial action is also required in these decades. These issues involve all of us and everyone has a part to play in conserving the natural environment.

Life on Earth may be robust for many species, even as far as withstanding the impact of global nuclear carnage. Many insects, for example, would survive, but it seems unlikely that higher species such as *Homo sapiens* would. The life span of humans is short relative to the age of the Earth or geological time, which is measured in thousands of millions of years – about 4,700 million years. The extinction of humans as a species may be an inevitable natural process, but such a conclusion is no reason for apathy and complacency over the consequences of polluting the environment so that the demise comes far quicker. Neither should entrepreneurs and businesses, or politicians, demand of scientists absolute proof of cause and effect before acting in a cautious way over pollution. Where a reasonable degree of doubt exists about the consequences of human actions, then there are, perhaps, sound reasons for taking a conservative approach. Those who are responsible for pollution often appear only too eager to employ scientists and engineers who are willing to bury their heads in the sand over environmental pollution. If human activities destroy the habitability of planet Earth, there will be no second chance. No opportunity will exist for those same scientists and technologists to undo the damage with a contrite heart. Apologies to future generations for our inept management of the environment are unacceptable when the opportunity for an attempt at sustainable development is imperative and possible now. Humankind must avoid the sins of commission and omission, but instead seek to be accused of only one thing – being over-cautious.

Who would survive?

If all the nuclear weapons in the world's arsenals were detonated (of which the USA and former Soviet Union possess more than 50,000), the Earth would continue and with it life in some form. Many species would survive and in time new species would evolve to occupy vacant ecological niches. But, in this doomsday scenario, one thing is virtually certain: human beings and other vulnerable species would be obliterated. Humans would not survive. 'To be, or not to be: that is the question.' This indeed is the question over our survival as human beings, together with the survival of many endangered species and fragile ecosystems. Shakespeare's simple and profound words spoken by Hamlet echo through all human actions on the environment. This is very much a book about the panoply of global environmental issues that confront human survival and the continuation of the natural world as it now exists, not the survival of the planet. Volcanic eruptions and earthquakes cannot be controlled yet, but humans can control the pollution of the atmosphere, oceans and land and, possibly, global climate change, at least within certain rather narrow limits.

Human activities appear to have exposed many parts of the natural environment to considerable risks (see Table 1.1, which shows some recent human-

Table 1.1   *Recent notable human-induced environmental disasters.*

Date	Event	Location	Consequence
1993	Break-up of tanker, the *Braer*, on the rocks of Pitful Head	Shetland, Scotland	Oil slick contained to 200–300 m from the shoreline but serious pollution of fishing grounds and fish farms, as well as sea animals and birds.
1992	Greek oil tanker, the *Aegean Sea*, runs aground and catches fire	La Coruña, Spain	Spillage of an estimated 16 million gallons of crude oil, creating a slick *c.* 18 by 1.5 km and causing contamination of *c.* 70 km of Spanish coastline. Serious pollution of sea life and clam and oyster fisheries.
1991	Oil fields set alight by Iraqi forces during the Gulf War	Kuwait	Spillage estimated at between 25,000,000 and 130,000,000 gallons of crude oil. Air pollution and potential increase in acid rain.
1991	Greek tanker, *Kirki*, breaks up	Cervantes, W. Australia	Spillage of 5,880,000 gallons of crude oil and pollution of conservation and fishing areas.
1989	Explosion in hull of Iranian supertanker, *Khark 5*	Atlantic Ocean, N. of Canary Is.	Spillage of 19,000,000 gallons of crude oil and 370 km oil slick, almost reaching Morocco.
1989	*Exxon Valdez* tanker is grounded on Bligh Reef	Prince William Sound, Alaska	Spillage of 10,080,000 gallons of oil and 1,170 km of Alaska coastline polluted. More than 3,600 km^2 contaminated. Thousands of birds and animals killed.
1988	Accident at water treatment works results in aluminium sulphate being flushed into local rivers	Camelford, Cornwall	Local people suffer from stomach and skin disorders. Thousands of fish killed.
1987	Abandoned radiotherapy unit containing radioactive materials leaks	Goiana, Brazil	Radioactive contamination affected 249 people.
1986	Explosion of nuclear reactor	Chernobyl, Ukraine	Official death toll 50. Radioactive cloud spreads across Europe contaminating farmland. Long-term effects on inhabitants of surrounding areas are not yet ascertainable.
1984	Union Carbide pesticide plant leaks toxic gas	Bhopal, India	Death of 2,352 people officially. Unofficially an estimated 10,000 died.
1983	Blow-out in Nowruz oil field	Persian Gulf	Spillage of 176,400,000 gallons of oil.
1980	Chemical spill due to Sandez factory fire	Basel, Switzerland	Rhine polluted for 200 km.
1979	Collision of the *Atlantic Empress* and *Aegean Captain*	Trinidad and Tobago	300,000 tonnes of oil spilled.
1979	Blow-out of Ixtoc oil well	Gulf of Mexico	600,000 tonnes of oil spilled.
1979	Release of radioactive stream after water pump breaks down	Three Mile Island	Pollution by radioactive gases. Partial core meltdown in reactor.
1979	Collision of *Burmah Agate*	Galveston Bay, Texas	Spillage of 10,700,000 gallons of oil.
1979	Uranium released from secret nuclear fuel plant	Erwin, Tennessee	Approximately 1,000 people contaminated.
1978	Cypriot tanker, the *Amoco Cadiz*, is grounded	Portshall, France	Spillage of 65,562,000 gallons of oil. Pollution of 160 km of French coast.
1977	Fire on the *Hawaiian Patriot*	N. Pacific	Spillage of 99,000 tonnes of oil.
1977	Well blow-out in Ecofisk oil field	North Sea	Spillage of 8,200,000 gallons of oil.
1976	The supertanker, the *Urquiola*, is grounded	La Coruña, Spain	Spillage of 100,000 tonnes of oil.
1976	Leak of toxic gas TCDD	Seveso, Italy	Topsoil had to be removed in worst-contaminated areas.
1975	Fire at Browns Ferry reactor	Decatur, Alabama	$100 million damage. Cooling water lowered significantly.
1974	Explosion of container of cyclohexane	Flixborough, UK	28 deaths.
1972	Collision of tanker *Sea Star*	Gulf of Oman	115,000 tonnes of oil spilled.
1971	Overflow of water storage space at Northern States Power Company's reactor	Monticello, Minnesota	50,000 gallons of radioactive waste dumped into the Mississippi River. Contamination of St. Paul water system.
1970	Collision of tanker *Othello*	Tralhavet Bay, Sweden	60,000–100,000 tonnes of oil spilled.
1957	Fire in Windscale plutonium production reactor ignited three tonnes of uranium	Cumbria, UK	Spread of radioactive material throughout Britain. Official death toll 39 but this is strongly contested.

After Crystal 1993.

induced environmental disasters). An important environmental question considered by scientists, engineers, policy-makers and other concerned citizens is the extent to which any natural variability in an ecosystem will be affected by human activities. The stability of many natural systems remains poorly understood, together with the amount of environmental stress a system can accommodate before rapidly changing to another state. Many human activities may involve a kick to the natural environment so hard that, like a line of collapsing dominoes, they destabilise with serious consequences.

Science and technology are used to understand and harness the world's resources but not always for the greater good of humankind. It is not uncommon to see that short-term economic gain tends to outweigh most other considerations. Wherever there is a conflict of interests the parties to any dispute can call on the vocal support of scientists, technologists and other experts to back rival claims and opinions. In most conflicts, embodied in the ancient Chinese yin–yang symbol, issues are rarely black or white, right or wrong, good or evil. There are many uncertainties, and it is in this very middle ground that much of societies' values, mores, customs and laws can be challenged. In the grey area of environmental issues, between what is called fact and supposition, certainty or possibility, scientific fact or irrational argument, the conflicting vested interests of various organisations, groups and individuals meet. There are those who would say that scientists should simply discover and state the facts and leave politicians and other decision-makers to take policy decisions which determine the application, if any, of that science and technology. There are those who promote a greater moral responsibility by scientists and technologists, encouraging them to discuss uncertainties and to voice ethical and moral issues raised by their work. The problems raised by discussing uncertainties are large. The general public often expects scientists to provide straightforward solutions, or it loses interest. Politicians with short-term agendas that owe more to the lifetime of a parliament or other term of office show little interest in any long-term commitment to solving environmental problems.

Environmental issues are inherently political but, paradoxically, politicians are not generally the best people to act as the custodians of such issues. Few politicians have a background as a professional scientist and so rely to a very considerable extent on the advice of other governmental bodies. A trend to be encouraged, however, is the establishment by some governments of scientific units with a remit to provide independent briefing papers on any issue of science, technology or medicine, which are separate from the official civil service, the administrative arm of government. Two examples of such units are the Office of Technology Assessment in the USA (OTA), and the Parliamentary Office of Science and Technology in the UK (POST).

Too many want too much

Too many humans want too much of the world's resources for themselves, whether it is food, land, power or influence. Over-population and waste are the two biggest problems facing the present generation. Other issues tend to stem directly or indirectly from these two problems. The ways in which global environmental issues are tackled will determine the legacy that is bequeathed to future generations. There are those who would not agree that over-population is a central problem; through religious and other beliefs they might claim that the real problem is the management of the resources on Earth, not the number of people. The issue of over-population is extremely contentious, but the Earth could be managed with much less risk if there were less demand for the limited, finite, global resources, and the natural environment were under less stress from planners, developers, industrialists, colonisers and others who exploit the land. Indeed, in February 1992 the US National Academy of Sciences and the Royal Society of London published a joint document on global problems in which world population growth is considered a central issue. The joint document, the first ever produced by these two academies, took two years to write and expresses 'deep concern' over the links between the estimated growth of the world's population of 100 million a year (based on the 1991 report of the UN Population Fund), and it also highlights the way in which human activities are causing 'major changes in the global environment'. Without a change in this growth of population and the present pattern of human activities, then, according to the document, 'science and technology may not be able to prevent either irreversible degradation of the environment or continued poverty for much of the world'. A direct corollary of these arguments concerns the fundamental issue of the root cause of environmental problems – is it a consequence of poverty or a result of affluence?

Humankind has the technical ability to explore Space, yet human suffering, starvation and disease seem as prevalent as they ever were. Over thousands of years, humans have developed a rich and diverse culture through many civilisations. Despite this technological age, with its enormous advances and achievements, the human species remains as aggressive as ever.

The twentieth century has witnessed two world wars and many regional conflicts, all of which have grown out of human greed and avarice. There were more than 10 million deaths in the First World War and more than 55 million in the Second World War. Artificially created radioactivity has been harnessed for peaceful use as an energy resource, but also used to kill tens of thousands of people in Hiroshima and Nagasaki. Nuclear weapons could have been used in 1948/49 over the Berlin blockade, or in 1963 during the Cuban missile crisis, or in Vietnam in the late 1960s. They could have been used on other occasions but were not; we cannot be sure that they will not be used at some future date. International diplomacy, while undoubtedly more sophisticated than in previous centuries, remains incapable of stopping wars in many parts of the world. Many of the issues addressed in this book can only be tackled in a climate of international diplomacy, confidence and good will. In the decaying Soviet Union, the momentous events of the third week in August 1991, with the abortive military coup to overthrow President Mikhail Gorbachev, followed by his resignation on Christmas Day 1991, with the handing over of power to Boris Yeltsin as President of Russia, and of the new commonwealth, symbolised the formal break-up of the Union. The events of the next few years in the dismembered Soviet Union may lead to greater superpower co-operation over the environment. It must be hoped that the danger of a nuclear war has receded somewhat since the death of Soviet communism, but this is by no means certain.

Studying Earth

Through studies and observations scientists have become increasingly aware of the relationships and interactions between the Earth and the Solar System, or with the Universe, the inorganic and organic. No matter how detailed these studies, there are always new principles and phenomena to be discovered. Some relationships are so complex that scientists are only just beginning to understand them, yet others seem very simple. The laws of mathematics, physics and chemistry permit a description of many natural phenomena, but most of the ideas about natural systems are simply reductionist models, commonly abstracted to a mathematical simplicity that does not adequately explain the real phenomena – but a start

has to be made somewhere. Many physicists concern themselves with a search for fundamental particles, the ultimate origin of matter and time. Yet the complexity of the living, organic world still defies such elegant mathematics. This point is well made by Richard Dawkins, a zoologist from Oxford University, in his book *The Blind Watchmaker*, in which he describes the 'sheer hugeness of biological complexity and the beauty and elegance of biological design'. Observations and experiments will always provide the essential link between theory and reality.

In order to make understandable sense of the natural world the various component parts need to be simplified into models of how things work – mathematical and conceptual abstractions from reality. In the environmental sciences, examples of such models include a model of atmospheric circulation, oceanic circulation, the internal heat engine of the Earth, and biological and chemical cycles. As the understanding of natural and artificial processes increases, so the need for a multi-disciplinary approach to these cycles or systems increases. For the future there will be an ever-increasing need to train people who have both a sound understanding of particular global environmental issues, for example through basic science, and also a broader appreciation of the societal context of these issues. Training people who are paradoxically both specialists and generalists is no easy task. Many multi-disciplinary subject areas are dealt with in this book, but it is perhaps worth singling out a few of these, such as the geological sciences, geography, climatology, **meteorology**, **hydrology**, oceanography, botany, zoology, **geodesy** and **pedology**.

Earth in Space

The Earth is one of nine planets that orbit the Sun. These heavenly bodies, together with their moons and the asteroids (a belt of fragmented planets between Mars and Jupiter), constitute the Solar System. Our Solar System containing planet Earth is just one of about 10^{11} (100,000,000,000) that form our galaxy, the Milky Way. This, in turn, is one of 10^{11} galaxies in the Universe, all with a similar number of planets and stars to our own galaxy. The Earth, therefore, is estimated to be just one of at least 10^{22} planets travelling in space, held in orbit by the gravitational forces that exist between the planets and stars. In the past couple of years there has been intense interest in the possibility of there being other planets like the Earth that contain liquid water and atmospheric-climatic conditions capable of supporting life. Based on the 'wobble' of two very distant stars, caused by the mutual gravitational attraction of an unseen planet and the star, astronomers in the last couple of years believe that they may have discovered two candidate planets that could have life.

Cosmic distances are large. The distance of the Earth from the Sun is a relatively small cosmic distance at about 150 million km. It takes around eight minutes for light to reach the Earth from the Sun. In most cases, however, cosmic distances are extremely large and so astronomers measure such distances in light years, which is the distance light travels in one year. In just one second, light travels 300,000 km. It would take 100,000 years for light to travel across the diameter of our galaxy. Humankind is travelling on the Earth as it revolves around the Sun at speeds of about 107,000 km hr^{-1}. The Sun travels around the galaxy at about 300 km s^{-1}, and the galaxy itself is travelling at enormous speed outwards from the centre of the Universe, which is still expanding after its creation in the **Big Bang**. The creation of the Universe probably occurred some 15,000 million years ago, with a consensus favouring a time approximately 14,700 million years ago. This contrasts with the date of 4004 BC for God's creation of the world proposed by the Irish archbishop of Armagh, James Ussher (1581–1656). The noise from the Big Bang is still reverberating through Space as a constant and measurable background level of radiation. So, humans are cosmic passengers on a journey at enormous speeds within the vastness of Space. The story of the creation of the Universe, and with it the Earth, is eloquently told in Stephen Hawking's best-selling book, *A Brief History of Time*.

Humankind is currently living through the Space Age, which really began in the 1960s. On 12 April 1961, Yuri Gagarin's historic Space flight aboard the *Vostok* capsule began the era of extraterrestrial human travel. The dream of countless earlier generations was fulfilled in this Soviet mission. The USA, with strong Presidential backing, especially from John F. Kennedy, and massive public investment, was the first to land astronauts on the Moon in the Apollo 11 mission in 1969, flown by Michael Collins, Edwin 'Buzz' Aldrin and Neil Armstrong. Upon landing, Armstrong spoke to Mission Control: 'Houston. Tranquillity Base. The Eagle has landed'; on 21 July 1969 at 3.56 a.m. BST, Armstrong was the first person to walk on the Moon, when he stated 'That's one small step for [a] man, one giant leap for mankind.'

On 12 April 1981, exactly twenty years to the day after Yuri Gagarin's flight, the space shuttle *Columbia* was launched by the USA. But tragedy was to strike the space shuttle programme when in 1986, just 73 seconds after lift-off, the space shuttle *Challenger* exploded, and with it the cosy myth of Space travel becoming routine into the 1980s. The accident also caused people to question the cost of Space travel, not only in terms of the lives of astronauts, but also in relation to broader human costs over the actual and perceived benefits. For example, is it morally defensible to spend billions of US dollars on a Space programme when so much of the world's population has pitifully inadequate food and shelter?

The outer layers of the Earth

Earth scientists divide the outer layers of the Earth into four main spheres or realms (Figure 1.1), which are the **lithosphere**, comprising the outer layers of the more solid Earth (uppermost mantle and crust), as rocks, sediments and soils; the **atmosphere**, the

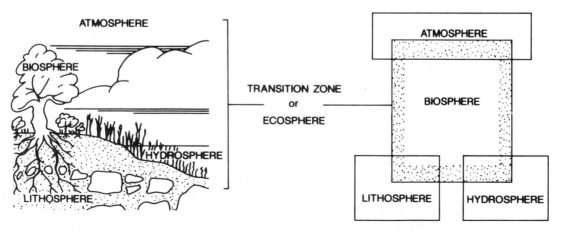

Figure 1.1 *The relationship between lithosphere, hydrosphere, atmosphere, biosphere and ecosphere. Adapted from White (1986).*

gaseous layers that extend from the Earth's surface up to about 100 km; the **hydrosphere**, the aqueous component that covers large parts of the planet, from a maximum depth of more than 11 km in the oceans to shallower and less extensive bodies of water such as shelf seas, lakes and rivers (the hydrosphere also includes snow and ice as glaciers and ice sheets, and the water found within the soils and rocks, such as that below the water table); and the **biosphere**, a term first extensively used by the Swiss geologist Suess to describe the thinnest layer, comprising organic matter, generally only up to a few metres thick and covering much of the land surface. This layer, at its thickest, reaches several tens of metres in the rainforests. It also extends into the atmosphere (because creatures fly and plant spores are blown by the wind), and deep into the oceans, seas and lakes. Human beings are part of the biosphere and interact naturally with the other three 'spheres'.

Although the outer layers of the Earth can be considered as comprising these four zones, they are inextricably linked and all are part of the ecosphere. **Ecology** is the study of the ecosphere, commonly loosely referred to as the study of the environment. Ecology is undoubtedly a difficult subject to study, because it utilises information and ideas from just about every other subject, from science to politics, from economics to culture.

Earth's energy sources

The landscape is fashioned by a wide variety of natural processes. These processes include volcanic eruptions, the slow and inexorable drift of continents and sea-floor spreading, earthquakes, and the formation of sediments from rock. Among the many Earth-surface processes are wind and ice action, ocean currents, tides, storms, tsunamis, the flow of water in streams and rivers, or through sediments and rocks as ground water, erosion, and landslides. Meteorological processes, such as wind, rain, tropical cyclones, and thunder and lightning, create the weather. Biological processes, such as the growth of plants and animals, death and the decay of organic matter, and the colonisation by species, occur within the global theatre of fair-weather processes and natural hazards – as well as catastrophes caused by humans.

The energy that drives the natural processes comes from three main sources (illustrated in Figure 1.2). The most important source of energy is the Sun, mostly reaching the Earth in the form of short-wave radiation. Some of this radiation is converted into long-wave radiation (towards the infrared end of the spectrum), which heats the Earth's surface and atmosphere. This heat energy is responsible for global and local variations in air temperature and pressure, which ultimately control the circulation of gases within the atmosphere and across the globe to give weather. Heat energy also controls the state of moisture (water) in the air and hence the form of precipitation. Short-wave radiation such as ultraviolet light is essential for providing the energy for life, for example in the processes of photosynthesis in green plants where carbohydrates are formed from the basic chemical building blocks of carbon dioxide and water in plant tissues. In the **food chain**, plants in turn provide the primary food source for animals.

The second major source of energy comes from within the Earth itself. This internal heat energy is produced mainly by the radioactive decay of elements such as uranium and thorium in the Earth's mantle and crust. The same heat energy allows rocks to behave plastically and to flow at depth or even partially melt to produce rock melts called **magmas**. Magmas may rise towards the surface of the Earth and pierce the crust to form **volcanoes**. Depending upon the chemistry of the parent magmas, particularly the silica and water content, volcanoes can be highly explosive and produce high eruptive columns, or behave in a more gentle (lower viscosity) manner. It is the explosive type of volcano that can eject large amounts of dust and aerosols high into the atmosphere to cause short-term global climate change.

The flow of rocks in the Earth's mantle at depths greater than 5–8 km below ocean floors and more than 35–50 km below continents provides a fundamental mechanism for the horizontal and vertical movement of the cooler surface layers of the Earth's lithosphere. The continents also can behave like extremely viscous liquids but at rates so slow that for most purposes they can be considered as solid and rigid. Earthquakes are the expression of the sudden release of stresses (force per unit area) built up within the lithosphere, especially the uppermost 12 km, as relatively rigid plates and blocks move past each other, with displacements typically measured in centimetres to metres.

The third main energy source responsible for many Earth processes exists because of the gravitational forces that mutually attract masses. Newton's **Law of Gravitation** states that the larger the mass of an object the greater the gravitational attraction it will exert on other masses. The mass of the Earth is considerable compared with bodies at its surface and hence objects are strongly attracted by the gravitational force towards the centre of the Earth, an observation appreciated in the anecdotal observation

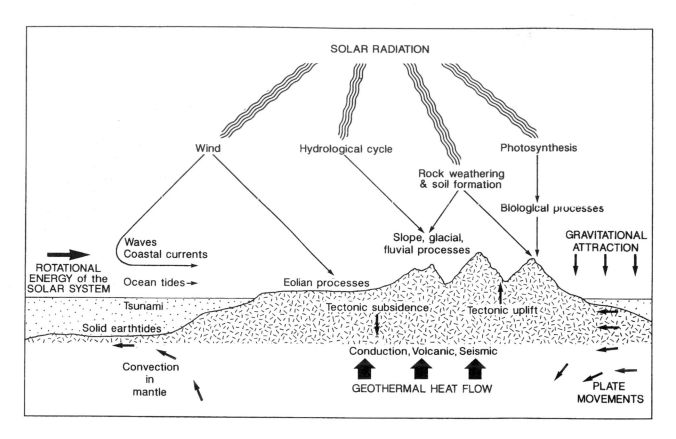

Figure 1.2   *The major energy sources and associated processes for planet Earth. Redrawn after White* et al. *(1986).*

of an apple falling on Newton's head! The Earth also attracts and is attracted by objects such as the Sun and the Moon. It is the gravitational force that maintains the planets in their orbits around the Sun, and the orbits of moons around their respective planets. Gravitational forces are important for Earth surface processes such as the surface run-off of water, rock and debris landslides, snow avalanches, and the movement of glaciers and ice sheets. The gravitational attraction of the Sun and Moon on the Earth is responsible for the daily rise and fall of the sea as tides, in most places observed as the twice-daily cycle between high and low tide. The changing position of the Earth relative to the Moon and Sun also controls the monthly tidal inequality from neap (smallest tidal range) to spring (greatest tidal range) tides, and the annual change from the summer and winter solstices to the spring and vernal (autumnal) equinoxes, respectively.

Tectonic processes and the Earth's interior

The outer layers of the Earth are continuously being modified by processes of weathering and erosion.

Physical and chemical weathering break down rocks and minerals to furnish new sedimentary particles into the Earth's surface environments – glacial, deserts, soils, rivers, lakes, and coastal and marine environments. At the same time, the Earth's surface is undergoing changes driven by its internal processes, which are responsible for producing the major morphological features of the Earth, the distribution of different rock types and mineral resources, and phenomena such as earthquakes and volcanoes. Geophysical evidence, mainly from seismological studies, shows that the Earth's internal structure and composition can be divided broadly into three major layers separated by two major discontinuities defined by changes in the velocity of **seismic waves** as they travel through the Earth (Figure 1.3). Extending from the centre of the Earth to approximately 3,740 km below the surface, the **core** comprises very dense material, probably mostly iron with lesser amounts of nickel and carbon. This has the properties of a solid in the inner part and of a liquid in the outer part. The Earth's magnetic field is created by the solid and metallic iron-rich inner core acting like a bar magnet. Periodically the Earth's magnetic field flips to a reverse polarity. These flips

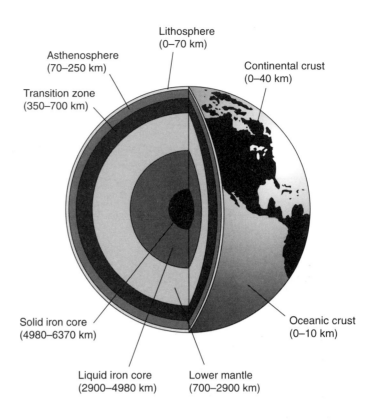

Figure 1.3 *Internal structure of the Earth. Distance of layer from the surface is given in km.*

...ice over periods ..., so the Earth's ...ens, but not on ...nderlying causes ...etic field are not ...siderable current ...nvective flow of ... reversals.

...extends to the ...), which separates ...Figure 1.4). The ...t 35 km beneath ...km beneath the ...ots of the largest ...ayas the Moho is ...mantle has most ...ut due to its high ... strain rates like ...erals that are rich ...per part of the ...with depth to ...elerate slightly ...wn as the **low-** ... a region of ...ibits its most ...e **asthenos-** ...heat energy ...ve elements such ...present in rocks at ...le has plastic-like ... flow when forces

...f the solid Earth. ...antle above the ...**osphere**, which is ...er. There are two ...here, capped with ...continental lithos- ...ighter continental ...here varies consid- ...ow some parts of ...efine. The highest ...es an altitude of ...eepest part of the ...Marianas Trench, ...maximum relative ...therefore nearly ...he Earth's polar ...re 12,756 km and ...mum difference in ...rface that supports the weight of mountains, plateaus and the ocean's deep trenches, reflecting the forces that maintain the relief and once removed cause the surface to

tend toward some equilibrium profile called **isostasy**.

The Earth's interior and outer layers are not passive. Partially melted rock within the asthenosphere flows and causes the lithospheric plates to move. In 1915 Alfred Wegener published his book, *Die Entstehung der Kontinente und Ozeane* (the first English edition, *The Origin of Continents and Oceans*, was in 1924), which provided the first scientifically argued case that the continents may have once been joined together into a supercontinent and then moved apart into their present positions. Wegener's theory of **continental drift** was based on:

- the similarity of fit of coastlines across opposite sides of the oceans, such as the jigsaw-like fit of the eastern coast of South America with the western coast of Africa, now separated by thousands of kilometres;
- the presence of ancient glacial deposits, **tillites**, now widely distributed throughout the continents, but which would make sense if the continents were fitted together so that the tillites were originally juxtaposed over one common South Pole; and
- palaeontological evidence in which species of distinctive non-migratory creatures are now widely dispersed.

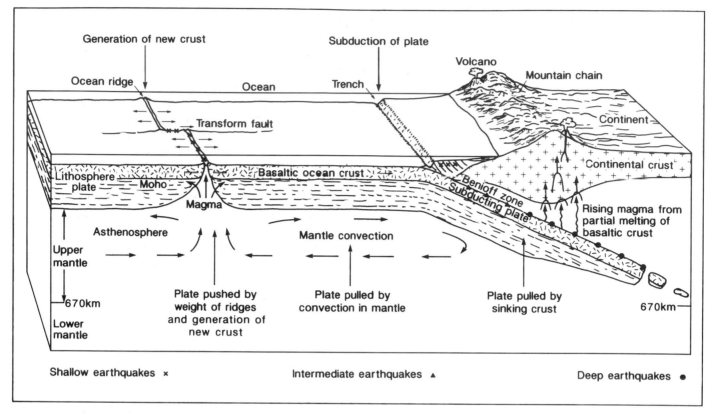

Figure 1.4   *The structure of the outer layers of the Earth, the major physiographic features, and the plate tectonic mechanisms responsible for the generation of new sea-floor crust, sea-floor spreading, the consumption or subduction of oceanic crust, earthquakes and vulcanicity. Redrawn after Selby (1985).*

His theory found little support amongst the scientific community because of the failure to provide a driving mechanism for continental drift. Arthur Holmes and others, however, suggested that the driving mechanism could involve the convection of molten rock at depth, heated by the decay of radioactive elements within the Earth, and proposed that hot rock could rise towards the Earth's surface, pushing the continents sideways. The molten rock would then cool and descend to be reheated again, and the convection process would continue.

Widespread support for Wegener's theory had to await the publication of a paper by Vine and Matthews in 1963, in which the overwhelming evidence supported the contention that the continents had indeed moved apart. Vine and Matthews had been examining data on **magnetic anomalies** observed along the mid-ocean ridges, from which they noted the symmetrical pattern of **palaeomagnetic** 'stripes' about the ridges and suggested that they were produced by magnetic minerals, which aligned themselves parallel to the Earth's magnetic field as lavas crystallised shortly after being extruded

from mid-ocean ridges. They further suggested that the Earth's magnetic field reversed periodically, producing the pattern of negative and positive anomalies symmetrically disposed about the ridge axes. Also, Vine and Matthews suggested that new ocean crust formed at the ridges and was pushed away from the ridge each time new lavas erupted to form additional oceanic crust. At last a plausible driving mechanism for the motion of the continents, backed up by clear scientific evidence, was proposed and the theory of **plate tectonics** rapidly became accepted by the wider scientific community.

As a result of this theory of **sea-floor spreading**, now subsumed within the theory of plate tectonics, considerable international scientific effort was put into establishing the history of the ocean floors, particularly through the activities of the Deep Sea Drilling Project, an international research programme that still continues today under the aegis of the second major phase of international drilling as the Ocean Drilling Program. Evidence in support of the theory of plate tectonics has also come from land-based work, such as palaeomagnetic studies on the continents, which show that the continents have drifted into their current positions. Seismological evidence shows that the vast majority of earthquakes are concentrated along very obvious linear zones, which turn out to define the boundaries of the plates making up the continents and oceans. In many oceanic regions, an inclined zone of seismicity, the **Wadati–Benioff zone**, was recognised as associated with **active continental margins**, where **volcanic island arcs** develop above regions of the mantle where slabs of oceanic lithosphere are being subducted back into the mantle – a means of accommodating the additional space required by the formation of new ocean crust at spreading ridges. As the subducting slabs descend back into the mantle at **subduction zones**, so the rocks undergo partial melting in response to the increasing temperature and pressure, their melting point being lowered by a veneer of downgoing wet oceanic sediments rich in sodium and potassium, to produce magmas that rise to form the volcanic island arcs.

Plate tectonic theory involves the recognition of seven major, and at least a dozen minor, present lithospheric plates, which are constantly in motion with respect to each other. These plates travel at velocities of up to 100 mm yr^{-1}, but average about 70 mm yr^{-1}. The zone of relative movement between plates, the plate boundary, is clearly defined by seismic activity. There are three main types of plate boundary. A **divergent plate boundary** at mid-ocean ridges is where new oceanic crust is being formed and the oceanic plates are moving laterally away from the **spreading ridge**. The second type is a **convergent plate boundary**, where two lithospheric plates are moving together and forcing one plate to be subducted beneath the other. If one plate comprises oceanic crust it will be subducted in preference to the more buoyant continental crust. If both plates are continental crust, as is the case between India and Asia, then neither can be subducted as they are too buoyant; therefore the thickness of the continental crust effectively doubles, not only to create a very high mountain chain – the Himalayas – but also to generate a high plateau – the Tibetan Plateau. Figure 1.5 shows reconstructions of the relative positions of the major continents in **Mesozoic** and **Cenozoic** times (see Figure 1.16), during the past *c.* 240 Ma.

At active convergent plate boundaries, for example off the eastern coast of the island arcs that constitute Japan and the Philippines, deep oceanic trenches parallel the plate boundary and are the surface expression of the subducting lithospheric plate. Where oceanic lithospheric plates collide with continental plates, the more dense oceanic plate is subducted beneath the continental lithospheric plate and creates mountain ranges of folded and faulted rocks, and volcanoes, near the edge of the continental plate. The mountains of the Western Cordillera and the Andes have formed in such a manner.

The third type of plate boundary develops where two plates slide past each other at so-called **transform plate boundaries** (conservative plate margins). One of the best-known examples of a conservative plate margin is defined by the San Andreas Fault System in California, where the North American continental plate is moving southeastwards and the Pacific oceanic plate is moving northwestwards along a complex fault system. Another example cuts along the Southern Alps of South Island, New Zealand, and is defined by the Alpine Fault Zone.

Today, the theory of plate tectonics forms the framework for much research in the Earth sciences. The theory is subject to continuous modification; for example, it is now appreciated that the continental plates are not as rigid and independent of one another as originally thought. Instead, the continental plates are best modelled as extremely viscous liquids with mountain belts maintaining their height by virtue of continuous 'push' or compressional forces. If these compressional forces are removed then the roots of the mountains will literally flow away under the force of gravity (gravitational collapse), while also being worn away through erosion, until the elevation of the continental crust is approximately at sea level (England 1992). Of course, this process operates at extremely slow rates measured in millimetres per year. Another aspect of research in plate tectonics concerns the forces involved in driving the plate motions – for example the slab pull forces as plates descend back into the mantle (e.g. Kerr 1995c) versus the ridge push forces as new oceanic crust is created at mid-ocean ridges. Amongst the more exciting recent developments in the theory of plate tectonics is the imaging of the structure of the mantle using seismic techniques in a field of study known as mantle tomography, or comparing plate tectonics on Earth with processes on other planets such as Mars or Venus.

Life on Earth

The Sun's rays provide the energy to drive the ocean currents and atmospheric processes – the weather that is so important in the global distribution of gases, water and heat. The chemical and physical breakdown

Present Day

50 Ma

100 Ma

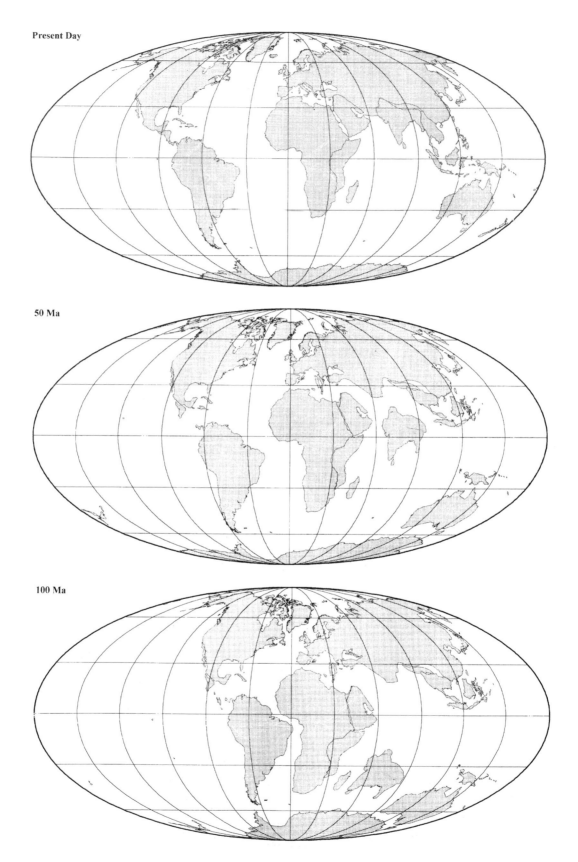

Figure 1.5 *Reconstructions of the relative positions of the continents from approximately 240 Ma to the present. Maps supplied by Cambridge Palaeomap Services Ltd (1996).*

150 Ma

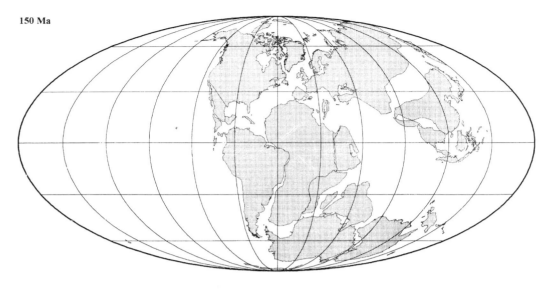

200 Ma

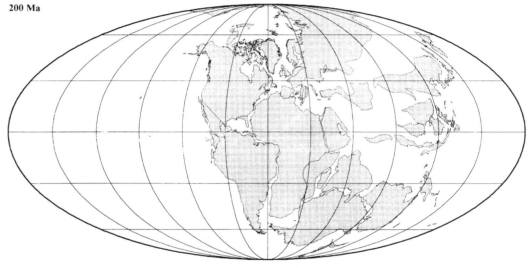

240 Ma

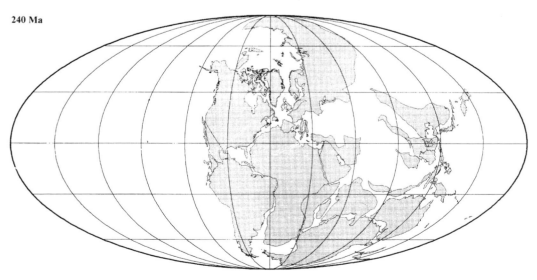

of rocks supplies the vital nutrients essential for life, and the rich variety of surface environments, from rocky desert to tropical soils, provides the substrate for life. Humans are short-stay passengers on the Earth and in terms of biodiversity represent a very small part of life on Earth, but human activities have led to the extinction of some species, pose a threat to many more, and may be profoundly altering the atmospheric and climate systems.

On cosmic scales, planet Earth seems insignificant, but it may be unique. It is the only planet that is known to be capable of supporting human life. Experiments continue in an attempt to find criteria to recognise life elsewhere in the universe. For example, Sagan *et al.* (1993) conducted an experiment during the 1990 fly-by of Earth with the Galileo spacecraft, whose principal aim was to characterise the main remotely sensed chemical and physical attributes of life on Earth, which can then be used to study other planets when they are identified outside the Solar System.

Chemical and physical arguments suggest that the Earth is about 4.7 billion years (4.7 Ga) old. Life on Earth is incredibly diverse. It has been estimated that the Earth contains as many as 1.4 million formally described species of animals and plants (Wilson 1989). Many more await detailed study and the conferment of a formal species name. New species are continually being recognised; most are plants and invertebrates, but occasionally higher creatures are discovered, such as the surprising discovery in Vietnam in 1993 of a new genus of bovid, *Pseudoryx nghetinhensis* sp. nov. (Dung *et al.* 1993). There are probably between 5 million and 30 million species, although most biologists regard 10 million as being the best approximation (Blum 1993). Biologically diverse groups that receive relatively little attention from scientists who study biodiversity include fungi, many insect species (e.g. mites), and organisms inhabiting the deep oceans. More than half of the total number of species of flora and fauna inhabit the rainforests, with their moist tropical climate.

The rainforests, where rainfall is in excess of 200 cm yr^{-1}, account for approximately 6 per cent of the land surface. The number of individuals of any species inhabiting the rainforests is truly amazing. In just one gram of soil, there may be as many as 100,000 algae, 16 million moulds and fungi, and several billion bacteria. Up to 5,000 species of organism can inhabit just one rainforest tree. In a single acre of rainforest in Panama, it is estimated that there are as many as 40 million animals, not counting the bacteria, fungi and moulds.

Other major ecosystems with extremely large biodiversity are the coral reefs, where a myriad of organisms occupy these **ecological niches**. Corals require a plentiful food supply, and well-oxygenated, warm waters that are essentially free of land-derived (terrigenous) sediments like mud and silt. These would otherwise make the waters cloudy and dilute the chemical factories which produce the calcium carbonate ($CaCO_3$) shells or tests that the coral polyps inhabit.

It is not just the rainforests and coral reefs that are teaming with life. Other **biomes** are surprisingly profuse and varied (see Box 1.1). Recently, for example, the Rockall Trough off western Scotland has been recognised for its diversity (Pearce 1995d). In fact, there are possibly more species living at depths of between 1,000 and 3,000 m in the Rockall Trough than in a tropical rainforest or coral reef. Most of the species present are microscopic mud-dwelling **nematodes**, whereas coral reefs have more vertebrates but in comparison less overall biodiversity. Brey *et al.* (1994) also showed that the Weddell Sea in the Antarctic Ocean has an incredible bottom-dwelling, or **benthic**, diversity, with over 300 invertebrate species, which is in the upper range for species diversity in tropical regions. An ignorance of the nature of such ecosystems and their biodiversity can lead to threats to ecosystems. This was well illustrated when the Shell Petroleum Company attempted to dump the *Brent Spar* in Rockall Trough. There are those who would argue that it was fortunate that the environmental pressure group Greenpeace was able through international pressure to dissuade Shell from dumping the *Brent Spar* at sea, thus saving a prolific biome and a potentially valuable genetic resource.

Biodiversity is a complex concept, involving a consideration of the number of genetic phenotypes and the actual number of species within a habitat, the abundance and dominance of species, and the diversity of habitats in a given area. Human activities throughout the world so frequently involve the profligate and thoughtless exploitation of natural ecosystems, which results in their destruction or at best marginalisation. This is particularly so in developing countries, where natural resources such as the forests are being exploited for short-term economic gain, often to provide developed countries with luxury items. Biodiversity as a concept has little currency with those who are most involved in the exploitation of natural resources, and it is difficult to protect environments without large economic incentives.

Studies have shown, however, that when considered in their entirety the economic value of the bio-

BOX 1.1 BIOTIC PROVINCES AND BIOMES

In 1876, A. R. Wallace divided the world into six biogeographical regions on the basis of the families or orders of animals that dominate particular regions, now known as **Wallace's realms**. Wallace recognised that animals filling the same ecological niche within each realm were of a different genetic stock from those in other niches. These basic concepts are still held true, but Wallace's realms are now extended to include vegetation in **biotic provinces**, which are defined by a characteristic set of taxa possessing a common genetic heritage, and are confined by barriers that inhibit the spread of the distinctive taxa into or from other biotic provinces.

The main biogeographic realms for animals include the *Australian*; the *Neotropical* of South America; the *Nearctic* of North America; the *Palaearctic* of Europe and Central Asia; the *Palaearctic* of European Africa; and the *Oriental* of Southeast Asia and the Indian subcontinent. The main vegetation realms include the *Australian floral region*; the *Antarctic floral region*; the *Neotropical floral region* of South America; the *Palaeotropical floral region* of Africa, the Middle East, the Indian subcontinent and Southeast Asia; and the *Boreal floral region* of North America, Europe and Central Asia. Biotic provinces can be explained by considering the dynamics and timing of continental drift as species were separated during continental break-up and the movement of the land masses into their present positions.

The Earth can also be divided into a series of comparable environments where organisms have evolved into similar forms and with like functions. Organisms have evolved to adapt to the climate, topography and available nutrients in a particular ecosystem or **biome**. It is convenient, therefore, to divide the biosphere into biomes. To some extent, the geographical distribution of certain types of organism can be predicted from a knowledge of the characteristics of the rainfall and temperature distribution for a particular region (Mather and Yoshioka 1968, Belsky 1990, Prentice *et al.* 1992; Figure 1.6).

The deserts of Africa and America, for example, have been geographically isolated for the past 180 Ma, yet the flora in each region looks remarkably similar. The plants in these like environments have evolved to adapt to similar stresses and ecological opportunities in a process known as **convergent evolution**. Alternatively, a population may be separated by a geographical barrier, and sub-populations may evolve independently, retaining similar characteristics in a process known as **divergent evolution**. When creatures adapt to a new environment and become highly specialised, as is the case on small island ecosystems, the process is known as **adaptive radiation**.

Biomes are named after the dominant organism, for example boreal forests rather than simply grasslands, and the dominant climatic conditions. The main biomes are shown in Figure 1.7A. The similarity between the distribution of biota and the world's climatic regions can be appreciated by comparing Figure 1.7A and Figure 1.11.

This illustrates the strong control that climate has on the distribution of biota. The distribution of vegetation is also strongly controlled by topography, as is illustrated in Figure 1.7B.

---

diversity within rainforests can be remarkably high. Peters *et al.* (1989) calculated the market value for timber, fruit and latex in one hectare of forest at Mishana, Rio Nanay, Peru, and showed that there are substantial profits to be made in conserving and sustaining the forest. Yet little is done to promote their development in favour of clearing for other activities. Biologists believe that biodiversity is critical in helping to support the ecological stability of regions as well as being vital for sustaining the **biogeochemical cycles**, for example the oxygen, carbon, nitrogen and sulphur cycles. Biomes also have great importance as potential genetic stores and as repositories for natural resources. Using controlled environmental chambers, Naeem *et al.* (1994) showed for the first time that declining biodiversity can profoundly alter the performance of ecosystems. They showed that reducing biodiversity results in a loss of biomass productivity, a loss of ecosystems, a decrease in the buffering against ecological perturbations, and a reduction in the ability of terrestrial ecosystems to sequester carbon dioxide (CO_2). Reducing biodiversity, therefore, will have profoundly deleterious effects on the ability of terrestrial ecosystems to absorb the recent increased atmospheric levels of anthropogenic greenhouse gases.

In a real-life situation in the grasslands of Minnesota, Tilman and Downing (1994) showed that the more diverse the grassland plant community the more resistant it is to drought. When the number of species of plants was reduced, the impact from drought was much greater. This reinforces the view that biodiversity helps support stability within an ecosystem.

In 1994 an international attempt to reduce the threat to biodiversity was signed under the United Nations Framework Convention on Biodiversity promoted at the Rio Conference and ratified in 1994. Under this UN convention, each signatory country where biodiversity is seriously threatened has to prepare a plan for conserving and sustaining the present biodiversity, and monitor its own genetic stock and provide financial support to aid in any necessary conservation programmes. Another means of conserving biodiversity is the World Conservation Strategy, established in an attempt to preserve bio-

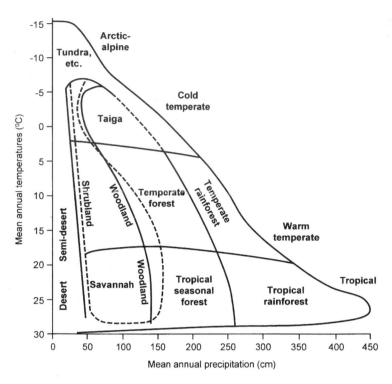

Figure 1.6 *The major terrestrial biomes based on temperature and rainfall. Redrawn after Belsky (1990).*

diversity, maintain ecological systems, ensure the sustainable use of ecosystems and initiate conservation schemes. These strategies have met with varying degrees of success, but they need enforceable legislation to be really effective.

The Earth provides the life-support system for this diverse and abundant array of organisms. The atmosphere filters out potentially lethal **radiation** from the Sun, yet at the same time allows some of the radiation to penetrate the atmosphere and provide the energy for plants to construct tissues of carbohydrates from **carbon dioxide** (CO_2) and water (H_2O) in the process of **photosynthesis** (see Box 1.2). These plants, in turn, are the food for the animal kingdom. In addition, the atmosphere provides the CO_2, O_2 and much of the water vapour needed for the basic functions of animal life.

Evolution and extinction

The study of fossils, past life and evolution, **palaeontology**, suggests that throughout the **Phanerozoic Era** (during the past 540 million years) of Earth history, there have been catastrophic extinction events when exceptionally large numbers of species became extinct. It is these major extinction events

and the radiation of new species that have been used to compartmentalise geological time. The causes of such extinction events remain controversial and there appears to be a range of different circumstances that brought about many of the major extinction events. Abrupt climatic shifts from greenhouse to icehouse conditions, meteorite impacts, and the configuration of the continents, are amongst the most commonly cited causal processes. Other major climatic influences on the evolution of organisms include the postulated atmospheric 'oxygen pulse' during the late Palaeozoic (mid-Devonian to late Permian periods) in which mainly biotically driven atmospheric O_2 levels are modelled to have reached a maximum of 35 per cent and then dropped to 15 per cent, compared with the present 21 per cent (Graham *et al.* 1995). Elevated O_2 levels would have accelerated diffusion-dependent metabolic processes such as respiration, increased air density and barometric pressure (e.g. 35 per cent O_2 compared with the present 21 per cent would give *c.* 21 per cent greater density), thereby promoting the radiation of certain species that were or became advantaged by an enhanced metabolic rate, turnover and resource accessibility (*ibid.*).

The initial idea that large-scale or mass extinction events are periodic was proposed by Fischer and Arthur (1977) and was based on their review of open-ocean, free-swimming or floating (**pelagic** or **planktonic**) fossil communities throughout the Mesozoic and Cainozoic eras (see Figure 1.16). The data led them to suggest an approximately 32 Ma periodicity in mass extinctions. Figure 1.8 shows Fischer's megacycles of extinction with global climate change, the major variations in global (**eustatic**) sea level, and intensity of volcanism (modified after Fischer 1982, in van Andel 1994). These data suggest that major extinction events appear to be related to periods of rapid and large-scale shifts in global climatic conditions – although there are exceptions. In contradistinction, other researchers recognised a 26 Ma cyclicity (Raup and Sepkoski 1984, 1986, Rampino and Stothers 1984, Hoffman and Ghiold 1985, Kitchell and Pena 1984, Sepkoski and Raup 1986). Such periodicity in mass extinction events has been used by some researchers to assert that it 'requires an astronomical explanation' (Whitmore and Jackson 1984). Others argue that extinction events are not unusual in the Earth's biotic evolution (e.g. Patterson and Smith 1987, 1989), or are perceived as more significant and devastating because of incomplete sampling and the way in which data are manipulated (e.g. Hoffman 1985, Patterson and Smith 1987, 1989).

Plate 1 *Earth rising above the surface of the Moon. The 1969 landing on the Moon provided a new perspective of our planet.*

Courtesy of NASA/Lunar and Planetary Institute.

Plate 2 *The space shuttle starting another mission.*
Courtesy of Comstock.

Plate 3 *Coral reefs are amongst the world's most diverse ecosystems. This plate shows some of this biodiversity on a small coral knoll.*
Courtesy of Comstock.

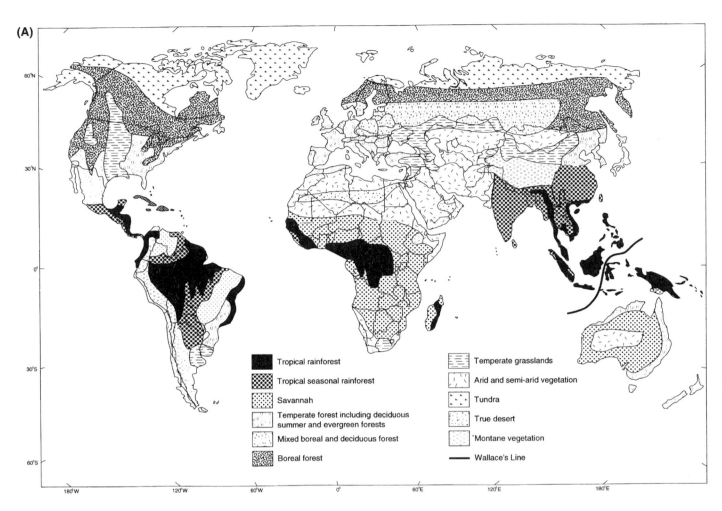

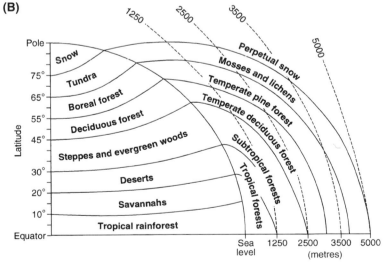

Figure 1.7 *(A) Generalised distribution of the world's vegetation types; and (B) the modification of the world's major vegetation zones by altitude. Redrawn after Goudie (1993a).*

Naturally, any debate about mass extinctions must rely on the central axiom that the fossil record is a representative and accurate record of past biota, something that is by no means resolved, as the preservation potential of organisms will depend upon factors such as any skeletal or hard parts that can more readily be fossilised, the conditions of fossilisation (or **taphonomy**), etc. Proponents of mass extinction events seek to explain such processes as due to catastrophic (e.g. extraterrestrial **bolide** impact or intensive and extensive terrestrial volcanism) or gradualistic in nature (e.g. racial senescence, ecological decline and/or significant changes in global sea level). Comet impacts may occur if the cloud of comets which orbit beyond Pluto, the so-called Oort cloud, is disrupted such that the orbits of comets are perturbed (and there are many thousands that have orbits which cut across that of the Earth), as for

example is caused by oscillations of the Sun perpendicular to the galactic plane (Schwartz and James 1984, Thaddeus and Chanan 1985). Alvarez and Muller (1984) show that most dated bolide impacts cluster with a 28.4-million-year periodicity.

BOX 1.2 OXYGENIC PHOTO-SYNTHESIS AND AN ATMOSPHERE RICH IN FREE OXYGEN

The critical role played by biological activities in releasing O_2 to the atmosphere is **oxygenic photosynthesis**, in which the water molecule is split to release pure oxygen:

$$CO_2 + H_2O \rightarrow CH_2O + O_2$$

It is the free oxygen released in this reaction that, over geological time, increased the levels of atmospheric oxygen to present concentrations. The burial of organic matter (shown in its simplest chemical formula, CH_2O, in the above equation) in sediments allows the release of free oxygen into the atmosphere. A corollary of the burial of organic carbon and biogenic sulphide is that electrically charged sulphate ions (SO_4^{2-}) and ions of iron in its ferric state (Fe^{3+}) increased at the Earth's surface and in the atmosphere. Oxidation of the Earth's atmosphere and surface environments was facilitated both by biological and sedimentary (geological) processes.

Although the Earth's atmosphere has changed to one in which free oxygen is present, the earliest life evolved in a very different atmosphere. There is a 3,800-million-year isotopic record of life on Earth, something that Earth scientists have discovered by examining the carbon contained in lithified sediments, or sedimentary rocks, at Isua in west Greenland.

The oxidation of the Earth's crust early in Earth history, and the associated increase in atmospheric oxygen, has been linked to the accumulation of reduced carbon in sedimentary rocks. By studying the carbon **isotope** composition of sedimentary organic carbon and carbonate, Des Morais *et al.* (1992) have shown that during the **Proterozoic** time period, 2.5–0.54 thousand million years ago (Ga), the organic carbon reservoir grew in size relative to the carbonate reservoir. They further showed that this increase and the transition to an oxidising atmosphere took place mainly during intervals of enhanced global sea-floor spreading, continental break-up and rifting, and **orogeny** in what is broadly referred to as **tectonic processes**. Around 3.0–2.4 Ga,

relatively small continental plates or cratons welded together to form the first relatively large and stable **continental plates**. These processes provided the templates for the accumulation of large amounts of sediments, and set the stage for the growth of carbonate platforms 2.6–2.3 Ga. Although there is evidence to suggest oxygenic photosynthesis in the algal mats called **stromatolites**, the net accumulation of atmospheric O_2 was virtually zero, because there was very little burial of organic carbon. Approximately 2.2–2.1 Ga, the large continental plates began to disintegrate by rifting apart and sea-floor spreading, and the break-up allowed the development of free O_2-deficient or **anoxic** basins in which organic matter could accumulate and be buried. Also, at this time, there appears to have been significantly enhanced erosion and continental run-off, inferred from the rise in sea water strontium or osmium isotope values (see Box 1.3).

Evidence from ancient soil profiles or **palaeosols** suggests that prior to about 2 Ga atmospheric O_2 levels were low, but such a situation would not have inhibited the efficient re-mineralisation of organic matter in microbial mats, as is the case today. A large part of the early free O_2 was probably consumed in reactions associated with the more voluminous **hydrothermal systems** on the sea floor associated with extensive **magmatic** and volcanic activity. Sea-water sulphate ion (SO_4^{2-}) levels, therefore, would have been much lower than in the modern oceans.

The views of Des Morais *et al.* (1992) challenge the widely held belief that major rapid changes in biological evolution controlled the long-term increase in oxygen levels in the atmosphere. The development of oxygenic photosynthesis took place at least 600 million years ago, prior to the accumulation of significant amounts of O_2 in the atmosphere. **Eukaryotic organisms**, those which require O_2 to **biosynthesise** the essential **lipids**, appear about 2.1 Ga.

Oxygenic photosynthesis must have provided a mechanism capable of sustaining a dramatic increase in atmospheric O_2 levels, but the timing and magnitude of the O_2 accumulation was regulated by tectonic processes controlling erosion and sedimentation (*ibid.*).

Another suggested explanation for mass extinctions is to link them to perturbations of the Oort cloud caused by the gravitational pull of a companion star to the Sun – the so-called 'Nemesis' scenario; but this hypothesis has been refuted by Carlisle (1995), who showed that the orbit of such a star would be intrinsically unstable (since it would travel between other neighbouring stars, which would perturb its orbit and cause the unstable orbit to decay within approximately 250 million years, leading to the separation of the 'pair'). Since binary star pairs tend to form simultaneously, it seems reasonable to suppose that any original companion star to the Sun, if it ever existed, would have separated thousands of millions of years ago. Also, the notion that mass

extinctions may owe their occurrence to the way in which the Solar System moves through the galactic plane has been challenged by Sepkoski (1990), who showed that the best-known mass extinction events are out of phase, and that the mass of the galaxy is insufficient to produce such periodic perturbations.

Five major mass extinction events are widely believed to have occurred in geological time: at approximately 435 Ma, the boundary of the Ordovician and Silurian periods when there was a major global glaciation; 375 Ma, late in the Devonian Period; 240 Ma, at the boundary between the **Permian** and **Triassic** periods; 210 Ma, in the Triassic Period; and at 65 Ma, the boundary between the **Cretaceous** and **Tertiary** periods (the so-called K–T

BOX 1.3 STRONTIUM AND OSMIUM ISOTOPES THROUGH GEOLOGICAL TIME

Strontium isotopes

In the Periodic Table of Elements (see Appendix 2), strontium (Sr) is an alkaline earth Group IIA element (included with Be, Mg, Ca, Ba & Rb). Sr has four naturally occurring stable isotopes: ^{88}Sr (82.53 per cent); ^{87}Sr (7.04 per cent); ^{86}Sr (9.87 per cent) and ^{84}Sr (0.56 per cent). The naturally occurring radiogenic isotope ^{87}Rb decays to the stable ^{87}Sr by β-decay (the loss of an electron), therefore the amount of ^{87}Sr increases through time in Rb-bearing rocks. Throughout geological time the ^{87}Sr/^{86}Sr ratio has increased.

The fractional crystallisation of magma tends to concentrate Sr in the mineral phase, particularly plagioclase feldspar, and leave any rubidium (Rb) in the liquid phase. Thus, the Rb/Sr ratio increases with increasing degrees of differentiation, i.e. chemical separation. Continental crust is enriched in Rb relative to oceanic crust and mantle, therefore continental crust has a significantly higher ^{87}Sr/^{86}Sr ratio, and thus is also enriched in ^{87}Sr with time.

Sr occurs as a trace element in most igneous, metamorphic and sedimentary rocks. The chemical weathering of continents releases Sr from the rocks into solution in rivers, lakes and ground water. The isotopic composition of such water is a function of:

- the age of the rocks and minerals that are being dissolved;
- the Rb/Sr ratio of these rocks and minerals, and
- the solubility of the constituent minerals.

Continental weathering is controlled by many variables, including global and local climate, and other geographic factors. The average ^{87}Sr/^{86}Sr ratio of continental crust is difficult to determine but has been estimated at a value of $c.$ 0.711 to 0.716. For fresh mantle-derived oceanic tholeiites, which have a typical composition for oceanic crust at mid-ocean ridges, ^{87}Sr/^{86}Sr is 0.709. Sea water Sr composition is a function of its interaction with these rocks.

Sr is supplied to the oceans by sources that display characteristic values of ^{87}Sr/^{86}Sr. The varying relative influence of these sources causes changes in the ^{87}Sr/^{86}Sr in the marine environment. Carbonate precipitates, such as the shells or tests of microscopic organisms, preserve the original ^{87}Sr/^{86}Sr isotopic ratio in sea water, since they are formed in isotopic equilibrium with the ocean water. Chemical isotopic analyses of marine carbonates show a temporal variation in ^{87}Sr/^{86}Sr ratio throughout geological time.

Present-day sea water has a mean Sr concentration of 8 ppmbv. The oceanic residence time for Sr is calculated to be about 1.9×10^7 yr to 5×10^6 yr. Since the residence time for Sr in the oceans is much longer than the mixing time for the oceans (1.6×10^3 yr), this leads to thorough homogenisation of the oceanic Sr isotopic composition – this was confirmed by Burke *et al.* (1982), who analysed 42 modern marine carbonates from oceanic basins and found that those with the same age have the same Sr isotopic signatures. Present-day sea water ^{87}Sr/^{86}Sr ratio is 0.7091, corrected for inter-laboratory bias to 0.70800 for the Eimer and Amend strontium carbonate (SrCO$_3$) isotope standard. Marine carbonates of the same age display similar (if not identical) Sr isotopic ratios anywhere on the Earth's surface (see above).

Oceanic Sr values are high compared with average river water concentrations of 0.068 mg/ml. As the Sr supplied to the oceans displays characteristics of its original rock/mineral sources, measurements of ^{87}Sr/^{86}Sr, taking into account such factors as the proportion of these isotopes having changed throughout the evolution of the oceans and continental crust, provide an indication of the changing relative importance of sediments supplied from various sources. The isotopic signature of Sr in marine fossil shells, therefore, is a powerful tool for interpreting past changes in the global rates of chemical weathering and the supply of river-derived (terrigenous) sediments to the world's oceans versus other geological processes such as the production rates of juvenile oceanic crust.

Osmium isotopes

The element osmium (Os) has many chemical similarities to Sr and provides another useful proxy for geological processes such as past rates of input of terrigenous sediments into the world's oceans. The residence time of Os in the oceans is about 10^4 years – much shorter than Sr, but still significantly longer than the mixing time for ocean water. Osmium isotopic values are expressed as the ratio of ^{187}Os to ^{186}Os in a sample, i.e. ^{187}Os/^{186}Os, or the ratio of rhenium (^{187}Re) to osmium (^{186}Os), i.e. ^{187}Re/^{186}Os.

The ^{187}Os/^{186}Os and ^{187}Re/^{186}Os values fall at the Cretaceous–Tertiary (K–T) boundary event (Peucker-Ehrenbrink *et al.* 1995) – something that could be due to the injection of cosmic, more radiogenic, material to oceans at that time (see Chapter 2).

boundary event), best known because it included the demise of the dinosaurs (see Chapter 2).

The biggest known extinction event in Earth history occurred at the end of the Permian Period – about 250 Ma. Unlike at the Cretaceous–Tertiary boundary, there is no iridium anomaly that can be ascribed to a meteorite impact, nor any other evidence for a bolide impact in sediments at the Permian–Triassic boundary; therefore, a different explanation is required.

At the boundary between the Permian and Triassic periods, much of the continental land mass was welded together in equatorial to low latitudes as the supercontinent of Pangaea. Pangaea was dominated

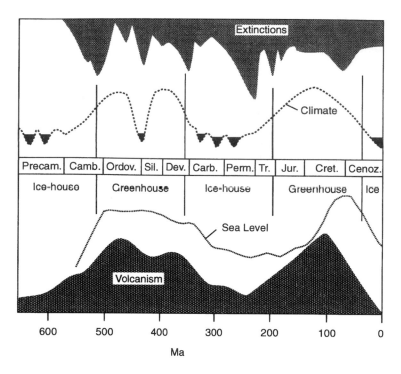

Figure 1.8 *Fischer's megacycles of extinction with global climate change, the major fluctuations in global sea level, and intensity of volcanism (modified after Fischer 1982, in van Andel 1994).*

by arid desert conditions with intense evaporation to produce extensive areas where thick accumulations of salt or evaporite minerals formed, because of the frequent evaporation and even desiccation of standing bodies of water. The amalgamation of many of the continental land masses into a single super-continent meant that the global amount of shelf-sea area available to support extensive shallow-marine and coastal ecosystems was greatly reduced. Consequently, the competition for suitable marine ecological niches by many organisms was intense, and the demand on available nutrients far exceeded that which was available. A crisis for life on Earth occurred and resulted in mass mortalities and extinctions of more than 95 per cent of all the existing species then living. The Permian–Triassic extinction event is an important example of how plate tectonics, coupled with climatic conditions, can provide an explanation for mass extinction events. Useful summaries of the Permo-Triassic extinction events and their possible cause are given by Wignall (1993) and Erwin (1994).

Other mass extinction events are known through-out the geological column, but compared with the five mass extinction events mentioned above, they were relatively small. The geological record shows that evolution is slow, at least by the yardstick of human longevity, and that environmental changes

can cause a dramatic reduction in the number of species (biodiversity), which only recovers in time spans measured in millions of years. Fossil evidence has shown that species inevitably become extinct over time and it has been estimated for example that marine invertebrate species evolve and become extinct over time intervals lasting about 10^6 to 10^7 years. Today, human activities threaten the bio-diversity, possibly on a scale that has not happened since other mass extinction events. This is illustrated in Figure 1.9, which shows the percentage of animals species known to be globally threatened in 1990 and the known causes of animal extinction since 1600. Pimm *et al.* (1995) have suggested that in well-studied but taxonomically diverse groups from widely different environments, recent extinction rates are 100 to 1,000 times greater than during their pre-human level. Even more disconcerting is their prediction that if all the species currently deemed threatened were to become extinct in the next century, the extinction rates will be 10 times greater than today's rates (*ibid.*).

The greatest threats to species have come in the relatively isolated environments such as islands and lakes. In these areas, perturbations to the environ-ment can cause rapid extinction. The South Atlantic island of St Helena was robbed of its unique plant flora in the nineteenth century because of defor-estation. The destruction of the tropical rainforests is probably the greatest crime against the diversity of species and fauna on Earth. It is estimated that about a half of the bird species have been obliter-ated from Polynesia because of hunting and the destruction of the rainforests. After the wholesale removal of large areas of rainforests, people are begin-ning to realise what damage has been done, but concerned individuals and organisations are still a long way from persuading the exploiters of the rain-forests to desist from destroying these ecosystems.

Myers (1988b, 1990) discusses the threat to bio-diversity by focusing on a series of '**hot-spot**' areas – regions with very large concentrations of species with high levels of **endemism** and which face clear threats of destruction. Myers (1990) identified eighteen hot-spots which supported 49,955 endemic plant species, 20 per cent of the total plant species identified on Earth, in an area of just 746,400 km², 0.5 per cent of the Earth's land surface. This is particularly alarming because this great number of plant species is confined to such a very small land surface area whose destruc-tion appears imminent. On a more optimistic note, however, Myers (1990) points out that if conserva-tion strategies are focused on these areas the pay-off will be considerable for relatively little effort.

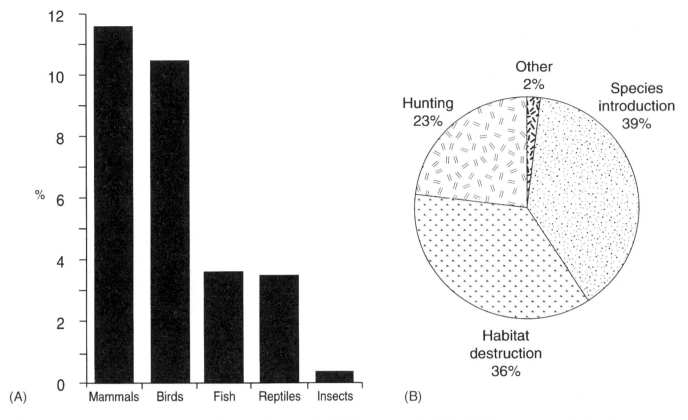

Figure 1.9 *(A) Percentage of animal species known to be globally threatened, 1990; (B) Known causes of animal extinction since 1600. Redrawn after World Resources Institute (1994–1995).*

There are scientists who believe that a potential threat to human life at some time in the future is posed by asteroids that approach the Earth regularly. Perhaps the main concern is over the irregularly shaped, 40-km-long asteroid Eros 433, which was discovered in 1898. In February 1996, NASA launched the NEAR (Near Asteroid Rendezvous) satellite probe to investigate the possibility of nudging such asteroids into orbits that could be less threatening.

Against this panoply of environmental issues, arguably the main problem is over-population of the planet. In the short term, the ingenuity of human endeavour is required to create a sustainable planet for the present and predicted near-future population levels. In the longer term, the world population must be reduced. This is not easy. Cultural, ethical, religious and socio-economic factors are inextricably interwoven into issues about maintaining biodiversity and conserving the Earth's rich variety of ecological niches.

Weather and climate

The atmosphere is a relatively thin layer of gases around the Earth, containing by volume approxi-

mately 78 per cent nitrogen, 21 per cent oxygen, 0.9 per cent argon, 0.03 per cent carbon dioxide and trace gases (Table 1.2). Since air is highly compressible, air pressure decreases with height above the Earth's surface. 50 per cent of the total mass of gases is concentrated into the lowest 5.5 km of the atmosphere whereas 100 km above the Earth's surface there is little more than 0.0000001 per cent of the gases present in the atmosphere, and the atmospheric pressure is < 0.0001 mb, as compared with atmospheric pressure at sea level, which is approximately 1,013 mb.

The atmosphere is divided into distinct horizontal layers, mainly on the basis of temperature (Figure 1.10). The evidence for this comes, for example, from radar wind-sounding balloons (RAWINSONE), radio-wave investigations, rocket flights and satellite sounding (Barry and Chorley 1992). The lower atmosphere, the **troposphere**, is heated by the surface of the earth and becomes cooler with height. The rate of change in mean air temperature with altitude is known as the **environmental lapse rate**, which is normally about 6.5°C km^{-1}. It is within the troposphere that most weather processes occur. The troposphere is capped by a layer of cold air – the

Table 1.2   *Composition of the atmosphere.*

Constituent	Chemical formula	Abundance by volume*
Nitrogen	N_2	78.08%
Oxygen	O_2	20.95%
Argon	Ar	0.93%
Water vapour	H_2O	variable (%–ppmv)
Carbon dioxide	CO_2	340 ppmv
Neon	Ne	18 ppmv
Helium	He	5 ppmv
Krypton	Kr	1 ppmv
Xenon	Xe	0.08 ppmv
Methane	CH_4	2 ppmv
Hydrogen	H_2	0.5 ppmv
Nitrous oxide	N_2O	0.3 ppmv
Carbon monoxide	CO	0.05–0.2 ppmv
Ozone	O_3	variable (0.02–10 ppmv)
Ammonia	NH_3	4 ppbv
Nitrogen dioxide	NO_2	1 ppbv
Sulphur dioxide	SO_2	1 ppbv
Hydrogen sulphide	H_2S	0.05 ppbv

* ppmv = parts per million by volume; ppbv = parts per billion by volume.
Source: Henderson-Sellers and Robinson 1986.

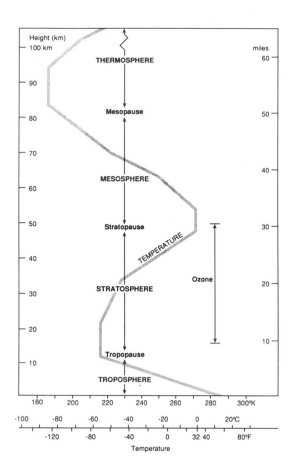

Figure 1.10   *Thermal structure of the atmosphere to a height of approximately 110 km (68 miles).*

lower **stratosphere**, which plays an important part in limiting the upper level of convection by gases within the troposphere and also acts as a ceiling on the weather.

Above the lower stratosphere **temperature inversion**, the stratosphere gradually increases in temperature upwards due to the absorption of the Sun's ultraviolet radiation by ozone and reaches a maximum temperature of about 0°C at the **stratopause**, above which is the **mesosphere**, where the temperature gradually decreases upwards to the **thermosphere**. Clouds have been observed in this zone, known as **noctilucent clouds**, and are believed to form as water vapour condenses around nuclei of meteoric dust or to be the result of increased anthropogenic atmospheric methane emissions, because they were not observed prior to the Industrial Revolution. In the thermosphere, temperatures again increase upwards because of the absorption of ultraviolet radiation by molecular and atomic oxygen. Above 100 km the atmosphere is affected by cosmic radiation, solar x-rays and ultraviolet radiation, which cause ionisation, for example to produce the **Aurora Borealis** (the northern lights) and **Aurora Australis** (the southern lights).

Weather can be thought of as a set of particular regional atmospheric conditions at any given time. Climate, however, is more difficult to define – it is the sum total of atmospheric conditions (weather) over a time period that permits a reasonable approximation of the more localised, regional, weather patterns.

A cursory study of Chapters 2 and 3 reveals that there are problems with this definition of climate. Climate has varied considerably over both short (10–10^2 years) and longer time periods (10^2–10^7 years) and the physical properties, and even the geographical position of a region and/or its topography change over geological time because of plate tectonic processes. For clarity, international convention loosely defines the climate of a region as the weather conditions experienced over a period of thirty years, usually taking climatic statistics from 1941–1970 (Lamb 1995). There are problems, however, even in using this time series (see Chapter 3). The world can be divided into different climatic regions as is shown in Figure 1.11. A particularly interesting question concerns the factors that control and drive the weather that characterises any region.

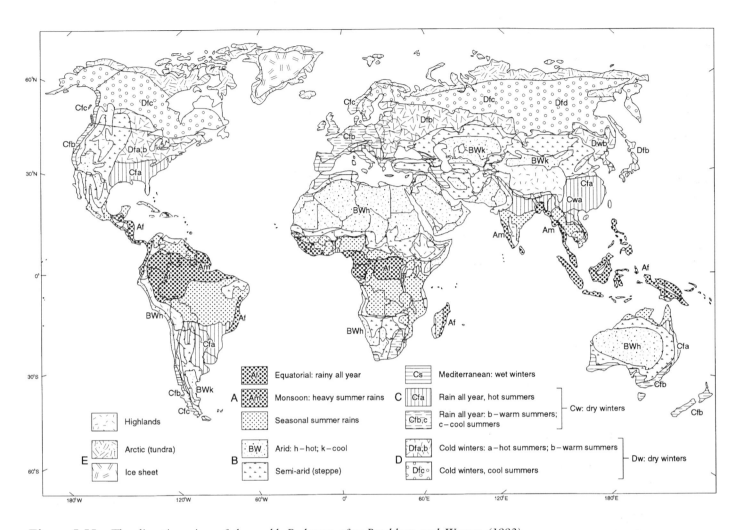

Figure 1.11 *The climatic regions of the world. Redrawn after Bradshaw and Weaver (1993).*

Ultimately, it is the Sun that is responsible for controlling weather and climate, leading to the alternate heating and cooling of the surface of the Earth, the atmosphere and the oceans. The Earth receives incoming solar radiation (**insolation**), mostly in the form of short-**wavelength** electromagnetic energy. Much of this energy is scattered, reflected and absorbed in the atmosphere. Insolation reaching the Earth's surface is absorbed, reflected or radiated back into the atmosphere as longer-wavelength electromagnetic energy, depending upon the properties of the surface. Dark surfaces, such as roads, will absorb much of the insolation, whereas light surfaces, such as snow, will reflect much of the insolation. This property of characteristic reflectivity is known as the **albedo**. Additionally, much of the insolation and radiated energy will be absorbed by greenhouse gases such as CO_2 and water vapour (see Chapter 3, which deals with the greenhouse effect).

The amount of insolation received at the Earth's surface will also vary as a function of latitude. Insolation can be likened to a series of approximately parallel rays of energy hitting a sphere (the Earth) from the Sun. The absorption of energy will be greatest at those surfaces perpendicular to the Sun's rays. This is because the energy per unit surface area is greater, whereas energy received at surfaces which are inclined at an acute angle to the incidence of the Sun's rays will have that energy distributed over a larger area – the energy per unit area will be less. From pole to equator, from high to low latitude, the Earth's surface becomes more perpendicular to the direction of travel of insolation and the total energy per unit area increases. Consequently, polar and mid-latitude regions receive less insolation perunit area than the tropical and equatorial regions and are thus generally cooler.

Variations in solar insolation values with latitude, and other geographical conditions, including alti-

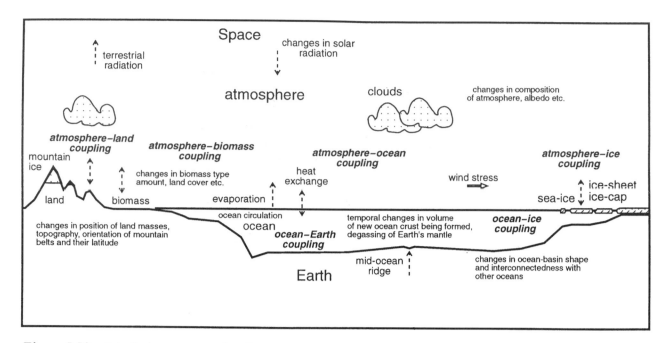

Figure 1.12 *Principal components of a climate system.*

tude, topography and the orientation of slopes, from north- to south-facing, all conspire to create differential warming and cooling of the Earth's surface. These temperature differences affect atmospheric pressure, so that in colder regions higher atmospheric pressures develop. **Pressure gradients** induce air to move from regions of high to low pressure, thereby creating winds. The rotation of the Earth deflects and accelerates the winds in a clockwise direction in the Northern Hemisphere and anticlockwise in the Southern Hemisphere. This deflecting force is known as the **Coriolis effect**. Ultimately, the winds will be deflected so that they tend to travel perpendicular to the pressure gradient, following lines of equal pressure, where they are known as **geostrophic winds** – some travel at velocities in excess of several hundred km h⁻¹. Winds will also be deflected or blocked by topography or other pressure systems. There are many other factors that influence weather and climate systems, such as the complex coupling between the atmosphere, oceans, biomass, land and even tectonic processes. Figure 1.12 attempts to illustrate these principal components and their interaction.

In order to characterise the weather and climate it is necessary to examine the world's wind circulation (Figure 1.13). In the mid-eighteenth century, George Hadley proposed a model for the general circulation of the atmosphere in which global atmospheric circulation was compared with a convective system in which air is warmed at the equator and cools at the poles. The warm air at the equator

becomes buoyant and rises vertically, and as it rises it cools and descends northwards and southwards away from the rising air at the equator. Eventually the cold dense air returns from the polar regions towards the equator. This model was eventually replaced by a three-cell model, in which the cells were modified because of the rotation of the Earth, i.e. the Coriolis effect. In this refined model the tropical cells were named after Hadley (southern and northern **Hadley cells**).

The Hadley cells meet at the **intertropical convergence zone** (ITCZ), which forms an irregular belt around the Earth and migrates seasonally between the latitudes of the two tropics. As the amount of atmospheric data increased in the late 1940s and 1950s it became increasingly apparent that a three-cell model was far too simple to explain the observed atmospheric circulation. According to the three-cell model the upper airflow in mid-latitudes should have been easterly, but observations showed that the winds are predominantly westerly. These **circumpolar westerlies** circle the poles. It became apparent that energy transfer in mid-latitude regions is dominated by horizontal cells rather than vertical cells. In the lower atmosphere this involves the development of low- and high-pressure systems, **cyclones** or **depressions** and **anticyclones**, respectively, and in the upper atmosphere there are wave-like wind patterns, which are described as **Rossby waves**. The Rossby waves meander, with the meander loops varying in amplitude such that where large-

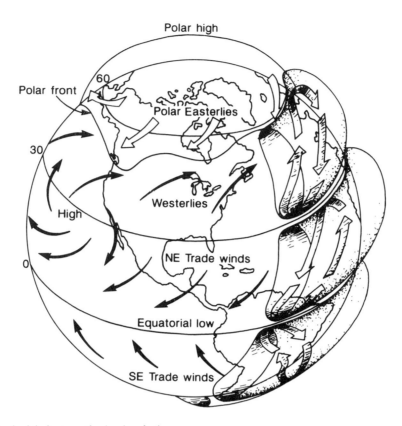

Figure 1.13 *Idealised global atmospheric circulation.*

amplitude waves form, this increases the migration of cold air into low latitudes and, conversely, warm air moves into high latitudes. The modern interpretation of global atmospheric circulation has retained the Hadley cells, but horizontal eddies are seen as dominating mid-latitudes and have even replaced the simple thermal cell of the polar latitudes.

Within the broad belts of the upper atmosphere flow, there are narrow bands of rapidly moving air known as **jetstreams**. These travel at speeds in excess of 160 km h⁻¹ and are associated with the zone of steep temperature and pressure gradients at the tropopause. The influence of these jetstreams extends into the lower atmosphere, affecting weather conditions. These include the polar-front jetstream, the subtropical jetstream above the subtropical high-pressure zone, and the tropical easterly jetstream above the ITCZ. Other weather systems such as **tropical cyclones** (see Chapter 8), the **El Niño southern oscillation** (see Chapter 3) and **monsoons** complicate the wind patterns and exchange of heat.

Given a basic understanding of atmospheric processes, it becomes easier to appreciate the classification of world climates as shown in Figure 1.11, since it is based on the variation of temperature and

precipitation under a given dominant atmospheric system. The similarity in distribution to vegetation in Figure 1.7A illustrates the importance of climate in controlling the distribution of biota. In addition, climate is also important in determining many other factors, such as the dynamics of geomorphological and soil-forming processes, and the hydrological regime within an area.

Hydrosphere

Most water is present on the Earth as the oceans and seas (97.41 per cent of all surface water volume), as rivers, lakes, within soil, animals and plants, and in the atmosphere as water vapour (0.014 per cent). The remaining water is stored as ice within the ice sheets and glaciers, and as ground water (2.576 per cent). Figure 1.14 illustrates the ways in which water is transferred continuously between these main components by evaporation, vapour transport, precipitation and flow across the surface of the land in what is termed the hydrological cycle.

Water is precipitated from the atmosphere as rain or snow, falling on the land and the oceans. Some of

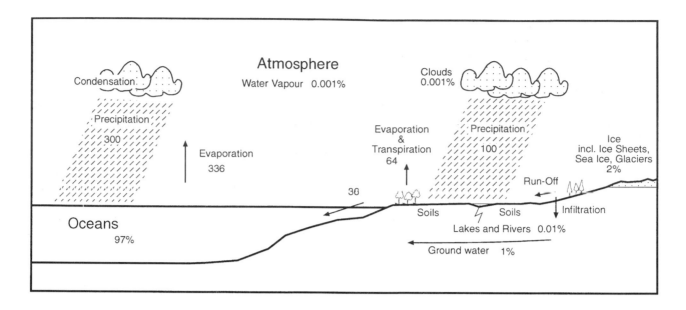

Figure 1.14   *The hydrological cycle, showing the movement of water through the atmosphere, lithosphere, hydrosphere and biosphere.*

this water will percolate into the soil and bedrock and flow as ground water, often towards the sea, and some of the water will flow via rivers to lakes and oceans. The amount of precipitation is counterbalanced by evaporation of water from the seas and lakes, and from the soil by direct evaporation or drawn up by plants and then released during **transpiration**. Once the water has been converted by evaporation into atmospheric water vapour it can then condense and returns to the Earth in the form of precipitation.

Many factors control the various components of the hydrological cycle, including the dynamics and variation in weather systems, the characteristic water storage and removal by biomes, the geology of a region, e.g. the type of **aquifer** present, the soil, and the type of surface drainage. In many regions, human activities have greatly modified the various components of the hydrological cycle, such as the withdrawal of ground water, the channelisation of rivers, the alteration of vegetational patterns and soils, changing the rates of percolation and evapo-transpiration, and also through modifying climate. Aspects of land use are discussed in depth in Chapter 9. Also, Chapter 5 deals with ways in which human activities have polluted water resources.

Interdependence

Interdependence involves the complex interaction between organisms in the biosphere and the in-

animate world, the lithosphere, atmosphere and hydrosphere. As an illustration of interdependence the following ostensibly simple system provides an insight into the inextricably interwoven aspects of the Earth's surface.

A plant such as a poppy growing in a field will anchor itself into the soil layers, a part of the lithosphere, by using its root system. The poppy will obtain most of its essential nutrients from minerals in the soil that have been derived from the chemical and physical weathering of rocks. At the same time, the poppy obtains carbon dioxide (CO_2) from the atmosphere to build up carbohydrates to form tissue. It obtains the water necessary for life from the ground water in the soil, and a very small amount directly from precipitation, which are part of the hydrosphere. When the poppy reproduces, atmospheric processes such as the wind help disperse the seeds and so facilitate propagation.

In life, the poppy is an integral part of the organic layer, the biosphere, yet at the same time it is in all four spheres, as part of the ecosphere. The poppy will also contribute to the atmosphere by producing oxygen during the process of photosynthesis. If, during life, one of the spheres is severely altered, for example, the hydrosphere becomes depleted of water or the soil (part of the lithosphere) becomes depleted of vital nutrients, then the plant wilts and dies. Upon death and decay the poppy remains in the biosphere during bacterial degradation and becomes part of the lithosphere through the addition of new nutrients to the soil, and it may even become fossilised in a rock to form a fossil fuel, such as **lignite** or coal.

From such extreme scenarios, the so-called domino effect of one deleterious action fuelling another can be appreciated. Many scientists refer to our planet as being in a state of delicate balance. If this 'balance of nature' is upset by altering the inputs to the natural systems, the consequences or outputs may be detrimental to many other dependent organisms, including human beings.

The Earth can be visualised as a system with many inputs and outputs. The systems concept was originally developed by the biologist Ludwig von Bertalanffy in the 1920s, and was later adopted in 1949 by the new science of **cybernetics**. Some of the outputs may become inputs again, that is they feed back into the system. These inputs may further enlarge or decrease the output, which in turn may feed back into the system again, and so on. Where the original effect is magnified or reduced, such loops are called **positive feedbacks**. When a feedback results in stability in the output, the feedback is said to have a countervailing effect as a **negative feedback**. Negative and positive feedback mechanisms are very important in understanding how the Earth's natural systems work.

All environmental systems are both open and in **dynamic equilibrium**, that is, there is an input of energy and matter and a corresponding output of energy and matter, which are in some way balanced. This balance is controlled by negative feedback mechanisms. Environmental systems are commonly resistant to positive feedback, which is evident by the time delay between the input and output or response. When positive feedbacks take effect, the response is usually in the form of major environmental change. Over geological time, the major stimulus causing positive feedbacks in the ocean–atmosphere system has been global climate change – many examples will be considered throughout this book.

The cycle of nutrients throughout the Earth's surface environments is essential to any consideration of global systems and feedbacks. There are six major nutrients that are essential for life – carbon, hydrogen, oxygen, nitrogen, phosphorus and sulphur. Figure 1.15 illustrates three of these cycles, showing the storage and transport mechanisms that distribute these chemical elements throughout the Earth. The quantities of each nutrient that are stored and being transported around the Earth are currently being altered by human activities, for example the increased emissions of carbon dioxide from the combustion of fossil fuels and deforestation have significantly altered the sources and sinks in the carbon cycle.

Thresholds

Most natural processes or events require a certain activation energy, which is commonly referred to as a **threshold**. A pain threshold, for example, is familiar to everyone. People with higher pain thresholds will tend to tolerate discomfiture longer than those with lower thresholds. There are many examples of natural thresholds where the consequences of some input process makes a sudden, abrupt, change of output. Though very important, thresholds for many events are not known or are poorly understood.

As an example, the discovery in 1985 of a significant depletion or 'hole' in the **stratospheric ozone layer** over Antarctica provides an illustration of the fact that the depletion of ozone has now crossed some sort of threshold set of conditions that had previously maintained a continuous ozone layer over this region. Changed environmental conditions precipitated by human activities should fuel a strong curiosity to discover which other thresholds in the ocean–atmosphere system may easily be reached to the detriment of the natural environment in order that preventive or ameliorative action can be taken. Myers (1995a) emphasised that the most important future environmental problems could be those that are still unknown to us – so-called environmental surprises. Myers describes one set of surprises, **environmental discontinuities**, as the result of ecological systems jumping over some threshold condition/s. Another set of environmental surprises, **synergisms**, result from two or more environmental processes interacting in such a way that the outcome is not simply additive but multiplicative. Clearly, it is important not only to supply solutions to environmental problems but also to raise appropriate new questions in an attempt to anticipate possible environmental surprises. Global warming could trigger a cascade of natural hazard effects, both directly though the meteorological processes associated with any climate change, and indirectly because of rising sea level (Figure 3.26 is a flow chart summarising the potential natural hazards).

Earth – a self-regulating organism

In the late 1960s, James Lovelock developed a hypothesis which he called the **Gaia Hypothesis**, named after the Greek word for the Earth Goddess. Lovelock and his colleagues suggested that the Earth is a self-regulating system, that is, one able to maintain the climate, atmospheric conditions, soil and ocean composition in a stable balance favourable to

(A)

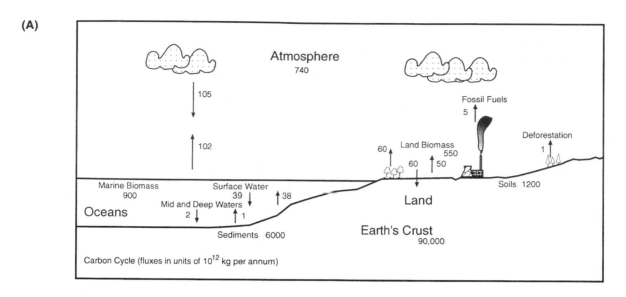

(B)

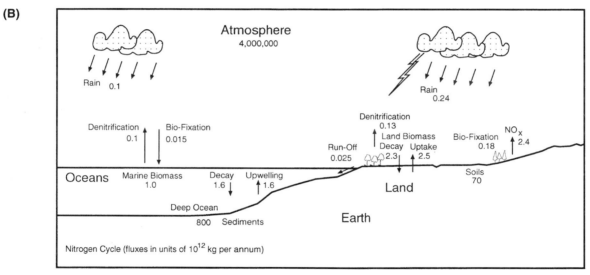

(C)

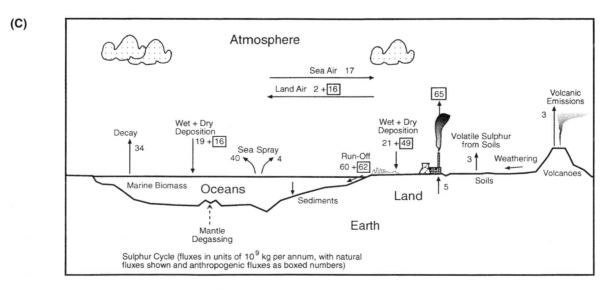

Figure 1.15   *Nutrient cycles for (A) carbon; (B) nitrogen; and (C) sulphur.*

life. In this hypothesis the inputs and outputs are perceived as delicately balanced and controlled by feedback mechanisms that maintain stability.

The Gaia Hypothesis proposes that the Earth's organisms collectively have an innate ability to self-regulate the external environmental conditions necessary to support and sustain life. The Gaia Hypothesis also explains the extinction of species as a consequence of their inability to continue to maintain the equilibrium conditions necessary to support life – the balance of Gaia.

The Gaia Hypothesis seeks to explain why the surface temperature of the Earth has remained relatively constant over the last four billion years since life first emerged from the primordial organic soups and gases of the planet, despite the fact that the Sun's heat has increased by about 25 per cent. Over the same period, the overall carbon dioxide level has dropped, reducing the heat-holding potential of the Earth. In the Gaia Hypothesis, these changes in the balance of gases in the atmosphere are explained exclusively as a consequence of biological activity, the fixation of carbon dioxide from the atmosphere by photosynthesising organic matter. Geological evidence, however, suggests that the amount of oxygen has remained essentially constant over the past 200 million years as a result of the balance of the complex interactions of organisms *and* the inorganic components of Earth, such as volcanic activity, etc. Lovelock and his colleagues believe that if human activity continues to disturb the **geosphere**, by disturbing the natural balance of Gaia, and if human activities are not harmonised with the natural processes of Gaia, then this life-support machine will no longer sustain *Homo sapiens* and therefore extinction will be inevitable. A new species will then evolve to occupy the vacated ecological niche.

To help illustrate the Gaia Hypothesis, Watson and Lovelock (1983) developed the 'daisy world model'. They imagined a world inhabited only by black and white daisies. In this scenario, as the Sun heated up, a lifeless world also warmed up because of the greater heat energy being emitted from the Sun. In daisy world, the black daisies absorbed more of the incoming solar radiation, and were thus favoured over the white daisies because of their more suitable survival strategy, at least during the early days of the faint Sun. As the Sun continued to heat up, however, the black daisies became less suited to the warmer world, and then the white daisies began to compete more successfully since they could maintain a better temperature balance brought about by their ability to reflect more sunlight – using a negative feedback to help cool the Earth's surface. In such a changed world, the white daisies could become more abundant than the black daisies. Eventually, the Sun would become so bright that all the daisies would die, unable to reflect the large amounts of solar radiation reaching the surface of the planet.

This simple daisy world model shows how evolving life on Earth could have modified global climate through both negative and positive feedback mechanisms. Lovelock and his fellow workers suggest that similar processes took place on Earth throughout geological time, and that the Earth will continue to regulate itself if human activities do not cause changes in global climate that are faster than any natural negative feedbacks that might operate to maintain the habitability of this planet.

An understanding of the way in which the Earth maintains overall global climatic stability is aided by considering the concept of **self-regulating mechanisms**, commonly referred to as negative feedback mechanisms. Some elementary chemistry is required in order to appreciate the long-term stability in atmospheric concentrations of CO_2, that is over time intervals greater than about 100,000 years, which is the residence time of carbon in the oceans – the time taken for an 'average' carbon atom introduced into the oceans to be removed, for example by being locked into a rock such as limestone. The long-term control of atmospheric CO_2 involves the 'carbonate–silicate geochemical cycle', which is a measure of the way in which gaseous CO_2 exchanges with CO_2 contained in carbonate rocks. The last mechanism involves the chemical weathering of silicate minerals and the accumulation of carbonates. Gaseous CO_2 is returned to the atmosphere as silicate minerals are formed or as carbonates are metamorphosed (subjected to intense heat and pressure) to release CO_2.

Silicate weathering depends on temperature because warmer conditions encourage the chemical reactions that break down the silicates. The rate of chemical weathering is increased with greater rainfall, also strongly influenced by surface temperature. A decrease of temperature at the Earth's surface should be accompanied by a reduced rate of silicate weathering, which in turn should induce an increased atmospheric concentration of CO_2 and an accompanying increase in surface temperature due to the greenhouse effect. The net result of such a scenario is that these feedback mechanisms cause the atmospheric CO_2 levels and climate to act as a self-regulating system.

Applying the same logic to the converse scenario, that is an increase in the surface temperature of the Earth, then the rate of silicate weathering increases and the removal of CO_2 from the atmosphere should cause a reduction in the Earth's surface temperature. This is the essence of an important self-regulating mechanism in the atmosphere brought about by negative feedbacks involving the greenhouse gas CO_2 and the greenhouse effect.

In some cases, diametrically opposed feedback mechanisms are proposed, for example there is disagreement over the possible implications for global climate change precipitated by a thawing of the entire Arctic permafrost. In such a scenario, some scientists postulate that the thawing will release CH_4 and CO_2 in sufficient quantities to make a substantial contribution to global warming. The contrary view invokes a negative feedback where under a warmer global climate with greater concentrations of atmospheric CO_2, there would be enhanced tree and other vegetational growth, which would act as a brake on or limit global warming. More research is needed in order to understand which sequence of events is more likely.

The rainforests provide an example of the importance that the world's flora play in the regional water balance and the distribution of clouds. Trees may even control the rate and timing of cloud nucleation by emitting a variety of cloud condensation nuclei to help produce local convective systems that may be as much as 5 km in diameter. The destruction of the rainforests could cause major perturbations in the global weather systems by disrupting regional and then global water balance, and by increasing atmospheric levels of CO_2. In recognition of the central role played by the rainforests in global climate, many environmentalists advocate increased afforestation as a way of sequestering the increased anthropogenic emissions of CO_2. Some scientists maintain that biological control of equilibrium conditions is actually far more important than the inorganic chemical reactions in maintaining an equable cocktail of atmospheric gases. Perhaps the best known and most ardent proponent of such a viewpoint is Lovelock, in his book *The Ages of Gaia*.

Not all scientists agree with the Gaia Hypothesis. A contrary view is that the Earth's atmosphere has evolved by chance chemical reactions and degassing from the mantle. Lovelock argues, however, that this and the traditional evolutionary theories (both Darwinian and **punctuated evolution**) are inadequate, because they invoke a passive role for **biota** throughout Earth history. Lovelock believes that the biota played, and continue to play, an active role in controlling their environment.

Today, the consensus of scientific opinion lies somewhere between a Gaian perspective and an appreciation of non-biological, often random, processes that collectively maintain the self-regulation of global climate through both negative and positive feedback mechanisms. A central problem with the Gaia Hypothesis is that it is untestable. The 'experiment' that has been run over about 4.7 Ga to create the world and life as it exists is a unique one-way sequence of chemical reactions. The Gaia Hypothesis cannot be used to predict specific future changes. These criticisms mean that it cannot become accepted as a theory, but remains a series of interesting speculations. The hypothesis, however, provides an interesting perspective on life on Earth, and a set of ideas for active debate amongst those concerned with the environment.

Long-term climatic stability?

Throughout much of the Earth's history global climate has shown a long-term stability. The 3,800-million-year-old metamorphosed sedimentary rocks at Isua, west Greenland, show that liquid water has existed on the Earth's surface at least from that time (Kasting 1989).

The concept of climatic stability merely carries the connotation of the continued presence of liquid water on the Earth's surface and the continued presence of life. Periods of substantial global cooling and major ice ages, and other times when the mean surface temperature was much warmer than today, can be traced back through the geological history of the Earth. In both extremes, however, liquid water existed as oceans, rivers and lakes, and life was sustained.

Throughout Earth history – geological time – the amount of solar energy flux reaching the Earth's surface has increased by about 25 per cent, and this has been associated with temperature changes, changes in the atmospheric proportion of gases, and the evolution of life (see Figure 1.16). Despite these changes during the past 540 million years, since the inception of organisms developing hard parts that have become fossilised as shells, tests, etc., oxygen and carbon dioxide levels have remained relatively constant. Also, the surface of the Earth, which it might seem reasonable to assume should be getting warmer, in fact has remained essentially constant. Any potential increase in the Earth's surface temperature has been offset by a decrease in the concentration of atmospheric greenhouse gases, particularly CO_2, by negative feedback mechanisms (see above section). Studies of theoretical changes in the amount of solar energy reaching the Earth, using reasonable ranges

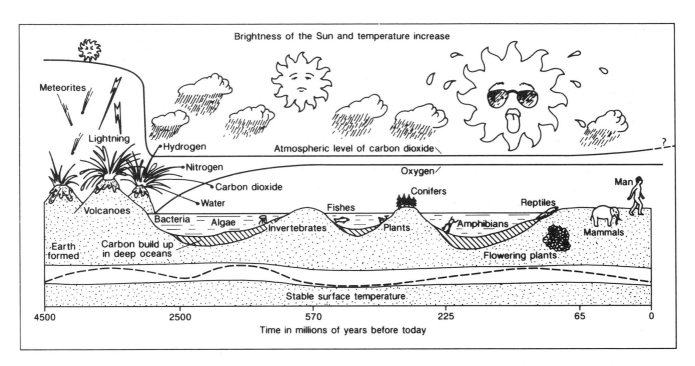

Figure 1.16 *Schematic development of the Earth's atmosphere and life, and the variability in the solar brightness and relative temperature throughout the Earth's history. Redrawn after Watson (1991).*

of values, suggest that such solar variability could not compete with the anthropogenic greenhouse gases as a more significant cause of global warming (Hansen and Lacis 1990). Thus, if the Sun were to radiate slightly less heat energy to the Earth, within the range of possible natural fluctuations, then the cooling that could result would be insufficient to offset the overall warming effect caused by anthropogenically generated emissions of greenhouse gases.

Simple energy-balance calculations (using the climate models referred to as the Budyko/Sellers type, published in 1969) predict that only a 2–5 per cent decrease in solar output could result in a runaway glaciation on Earth, yet solar fluxes 25–30 per cent lower early in the Earth's history (Gough 1981) apparently did not produce such an effect (Caldeira and Kasting 1992). A favoured explanation to circumvent this paradox is that the partial pressure of CO_2, as a result of higher rates of volcanic degassing, possibly associated with slower rates of silicate weathering in rocks, generated a large enough greenhouse effect to keep the Earth warm. Caldeira and Kasting (1992), however, argue that the oceans can freeze to form sea ice much more rapidly (< 1 year) than the rate at which CO_2 can accumulate in the atmosphere (> 10^5 years); therefore if such a transient global glaciation had occurred in the past

when solar luminosity was low, it may have been irreversible because of the formation of highly reflective CO_2 clouds. Had such a scenario occurred, argue Caldeira and Kasting, then the Earth might not be habitable today if it had not been warm during the first part of its history. As mentioned above, the presence of sedimentary rocks from Greenland shows that liquid water was present on the Earth's surface as early as 3.8 Ga, when solar luminosity was as much as 25 per cent less than at present. Large amounts of atmospheric ammonia (NH_3) and CO_2 could account for the warmer climate of the Earth back in the so-called **Archaean eon**.

Time and rates of change

The rates at which processes take place must be considered along with the magnitude and frequency of events. Many of the processes and events that shape the Earth and influence environmental change, such as global climate change, take place on time spans that are far outside the experience of a single human lifetime, for most a mere 70 or so years. Time in terms of the Earth history, or geological time, is measured typically in thousands of millions to millions of years.

Eon	Era	Period	Epoch	Million years before present	Geological events	Sea life	Land life
Phanerozoic	Cenozoic	Quaternary	Holocene	0.01	Glaciers recede. Sea level rises. Climate becomes more equable.	As now.	Forests flourish again. Humans acquire agriculture and technology.
			Pleistocene	2.0	Widespread glaciers melt periodically causing seas to rise and fall.	As now.	Many plant forms perish. Small mammals abundant. Primitive humans established.
		Tertiary	Pliocene	5.1	Continents and oceans adopting their present form. Present climatic distribution established. Ice caps develop.	Giant sharks extinct. Many fish varieties.	Some plants and mammals die out. Primates flourish.
			Miocene	24.6	Seas recede further. European and Asian land masses join. Heavy rain causes massive erosion. Red Sea opens.	Bony fish common. Giant sharks.	Grasses widespread. Grazing mammals become common.
			Oligocene	38.0	Seas recede. Extensive movements of Earth's crust produce new mountains (eg Alpine–Himalayan chain).	Crabs, mussels, and snails evolve.	Forests diminish. Grasses appear. Pachyderms, canines, and felines develop.
			Eocene	54.9	Mountain formation continues. Glaciers common in high mountain ranges. Greenland separates. Australia separates.	Whales adapt to sea.	Large tropical jungles. Primitive forms of modern mammals established.
			Paleocene	65	Widespread subsidence of land. Seas advance again. Considerable volcanic activity. Europe emerges.	Many reptiles become extinct.	Flowering plants widespread. First primates. Giant reptiles extinct.
	Mesozoic	Cretaceous	Late Early	97.5 / 144	Swamps widespread. Massive alluvial deposition. Continuing limestone formation. S. America separates from Africa. India, African and Antarctica separate.	Turtles, rays, and now-common fish appear.	Flowering plants established. Dinosaurs become extinct.
		Jurassic	Malm / Dogger / Lias	163 / 188 / 213	Seas advance. Much river formation. High mountains eroded. Limestone formation. N. America separates from Africa. Central Atlantic begins to open.	Reptiles dominant.	Early flowers. Dinosaurs dominant. Mammals still primitive. First birds.
		Triassic	Late Middle	231	Desert conditions widespread. Hot climate slowly becomes warm and wet.	Ichthyosaurs, flying fish, and crustaceans appear.	Ferns and conifers thrive. First mammals, dinosaurs, and flies.
			Early	243 / 248	Break up of Pangaea into supercontinents Gondwana (S) and Laurasia (N).		
	Paleozoic	Permian	Late Early	258 / 286	Some sea areas cut off to form lakes. Earth movements form mountains. Glaciation in southern hemisphere.	Some shelled fish become extinct.	Deciduous plants. Reptiles dominant. Many insect varieties.
		Carboniferous	Pennsylvanian / Mississippian	320 / 360	Sea-beds rise to form new land areas. Enormous swamps. Partly-rotted vegetation forms coal.	Amphibians and sharks abundant.	Extensive evergreen forests. Reptiles breed on land. Some insects develop wings.
		Devonian	Late / Middle / Early	374 / 387 / 408	Collision of continents causing mountain formation (Appalachians, Caledonides, and Urals). Sea deeper but narrower. Climatic zones forming. Iapetus ocean closed.	Fish abundant. Primitive sharks. First amphibians.	Leafy plants. Some invertebrates adapt to land. First insects.
		Silurian	Pridoli / Ludlow / Wenlock / Llandovery	414 / 421 / 428 / 438	New mountain ranges form. Sea level varies periodically. Extensive shallow sea over the Sahara.	Large vertebrates.	First leafless land plants.
		Ordovician	Ashgill / Caradoc / Llandeilo / Llanvirn / Arenig / Tremadoc	448 / 458 / 468 / 478 / 488 / 505	Shore lines still quite variable. Increasing sedimentation. Europe and N. America moving together.	First vertebrates. Coral reefs develop.	None.
		Cambrian	Merioneth / St. David's / Caerfai	525 / 540 / 590	Much volcanic activity, and long periods of marine sedimentation.	Shelled invertebrates. Trilobites.	None.
	Precambrian	Vendian		650	Shallow seas advance and retreat over land areas. Atmosphere uniformly warm.	Seaweed. Algae and invertebrates.	None.
		Riphean	Late / Middle / Early	900 / 1300	Intense deformation and metamorphism.	Earliest marine life and fossils.	None.
			Early Proterozoic	2500	Shallow shelf seas. Formation of carbonate sediments and 'red beds'.	First appearance of stromatolites.	None.
Archaean		Archaean (Azoic)		4600	Banded iron formations. Formation of the Earth's crust and oceans.	None.	None.

Geologists and cosmologists believe the age of the Earth is about 4,700 million years. The first bipedal **hominid** (*Australopithecus afarensis*) evolved about 3.75 million years BP, and *Homo sapiens* about 300,000–400,000 years ago (although Peking Man at 400,000 years BP was *H. erectus*) while true modern humans (*H. sapiens sapiens*) have been in existence for only about 40,000 years BP. In order to make sense of the history of the Earth, geologists divide time into a number of geological periods, mainly defined by global events that have had a profound effect on the biota during the past 540 million years (the Phanerozoic) but using essentially abiotic chemical and physical changes prior to this – in the **Precambrian** (Figure 1.17). The present period, for example, is called the **Quaternary** (see Chapter 2), with a beginning defined by evidence to suggest that it marks the start of the last major and abrupt global cooling at about 2.5 million years ago (2.5 Ma).

To appreciate the enormous extent of geological time, imagine that the entire Earth history is represented by a single calendar year – an analogy used by Stephen Gould in his book *Wonderful Life*, 'By the turn of the last century, we knew that the earth had endured for four billion years, and that human existence occupied but the last geological millimicro-second of this history – the last inch of the cosmic mile, or the last second of the geological year, in our standard pedagogical metaphors.'

Chaos theory: the unpredictability of events

In recent years the mathematics of **chaos theory** has been applied to many aspects of the Earth's natural systems. Many scientists believe that systems such as weather patterns and oceanic circulation contain inherently chaotic motions, for example associated with atmospheric and oceanic turbulence. There are regions of unpredictability with complex boundary conditions, but these may be contained within and/or be adjacent to regions with a good degree of predictability. In other words, whilst a system may be reasonably predictable at one scale many of its component parts may be very difficult to predict, or simply appear chaotic. Chaos theory has attracted considerable public interest because it attempts to explain in an aesthetically pleasing way the relation-

ships between randomness and predictability (orderliness), complexity and simplicity, and it has been widely applied to practical everyday experiences rather than merely mathematical abstractions. As a contrast to chaos theory, Newton's laws of motion epitomise classical scientific determinism, where the future is uniquely determined by the past.

Chaos theory endeavours to explain why determinism does not necessarily imply predictability. The minimal condition for the applicability of chaos theory to a situation is that the controlling equations must be 'non-linear' – otherwise the system is too simple for chaotic conditions.

Earth scientists need to be able to examine and appreciate the world at different scales, quite simply because humans exist and make use of and are influenced by natural and artificial processes at a variety of scales. Systems require an explanation in terms of the chemical reactions that take place on an atomic scale, and these need to be related to the effects on larger, mesoscopic to macroscopic scales. Scientists also need to understand just how well scale models of processes accurately reflect and mimic larger-scale phenomena, and what the potential amplifying consequences may be of a seemingly insignificant initial event – in other words the sensitivity of a system. Ultimately, there is the need to appreciate the consequences of the sum of small-scale processes on a global scale. The relatively new study of **fractal geometry**, the study of scale-invariant processes, is fast becoming a potential means of doing this.

The destruction of the ozone layer, a protective gaseous layer in the atmosphere that shields the Earth's surface from the harmful effects of the Sun's radiation, is a good example of the range of scales at which scientists can see processes operating that are interrelated. The chemical reactions that lead to the destruction of this protective ozone layer take place on the atomic scale, as compounds such as **chlorofluorocarbons** (**CFCs**) combine with ozone to break it down. This leads to regional effects such as the depletion of ozone over the Antarctic during the spring and late summer, which in turn, allows more radiation to reach the surface of the Earth. Radiation at the short-wavelength part of the **electromagnetic spectrum** can be harmful to organisms, and may destroy animals and plants, especially important bacteria, and can cause mutations and cancers. The equations that govern the rate of chemical reactions are non-linear, depending upon such factors

---

Figure 1.17 *The geological time scale. Time is divided up in a way that reflects the major events in the evolution and/or extinction of species of animals and plants. Redrawn after Harland* et al. *(1990).*

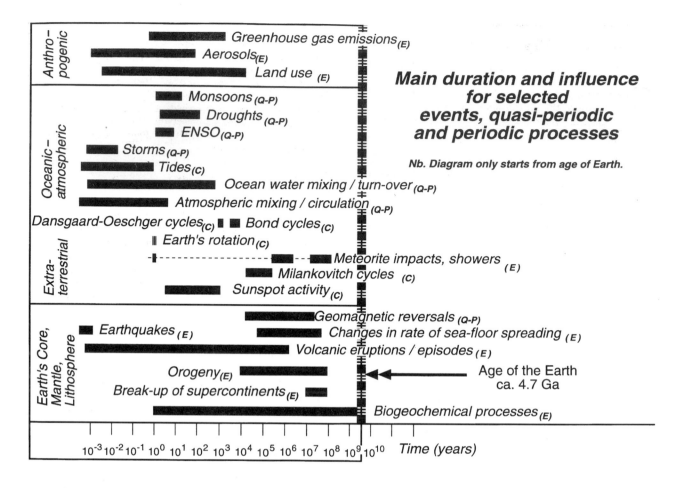

Figure 1.18 *Processes involved in environmental change, especially global climate change, and their time scales.*

as the concentration of the reactants (typically raised to some power, e.g. the square of the concentration where two molecules are involved), the concentration of the reaction products, temperature, pressure and the presence of catalysts. Chaos theory, therefore, may be applied to chemical processes such as stratospheric ozone destruction, or the chemical effects of the anthropogenic emission of **greenhouse gases**.

Figure 1.18 summarises the main processes involved in environmental change and their scales, providing a useful overview and reference point for much of what is discussed throughout this book. Whilst understanding the causes and effects of global environmental change, it is not as easy to predict the actual timing of any abrupt, and possibly catastrophic changes – due, at least in part, to the inherent chaos in any system.

Chapter 1: Key points

1 Although the Earth is one of an estimated 10^{22} planets in the Universe, it may be unique in supporting life as we know it. The outer layers of the Earth comprise the atmosphere, biosphere, hydrosphere and lithosphere. These are interrelated as the ecosphere.

2 The Earth's interior comprises the inner and outer core, upper and lower mantle, and crust. The upper part of the mantle and crust, the lithosphere, can be considered as essentially rigid and resting on the asthenosphere, which has the capacity to flow. Plate tectonic theory explains the movement of rigid/semi-rigid lithospheric continental plates over the extremely low strain-rate upper mantle; the construction of oceanic plates at oceanic spreading centres; and their destruction by subduction back into the mantle at subduction zones. It can account for many Earth surface features, earthquake zones and volcanoes. Plate tectonic theory, however, is not adequate for explaining some aspects of the mechanical behaviour of continental crust, which can be considered as behaving not strictly as rigid plates but as extremely viscous fluids. The models adopted to explain the mechanical behaviour of continental crust, fluid versus solid, depend very much upon the features and processes that are being evaluated.

3 On Earth, biological diversity – biodiversity – is enormous and is sustained by energy from the Sun and the Earth's internal energy systems. Organisms inhabit particular ecological niches, biomes and biotic provinces, which are dominated by the climatic conditions, nutrient supply and competition for living space. Biodiversity is essential to maintaining the stability of ecosystems and biogeochemical cycles. Biological activity plays a critical role in releasing free O_2 to the atmosphere and hydrosphere by oxygenic photosynthesis. Evidence from chemical isotopes in sedimentary rocks from Isua in west Greenland suggests that life existed on Earth approximately 3,800 million years ago. A study of carbon and strontium isotopes in sedimentary rocks suggests that free O_2 started to accumulate in substantial quantities in the Earth's atmosphere about 2,000 million years ago as oxygen-deficient (anoxic) basins began to form, which allowed organic carbon to be buried. Prior to this, oxygen was held in carbonate rocks as the so-called 'carbonate reservoir'. Oxygenic photosynthesis took place at least 600 million years ago and provided a mechanism capable of sustaining atmospheric free O_2 levels. The Sun provides the energy to drive photosynthesis, and the atmospheric and hydrological systems.

4 The atmosphere is divided into layers on the basis of temperature. Most weather processes are restricted to the lowermost layer, the troposphere. The overlying stratosphere has an important part to play in maintaining the habitability of this planet, for example in the ozone layer. The world can be divided into climatic regions, which are influenced and partially defined by the general atmospheric circulation, and are driven by differential heating and cooling of the Earth's land surface, oceans and atmosphere, and the Earth's rotation.

5 The hydrological cycle involves the storage and transfer of water throughout the world by hydrological, atmospheric, biological and geological processes.

6 A 'systems approach' allows the various components of the ecosphere to be studied independently, from which it is possible to appreciate both 'negative' and 'positive' feedbacks. The Gaia Hypothesis describes the Earth as a self-regulating organism, able to sustain itself in equilibrium with any major long-term environmental changes, thereby maintaining an optimum global climate conducive to survival by the successful organisms. Processes and events within these 'spheres' or systems may change from one level or condition to another when an input has reached a 'critical threshold'. Chaos theory proposes that natural systems have at least some components that are fundamentally unpredictable.

Chapter 1: Further reading

Botkin, D. and Keller, E. 1995. *Environmental Science: Earth as a Living Planet*. Chichester: John Wiley & Sons Ltd, 627 pp.
A colourful and well-illustrated introductory book on the principles of environmental science for high school and university students. The text is simple to follow, aided by case studies and explanations in boxed text, and a series of appendices. It is divided into eight sections, which include environment as an idea; Earth as a system; life and the environment; sustainable living resources; energy; water environment; air pollution; and environment and society.

Bradshaw, M. and Weaver, R. 1993. *Physical Geography: An Introduction to Earth Environments*. London: Mosby, 640 pp.
A comprehensive and well-illustrated textbook outlining the principles of Earth systems at an introductory level suitable for high school students and first year undergraduates. Atmosphere–ocean systems are described in terms of their dynamics; plate tectonics is introduced; processes of geomorphology are described; aspects of human interaction with the natural environment are discussed; and ecological systems are outlined in which there is a useful emphasis on soil dynamics and the characteristics of biomes.

Broecker, W.S. 1987. *How to Build a Habitable Planet*. Palisades, New York: Eldigio Press, 288 pp.
An extremely readable introduction to the origin and evolution of the Earth. Broecker manages to make seemingly complex scientific arguments simple and interesting. This book is highly recommended as an introductory book for both students and teachers wishing to understand some basic geochemical arguments about the Earth. Obtaining copies can be difficult.

Brown, G.C., Hawskesworth, C.J. and Wilson, R.C.L. (eds) 1992. *Understanding the Earth: a New Synthesis* Cambridge: Cambridge University Press.
This is an excellent textbook written for the British Open University. It is a compilation of authoritative chapters by Earth scientists and summarises a selection of important geological problems. It is easy to read and well illustrated, and the use of boxed text helps to highlight important points. It is essential reading for anyone studying Earth/geological sciences.

Dawkins, R. 1986. *The Blind Watchmaker*. Longman Scientific and Technical, 332 pp.
An examination of the evolution of life, which inspires the reader with a vision of existence and the elegance of biological design and complexity. Dawkins argues for the truism of Darwinian theory and shows for example how modern views such as punctuated evolution are part of neo-Darwinian theory. An excellent supplementary book for many courses in the natural sciences and environmental studies.

Gleick, J. 1987. *Chaos*. USA: Viking Press, 352 pp.
A readable account of the historical development and the elementary principles of the science of chaos.

Goudie, A. 1993. *The Nature of the Environment* (third edition). Oxford: Blackwell, 397 pp.
A comprehensive introduction to the world's natural environments. It examines the dynamics of the processes acting on the landscape and environment, past, present and future. This book integrates the study of landforms, climate, soils, hydrology, plants and animals to provide a good understanding of the nature of environments on both a global and a regional scale.

Hall, N. (ed.) 1992. *The New Scientist Guide to Chaos*. London: Penguin Books, 223 pp.
A well-written, easy-to-follow introduction to the essentials of chaos theory.

Huggett, R.J. 1995. *Geoecology: an Evolutionary Approach*. London: Routledge, 320 pp.
A useful text that examines the dynamics of geo-ecosystems. Huggett develops a simple dynamic systems model for geo-ecosystems as entities constantly responding to changes within themselves, their near-surface environments – the atmosphere, hydrosphere and lithosphere – and external influences, both geological and cosmic. This book

is highly recommended to students and teachers as supplementary reading, since it will give the reader an appreciation of the complex interdependence of animals, plants and soils, and their interaction with other terrestrial spheres.

Jackson, A.R.W. and Jackson, J.M. 1996. *Environmental Science: the Natural Environment and Human Impact*. Harlow: Longman Scientific & Technical.
This is a useful introductory text on environmental science. The first section explores the fundamental concepts of the natural environment, the interactions between the lithosphere, hydrosphere, atmosphere and biosphere. The second part looks at the environmental consequences of human activity as a result of natural resource exploitation.

Lovelock, J.E. 1988. *The Ages of Gaia: a Biography of our Living Earth*. Oxford University Press, 252 pp.
The follow-up book to *Gaia: a New Look at Life on Earth* (1982), which elaborates on the Gaia view of Earth. This book examines the interaction between the atmosphere, oceans, the Earth's crust, and the organisms that evolve and live on Earth. Lovelock discusses recent scientific developments, including those on global warming, ozone depletion, acid rain and nuclear power. This book provides a thought-provoking look at interdependence, and the role of negative and positive feedbacks in controlling the evolution and adaptability of life.

Manahan, S.E. 1993. *Fundamentals of Environmental Chemistry*. Michigan: Lewis Publishers, 844 pp.
A comprehensive and well-written textbook aimed at students having little or no background in chemistry. This book gives the fundamentals of chemistry and environmental chemistry needed for a trade, profession, or curriculum of study requiring a basic knowledge of these topics. It also serves as a general reference source. This book will appeal to those involved in college and university studies where the environmental course has a relatively strong science base, and is unlikely to appeal to those in the social sciences and geography.

Meadows, D.H., Meadows, D.L. and Randers, J. 1992. *Beyond the Limits: Confronting Global Collapse, Envisaging a Sustainable Future*. Post Mills, Vermont: Chelsea Green Publishing.

Nebel, B.J. and Wright, R.T. 1993. *Environmental Science: The Way the World Works* (fourth edition). Englewood Cliffs, New Jersey: Prentice Hall, 630 pp.
An undergraduate environmental textbook with a central theme of sustainability. There are four sections in this book: Part I, what ecosystems are and how they work; Part II, finding a balance between population, soil, water and agriculture; Part III, pollution; Part IV, resources: biota, refuse, energy and land. The text has various elements that provide teaching aids, e.g. learning objectives, review questions, etc. While this book is useful, it has the somewhat irritating presentation style of very well-drawn and sophisticated diagrams alongside over-simplistic, naive artwork.

The book is aimed at college students taking environmental courses.

Summerfield, M.A. 1991. *Global Geomorphology*. Harlow: Longman Scientific & Technical, 537 pp.
An excellent comprehensive textbook on geomorphology, ideal for everyone interested in the Earth's surface and internal processes. It is beautifully illustrated, with useful tables and boxed text.

Yearley, S. 1992. *The Green Case: A Sociology of Environmental Issues, Arguments and Politics*. Routledge, London, 197 pp.
A comprehensive account of the basis of 'green' arguments and of their social and political implications. Yearley examines the reasons for the success of leading environmental campaign groups (such as Greenpeace), and analyses developments in green politics and green consumerism. The book explores many of the major ecological issues in the developing world, and Yearley argues that these problems are inextricably linked with debt and their need for development. A well-written sociological perspective, and a recommended supplementary book for those interested in the broader aspects of global environmental issues.

Blow, winds, and crack your cheeks! rage! blow!
You cataracts and hurricanoes, spout
Till you have drench'd our steeples, drown'd the cocks!
You sulphurous and thought-executing fires,
Vaunt couriers to oak-cleaving thunderbolts,
Singe my white head! And thou, all-shaking thunder,
Strike flat the thick rotundity o' the world!
Crack nature's moulds, all germens spill at once
That make ingrateful man!

Spoken by Lear.
William Shakespeare, *King Lear*, Act III, Scene ii.

The night has been unruly: where we lay,
Our chimneys were blown down; and, as they say,
Lamentings heard i' the air; strange screams of death,
And prophesying with accents terrible
Of dire combustion and confus'd events
New hatch'd to the woeful time. The obscure bird
Clamour'd the livelong night: some say the earth
Was feverous and did shake.

Spoken by Lennox, Nobleman of Scotland
on the night Macbeth dies.
William Shakespeare, *Macbeth*, Act II, Scene iii.

The Earth's climate has not always been as it is today. There have been times in the geological past when the global climate was warmer or considerably colder than at present. The geographic and temporal distribution of organisms, preserved as fossils, and the particular chemical signatures and sediment types available for study, show that the Earth's climate has fluctuated over geological time. As an example, 4.5–3.5 million years ago, parts of eastern Antarctica were a lot warmer. During the **Tertiary Period** of Earth history, from about 65 million years ago (65 Ma), but prior to 1.64 Ma, boreal forests were growing in the Canadian High Arctic as far north as 78°N, now preserved as fossil forests (Plate 2.1). Although it is now known that there have been substantially different climates in the past, the exact causes of such variations remain unclear.

Beside the intellectual curiosity that drives humankind in search of knowledge about past climates on Earth, about how major climatic change may come about, and the rates at which such changes could occur, it is possible to begin to make sensible predictions and models about negative and positive feedback processes in controlling global climate change. Put more simply, the geological record provides an unprecedented insight into the circumstances in which greenhouse and icehouse conditions occur, and the opportunity to assess the potential impact of human activities in controlling climate change.

Climates, both past and present, are studied by many people; meteorologists trying to improve weather prediction and construct climatic models for the future, archaeologists wishing to understand the climatic conditions that prevailed during the early development of human life around the globe, geographers and agro-economists involved in predicting the potential impact of climate change on world and regional food supply, and Earth scientists endeavouring to unravel the history of our planet and the dynamics of Earth surface processes. Global warming, acid rain, the potential effects of a nuclear winter, and how other forms of chemical pollution in the atmosphere or oceans affect climate have all contributed to a resurgence of interest in past climates, primarily as a key to predicting future climatic change.

Earth scientists frequently find themselves at the centre of media attention, with large sums of money

Plate 2.1 *Fossil tree stump preserved at 79°N on Axel Heiberg Island in the Canadian High Arctic. This provides evidence for the existence of high-latitude boreal forests in polar regions during Tertiary times.*

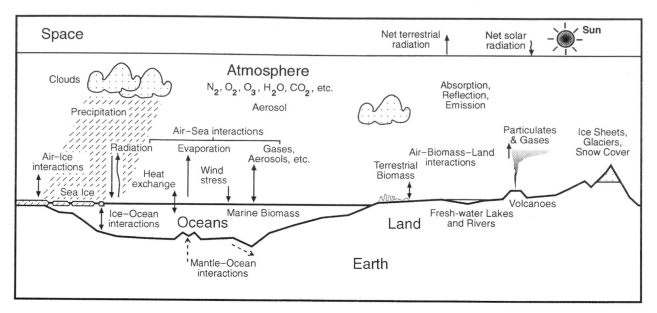

Figure 2.1 *Generalised climate system to show interaction between principal components in the ocean–atmosphere–lithosphere systems.*

more readily available for research into climate change and past climates. Computer-based climatic models, commonly referred to as **general circulation models** (**GCMs**), are in vogue. The past few years have witnessed a concerted effort to understand causal factors that contribute to global climate change. **Palaeoclimatology** as a scientific subject is truly inter-disciplinary, regularly and necessarily involving many different Earth scientists, chemists, biologists, physicists, astronomers and mathematicians. It is, perhaps, more than any other current scientific pursuit, the youngest science looking for universal recognition.

Earth scientists have now established that global climatic changes occur on many time scales up to hundreds of millions of years, but they have not yet developed well-constrained cause-and-effect models for global changes in climate. One of the main ways to understand past climates and the nature of climate change over the past few hundred thousand years is through the study of ice cores and sediment cores; therefore increasing attention is being focused on the climatic signatures preserved in such cores.

This chapter considers some of the major, sometimes abrupt, changes in the Earth's climate at a few selected time intervals. In terms of climate change, humanity is currently in a particularly interesting period of geological time, the **Quaternary Period**, often referred to as the present Ice Age. During this period, which extends back for over 1.64 million years (Harland *et al.* 1989), the Earth's climate has cooled down and undergone a series of rapid

fluctuations between warm and cold phases. It is important for Earth scientists to understand the nature of these changes if they are to resolve the effects of human activities and natural variation in the climatic system. Particular attention, therefore, is given to the nature and study of the Quaternary Period in this chapter. Whatever the exact cause, or causes, of the sudden past shifts in the Earth's climate, the one thing that Earth scientists are certain of is the catastrophic consequences for life on Earth at such times. Clearly, just as current political thought and, hopefully, action is built upon the lessons that history teaches, so humankind should attempt to understand Earth history better in order to appreciate the potential that exists, either natural or human-made, for destroying various types of animal and plant life on this planet. Human activities may be exerting a forcing effect on world climate (see Chapter 3).

A simplified climate system is shown in Figure 2.1, from which it can readily be appreciated that controls on climate are either external (e.g. the solar flux) or internal to the Earth (e.g. heat, gas and fluid flux from the Earth's core and mantle). For truly global shifts in climate, the Earth's surface heat energy (in the biosphere, principally the atmosphere and oceans) must be dissipated worldwide. Although the atmosphere, as a mixture of gases, will respond most rapidly to any major temperature change (see Chapter 1), it is the oceans that act as the main conveyor belt for heat energy worldwide. Any rapid and abrupt alterations in the nature of the oceanic conveyor belt

Plate 4
*Biodiversity: (A)
Tundra, Arctic
poppies, Northern
Ellesmere Island,
Canadian High
Arctic; (B)
Joshua trees,
Joshua National
Monument, USA;
(C) Koala bear,
eastern Australia;
(D) Spider,
Japan.*

Plate 5 *Meteorite impact crater, at Wolf Creek Crater reserve, Australia, is 853 m in diameter and the fourth largest of meteoric origin discovered on Earth. Courtesy of Ian Oswald-Jacobs Aerial Photography, Apsley, Australia*

Plate 6 *Gosses Bluff, Australia, the inner 5-km wide 'halo' of an approximately 130 Ma comet impact. The original crater was about 20 km in diameter but has been eroded to remove more than 2,000 m of overlying rock and sediment. The comet consisted of carbon dioxide, ice and dust. The impact was about one million times more powerful than the Hiroshima atomic bomb. Courtesy Ian Oswald-Jacobs Aerial Photography.*

Plate 7 *Large-scale diamond mining operations in the Kimberley region of Western Australia. Courtesy Ian Oswald-Jacobs Aerial Photography.*

will have profound effects on the distribution of this heat energy. The following section is a brief look at the role of the oceanic conveyor belt in global climate change.

The role of the oceans in climate change

The oceans play a fundamental part in controlling and changing global climate. Ocean circulation is driven essentially by solar energy and this circulation acts like a giant conveyor belt moving heat or thermal energy around the Earth. The mixing time of the ocean waters is about 1,500 years, which means that any climate change on a millennial or longer time scale has the potential to have the atmosphere and oceans in some degree of thermal equilibrium. Global climate change measured on a century to decadal scale is very unlikely to be a consequence of global oceanic circulation but rather high-frequency fluctuations in global mean air temperature.

The oceans act as a giant conveyor belt for the global distribution of thermal energy. Surface currents are warmed in low latitudes and carry heat polewards, whereas the surface currents lose heat at high latitudes and flow equatorwards. Predictably, there is a general similarity in the pattern of surface winds and ocean surface current directions, because of the frictional coupling between these systems, but this is only an average as the distribution of land tends to constrain oceanic circulation to a much greater effect than for surface winds. The Earth's rotation causes a significant deflecting force to act on wind and ocean currents – to the right in the Northern Hemisphere and to the left in the Southern Hemisphere, known as the **Coriolis force** (mathematically expressed as the product of mass × its speed × twice the angular velocity of the Earth × the sine of the latitude). The Coriolis force and the frictional forces resulting from wind shear across the ocean surface waters set up complex vortices and eddies, the best-known of which is the so-called **Ekman motion**, which leads to objects such as boats and icebergs moving at about 20–40° to the prevalent wind direction rather than parallel to it. Actually, the angular deflection increases with depth, so that the surface currents begin to have a spiral pattern with current speed decreasing with depth – the **Ekman spiral**.

In certain parts of the oceans at high latitudes the sea water is cooled sufficiently to sink and flow equatorwards as **thermohaline circulation**, driven by virtue of temperature and salinity differences. In three dimensions, thermohaline circulation in the oceans is extremely complex and remains poorly understood. The actual shape of the ocean basins, sea-floor gradients and topography, and the physical nature of the interconnections between ocean basins, all affect the thermohaline circulation patterns.

Many research workers now advocate a mutual interaction between global climate and ocean-current circulation. Broecker and Denton (1990) suggest that warming in the Northern Hemisphere prompts biological activity, and the consequent production or release of CO_2 from the oceans to the atmosphere. In turn this changes the ocean circulation, together with the way in which heat energy is transferred through the oceans. Such changes in the thermal structure of the oceans induce the formation of the **North Atlantic deep water** (**NADW**), a deep-ocean current that is currently active but did not flow as strongly during glacial times. The formation of the NADW involves the upwelling of north-flowing waters of high salinity from depths of about 500 m, and as these cold waters rise to the surface they replace the warmer surface waters that flow southwards, aided by the strong winter winds (Figures 2.2 and 2.3). As the NADW travels northwards, it loses heat energy and cools, which together with its high salinity, leads to an increase in water density, and it therefore begins to sink to abyssal depths (in the vicinity of Iceland) and then flow south, across the equator, towards Antarctica and into the Pacific Ocean. This 'Atlantic Conveyor', as it has become known, releases vast amounts of heat energy during this process, approximately equivalent to about a quarter to one third of the direct input of solar energy to the surface of the North Atlantic. The volume of flowing water is immense, roughly equivalent to 20 times the combined flow of all the world's rivers. Scientists now believe that towards the end of a glacial period, when the NADW begins to form it fashions a different pattern of global oceanic circulation and redistributes the heat energy in a manner different to that of the present day. Such changes in ocean circulation and heat exchange between the oceans and atmosphere may have had a profound effect on global climate and help drive the rapid climatic changes (see also the summary in Street-Perrott and Perrott 1990).

Oceanic circulation patterns are associated with the exchange of thermal energy (heat) between the ocean–atmosphere system and therefore exert a fundamental control on climate. A major problem which hinders a good understanding of oceanic circulation is that any models have been highly schematic in treating the circulation as an essentially laminar

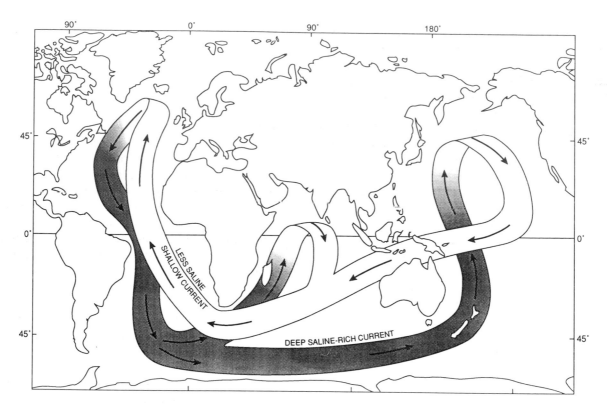

Figure 2.2 *The thermohaline (salt–heat) conveyor belt in the oceans. Solid arrows show the flow of deep, cold and salty water; open arrows show the return flow. Notice that the deep currents begin in the North Atlantic, in the East Greenland Sea, then move southwards from the Atlantic into the Pacific Ocean. The upper, warmer, current may begin in the tropical seas around Indonesia, and includes the strong flow out of the Gulf of Mexico. Redrawn after Street-Perrott and Perrott (1990).*

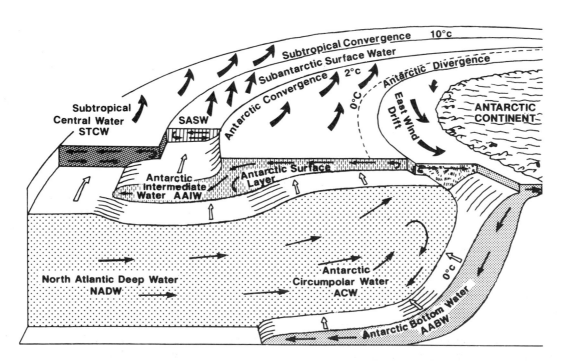

Figure 2.3 *Schematic illustration to show the principal water masses in the Southern Ocean in proximity to Antarctica. The water masses have different temperatures and densities, and move as discrete currents. There is upwelling of cold, nutrient-rich water where the surface currents diverge, whereas 'downwelling' takes place where currents converge. The Antarctic Bottom Water (ABW) flows into the Atlantic Ocean. Redrawn after Williams et al. (1993).*

flow phenomenon that ignores the actual turbulent flow conditions. Recently, however, more actualistic computer models have been developed to simulate oceanic thermohaline circulation, for example by utilising hydrographic velocity data over the rapid spatial variations actually exhibited by ocean currents (e.g. Macdonald and Wunsch 1996). The model results of Macdonald and Wunsch suggest that global oceanic circulation is best treated as comprising two nearly independent cells – one connecting overturning in the Atlantic Ocean to other ocean basins through the Southern Ocean, and the other connecting the Indian and Pacific Ocean basins through the Indonesian archipelago. The resultant heat flux estimates from this model suggest net heat losses in the North Atlantic and Pacific Oceans, heat gain in equatorial regions, and heat loss throughout most of the Southern Ocean (*ibid.*).

Techniques for studying past climates

Palaeoclimatologists looking back in time on a scale of hundreds of years have historical records as well as an enormous range of sophisticated scientific techniques to probe past climates. Many techniques are available, and their applicability depends upon the age of the sediments and fossils; each is associated with varying degrees of confidence, or error bars.

To interpret the record over hundreds of thousands of years, scientists have to rely on various subtle techniques, and obviously without recourse to human records. Looking even farther back through geological time, on a scale of millions to hundreds of millions of years, the available data for confident climatic reconstructions become more uncertain, the techniques utilised become more subtle, and the assumptions made become critical. Despite the apparently impossible odds, Earth scientists are able to use a vast range of different data and techniques to interpret ancient climates (palaeoclimates).

Just what are the tools of the trade for deciphering past climates? Careful study of ancient sediments, which are now lithified, can show the type of environment that they accumulated in, for example desert, glacial, river, lake, coastal, shallow or deep marine setting. Particularly diagnostic sediments include coals, minerals formed by evaporation of saline water (**evaporites**) such as rock salt (**halite**) and gypsum (calcium sulphate $CaSO_4.2H_2O$), glacial sediment (**till**), carbonate reefs, and sedimentary ironstones. If the sediments have a good magnetic record locked into the microscopic iron oxide mineral grains, then it may be possible to unravel their lati-

tudinal position on the surface of the Earth when they accumulated, for example whether they were deposited in the equatorial, temperate or polar regions.

Fossils

The remains of dead organisms (**fossils**) are extremely important in understanding ancient environments and past climates. Large colonies of reef corals, for example, suggest low-latitude/equatorial, warm, clear waters as off the Bahamas or Great Barrier Reef today. Fossils are also vitally important in helping to date ancient sediments accurately, something that is essential in any discussion of what the Earth's climate was like at various times in the geological past.

The analysis of pollen as an aid in the interpretation of palaeoenvironmental change is one of the most widespread methods adopted by palaeoclimatologists. Pollen grains extracted from ancient organic deposits such as peat provide information regarding changes in vegetation through time. Pollen grains are easily preserved because they are protected by a highly resistant waxy coat called **sporopollenin**. Pollen grains are identified under optical and scanning electron microscopes to determine the species by examining their shape and surface textures. The percentages of different pollen grains are estimated under the optical microscope and plotted graphically on pollen diagrams. This allows the scientists who study pollen, called **palynologists**, to determine the changes of vegetation down through a section of a sedimentary deposit, i.e. through geological time.

Care must be taken when interpreting pollen data, as pollen may have travelled large distances and it does not necessarily represent the climatic conditions at the location it occurs in. Also, pollen may be derived from older sediments, eroded out and then redeposited with younger sediments and younger fossils. Nevertheless, with all these caveats, a careful pollen analysis provides an excellent tool for interpreting palaeoenvironments, especially when many pollen sites are compared along with other kinds of palaeontological information, which includes fossil mosses, diatoms and insects.

Over the last twenty years, fossil insects have provided an exciting new method of studying environmental change throughout the Quaternary. These include the study of bugs, flies, bees, dragonflies and beetles. Beetles (*Coleoptera*) provide the best value, because they have very robust chitinous exoskeletons that tend to survive with their original chemical signature. They are well preserved in a wide variety of deposits, and they can often be identified from isolated fragments of the body, including head, thorax, wing covers and genitalia.

Unlike pollen, fossil beetles are commonly preserved at or in very close proximity to where they lived. They are the best-studied and collected group of insects, colonising almost every terrestrial, freshwater and intertidal environment. Many species show a marked preference for a particular environment, where humidity, temperature, vegetation, water conditions and substrate satisfy a rather limited range. They are, therefore, good indicators of palaeoclimate and particularly **palaeotemperatures**. It has been shown that subtle variations in temperature over the last 50,000 years, particularly the cold phases (**stadials**) and warm phases (**interstadials**) in northwest Europe, can be picked out by the dominance and presence of various beetle species (Coope 1986).

Past climates can be interpreted simply on the basis of the types of vegetation and animals that lived in certain geographical areas, or using the sediments that were laid down in particular areas. Most studies of past climates have focused on rock types that contain abundant fossils, that is ancient environments where many organisms lived, died and were preserved. Such continental environments include swamps, lakes and rivers. There has been a tendency to neglect the ancient dry (arid) regions simply because they yield less data.

There is a more subtle signature locked into the geological column. It involves the use of sophisticated chemical techniques, and this is very much where the study of palaeoclimates has reached today.

Tree rings and recent changes in climate

Studies of tree rings can be used to infer past climates. An example of this approach is the work undertaken by Earth scientists examining west European oaks and their tree-ring characteristics back to 1851 (Kelly *et al.* 1989). Temperature, barometric pressure and precipitation (rainfall) data are available for the last 150 years or so from the study area.

The width of tree rings is related to the rate of growth, which in turn tells us something about the overall climatic conditions in any particular year. By studying many trees across a wide area, such as northwest Europe, it is possible to see if there were years in which a significant proportion of trees show similar changes in growth-ring width. Using these techniques on west European oaks, it was shown that the years in which there was greater growth of tree

rings tended to be associated with enhanced cyclonic activity over the middle latitudes of Western Europe, accompanied by an increase in precipitation (rain). Temperature variations appear not to have played a significant role in the growth of the tree rings.

Changes in the growth rates of tree rings can be related to past climate. By studying the chemical isotopes of the cellulose in the tree rings, it is possible to interpret the past composition of the atmosphere and the hydrosphere. As a reliable and absolute time scale is developed, so this technique is becoming a very powerful means for understanding the changes in global climate brought about by the change from the last major glaciation (**Pleistocene**) to our present warmer (**Holocene**) period.

Tree-ring time scales are now being established that go back nearly 10,000 years. By using tree remains from the oak (*Quercus robur*, *Quercus petraea*) and pine (*Pinus sylvestris*) that have accumulated in the river terraces of south central Europe, Becker *et al.* (1991) have compiled a 'dendrochronological' (tree ring) record of the last 9,928 years and 1,604 years, using the oak and pine, respectively. By calibrating these dendrochronologies, an absolute time scale can be established. Such correlations have led Becker and his colleagues to suggest that the last significant cold phase (commonly referred to as the **Younger Dryas**) must have ended at a minimum of 10,970 years BP.

The effects of volcanic eruptions on global climate are recorded in tree-ring signatures. Detailed explanations of the methodologies and examples of chronologies can be found in Fritts (1976) and Schweingruber (1989). LaMarche and Hirschboech (1984) were able to correlate frost rings in bristlecone pines in the western United States with major volcanic eruptions on a global scale. Baillie and Munroe (1988) correlated exceptionally narrow tree rings in Northern Ireland and California dating to 1627/8 BC with the eruption of Santorini in the Aegean Sea. This eruption was originally dated using Late Minoan Stage 1a pottery at about 1500 BC, although radiocarbon dating suggested a slightly earlier date (Bell and Walker 1992). The tree-ring date is further confirmed by an acidity peak in the Dye 3 Greenland ice core (Hammer *et al.* 1987). It is argued that the massive collapse of the Minoan civilisation on Crete, 120 km away, was related to this eruption (Watkins *et al.* 1978).

Using ice cores from Crete in Greenland, Hammer *et al.* (1980, 1981) have shown evidence for volcanic activity over the past 1,500 years. Their studies were based on the acidity levels in annual ice layers as established by electrical conductivity measurements, which reflect the amount of sulphuric acid washed out of the atmosphere in any year – a function of the amount of volcanic aerosols present in the atmosphere at that time. By comparing their data with tree rings and isotope data, Hammer *et al.* (1980, 1981) were able to correlate the acidity with records of temperature variations in the Northern Hemisphere, and the close correlation between ice core acidity and Late Holocene glacier variations led Porter (1986) to suggest that sulphur-rich aerosols emitted by volcanic eruptions are one of the main driving forces for global cooling.

Extent of glaciers, ice caps, landforms and sediments

Particularly important in the study of palaeo-environmental change is the reconstruction of the former extent of ice bodies such as valley glaciers and ice sheets. During glaciations, when the Earth's climate was much colder, precipitation was dominated by snowfall. Over years, the compacted and buried snow became thick enough to change its structure and form glacier ice. As a result, valley glaciers and ice sheets formed, increased in size and flowed across the continents. These glaciers eroded the landscape and deposited glacial debris to form a rich variety of landforms. In response to the changing global climate, there have been many advances and retreats of glaciers, some of which may be globally synchronous, but others appear to have been more localised. In some areas, ice sheets were very extensive. During the last glaciation, for example, the **Laurentide ice sheet** stretched from Banks Island southwards, flowing from three main ice domes,

which were located (i) southeast of Hudson Bay, (ii) north of Hudson Bay and (iii) over Keewatin. The glaciers in the Arctic were constrained at their high latitudes by severe aridity and actually advanced only about 20–30 km southwards. During the last major glaciation, ice covered most of Northern Europe, extending south to the North German Plain from the Fenno-Scandinavian ice sheet, and south to the English Midlands from the British ice sheet (Figure 2.4A, B, C). It was from evidence such as this for the former extent of continental ice during past glaciations, particularly on the continents of South America, Africa, Australia and India, that led Alfred Wegener, in 1915, to propose that the continents had drifted around the surface of the Earth. Wegener used such information to reconstruct the supercontinent of **Gondwana**. These ideas were embodied in Wegener's theory of continental drift, which provided many of the early ideas that were incorporated into the present theory of plate tectonics.

Mapping and geochronological dating of glacial landforms provides information on the former extent and temporal variation relating to past climate. Research has shown that several periods of ice advance can be identified for most high- and mid-latitude regions of the world (Plate 2.2). Many of these occurred at the same time, suggesting global changes in climate. Figure 2.4 shows the expansion of glaciers from selected parts of the world, and shows the degree to which glacial advance can be correlated. Of particular interest are the fluctuations during the past few centuries, especially during the seventeenth century, which was a cold period known as the **Little Ice Age** (see Grove 1988). Christmas cards that use paintings from this time show a great deal of snow and ice – the picturesque white

Plate 2.2 *Yosemite National Park, USA, illustrating the evidence for former glaciations. The deep U-shaped and hanging valleys were once filled with glacial ice, which helped to erode and carve them into their present form.*
Courtesy of K.C.G. Owen.

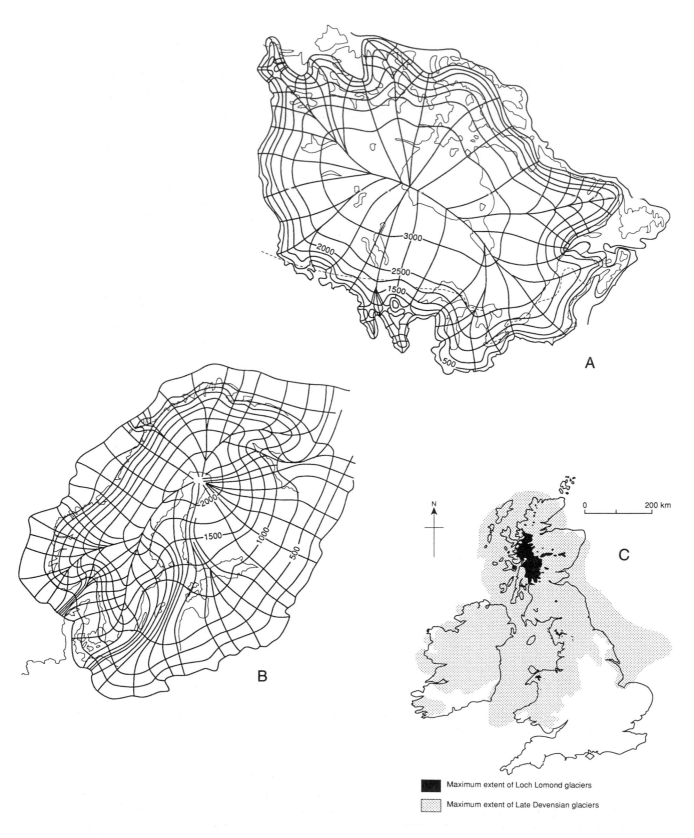

Figure 2.4 *Computer model of (A) Laurentide ice sheet at its maximum extent, and (B) the Fennoscandian ice sheet at its maximum extent. Redrawn after Boulton* et al. *1985, reproduced with permission of The Geological Society. (C) The last glaciers in Britain and Ireland. Redrawn after Bowen* et al. *(1986).*

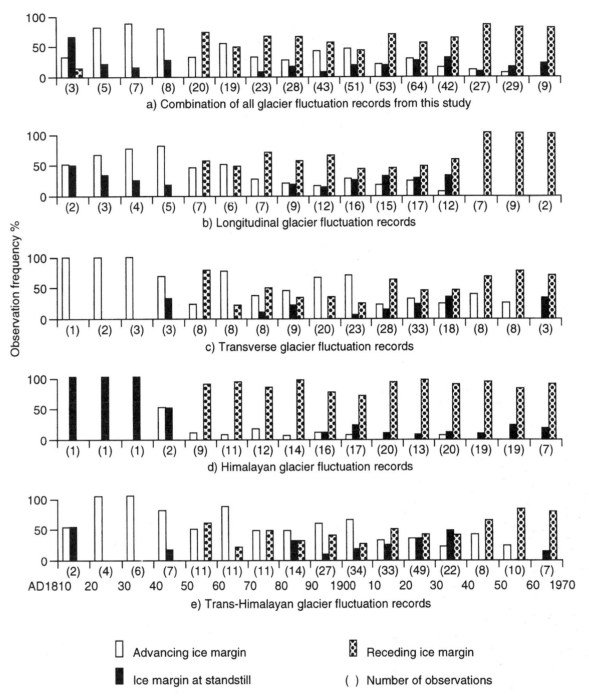

Figure 2.5 *Frequency of glacier activity for selected Himalayan and Trans-Himalayan glaciers from 1810 to 1970. See text for explanation. Redrawn after Mayewski and Jeschke (1979).*

Christmas. Even the River Thames in London froze over.

The response of glaciers to climate change is complex. Mayewski and Jeschke (1979), for example, showed how Himalayan and Trans-Himalayan glaciers fluctuated between 1810 and 1970 (Plate 2.3). They considered a sample of 112 glaciers in

Pakistan, India and Nepal. Figure 2.5 shows their composite record for all glacier fluctuations in the Himalayas and Trans-Himalayas from 1810 to 1970, plotted as 10-year periods. The data are presented as percentages of the number of glaciers in the sample (indicated in parentheses) that are advancing, receding or in equilibrium. (a) is a composite of all

Plate 2.3 *View looking southwards towards Shivling showing the characteristics of a typical glaciated Himalayan environment. The large moraines in the middle ground were formed during the last major glaciation, while the small moraines in the foreground were formed during the retreat of the Gangotri Glacier during the last century. Note the tents for scale. The Gangotri Glacier is regarded as the source of the Ganges, and it and other glaciers in the Himalayas are responsible for providing much of the waters to the Indo-Gangetic Plain. An understanding of the dynamics of glaciers like this is essential for the management of mountains and their adjacent forelands.*

glaciers studied throughout the Himalayas and Trans-Himalayan mountains, while (d) and (e) divide the glaciers into the monsoon-influenced Himalayas and the drier, higher, Trans-Himalayan ranges. (b) and (c) represent glaciers of different types: the transverse glaciers are small and steep, and generally flow perpendicular to the trend of the range, while the longitudinal are relatively long and wide, flowing in east–west trending valleys. The data show that glacier advance was dominant until 1850, since which time retreat has dominated. The data also show that the glaciers do not all respond in the same way and this may be due to different climatic changes throughout the mountain ranges. Since 1850, for example, retreat has been more dominant in the Himalayan glaciers (d) as compared with the Trans-Himalayan glaciers (e). This may be the result of changes in the intensity of the Indian monsoon, which has profound effects on the Himalayas but has little effect on the Trans-Himalayas. In addition, glacier type is important, for example retreat has been more dominant in the longitudinal glaciers than the transverse glaciers. These data, therefore, illustrate the complexity of predicting regional changes in glaciation in response to fluctuating climate.

From this and other types of data discussed earlier, it is possible to reconstruct, with a relatively high degree of accuracy, estimates of temperature changes over the last several hundred thousand years, from which it is possible to begin to understand the nature of changes in global climate.

Many studies of past global climate change rely upon using deep-sea sediment cores with the preserved planktonic and benthic faunas. A number of basic assumptions or criteria must be met, the most important being:

- There is an essentially continuous preserved record in the sediments and fossils of past climatic episodes. Ideally, the annual seasonal events are preserved.
- There is a direct link or response between any changes in sea surface temperature (SST) and biomass production (bio-productivity) in surface waters and benthic events and/or the preserved record in the benthic realm.
- The observed and/or inferred climatic events in one location can be correlated to similar events worldwide, e.g. from the Arctic to Antarctic, from ocean basin to ocean basin, or between land and ocean.

The second criterion above is, perhaps, the most difficult to establish. In recent years, however, scientists are increasingly demonstrating that the deep-sea environment is subject to rapid, abrupt, seasonal changes (Deuser and Ross 1980, Billett *et al.* 1983, Smith 1987, Sayles *et al.* 1994), and that the deep-sea floor environment is coupled to the rapid vertical transport of particulate matter through the water column, with only limited time for degradation prior to arriving at the sea floor (Sayles *et al.* 1994).

Another important means of examining past climates is through the study of soils, since they form at or very close to the Earth's surface, where the atmosphere has a direct effect on their physical and chemical nature. The study of soils, **pedology**, has proved particularly useful in reconstructing Quaternary climate change (e.g. Catt 1991, 1993). In China, for example, ancient soil profiles, called **palaeosols**, are interlayered with silt-rich layers formed mainly from wind-blown sediments known as **loess** and, together, these show quasi-periodic changes in proxy climate indicators, which include the following:

- Particle size distribution: during glacial and stadial intervals, the colder and stronger northerly winds brought coarser-grade silt as loess, whereas interglacials and interstadials are associated with greater clay content in the soils.
- %CaCO$_3$: during the more humid interglacials and interstadials, %CaCO$_3$ shows a reduction in the soils due to the enhanced chemical weathering under warmer climatic conditions, whereas during glacials and stadials %CaCO$_3$ increases in the loess.
- Magnetic susceptibility: during warmer intervals (interglacials and interstadials), there is a tendency towards increased rates of chemical weathering to release more iron oxides into the soils, thereby increasing the magnetic susceptibility of the soil layer compared with that of the loess.

Sea level change

During times when the Earth's surface, particularly the continents, have hosted substantial ice sheets, rapid and abrupt changes in the global ice volume appear as sea level changes, which occur at frequencies of 10^3–10^5 years, and with amplitudes from centimetres to more than 100 m, resulting from the expansion and contraction of continental ice sheets.

A puzzle, however, has been to explain such fluctuations in global (eustatic) sea level even at times during Earth history (e.g. the **Triassic, Jurassic** and **Cretaceous** periods) when there appears to have been no significant continental ice. Jacobs and Sahagian (1993) argue that these latter sea level fluctuations, producing smaller rises and falls in sea level (up to about 10 m), result from periodic (Milankovitch frequency – see below) climate-induced changes in lake and ground water storage.

Raised beaches and coral reefs provide important information regarding sea level changes throughout the Quaternary, and reflect the amount of water stored as glaciers during a glaciation, and the volume of water released into the oceans when ice sheets melted (see Figure 2.6). If the entire Greenland ice sheet (with an estimated 2.82×10^6 km^3 of ice) melted, global sea level would rise by about 6 m. If the entire Antarctic ice sheet melted, global sea level would rise by approximately 60 m. Additionally, raised shorelines may allow reconstructions of former ice thickness, because the growth of ice sheets and glaciers depresses the Earth's crust due to their extra weight. When the ice melts, the Earth's crust responds to the released stress by rebounding upwards, in a process known as **glacio-isostatic rebound**. In coastal areas, as the crust rises, coastal regions and raised beaches are uplifted to form raised shorelines. These raised shorelines can be dated by radiometric ages on fossil shells and other organic matter (Plate 2.4).

It transpires that the amount of uplift is directly proportional to the thickness of ice. The uplift history, however, is complex because as the ice sheets melt, sea level also rises. To determine the absolute amounts of uplift, curves for global sea level (or

Plate 2.4 Rapid changes in sea level can result in the development and preservation of coastal features such as spits, barrier islands and lagoons, for example as seen here at Chesil Beach in the UK.

'eustatic') changes have been constructed using shorelines and coral reefs in geologically stable areas that were not glaciated (Jelgersma 1966, Fairbanks 1989). The major problem in determining the nature of any change in sea level lies in the difficulties in discriminating between global glacio-eustatic signals from local or regional crustal movements, and the unknown effects on tides caused by changes in the coastline and bathymetry during times of changing sea level (see discussion by Scourse 1993). Curves for sea level change at any location show that the rate of change is not linear, but rather asymptotic in nature, displaying an accelerating then decelerating trend between periods of apparent stasis.

A study of the altitudes and ages of raised beaches from the Ross embayment, Antarctica, and east Antarctica suggests that during the **Last Glacial Maximum** (**LGM**), the ice margin was thinner and less extensive than previously thought, and that its contribution to the fall in sea level was only 0.5 m to 2.5 m (Colhoun *et al.* 1992). Until this latter study, most models indicated widespread thickening of the ice sheet margins of between 500 m and more than 1 km, sufficient to induce a fall in sea level of around 25 m, whereas geological data support a more limited ice expansion and corresponding fall in sea level of approximately 8 m (*ibid.*). These recent studies imply that during the LGM, the drop in sea level was less than previous estimates suggest, or that the Northern Hemisphere ice volume was much greater than current estimates suggest.

Further complications have to be taken into account in constructing global sea level curves, such as secular variations in the global **geoid** due to subtle changes in the Earth's gravitational field induced by plate tectonic processes, and the growth or decay of ice sheets.

One of the most common perceptions held by many scientists and non-scientists is that global warming will lead to the melting of polar ice sheets, with a concomitant rise in global sea level. Moderate temperature rises, however, could cause increased precipitation in high latitudes, resulting in greater amounts of water being locked up as snow on the polar ice caps. In the latter scenario, there would be a global, or 'eustatic', fall in sea level. Snow accumulation rates in Antarctica are known to be dependent upon the mean annual air temperature above the surface inversion layer (Robin 1977), something that is consistent with the lower accumulation rates during the LGM (Lorius *et al.* 1985). The total annual water budget of Antarctica is several times greater than that of Greenland, with the snow that falls on the grounded ice being equivalent to approximately 5 mm per annum of global sea level change

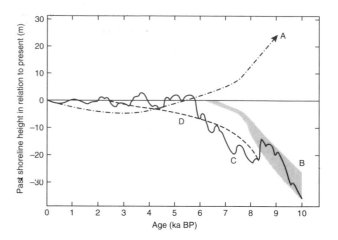

Figure 2.6   *Holocene sea level records for (A) Baffin Island; (B) eastern Australia; (C) a synthesis of several areas; and (D) the Netherlands. These data are associated with error bars that are not shown, but emphasise local rather than global (eustatic) changes in sea level, although many areas show a rise in sea level due to post-glacial melting of ice and thermal expansion of the ocean waters. High-latitude areas, such as Scandinavia and Arctic Canada, show a relative fall in sea level (e.g. Baffin Island) associated with the isostatic rebound of the continents after removal of considerable thicknesses of ice by melting. Redrawn after Williams et al. (1993).*

(Jacobs 1992). Over the past century, the observed rise in global sea level has been in the range 1.0–2.4 mm per year, with a 'best guess' estimate of about 1.5 mm per year (IPCC report, *Climate Change*, 1990). At the lower limit, most observed sea level rises could be explained by the thermal expansion of the oceans, together with the melting of temperate and Greenland margin glaciers (*ibid.*).

Two lines of evidence may suggest the growth of polar ice sheets, that is satellite altimeter measurements over Greenland (Zwally *et al.* 1989), and positive correlations between net snow accumulation and increased air temperature (Morgan *et al.* 1991). Satellite altimetry measurements are limited in duration, and can be compromised by a number of factors, including the changing distance from moisture sources. Jacobs (1992) concluded that it is too early to say whether the Antarctic ice sheet is shrinking or growing.

Chemical methods

The chemistry, including isotope studies, of sediments and fossils as a tool for trying to understand

BOX 2.1 CHEMICAL ISOTOPE METHODS IN PALAEOCLIMATOLOGY

Oxygen isotopes

The ratio of the heavier ^{18}O to the lighter ^{16}O isotope in the remains of planktonic micro-organisms and benthic organisms, such as foraminifera, reflects the isotopic composition of sea water at the time their tests formed, assuming that the shells have not undergone any chemical alteration after reaching the sea floor and been buried. With some caveats, changes in isotopic composition reflect changes in the relative proportions of the lighter to the heavier isotope of oxygen locked up in ice sheets and glaciers, giving a measure of global ice volume, which can be used to infer sea water (generally surface water) temperatures, and past global climate. Thus, the secular change in oxygen isotopes in fossils and sedimentary rocks can be used to infer past water temperature. During glacial periods, large volumes of sea water are locked up in polar ice caps. The lighter ^{16}O isotope is preferentially incorporated into the ice crystals because water vapour formed by evaporation of liquid water is enriched in ^{16}O, so the global sea water becomes relatively enriched in ^{18}O. The marine organisms that secrete calcium carbonate ($CaCO_3$) shells using oxygen atoms from sea water will have varying ratios of ^{16}O to ^{18}O, which reflect changing polar ice volume or climate.

The isotopic composition of oxygen is expressed in terms of differences in $^{18}O/^{16}O$ relative to a standard called SMOW (standard mean ocean water), with reference to a large volume of distilled water distributed by the US National Bureau of Standards (NBS), such that: $^{18}O/^{16}O$ (SMOW) $= 1.008$ $^{18}O/^{16}O$ (NBS-1). The isotopic composition of oxygen in a sample is expressed as per mil (‰) differences relative to SMOW such that:

$$\delta^{18}O = \frac{[(^{18}O/^{16}O)_{sample} - (^{18}O/^{16}O)]}{(^{18}O/^{16}O_{SMOW})} \times 10^3$$

Positive values of $\delta^{18}O$ indicate enrichment of a sample in ^{18}O, whereas negative values indicate depletion. The SMOW standard tends to be used for $\delta^{18}O$ values in waters and silicates, whereas for carbonate oxygen the PDB (Upper Cretaceous Peedee Formation belemnite fossil, South Carolina) standard is commonly used ($\delta^{18}O_{SMOW} = 1.03086$ $\delta^{18}O_{PDB} + 30.86$). The $\delta^{18}O$ in polar snow and ice depends principally upon the temperature of formation of the precipitation. The isotopic composition of oxygen in a carbonate sample is determined from the CO_2 gas obtained by reaction with 100 per cent phosphoric acid, normally at 25°C. Using the oxygen isotopes, ^{16}O and ^{18}O, for palaeotemperature studies is also a well-tried and tested technique. The $\delta^{18}O$ values from marine shelly material made of calcium carbonate ($CaCO_3$) are routinely used to infer palaeotemperatures and palaeoclimates. Oxygen isotope composition of pre-Carboniferous (> 360 Ma) normal marine carbonates, cherts and phosphates (including fossil brachiopod shells), for example, suggests that early

Devonian (*c.* 390 Ma) low-latitude sea water was at 25 ± 7°C (Gao 1993), somewhat similar to modern oceans, at least for some of this time period. Similar high $\delta^{18}O$ values have also been obtained for older Ordovician and Silurian samples (Wadleigh and Velzer 1992).

In palaeoclimatology, past near-sea-surface temperatures are calculated from isotopic data in carbonates, but the relationship between both is dependent upon the vital effects of individual species, such that any equation linking temperature and isotopic composition must be derived for individual species and cannot form the basis of a generally applicable equation. An example of such an empirical equation, proposed by Anderson and Arthur (1983), is as follows:

$$T°C_{water} = 16.9 - 4.2\,(\delta^{18}O_{calcite\ PDB\ scale} - \delta^{18}O_{water\ SMOW\ scale}) + 0.13\,(\delta^{18}O_{calcite\ PDB\ scale} - \delta^{18}O_{water\ SMOW\ scale})^2$$

Carbon isotopes and changes in biomass productivity

Carbon is the key element for life, and it occurs as a mixture of two stable isotopes, carbon-12 (^{12}C) and the heavier carbon-13 (^{13}C), along with a relatively short-lived radioactive nuclide of carbon-14 (^{14}C). In total, carbon occurs as seven isotopes (^{10}C, ^{11}C, ^{12}C, ^{13}C, ^{14}C, ^{15}C, ^{16}C), two of which are stable, ^{12}C and ^{13}C. ^{12}C makes up 98.89 per cent of the total carbon budget, with ^{13}C accounting for 1.11 per cent. Carbon isotopes are used for interpreting the photosynthetic strategy that fixes fossil organic matter. Geologically important carbon reservoirs include carbonate rocks, which contain no radiocarbon, as the residence time is much greater than the radiocarbon half-lives (0.74 s for ^{16}C, to 5,726 yrs for ^{14}C).

Oceanic carbon exists mainly in four forms:

- DIC = dissolved inorganic carbon
- DOC = dissolved C_{org} (organic carbon)
- POC = particulate C_{org}
- Marine biota

The most commonly used standard is with reference to the Peedee Formation belemnite fossil (PDB standard) or the University of Chicago standard, which was the first material analysed by H.C. Urey *et al.* in 1951. The isotopic composition of carbon in a sample such as a fossil shell is expressed as the $\delta^{13}C$ value per mil. (‰)

$$\delta^{13}C = \frac{(^{13}C/^{12}C)_{spl} - (^{13}C/^{12}C)_{standard}}{(^{13}C/^{12}C)_{standard}} \times 10^3$$

If a sample is enriched in ^{12}C relative to the standard, then the $\delta^{13}C$ value is negative. If a sample is enriched in ^{13}C relative to the standard, then the $\delta^{13}C$ value is positive.

All the common photosynthetic pathways discriminate against ^{13}C in favour of ^{12}C, therefore living oganisms show a very strong preference for the lighter carbon isotope, ^{12}C. Consequently, the heavier isotope, ^{13}C, tends to remain in the Earth's surface reservoir of oxidised carbon, mainly as dissolved bicarbonate in sea water. The increased $^{12}C/^{13}C$ ratio, a proxy for the principal carbon-fixing chemical reactions associated with photosynthesis, occurs in sedimentary organic matter as far back in the geological record as almost 4 Ga (4×10^9 years ago) and suggests that there was prolific microbial life on Earth (Schidlowski 1988). The implication of this very early microbial life is that there was at least partial biological control on the terrestrial carbon cycle at a very early stage in the evolution of the hydrosphere, something that could allow life itself to modify its evolutionary environment in agreement with the Gaia Hypothesis.

The carbon delta ($\delta^{13}C$) value can even be used to study herbivore diets since the isotope ratio is passed on to the grazing animal and is deposited in the animal's bone collagen, which has a greater preservation potential than the softer organic matter. Small organisms with shelly matter, such as snail shells, contain sufficient organic material to analyse their palaeoclimatic signature. Indeed, snail shells have been used to extract carbon isotope signatures for understanding the climate in the Negev Desert, Israel, 3,000–4,000 years ago. Ancient, well-preserved, bone material in fossil vertebrates makes it possible to interpret the climatic conditions under which that animal lived.

Measurements of $\delta^{13}C$ values from the CO_2 trapped in air bubbles in an ice core from Byrd Station, Antarctica, have shown that during the Last Glacial Maximum atmospheric concentration of CO_2 was 180–200 ppmbv, much lower than the pre-industrial values of about 280 ppmbv.

Nitrogen isotopes

The isotope ratios of nitrogen are just beginning to be utilised. Nitrogen which is fixed, for example, by symbiotic bacteria in leguminous plants contains about the same $^{15}N/^{14}N$ ratio as the ambient atmosphere. Most non-symbiotic plants, however, possess up to five or more parts per thousand (ppt) ^{15}N. Thus, the nitrogen isotope signature in fossil organic matter allows some insight into the contribution by nitrogen-fixing organisms to its decay or preservation. Nitrogen ratios may prove useful to palaeoclimatologists because the biological fixation of nitrogen described here tends to decrease as soils become drier.

An example of the use of nitrogen isotopes in studying past climatic–oceanographic conditions is in determining the causal factors for the formation of deep-marine (from cores collected in *c.* 1,375 m water depth) layers of organic-rich sediments (sapropels, with up to 4.5 per cent by weight organic carbon) in the eastern Mediterranean, from the mouth of the Nile, during the Holocene and Upper Pleistocene to about 450,000 years BP (Calvert *et al.* 1992). The $\delta^{15}N$ record, which closely follows those of the organic carbon trends but as an inverse relationship, displays large and systematic variations, with an amplitude up to 9‰, and with the systematically lighter values in the sapropels reaching 0.3‰ and the heaviest values being confined to the glacial

stages (*ibid.*). Amongst the possible explanations for the accumulation of the sapropels, the more plausible include:

1 enhanced preservation of organic carbon in anoxic bottom waters with reduced rates of renewal of the deep water, possibly due to lower sea levels associated with the Last Glacial Maximum, and/or because of reduced salinity in surface waters linked to increased run-off of surface waters;
2 a greater flux of organic matter to the sea floor associated with increased primary production related to increased surface-water run-off.

Calvert *et al.* found that the sapropels contain significantly lower nitrogen isotope ratios ($^{15}N/^{14}N$) than the intercalated marls (calcareous muds). They concluded that the large differences could not be due to either variable mixtures of marine and terrestrial organic matter with different isotopic compositions, or to differences in the type and extent of post-depositional alteration. A terrestrial contribution to the sapropels is minor, since the $\delta^{13}C_{organic}$ values (mean −21.0 ± 0.82‰) are identical to those in plankton from the present Mediterranean, and there is no gradient in the isotope values in cores recovered at varying distances from the Nile, the main source of any terrestrial sediment input (*ibid.*). The variation, however, is consistent with a greater utilisation of dissolved nitrogen during the accumulation of the sapropels, that is, the formation of the sapropels was associated with high productivity of plankton in surface waters causing a higher flux of organic matter to the sea floor (*ibid.*).

Cadmium/calcium ratios and sea water temperatures

Studies of deep-ocean benthic (bottom-dwelling) foraminifera have demonstrated that there is a relationship between the amount of dissolved cadmium (Cd) in sea water and the Cd/Ca ratio in biogenic calcium carbonate (Boyle 1988), something that has also been shown for scleractinian corals from the Galapagos Islands (Shen *et al.* 1987). Other studies have confirmed that Cd/Ca ratios in fossil shell material can provide insights into past oceanic circulation and, therefore, palaeoclimates. Upwelling of nutrient-rich waters in the oceans is driven by temperature differences between air masses over the land and oceans. These relationships have been used by van Geen *et al.* (1992) in a study of the Cd/Ca ratio in the shell of the benthic foraminifera *Elphidiella hannai* (from sediment cores in the mouth of San Francisco Bay), which is proportional to the Cd concentration in coastal waters, in order to calculate the past changes in mean upwelling intensity along the west coast of North America. *E. hannai* inhabits waters shallower than about 50 m along this coast. This study revealed that the foraminiferal Cd/Ca ratio has decreased by about 30 per cent from 4,000 years ago to the present day, probably because of a reduction in coastal upwelling. Van Geen *et al.* interpret these changes to reflect the weakening of the northwesterly winds that drive upwelling, associated with the decreased summer insolation of the Northern Hemisphere by about 8 per cent over the past 9,000 years as a consequence of systematic changes in the Earth's orbit around the Sun.

Natural variability in stable isotope systems in sedimentary environments

The range of variation in sedimentary systems for selected, commonly used stable isotope systems can be summarised as follows:

D/H (deuterium/hydrogen)	δD –430 to +50 ‰
$^{13}C/^{12}C$ (carbon)	$\delta^{13}C$ –90 to +20 ‰
$^{18}O/^{16}O_{SMOW}$ (oxygen on SMOW scale)	$\delta^{18}O$ –45 to +40 ‰
$^{34}S/^{32}S$ (sulphur)	$\delta^{34}S$ –40 to +50 ‰

BOX 2.2 ICE CORES

Ice cores provide a unique archive of past climatic conditions, including atmospheric chemistry. Complete ice cores record the annual, seasonal changes in atmospheric gases, chemicals such as acids, trace metals and wind-borne dust, which were sealed into the falling snow and buried to form ice. The stable isotopic composition of the ice (see Box 2.1) depends on the air temperature at the time the snow formed and accumulated, thereby providing a means of calculating past atmospheric temperatures.

Increasingly, scientists wishing to document past climatic conditions, and understand the causes and effects of climate change, are analysing the chemical and physical nature of ice cores. Ideally, ice cores are drilled in parts of the world where there is likely to be an undisturbed and continuous signature of past climates, for example in the Greenland ice sheet and in Antarctica. Examples of ice cores include:

- the American 'Thule' core, drilled to the bottom of the Greenland ice sheet between 1963 and 1966 in northwest Greenland, near Thule, retrieving a 120,000-year record, and a 100,000-year record, also reaching the underlying bedrock, drilled near a radar station in southeast Greenland between 1979 and 1981;
- the American Byrd core from west Antarctica, drilled in 1968, and giving a record of the past 70,000 years;
- the 2,083 m long Vostok ice core from east Antarctica, drilled by the Soviets in the early 1980s and analysed jointly with French scientists, recording the past 160,000 years;

- a 3,028 m long core drilled from 1990 to 1992 at Summit (72° 34'N, 37° 37'W) on the Greenland ice sheet to its base, by the Greenland Ice Core Project (GRIP), under the aegis of the European Science Foundation (with researchers from Belgium, Britain, Denmark, France, Germany, Iceland, Italy and Switzerland), going back 250,000 years – the first to contain information from two ice ages and the three intervening warm interglacials. The GRIP cores have been particularly useful in providing a high-resolution record of atmospheric CO_2 and CH_4 budgets. About 30 km away from Summit a complementary Greenland Ice Sheet Project 2 (GISP2), run independently by the USA, has also been undertaken. GISP2 reached bedrock in the summer of 1993.

After the burial and compaction of snow and its transformation to ice, the layers of ice may be subjected to disturbances because of ice flow, tensional stresses in the ice, and exhumation by the stripping away of younger layers to form an ice surface. Thus, the dating of ice cores requires considerable care. Ice cores are dated using various techniques. The latest GRIP core was dated back to 14,500 years ago by counting the annual layers. The counting was made possible by the acid and dust content of the ice core. Summer snow contains peak amounts of acid, whereas dust content peaks during the winter and spring seasons. For the GRIP core, calculations using two well-dated 'fixed points' were employed to calibrate the rest of the ice core record, i.e. the cold period about 11,500 years ago that followed the last glaciation – the Younger Dryas – and the very cold interval 113,000 years ago, after the 'Eemian interglacial'.

past climates, and estimating palaeotemperatures, and oceanographic and atmospheric conditions, is coming of age. Many chemical techniques are now available, and their use and interpretation is the subject of considerable current research. Box 2.1 summarises the underlying chemical rationale behind some of the most commonly employed isotopic techniques.

Amongst the chemical methods for gathering a high-resolution record of past climate change through the Quaternary Period is the recovery of

continuous ice cores from ice caps (Box 2.2). Earth scientists can measure the chemical properties of trapped air bubbles, oxygen isotopes and deuterium (a heavy isotope of hydrogen), and dissolved and particulate material in the ice. Perhaps the best known of these cores is the Vostok ice core (Box 2.2), which was drilled in east Antarctica and recovered over several years from the Soviet station, Vostok. This ice core totalling 2,083 m in length, extends back in time 160,000 years (Barnola *et al.*

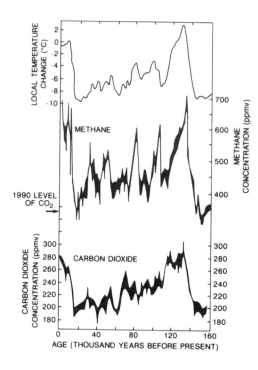

Figure 2.7 *Analysis of trapped air in the Vostok ice core to show the methane and carbon dioxide concentrations during the last 160,000 years. Notice the close correlation between methane and carbon dioxide with local temperatures over this period. Redrawn after Lorius et al. (1988).*

1987, Genthon *et al.* 1987). A study of the CO_2 in air bubbles trapped within the ice core has shown that during the last interglacial period, about 125,000 years ago, average atmospheric temperatures were probably around 2°C higher than at any period since the ice sheets started melting approximately 18,000 years BP (Figure 2.7). During the last interglacial, it seems that the peak mean global temperatures could have been similar to those of the projected anthropogenically created greenhouse period.

Earth scientists have applied similar techniques to the shells of microfossils going much further back in time to produce palaeotemperature curves stretching back 100 million years and 300 years into the future (Figure 2.8). The curves are derived from the data obtained from the shells of planktonic, near-surface organisms and deeper-water species. From this graph, it can readily be seen that sea water, and therefore mean Earth surface temperatures, were somewhat warmer 80–140 Ma, during the Cretaceous Period of Earth history.

The calcareous skeletons of planktonic foraminifera are commonly chosen for isotopic analysis because

these organisms live in surface waters and, therefore, they provide one of the best measures of surface water temperature: in turn, sea-surface temperatures can be linked to global temperatures. Determination

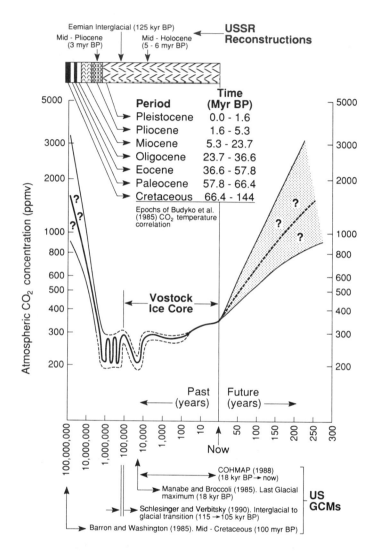

Figure 2.8 *Variations of atmospheric carbon dioxide concentration for the past 100 million years and the projected next 300 years. The upper scale (top left) shows only palaeoclimates as reconstructed by Russian researchers, keyed against a table showing the name and duration of each period. The lower scale shows periods simulated by general circulation models, both for the past (to the left of 'Now') and for the future (shown to the right of 'Now'). Note that the future scale is linear in contrast to the scale for the past, which is exponential. Palaeoclimatic changes were at least partly due to the greenhouse effected by fluctuating carbon dioxide levels. Human activities could create global greenhouse conditions similar to those that occurred naturally in the past. Redrawn after Hoffert (1992).*

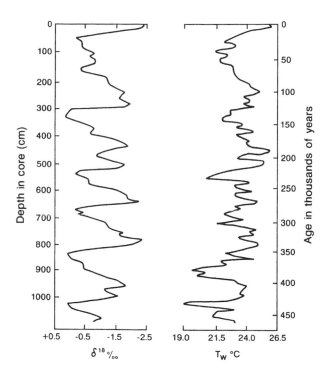

Figure 2.9 *Variations in sea-surface water temperature inferred from the oxygen isotopes recovered from the shells or tests of fossil microscopic floating, or free-swimming (planktonic), marine micro-organisms, expressed in parts per thousand and, by convention, expressed as $\delta^{18}O$ values based on the reference Caribbean core V12–122. Redrawn after Imbrie et al. (1973).*

of past ocean temperatures, using isotopes, also involves an estimate of the volume of water stored in ice sheets and the oceans, and the volume change per millionth part change in isotopic composition of a shell that is precipitating out of sea water, and in isotopic equilibrium with the sea water. The calculations of ice volumes, however, are prone to many errors. Calculations of the water stored in ice sheets during the Last Glacial Maximum, for example, range from 47×10^6 km^3 (Dansgaard and Tauber 1969) to 100×10^6 km^3 (Craig 1965). The principal factor controlling the isotopic composition of the oceans is the oceanic water volume and, therefore, the oxygen isotope curves predominantly represent fluctuations in the global ice and ocean volumes. Work on benthic foraminifera may be more truly representative of oceanic volume changes, since waters near the ocean floors remain relatively unaffected by global temperature changes, so that temperature-dependent variations in oxygen isotopes can be disregarded (Dansgaard 1984, Shackleton 1987).

Detailed studies of $\delta^{18}O$ values (see Box 2.1) from Quaternary marine microfossils dating back over the past 450,000 years have revealed fluctuations in climate over time scales of tens to hundreds of thousands of years (Shackleton and Opdyke 1973, Chappell and Shackleton 1986). Figure 2.9 shows the variations in the sea-surface temperature calculated from the $\delta^{18}O$ values measured from a core collected from the Caribbean. The last Ice Age can be seen as higher ^{18}O values from just over 110,000 to 20,000 years ago. This isotopic signal thus provides a record of glacial and interglacial stages. By convention, odd-numbered stages represent interglacials and even numbers glacials. The record shows that there have been more cycles than have so far been identified from other lines of evidence on the continents. It also shows that glacial stages are about five times longer than interglacials, and their termination is rapid. Furthermore, the record shows small perturbations in the average climate during glacials and interglacials, i.e. stadials and interstadials, respectively (discussed in more detail later in this chapter). For the last 15,000 years, there have been dramatic climatic changes on a scale from a few hundred to a few thousand years, spanning the deglaciation from the last glacial phase into the present interglacial.

There are other ways of studying past climates besides using the isotopes of various elements. Earth scientists have used the concentration of toxic metals such as copper, zinc and cadmium (Box 2.1) in cores from the Antarctic ice to assess the past, natural atmospheric conditions in the troposphere during the last 40,000 years (Batifol et al. 1989). The highest values of these toxic metals occurred during the Last Glacial Maximum of the last ice age some 25,000–16,000 years ago.

The source of these higher levels of copper, zinc and cadmium during the Last Glacial Maximum are believed to be wind-borne dust, which would be preferentially concentrated in the troposphere during drier climatic conditions associated with glacial phases. There are, however, increased concentrations of these metals since about 13,500 years ago, which may be, at least in part, due to volcanic and/or biogenic activity, and increased metal contents in sediments over the last few centuries due to increased industrialisation and pollution.

Causes and rates of global climate change

In order to consider present and past climate, it is important to have at least a rudimentary understanding of the principal components of any climate system (see Figure 2.1), the structure of the Earth's atmosphere (see Figure 1.10), an idealised global wind circulation (see Figure 1.13), and the ocean conveyor belt (see Figures 2.2, 2.3), which distributes heat around the Earth's surface. Up to about 100 km above sea level, the Earth's atmosphere comprises an essentially uniform mixture of gases but with substantially varying proportions of water vapour, concentrated mainly in the troposphere (see Table 1.2). Also, it is important to understand the principal chemical cycles and fluxes that control climate and climate change (see Chapter 1).

Global climate change is driven by both external and internal controls on the Earth's ocean–atmosphere–biosphere system. External controls include the Sun, which has a direct and important influence. Short-term changes in global climate, on a scale from tens of thousands to hundreds of thousands of years, appear to be a result of slight changes in the distribution and amount of solar radiation, or solar flux, reaching the surface of the Earth. Such changes in solar flux result from variations in the orientation and proximity of the Earth to the Sun. These factors can be thought of as external controls on climate. The astronomical factors that control the actual movement of the Earth around the Sun play a major role in global climate change. Internal controls involve the heat flux and gaseous emissions from the Earth's mantle into the ocean–atmosphere–biosphere system, together with volcanic activity, the position and latitudinal distribution of the continents, and the topography of the Earth's surface.

Decadal- to century-scale global climate change is commonly explained as due to one or several of the following:

- random atmospheric variability;
- solar variability;
- inherent or forced fluctuations in the production rate of the North Atlantic Deep Water (NADW);
- natural variations in the atmospheric concentrations of trace gases; and
- natural variations in volcanic aerosols.

The following sections examine the various controls on global climate change, at a variety of temporal scales, and moving from short- to long-duration events and cycles. The first section considers the role played by micro-organisms – a major biotic factor – in controlling and responding to global climate change over a wide range of time scales.

Micro-organisms in the world oceans and seas

There are scientists who believe that as global warming commences, marine plankton (microscopic plants and animals) may show a multiplying effect. As a counterpoint, there are also equally eminent scientists who believe that as atmospheric CO_2 levels begin to rise, the rate at which the marine plankton absorb this greenhouse gas may actually decrease, with the result that the rate of warming increases. This latter scenario is an example of a positive feedback mechanism. Since the oceans contain about 20 per cent more carbon than the total land plants, animals and soil, the oceans with their biota probably represent the principal factor in controlling global atmospheric CO_2 levels. At present, CO_2 released by human activities adds about 7 ± 1.2 gigatonnes of carbon per year (GtC yr^{-1}) to the atmosphere, about 2 GtC yr^{-1} of which is believed to be sequestered in the oceans, and in a steady state; phytoplankton fix about 35–50 GtC yr^{-1}, representing a significant part of the natural carbon cycle (Falkowski and Wilson 1992). Considerable scientific debate is focused on the potential ability of changing ocean productivity to sequester, or 'draw down', any increased (anthropogenically-created) CO_2 in the surface waters and, therefore, act as a buffer on global climate change. Records of mainly coastal water data, spanning the period 1900 to 1981 for the North Pacific, indicate that although very minor changes in phytoplankton biomass have occurred over the 70-year time interval, they are too small to have a significant effect on the rise in atmospheric CO_2 concentrations (*ibid.*).

Unfortunately, this 'multiplier' effect is poorly researched and, in past GCMs, has tended not to be an important part of most computer models. Indeed, in 1989, the five principal computer-based models for predicting global climates did not take account of the positive feedback mechanism due to plankton, i.e. four programs in the USA and one in the UK at the Meteorological Office, Bracknell. Current models assume that 50 per cent of the CO_2 injected into the atmosphere as a consequence of the burning of fossil fuels is 'drawn down' into the oceans by marine plankton where it is stored. Clearly, the significance of plankton in controlling climate may well invalidate this assumption and lead to underestimates of global warming rates.

It is now believed that the past glaciations during the Pleistocene Period ended with slight changes in the solar flux to the Earth's surface caused by variations in the Earth's orbit, known as **Milankovitch cyclicity** after the Yugoslav astronomer who cata-

BOX 2.3 TESTING THE IRON PUMP IN THE OCEANS

More than 20 per cent of the surface waters in the open oceans contain major plant nutrients such as nitrate, phosphate and silicate, and receive sufficient light energy to support phytoplankton blooms yet support only low abundances. Such zones are commonly referred to as high-nitrate, low-chlorophyll (HNLC) areas. In oceanic areas far from continental and/or shallow-marine sedimentary sources, Martin (1990) and Martin *et al.* (1990, 1991) noted that HNLC regions appear to coincide with areas of particularly low concentrations of wind-blown terrestrial dust, the main source of biogenically available iron. Understanding these HNLC areas is important, because it has been suggested that if they support increased biomass production significant amounts of atmospheric carbon dioxide can be sequestered and, therefore, they will exert a major control on global climate change. Research suggests that availability of iron (Fe) in the surface waters in many parts of the oceans may limit phytoplankton growth: open-ocean surface-water Fe concentration is about 10^{-12} Moles (picomoles). During glacials, the increased aridity and atmospheric dustiness should lead to greater amounts of wind-blown Fe-rich dust reaching the surface waters in the oceans, a process that may have stimulated enhanced oceanic surface-water biomass production, a greater draw-down of atmospheric CO_2, and further global cooling to sustain the cold interval. Paradigms such as this require testing.

In mid-October 1993, a test of the iron-limiting hypothesis by Martin *et al.* (1994) was carried out over *c.* 64 km² in the open equatorial Pacific Ocean 500 km south of the Galapagos Islands by seeding the surface waters with iron filings, resulting in a change in Fe concentration from *c.* 0.06 nM to 4 nM (1 nanomole $= 10^{-9}$ M). Because in bottle experiments such high concentrations of Fe are sufficient to produce large increases in chlorophyll and lead to a total depletion of the available major nutrients within five to seven days, the open-ocean experiment was monitored for ten days (*ibid.*). The Fe-rich surface waters were tracked using a harmless chemical tracer, sulphur hexafluoride (SF_6), which was mixed with the iron.

In the study area, primary biological productivity within the surface waters showed a three- to four-fold increase in all the size fractions, with chlorophyll increases to nearly three-fold, demonstrating 'a direct and unequivocal biological response of the equatorial Pacific ecosystem to added iron' (*ibid.*). The results were inconclusive, however, in regard to showing clearly if phytoplankton growth was finally limited by the availability of other trace metals, increased grazing pressure and/or the sinking of larger phytoplankton. Also, significant amounts of Fe may have been lost from the experiment by sinking into deeper water. Surprisingly for the researchers, the phytoplankton growth was not associated with a significant draw-down of atmospheric CO_2 by the algae – only about 10 per cent of the predicted amount that would have been sequestered if the Fe had allowed the phytoplankton to grow until total depletion of the available nitrate and phosphate. The conclusion of this experiment was that it is possible to enrich an area of open ocean with Fe and stimulate significantly increased phytoplankton production, but that in itself this enhanced biomass production does not directly influence atmospheric CO_2 levels (*ibid.*).

A contrasting view about the link between available Fe in surface waters and atmospheric CO_2 levels is taken by Kumar *et al.* (1995). During glacials, when there is enhanced atmospheric aridity, there were significantly greater supplies of wind-blown dust to the oceans, and a corresponding increase in the amount of wind-blown iron, more than five-fold in the glacial sediments of the Atlantic sector of the Southern Ocean from the Patagonian deserts (*ibid.*). The use of radionuclide proxies ($^{231}Pa/^{230}Th$, $^{10}Be/^{230}Th$ and authigenic U) shows that glacial sediments in the southernmost Atlantic Ocean (part of the Southern Ocean) over the past 140,000 years have substantially enhanced fluxes of biogenic particulate matter from the surface waters to the sea floor, providing a plausible explanation for the sequestration of atmospheric CO_2 during glacial periods, which supports the hypothesis that the Fe limitation in today's Southern Ocean was relieved during glacials by a much increased supply of Fe from wind-blown dust (*ibid.*).

---

logued these changes. Such small changes in the amount of solar energy reaching the Earth's surface were multiplied by the decreased ability of the marine plankton to absorb CO_2.

The glacial events, which lasted as long as 100,000 years, therefore switched off rather rapidly – for example, ice cores from south Greenland revealed a 7°C rise in just 50 years following the last major glaciation. Indeed, the idea that past increases in atmospheric CO_2 levels might be responsible for global warming was suggested by the research results of Shackleton *et al.* (1983), who showed that in cores from the Pacific Ocean, CO_2 levels increased after slight changes in the Earth's orbit but prior to

the start of an increase in global temperatures. A major source of this CO_2 appears to be the marine plankton.

In central Antarctica, samples of ice taken from the Vostok ice core have provided one of the longest palaeoclimatic records, for example, including variations in atmospheric CO_2 concentrations that are set within a much longer geological time framework, and projected 300 years into the future (see Figure 2.8). It is also from the Vostok ice core that the first historical record of biogenic sulphur emissions from the Southern Hemisphere oceans has been gleaned (Legrand *et al.* 1991). It has demonstrated that at the end of the last ice age, levels of methyl sulphonic acid, produced by marine plankton, decreased significantly at the same time as atmospheric CO_2 levels increased (*ibid*.). These findings strongly support the role of plankton as a major factor in controlling atmospheric CO_2 levels and, therefore, climate. Legrand *et al.* have shown that the concentrations of methyl sulphonate and non-sea-salt sulphate, products of the atmospheric oxidation of dimethyl sulphide from plankton in the oceans, vary systematically over a complete 160,000-year glacial–interglacial cycle. During the later stages of the glacial period, there was increased oceanic emission of dimethyl sulphide compared with the present day. At around 13–14 ka, the end of the last glaciation, mean methyl sulphonate levels changed from about 31 to 5 parts per billion by volume (ppbbv), and non-sea-salt sulphate dropped from 222 to 102 ppbbv. The enhanced productivity from the biota in the oceans, and correspondingly increased emissions of dimethyl sulphide from the plankton, appears to have taken place between 18,000 and 70,000 years ago. So, the ocean–atmosphere sulphur cycle, linked to marine plankton, is extremely sensitive to global climate change. The biogenic aerosols play an important part in forcing global climate change by altering the cloud albedo (cover and ability to insulate the Earth's surface) and distribution, or because of their direct effects on absorbing and re-radiating solar radiation.

El Niño events

El Niño events are relatively large perturbations of a climatic process that occurs annually in the Pacific Ocean. The Japanese Meteorological Agency (JMA) recognises a warm extreme in the ENSO cycle (El Niño Southern Oscillation) – an El Niño – as being under way when sea-surface temperature (SST) in the tropical Pacific Ocean reaches a minimum 0.5°C above normal for at least six consecutive months. The underlying cause of El Niño events is the eastward propagation of a downwelling **Kelvin wave** across the equatorial Pacific Ocean (Busalacchi and O'Brien 1981). These Kelvin waves are confined to a narrow belt by the Coriolis force. Also, the consequence of such a wave propagation is a small change in sea level that can be detected by satellite altimeter. A revised definition of an El Niño event has been proposed that takes into account this change in sea level such that an El Niño event is under way when 'sea level at Galapagos is 2 cm above its normal height for six or more consecutive months, corresponding to a thermocline **downwelling** of 40–60 cm' (Meyers and O'Brien 1995).

In a 'normal year', the variations in the atmosphere–ocean system produce a fairly predictable pattern of ocean currents in the southern Pacific Ocean, and in which the sea-surface temperature is highest in the west (> 28°C) which helps to induce the movement of strong warm maritime Southeast Trade winds into Indonesia, and with them heavy rainfall. A corollary of this is that cold, nutrient-rich bottom waters up-well to replenish surface waters off the western coast of South America. In contrast to such normal years, during an El Niño, also referred to as an El Niño Southern Oscillation event, surface water temperatures greater than 28°C develop much farther eastwards and allow the intertropical convergence zone (ITCZ) to migrate southwards and suppress the Southeast Trades, or even reverse them. The result of these changes is that rainfall is heaviest in the central-east Pacific and upwelling of cold, nutrient-rich, bottom waters is weakened. The decreased upwelling leads to a reduction in marine productivity. With less bioproductivity, less CO_2 is sequestered from the atmosphere–ocean system by organisms and this can lead to greater concentrations of CO_2, a greenhouse gas, in the atmosphere. The El Niño effect is illustrated in Figure 2.10. El Niño events appear to be associated with enhanced atmospheric convection, inferred from the associated increase in atmospheric water vapour – the water vapour anomalies being detected slightly east of the wind anomalies – where surface winds converge – and slightly to the west of the sea-surface temperature anomalies (Liu *et al.* 1995). A series of westward-moving tropical instability waves have been observed in the eastern Pacific Ocean travelling at a speed of about 50 km/day (Legeckis 1977).

El Niño events result in the release of large amounts of CO_2 into the atmosphere. At the meteorological observatory on the Hawaian peak of Mauna Loa, Keeling *et al.* (1989) have documented an increase in the rate of release of CO_2 into the atmosphere and showed that it rose by more than two-thirds over their last two-year observation period: the result has been an increase in atmospheric CO_2 from pre-industrial levels of 270 to 350 ppmbv.

Observations during the 1987 El Niño showed that for the upper range of sea-surface temperatures, the greenhouse effect increases with surface temperature at a rate exceeding the rate at which radiation is emitted from the surface. In computer models, the atmospheric response to the so-called 'super greenhouse effect' is the formation of highly reflective cirrus clouds, which shield the ocean from the solar radiation (Ramanathan and Collins 1991). In effect, they may act like a thermostatic umbrella around the

Earth to regulate the temperature of the sea surface to less than 305 Kelvin. This model involves a negative feedback to regulate the surface temperature.

Figure 2.11 shows the occurrence of ENSO events compared with the variation in atmospheric CO_2 recorded at Mauna Loa, Hawaii, after removing the overall anthropogenic trend and seasonal signal (Meyers and O'Brien 1995). The graph shows in general a good correlation between atmospheric CO_2 levels, sea level changes and El Niño events, e.g. the El Niño events of 1965, 1972, 1976, 1982–1983, 1991 and 1992–1993 correspond to a local minimum in the atmospheric CO_2 anomaly (*ibid.*). It has been suggested that the initial decline in atmospheric CO_2 levels is due to the suppression of CO_2 outgassing as a consequence of downwelling, and the subsequent increase is commonly related to the response of terrestrial vegetation (Keeling *et al.* 1989).

An El Niño event occurred in 1987 and 1988 and was associated with a change in the wind patterns and ocean currents in the Pacific Ocean, leading to severe droughts. This El Niño event ended in June 1989 with a decrease in the observed surge of CO_2 levels. The El Niño of 1987 caused the equatorial Pacific Ocean to warm by as much as 3°C, believed by some scientists (e.g. Ramanathan and Collins 1991) to be sufficient to cause a potential atmospheric warming. Such predictions are, of course, only as good as the computer models themselves and the data which go into them. Until the models are adequately tested, we have to be cautious in assuming that the ocean–atmosphere system will operate like a giant thermostat to regulate the mean global temperatures and global climate within relatively narrow limits.

In contrast to the El Niño events recorded in the 1980s, the equatorial warming events of the 1990s have been more frequent, less intense and of shorter duration.

An anomalous warming event in the tropical Pacific recorded between July and December 1994 may have been due to an El Niño (Liu *et al.* 1995). These intra-seasonal episodes involved four distinct groups of equatorial westerly wind anomalies observed by scatterometer, which initiated eastward-propagating, downwelling Kelvin waves that, in turn, caused a rise

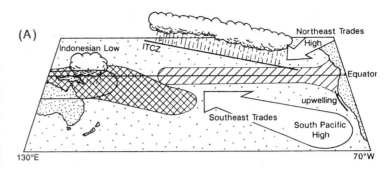

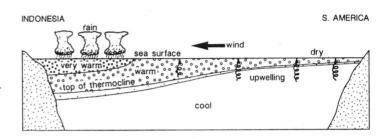

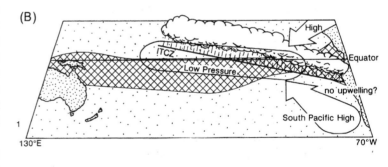

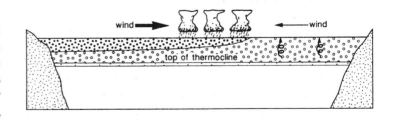

Figure 2.10 *Variations in the atmospheric systems, ocean temperature profiles and ocean currents in the southern Pacific Ocean during (A) a normal year, and (B) an El Niño Southern Oscillation (ENSO) event. In a normal year (A), the sea-surface temperature is highest in the west (> 28°C), helping to induce the movement of strong, warm, maritime trade winds into Indonesia, which creates heavy rainfall. On the western coast of South America, cold bottom waters upwell to provide fresh nutrients to surface waters. During an ENSO event, the surface water temperatures (> 28°C) develop much further eastwards, allowing the intertropical convergence zone (ITCZ) to migrate southwards and suppress the Southeast Trade winds or even reverse them. As a result, rainfall is heaviest in the east central Pacific; upwelling of cold, nutrient-rich bottom waters is weakened and marine productivity is reduced. After Open University Case Studies in oceanography and marine affairs (1991).*

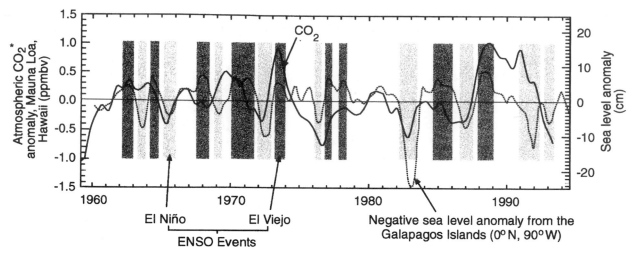

Figure 2.11   *The occurrence of ENSO events compared with variations in CO₂. The light-shaded regions indicate El Niño and the dark-shaded regions show El Viejo: both are designated by Galapagos sea level datums. The solid line indicates the atmospheric CO₂ concentration anomaly from Mauna Loa, Hawaii, after removing the trend and seasonal signal. The dotted line indicates the negative sea level anomaly from the Galapagos Islands at 0°N, 90°W. Redrawn after Meyers and O'Brien (1995).*

in sea level detected by space-borne altimeter in the TOPEX/Poseidon spacecraft (*ibid.*). The surface Kelvin waves were detected by a visible-infrared radiometer.

Monsoons

The word **monsoon** is derived from the Arabic word *mausim*, which means season. In the Indian subcontinent monsoons involve the northward movement of cloud and rain over this area early each summer, brought about by seasonal variations in solar energy and topography, occurring with great regularity but with extreme variations in their intensity from one year to the next. The monsoons involve an essentially continuous southerly flow of warm and moist surface air, while the High Himalayas block the cold northerly winds that would otherwise cool the subcontinent, which instead heats up over the summer months to a much greater degree. The already moist air gains even greater amounts of water vapour from the Arabian Sea, which has been heated to 28–29°C by the strong sunshine prior to the onset of the monsoon. These southwest monsoon winds reach the coast of India, where the forced ascent of the air masses along the western edge of the Himalayan range causes heavy and large rainfall.

Sediments from the deep ocean floor show that the monsoon cycle has existed for the past 12 million years, having commenced as a consequence of the uplift of the Himalayas.

Sunspot activity

From the time when gravitational attraction contracted the pre-solar nebula to the initiation of thermonuclear reactions that defined it as a star (taking *c.* 40 million years), the Sun has evolved over approximately 4.7×10^9 years, and standard cosmological theory suggests that its luminosity has steadily increased from an initial value of about 70 per cent of the present level (Gilliland 1989). The mass of the Sun is 1.99×10^{33} g, assuming that the gravitational constant, G, is 6.67×10^{-8} c.g.s. The solar radius is 6.96×10^{10} cm, and surface gravity is about 30 times that on Earth. The mean rate at which energy is generated is 1.94 ergs g^{-1} s^{-1}, and the human body generates energy per unit mass at a rate 1,000 times that of the solar core (*ibid.*). The Sun is bright only by virtue of its size and not because of the rate of nuclear fusion.

It has been suggested, although this remains extremely controversial, that the Sun's core may undergo episodic mixing every few hundred million (2.5×10^8) years due to a process known to astrophysicists as 'overstability' (Dilke and Gough 1972), causing a significant change in the flux of **solar neutrinos**. This process has been invoked as a possible explanation of geological ice ages (Mitchell 1976). On much shorter time scales, from decades to thousands of years, the Sun exhibits sunspot activity.

Sunspots are areas of cooler gas and stronger magnetic fields in the Sun's surface, or photosphere. Typically, the observed temperature of a sunspot is about 3,900 K, compared with the background normal photosphere at 5,600 K. Other bright features in the photosphere are known as plages. Using a technique of **helioseismology**, the time it takes acoustic waves to travel inside the Sun, Duvall *et al.* (1996) have shown that below sunspots there are very powerful downflows with velocities in the order of about 2 km s^{-1}, persisting to depths of around 2,000 km. Records of sunspot activity since about 1700 show a cyclicity of roughly 11 and 100 years. By dating samples of wood using the radioactive isotope of carbon, ^{14}C (produced in the atmosphere by the interaction of cosmic rays with atoms of the nitrogen isotope ^{14}N), a 9,000-year record of solar activity has become available to us (*New Scientist*, 1989). During periods of increased solar activity, more particles are emitted from the Sun as a solar wind, which effectively holds back more of the cosmic rays and, therefore, less ^{14}C is produced in the Earth's atmosphere. Data gathered during the last 200 years show that variations in sunspot activity correlate closely with the ^{14}C record. Correlating sunspot cycles with historical data has led to uncertainties and conflicting views about the cause of short-term fluctuations in global climate. Tropical temperature records, for example, show a positive correlation with sunspot activity for the period 1930 to 1950, but a negative correlation between 1875 and 1920. It has been suggested that there could have been a correlation for this latter time interval, but that it is masked by variations in stratospheric ozone concentrations. Ozone appears to be more abundant about two years before sunspot minima, resulting in stratospheric warming, which in turn weakens the subtropical anticyclones and mid-latitude westerlies. Cool and dry weather then follows, slightly out of phase with the sunspot cycle.

The cold winters of the Little Ice Age have been correlated with 100-year sunspot cycles, corresponding with a so-called 'quiet Sun', or 'sunspot minima'. Similar low winter temperatures occurred during the nineteenth century. Sunspot maxima correlate with high annual temperatures. It is predicted that the twenty-first century will be in a sunspot minimum, whereas the twentieth century is presently in a sunspot maximum. The Earth may, therefore, return to Little Ice Age conditions during the next century, if this is not offset by human-induced global warming (Thompson 1992).

At the first ever joint meeting between the (British) Royal Society and the French Academie des Sciences in London in February 1989, Sonett (University of Arizona, USA) suggested that the ^{14}C record shows a dominant 200-year cycle, modulated by shorter 80- to 90-year (**Gleissberg**) cycles and longer 1,000-year and 2,300-year cycles. The 200-year cycle in ^{14}C may well account for the Little Ice Age recorded throughout Europe in the seventeenth century (linked to a quiet period of sunspot activity). Other element isotopes produced by the bombardment of cosmic rays with particles in the Earth's atmosphere show cyclic variations in abundance. Beryllium, as the isotope beryllium-10 (^{10}Be), forms in this way and settles to the ground unabsorbed by living organisms, and in cores from the Antarctic it shows a cyclic variation in abundance of about 194 years – close to the 200-year cycle interpreted from the ^{14}C record.

In summary, apart from the three principal short-term modes of sunspot cycle activity, i.e. 11, 22 and 33 years, other dominant modes are the 80–90-year Gleissberg cycle and the *c.* 200-year cycle. Minor modes of sunspot activity occur at 44, 52, 57, 67, 105, 130, 140, 180, 222 and 420 years (Glenn and Kelts 1991). The longer-term modes, e.g. 1,000 and 2,300 years, are less well documented, because of a lack of historical records.

There is some debate amongst cosmologists about the true nature of solar flares and sunspot activity, also referred to as coronal mass emissions (CMEs). Basically, the argument centres around whether solar flares are the expression of *or* cause of CMEs.

Milankovitch cyclicity

The Yugoslavian astronomer, Milutin Milankovitch calculated how summer radiation at latitudes 55°N, 60°N and 65°N varied during the past 650,000 years, then mailed his graphical results to the great German climatologist, Wladimir Koppen. Koppen immediately wrote back to Milankovitch to say that the data could reasonably be matched to the periodicity of the Alpine glaciations that had been reconstructed by Penck and Bruckner some 15 years earlier. In 1924, the 'Milankovitch curves' were published in Koppen's and Alfred Wegener's book *Climates of the Geological Past*, which allowed Milankovitch's work to reach a wide scientific audience. Milankovitch then began work on calculating radiation curves for eight latitudes ranging from 5°N to 75°N, and published

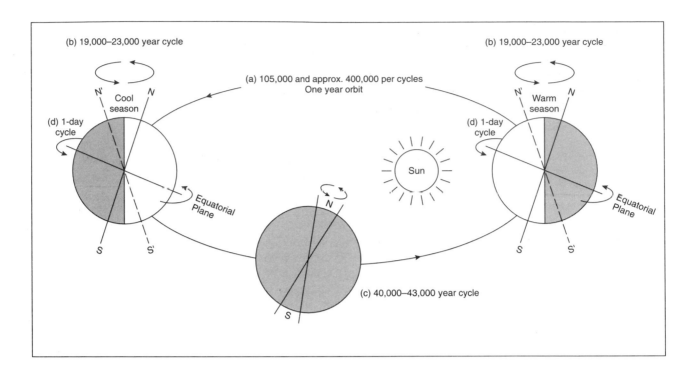

Figure 2.12   *The variability in the Earth's orbit around the Sun at various time scales measured in tens of thousands of years, and referred to as orbital parameters. The temporal variation in orbital parameters causes long-term changes in the amount of solar energy reaching the surface of the Earth, which in turn can result in significant changes in global climate, referred to as Milankovitch cyclicity, so named after one of the first people to propose a link between changes in the Earth's orbit and global climate change. Adapted from Peltier (1990).*

this work in 1930 in a volume entitled *Mathematical Climatology and the Astronomical Theory of Climate Change.* Milankovitch was not to rest there. He set about the task of calculating just how much the ice sheets would respond to a defined change in solar flux or solar radiation, which was published in 1938 in a volume called *Astronomical Methods for Investigating Earth's Historical Climate.* In 1941, Milankovitch published the comprehensive results of his life's work as a unifying theory linking the astronomical control on variations in the amount of solar radiation reaching the surface of the Earth and climatic change, in his book *Canon of Insolation and the Ice Age Problem.*

The work of Milankovitch and others has provided a major and fundamental contribution to the Earth sciences, where in an attempt to understand the forcing mechanisms for global climate change better, Earth scientists are utilising various astronomical studies that reveal three scales of global climate change caused by temporal variations in the nature of the Earth's orbit around the Sun (Figures 2.12 and 2.13). These external controls or orbital factors are:

- changes in the Earth's **precession** on a scale of about 19,000–23,000 years caused by the slow variation in the annual position of the perihelion (Earth's closest distance to the Sun). Precession is caused by the gravitational pull of the Sun and Moon on the Earth's equatorial bulge such that the Earth's axis of rotation describes a circular path where it is said to precess. The Earth is at present nearest to the Sun in the Northern Hemisphere winter but in *c.* 10,000 years it will be farthest from the Sun at that season;
- changes on a time scale of about 41,000 years caused by variations in the **obliquity** of the Earth (tilt of the Earth's axis of rotation); and
- changes on a scale of 100,000 and 400,000 years caused by the Earth's **eccentricity** (the shape of the Earth's orbit, cyclically changing from more circular to more elliptical and back again).

Collectively, these three orbital parameters are known as **Milankovitch cyclicity**.

Although the eccentricity varies with the 100-ka period, the variation in the incoming solar radiation is relatively weak compared with that of the precession and obliquity periods; perhaps, therefore, it is surprising that the 100-ka signal appears to dominate many past climatic records, at least to about 1 Ma, prior to which, over the preceding million years or so, it was much less important. Thus, it seems that the 100-ka glacial cycles may not be due to eccentricity as the forcing mechanism. Mathematical arguments using signal processing methods applied to the variation in orbital parameters of the Earth have been proposed to explain how variations in the frequency of the obliquity, not 100-ka eccentricity, cycle can produce a 100-ka period (Liu 1992). In effect, variations in the obliquity of the Earth's orbit occur because of the coupling between the motion of the Earth's orbital plane (due to the gravitational perturbations caused by the other planets) and the precession of the spin axis resulting from the solar torque exerted on the Earth's gravitational bulge. In actuality, variations in the magnitude of the solar torque are governed by the instantaneous distance of the Earth from the Sun, which is controlled by the eccentricity – thus, in a more complex manner, the variations in the Earth's obliquity are indeed partially determined by the eccentricity (*ibid.*). Only when the eccentricity is large will the maximum rates of frequency variation of the obliquity occur: minimum rates of frequency variation can occur irrespective of the magnitude of the eccentricity (*ibid.*).

Rapid sub-Milankovitch climate change

The rates at which global climate change occur, together with their abruptness, is now well established, e.g. in the GRIP ice core (Figure 2.14). Rapid fluctuations in $\delta^{18}O$ values have been recognised for some time as typical of many parts of the Quaternary, e.g. from the Dye 3 and Camp Century Greenland ice cores for between 80,000 and 30,000 years ago (Figure 2.15), interpreted as indicative of rapid changes in ice volume and, possibly, temperature. Isotope and chemical analyses from the GRIP (Greenland Ice Core Project) ice core from Summit, central Greenland, suggest that in Greenland, between approximately 135,000 and 115,000 years ago, during the last interglacial (known as the Eemian interglacial in Europe and correlated with the Sangamon in North America, which was warmer overall than the present case), there were intervals of severe cold conditions, which began extremely rapidly and lasted from decades to centuries (Dansgaard *et al.* 1993, GRIP Members 1993). The past 10,000 years have witnessed a relatively stable, interglacial climate, but prior to this during the last ice age, which lasted about 100,000 years, and in

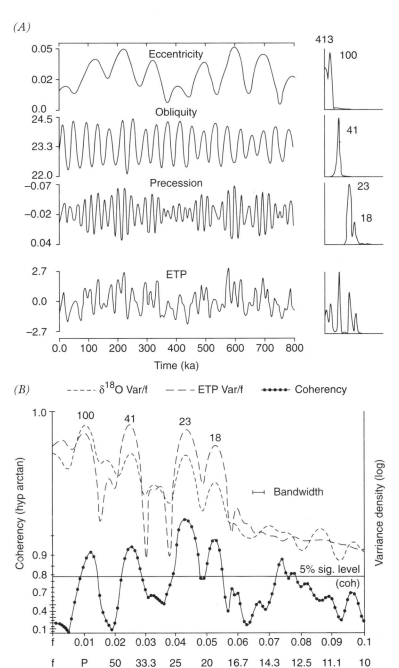

Figure 2.13 (A). *Numerical simulations of variation in the precession ($\Delta e \sin \varphi$), obliquity and eccentricity (degrees) during the past 800,000 years. The curve labelled ETP represents a normalised and summed combination of the above quantities. Shown on the right-hand side are the power spectra of each curve with the dominant periods in thousands of years indicated. (B). Power spectra comparison between ETP and $\delta^{18}O$ variations for the past 780,000 years, showing good agreement between calculations and the geological record. The lower curve shows the coherency. Redrawn after Imbrie* et al. (1984) *in Torbett (1989).*

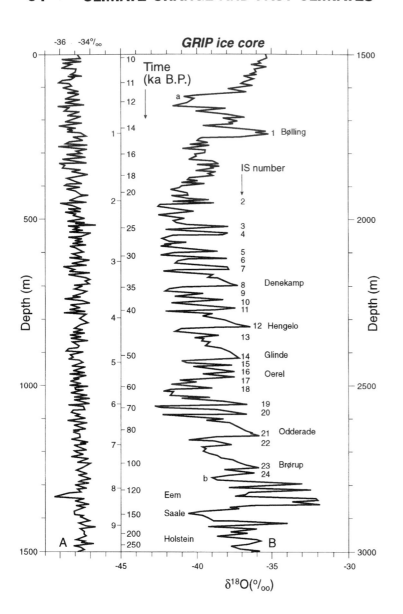

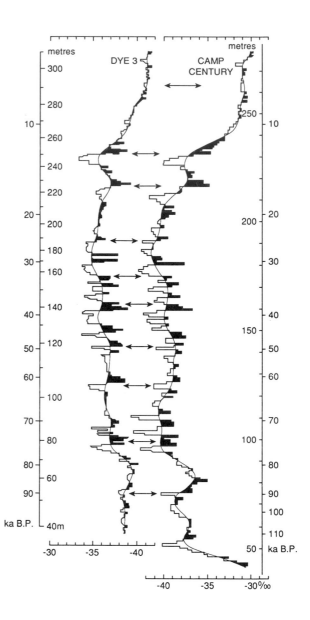

Figure 2.14 *δ¹⁸O record from the GRIP Summit ice core. Note that the sections are plotted in two on a linear scale (A) and a log scale (B). Each point represents 2.2 m of core increment. Glacial interstadials are numbered to the right of the B curve. The time scale in the middle was obtained by counting annual layers back to 14.5 ka BP, and beyond that by ice flow modelling. The glacial interstadials of longest duration are reconciled with European pollen horizons. Redrawn after Dansgaard et al. (1993).*

Figure 2.15 *Rapid fluctuations of ice volume during the last glacial period, from about 80,000 to 30,000 years ago, derived from δ¹⁸O profiles obtained from two Greenland ice cores, Camp Century and Dye 3. Note rapid bimodal fluctuations. Carbon dioxide measurements also reveal variations between two states, in general agreement with the δ¹⁸O data. Redrawn after Oeschger and Mintzer (1992), originally published by Dansgaard et al. (1982).*

the transitional period, global climate change was abrupt and erratic. The GRIP team has shown that changes of up to 10°C occurred within a couple of decades, possibly even less than a decade.

   Other examples of rapid climate change come from uranium/thorium dating of carbonate lake (lacustrine) sediments in the dry valleys along the western margin of the West Antarctic ice sheet, which shows

that there were rapid and marked retreats of grounded ice 130,000–98,000 years ago (Denton *et al.* 1989). The apparently sudden and sporadic, possibly chaotic, collapse of the West Antarctic ice sheet over the past million years led MacAyeal (1992) to develop a 'finite-element' computer model of ice sheet flow and mass balance that reproduces the present-day flow regime of the ice sheet. He pointed

BOX 2.4 HEINRICH EVENTS AND RAPID CLIMATE CHANGE

Evidence of repeated rapid climate change on a time scale of about 10,000 years (actually changing from about 13,000 years to 7,000 years spacing over the last glacial cycle) has come from deep sediment cores from the Dreizack seamounts in the eastern North Atlantic in what have become widely known as Heinrich layers (Heinrich 1988). The *c.* 10,000-year periodicity may be associated with alternations in the relative strength of the Northern and Southern Hemisphere polar seasonality caused by the precession of the Earth.

Heinrich layers differ from more typical ice-rafted debris in four main ways:

- *c.* 20 per cent of the sand-sized material is detrital limestone, whereas the surrounding glacial sediments have virtually none;
- the clay size fraction contains *c.* 1 Ga-old rock fragments, more than twice the age of the ambient glacial sediments;
- the Heinrich layers, in contrast to the ambient glacial sediments, do not contain clay minerals derived from basal tills; and
- they contain relatively few foraminifera tests, typically an order of magnitude less than the surrounding glacial sediments.

These characteristics of Heinrich layers, and the overall eastward thinning by more than an order of magnitude away from the Labrador Sea to the termination of the iceberg route at about 46°N, show that the Heinrich layers originated as sediment derived from Canada. The widespread distribution, to 40°N, of the polar foraminifera *Neogloboquadrina pachyderma* (left-coiling) suggests that the Heinrich events occurred when the North Atlantic was at its coldest, and the lower $\delta^{18}O$ values in the associated foraminifera in these layers suggest that there was a low-salinity water mass above the site where Heinrich layers accumulated. All this evidence points to time intervals when the North Atlantic was covered by extensive sea ice as in the present Arctic Ocean.

Heinrich events are demonstrably coincident with rapid and major changes in the thermal conditions in the North Atlantic region, e.g. at the transition from the relatively warm interglacial marine stage 5 to the cold last glacial marine stages 4, 3 and 2 (Heinrich event number 6). Evidence such as this led Broecker (1994) to speculate that Heinrich events resulted from the periodic release and melting of massive icebergs into the North Atlantic from the Canadian margin, to input large volumes of fresh water into the oceanic conveyor belt and disrupt the formation of deep water masses. As a consequence of the release of these massive icebergs, Broecker postulated that the catastrophic disruption of deep-water formation in the North Atlantic forced a switch between glacial and interglacial patterns of thermohaline circulation.

out that the distribution of basal till, which helps lubricate ice sheet movement, possesses inherently irregular behaviour.

Bond and Lotti (1995) have offered an explanation for the **Dansgaard–Oeschger cooling cycles (D–O cycles)** (Dansgaard *et al.* 1993; Figure 2.16) by showing that the amount of glacial ice discharged in the North Atlantic increased suddenly every 2,000 to 3,000 years coincident with these cycles. Thus, the D–O cooling cycles occur at a greater frequency than the 7,000–10,000-year cycles related to the massive discharge of icebergs into the North Atlantic associated with the **Heinrich events/layers** (see Box 2.4). However, each Heinrich event was followed by a pronounced global warming and then a package of higher-frequency D–O cycles in a progressive cooling trend (Bond *et al.* 1992, 1993) – referred to as **Bond cycles** (Figure 2.16). The minima in CH_4 concentrations during the cold intervals of the D–O cycles and Younger Dryas are thought to be related to the storage of CH_4 in tropical wetlands. As yet, there is no satisfactory cause-and-effect explanation for the D–O and Bond cycles, but they appear to be linked in some very profound way to the release of enormous quantities of fresh water into the North Atlantic and its interference with the production of

deep water at the start of the ocean conveyor belt (Broecker 1994).

Although studies of Greenland ice cores and North Atlantic deep-sea sediments suggest rapid climatic changes during the last glaciation and preceding interglacial, in agreement with observations from lacustrine sediments along the Californian margin and in France, the first evaluation of such data from the high-latitude North Pacific region was with two high-resolution records of input of ice-rafted debris to the sub-Arctic Pacific Ocean preserved at ODP Sites 882 and 883 (Kotilainen and Shackleton 1995). They examined climatic variability in the North Pacific Ocean during the past 95,000 years using a γ-ray attenuation tool (GRAPE, Gamma Ray Attenuation Porosity Evaluator) over unsplit core to provide a measure of the ratio of biogenic opal to terrigenous material (based on chemical analyses of the sediments). Biogenic opal and terrigenous material have large differences in wet-bulk density: biogenic opal is associated with high porosity and low GRAPE density, whereas the terrigenous material from ice-rafting has low porosity and high GRAPE density. Kotilainen and Shackleton found a good agreement between the GRAPE data from ODP Sites 882 and 883 and GRIP $\delta^{18}O$ record from Summit

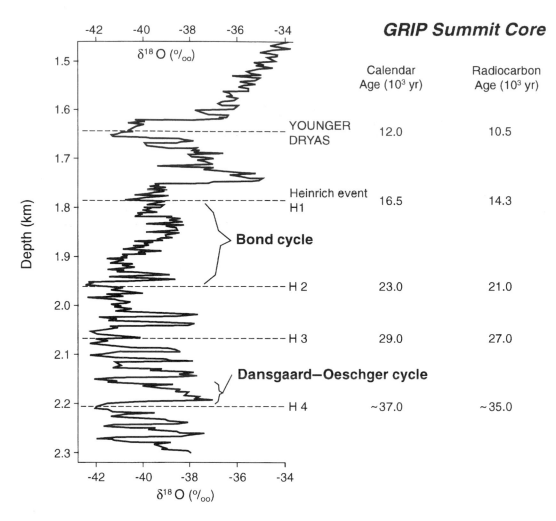

Figure 2.16 *Placement of Heinrich events in the GRIP ice core oxygen isotope record from Summit, Greenland, after Bond et al. (1993). Heinrich events occur in the last cold phase of a series of Dansgaard–Oeschger cycles and precede a major interstadial warm pulse. Redrawn after Broecker (1995).*

in Greenland, with the higher GRAPE values correlated with lower $\delta^{18}O$ values in GRIP (Figure 2.17). The conclusion is that during the many brief cold events associated with the last glacial (stages 2, 3 and 4) there were pulses of ice-rafted material, probably caused by the increased discharge of icebergs, into the North Pacific (cf. Heinrich events in North Atlantic), i.e. high-frequency climatic variability as characteristic of the entire north high latitudes (*ibid.*).

A high-resolution palaeoclimate and palaeogeographic record for the past 20,000 years, gleaned from benthic and planktonic foraminifera and sediments in the Santa Barbara basin on the eastern margin of the North Pacific Ocean, shows rapid oscillations (*c.* 1,000–3,000-year variation) in the benthic environment between low-O_2 conditions producing laminated sediments associated with warmer climatic

conditions, and higher O_2 conditions during which bioturbated, non-laminated sediments accumulated during cooler climatic conditions (Kennett and Ingram 1995). It appears that during cooler climatic intervals, relatively young bottom waters form as a consequence of the enhanced production of intermediate waters derived from nearby sources, whereas the warmer periods are associated with older bottom waters derived from more distal sources (*ibid.*). Furthermore, Kennett and Ingram found that the climate-controlled changes in ocean circulation operating in the Santa Barbara basin were synchronous with those documented from the North Atlantic, suggesting a tight coupling mechanism between the Atlantic and Pacific Ocean basins. The actual cause of this coupling remains poorly understood and may have resulted from changes in the strength of the thermohaline circulation controlled by the produc-

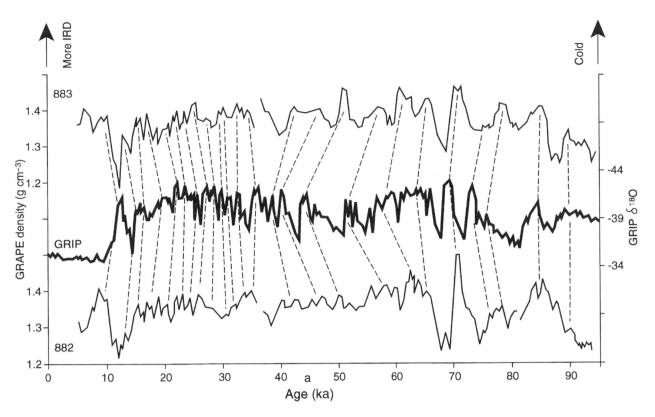

Figure 2.17 *Age plotted against GRAPE (Gamma Ray Attenuation Porosity Evaluator) tool density records from Ocean Drilling Program Sites 882 and 883 (thin lines), and oxygen isotope record from the GRIP Greenland ice core in Summit, Greenland (thick line). Possible correlation between GRAPE records and GRIP oxygen isotope record shown by dotted lines. Redrawn after Kotilainen and Shackleton (1995).*

tion of North Atlantic Deep Water (NADW), i.e. changes in the ocean conveyor belt, or possibly as a direct consequence of global climate change transmitted through the atmosphere (*ibid.*).

In contradistinction to the latter viewpoint, geochemical data have challenged the perception that rapid climate change in the North Atlantic at the end of the last glacial was due to the switching 'on' and 'off' of the thermohaline circulation (Lehman and Kelgwin 1992, Veum *et al.* 1992), but rather that the oceanic circulation oscillated between a warm, deep mode and a cold, shallow mode (Rahmstorf 1994). Computer model simulations in an idealised coupled ocean–atmosphere system for the North Atlantic are capable of reproducing such rapid climatic fluctuations as a response to the abrupt input of fresh water, resulting in a fall in sea-surface temperature by up to 5°C within less than ten years (*ibid.*). Also, these models suggest that the rate of production of North Atlantic Deep Water (NADW) is identical in a cold or warm climate, although in a cold climate the NADW sinks to intermediate depths only, and Antarctic Bottom Water (ABW) pushes northward to cover the entire abyssal Atlantic Ocean (*ibid.*).

Proxy temperature records, $\delta^{18}O$, from Greenland Summit ice cores and North Atlantic sediment cores (DSDP Site 609, ODP Site 644, and V23–81) have revealed a high degree of climatic instability during the last glacial period. This climatic instability has been shown to have been either in phase with, or phase-locked, with air temperature changes over Greenland (Fronval *et al.* 1995). Such a relationship suggests that the rapid changes in heat fluxes in the North Atlantic region were due not only to rapid and abrupt release of large volumes of fresh water from the North American and Greenland but also the Fennoscandian ice sheet (*ibid.*).

Wilson (1964, 1969) suggested that a glacier surge of the Antarctic ice sheet could have considerable effects on global climate. A surge would increase considerably the aerial extent of **ice shelf**, thereby increasing the Earth's albedo (reflectivity), with a consequent global cooling and increased formation of ice sheets in the Northern Hemisphere, in turn initiating renewed glaciation. Break-up of an ice shelf would decrease the albedo, favour rapid melting of ice sheets and, therefore, the termination of a glaciation – associated with a rapid rise in global sea level.

Wilson (1969) argues that if surges occurred, they would cause a rapid rise in sea level (100 years or less) as the ice melts, with renewed ice storage being associated with much slower falls in global sea level, of the order of 50,000 years. Interglacial pollen profiles should, therefore, record a rapid but tempo-

rary marine transgression beginning at the break of climate, and although some evidence exists for such profiles in the UK and USA, these cannot be linked unequivocally to surging but may be the result of other factors such as localised tectonic subsidence.

There are increasing data to show the global correlation and synchronous nature of high-frequency climate change, for example from the Arctic (Greenland) to the Antarctic (Figure 2.18), between the major ocean basins, and between the record from the oceans and nearby ice sheets (Figures 2.19, 2.20 and 2.21). The correlations, however, are not always on a one-to-one basis. For example, isotopic correlations by Bender et al. (1994) between Greenland and Antarctica for the past 140 ka, as a proxy for past global climate change, have resulted in the identification of 22 interstadials from the GRIP and GISP2 ice core records, Greenland, during the part of the last glacial spanning the interval 105–20 ka, compared with only nine interstadials in the Vostok ice core from Antarctica for the same period. Bender et al. show that only for interstadials identified from the Greenland GISP2 ice core lasting longer than 2,000 years is there a corresponding warm interval over Antarctica (Figure 2.18). Warm interstadials can be related to local minima in the $\delta^{18}O_{foraminifera}$ record from the V19–30 deep-sea sediment core, interpreted as times of relatively high sea level. Tie-points used as the basis for correlations on the ice core age versus depth profile for the GISP2 ice core were identified by Bender et al. at 52.4 ka (2,450 m), 57.5 ka (2,500 m), 63.8 ka (2,550 m), 69.5 ka (2,589 m), 75.6 ka (2,628 m), 82.2 ka (2,667 m), 87.6 ka (2,693 m), 93.8 ka (2,719 m), 100.1 ka (2,745 m), 107.4 ka (2,784 m) and 111.0 ka (2,808 m). Below about 2,400 m in the GISP2 ice core (50 ka), the two age models used begin to diverge with increasing depth to 2,800 m (26 ka), where there is the largest discrepancy, the cause of which is not understood but is probably due to a combination of errors in the absolute chronology to which the $\delta^{18}O_{atmosphere}$ curve was referenced and because of the loss of

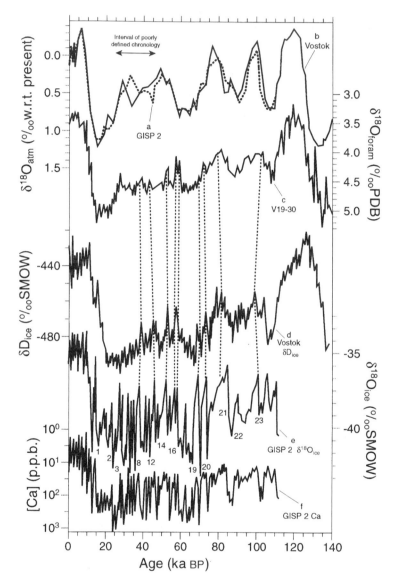

Figure 2.18 *Correlations between the Greenland and Antarctic climate records for the last glacial and interglacial to c. 100 ka, showing the generally good global synchroneity of climate change. Redrawn after Bender et al. (1994). $\delta^{18}O$ PDB or SMOW scale is indicated (see Box 2.1 for explanation). For hydrogen/deuterium (D), $\delta D = [((D/H)_{sample} / (D/H)_{standard}) - 1]$, expressed as per mil. (‰). Time scales used are (1) GISP2 ice core to 2,250 m depth, layer-counting chronology of Meese et al. (1994) and Alley et al. (1993); (2) GISP2 ice core between 2,250 and 2,800 m depth, gas age by correlation of the GISP2 $\delta^{18}O_{atmosphere}$ record with the Antarctic Vostok $\delta^{18}O_{atmosphere}$ of Sowers et al. (1993) between 37.9 and 11.0 ka, labelled **a** (GISP2) and **b** (Vostok). Also marked is interval over which Vostok ice core age control is poor (25–49 ka). **c**. Benthic $\delta^{18}O_{foraminifera}$ record from deep-sea sediment core V19–30 (3° 21'S, 83° 21'W, 3,091 m Uvigerina senticosa) as a proxy for variations in the volume of continental ice. **d**. Vostok ice core δD_{ice} as proxy for temperature with dashed lines tying the interstadial events at Vostok with the longer Dansgaard-Oeschger cycles in GISP2 ($\delta^{18}O_{ice}$). **e**. Enumerated interstadial events below the GISP2 record as identified by Dansgaard et al. (1993). **f**. Calcium (Ca) data from GISP2 on an inverted log scale. See text for explanation. Redrawn after Bender et al. (1994).*

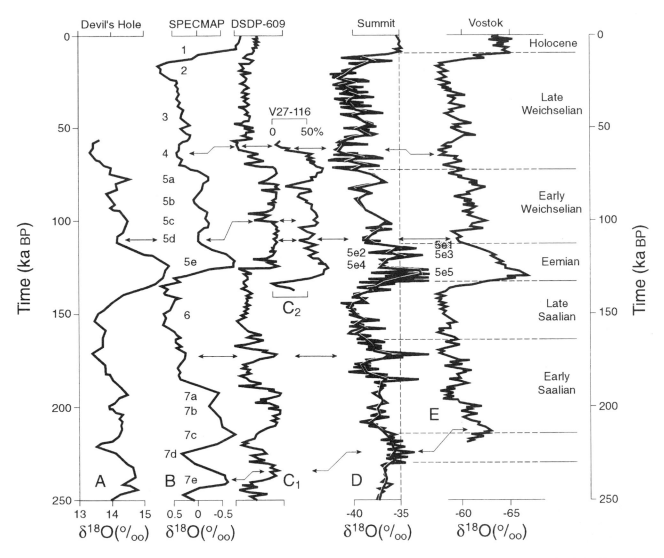

Figure 2.19 *Four climate records spanning the last glacial cycle plotted on a common linear time scale. (A) δ¹⁸O variation in vein calcite from the Devil's Hole, Nevada, dated by U/Th methods. (B) The SPECMAP standard isotope curve with conventional marine isotope stages and sub-stages, dated by orbital tuning. (C1) Grey-scale measurements along 14.3 m of ocean sediment cores from DSDP Site 609. (C2) per cent CaCO₃ in Atlantic sediment core V27–116 (through isotope stage 5) from locations WSW and W of Ireland: scale in arbitrary units on top and dated by orbital tuning. (D) δ¹⁸O record along the upper 2,982 m of the GRIP Summit ice core. Each point represents a 200-year mean value. The heavy curve is smoothed by a 5-ka gaussian low-pass filter. Dating by counting annual layers back to 14.5 ka BP and beyond that by ice flow modelling. Along the vertical line, which indicates the Holocene mean δ value, is added an interpretation in European terminology. (E) δD record from Vostok, East Antarctica, converted into a δ¹⁸O record by the equation δD = 8 × δ¹⁸O + 10‰. Dating by ice flow modelling. Redrawn after Dansgaard* et al. *(1993).*

annual layers in the ice as a result of thinning (*ibid.*). A fundamental inference from this study is that there are times when the oceanic and atmospheric records are not coupled in a simple way, and that there are times when the Northern and Southern Hemispheres, and probably different continents, experience the effects of stadials and interstadials to varying degrees. Bender *et al.* propose that such climatic

differences between the Arctic and Antarctic may be a consequence of periodic suppression in the production of cold North Atlantic Deep Water (NADW), etc.

Any analysis of deep-sea cores that attempts to link the sedimentary geochemical signature and isotopic record, particularly for benthic microfossils such as foraminifera (e.g. the δ¹³C record), to changes in sur-

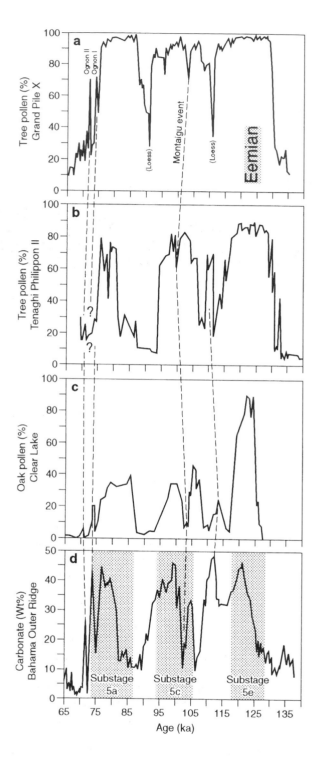

face-water biological productivity (biomass) assumes that there is a direct link. While this assumption is intuitively likely, it was not until studies such as those by Ganeshram *et al.* (1995) from the eastern tropical North Pacific, for the past 140 ka, that a clear link between large changes in the nutrient supply and various geochemical proxy data was firmly established (Figure 2.22).

Amongst the innovative ways of looking at global climate change in the ancient sedimentary record, in addition to geochemical and palaeontological techniques, the colour of very finely laminated deep sea sediments can be used from basins with as continuous a depositional record as one is likely to encounter anywhere. Studies using such colour variations in deep-sea cores from the offshore Californian basins, with digitised data and analysis by power spectra methods, has revealed annual, sunspot, ENSO and Milankovitch changes (Figure 2.23).

Examples such as these emphasise the hierarchy and complexity of rapid global climate change and emphasise the need for even more proxy data, such as that obtained from sediments, their chemistry, and landforms, in order to test various postulates and models proposed to explain such rapid and/or abrupt changes in global climate.

Volcanic activity

Volcanic activity influences both long- and short-term global climate (Plate 2.5). On a scale of many millions to tens of millions of years, increased igneous activity can emit enormous volumes of greenhouse gases and increase the rate at which new oceanic crust is generated at spreading centres, such as the present-day Mid-Atlantic Ridge or East Pacific Rise. Increased emissions of greenhouse gases can lead to substantial global warming. The enhanced production of thermally warm and buoyant oceanic crust causes a shallowing in the mean water depth in the oceans, which in turn leads to a flooding of the land surface, seen as a rise in sea level. Both these effects occurred together during the Cretaceous Period of Earth history, with the result that during that greenhouse phase global or

Figure 2.20 *Correlation of millennial-scale climatic events in the North Atlantic, using weight per cent (wt. %) CaCO₃ in the western North Atlantic core GPC9 (28° 14.7'N, 74° 26.4'W) as a proxy for deep ocean circulation (d), with pollen records from La Grande Pile in France (a), from Tenaghi Philippon in Macedonia (b), and from California (c). The SPECMAP age model was applied to the pollen data. Shaded intervals in the GPC9 panel denote warm substages of interglacial isotope stage 5. Dashed lines correlate the Ognon I and Ognon II warm events of c. 70–75 ka, the cold Montaigu event, which occurred c. 103 ka, and a warm event within stage 5d (c. 113 ka). Kelgwin et al. (1994) propose that these and other short-duration events in palaeoclimatic proxy data are global, and may be related to brief changes in North Atlantic thermohaline circulation. Redrawn after Kelgwin et al. (1994).*

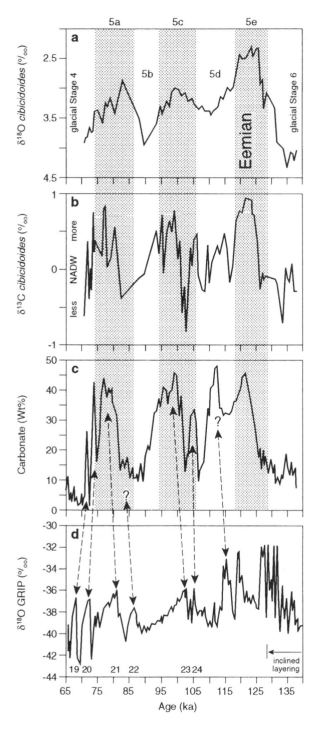

Plate 2.5 *Mount Fuji, Honshu, Japan. Volcanic eruptions may cause global climate change over durations of up to a few years if they eject sufficient aerosols high enough into the atmosphere to alter the Earth's albedo.*

eustatic sea level was up to a couple of hundred metres higher than at present.

In the shorter term, down to periods of a year, volcanic eruptions can eject large volumes of gases and ash which have relatively short-term effects on climate. Large eruptions can pump enough ash into the higher levels of the atmosphere to cause a reduction in the solar flux to the Earth's surface. The 1991 eruption of Mount Pinatubo caused a cold-air temperature anomaly throughout the Middle East during the winter of that year, which in turn appears to have led to unusually deep vertical mixing of the waters in the Gulf of Eilat (Aqaba) to depths greater than 850 m (Genin *et al.* 1995). This deep vertical mixing generated the increased supply of nutrients to the surface waters and, therefore, large phytoplankton and algal blooms which, by the following spring, formed a thick mat of filamentous algae over large areas of the coral reefs, with the result that there was extensive coral mortality, especially of the

Figure 2.21 *δ¹⁸O and δ¹³C isotope data from the benthic foraminifera Cibicidoides spp. in the western North Atlantic core KNR31-GPC9 (28° 14.7'N, 74° 26.4'W) compared with the δ¹⁸O of the GRIP ice core from Summit, Greenland (d). GPC9 data are plotted versus age using δ¹⁸O stratigraphy and the Martinson et al. (1987) chronology. Warm interglacial sub-stages in the sediment core are shaded, and interstadials in the core are numbered. The ice core chronology is pinned at the 110-ka level, denoted by the solid vertical line. Both GRIP and GISP2 records correlate, and are thought to be reliable back to interstadial 23, but the deeper occurrence of inclined layering at GRIP suggests that it may have a record that is reliable as far back as c. 129 ka. Variability in the δ¹³C record in GPC9 probably reflects changes in the relative proportion of North Atlantic Deep Water (NADW) and Antarctic Bottom Water (ABW). CaCO₃ variability, likewise, is interpreted by Kelgwin et al. (1994) as due to changes in the thermohaline circulation, which affects meridianal heat flux in the surface North Atlantic, probably linked to atmospheric temperature over Greenland for events indicated by the short dashed arrows. Redrawn after Kelgwin et al. (1994).*

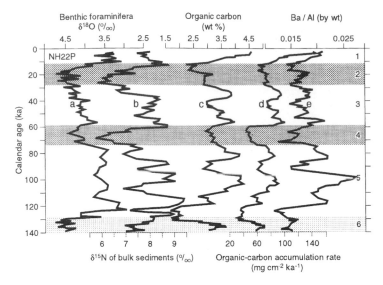

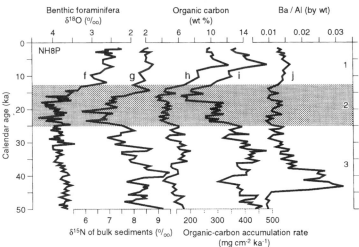

Figure 2.22 *Graphs to show correlation between large changes in oceanic nutrient supply from glacial to interglacial periods for the eastern tropical North Pacific during the past c. 140 ka, measured as (**a, f**) benthic foraminifera δ¹⁸O record (‰); (**b, g**) δ¹⁵N of bulk sediments (‰); (**c, h**) organic carbon (wt. %); (**d, i**) organic carbon accumulation rate (mg cm⁻² ka⁻¹); (**e, j**) Ba/Al ratio by wt. (high Ba concentrations in marine sediments as a proxy for high productivity). Redrawn after Ganeshram et al. (1995).*
Organic-carbon accumulation rate (mg cm⁻² ka⁻¹) = 2,400 (mg cm⁻³) × [1 − porosity] × [sedimentation rate (cm ka⁻¹)] × [fraction organic carbon], where assumed grain density is 2,400 mg cm⁻³, and porosity was calculated from the measured water content of the sediments. The Ba concentration was normalised to the Al to minimise the variability in the profile that could be attributed to aluminosilicate-hosted Ba, e.g. Ba locked up in clay minerals rather than the organic matter.

branching colonial corals and solitary mushroom corals (*ibid.*). This example shows how volcanic activity can cause dramatic very short-term changes in an ecosystem, permissible because of weak water-column stratification.

Figure 2.24, based on data from the North Atlantic sediment cores V23–82 and on oxygen isotope stages, summarises the major volcanic eruptions during the Late Quaternary in relation to summer sea-surface temperatures. There is also an expanded part of Figure 2.24 for the past 100 years, showing the relationship between major historic volcanic eruptions and the Northern Hemisphere mean annual temperature anomaly. From this figure, it is possible to infer that immediately following some major volcanic eruptions, there is a drop in mean annual temperature, for example associated with Krakatau and Mont Pelée. Lamb (1972) noticed that the wettest and coldest summers over the past three centuries coincided with time intervals of enhanced volcanic activity and, also at such times, Arctic sea ice appears to have been more extensive and persistent.

The increased volcanic activity in the late 1940s and mid-1960s could account for the cold winters during this period. The eruptions of Mount St Helens in the USA (June 1980) and El Chichon in Mexico (April 1982) appear to have caused only a short-term reduction in solar radiation, thus not all volcanic activity causes significant climatic change. The increased use of motor vehicles from the late 1940s onwards may have also increased atmospheric dust by combustion–dust loading. This may have been more important in cooling the winters during the middle of this century. Today, motor vehicle emissions have been greatly reduced in developed countries by improved legislation, but as anyone who has travelled in the developing world knows, there is little or no pollution control in big cities such as Delhi, Bangkok and Beijing.

During the Late Quaternary, the Toba eruption in northern Sumatra, dated at c. 73,500 years BP, was probably, by order of magnitude, the largest volcanic eruption (Chesner *et al.* 1991). The eruption has been correlated with the oxygen isotope stratigraphy (Ninkovich *et al.* 1978). Ash from the Toba eruption was transported up to 2,500 km west of Sumatra and deposited on land as far away as India (Stauffer *et al.* 1980, Ninkovich *et al.* 1978). 10¹⁵ g each of fine ash and sulphuric acid were believed to have been emitted (Rampino and Self 1992). It is argued that the eruption of such large amounts of ash led to an increase in atmospheric turbidity and global cooling of the order of 3–5°C over a period of several years. This may have initiated rapid ice growth and correspondingly lowered global sea levels, which in turn could have enhanced global cooling and greater sea level falls attributed to the transition from oxygen isotope stage 5a to stage 4 (*ibid.*). Rampino and Self emphasise, however, that the Toba eruption occurred after the start of global sea level fall in the transition of stage 5a to 4, suggesting that other factors were important in initiating the global climatic shift to cooler conditions. The Toba eruption, however, appears to have at least provided a contributory causal factor that probably helped to drive global cooling.

Ninkovich *et al.* (1978) and Fisher and Schmincke (1984), have postulated that the column height of the **tephra** from the Toba eruption may have reached 50–80 km, although others have suggested more modest heights of 27–37 km (Rampino and Self 1992). Wood and Kenneth (1991) have argued that the eruption may have produced a co-ignimbrite eruption column that could have reached heights of 23–32 km. These lower eruption height estimates support the argument that the mass of sulphuric acid injected into the atmosphere may be more important in influencing global climate change than the actual physical power of the eruption. In order to quantify the role of sulphuric acid aerosols in influencing global climate change, further research is necessary.

Large igneous provinces

Over time intervals measured in tens of millions of years, global climate is strongly influenced by the amount of new oceanic crust being produced at oceanic spreading centres (such as the Mid-Atlantic Ridge and the East Pacific Rise, linear, mainly submarine, mountain chains and associated central depressions or graben formed by the extrusion of new and hot basaltic lavas), and also from so-called mantle plumes. Mantle plumes rise diapirically through the Earth's mantle, and are caused by the detachment of mantle melts or magmas from depths in the Earth of 650–670 km (the transition between the lower and upper mantle), and possibly even from sources as deep as the core–mantle boundary to produce so-called 'super-plumes'. At the Earth's surface, the expression of such mantle plumes is the eruption of large volumes of basaltic igneous rocks to produce so-called 'large igneous provinces' with diameters of up to about 1,400 km. Mantle plumes are in the order of 200°C hotter than the surrounding mantle through which they rise, and therefore are commonly associated with large-scale uplift or doming of the Earth's crust.

An ancient example of a large igneous province produced by a mantle plume acting like a blow torch to the base of the Earth's crust is the 'Tertiary North Atlantic Igneous Province', represented above sea

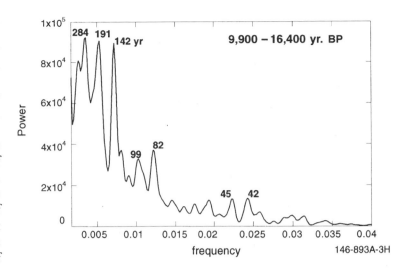

Santa Barbara Basin sedimentary climatic cycles. Power spectra from Fourier transformations of colour density.
ODP Site 893

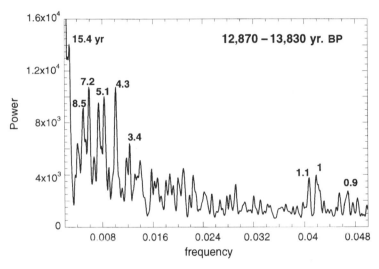

Figure 2.23 *Examples of power spectra based on colour variations in finely laminated sediments obtained from deep-sea drilling sites in the eastern Pacific basin off California. These power spectra for time intervals shown in years BP reveal annual (seasonal) changes, ENSO, sunspot, and Milankovitch cycles and quasi-periodic cycles. Although the mathematical treatment of the original digitised data inevitably retains some harmonic frequencies, many of the numbered power spectra peaks are believed to reflect real changes in climate that, excluding the annual and ENSO events, probably were truly global. Redrawn after Schaaf and Thurow (1995).*

level by parts of Iceland and northwest Scotland, and which was extruded over a very short geological time interval approximately 55 million years ago. Other examples include the Ontong–Java Plateau in the western central Pacific Ocean, where an estimated 12–15 km^3 of igneous rock was extruded annually, or the approximately 65 Ma Deccan Plateau basalts, India, where an estimated 2–8 km^3 of igneous rock

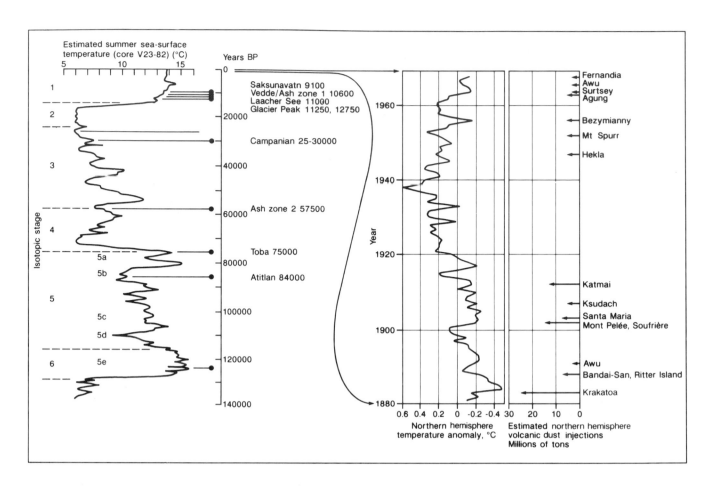

Figure 2.24 *Timing of major volcanic eruptions in the Late Quaternary in relation to summer sea-surface temperatures from North Atlantic core V23–82 and oxygen isotope stages, and in relation to historical records of the Northern Hemisphere temperature anomalies during the last 100 years. Estimates of the amounts of ejected volcanic dust are given for eruptions during that period. Notice the close coincidence between major volcanic eruptions, rapid changes of sea-surface temperatures and negative temperature anomalies. Adapted from Dawson (1992), and Decker and Decker (1989).*

was extruded annually (Coffin and Eldholm 1993). Given that the estimated global network of mid-ocean ridges has produced 16–26 km³ of new oceanic crust each year over the past 150 million years, these large igneous provinces have created new crust at rates comparable with, or greater than, that of sea-floor spreading. It has been estimated that a single flood basalt event that generates 1,000 km³ of lava, typical of the 16 Ma Columbia River igneous province in the western USA, is associated with the emission of 16×10^{12} (trillion) kg CO_2, 3×10^{12} (trillion) kg of sulphur and 30×10^9 (billion) kg of halogens (F, Cl, Br) (*ibid.*). Since large volumes of gases such as CO_2 and SO_2 are emitted from the Earth's mantle, any dramatic increase in the rate of generation of oceanic crust and associated mantle degassing (and/or accelerated global igneous activity) over short time

intervals will have a profound forcing effect on global climate. A good example of this effect occurred during the Cretaceous Period of Earth history, when igneous activity peaked around 120 million years ago with very large-volume volcanic activity centred in the Pacific Ocean basin. This Cretaceous igneous activity appears to have been associated with a greenhouse period of Earth history, when global mean annual temperatures were much higher than today (of the order of 10°C higher), global sea level was higher (by more than 100 m), and organic-rich black muds accumulated in many parts of the world's oceans in oxygen-poor waters created by the decreased rate of ocean-current circulation in the warmer climate and, therefore, its reduced ability to dissolve oxygen and ventilate the world's oceans.

Many of the large igneous provinces appear to be

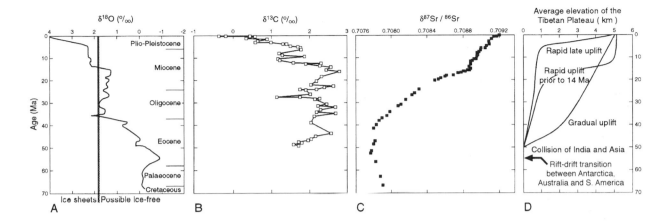

A = deep-sea Atlantic Ocean cores
B and C = marine carbonates
NB. broad correlation with uplift prior to 14 Ma

Figure 2.25 *Comparison between isotope curves and models for Tibetan uplift. (A) Simplified compilation of δ¹⁸O measurements from deep-sea cores in the Atlantic Ocean. (B) δ¹³C curve for marine carbonates over the past 70 Ma. (C) ⁸⁷Sr/⁸⁶Sr curve for marine carbonates for the last 70 Ma. (D) Contrasting models for the Tibetan uplift. Note the broad correlation between isotopic curves (see text for discussion), and between the model for rapid Tibetan uplift before 14 Ma. A, B and C redrawn after Raymo and Ruddiman (1992). We believe it is more likely that the Tertiary deterioration in global climate was forced by the continental separation of Antarctica from Australia-Tasmania and from South America to cause the circum-polar Southern Ocean circulation which helped thermally to isolate Antartica with its ice build-up.*

associated with large-scale or mass extinction events in Earth history. For example, the biggest extinction event known throughout Earth history occurred 248 Ma, when about 95 per cent of all marine species were wiped out in an event that coincided with the eruption of the voluminous Siberian Traps, a major igneous province. While a large meteorite impact may have been the principal cause, the eruption of the Deccan Traps about 65 Ma may have contributed to the major extinction event that witnessed the demise of the dinosaurs.

Continental positions and mountain-building

The very long-term changes in global climate, over hundreds of millions of years, are strongly controlled by the position of the continents. As the plates that make up the outer surface of the Earth relentlessly move around, at speeds typically measured in mm to cm per year, so the size and position of continents or land areas change. The theory that explains the movement of these plates is well known as **plate tectonics**. At times in Earth history, there have been supercontinents (e.g. with names such as **Pangaea**

and **Gondwana**), when many continental plates were locked together. At other times, the distribution of continents has been more like it is today, with many large continents separated by large oceans. The size and distribution of these continents, for example centred over polar or equatorial latitudes, profoundly affects global climate. Also, the rate at which ocean basins floored by oceanic crust are created has varied on a time scale measured in tens of millions of years. At times when there was fast production of new oceanic crust at mid-ocean ridges (or spreading centres), greater amounts of heat energy were released from within the Earth together with more greenhouse gases. The result of this enhanced heat exchange between the solid Earth and hydrosphere–atmosphere–biosphere is that it could have caused past greenhouse periods in the Earth's history. These factors can be thought of as internal controls that are entirely a consequence of processes within the Earth's heat engine.

Some scientists believe that mountain-building episodes can give rise to ice ages. Ruddiman and Kutzbach (1991), and more recently Raymo and Ruddiman (1992), for example, have proposed that the uplift of Tibet, the Himalayas and the American southwest caused large areas of land in low latitudes

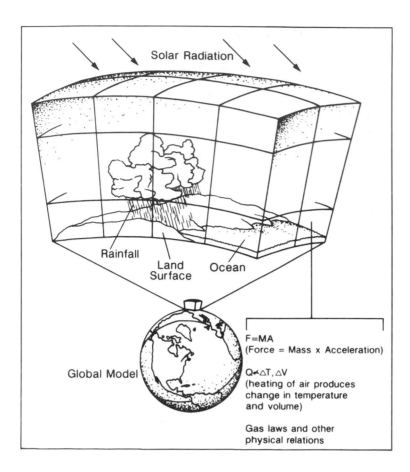

Figure 2.26 *Schematic diagram of global climate system, to illustrate the way in which the Earth's atmosphere–ocean system, and land surface area, is divided into thousands of boxes with sides typically extending several hundred km in latitude and longitude, and with altitudes of a few km. In a general circulation model (GCM), the computer treats each box as a single element as it calculates the evolving global climate. The GCM imposes seasonal and latitudinal changes of incoming solar radiation, the height and shape of the continents, and other external conditions that affect the behaviour of the atmosphere. In GCMs, the equations may be solved in hourly increments over at least 20 years of simulated time to generate an output that is statistically 'accurate'. Such large and time-consuming calculations require the use of supercomputers. Redrawn after Ruddiman and Kutzbach (1991).*

to reach a height that altered global atmospheric circulation patterns, which helped induce global atmospheric cooling (see Figure 2.25). In addition, they argue that increased uplift exposed more rock, which then underwent accelerated rates of chemical and physical weathering. During many weathering reactions, CO_2 is extracted from the atmosphere to react with the decomposing minerals and form bicarbonates. These bicarbonate compounds are soluble in water and are carried in solution to be deposited finally as sediments in the oceans. Also, the uplift increased river gradients, causing the rivers to erode more deeply and carry sediment to the sea at greater rates, and the uplift could have increased storminess along the mountain front, leading to more rainfall and faster-flowing rivers. In essence, there is a net removal of CO_2 from the atmosphere during the chemical reactions associated with the breakdown of rock-forming minerals, a process that can therefore reduce any potential greenhouse warming, and hence encourage a global cooling. Such 'tectonic' processes of mountain building, or orogeny, could provide a negative feedback to the ocean–atmosphere system.

Modelling global climate and climate change (GCMs)

In attempts to understand the nature of global climate change better, scientists are developing computer models to replicate present climatic conditions and to predict future changes in climate. Climate models are used by many research groups to evaluate the effects of the various positive and negative feedbacks that can influence climate change. In effect, such models are less sophisticated versions of the weather forecasting models that appear on the world television networks. The various computer-based 'general circulation models', or GCMs, represent the atmosphere as a finite number of stations both in geographic locations around the world, and three-dimensionally as vertically stacked points in the atmosphere (Figure 2.26). In many GCMs, the oceans tend to be represented as stations with a defined sea-surface temperature, although more sophisticated models are beginning to divide the ocean into vertical slices. From all these atmosphere–ocean stations, a three-dimensional grid of points is fed into a computer program, whose physical states are mathematically linked to neighbouring points.

The program is then run, and the numerical relationships are allowed to evolve in discrete temporal steps until predetermined conditions are satisfied, e.g. a certain time period has elapsed. Because the more sophisticated computer programs require very large amounts of memory and relatively lengthy running times, supercomputers are well suited to GCMs.

An important aspect of GCMs is that they are only models, and the output can only be as good as the data that is input – they are approximations of what may actually happen. For example, the atmosphere and oceans are continuous fluids, but they are represented as finite points in the model. In most GCMs, grid points typically involve horizontal separations of 500 km (100 km in more refined models), and with time steps of say 30 minutes. Cloud cover and cloud type, for example, are parametrised so that their evolution is described by substantial approximations to the physical and chemical processes affecting them, ideally in a manner that preserves the important spatially averaged properties of the variable. Ocean circulation and the way in which heat is transferred within the ocean–atmosphere system is a current area of research, therefore not included in most GCMs, although the most sophisticated models now include a layered ocean and heat transfer. To improve GCMs, much more research is required, especially sensitivity analyses of GCMs to many poorly understood variables, e.g. cloud types and cloud-forming processes, and heat transfer in the oceans.

Although GCMs are being developed mainly for predicting future potential climate, Earth scientists are beginning to make use of such models to try and understand past climates. Probably the three best-known main GCMs are the Canadian Climate model, the US Geophysical Dynamics Laboratory model and the UK Meteorological Office (UKMO) model. Figure 2.27 summarises some of the results from several of these GCMs. Much of the variation between the GCM results is due to the different weightings given to various assumptions.

Quaternary climates

Historical perspective

Towards the end of the eighteenth century, Earth scientists such as the Scottish geologist James Hutton (often referred to as the father of the science of geology) and John Playfair were among the first to develop a theory of glaciation to explain many of the geological phenomena that were then ascribed to the biblical Flood (or diluvial theory). The glaciation

Plate 2.6 *Portraits of selected scientists who have made fundamental contributions to the development of the science of palaeoclimatology: (A) James Hutton; (B) Louis Agassiz; (C) Charles Lyell; (D) Archibald Geike; (E) Milutin Milankovitch.*
B, C and D courtesy of the Royal Geographical Society. A courtesy of the Department of Geology and Geophysics, University of Edinburgh. E courtesy of Vlaso Milankovitch.

theory had already been presented to the Swiss Society of Natural Sciences in 1837 by its young president, Louis Agassiz but was not destined to become widely accepted until the 1860s. There were a number of competing theories besides the diluvial explanation. In 1833, the English geologist Charles Lyell explained the features now known as being of glacial origin, such as erratics and drift deposits, as the products of floating icebergs. In 1840 the Reverend William Buckland, Professor of Mineralogy and Geology at Oxford University, and Charles Lyell eventually accepted the arguments by Agassiz for glaciation. Until then, Buckland had been a committed catastrophist. Indeed, in 1863 Archibald Geike proposed a multiple glaciation hypothesis to explain the superficial glacial deposits in Scotland, a view that is generally accepted to this day (Plate 2.6).

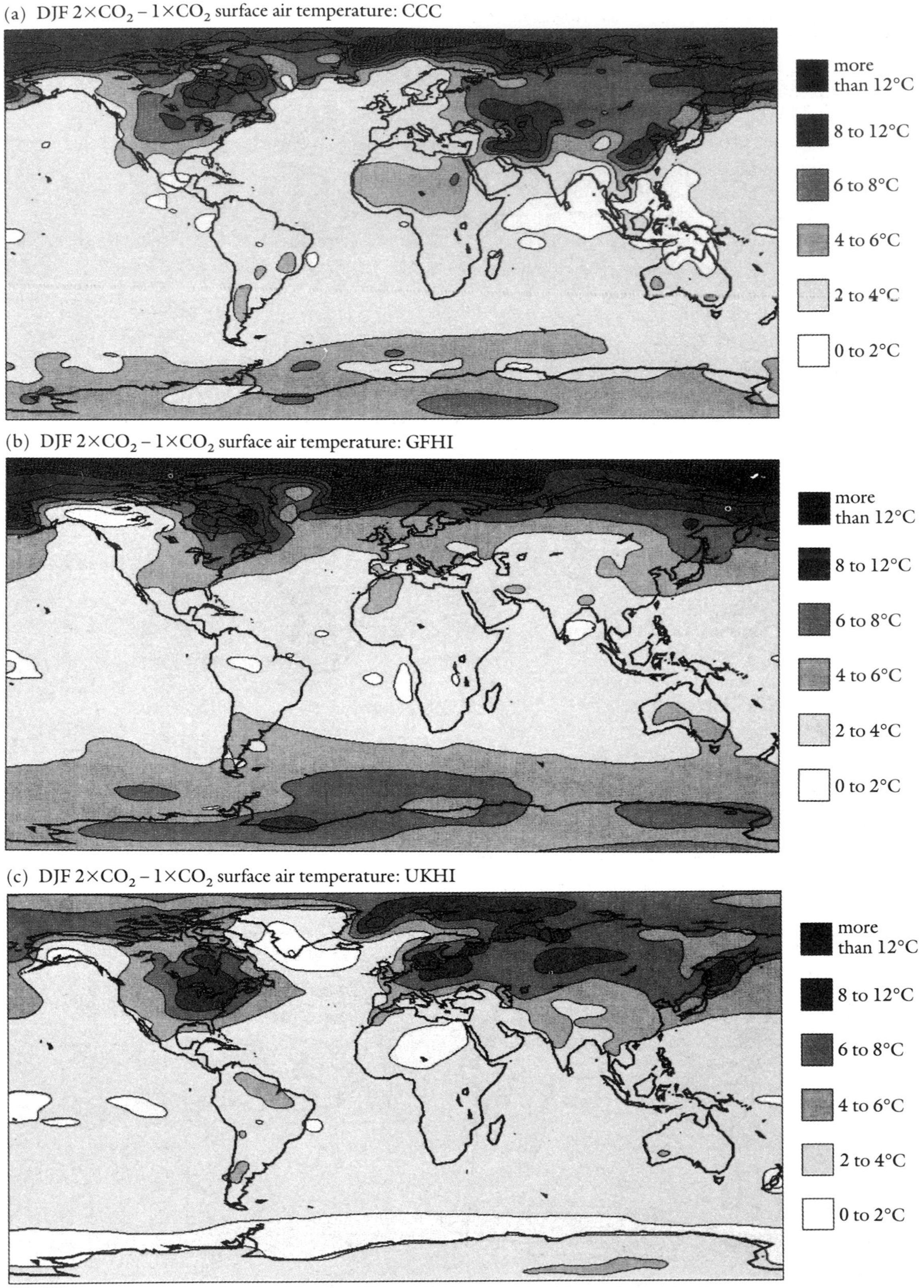

(a) DJF $2\times CO_2 - 1\times CO_2$ surface air temperature: CCC

(b) DJF $2\times CO_2 - 1\times CO_2$ surface air temperature: GFHI

(c) DJF $2\times CO_2 - 1\times CO_2$ surface air temperature: UKHI

Figure 2.27   *GCM output. The change in surface air temperature (10-year means) due to doubling carbon dioxide, for (a, b, c) December–February, and (d, e, f) June–August, respectively, as simulated by three high-resolution models: (a and d) CCC: Canadian Climate Centre; (b and e) GFHI: Geophysical Fluids Dynamics Laboratory; (c and f) UKHI: United Kingdom Meteorological Office. See legend for contour details. After IPCC (1992).*

(d) JJA $2 \times CO_2 - 1 \times CO_2$ surface air temperature: CCC

(e) JJA $2 \times CO_2 - 1 \times CO_2$ surface air temperature: GFHI

(f) JJA $2 \times CO_2 - 1 \times CO_2$ surface air temperature: UKHI

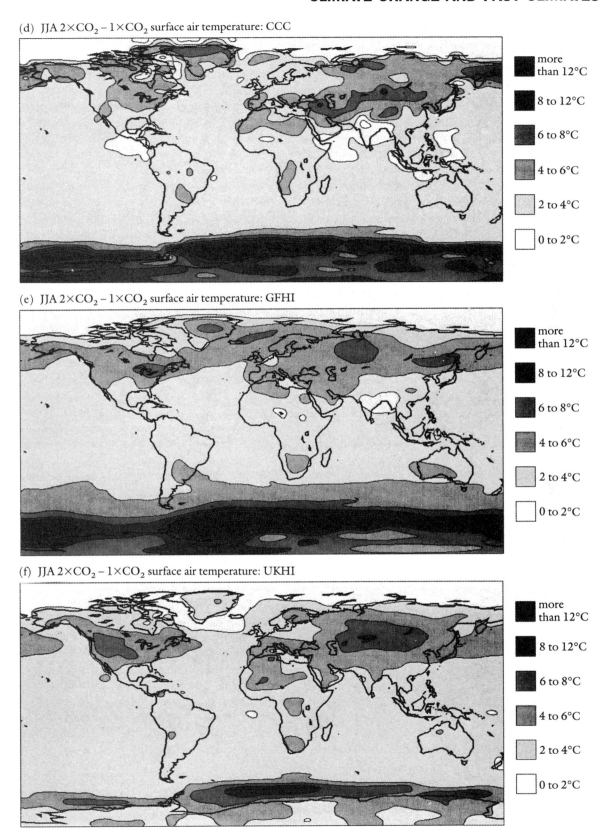

more than 12°C

8 to 12°C

6 to 8°C

4 to 6°C

2 to 4°C

0 to 2°C

As far back as 1909, the Alpine glaciations and interglacial periods were interpreted as alternating warm and cold stages by the German geographers Albrecht Penck and Eduard Bruckner. In 1842, the French mathematician Adhemar invoked changes in the orbit of the Earth around the Sun as the main reason for such climatic change, while in 1864 the Scottish geologist James Croll hypothesised that changes in the Earth's orbital eccentricity could be the cause of the ice ages, a theme he elaborated upon in his book *Climate and Time,* published in 1875. Without a very precise means of dating the climatic changes and linking them to orbital parameters, these ideas lay dormant. It was not until well into the twentieth century, between 1920 and 1940, that these astronomical interpretations for climatic changes on Earth found support and widespread acceptance throughout the scientific community.

A brief history of the Quaternary

The Quaternary Period is defined by Earth scientists as the relatively recent period of geological time spanning the last 1.64 million years of Earth history: many scientists studying the Quaternary would argue that this period should extend back to 2.5 Ma (discussed in detail by Shackleton *et al.* 1990). Floral and faunal evidence suggests that there was an abrupt change from warm to cold climatic conditions anywhere between 2.5 Ma and 1.64 Ma, depending upon which data are used. In 1985, at Vricia in Calabria, Italy, the *International Commission on Stratigraphy* formally defined the base of the Quaternary Period as being where a claystone horizon containing the first appearance of a cold-loving, or thermophobic, foraminifera directly overlies a black mud rich in organic calcium carbonate (called a **sapropel**). The identification of ice-rafted debris in cores from the Antarctic deep sea, however, has placed the onset of glaciation as far back as 3.5 Ma (Opdyke *et al.* 1966), together with other lines of evidence, although dating of marine diatom-bearing glacio-marine strata in east Antarctica suggests that there was an extensive deglaciation of Antarctica during the mid-Pliocene Period *c.* 3 Ma (Barrett *et al.* 1992). Actually, Antarctica supported a continental ice sheet at least as far back as about 35 Ma – earliest Oligocene time (see review of Quaternary by Boulton 1993). In the North Atlantic region, a study of foraminifera linked to oxygen isotope data recovered from a deep-sea drilling site revealed evidence for the onset of glaciation associated with progressively deteriorating climatic cycles,

and ice sheet initiation, at about 2.5 Ma (Shackleton *et al.* 1984). The last 10,000 years of this time interval, defined as the period following the last glaciation, is referred to as the Holocene, and from 1.64 million to 10,000 years ago as the Pleistocene (see time chart in Chapter 1). Further back in geological time, at least five other major global ice ages are known, two in the late Precambrian and three in the Phanerozoic.

The last glaciation was a period of extreme cold on Earth, when the polar ice caps were more extensive than today, much of the continents were covered by continental glaciers and ice caps, and sea level was much lower than at present. Following the last glacial maximum (LGM), deglaciation in Antarctica was well advanced by about 10,000 years BP, and by 6,000 years BP was complete (Colhoun *et al.* 1992). Figure 2.28 shows the variation of relative temperature during the last 20,000 years and the advance of glaciers from selected regions of the world. Also shown is the variation in solar radiation as a precession-related Milankovitch cycle for a latitude of 65°N (Pielou 1991, Grove 1979). Perhaps it is important to emphasise that we are currently still in an icehouse world, experiencing a warm interglacial, something that is apparent from Figure 2.28.

During past glaciations, global sea level was lower because large volumes of sea water were frozen as ice. Colhoun *et al.* (1992) have suggested that the role of Antarctic ice in contributing to global sea level fall at the LGM was dependent on the thickness and extent of peripheral ice. Models suggest that there was probably a thickening of 500–1,000 m, which induced a sea level drop of about 25 m. Evidence from raised beaches in the Ross embayment and East Antarctica shows that sea level dropped by only 0.5–2.5 m. This discrepancy suggests that either sea level fell less than present estimates suggest, or that ice volumes in the Northern Hemisphere must have been considerably larger to account for a global sea level lowering during the LGM. The last major glaciation, when ice cover was at its maximum extent, about 18,000 years ago ('Last Glacial Maximum'), ended fairly abruptly as the mean surface temperature of the Earth increased. Large-scale melting of polar and continental ice ensued and the return of this water mass to the oceans and seas, together with its thermal expansion, led to a rise in sea level of up to 120 m in some parts of the world. Ice core studies at Vostok and Dome C, Antarctica, suggest that during the LGM, the surface of central Antarctica was 200–300 m lower than at present (*ibid.*).

The last major glaciation is known by different names throughout the world, for example the

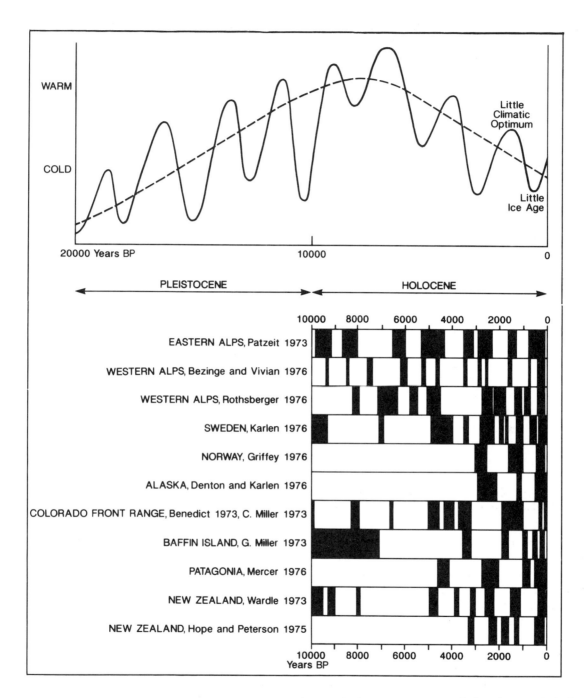

Figure 2.28 *Schematic variations in relative temperature during the last 20,000 years and the advance of glaciers from selected regions of the globe (shaded). The dashed curve shows how solar radiation varied as a precession-related Milankovitch cycle for a latitude of 65°N. Redrawn and adapted after Pielou (1991) and Grove (1979).*

Devensian in Britain, the **Wisconsin** in North America and the **Weichselian** in mainland Europe. The Ice Age was not a single event, but a number of closely spaced cold-period glaciations (glacials) with durations of the order of 100,000 years separated by intervening warmer periods referred to as interglacials (as opposed to brief warm intervals within glacial stages, known as interstadials), lasting 10,000–20,000 years. During the Quaternary, this pattern of glacial and interglacial periods seems to have repeated itself at least ten times. Indeed, a chronology of glaciations in the USA (Figure 2.29)

Figure 2.29
The chronology of
glaciations in the
USA. The main
glacial advances
are shaded. Notice
that the glaciations
occurred as early as
the Pliocene Period.
Redrawn after
Goudie (1992).

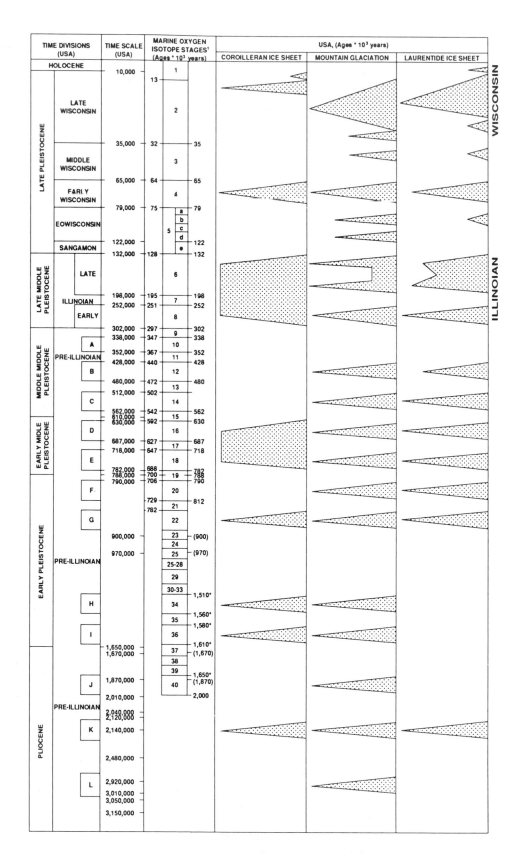

Table 2.1 *Sequence of Pleistocene phases in the Northern Hemisphere.*

Rhine estuary	Britain	Alpine foreland	European Russia	North America
WEICHSELIAN	DEVENSIAN	WÜRM	VALDAI	WISCONSIN
Eemian	Ipswichian	Riss–Würm	Mikulino	Sangamon
SAALIAN	WOLSTONIAN	RISS	MIDDLE RUSSIAN	ILLINOIAN
Holsteinian	Hoxnian	Great Interglacial	Likhvin	Yarmouth
ELSTERIAN	ANGLIAN	MINDEL	WHITE RUSSIAN	KANSAN
Cromerian	Cromerian	Günz–Mindel	Morozov	Aftonian
MENAPIAN	BEESTONIAN	GÜNZ	ODESSA	NEBRASKAN
Waalian	Pastonian	Donau-Günz	Kryshanov	
EBURONIAN	BAVENTIAN	DONAU		
Tiglian	Antian			
PRETIGLIAN	THURNIAN			
	Ludhamian			
	WALTONIAN			
'Pre-Glacial'				

Glacials in capitals; Interglacials in lower case.
Source: Compilation from various sources in Goudie 1992.

reveals many more glaciations, even extending back into the **Pliocene Period** at about 3 Ma. Also, it is now well established that each glaciation involved several, multiple advances and retreats of the ice sheets.

Cave sediments at Skjonghelleren, western Norway, provide good evidence for multiple glaciation during the Weichselian. Larsen *et al.* (1987) have identified evidence for three glaciations over the past 70,000 years. The caves were formed at a high sea level by wave action some time during the Weichselian. These comprise three beds of glacio-lacustrine sediments, which formed subglacially, and which are interbedded with structureless blocky deposits (called diamictons) formed by the collapse of the roof during ice-free periods. The entire sequence was deposited during the Weichselian stage (Figure 2.30). Some of the diamictons contain bones and teeth of birds, mammals and fish, which have been dated using radiocarbon methods, along with dates on speleothems using uranium isotope dating techniques. These radiometric dates cluster around 30,000 BP, the end of the Alesund interstadial, and between 12,000 and 10,000 BP.

Figure 2.31 shows the continental and sea ice at its maximum extent during the LGM. Ruddiman and McIntyre (1981) have discussed the changes in the position of the Polar Front, and the limit of sea ice, as a response to global climate change during the Late Pleistocene to early Holocene (Figure 2.32). From such data, it appears that there was a northward migration (retreat) of the Polar Front *c.* 20,000–11,000 BP, followed by a re-advance at approximately 11,000 BP, attributed to the Younger Dryas.

Table 2.1 shows, for comparative purposes, the correlations of synonymous names for the various Pleistocene phases in the Northern Hemisphere. Actually, between about 11,000 and 10,000 years ago, there was a brief return to near-glacial conditions in an event called the Younger Dryas. This event interrupted the change from the Pleistocene glacial to warmer Holocene climates. Indeed, a study of high-resolution ^{18}O isotope records from benthic (bottom-living) and planktonic microfossils in two radiocarbon-dated cores from the Sulu Sea, western Pacific, has shown that the Younger Dryas was a global event that occurred synchronously and as far afield as in the surface and deep waters of the North Atlantic and the Sulu Sea in the western Pacific, and was associated with low atmospheric CO_2 concentrations (Kudrass *et al.* 1991).

High-resolution reconstructions of past atmospheric $^{14}C/^{12}C$ ratios from annually laminated lake sediments in Lake Gosciaz in central Poland, which may provide important information on the mechanisms of abrupt climate change, show abnormally high ^{14}C concentrations during the Younger Dryas and early Holocene (Goslar *et al.* 1995). Any changes in the size of the various global carbon reservoirs or the exchange rates between them is likely to be manifest in the most robust appropriate atmospheric tracer, the natural radiocarbon isotope ^{14}C. The anomalously high ^{14}C concentration (obtained from plant macrofossils of terrestrial origin synchronised to the younger German oak and older German pine chronologies using a 'wiggle-matching' procedure) in the lake sediments has been interpreted as an expression of a reduced rate of ventilation in the

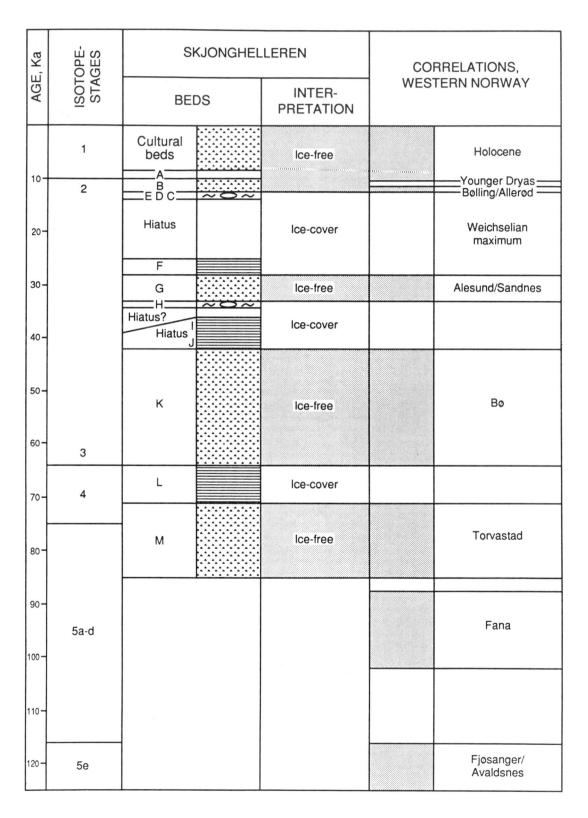

Figure 2.30 *Summary of the stratigraphy in Skjonghelleren and its correlation with western Norway for the past 120,000 years. Redrawn after Larsen et al. (1987).*

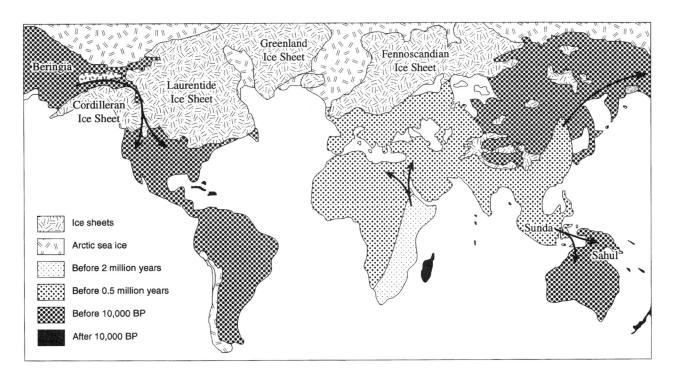

Figure 2.31　*World map to show the migration of humans at various times during their evolution. Redrawn after Roberts (1989).*

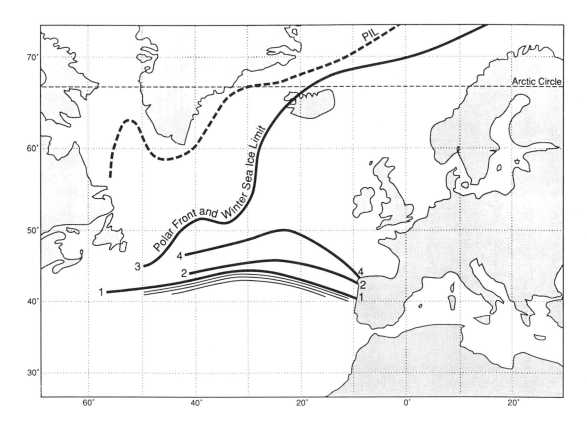

Figure 2.32　*Position of the Polar Front, and limit of winter sea ice, during the period c. 20,000–10,000 years BP. 1 = 20,000–16,000 years BP; 2 = 16,000–13,000 years BP; 3 = 13,000–11,000 years BP; 4 = 11,000–10,000 years BP. Thin lines represent the pronounced thermal gradient to the south of the Polar Front. PIL = approximate southern limit of pack ice at present day. Redrawn after Ruddiman and McIntyre (1981).*

deep ocean, probably resulting from the decreased intensity of the North Atlantic Deep Water (NADW) formation (*ibid.*). This research also led to the Younger Dryas/Preboreal boundary being fixed at 11,640 ± 250 years BP (*ibid.*).

The Younger Dryas was probably the result of the sudden increased rate of melting of the Laurentian ice sheet, with large volumes of cool melt water entering the oceans and affecting atmospheric temperatures. Recent studies on oxygen isotopes in an ice core from Camp Century in Greenland suggest that the Younger Dryas terminated very abruptly, possibly even within a few decades (Johnsen *et al.* 1992). In effect, the Younger Dryas was a brief cool interval in a warmer period, and is referred to as a **stadial**. Evidence is now emerging for abrupt and rapid changes in global climate during the Quaternary, during periods of climatic instability (see Boxes 2.5 and 2.6). These climatic changes are related to glacial–interglacial cycles, which in turn are related to changes in the global carbon cycle. This supports the view that there are strong links between climate, biogeochemical cycles and metabolic processes in organisms.

CO_2 and CH_4 concentrations in the atmosphere have also changed considerably during past glacials and interglacials. During interglacials, there is approximately 25 per cent more CO_2 and 100 per cent more CH_4. These changes in the concentrations of atmospheric gases have important implications for understanding the global carbon cycle. They suggest, for example, that organic productivity and carbon storage was greatest during glacial periods, thereby providing a sink for carbon, for example in the oceans. Such changes in CO_2 and CH_4 concentrations from glacial to interglacial periods appear to have taken place suddenly, that is within a few hundred years (Jouzel *et al.* 1987, Stauffer *et al.* 1988). The precise causes of these changes in atmospheric gas concentrations, and the threshold conditions that precipitated a switch from glacial to interglacial period, remain poorly understood. The release of CH_4 stored as methane-gas hydrates in permafrost may have provided a significant contribution to the rapid rise in atmospheric CH_4 and CO_2, leading to the global temperature rise at the end of the last major glaciation about 13,500 years ago. The release of CH_4 would have led to a strong positive feedback, which could have had the net effect of amplifying the emission of greenhouse gases. This warming, driven by methane release from various reservoirs, may have induced the release of CO_2 from the oceans to the biosphere, thereby stabilising the interglacial carbon cycle at a different level of produc-

tivity. The study and understanding of these changes are important, because a small anthropogenically induced warming could thaw permafrost and release CH_4 from methane-gas hydrates.

Data from the Vostok ice core for the past 160,000 years suggest that tropical wetlands are a leading influence on variations in atmospheric CH_4 levels (Petit-Maire *et al.* 1991). During glacial maxima, CH_4 levels have fluctuated naturally around 350 parts per billion by volume (ppbv), compared with 650 ppbv during warm interglacial periods. The CH_4 record from the Vostok ice core shows four significant temporal periodicities at 110, 38, 24 and 19 ka, in agreement with the orbital parameters of the Earth, i.e. eccentricity (100 ka), obliquity or tilt (41 ka) and precession (23 and 19 ka). This correlation led Petit-Maire *et al.* to propose that orbitally driven changes in monsoon rainfall exert a crucial role in controlling CH_4 emissions from low-latitude tropical wetlands. The precession and eccentricity of the Earth are the principal controls on long-term variations in insolation in the tropics, whereas obliquity or tilt becomes increasingly important with higher latitudes.

Three deep ice cores recovered from the Greenland ice cap show $\delta^{18}O$ profiles that reveal irregular but well-defined episodes of relatively mild climatic conditions, or interstadials, that occurred during the middle and later parts of the last glaciation (Johnsen *et al.* 1992). The oxygen isotope record from these cores suggests that the interstadials lasted from 500 to 2,000 years, and their irregular development has been interpreted in the context of complex behaviour of the North Atlantic Ocean circulation (*ibid.*).

During the past 200,000 years up until the LGM, there was increased global aridity, which led to the most extensive spread of deserts and sand dunes in low latitudes (Sarnthein 1978). Regions such as the Western Sahara and the Sahel were, therefore, once much more extensive. This conclusion is supported by the work of Hovan *et al.* (1989), who examined the influx of wind blown (aeolian) sediments in a deep-sea core from the northwest Pacific Ocean, at a site about 3,500 km downwind from central China. They were able to show increased quantities of wind-blown sediment in the core and relate this to enhanced wind action during a more arid climatic period, which was linked to glacial stages as determined by the oxygen isotope curve. In addition, they were able to relate this influx of aeolian sediments to a sequence of wind-blown silts in Xifeng, China. These wind-blown silts, known as loess (Plate 2.7), contain fossil soils (palaeosols). Palaeosols are thought to develop mainly during interglacials, and

Plate 2.7 *Loess exposed at Luochuan, Shannxi Province, central China, representing 900,000 years of deposition in this 70m thick exposure. Loess sequences such as this provide the most continuous continental record of Quaternary climate change.*
Courtesy of E. Derbyshire.

correspond to times of decreased aridity (wetter intervals), and faster rates of accumulation of aeolian sediments. The evidence from loess supports the view of increased aridity during glacial periods.

Figure 2.33 is a summary of the loess–palaeosol stratigraphy and magnetic susceptibility (MS) of selected loess sequences in China in comparison with deep-sea oxygen isotope curves and aeolian flux into the Pacific Ocean (see Box 2.5). A comparison of the loess stratigraphy and MS at Xifeng with the deep-sea oxygen isotope curve from the equatorial Pacific Ocean reveals a good degree of similarity. There is a correlation between times of loess accumulation, cold climatic intervals, decreased MS values and increased aeolian flux (Hovan *et al.* 1989, Kukla *et al.* 1990, Liuxiuming *et al.* 1992).

There are many ways in which Earth scientists can read the history of the dramatic and cyclic changes in the Earth's climate over the Quaternary. One method is to study the type and relative abundance of plant spores or pollen in ancient sediments.

Recently, French scientists have used the pollen record from sediment cores in eastern France (La Grande Pile and Les Echets) to reconstruct a 140,000-year continental climate (Guiot *et al.* 1989). Mook and Woillard's (1982) work on a core of pollen-rich laminated sediments at La Grande Pile is particularly important as it provides a continuous continental pollen record reflecting climatic change over the past 140,000 years. Sixteen radiocarbon dates help to provide a detailed chronology. Mook and Woillard recognised the onset of a cold period at 70,000 BP, marked by the disappearance of deciduous forests, which they correlate with the transition from oxygen isotope stage 5a to stage 4, i.e. the Early Weichselian–Middle Weichselian transition.

The fundamental assumption behind this, and similar palaeoclimatic studies, is that corresponding vegetation and pollen types existed in similar ecological niches to their counterparts today. The validity of such assumptions needs much more research before scientists can feel confident about the interpretations, but they represent reasonable criteria from which to begin palaeoclimatic studies. Present-day plants may have different climatic requirements than those ancient plants, the variability of past climates may not exist today, and human activities have undoubtedly made a unique impact on modern plant life. With these provisos in mind, it is possible to begin to look at some palaeoclimatic modelling based on fossil plant material in the geological record.

The pollen records for the last 140,000 years, based on data from eastern France, suggest that the Holocene and the last interglacial (known as the Eemian Interglacial) were the warmest and most humid climates of the last 140,000 years (Guiot *et al.* 1989). The main period of global ice growth commenced before 110,000 years, which is defined as the end of the Eemian Interglacial. If the growth of continental ice sheets between latitudes 50° and 60°N implies a cold and humid climate, as suggested by climatic models, then the pollen data indicate three major periods of ice development in Europe during this time interval. The oldest occurred as a humid and markedly cold climate towards the end of the Eemian (approximately 110–115 ka) which immediately pre-dated the even colder and drier Melisey I Stadial (approximately 103–110 ka).

The next period of major ice development in Europe occurred towards the end of the St-Germain I Interstadial, which was very humid and moderately cold, and which was succeeded by the cold, dry Melisey II Stadial (approximately 83–92 ka). The third major ice growth occurred at the end of the St-Germain II Interstadial and into the start of

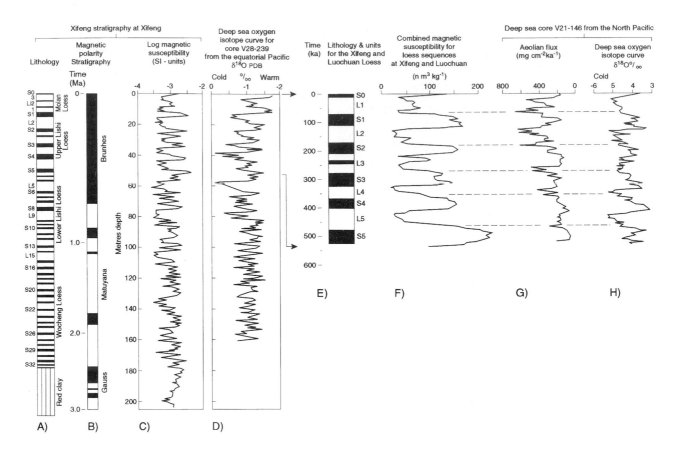

Figure 2.33 *Summary of the loess–palaeosol stratigraphy and magnetic susceptibility (MS) of selected loess sequences in China in comparison with deep-sea oxygen isotope curves and aeolian flux into the Pacific Ocean. A, B and C compare the loess stratigraphy and MS at Xifeng with the deep-sea oxygen isotope curve from the equatorial Pacific Ocean. Note the similarity between the curves. The age of the loess–palaeosol sequence at Xifeng was determined using palaeomagnetic dating (B). (E) and (F) show the combined stratigraphy for three sections at Xifeng and Luochuan. (G) and (H) show the aeolian flux and oxygen isotope curve for a deep-sea core from the North Pacific Ocean. Note the good correlation between loess, cold intervals, decreased MS values, and increased aeolian flux. Adapted and redrawn after Liu* et al. *(1992), Kukla* et al. *(1990) and Hovan* et al. *(1989).*

the substantially colder (and moderately humid) Lower Pleniglacial, prior to the second very cold, dry part of this major stadial (approximately 45–72 ka).

Temperate conditions, not unlike those of the present day, especially in terms of temperature, appear to have existed during the St-Germain I and II Interstadials, approximately 72–84 ka and 93–104 ka, respectively. Perhaps, the most surprising findings from these pollen data are that these temperate climatic phases during the St-Germain I and II Interstadials have not been recognised in sediment cores from the Antarctic ice cores, Pacific Ocean records, Atlantic Ocean deep-water temperature estimates or the northern European record. Guiot *et al.* (1989) suggest that this apparent discrepancy could be due to steeper thermal gradients than occur today between the poles and equator.

There is growing evidence to suggest rapid advances of the Laurentide ice sheet in North America, with the 5,000–10,000-year intervals between the events being inconsistent with Milankovitch orbital frequencies. This evidence comes from layers of ice-rafted sediments, known as 'Heinrich layers', in the North Atlantic (see Box 2.4). The six most recent of these layers, which accumulated 70,000–14,000 years ago, indicate marked decreases in sea-surface temperature and salinity, reduced fluxes of foraminifera to the sea floor, and enormous discharge of icebergs from eastern Canada as glaciers entered the sea and began to break up (calve) over short time intervals (Bond *et al.* 1992). Melting of very large volumes of icebergs drifting across the North Atlantic must have been a major factor in reducing the salinity of the surface waters,

Plate 8 *Icebergs frozen into sea ice in Otto Fjord, northern Ellesmere Island, Canadian High Arctic. These icebergs form as glaciers calve when they enter the sea. Global warming may lead to widespread melting of sea ice which in turn may lead to a decrease in aridity in the Arctic. Increased precipitation, primarily as snowfall, will lead to the growth of glaciers, rather than their melting.*

Plate 9 *Glaciated peaks in the Karakoram Mountains (Khunjerab, northern Pakistan) on the southwest edge of the Tibetan Plateau. Uplift of this region was probably an important factor in contributing to global cooling during the late Tertiary and for the onset of the Quaternary Ice Age.*

Plate 10 *The research vessel* Polar Duke *in the Le Maire Channel beneath ice cliffs of the Antarctic ice sheet.* Courtesy of Professor B.F. Windley.

BOX 2.5 LOESS DEPOSITION AND PALAEOCLIMATE

The most complete terrestrial sedimentary record for palaeoclimate change is provided by thick loess sequences in central China. Loess formation, transportation and deposition is strongly controlled by climatic conditions. During cold glacial times, desert regions become more arid and the production of silt increases and its transportation is enhanced, while in glaciated areas glacial grinding increases, and wind systems intensify, producing more silt, which is transported and deposited further away from the source than during interglacial times. As a consequence, loess deposition increases. During interglacial times, soil-forming processes dominate and silt deposition is greatly reduced, and as a result loess successions have distinct palaeosol horizons. The grain size characteristics and mineralogy of the loess should, therefore, reflect past climatic conditions.

The thickest loess occurs near Lanzhou on the Loess Plateau in central China, where it exceeds 330 m in thickness and has a palaeomagnetic age at its base of 2.48 Ma (Heller and Lui 1984). The onset of loess deposition is approximately contemporaneous with the start of the Quaternary Ice Age, and this suggests that the build-up of ice sheets and the intense rapid global climate change may have resulted in loess-forming processes becoming more dominant. It is also argued that the uplift of the Tibetan Plateau during Late Tertiary times may have led to increased aridity in Central Asia and the onset of loess deposition. Up to 37 identifiable palaeosols (S1 to S32, S1 being the youngest) alternating with loess units (L1 to L33, L1 being the youngest) have been identified in the Chinese loess (see Figure 2.33). These are thought to represent alternations from cold, dry periods, with high rates of loess deposition, to warm, wet periods, with lower rates of deposition and the formation of soils. At the time of deposition, magnetic minerals within the loess align themselves parallel to the Earth's magnetic field and thus palaeomagnetism provides a useful method for dating the loess. Magnetic susceptibility (MS) has been used to detect palaeoclimatic variations, to correlate palaeosols and to correlate with the oxygen isotope record. MS is generally higher in palaeosols than in loess. Although the reasons for this variability are not fully understood, it may be attributed to the enrichment of detrital magnetic minerals in soils during interglacials, due to concentration by decalcification and soil-compaction processes. It may also be the result of sub-aerial deposition of ultrafine magnetic minerals from distant sources, the concentrations of which are diluted during the higher rates of silt deposition associated with cold times. Alternatively, the MS may be the result of

in situ formation of magnetic minerals by soil-forming processes. On the basis of the types of magnetic minerals present within the loess, it has been suggested that *in situ* formation of magnetic minerals by soil-forming processes is the most important control on the MS. Therefore the MS may be broadly considered to be a function of palaeo-precipitation. One of the most intensive MS studies on loess was undertaken at key sections on the Loess Plateau (Xifeng, Luochuan: Figure 2.33) (Kukla 1987, Kukla *et al.* 1988, 1990). The sediments at these locations represent a time span of about 2.5 Ma. The combined magnetic susceptibility results from these sections showed a general agreement with the astronomically tuned oxygen isotope deep-sea chronology in the upper part of the succession, but less of an agreement prior to 0.5 Ma (Figure 2.33). Gross correlations are possible between other sections throughout the Loess Plateau, e.g. Lui Jia Po and Baoji near Xian in the warm, humid south and at Lanzhou in the semi-arid west (Luixiuming *et al.* 1992). Minor variations are more difficult to correlate, probably because of regional variations in climate and soil-forming processes. Broad correlations have also been made between the MS results and aeolian sediment present in deep-sea cores from the Pacific Ocean (Hovan *et al.* 1989). High MS values correspond well with high concentrations of aeolian sediment in deep-sea cores (Figure 2.33E, F and G). This probably indicates glacial times when stronger westerly winds carried sediment from China into the Pacific Ocean. A better understanding of the controls on MS will help to refine its use as a very detailed proxy measure of climate change.

In the Chinese loess, the median grain size is essentially a measure of the vigour of the northwesterly (winter) monsoon. Coarse median values probably represent cold and dry glacial times. Some loess horizons are particularly sandy, such as the L9 and L15 sandy loess units (Figure 2.32A), and these are thought to represent extensive advances of the desert margins. Median grain sizes match closely with MS, but recent results show a much more complex pattern than that obtained from the MS analysis. This technique has great potential for detailed interpretation of past climate.

Study of the microscopic structures (micromorphology), clay mineralogy, organic carbon and faunal excrement within palaeosols has recently helped to determine the nature of soil-forming processes across the plateau and between palaeosols of different ages (Derbyshire *et al.* 1991). The abundant molluscan fauna within the loess is also beginning to be used to help provide information on past humidity and temperature. Early results show that the abundance of molluscs closely parallels the MS, further supporting the idea that these were times of thermal and humidity maxima (*ibid.*).

implied by the δ18O values: Bond *et al.* also noted that the salinity drop would have been sufficient to shut down the thermohaline circulation of the North Atlantic. The ice-rafted sediments on the sea floor, including detrital carbonate (limestone and dolomite with a provenance in eastern Canada), delineate the

path of the icebergs and show that they must have travelled more than 3,000 km, a distance that in itself suggests extreme cooling of the surface waters, and substantial volumes of drifting ice (*ibid.*). Indeed, the Heinrich layers all show a dominance of the left-coiled planktonic foraminifera species *Neoglobo-*

BOX 2.6 MINERALOGY AND CLIMATE CHANGE

Temporal variations in the type of minerals being fed through rivers into large deltas can be used to determine climatic changes. By looking at the changing ratios of two groups of heavy minerals, pyroxenes and amphiboles, Foucault and Stanley (1989) have elucidated palaeo-climatic changes in East Africa during the last 40,000 years or so, in the time interval referred to as the Late Quaternary. The Nile river system, formed mainly by the drainage from three large rivers, the White Nile, Blue Nile and Atbara, flows across nearly 35° of latitude from south of the Equator to the Mediterranean. The river drains a vast area of mixed climates, from humid tropical to warm, arid conditions. During the Quaternary, changes in global climate caused the climatic belts to migrate large distances, with the effect that there were changes in the sediment yield of the river, as well as the mineralogy and grain size of sediments reaching the Nile Delta. Detailed studies of sediment mineralogy in age-dated cores from the three tributary rivers and the main Nile were made in the context of the drainage basins, with their different geology and climate/vegetation cover.

Decreased amounts of pyroxenes relative to amphiboles in the sediments of the Nile Delta and main Nile, eroded from volcanic rocks on the Ethiopian Plateau, suggest increased vegetation cover with a more humid climate in the Blue Nile and Atbara drainage basins. More humid conditions would probably have led to a longer rainy season and a greater cover of vegetation, which in turn would have reduced erosion of sediments. Now, even if the wetter conditions led to an *increase* in river discharge, the sediment load carried by the Blue Nile and Atbara would have *decreased*. The decreased sediment load would result in a reduced supply of pyroxenes to the main Nile. On the other hand, increased proportions of pyroxenes supplied to the main Nile and the delta probably indicate reduced vegetation cover, accelerated rates of erosion of the Ethiopian Plateau and a more arid climate.

Measurements of African lake levels in the Ethiopian Rift and Plateau, tied to this mineralogical data, suggest that high lake levels, lower pyroxene values and a more humid, wetter climate prevailed in northeast Africa about 40,000–17,000 and 7,000–4,000 years ago, and from 1,500 years ago to the present day. Low lake levels, increased abundance of pyroxenes and a more arid climate existed about 17,000–7000 and 4,000–1,500 years ago. The significance of changing lake levels in response to fluctuating global climate has been well discussed by Street-Perrott and Perrott (1990).

---

quadrina pachyderma, which indicates a deep southward penetration of polar water. The actual cause of the ice sheet surging remains unclear, but Bond *et al.* proposed that shortly after sea-surface temperatures and foraminifera fluxes to the sea floor began to decline, ice streams in eastern Canada and possibly in northwestern Greenland advanced rapidly, leading to massive calving as ice fronts reached maximum seaward positions. The lower sea surface-water temperatures, created by the release of large volumes of ice, would have slowed melting rates and facilitated the long-distance transport of ice-rafted sediments.

Other lines of evidence also suggest sub-Milankovitch (short-term) climatic shifts. For example, sediment cores from the eastern equatorial Pacific have revealed vast 1.5 to 4.4 Ma laminated diatom mats, which accumulated rapidly at rates exceeding 10 cm per year over distances of more than 2,000 km (Kemp and Baldauf 1993).

In East Africa, there is a good correlation between lake sediment stratigraphy, geochemistry and lake water levels, all of which can be correlated with Late Quaternary global climatic fluctuations. Indeed, there are good case studies of links between mineralogy and climate change (Box 2.6). Street-Perrott and Perrott (1990) showed that periods of low lake levels generally occurred at about 13,000 BP, 11,000–10,000 BP, 8,000–7,000 BP and 4,500–2,500 BP. They attribute the last two low stands in lake level to prolonged periods of aridity produced during times of anomalously low sea-surface temperatures in the North Atlantic. These low temperatures may have been caused by large volumes of glacial melt water entering the North Atlantic during deglaciation and increasing the ocean salinity stratification. Such changes could then suppress the formation of North Atlantic Deep Water (NADW) and further lower the sea-surface temperature, leading to decreased rainfall and, therefore, lower lake levels. During periods when Laurentide melt waters flowed into the Gulf of Mexico, the production of NADW would return and lake levels would show a corresponding rise. During the Last Glacial Maximum, much of the North Atlantic would have had a cover of ice, the production of NADW would have been impeded, and arid conditions would have prevailed over much of Africa and America.

The postulated effects on the oceanic conveyor belt (the thermohaline circulation) caused by the abrupt release of enormous volumes of fresh water as melt water from continental ice sheets into the North Atlantic – the ice armadas of Broecker (1990) – have been modelled using a coupled ocean–atmosphere computer simulation (Manabe and Stouffer 1995). In the computer model, Manabe and Stouffer showed that in response to a massive surface flux of fresh water into the northern North Atlantic, the thermohaline circulation weakens abruptly, intensifies and then

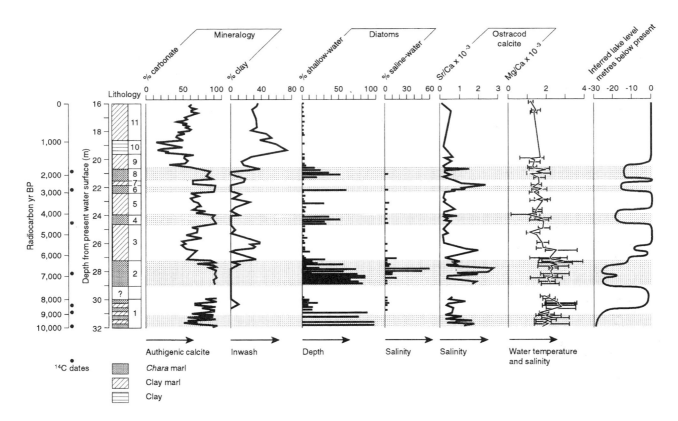

Figure 2.34 *Mineralogical, biological and chemical data from tropical African lake sediments in Lake Tigalmamine, Middle Atlas Mountains, Morocco (32° 54'N, 5° 21'W), to show century-scale Holocene arid intervals in tropical and temperate zones. Lithology, radiocarbon chronology and biostratigraphy of the Tigalmamine C86 core, tropical Africa. Sr/Ca and Mg/Ca element concentrations were measured by inductively coupled plasma mass spectrometry (ICPMS), and each element value represents the mean value of three separate measurements on individual valves: the ranges are shown for Mg/Ca. Inferred shallow-water lake-level phases are indicated by shading. Redrawn after Lamb* et al. *(1995).*

weakens again, followed by a gradual recovery to create events that resemble the abrupt changes in the ocean–atmosphere system recorded from ice and deep-sea cores. Furthermore, the model simulation suggests that these high-frequency and abrupt climatic variations appear to be associated with particularly large changes in surface air temperature in the northern North Atlantic Ocean and vicinity but relatively small changes throughout the rest of the world (*ibid.*). The sensitivity of the North Atlantic thermohaline circulation to the input of fresh water has also been computer modelled by Rahmstorf (1995), who came to similar conclusions to those of Manabe and Stouffer, i.e. that relatively local changes in fresh-water flux can induce transitions between different equilibrium states, and may trigger convective instability in the oceans with temperature changes of several degrees on time scales of only a few years.

The Holocene is marked by rapid shifts in global and regional climate, with the global changes reflecting sunspot maxima and minima, ENSO events

and other poorly understood decadal- to millennial-scale changes. A study by Lamb *et al.* (1995) of the mineralogical, biological and chemical data from subtropical African lake sediments in Lake Tigalmamine, Middle Atlas Mountains, Morocco (32° 54'N, 5° 21'W), has revealed century-scale Holocene arid intervals in tropical and temperate zones (Figure 2.34).

The shift from glacial to interglacial, from Pleistocene to Holocene, permitted humans to colonise hitherto inaccessible and frozen landscapes (Figure 2.31). As an example of the changing pattern of vegetation following the LGM and into the Holocene, Figure 2.35 shows the situation in eastern North America 18 ka, 10 ka, 5 ka and 200 years ago (Delcourt and Delcourt 1981). It is against a background of major global climatic amelioration that human activities should be placed. The following section considers the evolution of *Homo sapiens* and human colonisation, particularly during the Holocene.

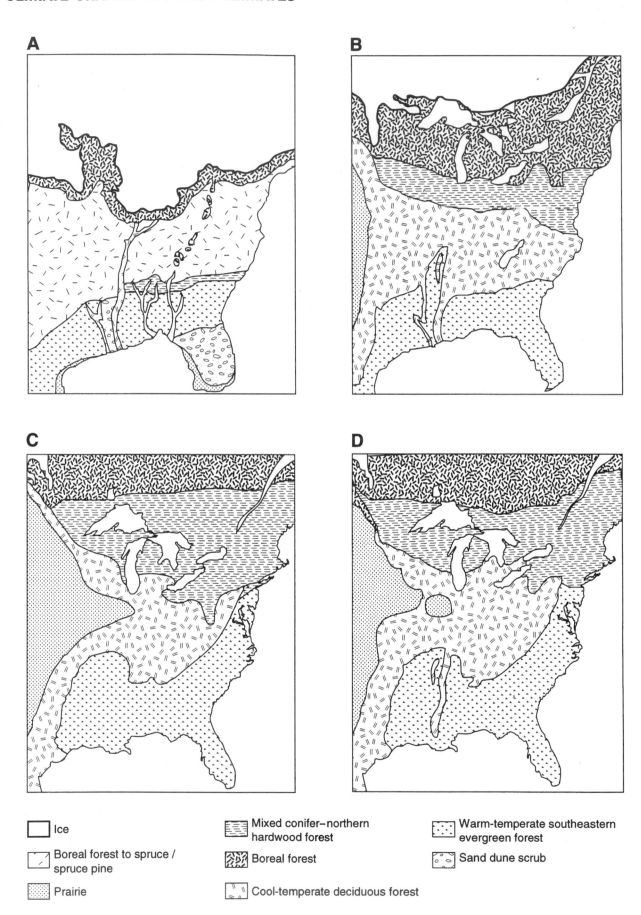

Ice

Boreal forest to spruce / spruce pine

Prairie

Mixed conifer–northern hardwood forest

Boreal forest

Cool-temperate deciduous forest

Warm-temperate southeastern evergreen forest

Sand dune scrub

Human evolution in the Quaternary Period

A study of the Quaternary Period is important in understanding human evolution and colonisation, including the human impact on the natural environment. It was a little time prior to the onset of the Ice Age that the first bipedal hominids evolved (3.75 million years BP). These were known as *Australopithecus afarensis*, the most famous fossil of which was unearthed by Louis Leakey in Ethiopia in the mid-1970s and became known as 'Lucy'. It is believed that the genus *Homo* evolved from *A. afarensis* about 2 million years ago. The first species was *H. habilis*, but within another 500,000 years *H. erectus* evolved. *H. erectus* probably organised themselves into groups for hunting and gathering food, as well as making tools and utilising fire. These were the forerunners of our modern, organised society. Many scientists believe that *H. erectus* was the ancestor of modern humans (*H. sapiens*), who evolved about 200,000 years ago. Neanderthals (*H. sapiens neanderthalensis*) are also believed to have evolved from *H. erectus* but became extinct about 30,000 years BP.

By 50,000 years BP, *H. sapiens sapiens* had spread to Australia. They arrived in the Americas between 14,000 and 12,000 years BP and by the start of the Holocene had colonised every continent except Antarctica. Their migration was undoubtedly influenced by climatic change, often aided by the extensive coastal regions that were created as a result of the fall in sea level caused by water being locked up in the ice sheets during the last glacial.

Towards the close of the last glacial, about 15,000 years BP, hunter-gathering communities began to develop, and these groups began to clear land for farming and settlements. They were the first humans to initiate the process of deforestation. This practice of forming organised settlements and land-clearing began in many regions, particularly in the Near East, Central Asia and South America. In the Near East, by 10,000 years BP, the domestication of plants and animals was well established. By about 9,000 years BP, Jericho, one of the earliest permanent settlements, was established, associated with cultivating cereals: wheat and barley. The domestication of animals became more sophisticated. Approximately 8,000 years BP, chickens, for example, were domesticated from the red jungle fowl of Southeast Asia, and horses were first domesticated in the Ukraine about 4,000 years BP.

Such changes led to the modification of the landscape, vegetation, soil and water courses as modern humans cleared more forest for farming and the establishment of permanent settlements. The need for tools also had a profound effect on the environment, as more trees were required for charcoal to aid in the smelting of metal ores. In the Near East by 7,000 years BP, copper was being smelted, which then gave way to arsenic bronze by 5,000 years BP, and eventually iron about 4,000 years BP.

This societal organisation provided a more secure environment for human survival, and even produced surplus food. Improved living conditions meant that humans could spend time in pursuits other than basic survival, for example in making jewellery and ornaments. Barter thus became possible. Religious activities also developed. Between 5,000 and 4,000 years BP, large monuments were being constructed. These included the pyramids in Egypt (*c.* 4,500 years BP), the Minoan palaces of Crete (4,000 years BP), and the construction of Stonehenge in England (*c.* 4,000 years BP).

Wood (1992) reviewed the evolution of *Homo* in the light of recent advances in techniques for absolute dating (e.g. Grun and Stringer 1991), and also reassessed some of the evidence from fossils. He argued that a simple unilineal model for the evolution of humans, where *H. habilis* succeeded the australopithecines and then evolved via *H. erectus* into *H. sapiens*, is untenable. Despite such arguments amongst the experts, no clear consensus on human evolution has emerged, therefore the actual pattern of human evolution and colonisation of the Earth remains unclear.

One controversial aspect of hominid evolution is whether it led to the extinction of many large mammals. Stuart (1993) has pointed out that by the beginning of the Holocene, 10,000 years BP, much of the 'megafauna', defined as mammals exceeding 40 kg mean adult body weight by Martin (1984), was extinct, with the estimated losses as follows: 46 out of 58 genera in South America (80 per cent), 33 out of 45 in North America (73 per cent), fifteen out of sixteen – leaving only the red kangaroo – in Australia, seven out of 24 in Europe (29 per cent), and in contrast to these high losses, one out of 44 in North Africa (2 per cent). The most viable explanation for these extinctions is global climate change and/or human predation on the largest, slowest-breeding species with relatively small populations (Stuart 1993). Currently, there simply is

---

Figure 2.35 *Palaeo-vegetation maps for eastern and central North America at about (A) 18 ka; (B) 10 ka; (C) 5 ka; (D) 200 years ago. After Delcourt and Delcourt (1981), reproduced in Gates (1993).*

insufficient data to resolve the principal cause of these extinctions.

The important point through this deviation into human history is that throughout the Quaternary Period the human impact on the natural environmental is inextricably linked to changes in the landscape, particularly vegetation patterns, as far back as the Late Pleistocene and early Holocene (see Figure 2.35). Currently, there is great debate regarding the extinction of many species of animals, as well as major changes in natural vegetation, that occurred near the end of and after the last glacial stage. The fossil record for the last interglacial shows a decline in diversity of species (see previous paragraph). In Europe, during the last interglacial, abundant elephants, rhinos, bison and giant deer were present. In Australia, a more diverse marsupial fauna existed, including giant wombats, giant kangaroos and a diprotodont (a marsupial somewhat like an hippopotamus), and in New Zealand giant birds were abundant. In each continent outside Africa, these faunas disappeared as complex human societies evolved. In Australia, the marsupials were greatly reduced by 30,000 years BP, while in North America three-quarters of the genera disappeared by about 11,000 years BP. The most recently colonised regions of the world, such as Madagascar (c. 1,500 years BP) and New Zealand (1,000 years BP), saw the extinction of large flightless birds such as the rocs and moas, respectively.

A detailed study of the Quaternary Period of Earth history allows us to assess the possible relationship between the growth of human society and the extinction of various species, together with any environmental changes, around the end of the last glacial stage. It may be that the extinctions and changes in the natural environment occurred entirely independently of human activities, because of natural processes that exerted a more profound influence, for example the changes in the ocean–atmosphere system brought about by the end of the last glaciation.

Meteorite impacts on Earth and global climate change

The collision of large meteorites (bolides) with the Earth may cause global climate change and the extinction of species. It has been estimated that the Earth's global climate will only be significantly affected by the impact of meteorites greater than 1 km in diameter – the size required to inject sufficient dust into the atmosphere to perturb global climate (Bland *et al.* 1996). In order to penetrate the lower atmosphere, a meteorite must be greater

Table 2.2 *Meteorite impact craters and age.*

Crater	Diameter (km)	Age (Ma)
Vredefort, South Africa	300	2,006
Sudbury, Canada	250	1,850
Beverhead, USA	60	c. 600
Acraman, Australia	90	> 570
Charlevoix, Canada	54	357
Manicougan, Canada	100	214
Puchezh-Katunki, Russia	80	175
Kara, Russia	80	175
Tookoonooka, Australia	55	128
Chicxulub, Mexico	170	64.98
Chesepeake Bay, USA	85	35.45
Popigai, Russia	100	35
Meteor Crater, Arizona, USA	1.2	0.05

Source: Bland *et al.* 1996.

than the threshold of 50 m in diameter upon entry: Meteor Crater, Arizona, was formed by the impact of a *c.* 60 m diameter object 50,000 years ago. Meteorites with an impact crater diameter greater than 50 km, with their approximate age of impact, are given in Table 2.2.

Shoemaker *et al.* (1990) estimate that there are about 1,000 Earth-crossing asteroids (ECAs) with a diameter greater than 1 km, suggesting that a K–T boundary sized impact (i.e. producing an impact crater > 150 km in diameter) occurs once every 100 million years.

The Cretaceous–Tertiary (K–T) boundary event, 65 Ma

An example of abrupt global climate change occurred about 65 million years ago, when a giant meteorite impacted on the Earth's surface. This is particularly interesting because it provides Earth scientists with information on how external, cosmic processes may lead to major climate change and extinctions of fauna and flora.

Approximately 65 million years ago (64.5 ± 0.1 Ma) as dated using an argon laser probe technique at the US Geological Survey in Denver, Colorado, on Haitian tektites – spherules of glass generated by the meteorite impact (see below) – a phenomenal catastrophe hit the Earth, the consequences of which were fatal for many organisms. An estimated 70 per cent of the flora and fauna on Earth became extinct around or at the K–T boundary event. Such is the significance for the evolution of life on Earth that Earth scientists define the time era after 65 million years ago as the Tertiary, and the immediately preceding time interval as the end of the Cretaceous Period, hence the K–T boundary event. The 'K' is from the German spelling 'Kretaceous'.

Detailed palaeontological research is now suggesting that a number of the extinction events associated with the K–T boundary were actually well under way in many species prior to any possible meteorite impact – even allowing for inaccuracies in dating the event. Such evidence suggests that, worldwide, many environments were already under severe stress, and perhaps the meteorite impact merely acted like the proverbial nail in the coffin for many organisms.

The most widely known event at the K–T boundary was the extinction of the dinosaurs. Their demise allowed the humble mammals to inherit the role of dominance from the dinosaurs, and paved the way for human beings. There have been many theories to explain their extinction, but here only the most plausible event is presented, an explanation subscribed to by most Earth scientists, which is the impact of a massive meteorite.

Towards the end of the 1970s, an Earth scientist named Walter (L.W.) Alvarez was researching the rates at which ancient clay-rich marine sediments were laid down around the Cretaceous–Tertiary boundary near Gubbio in Italy. Chemical analyses of these clays revealed an unexpected abundance in a chemical element called iridium, now known as an **iridium anomaly**. Alvarez and his co-workers interpreted this anomaly as a result of an enormous meteorite impacting onto the Earth at the end of the Cretaceous Period. This hypothesis was published by Alvarez in 1980 in the American journal *Science*. This meteorite may have been about 10 km across, and upon impact had an estimated explosive energy equivalent to 100 million megatons of TNT, or roughly 10,000 times the world's total nuclear arsenal (Ravven 1991).

The iridium anomaly was discovered in other rocks of the same age from around the world, but always in marine sediments. A popular interpretation, therefore, was that it was caused by chemical reactions in sea water, which preferentially extracted iridium into the sediments. This notion was shattered in 1981 with the discovery of the same iridium anomaly in terrestrial (land) sedimentary strata dated at 65 Ma in New Mexico. However, the geochemical iridium spikes at extinction horizons, commonly associated with spikes in the other platinum-group elements (Ru, Rh, Pd, Re, Os, Pt, Au), can be the result of post-depositional redistribution in the sediments because of changes in redox conditions at or near the sea floor (Colodner *et al.* 1992). Such geo-chemical spikes and ratios may, therefore, be characteristic but not diagnostic of a cosmic source.

Other elements, beside the platinum-group elements, were found to be enriched in sediments occurring at the K–T boundary, for example nickel, chromium, cobalt, gold and iron, all of which have been interpreted as the result of a large meteorite impact. Although these siderophile elements (those soluble in iron), occur in varying abundances on Earth, their relative abundances and concentrations at the K–T boundary are quite unlike those of typical terrestrial rocks, but similar to those encountered in certain types of meteorites.

Another line of evidence in favour of an impact event at the K–T boundary is the presence of highly deformed or 'shocked' quartz in which the crystal structure is believed to have suffered very rapid strain during a meteorite impact (Plate 2.8). Also, varieties of silica that form only at extremely high pressures, such as caused by a meteorite impact, are found in the sediments of the K–T boundary at Raton Pass, Mexico. These silica minerals are coesite and stishovite, which require respectively 20 kilobars cm^{-2} and 110 kilobars cm^{-2} of pressure to form (1 bar is equivalent to 1 atmosphere, which equals 1 kg cm^{-2}). Krogh *et al.* (1993) undertook a U–Pb date on the shocked zircons from distal ejecta (at the Berwind Canyon site in the Raton Basin, Colorado) which yielded a date of 65.5 ± 3 Ma – shocking has reset isotopic clocks: degree of isotopic resetting correlates well with amount of shock-induced textural change in zircons.

Chondritic meteorites also contain abundant 3–5 μm-sized diamonds, something that prompted the search for similar small diamonds in the sediments at the K–T boundary. The boundary clay from Red Deer Valley, Alberta (known as the 'Knudson Farm' locality), has indeed yielded a white fraction containing 97 per cent more carbon, which is absent from the surrounding layers. Two Canadian scientists have demonstrated that this carbon-rich material is almost certainly very small diamonds (Carlisle and Braman 1991) and, therefore, provided additional supportive evidence for the meteorite hypothesis.

There are a number of other lines of evidence pointing towards an extraterrestrial, meteorite impact, cause for the event at the K–T boundary, including the presence of so-called 'spheroids' or droplet-shaped amorphous minerals in the sediments. These sand-size spheroids are believed to result from the crystallisation at high temperatures of material melted by a meteorite impact and rapidly ejected into the air and water. Sites where these spheroids, mainly of the mineral feldspar, occur include the K–T

Plate 2.8 *Cretaceous–Tertiary boundary interval exposed at Risks Place, Montana, showing the meteorite impact layer, which contains shocked quartz and high concentrations of iridium, together with other chemical anomalies.*
Courtesy of M. Collison.

boundary clays at Caravaca in southern Spain, Petriccio in Italy, El Kef in Tunisia and the central Pacific Ocean (see Box 2.7).

An intriguing aspect of the K–T boundary event is the evidence that is emerging for global fires. The percentage of carbon in sediments at the K–T boundary is much greater than expected, with the carbon occurring as fluffy aggregates of 0.1–0.5 μm graphite. Fluffy graphitic carbon is similar to charcoal that is produced from forest fires today. Analysis

of clay samples from the K–T boundary at Woodside Creek, New Zealand, Stevn's Klint, Denmark, and Caravaca, Spain, led Wendy Woolbach *et al.* (1985) to suggest a worldwide flux of carbon about 10,000 times greater than the present day and 1,000 times greater than in the underlying Cretaceous and overlying Tertiary sediments. The source of this graphitic carbon is unlikely to have been the meteorite, but a massive impact event could have caused devastating fires that raged throughout enormous areas of land.

BOX 2.7 K–T BOUNDARY METEORITE IMPACT SITE

Geochemical analyses of the K–T boundary clays suggest that the site of meteorite impact was in the deep oceans, penetrating 3–5 km into the oceanic crust. The shocked quartz, however, indicates at least a thin cover of land-derived, continental material overlying the oceanic crust. Few people have suggested a precise site for the enormous meteorite impact, but potential sites, based on age, dimensions and shape, that have been proposed include the Amirante Basin, west Indian Ocean, the Nicaragua Rise in the Caribbean Sea, and the 65 Ma, 35-km diameter, Manson impact crater in Iowa.

In March 1991, new evidence was presented to the Lunar and Planetary Science Conference in Houston, Texas, in favour of an impact site on the Yucatan Peninsula in the southern Gulf of Mexico (Ravven 1991). A particularly thick layer rich in spherules was interpreted as resulting from the ejection of vaporised and melted material from the meteorite impact, which was spread over a very large area. Similar spherule layers, albeit much thinner, have been identified at many K–T boundary sites. In northern Yucatan, Mexico, the prime candidate for the site of the meteorite impact is the *c.* 200-km diameter Chicxulub impact structure (Hilderbrand *et al.* 1991; Plate 2.9), which contains deformed or 'shocked' rock fragments that are similar to those found worldwide at the K–T boundary, an observation that may favour a single meteorite impact rather than a comet shower (Sharpton *et al.* 1992). The impact structure is associated with igneous rocks (andesites), produced by the impact, which have been radiometrically dated by $^{40}Ar/^{39}Ar$ techniques as 65.2 ± 0.4 Ma (*ibid.*), in good agreement with the recently reported date of 64.98 ± 0.05 Ma (Swisher *et al.* 1992).

It may be that there was not one but several meteorite impact sites at the K–T boundary, something proposed in 1988 by Eugene Shoemaker of the US Geological Survey at Flagstaff, Arizona. A comet passing close to the Sun could have fragmented and caused several meteorites to impact on Earth at several locations. The multiple impact hypothesis might explain why some Earth scientists now recognise a number of impact sites; the most plausible candidate sites include the Caribbean, the Manson crater (Iowa), and the 105-km diameter Popigai crater (Siberia), all dated to approximately 65 Ma.

In the Brazos River, Texas, and the New Mexico sites, there is also evidence of tsunami (Japanese for 'habour wave') activity, possibly caused by a meteorite impact, but this event is about 230,000–330,000 years *after* the principal K–T boundary extinctions (Montgomery *et al.* 1992). If, as some scientists suspect, the K–T boundary meteorite impact was not actually a single event, but perhaps many smaller impacts associated with a very large, main impact, then the Brazos River section, although slightly younger in age, may represent a part of the K–T meteorite shower events.

These fires could have ignited various shallow deposits of fossil fuels such as coal to release even more carbon into the land and atmosphere at that time. The meteorite impact need not have been on land to cause such catastrophe. An oceanic impact could still have led to enormous fireballs and expanding clouds of rock vapour.

Many suggestions exist as to just how the dinosaurs became extinct, something that is now known to have occurred over a few million years. Perhaps the most reasonable interpretation is that the meteorite impact ejected huge volumes of very fine material into the upper atmosphere, together with the soot and other materials contributed by global fires. Such clouds would have been very effective in absorbing sunlight and solar energy to stop it reaching the surface of the Earth. The atmosphere would also have become extremely polluted by the emission of very large amounts of gases from the wildfires to produce poisonous chemicals called pyrotoxins.

Evidence from the remains of plants that were living at the time of the impact event can even give us a clue as to the season and month when the devastating meteorite hit the Earth. A study by Jack Wolfe, at the US Geological Survey in Denver, Colorado, of aquatic leaves in the K–T boundary section near Teapot Dome, Wyoming, shows the preservation of

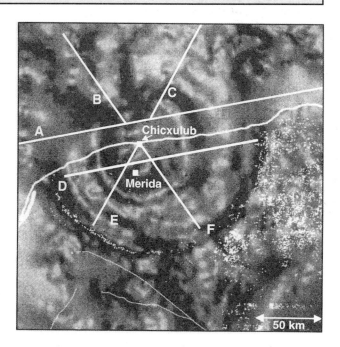

Plate 2.9 *Chicxulub meteorite crater off western Yucatan Peninsula, Gulf of Mexico. Horizontal gravity gradient across the crater from Hilderbrand* et al. *(1995). White dots represent cenotes (sink-holes in the limestone); the letters denote seismic lines.*
Courtesy of British Institutions Reflection Profiling Syndicate (BIRPS) (1995).

detail that can only be achieved experimentally in aquatic leaves by freezing. The impact of a huge meteorite would be expected to produce large amounts of light-attenuating debris in the atmosphere and, therefore, an 'impact winter'. Reproductive stages reached by the fossil aquatic plants at the time of death suggest that the freezing took place in approximately early June, that is in the early summer in the Northern Hemisphere (Wolfe 1991).

The other line of evidence for a protracted time interval of meteorite falls around the K–T boundary comes from the geochemistry. There are amino acids that are extremely rare on Earth but known to occur in meteorites. Research by Zahnie and Grinspoon (1990), into the K–T boundary site at Stevn's Klint, Denmark, has demonstrated that the concentration of these organic compounds shows an increase over about 50,000 years prior to the K–T boundary, followed by a fall-off afterwards, again over approximately 50,000 years. They suggested that if the amino acids came only with the big meteorite impact, then they would have been incinerated at the K–T boundary. Interestingly, the amino acids occur only in the few centimetres above and below but not in the boundary clay itself. In order to explain this anomaly, Zahnie and Grinspoon have suggested that the amino acids were deposited with the dust from a giant comet trapped in the inner Solar System, a fragment of which formed the K–T impactor. The amino acids would have been swept up by the Earth prior to and following the meteorite impact and therefore preserved in sedimentary layers, but those deposited at the K–T event would have been destroyed.

At the K–T boundary, there is also a change in the strontium isotope composition of sea water, recorded from foraminifera in an exceptionally thick, palaeontologically well-characterised K–T section exposed at Bidart in southwest France. Here, there is a rapid increase in $^{87}Sr/^{86}Sr$ (strontium isotopes) of ocean water about 1.5–2.3 million years before the boundary (Nelson *et al.* 1991). Bioturbation, the burrowing activity of organisms, cannot explain these changes before the K–T boundary, because vertical mixing by such processes typically involves up to about 10 cm of uncompacted sediment, equivalent to about 5 cm of compacted sediment. The studied section at Bidart is two to three orders of magnitude thicker, with the chemical anomaly appearing at approximately 90 m, reaching a maximum 40–50 m below the K–T boundary. The actual change in strontium isotopic signature of the ocean water is explained by a 10 per cent increase in strontium supply to the oceans from the continents over about

one million years. Such a change could be brought about by increased erosion of sediments from the land, induced by a major change in global climate (*ibid.*).

In effect, the Earth could have suffered many of the effects of a nuclear winter at the K–T boundary. A few years of darkness and freezing temperatures at the Earth's surface would have ensued. Plants would have been unable to photosynthesise the green pigment chlorophyll and would have died. Animals, particularly the 'higher' species, would have died both as a direct consequence of the meteorite impact and through starvation. Much of the complex food chains may have broken down as plants died. Using the scenario of a huge meteorite impact, it has been calculated that light levels would certainly have been too low for photosynthesis to occur for about 200 days, and that global temperatures at the surface of the Earth would have been below freezing because of the lack of sunlight penetrating the dense and poisonous atmosphere.

Even life in the oceans appears to have been killed by the meteorite impact. Microscopic organisms that secrete shells or plates of calcium carbonate (calcite) suffered extinction. A group of such organisms to become extinct at the K–T boundary were various species of calcareous plankton. The shells of dead organisms on the sea floor appear to have been subjected to dissolution in shallow marine waters, where such processes would not be expected. Geochemical evidence suggests that at the K–T boundary, the oceans suffered an unprecedented depletion of calcium, so essential for life. The actual cause of this decalcification of the ocean waters remains puzzling.

A possible cause may have been a dramatic shallowing of the depth at which material made of calcite dissolves in the world's oceans, known as the carbonate compensation depth, or CCD. At present, particles of calcite such as the tests or shells of dead microscopic organisms that are settling slowly through the water column begin to dissolve in the open oceans at depths of about 5.4 km in the Atlantic Ocean and 4.3 km in the Pacific Ocean. It has been suggested that at the K–T boundary, the position of the CCD rose to within the photic zone, less than a couple of hundred metres of water depth, with the result that organisms such as the calcareous plankton were unable to secrete their calcite shells. The result would indeed have been catastrophic with mass extinctions.

One possible reason for this decalcification is that as the huge meteorite travelled through the atmosphere and became very hot, high-temperature shock waves were generated and nitric oxide (NO) was formed, just as occurs today when lightning causes shock heating. Nitric acid (HNO_3), along with other acids, would form and cause rain with an estimated pH of 0–1 (strongly acidic) to fall. Such acid rain would rapidly cause a critical decalcification of the upper ocean waters and the CCD would rise substantially.

The acid rain hypothesis finds additional support in the high levels of nitrogen found in many of the sediment samples analysed from the K–T boundary. Furthermore, the ratio of the strontium isotopes $^{87}Sr/^{86}Sr$ show a sharp increase at the K–T boundary, something that is predicted by very acidic rain water dissolving large quantities of continental granites and releasing the abundant ^{87}Sr isotope into the water cycle from these rocks. Of course, the very acidic rain water could have been a direct cause of enormous fatalities and mass extinctions of species.

The possibility that the impact of a huge meteorite at the K–T boundary generated NO is supported by data from a much smaller meteorite shower in 1908 called the Tunguska meteor fall. This meteorite fall is estimated by Turco (1981) to have caused a substantial depletion in ozone from the ozone layer. It has been calculated that as much as 30 million tonnes of NO could have been produced, and that approximately 45 per cent of the ozonosphere in the Northern Hemisphere was destroyed. Clearly, the much greater magnitude of a meteorite impact at the K–T boundary would have led to devastating consequences compared with the 1908 Tunguska meteor fall.

If the ozone layer was destroyed by the K–T boundary event, then the immediate result of the meteorite impact would have been that lethal doses of ultraviolet radiation and heat from the Sun would have struck the Earth's surface. Such radiation levels could have contributed to mass extinctions.

Not all Earth scientists believe in the impact theory. Archibald (1993) argues that the commonly quoted mass extinctions at the K–T boundary are misleading, and that the actual number of species that survived were 52–72 per cent as compared with the commonly quoted 75 per cent figure that became extinct. He suggests that many species did not actually become extinct in the true sense that their entire gene pool was wiped out, but rather that species disappeared locally. Anderson (1993) believes that many dinosaurs could survive the darkened skies and global cooling associated with a nuclear winter brought about through an asteroid impact, based on current work being undertaken on dinosaur fossils from Dinosaur Cove, in the Otway Range 220 km west of Melbourne, Australia. During the Cretaceous Period, when the dinosaurs of Dinosaur Cove lived,

the palaeo-latitude lay between 70°S and 80°S, a region that would have experienced between six weeks and four and a half months of continuous darkness. Anderson argues, therefore, that dinosaurs may have been much more adaptable to environmental stress than has previously been thought. Officer (1993) has suggested that there is evidence to show that dinosaurs actually died out *before* the iridium anomaly, and that the impact could not, therefore, have been the principal cause of their extinction. Alternatively, Officer proposes that volcanic eruptions and global sea level changes may have been more important in causing the extinction of the dinosaurs. Volcanic eruptions can cause significant climatic changes, although some of the largest known eruptions, such as Toba in 75,000 BP, did not cause any species extinctions. Volcanic eruptions may also produce large amounts of iridium, although detailed studies of the amounts which can be produced by volcanic activity remain poorly quantified. Swinburne (1993) also argues that other fossil groups that are lumped into the total number of species which became extinct at the end of the Cretaceous Period, such as inoceramid and rudist bivalves, actually died out two and ten million years, respectively, prior to the K–T boundary event; thus, their extinction cannot be attributed to a single impact event at the K–T boundary. The problem with arguments such as these against a meteorite impact is that they ignore the cumulative evidence for an impact, merely showing that any one aspect of the evidence can be interpreted in other ways. Furthermore, these arguments commonly involve exploiting the uncertainties in precise dating of events.

The K–T boundary event was indeed a catastrophe for life on Earth, and something that only the most hardy and fortunate species survived. The Earth's ecosystems were stressed almost to the limit. The meteorite impact event brought immediate devastation with acid rain and an 'impact winter' of prolonged freezing temperatures because of the dust particles blocking out much of the sunlight, together with a possible depletion of stratospheric ozone to contribute further to global cooling. In short, at the K–T boundary, for the survivors of the actual impact, life on Earth experienced global acid rain, an impact winter and an ensuing period with perhaps little suitable vegetation as part of any diets. Not surprisingly, this chain of events was more than could be borne by 70 per cent of the species of flora and fauna.

Other geological mass extinction events caused by meteorite impacts

Iridium anomalies have been identified associated with other mass extinction events at the Precambrian–Cambrian boundary (570 Ma), Ordovician–Silurian boundary (435 Ma), within the Devonian (Frasnian–Famennian Stage) (365 Ma), and within the Carboniferous (Mississippian–Pennsylvanian Stages) (325 Ma) periods of Earth history.

Throughout the geological column, other examples of meteorite impact events are being identified. Beneath Chesapeake Bay and the adjacent Middle Atlantic Coastal Plain, US east coast, there is a 60 m thick boulder bed interval containing a mixture of sediments of different ages, distributed over an area of > 15,000 km^2, which is matched to a layer of equivalent-age impact material recovered from a deep-sea drilling site on the New Jersey continental slope (Deep Sea Drilling Project Site 612), and is interpreted as the result of a meteorite impact in the Late Eocene (Poag et al. 1992); the tektite glass (part of the impact ejecta, including shocked quartz) from DSDP Site 612 has been radiometrically dated by $^{40}Ar/^{39}Ar$ methods to be 35 ± 0.3 Ma (Obradovich et al. 1989). The candidate impact site for this Eocene event has also been identified by seismic reflection profiling across the continental shelf, and is represented by a 15–25 km wide impact crater with a central 2–3 km wide zone of disturbed sediments about 40 km north-northeast of DSDP Site 612, also extending several kilometres down (Poag et al. 1992).

Younger rocks with iridium anomalies, possibly caused by meteorite impact, include an 11 Ma event in the Miocene Period, where iridium levels are 15 times greater than the background values. More work needs to be undertaken on this latter event to see what caused it. Not all iridium anomalies, or mass extinctions of species, have to be caused by meteorite impacts. Whatever the trigger for these mass extinction events, they would have been associated with changes in global conditions so severe as to make sustainable existence impossible for the species of fauna and flora that became extinct.

Drying of the Mediterranean Sea, 5–6 Ma

The discovery that the Mediterranean ocean basin dried up to become a desert came as a dramatic and fascinating discovery to a team of Earth scientists drilling and recovering cores in the Mediterranean Sea in 1970. It was an international team on board the scientific drill ship *Glomar Challenger* taking part in the Deep Sea Drilling Project (DSDP), aimed at understanding more about the world's ocean basins. This is an example of how plate tectonic processes and climatic conditions conspired to exert a dramatic effect on the climate of a very large region, the Mediterranean, and in this important respect it differs from the previous case studies of global climate change.

How did the Mediterranean ever become an enclosed ocean basin that could dry up? Some 20 million years ago, the plate containing Arabia (the Arabian Plate) impinged against the Eurasian Plate, to the north, to cut off the Mediterranean from a closing ocean to the east named Tethys. Once the Mediterranean became landlocked (enclosed), the only connection to the large oceans was to the west between the narrowing seaway that separates North Africa from Europe at the Straits of Gibraltar. The climate became drier, and without an open, wide marine seaway connecting it to other oceans, over a period of about one million years, the Mediterranean virtually dried up.

Today, the evidence of this desert lies up to 3,000 m below the surface of the sea. This incredible discovery is in the layers of an evaporite mineral called gypsum, or calcium sulphate ($CaSO_4.2H_2O$), which was formed by evaporation of the Mediterranean sea water under desert conditions about 5–6 million years ago. The extreme evaporation of such a large volume of saline water led to the accumulation of more than 1,000 m of evaporite salt deposits. Seismic surveys of the sedimentary layers below the Mediterranean reveal a bright reflecting surface known as the 'M' reflector, which is this layer of salt.

The present sea floor of the Mediterranean 5–6 Ma lay some 2,000 m below the then sea level west of Gibraltar in the Atlantic Ocean. Such an enormous difference in sea level led to the rivers draining into the Mediterranean excavating deep, steep-sided, valleys or ravines into the underlying sediments and rocks. Using sophisticated geophysical techniques to look at the subsurface rock strata, Earth scientists have identified buried river gorges up to about 1 km below the present land surface containing ancient river gravels and sands, and dating from 5–6 Ma associated with rivers such as the Nile and Rhône.

As the Mediterranean evaporated, the waters became stagnant and extremely salty, a condition known as hypersalinity. Most organisms simply could not cope with the hostile environment and died. The Mediterranean basin became, in effect, a death valley with a series of salt lakes that periodically dried up completely. Calculations of the volume of evaporite minerals compared with the typical 35 grams of dissolved salts in every litre of sea water suggest that something like 30–35 times the volume of water in the present Mediterranean would have been necessary to form the 1 km thick salt deposits. The only way to do this would have been for periodic flooding of the Mediterranean by incursions of salty sea water, which then evaporated to leave yet more evaporite minerals. So, the Mediterranean cannot have been completely isolated from the world's oceans.

About five million years ago, the dam that separated the Mediterranean from the Atlantic Ocean was finally breached. Sea water cascaded down the world's most impressive waterfall at a rate of approximately 40,000 km^3 per year, taking about 100 years to fill the Mediterranean. The waterfall at Gibraltar was 100 times larger than the Victoria Falls on the Zambesi River. The salinity crisis (called the Messinian salinity crisis after the geological time interval when it occurred) thus came to an end. Very deep water again covered the sea bed, which had been dry land. The hot desert climate at the bottom of the Mediterranean was reclaimed by the sea.

Conclusions

Throughout this chapter, the evidence for past changes in global climate has been considered. Periodic or quasi-periodic global climate change occurs on all temporal scales from decadal, through century and millennial, up to millions of years. The evidence is multi-faceted and extensive, varying in the amount of information, type of data, and the confidence with which the interpretations are made. Furthermore, whilst the causal factors and rates of global climate change still require much more research, it is clear that evidence from the geological record reveals climatic conditions that were much more extreme than those experienced by humans. However, a concern for the natural environment that currently exists, together with attempts to make better predictions for future climate change, can only be made with continued research, both into past climates and by gathering detailed observations of present atmospheric, ocean and land physio-chemical conditions.

Chapter 2: Key points

1   The Earth's climate has changed throughout geological time and is still undergoing change. Palaeoclimatology is the study of past climate. There have been at least six major cold periods, or Ice Ages, throughout geological time. Since about 2.5 million years ago, global climate has cooled considerably, and the Earth entered the present Ice Age, referred to as the 'Quaternary Period', during which global climate has fluctuated between cold (glacial) stages and warm (interglacial) stages, with less intense warm (interstadial) and cold (stadial) periods.

2   Natural causes of global climate change include:

●   internal Earth processes such as plate tectonic processes, which lead to a redistribution of land masses and altitude, which in turn influence global atmospheric, hydrological and biological systems, together with volcanic activity, which may cause changes in atmospheric aerosols and gases;
●   processes external to the Earth, such as sunspot activity and Milankovitch cyclicity resulting from variations in the Earth's orbital parameters around the Sun, all of which lead to variations in the amount of solar insolation to the Earth's surface, thereby causing changes in the atmosphere–ocean system, e.g. changes in biomass production and burial;
●   catastrophic events such as large meteorite impacts, which may cause large-scale extinction events and thereby open up ecological niches for existing or new species to inhabit and evolve within.

3   Palaeoclimatology is studied using many different methods and techniques. Petrological techniques use characteristic sediment and rock types to interpret past climates (evaporite minerals, glacial deposits, etc.). Palaeontological techniques, such as the use of pollen spores, provide proxy data on global climate. Chemical methods include the study of stable oxygen isotopes, e.g. from foraminifera in deep-sea sediment cores, and air bubbles trapped in glacial ice, which provide an indication of sea-water temperature and, indirectly, an estimate of the relative amounts of sea water stored as glacial ice. Stable carbon isotopes in fossil organic matter can be used to evaluate changes in biomass production, which is a function of both regional and global climate. Stable nitrogen isotopes in fossils may be used as a proxy indicator of the contribution of nitrogen fixation by leguminous plants, again strongly influenced by global climatic conditions. Concentrations of various trace metals such as cadmium (Cd) in fossils (commonly expressed as a cadmium:calcium ratio) provide an insight into sea-water temperatures, and by extrapolation oceanic circulation patterns and global climate. The distribution of fine, wind-blown sediment, or loess, is an indicator of global aridity. Variations in the thickness of tree rings provide important information on past changes in climate, at least on an annual basis for the past 9,928 years. Glacial erosional and depositional landforms provide evidence for the extent of former ice sheets, a proxy for global climate. Raised beaches indicate the extent and position of former sea levels, which are a function of both global climate and tectonics.

4   The Quaternary Period is most often used in the prediction of future global climate change, because most data remain available from all the geological periods for study. The start of the Quaternary Period and the onset of the last Ice Age is debated but probably occurred about 2.5 million years ago. Glacials were periods of extensive ice cover lasting between 100,000 and 200,000 years, whereas interglacials, lasting 10,000 to 20,000 years, were much warmer periods, some being warmer than the present interglacial. These fluctuations in global climate are probably controlled mainly by Milankovitch cyclicity – the orbital characteristics of the Earth around the Sun. The glacial–interglacial cycles were complex with rapid transitions and perturbations in climate. The best-studied transition is the last glacial (Devensian/Wisconsin/Weichselian) to the present interglacial (Holocene). During this transition, there was a brief return to near-glacial conditions (Younger Dryas Stadial). Studies of the isotopes in ice cores (e.g. Vostok ice core), palaeontology, sedimentology and geomorphology provide important information on the rates of change of global atmospheric conditions and their resultant effects, including increased biological productivity, lower global sea levels, and increased aridity during the last glacial. Humans evolved during the Quaternary Period, the first bipedal hominid (*Australopithecus afarensis*) 3.75 million years ago, the first *Homo* (*H. habilis*) two million years ago and modern humans (*H. sapiens*) about 200,000 years ago. The development of human culture has affected the global biota and climate.

5   Global climate has been influenced by meteorite impacts. A major extinction event, which included the dinosaurs, took place about 65 million years ago at the Cretaceous–Tertiary (K–T) boundary. This major event is believed to be the result of one or more meteorites colliding with the Earth. Evidence for one or more meteorite impacts is provided by the iridium anomaly, which is present in rocks at the K–T boundary, along with other platinum-group elements showing concentrations characteristic of extraterrestrial bodies or meteorites, the common occurrence of spherules of molten glass and very high-pressure minerals (coesite and stishovite) atypical of conditions at or close to the Earth's surface, high concentrations of burned organic carbon (charcoal or fluffy carbon), and concentrations of rare amino acids that are more common in meteorites. Several locations have been suggested for the impact crater(s), with the most

favoured site being near the Yucatan Peninsula in the Gulf of Mexico. The meteorite impact(s) caused global fires, enhanced levels of atmospheric aerosols and reduced sunlight, which in turn led to global cooling, and poisonous chemicals called pyrotoxins having extremely serious effects on the most evolved life forms such as the dinosaurs. Other effects of the meteorite impact appear to have included very acidic rain, a depletion of the stratospheric ozone layer, and decalcification of the oceans.

6   Other mass extinctions have occurred throughout geological time, some of which may also be due to meteorite impacts, but at least some of which were caused by other processes leading to global climate change. The greatest extinction event known in Earth history, which occurred 250 million years ago, at the close of the Permian Period and the start of the Triassic Period, and involved the extinction of about 95 per cent of all living species, does not appear to have been associated with a meteorite impact but, rather, the growth of a supercontinent in low/equatorial latitudes, which caused a dramatic reduction in the area of favourable ecological niches, an unquenchable demand for nutrients and the exhaustion of sufficient nutrients to sustain the biomass. These circumstances conspired to lead to a crisis for life on Earth and mass extinctions.

7   There are examples of spectacular regional changes in climate caused by plate tectonic processes. About 5–6 million years ago, the Mediterranean Sea became landlocked as a result of plate tectonic processes, with the result that the Atlantic Ocean waters were sealed off from those of the Mediterranean in the region of the Straits of Gibraltar. The Mediterranean Sea evaporated and changed the regional climate to desert conditions; the evaporation of the sea water, probably periodically replenished by catastrophic flooding from the Atlantic Ocean, caused the accumulation locally of up to about 1 km in thickness of salts or evaporite minerals. This event is referred to as the Messinian salinity crisis. About five million years ago, the Straits of Gibraltar were breached by the Atlantic Ocean waters, which then flooded back into the Mediterranean.

8   An understanding of past global and regional climate change, the causes, processes and effects, is important to humankind in order to distinguish natural from human-induced climate change.

Chapter 2: Further reading

Barry, R.G. and Chorley, R.J. 1992. *Atmosphere, Weather & Climate.* London: Routledge, 392 pp.

Bell, M. and Walker, M.J.C. 1992. *Late Quaternary Environmental Change.* Harlow, UK: Longman Scientific & Technical, 273 pp.

Bradley, R.S. 1985. *Quaternary Paleoclimatology – Methods of Paleoclimate Reconstruction.* London: Unwin Hyman, 472 pp.
A comprehensive textbook suitable for undergraduates and researchers wishing to appreciate the various methods used in the reconstruction of past climates. Topics covered include the nature of global climate change; dating methods; ice core studies; the study of marine sediments; non-marine geological evidence; non-marine biological evidence; pollen analysis; dendroclimatology; and historical data.

Bradley, R.S. and Jones, P.D. 1995. *Climate Since A.D. 1500.* London: Routledge, 706 pp.

Dawson, A.G. 1992. *Ice Age Earth.* London: Routledge, 293 pp.
A detailed review of the fluctuations in the Earth's climate during Late Quaternary time. Suitable for undergraduate students and researchers interested in the complex and dynamic changes that affected the Earth's surface and atmosphere during this period. Topics considered in depth include ocean sediments and ice cores; general circulation models for the Late Quaternary; glaciation and deglaciation during Late Quaternary time; Late Quaternary environments; Ice Age aeolian activity; Late Quaternary volcanic activity; crustal and subcrustal effects; Late Quaternary sea level changes; and Milankovitch cyclicity in exerting a control on global climate.

Gates, D.M. 1993. *Climate Change and Its Biological Consequences.* Sunderland, Massachusetts: Sinauer Associates, Inc., 280 pp.
An extremely readable textbook on climate change and its biological consequences, with clear diagrams. The book is aimed at college/undergraduate students, and is in eight chapters: Chapter 1, Climate change: cause and evidence; Chapter 2, Past climates; Chapter 3, Plant physiognomy and physiology; Chapter 4, Past vegetational change; Chapter 5, Forest models and the future; Chapter 6, Ecosystems; Chapter 7, Agriculture, droughts, and El Niño; and Chapter 8, What to do?

Hsü, K.J. 1983. *The Mediterranean was a Desert: A Voyage of the Glomar Challenger.* New Jersey: Princeton University Press, 197 pp.
Written by one of the co-chief scientists on the deep-sea drilling vessel *Glomar Challenger*'s voyage to the Mediterranean in 1970, which first showed the Messinian salinity crisis, when the ocean basin dried up. This very readable book describes the evidence that led to the proposal that the Mediterranean Sea had evaporated. It introduces geological concepts with a minimum of terminology to explain the significance of the discovery and describes the technical problems encountered in undertaking such work. An interesting introduction to the excitement associated with discoveries made by Earth scientists who are involved with drilling into the sediments and rocks in the deep oceans.

Imbrie, J. and Imbrie, K.P. 1979. *Ice Ages: Solving the Mystery.* London: Macmillan, 229 pp.
A very readable, if somewhat dated, historic account of the causes and effects of Ice Ages. Strongly recommended to any student and teacher who wants a good historical background in global climate change.

Lamb, H.H. 1995. *Climate History and the Modern World* (second edition). London: Routledge, 433 pp.

Lowe, J.J. and Walker, M.J.C. 1997. *Reconstructing Quaternary Environments* (second edition). Harlow: Longman, 446 pp.
This is an essential text for both students and researchers who are involved in reconstructing Quaternary palaeoenvironments. It is comprehensively referenced and illustrated with up-to-date examples ranging from biological, geochemical, geomorphological and geochronological techniques.

McIlveen, R. 1992. *Fundamentals of Weather and Climate.* London: Chapman & Hall, 497 pp.

Parry, M. and Duncan, R. (eds) 1995. *The Economic Implications of Climate Change in Britain.* London: Earthscan Publications., 133 pp.

Williams, M.A.J., Dunkerley, D.L., Deckker, P. De, Kershaw, A.P. and Stokes, T. 1993. *Quaternary Environments.* London: Edward Arnold, 329 pp.
A comprehensive and well-illustrated text which examines the environmental changes that have taken place throughout Quaternary time. Useful for undergraduate students as well as a reference source for teachers and researchers. Emphasis is placed on the interactions between geological, biological and hydrological processes that have caused environmental change throughout this period and have resulted in the present environments.

'In the cities
there is even no more any weather
the weather in town is always
benzene, or else petrol fumes,
lubricating oil, exhaust gas.

As over some dense marsh, the fumes
thicken, miasma, the fumes of the automobile
densely thicken in the cities ...

In London, New York, Paris
in the bursten cities
The dead tread heavily through the muddy air
through the mire of fumes
heavily, stepping weary in our hearts.'

D.H. Lawrence, 'In the Cities'

This chapter examines the two main issues relating to global atmospheric change of ozone depletion and emissions of greenhouse gases and, therefore, provides a contrast to the generally more local atmospheric pollution caused by acidic deposition, or acid rain (see Chapter 4). Although the impact on global atmospheric change caused by human activities is emphasised, natural processes are also discussed. Central to any consideration of global atmospheric change is an appreciation of the radiation balance of the Earth's atmosphere (Figure 3.1).

This chapter is divided into four parts: ozone depletion and global cooling; the greenhouse effect and global warming; natural phenomena and global climate change; and finally a part on international action to control atmospheric pollution that may contribute to global climate change.

Stratospheric ozone depletion

Ozone was discovered by the Austrian chemist Schonbein in the 1840s. Studies of atmospheric ozone (O_3) go back into the early part of the twentieth century because it was seen as a potentially useful tracer of what was happening in the atmosphere. The recognition of a substantial depletion in the concentration of stratospheric O_3 had to wait until the 1970s. Also the role of CFCs in stratospheric O_3 depletion was not appreciated until the 1970s, when the so-called hole in the ozone layer was discovered, and first published in 1985 in the international scientific journal *Nature* by the British Antarctic Survey (BAS). Over Antarctica, O_3 depletion occurs during the boreal autumn (September) when the Antarctic polar vortex is isolated from other wind systems.

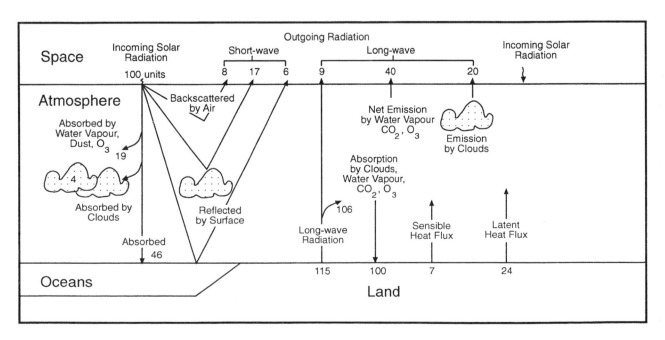

Figure 3.1 *Radiation balance of the Earth. Numbers refer to arbitrary units of radiation. Redrawn after* Our Future World: Global Environmental Research *(1989).*

The large fleets of aircraft that emit nitrogen oxides – which play a role in stratospheric ozone depletion – were viewed as the principal cause of ozone depletion, but subsequently the main culprits were identified as **chlorofluorocarbons (CFCs)**. The perceived safety of CFCs as chemicals in the manufacture of various products, such as refrigerants, made them appealing, but later work showed the role of atmospheric aerosols containing CFCs in releasing reactive chlorine, which breaks down the ozone molecule.

Ozone in the upper atmosphere, the stratosphere, is part of an important naturally occurring shield around the Earth. The ozone layer is involved in controlling the thermal structure of the stratosphere by absorbing incoming ultraviolet solar radiation and the outgoing longer-wavelength radiation from the Earth's surface.

Ozone forms naturally in the stratosphere by the action of sunlight splitting an oxygen molecule (O_2) into two separate oxygen atoms. These oxygen atoms then react with other oxygen molecules in the presence of a catalyst (e.g. hydroxyl radical, OH^-; water; hydrogen peroxide, HO_2) to produce ozone (O_3). Reactions between ozone molecules and sunlight can also lead to the destruction of the O_3 molecules.

In polar regions, stratospheric ozone depletion during the winter months occurs mainly through the catalytic action of chlorine, which is freed by chemical reactions that take place on polar stratospheric cloud (PSC) particles. In contrast, at middle to low latitudes, where the solar illumination is more intense, and because PSCs are absent, the rate of ozone destruction is influenced by a combination of different catalytic reactions. The relative importance of the possible chemical reactions that lead to stratospheric ozone depletion, and the precise controls on influencing such depletion, remain controversial. For example, gas phase models of the atmosphere suggest that nitrogen oxides, rather than chlorine and associated chemical species, are more important in destroying stratospheric ozone (Fahey *et al.* 1993). *In situ* measurements of stratospheric sulphate aerosol, reactive nitrogen and chlorine concentrations at middle latitudes by Fahey *et al.* confirm the importance of aerosol surface reactions that convert active nitrogen to a less reactive, reservoir form, resulting in mid-latitude stratospheric ozone being less vulnerable to active nitrogen but more vulnerable to chlorine species. The effect of aerosol reactions on active nitrogen depends on the rates of gas phase reactions, therefore following volcanic eruptions aerosol concentrations will have only a limited effect on ozone depletion at these latitudes (*ibid.*). Recently, it has been proposed that the chemical reactions in PSCs that lead to O_3 deple-

tion are more complex than originally thought, with the extent of O_3 loss being dependent on the ability of PSCs to remove NO_x permanently through deposition, which in turn depends upon PSC particle size, controlled by the composition and formation mechanisms for such particles (Toon and Tolbert 1995).

Ozone is an effective greenhouse gas, particularly in the upper and middle troposphere. It is formed in the atmosphere, where a series of complex chemical reactions are catalysed by the action of sunlight on carbon monoxide (CO), methane (CH_4), nitrogen oxide radicals (NO_x) and non-methane hydrocarbons (Figure 3.2).

A reduction in the amount of ozone in the upper atmosphere means that more solar radiation reaches the troposphere and Earth's surface, which in turn leads to greater surface warming. Reduced O_3 levels in the stratosphere, however, also mean that this part of the atmosphere becomes cooler, since it now absorbs less long-wavelength and solar radiation, and emits less to the troposphere – the result is that the Earth's surface will tend to cool. It so happens that the warming due to incoming solar radiation, related to the ozone column in the atmosphere, and the cooling because of the long-wavelength radiation, related to the actual vertical distribution of the ozone, are similar in magnitude. So, the juggling act between the magnitude of the ozone-related cooling or warming of the atmosphere and Earth's surface is critically affected by the magnitude of any change in the ozone concentration and distribution – obviously strongly influenced by latitude, altitude and the seasons. Furthermore, the creation and destruction of O_3 in the stratosphere is affected by the reactive chemical elements of oxygen, hydrogen, nitrogen and the halogens (e.g. chlorine and bromine). Elevated levels of incoming solar ultraviolet-B (UV-B, with wavelengths between 280 and 320 nm) radiation due to the destruction of the stratospheric ozone layer could lead to reduced bacterial activity in the surface layers of the world's oceans, with an accompanying increase in the concentrations of labile dissolved organic matter because bacterial uptake of this is suppressed (Herndl *et al.* 1993). UV-B radiation (Box 3.1) probably influences the recycling of organic matter in the surface layers of the oceans, because the processes are mediated by bacterioplankton, which are affected by solar radiation.

Early in 1992, a combination of anthropogenically created pollutants and a cocktail of chemicals from volcanic eruptions caused an unprecedented problem in the upper atmosphere. The news was released on 3 February 1992 by both the European Ozone Research Co-ordinating Unit and US government

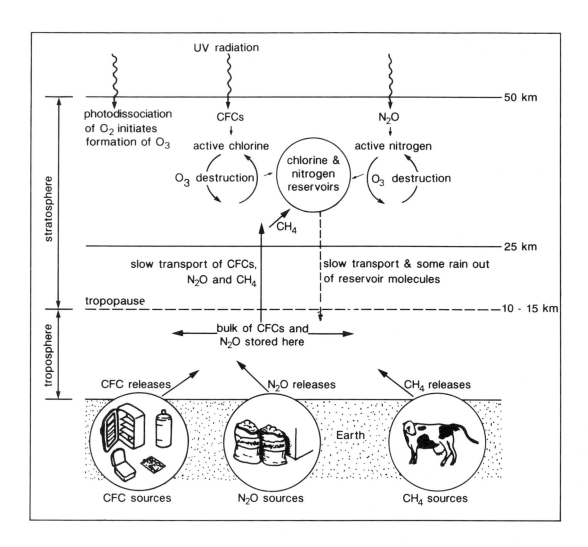

Figure 3.2 *Schematic diagram to show the principal sources of atmospheric ozone, and the main reactions that cause ozone depletion in the stratosphere. Redrawn after Smith and Warr (1991).*

scientists. Measurements revealed that chlorine-based chemicals were present in concentrations great enough to cause a complete depletion of O_3, or what has become known as a hole in the O_3 layer that protects people from being sunburnt by the ultraviolet radiation; skin cancer and eye cataracts can be caused by excessive exposure to UV radiation. On 11 January 1992, high levels of O_3-destroying chlorine chemicals were recorded over Moscow, Amsterdam and London.

A hole in the Earth's protective ozone layer

In 1977, the British Antarctic Survey observed and recorded a zone of stratospheric O_3 depletion, which is commonly referred to as a 'hole', in the naturally occurring ozone layer between 20 and 30 km above the Earth's surface. It was not until a decade later, however, that concern arose about the possible implications of this observation. Farman *et al.* (1985) were the first to show that the springtime values for total stratospheric ozone concentrations at the British Antarctic Survey stations, Argentine Islands at 65°S 64°W, and Halley Bay at 76°S 27°W had fallen significantly since 1957. They emphasised that lower stratospheric circulation had not changed and, therefore, the decreased stratospheric ozone levels were attributable to a chemical cause. They further suggested that the very low temperatures that prevail in midwinter, until after the spring equinox, make the stratosphere over the Antarctic region uniquely

---

BOX 3.1 UV-B AND THE OZONE SHIELD

The ozone layer absorbs part of the outgoing long-wave radiation and re-radiates it back to the troposphere below, and to the Earth's surface. The importance of stratospheric ozone is its role in controlling the UV-B reaching the Earth's surface. UV-B is a normal component of sunlight, with up to about 0.5 per cent of the energy reaching the Earth's surface under a clear sky at noon comprising biologically active UV-B radiation, but any significant increase in UV-B radiation above natural levels is potentially harmful to human health and the environment. Naturally, the absolute intensity of UV-B radiation reaching the Earth's surface is influenced by many factors, including the angle of the incident sunlight, principally controlled by the seasons and the time of day. Estimating UV-B intensity at the Earth's surface cannot be done from measuring stratospheric ozone levels alone, therefore it is important to obtain accurate UV-B data to establish long-term trends and causal factors. UV-B that passes through the stratosphere may be absorbed and scattered by air pollution, including ozone, in the lower atmosphere

---

sensitive to the destruction of O_3 in chemical reactions involving chlorine molecules. Antarctic ozone depletion generally occurs between altitudes of 12 and 22 km, the main region of stratospheric cloud formation. The size of this zone or hole of depleted O_3, which has fluctuated over the years, is increasing. The hole exists because the ozone-producing reactions have been inhibited or reduced in activity, possibly as a result of excessive anthropogenic emissions of certain ozone-destroying CFCs and other chemical species.

In the stratosphere, anthropogenic chlorine is converted to chemically reactive forms that lead to a depletion of the ozone (Figure 3.3), with particularly large O_3 losses during the springtime in Antarctica. Heterogeneous chemistry in stratospheric clouds, followed by the action of sunlight, converts the stratospheric chlorine from relatively inert forms to the much more reactive forms, of which ClO is dominant. Enhanced ClO is now known to precede the Antarctic and Arctic O_3 depletion (Waters *et al.* 1993, and references therein). Waters *et al.* have suggested that the O_3 loss in the south, long before the development of the Antarctic O_3 hole, can be masked by the influx of O_3-rich air. Although there is a decline in the absolute amount of anthropogenic emissions of gases that put chlorine into the stratosphere, these emissions will continue to have an increasing effect over the next decade, and remain for about a century at levels higher than those that were initially responsible for the Antarctic O_3 depletion because of their lifetime in the stratosphere.

The British research base on Antarctica, Halley, has monitored the meteorological conditions in this region since 1957, and up until 1977 there appeared to be no cause for alarm: climatic conditions appeared stable and the O_3 seemed intact. In 1979, however, a thinning of the O_3 layer was noted but its significance went unappreciated, probably because the British base was the only meteorological station in the world to record these changes and, at the time,

the results were considered anomalous, probably because of outdated instrumentation giving erroneous results! The stage was set for the dramatic discoveries of the 1980s.

On 7 October 1987, the American *Nimbus 7* satellite, which was monitoring the O_3 layer over Antarctica, recorded a substantial depletion of the O_3 layer at a height of 16.5 km, with a 97.5 per cent destruction of the amount of ozone measured on the 15 August 1987 (Farman 1987). This depletion, equivalent to more than the area of the USA, had developed from the Antarctic spring, with more than 50 per cent of the O_3 over Antarctica being destroyed within 30 days.

In the Antarctic spring of 1991, balloon-borne observations showed local ozone reductions approaching 50 per cent in magnitude which were observed at altitudes of 11–13 km (lower stratosphere) and 25–30 km (upper stratosphere) above the South Pole and McMurdo Station – these reductions being in addition to the normal springtime reductions at altitudes between 12 and 20 km (Hofmann *et al.* 1992). Until then, ozone depletion had not been observed at these altitudes, and by September 1991, the net result was an ozone column 10–15 per cent less than had been recorded in previous years. Hofmann and his colleagues also observed that this depletion coincided with penetrations into the lower stratospheric polar vortex of increased concentrations of sulphate aerosol particles (for significance, see section on sulphate aerosols) from the volcanic eruptions that took place in 1991, such as the eruption of Mount Hudson, Chile, at 46°S on 12–15 August, and from Mount Pinatubo, the Philippines, at 10°N on 15–16 June. The most plausible explanation for this ozone depletion, observed for the first time in the 11–13 km altitude layer, is that it occurred because of 'heterogeneous reactions' in the poleward-drifting volcanic cloud (*ibid.*).

Attention has naturally turned from the Antarctic to include the Arctic. Are there signs of a hole in the O$_3$ layer there? It was not until 1989 that a clear affirmative came (Hofmann *et al.* 1989). Not only were scientists able to detect the type of stratospheric clouds that allow the O$_3$-destroying reactions to occur, but they were able to measure the beginning of ozone depletion at a height of 22–26 km. This followed the coldest January in the North Pole stratosphere for at least 25 years.

A major problem to be solved by further research is to establish exactly where in the 20 to 40 km zone above ground level the thinning of the ozone layer is most dramatic. Heterogeneous reactions in the lower stratosphere present the greatest risks to the ozone layer, since it is here where most of the protective ozone is concentrated.

Ozone that is present in the stratosphere is known as stratospheric or high-level ozone, while O$_3$ that is present in the troposphere is known as tropospheric or low-level ozone. Tropospheric O$_3$, which controls the chemical cycling of atmospheric trace gases and exerts an important effect on global climate, is supplied naturally by downward transport from the stratosphere and, depending upon the local levels of NO$_x$, is produced by photochemical reactions. Stratospheric O$_3$ is decreasing, whereas above polluted regions in the Northern Hemisphere, tropospheric O$_3$ is increasing and often rises above the natural background levels. Over Europe, tropospheric O$_3$ concentrations may have increased by more than a factor of 2 in the last 100 years (Volz and Kley 1988). At sufficiently high concentrations, tropospheric O$_3$ is damaging to life and is probably partially responsible for forest die-back near industrialised centres. The potentially harmful effects of tropospheric O$_3$ were not appreciated until the 1950s, when it was identified as a photo-toxic component in the creation of photochemical smogs in Los Angeles and, subsequently, in many other urban areas.

Tropospheric O$_3$ is a waste product from automobile exhausts and many industrial processes. As a consequence of this, the concentration of tropospheric O$_3$ is at its greatest around large industrial cities, and poses a major threat to public health. Tropospheric O$_3$ is also a greenhouse gas and may be an important contributor to global warming if it is produced in large quantities. The global increases in tropospheric O$_3$ are a cause of worldwide concern. Studies of forest trees are leading to the definition of O$_3$ response thresholds. In a five-year study of the serial changes in the circumference of 28 mature lobiolly pine (*Pinus taeda* L.) trees, it was found that O$_3$ concentrations of ≥ 40 nl l^{-1} interacted with low soil moisture and high air temperatures to reduce short-term rates of stem expansion (McLaughlin and Downing 1995). Annual growth rates in this pine

were found to be inversely related to seasonal O$_3$ exposure and soil moisture stress (*ibid.*). If future predicted O$_3$ levels (IPCC Report 1992, 1994) are achieved in an anthropogenically enhanced greenhouse world, then the combined effect of greater tropospheric O$_3$ concentrations will be to alter the growth rates in plants, in some cases, as in the lobiolly pine, by reducing them.

Ozone loss is most pronounced during the Northern Hemisphere winter months: the Antarctic spring. Over the Antarctic, the ozone layer is destroyed by so-called heterogeneous reactions. These are reactions of chemicals in different states, for example as between gas and liquid, gas and solid, or solid and liquid. Such heterogeneous reactions take place on the surface of crystals in freezing clouds in the stratosphere. The catalysts for these reactions are chlorofluorocarbons (CFCs) produced by human activities. Since these reactions were not predicted, their discovery came as a surprise in the 1980s. The reactions are so rapid that 95 per cent of the destruction of the ozone layer in any year occurs in the first few weeks of the beginning of each Antarctic spring. A simplified series of reactions that lead to the breakdown of atmospheric ozone are illustrated in Figure 3.3.

The particular clouds in the stratosphere where the ozone is destroyed over Antarctica apparently form only at temperatures below about –80°C, although there is another cloud type responsible for ozone depletion that forms at –72°C. This latter cloud type is confined to polar air and is nine times more abundant than the colder clouds. Furthermore, the warmer clouds are widespread over the Arctic, whereas the cold types occur only over Antarctica. The warmer clouds are believed to contain fewer reactive chemicals than the cold types. For example, the warmer clouds do not contain hydrochloric or sulphuric acids, but they do contain nitric acid, which can trigger the heterogeneous reactions so harmful to the ozone layer.

Conventional wisdom puts the blame for the depletion of the ozone layer over Antarctica almost entirely on the accelerated anthropogenic emissions of certain greenhouse gases, such as the chlorine compounds, CFCs. But in the USA, at Boulder, Colorado, a group of scientists from the National Oceanic and Atmospheric Administration (NOAA) suspects that natural fluctuations in the sea-surface temperature in the eastern equatorial Pacific may be a major control on the concentrations of O$_3$ in the atmosphere. Their research in the eastern Pacific over the past 25 years (Joyce 1991) has shown that between 1962 and 1975, when the eastern equatorial

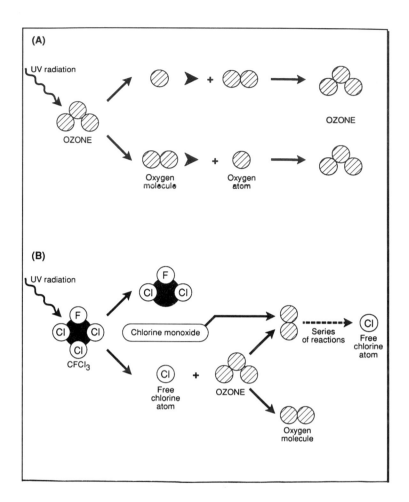

Figure 3.3 *(A) The naturally occurring chemical processes leading to the formation and decomposition of ozone in the atmosphere in the presence of ultraviolet radiation. (B) The decomposition of ozone initiated by chlorine atoms released during the breakdown of a commonly occurring, anthropogenically generated CFC believed to be harmful to the atmosphere ($CFCl_3$). Not all the two-atom (diatomic) molecules of oxygen combine to form ozone, and the free chlorine atoms that are liberated are potentially capable of initiating further reactions that lead to the breakdown of ozone.*

Pacific cooled, the global atmospheric O_3 budget increased. Then, between 1976 and 1988, when the eastern equatorial Pacific waters warmed, the global atmospheric O_3 budget decreased. So there may be a good correlation between sea-surface temperature and atmospheric ozone levels, but the mechanism by which they are linked remains unclear. Of course, finding natural, non-anthropogenic, cause-and-effect relationships between the levels of ozone and sea-surface temperatures is not a recipe for complacency in controlling human emissions of various gases.

In the stratosphere over Europe, the concentration of O_3 is decreasing at a rate approximately twice as fast as previously thought (Brown 1991). In a UK report published in July 1991 by the Stratospheric Ozone Review Group, it was stated that the concentration of O_3 in a wide band from the latitude of southern England to about latitude 30°N decreased by 8 per cent between 1979 and 1990. The potential problem of O_3 depletion is not confined just to the Antarctic and Arctic – the effects may be greatest

at the poles, but the knock-on effects of O_3 depletion over other parts of the globe, such as Europe, are now being appreciated. While the additional CO_2 would warm the lower atmosphere, it could cool the lower stratosphere and increase the formation of clouds that convert the potential O_3-depleting species to their active forms (Austin *et al*. 1992), i.e. enhance the stratospheric cloud chemistry that leads to the destruction of O_3 by chlorine from anthropogenically produced CFCs. In a numerical 3-D simulation of the Northern Hemisphere winter stratosphere, Austin *et al*. show that a doubling of the atmospheric CO_2 concentration, something that is likely to happen in the next century if steps are not taken to avert global warming, could lead to the formation of an O_3 hole in the Arctic and over northern Europe comparable with that observed over Antarctica, with almost 100 per cent local depletion of the O_3 in the lower stratosphere. The upper stratosphere will be affected to a lesser degree, and the Arctic will still have greater protection each spring compared with the Antarctic. But, since there are

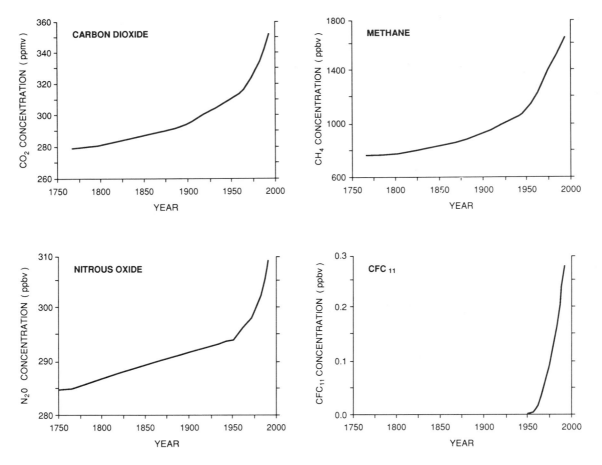

Figure 3.4 *Changes since the middle of the eighteenth century in the atmospheric concentration of carbon dioxide, methane, nitrous oxide and the commonly occurring CFC, CFC-11. Over the last few decades there has been a very large increase in the atmospheric concentrations of CFCs, which were absent before the 1930s. After IPCC (1990).*

many more people living at high latitudes in northern Europe and Canada, the risk of skin cancer, cataracts and other hazards will be enhanced. Austin *et al.* estimate that only about 20 per cent of the winters might produce an ozone hole over the Arctic, but persist into April or even May of such years.

Surface-based global measurements of atmospheric CH_4 and carbon monoxide (CO) show a significant decrease in their growth rates in 1991 and 1992, particularly in the Northern Hemisphere. The unprecedentedly large depletion of stratospheric O_3 in 1991 and 1992, thereby exposing the troposphere to additional UV radiation, leading to increased concentrations of the hydroxyl radical (OH^-) (see Box 3.5), the major atmospheric sink for CH_4 and CO, has been invoked as a plausible explanation (Bekkl *et al.* 1994). Indeed, Bekkl *et al.* have produced 2-D model simulations that show that almost 50 per cent of the 1992 decrease in CH_4 and CO growth rates can be accounted for by the observed reduced stratospheric O_3 concentrations.

CFCs, HCFCs, HFCs, halons

Human activities involve the use of aerosols in agriculture, industry and domestic situations that release chlorofluorocarbons, or CFCs. Chlorine in CFCs has been linked to stratospheric ozone depletion. The chemical stability of CFCs gives them long atmospheric lifetimes, and because they provide a long-term source of chlorine in the stratosphere, CFCs are seen as being a serious source of atmospheric pollution that could contribute to global climate change. Peak chlorine loading on the atmosphere will be reached over the next five years and, depending upon the exact date for phasing out CFCs, the loading should return to present-day levels some time between the years 2000 and 2010. Scientists participating in the United Nations Environment Programme (UNEP)/World Meteorological Organisation to assess the role of CFCs in contributing to stratospheric ozone depletion have given a high priority to minimising the future risks of ozone depletion by phasing out such harmful anthropogenically created chemicals.

CFCs are widely used in the electronics industry, where, for example, CFC-113 is a solvent used in more than 100 specialised applications. Pre-Industrial Revolution levels of CFCs were zero, so the emission of these molecules into the atmosphere is entirely due to human activities. While large parts of industry have attempted to develop alternative substances, there are many who believe that the electronics industry has been particularly slow in responding to the need for considerable research and development into replacement chemicals. Human activities, however, still result in the current annual production of 10^6 tonnes of CFCs, but the world consumption of CFC-11, -12 and -113 is now 40 per cent less than their 1986 levels, which is considerably less than the quantities permitted under the Montreal Protocol: the 1990 London Amendments to the Montreal Protocol require further reductions.

Other chemical compounds that are believed to be destroying the stratospheric ozone layer include the oxides of bromine, which are much more potent than the equivalent quantities of chlorine compounds. Reactions of bromine monoxide (BrO) and chlorine monoxide (ClO) can destroy ozone even in the absence of sunlight, which generally initiates such destructive reactions. Another set of reactions with ClO and BrO produces OClO, believed to be one of the gases responsible for the destruction of the ozone layer over Antarctica and the Arctic in the spring. The fumigant methyl bromide is a major ozone depleter in the upper atmosphere, and worries over its adverse effects on health and safety (toxic by inhalation, and it can cause pulmonary oedema and disorders of the central nervous system) led the Netherlands to drastically cut back its use between 1981 and 1989. Of the total annual global production of about 67,000 tonnes of methyl bromide, the USA uses about 43 per cent (26,000 tonnes), 22,300 tonnes of which is used as a soil fumigant. In November 1992, an international agreement was reached at a meeting of the Montreal Protocol held in Copenhagen to freeze the production and consumption of methyl bromide at 1991 levels, to take effect from 1 January 1995. Besides the natural emissions of methyl bromide, anthropogenic emissions may account for 0.05–0.01 of the observed annual global ozone depletion of 4–6 per cent and could increase to about one-sixth of the predicted ozone loss by the year 2000 if annual methyl bromide production increases at the current rate of 5–6 per cent (Buffin 1992).

Two of the halons that are particularly responsible for destroying the ozone layer are $CBrClF_2$ and $CBrF_3$. The concentration of $CBrClF_2$ is now 2 parts per trillion by volume of the atmosphere, and since the early 1980s it has increased at a rate of 12 per cent per annum, while that of $CBrF_3$ is now 1.3 parts per trillion and increasing at 5 per cent per annum (Singh *et al.* 1988). $CBrF_3$, amongst all the chemicals that are destroying the protective ozone layer, is perhaps the most effective and efficient of all the CFCs known at present. The British Antarctic Survey scientists now believe that the principal chemical culprits that are destroying the ozone layer are two particularly widely used compounds of bromine (1211 and 1301) which have a long residence time in the stratosphere (Thompson 1992).

Alternatives to CFCs are being developed and marketed. HCFCs and HFCs, for example, contain hydrogen in the structure and, unlike CFCs, have short atmospheric lifetimes and tend to be destroyed in the lower atmosphere by natural processes. HFCs contain no chlorine and, therefore, do not contribute to stratospheric ozone depletion, whereas HCFCs contain relatively small amounts of chlorine, which provides some contribution to stratospheric ozone depletion, but HCFCs are greenhouse gases. As examples of potential substitutes for various CFCs, HFC-134a could replace CFC-12 in refrigeration, air-conditioning, certain foams and medical aerosols, HCFC-123 could replace CFC-11 in refrigeration and air-conditioning, HCFC-141b could replace CFC-11 in energy-efficient insulating foams and solvent cleaning, HCFC-124 could replace CFC-114 and HFC-125 could replace CFC-115 in certain refrigeration uses, and HCFC-225ca/cb could substitute for CFC-113 in solvent cleaning (in the precision engineering and electronics industries).

The relative **ozone depletion potentials** (**ODPs**) of various CFCs, HCFCs and HFCs, calculated over their full lifetimes in the atmosphere, are compared in Table 3.1. Although HCFCs and HFCs appear to break down relatively easily in the lower atmosphere, the ultimate breakdown products are acidic compounds that will contribute to acid rain at minimal levels, but will not contribute to the formation of photochemical smog in urban areas. The hydrogen, chlorine and fluorine released by the breakdown products of HCFCs and HFCs should be removed from the atmosphere, by dissolution in cloud water followed by precipitation as rain, within an average of around two weeks. Trace amounts of other potentially harmful breakdown products, such as carbonyl and trifluoroacetyl halides, are expected to remain in the atmosphere for a few months, where they should be incorporated into cloud water, the oceans and land surface, and hydrolysed to CO_2 and

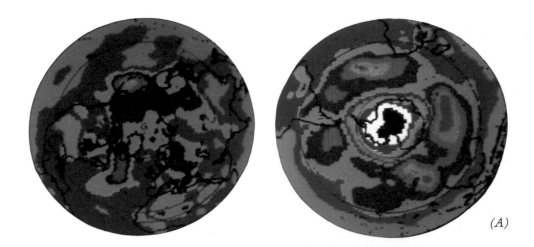

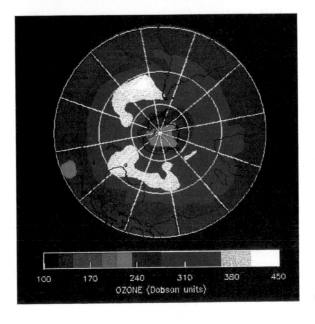

Plate 11 Maps of the 1989 to 1991 ozone column from the Total Ozone Mapping Spectrometer (TOMS) on board NASA's Nimbus 7 Satellite. (A) Maps for the Antarctic and Arctic for 1989. These show greater depletion of the ozone column in the Antarctic. (B) and (C) the variation in ozone depletion in the Antarctic summers for 1990 and 1991. Note that the amount of ozone present is measured in Dobson units.

If the atmosphere was compressed under a pressure of 1000 millibars, it would be 8 km thick, the thickness of oxygen would be about 1.5 km and ozone about 3 mm.

One Dobson unit is a hundredth of a millimetre of ozone in such a compressed atmosphere at standard pressure and temperature. Courtesy of NOAA/NESDIS/ NCDC/SDSD.

Plate 12 Atolls comprising the Maldives. The highest natural point on the islands is 2 m above sea level. These islands are under real threat from sea level rise induced by global warming. Courtesy of D. Sansoni/ Panos Pictures.

Plate 13 A sub-polar glacier flowing into a fjord on northern Ellesmere Island. Blocks of glacier calve and form icebergs as the glacier enters the sea. This process constrained the extent of glaciation in high latitudes during the Last Glacial.

Table 3.1 *Ozone depletion potentials (ODPs) of the principal CFCs, HCFCs and HFCs.*

	Chemical	ODP
CFCs	11	1.0
	12	1.0
	113	0.8
	114	1.0
	115	0.6
HCFCs	22	0.055
	123	0.02
	124	0.022
	141b	0.11
	142b	0.065
	225ca	0.025
	225cb	0.033
HFCs	32	0
	125	0
	134a	0
	152a	0

Source: AFEAS (Alternative Fluorocarbons Environmental Acceptability Study) and PAFT (Programme for Alternative Fluorocarbon Toxicity Testing) Member Companies 1992.

trifluoracetic acid, respectively, and form the corresponding hydrochloric and hydrofluoric acids (AFEAS 1992). Further independent research is needed to evaluate any potentially harmful environmental impacts from these breakdown products of the alternative fluorocarbons.

The political sensitivity of attributing lower stratospheric ozone depletion to the anthropogenic emissions of CFCs has led to a re-examination of the evidence. A four-year global time series of satellite observations of hydrogen chloride (HCl) and hydrogen fluoride (HF) in the stratosphere has shown conclusively that CFCs rather than other anthropogenic or natural emissions are indeed responsible for the recent global increases in stratospheric chlorine concentrations (Russell *et al.* 1996). It was also found in this study that all but a few per cent of the observed stratospheric chlorine can be accounted for from known anthropogenic and natural tropospheric emissions (*ibid.*).

The greenhouse effect – global warming

Most of this chapter examines global warming, the so-called **greenhouse effect**, a phenomenon that has become widely reported over the last few years. It was first observed in 1896, independently, by both the Swedish chemist Arrhenius and the American

geologist Thomas C. Chamberlain. In 1861, John Tyndall of Manchester was certainly amongst the first people to suggest that the large amount of carbon dioxide produced by combustion could affect the radiation balance to the Earth. It is interesting to note that Arrhenius suggested that by doubling the natural atmospheric levels of CO_2, average temperatures would rise by about 5–6°C. This phenomenon has been termed the greenhouse effect because it was originally thought that greenhouses are heated in a similar manner. The Sun's rays passing through the glass of a closed greenhouse include shorter wavelength (ultraviolet) radiation, which is absorbed by objects inside, which in turn re-radiate the heat but at longer wavelengths (infrared) to which the glass is nearly opaque. The heat is therefore trapped in the greenhouse with the net result that there is a sharp rise in temperature, together with more condensation. The condensation of water particles on the glass then leads to some cooling, but without ventilation and in bright sunlight the greenhouse can reach intolerable temperatures. The commonly cited analogy is not perfect, because the warming of air in a greenhouse is mainly due to the trapped air inside, which is unable to mix with the cooler air outside, but it represents a crude way of looking at the global greenhouse effect.

To investigate the extent to which human activities have begun to affect global climate and warm the planet, in 1990 a major review of the scientific evidence was published by the Intergovernmental Panel on Climate Change or IPCC (IPCC report, *Climate Change*, 1990), in preparation for the World Climate Conference, which took place in November of that year. This report was followed by an update in 1992 in which, although some of the earlier predictions were revised downwards, the findings remained essentially the same – that anthropogenic emissions of greenhouse gases are contributing to global warming. Perhaps the most significant shift in perspective by the IPCC between its 1990 and 1992 reports concerns the rate at which greenhouse gas concentrations are increasing (Figure 3.4), which is the principal control on how fast the world might be warming. Figure 3.5 shows the evidence for increased CO_2 levels from pre-industrialised times to the present, based on the analysis of air trapped in ice cores and, since the late 1950s, from precise, direct measurements of atmospheric concentration. The long-term rise in atmospheric CO_2 closely follows the increase in anthropogenic CO_2 emissions (Figure 3.5a). Under the IPCC 1990 'business-as-usual' (BAU) scenario, they estimated that the CO_2 doubling milepost could be reached as early as 2025, but the more recent forecasts predict

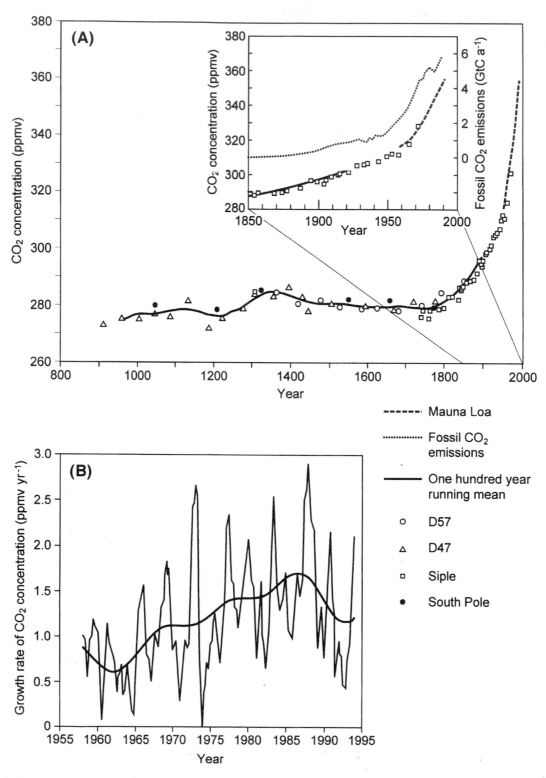

Figure 3.5 *(A) CO$_2$ concentrations over the past 1,000 years from ice core records (D47, D57, Siple and South Pole) and (since 1958) from Mauna Loa, Hawaii, measurement site. The smooth curve is based on a 100-year running mean. The rapid increase in CO$_2$ concentrations since the onset of industrialisation is evident and has closely followed the increase in CO$_2$ emissions from fossil fuels (see inset period from 1850 onwards). (B) Growth rate of CO$_2$ concentrations since 1958 in ppmbv yr^{-1} at the Mauna Loa station showing the high rates of the early 1990s and the recent increase. The smooth curve shows the same data but it has been filtered to suppress any variations on time scales < c. 10 years. Redrawn after IPCC report 1994 (Houghton* et al. *1995).*

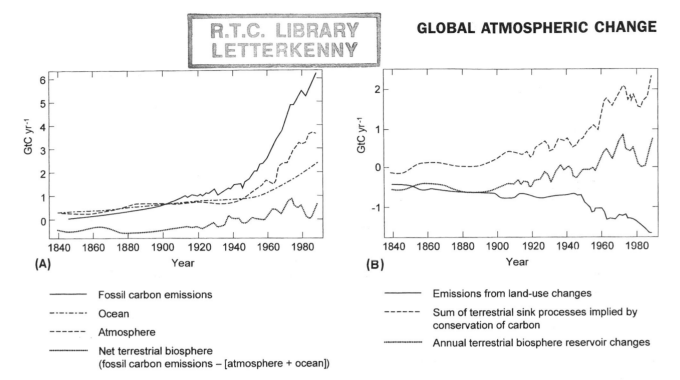

Figure 3.6   *(A) Fossil carbon emissions (based on statistics of fossil fuel and cement production), and representative calculations of global reservoir changes: atmosphere (deduced from direct observations and ice core measurements); ocean (calculated with the GFDL ocean carbon model); and net terrestrial biosphere (calculated as residual imbalance). The calculation implies that the terrestrial biosphere represented a net source to the atmosphere prior to 1940 (negative values) and a net sink since about 1960. (B) The carbon balance of the terrestrial biosphere. Annual terrestrial biosphere reservoir changes from (A), land use flux (plotted negative because it represents a loss of biospheric carbon) and sum of the terrestrial sink processes (e.g. Northern Hemisphere regrowth, CO$_2$ and nitrogen fertilisation, climate effects) as implied by conservation of carbon mass. (A) and (B) redrawn after IPCC report 1994 (Houghton et al. 1995).*

that this doubling will be delayed until 2050 or beyond. Figure 3.6 illustrates the changes in the global carbon reservoir and the balance within the terrestrial biosphere. Initial estimates of global warming and the rise in sea level (caused both by an expansion of the world's oceans because warmer water occupies a greater volume, and through melting of polar ice) suggested a rise of between 10–30 cm by 2030, and 33–75 cm by 2070, compared with present sea level. The most recent estimates, however, have revised these figures downwards, and suggest a global sea level rise of 2–4 cm per decade due to thermal expansion of ocean waters alone, and an additional, current 1.5 cm per decade contributed by melting ice caps and glaciers.

Although the essential conclusions of the IPCC 1990 and 1992 scientific assessments remain unchanged, the IPCC 1994 assessment included:

- revised values of global warming potentials (GWPs), with most values 10–30 per cent greater and uncertainties in GWPs typically ± 35 per cent;
- a revised GWP for methane to include both direct and indirect effects;
- stabilisation of atmospheric CO$_2$ levels between

one and two times present concentrations, i.e. 350 and 750 ppmv, only if anthropogenic CO$_2$ emissions fall substantially below 1990 levels;

- an improved estimation of radiative forcing by aerosols with the appreciation that aerosol radiative forcing values are highly uncertain and with substantial regional effects that cannot be viewed as a simple offset to greenhouse gas forcing;
- a recognition of the reduced rate in growth of atmospheric CO$_2$ emissions between 1991 and 1993 compared with the average rates over the past decade, comparable with similar temporary reductions since the 1950s but also noting the increased CO$_2$ emission rates since the latter part of 1993;
- the rate of increase in abundance of atmospheric methane has declined over the past decade, showing a substantial decrease during 1991–1992 but with an apparent increase in growth rate late in 1993;
- new estimates of the global carbon budget, particularly relating to terrestrial carbon uptake during the 1980s by forest regrowth in the Northern Hemisphere; and
- the climatic impact of the Mount Pinatubo

BOX 3.2 CLIMATE SENSITIVITY

The concept of climate sensitivity has been developed to provide an indication of the amount of global warming that could result from the increased concentration of greenhouse gases. Climate sensitivity is commonly expressed as the global temperature rise that would result from a doubling of CO_2 concentrations relative to pre-industrial levels. Climate sensitivity is estimated from general circulation models or GCMs, which are computer models that simulate the actual or real behaviour of global climate systems. In order to deal with the complexity of real climate systems over land and the oceans, GCMs need to be run on the fastest available supercomputers. Current IPCC estimates of climatic sensitivity are that it is 1.5–4.5°C, with a best estimate of approximately 2.5°C.

Recently, Working Group I of the IPCC evaluated various scenarios for stabilising the atmospheric CO_2 concentrations at 350, 450, 550, 650 and 750 ppmbv, as in the IPCC 1994 report but in this case explicitly, albeit quantitatively, incorporating global economic considerations (Wigley *et al.* 1996). Its results utilised the standard IPCC-recommended estimate of climate sensitivity of 2.5°C equilibrium global-mean warming for a doubling of atmospheric CO_2 levels, and best-estimate ice-melt model parameters. Where only greenhouse gases are considered, the numerical models suggest that sea level rise and temperature change are noticeably affected by the particular choice of route toward stabilisation at 550 ppmbv (*ibid.*).

Inferences such as these about climatic sensitivity are critically dependent on perceived future SO_2 emissions. Reduced CO_2 emissions lead to global cooling, but reduced SO_2 emissions lead to enhanced global warming. There is an increasing need to develop climate sensitivity models that evaluate scenarios where there is some degree of CO_2/SO_2 coupling.

eruption in June 1991, which led to a large transient pulse in stratospheric aerosols, causing a *c.* 0.4°C surface cooling over about two years, consistent with the model prediction of a global cooling of 0.4–0.6°C.

Manabe and Stouffer (1993) have examined the century-scale effects of increased atmospheric CO_2 on the ocean–atmosphere system. Using computer models to investigate climate sensitivity (see Box 3.2), they concluded that by doubling and quadrupling the concentration of atmospheric CO_2, global mean surface air temperature may increase over 500 years by 3.5°C and 7°C, respectively, with corresponding global sea level rises of 1 m and 2 m, due to thermal expansion of ocean waters alone. Any melting of ice sheets could make these values much larger. Also, under such a scenario, suggest Manabe and Stouffer, a quadrupling of atmospheric CO_2 levels could force the oceans into a new stable state in which thermohaline circulation would have ceased entirely, and the thermocline deepened substantially. Such changes in the structure of the oceans would prevent the ventilation of the deep ocean, with a likely profound effect on the carbon cycle and the biogeochemistry of the coupled system. Deep-ocean anoxia would prevail, and considerable amounts of the carbon budget could be locked up in organic-rich muds, in turn severely stressing biomass production in surface waters, and thereby leading to mass mortalities and, possibly, extinction events in marine organisms. According to the IPCC (IPCC report, *Climate Change*, 1990), a quadrupling of greenhouse gases could occur under its business-as-usual scenario if greenhouse gas emissions continued unchecked until the end of the twenty-first century, therefore these arguments are potentially realisable.

Computer simulations with increased concentrations of greenhouse gases suggest that in a warmer global climate there would be a decrease in high-frequency temperature variability and an increase in the proportion of precipitation occurring in extreme events (Karl *et al.* 1995). A detailed analysis of high-frequency temperature and precipitation data from hundreds of widely separated sites in the USA, former Soviet Union, China and Australia over the past 30–80 years supports this conclusion, as day-to-day temperature variability has decreased in the Northern Hemisphere, and at least in the USA the proportion of total precipitation during one-day extreme events shows a significant increase (*ibid.*).

Some scientists believe that the principal control on global temperatures is not the amount of greenhouse gases in the atmosphere but, rather, the magnitude of the heat transfer between the tropics and the poles. GCMs currently seem unable to produce a scenario based on atmospheric CO_2 concentration alone that can explain why about 60 Ma there were giant forests like those of Florida today covering the Arctic almost to the North Pole. The remains of these forests have been uncovered on Ellesmere Island and Axel Heiberg Island in northern Canada. Whilst these polar forests were in virtual darkness for three months of each year, temperatures appear never to have dropped below 0°C.

Measurements of atmospheric turbidity, for most practical purposes considered as an indication of atmospheric dustiness or dirtiness, have shown enormous increases (30 to 57 per cent) since the beginning of this century. Bryson (1968) estimates that these increases may lead to a 3.5–6.5°C lowering of the Earth's surface temperature. During the 1960s, it was from evidence such as this that the view that the Earth may return to glacial conditions within the near future was fostered. Ironically, any cooling this might initiate will probably be offset by warming induced by anthropogenically produced greenhouse gases. Today, however, there is more concern about global warming than cooling. Figure 3.7 shows the annual average values of atmospheric optical depth over the USA. The values increase dramatically over the industrialised areas.

The natural balancing act

The air around the planet provides a balanced mix of gases and water vapour to sustain life at a comfortable level. Amongst the planets in the solar system, the Earth is unusual because it contains substantial amounts of methane (CH_4) and ammonia (NH_3), which should not coexist with oxygen. CH_4 burns in oxygen, and this therefore means that some mechanism must be present that maintains CH_4 in the atmosphere; this mechanism is the metabolic processes of life itself. Scientists have looked for the presence of methane on distant planets as an indicator of extraterrestrial life. The atmosphere is largely responsible for keeping the Earth's surface temperature range at levels that generally keep water, so essential for life, in a liquid state and provide a climatic range conducive to existence as at present.

A comparison of the Earth and the Moon shows that the Moon, which lacks an atmosphere and is at roughly the same distance from the Sun as the Earth, has a much greater range of surface temperatures. The side of the Moon facing the Sun can reach 100°C, while on the opposite side night temperatures drop to a staggering −150°C. In fact, the average lunar temperature is −18°C, at which level the incoming heat energy, or solar flux, roughly balances the energy radiated from the Moon's surface. In contrast, the average temperature at the Earth's surface is 15°C.

The much higher surface temperature of the Earth compared with that of the Moon is a product of the atmosphere acting as a blanket to keep a large amount of the solar energy trapped on Earth. Without the atmosphere, the Earth's surface temperatures would be similar to that of the Moon.

The Earth's atmosphere provides the all-important insulation to sustain the rich diversity of life on Earth. It is easy, however, to overlook the fact that organisms have evolved to survive best in the Earth's climate as it exists without the intervention of humans to change this equilibrium. If human activities exert a forcing effect on the global climate to cause an irreversible and rapid climatic change, either through a warming or cooling, then many organisms, including people, may not survive. This is the essence of human interest and concern over the greenhouse effect.

The surface of the Earth emits radiation, mostly in the longer-wavelength or infrared part of the spectrum, and at a power of about 390 watts per square metre, which represents about one-half of the total incoming solar energy (or flux). The weather in the lower atmosphere, also referred to as the troposphere, is the consequence of this energy exchange, or balancing act, between incoming and outgoing radiation. Figure 3.1 illustrates the radiation and heat balance of the atmosphere and the major components that contribute to the climatic system.

Certain gases, the greenhouse gases, such as water vapour (H_2O), carbon dioxide (CO_2), methane (CH_4), ozone (O_3), nitrous oxide (N_2O), chlorofluorocarbons (CFCs) and halons absorb and emit infrared radiation at wavelengths that are longer than for visible light, that is at wavelengths greater than 12 μm. CFCs tend to absorb infrared radiation on the shorter-wavelength side of 12 μm where, without their existence, there would be a crucial window for infrared radiation to escape from the Earth back into space.

The temperature of the troposphere decreases with height above the ground at a rate of about 6.5°C per km, known as the environmental lapse rate. The overall pattern is for the lower, warmer layers of the atmosphere to absorb radiated energy from the ground and the rising air, and to prevent a lot of this energy being re-radiated upwards, with the net result that there is a reduction in the amount of infrared radiation radiated back into Space from the Earth. Without this layer of greenhouse gases, the Earth would be much like Mars, cold with an average surface temperature of about −55°C, and also lifeless. Another interesting comparison is between the Earth and Venus. Venus is approximately the same size as Earth, and probably has a very similar composition, but the atmosphere is different. The atmosphere of Venus is nearly all CO_2, with an atmospheric pressure at its surface of about 90 Earth atmospheres and temperatures of around 470°C, and it is also devoid of water. In the past, however, Venus had water on its surface, but early in its history, as the Sun began to brighten, the surface of Venus became hot and water vapour rose to the upper layers of the atmosphere, where it broke down to liberate

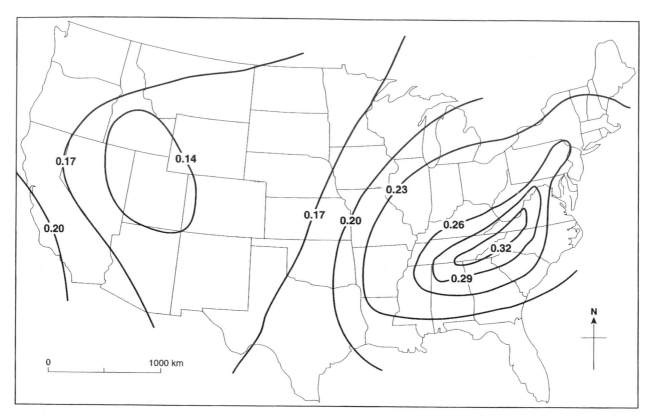

Figure 3.7 *Annual average values of atmospheric optical depth over the USA, showing a pronounced maximum over the industrialised areas. Redrawn after Flowers et al. (1969).*

hydrogen from the water, which was then lost into Space. CO_2 increased in the atmosphere of Venus and produced a runaway greenhouse effect, warming the surface to its present-day temperature. In contrast, on Earth, the runaway greenhouse effect was averted, probably through the evolution of life. There are, however, similar amounts of CO_2 on Earth and Venus, but on Earth much is stored in limestones and other carbonate sediments, oil, and coal. The comparison is important because human activities have been releasing the CO_2 stored in the rock reservoirs over the last few decades at ever-increasing rates. In a worst-case scenario, Venus may be a celestial warning of things to come.

To maintain equilibrium temperatures at the surface of the Earth, the incoming thermal energy from the Sun must balance that emitted from the Earth – in this case, there must be some balance between the total absorbed and emitted infrared radiation. There is a natural greenhouse effect, resulting from non-anthropogenic causes, and one of the biggest challenges for those working on this problem is to appreciate the balance between the natural and anthropogenic controls on global climate change. Amongst the chemical processes conspiring to main-

tain equilibrium temperatures, sulphate ions in the atmosphere play a role by creating a negative feedback to any warming (see Box 3.3).

Clouds are probably the most important self-regulating mechanism in controlling feedback mechanisms in the ocean–atmosphere system. Clouds prevent a runaway greenhouse or icehouse world. The warming effect of clouds is many times greater than, for example, that which is ascribed to doubling of the atmospheric CO_2 levels. Also, the overall effect of clouds is to make the Earth 10–15°C cooler than it would otherwise be if the planet were cloudless. There are two major sets of clouds. First, there are tropical clouds, which reflect sunlight back into the atmosphere to cool the global climatic system, but they also exert a greenhouse warming effect. Second, there are middle- to high-latitude clouds of the Northern and Southern Hemispheres, which have a net cooling effect. If these two sets of clouds were induced to behave differently, then this could have dramatic consequences on global heat budgets and regional weather systems, and lead to changes in regional and global precipitation patterns. Clouds also provide extra hydroxyl ions, which are capable of oxidising CH_4 and NO_x (not a greenhouse gas),

BOX 3.3 SULPHATE IONS, NEGATIVE FEEDBACK AND CLIMATIC COOLING

The climatic significance of **sulphate aerosols** derived from anthropogenic sulphur dioxide (SO_2) emissions comes from three main effects:

1 a possible increase in longevity of individual clouds;
2 a reduction in incoming short-wavelength solar radiation under clear sky conditions; and
3 a possible increase in cloud reflectivity because the sulphate aerosols act as cloud condensation nuclei (Wigley and Raper 1992).

Sulphate aerosols are very short lived, therefore increases or reductions in anthropogenic emissions have a rapid effect on global warming. There are large uncertainties associated with estimating the radiative forcing caused by sulphate aerosols, for example the quantitative links between changes in the mass of sulphate aerosols and changes in the number of cloud condensation nuclei are poorly understood. Such uncertainties mean that arguments about the exact consequences of atmospheric sulphate aerosols derived from anthropogenic SO_2 emissions should be treated speculatively, with more research being necessary in order to improve the understanding of cause and effect.

Sulphate ions (SO_4^{2-}) represent the dominant aerosol species in the Antarctic atmosphere, as well as being a significant part of the ice and snow. The sulphate ion is the end-member of the oxidation processes in the atmospheric sulphur cycle. On a global scale, anthropogenic emissions of SO_2 make the largest contribution to the amount of atmospheric sulphur (see Chapter 4). When anthropogenic emissions are excluded, the greatest concentration of tropospheric sulphate is in tropical regions, where the emission of biogenic dimethyl sulphide is highest. The contribution of volcanic eruptions to increased levels of tropospheric sulphate is greatest over regions such as Indonesia and the western part of the North Pacific. Increased levels of tropospheric sulphate may lead to global climate change because of changes in cloud albedo, their backscattering of incoming sunlight. As the mass concentration of aerosol sulphate increases, the number of cloud condensation nuclei also increases, but not in a simple way. The SO_4^{2-} produced through aqueous-phase oxidation becomes associated with pre-existing particles and does not appear to produce more particles directly. The size distribution, however, of pre-existing aerosol particles will be altered, which may lead to the generation of additional particles. The part of the emitted SO_2 available for creating new particles is that which is oxidised by hydroxyl radicals in the gas phase. Increasing tropospheric sulphate, therefore, should lead to a negative feedback, or negative forcing, of global climate, that is a cooling effect. A large amount of current research with climatic models tied to real measurements is aimed at attempting to understand the relative importance of positive climate forcing due to increases in CO_2 and other greenhouse gases versus negative feedback caused by SO_2 emissions.

Over the past hundred years, human activities have increased the global emissions of sulphur gases by a factor of about 3, with the result that there are enhanced sulphate aerosol concentrations, up to a factor of 100 in the Northern Hemisphere, but very small over the Southern Hemisphere oceans (Langer *et al.* 1992). Computer modelling of the distribution of tropospheric sulphate concentrations with actual mean monthly meteorological data, by Langer *et al.*, suggests that at most 6 per cent of the anthropogenic emissions of sulphur is available for the formation of new aerosol particles because about one-half of the sulphur dioxide is deposited on the Earth's surface, while most of the remainder is oxidised in cloud droplets, with the result that the rest of the SO_4^{2-} becomes associated with preexisting particles. Langer *et al.* calculate that the net effect is that the rate of formation of new sulphate particles may have doubled since pre-industrial times.

The second major source is part of the natural rhythm of life on Earth – sulphur released into the atmosphere from the oceans through biological activity. This organic (biogenic) sulphur is produced by the activity of the countless millions of sulphate-reducing plankton and algae. Another important source of sulphur is volcanic emissions, when it is injected high into the stratosphere. The biogenic sulphur source produces a chemical known as dimethyl sulphide, which is then oxidised to methyl sulphonate and the non-sea-salt sulphate ion (SO_4^{2-}). The importance of the biogenic sulphate source in controlling global climate has been mooted for several years, but it is only recently that strong scientific evidence for this has been demonstrated. The evidence comes from both present-day atmospheric sampling over Antarctica and also from ice cores that reveal past climatic events (see Chapter 2). A study of aerosol samples that were collected continuously at Mawson, Antarctica, between February 1987 and October 1989 has shown that the concentrations of both methyl sulphonate and non-sea-salt sulphate have a strong seasonality peaking in the austral summer, which parallels the cycle of oceanic biogenic sulphur production (Prospero *et al.* 1991). So, there appears to be sound scientific evidence to link the Antarctic atmospheric sulphur cycle with biological processes in the Southern Hemisphere oceans.

The 1992 IPCC report suggests that the cooling effect of sulphur emissions may have offset a significant part of the greenhouse warming in the Northern Hemisphere during the past several decades. Figure 3.8 shows the annual direct radiative forcing from anthropogenic sulphate aerosols in the troposphere. Note that the forcing is greatest over or close to industrialised areas, where sulphate aerosols are emitted from anthropogenic sources.

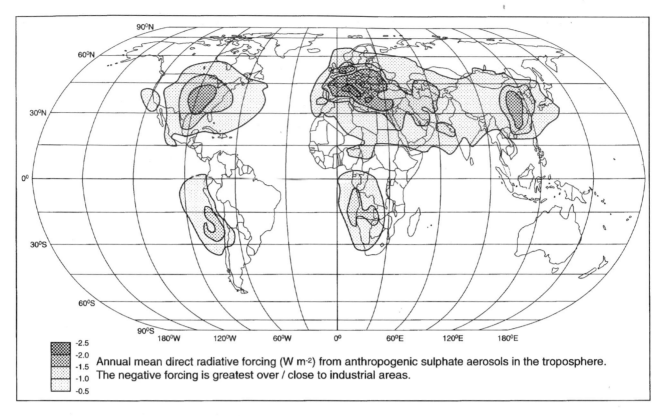

-2.5
-2.0
-1.5
-1.0
-0.5

Annual mean direct radiative forcing (W m-2) from anthropogenic sulphate aerosols in the troposphere. The negative forcing is greatest over / close to industrial areas.

Figure 3.8   *Annual mean direct radiative forcing (W m⁻²) resulting from anthropogenic sulphate aerosols in the troposphere. The negative forcing is greatest over or close to industrial areas. Redrawn after Houghton et al. (1995).*

thereby removing the greenhouse gases from the atmosphere and, hence, reducing any greenhouse effect (see Box 3.3).

It appears that wetter clouds can dampen the greenhouse warming (Slingo 1989). By altering the variables that go into computer-simulated climatic models, research suggests that near their freezing point, liquid-water clouds replace ice-crystal clouds. Liquid-water clouds dissipate more slowly than ice-crystal clouds, and they reduce the amount of solar energy absorbed, thereby reflecting more of this energy back into Space. The net result is a smaller degree of global warming than might otherwise occur

if there were no change in cloud type. Some computer runs with cloud type taken into account in a scenario where there is a doubling of atmospheric CO_2 emissions lead to an estimated equilibrium surface warming being reduced from 5.2 K to 2.7 K – a decrease of almost 50 per cent.

Seeing the greenhouse effect

Since detailed records started in 1973, ice and snow cover has decreased by about 8 per cent in the Northern Hemisphere, with a 2 per cent drop

BOX 3.4 ARCTIC GEOTHERMS

Measurements of a series of vertical temperature profiles in boreholes drilled into the permafrost in northern Alaska have revealed anomalously high temperature gradients in the upper 100 m of the permafrost (Lachenbruch and Marshall 1986). Since heat transfer in permafrost is almost exclusively by conduction, because there is no flow of ground water to transfer heat by other means, Lachenbruch and Marshall suggested that these anomalous temperature gradients indicate that global

warming is already under way. Analysis of the data, using heat-conduction theory, suggests that warming in the order of 2–4°C has occurred this century, probably over the past few decades. This evidence is particularly convincing because it provides a long-term signal of major climatic change and measures the direct thermal consequence of any change in atmospheric temperature conditions. Data of this type, gathered by drilling and coring ice, is extremely useful since the information can be recovered when and where desired, but it is costly.

between 1983 and 1992 in the amount of Arctic sea ice cover. Scientists from the British Antarctic Survey (BAS), based in Cambridge, have been documenting somewhat disturbing indications of possible global warming. On the Antarctic Peninsula below the King George VI Ice Shelf, field parties drilling into what should be solid ice have discovered many pockets of sea water at almost 0°C; also, the ice in this area has retreated about 30 miles in the last 40 years. Clearly, if this situation is repeated around Antarctica, then it could have a devastating impact on the polar ice cap, with rapid retreat and melting taking place. Furthermore, US and UK scientists estimate that the Antarctic ice sheet is shrinking annually by about 470,000 tonnes; also in Antarctica, since 1966, the Wordie Ice Shelf has disintegrated and decreased in area by about 1300 km². Other early warnings of a change in global climate caused by the greenhouse effect include measurements by American satellites of an increase in surface temperature of the North Atlantic Ocean by 1°C. Widespread bleaching of tropical corals has been documented, and is regarded as arising from exceptionally warm ocean waters (Gleason and Wellington 1993).

If the entire ice cap melted, then sea level around the globe could be expected to rise by 50 to 55 m. This scenario is extremely unlikely. If, however, as seems more probable, CO_2 levels double in the atmosphere into the next century, then a rise in sea level of about 1 m can be expected (IPCC report, *Climate Change*, 1990). Whole areas of low-lying coastal and estuarine environments will be flooded and many islands would become totally submerged, for example some of the Maldive Islands. It could be that the magnitude of the sea level changes will not be constant throughout the world, because the Earth's rotation causes greater effects in the low latitudes/tropics.

Many uncertainties remain as to whether global warming has actually started, with some evidence, for example from Arctic geotherms, suggesting that it is already under way (see Box 3.4). Indeed, the IPCC emphasises uncertainties in predicting the timing, magnitude and regional patterns of climate change, because there are important areas of ignorance concerning the sources and sinks of greenhouse gases, together with their control on interactions in the ocean–atmosphere system, e.g. in affecting cloud type and cloud cover, polar ice sheet development, ocean circulation, etc. Such large areas of ignorance and uncertainty, and the potentially very serious implications for the natural environment that might result from the anthropogenic emissions of greenhouse gases, have created universal interest both in

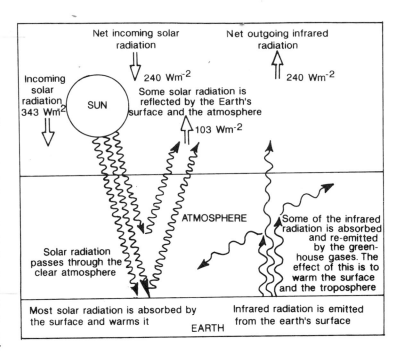

Figure 3.9 *Simplified diagram to illustrate the long-term global radiative balance of the atmosphere. Redrawn after Houghton et al. (1995).*

detecting any actual global warming and in the causal factors. With these factors in mind, the following section considers the principal chemical culprits that contribute to global warming.

Measuring radiative forcing

The 1994 IPCC report paid considerable attention to **radiative forcing**, which it defined as a change in average net radiation at the top of the troposphere (tropopause) due to a change in either solar or infrared radiation. Figure 3.9 shows the long-term global radiative balance of the atmosphere. The net input of solar radiation is 240 W m⁻², and this must be balanced by the net output of infrared radiation. Positive radiative forcing tends on average to warm the surface, whereas negative radiative forcing will tend to cool the surface.

Radiative forcing for the various chemical species that contribute to global warming (see next section) can be determined using the optical properties of the atmosphere. Two complementary methods are:

- Measurements of the chemical composition and size distribution of the chemical species and particles (e.g. Kiehl and Briegleb 1993). The optical properties and radiative forcing are calculated by

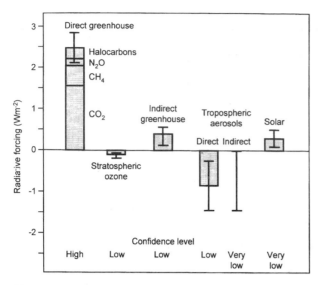

Figure 3.10 *Estimates of the globally averaged radiative forcing due to changes in greenhouse gases and aerosols from pre-industrial times to the present day, and changes in the solar variability from 1850 to the present day. The height of each bar shows the mid-range estimate of the radiative forcing and the lines indicate the possible range of values. An indication of relative confidence in the estimate is given below each bar. The contributions of individual greenhouse gases are indicated on the first bar for direct greenhouse gas forcing. The major indirect effects are depletion of stratospheric ozone (caused by CFCs and other halocarbons) and an increase in the concentration of tropospheric ozone. The negative values for aerosols should not necessarily be regarded as an offset against the greenhouse gas forcing, because of doubts over the applicability of global mean radiative forcing in the case of non-homogeneously distributed species such as aerosols and ozone. Redrawn after IPCC report 1994 (Houghton et al. 1995).*

making certain assumptions concerning particle shape, morphology and the distribution of particle properties with size.

- Direct measurement of the key properties at representative sites, and the extrapolation of these

Table 3.2　*Characteristics of greenhouse gases.*

Gas	Major contributor?	Long lifetime?	Sources known?
Carbon dioxide	yes	yes	yes
Methane	yes	no	semi-quantitatively
Nitrous oxide	not at present	yes	qualitatively
CFCs	yes	yes	yes
HCFCs, etc.	not at present	mainly no	yes
Ozone	possibly	no	qualitatively

After IPCC 1990.

results to other areas (e.g. Charlson *et al.* 1991). The scattering and absorption coefficients are then used to calculate the aerosol concentration and composition.

Figure 3.10 shows the 1994 IPCC estimates for the globally averaged radiative forcing due to changes in greenhouse gases and aerosols from pre-industrial times to the present day and changes in solar variability from 1850 to the present day.

The chemical culprits

Human activities release gases that have a warming effect such as carbon dioxide (CO_2), methane (CH_4), nitrous oxide (N_2O) and tropospheric (ground-level) ozone (O_3); with other potentially harmful chemicals including carbon tetrachloride, which was used extensively in dry cleaners in North America and Europe. In its 1990 report (IPCC report, *Climate Change*, 1990), the IPCC listed the main greenhouse gases (Table 3.2), those influenced by human activities (Table 3.3), and their radiative effects, together with the indirect trace gas chemical–climate interactions (Table 3.4).

These gases absorb infrared radiation in the range 7–19 μm, which is part of the 'window' through which more than 70 per cent of the radiation emitted from the surface of the Earth escapes into Space. The IPCC Report, *Climate Change*, published in 1990, showed the dramatic increase in the principal greenhouse gases that have resulted from human activities (Figure 3.4). Using the 1992 IPCC update, the greenhouse gases are believed to show the following annual increases: 0.5 per cent for CO_2, 0.5 per cent for CH_4, 4 per cent for CFCs and 0.25 per cent for nitrous oxide; their molecule-for-molecule global warming potential, normalised to CO_2 (= 1), remaining unchanged, is (with the 1990 IPCC estimates in parentheses): CH_4 = 11 (21); CFCs = 0 (> 1,000); N_2O = 260 (290). CO_2 emissions are commonly measured according to the carbon content, in millions of tonnes of carbon (MtC), where 1 tonne of carbon is equivalent to 3.67 (or 44/12) tonnes of carbon dioxide.

These greenhouse gases cause radiative forcing, a measure of their ability to perturb the heat balance in a simplified model of the Earth–atmosphere system. The concept of **global warming potentials (GWPs)** provides a simple way to describe the potential of greenhouse gas emissions to influence future global climate by radiative forcing, which is controlled by various parameters such as the amount of gas emitted,

Table 3.3 *Atmospheric concentrations of key greenhouse gases influenced by human activities.[1]*

Parameter	CO_2	CH_4	CFC-11	CFC-12	N_2O
Pre-industrial atmospheric concentration (1750–1800)	280 ppmv[2]	0.8 ppmv	0	0	288 ppbv[2]
Current atmospheric concentration (1990)[3]	353 ppmv	1.72 ppmv	280 pptv[2]	484 pptv	310 ppbv
Current rate of annual atmospheric accumulation	1.8 ppmv (0.5%)	0.015 ppmv (0.9%)	9.5 pptv (4%)	17 pptv (4%)	0.8 ppbv (0.25%)
Atmospheric lifetime[4] (years)	(50–200)*	10	65	130	150

[1] Ozone has not been included in this table because of lack of precise data.
[2] ppmv = parts per million by volume; ppbv = parts per billion by volume; pptv = parts per trillion by volume.
[3] The current (1990) concentrations have been estimated based upon an extrapolation of measurements reported for earlier years, assuming that the recent trends remained approximately constant.
[4] For each gas in the table, except CO_2, the 'lifetime' is defined here as the ratio of the atmospheric content to the total rate of removal. This time scale also characterises the rate of adjustment of the atmospheric concentrations if the emission rates are changed abruptly. CO_2 is a special case since it has no real sinks, but is merely circulated between various reservoirs (atmosphere, ocean, biota). The 'lifetime' of CO_2 given in the table is a rough indication of the time it would take for the CO_2 concentration to adjust to changes in the emissions.
* The way in which CO_2 is absorbed by the oceans and biosphere is not simple and a single value cannot be given.
Source: IPCC 1990.

Table 3.4 *Direct radiative effects and indirect trace-gas chemical–climate interactions.*

Gas	Greenhouse gas	Is its tropospheric concentration affected by chemistry?	Effects on tropospheric chemistry?*	Effects on stratospheric chemistry?*
CO_2	Yes	No	No	Yes, affects O_3 (see text)
CH_4	Yes	Yes, reacts with OH	Yes, affects OH, O_3 and CO_2	Yes, affects O_3 and H_2O
CO	Yes, but weak	Yes, reacts with OH	Yes, affects OH, O_3 and CO_2	Not significantly
N_2O	Yes	No	No	Yes, affects O_3
NO_x	Yes	Yes, reacts with OH	Yes, affects OH and O_3	Yes, affects O_3
CFC-11	Yes	No	No	Yes, affects O_3
CFC-12	Yes	No	No	Yes, affects O_3
CFC-113	Yes	No	No	Yes, affects O_3
HCFC-22	Yes	Yes, reacts with OH	No	Yes, affects O_3
CH_3CCl_3	Yes	Yes, reacts with OH	No	Yes, affects O_3
CF_2ClBr	Yes	Yes, photolysis	No	Yes, affects O_3
CF_3Br	Yes	No	No	Yes, affects O_3
SO_2	Yes, but weak	Yes, reacts with OH	Yes, increases aerosols	Yes, increases aerosols
CH_3SCH_3	Yes, but weak	Yes, reacts with OH	Source of SO_2	Not significantly
CS_2	Yes, but weak	Yes, reacts with OH	Source of COS	Yes, increases aerosols
O_3	Yes	Yes	Yes	Yes

* effects on atmospheric chemistry are limited to effects on constituents having a significant influence on climate.
Source: IPCC 1990, based on Wuebbles *et al.* 1989.

its infrared energy absorption properties, and the amount of time (residence time) of each gas in the atmosphere. The 1992 IPCC report, using the revised GWPs, estimates that the contribution made by the main greenhouse gases to global warming breaks down as follows: 72 per cent due to carbon dioxide, 18 per cent due to methane and 10 per cent due to nitrous oxide. The following sections review the principal greenhouse gases.

CFCs, HCFCs and HFCs

Global warming potentials (GWPs) relative to a CO_2 molecule have been calculated for the principal CFCs, HCFCs and HFCs, and are given in Table 3.5 (from AFEAS 1992). Initial research suggested that CFCs, because they are greenhouse gases, are important contributors to global warming, but their ability to destroy stratospheric ozone and thereby contribute

Table 3.5 *Global warming potentials (GWPs) of principal CFCs, HCFCs and HFCs compared with CO_2 and CH_4.*

Compound	Estimated atmospheric lifetime	GWPs for various integration time horizons*		
		20 yrs	100 yrs	500 yrs
CO_2	**	1	1	1
CH_4***	10.5	35	11	4
CFC-11	55	4500	3400	1400
CFC-12	116	7100	7100	4100
CFC-115	550	5500	7000	8500
HCFC-22	15.8	4200	1600	540
HCFC-123	1.7	330	90	30
HFC-125	40.5	5200	3400	1200
HFC-141b	10.8	1800	580	200
HFC-225ca	2.7	610	170	60
HFC-225cb	7.9	2400	690	240
HFC-134a	15.6	3100	1200	400
HFC-152a	1.8	530	150	49

* 'integration time horizon' is the timespan over which GWPs are calculated for this study from the cumulative radiative forcing over a given integration time horizon.
** The decay of carbon dioxide concentrations cannot be reproduced using a single exponential decay lifetime, thus there is no meaningful single value for the lifetime that can be compared directly with other values in this table.
*** GWP values include the direct radiative effect and the effect due to carbon dioxide formation, but exclude any effects resulting from tropospheric ozone or stratospheric water formed as methane decomposes in the atmosphere.
Source: AFEAS (Alternative Fluorocarbons Environmental Acceptability Study) and PAFT (Programme for Alternative Fluorocarbon Toxicity Testing) Member Companies 1992.

to global cooling, suggests that CFCs do not provide a net contribution to global warming, i.e. their global warming and cooling potentials more or less cancel out.

Plate 3.1 *Heavy industrialisation along the Yellow River in central China, emitting large quantities of greenhouse gases. The growth of China's industry and its poor environmental legislation poses one of the greatest threats in terms of reducing greenhouse gas emissions.*

Water vapour

Water vapour is the principal greenhouse gas. It absorbs light waves strongly in the range 4–7 μm, whereas CO_2 absorbs in the band 13–19 μm. The concentration of water vapour in the troposphere is determined internally within the global climate system, and on a global scale it is unaffected by anthropogenic sources and sinks. The coldest and hottest places on Earth are also the driest, such as the deserts, with central Asia being the coldest and driest and central Australia the hottest and driest. At night, energy escapes into Space to make these places cold, while during the day the lack of cloud cover allows more solar radiation to reach the ground and thereby make it hot. These places have least water and are therefore least able to maintain an equable climate throughout the day.

Carbon dioxide

Carbon dioxide (CO_2) is one of the main greenhouse gases, and is of greatest concern as a controllable gas emission caused by human activities. Following the Industrial Revolution, the combustion of fossil fuels, together with deforestation, has caused an increase in the concentration of atmospheric CO_2

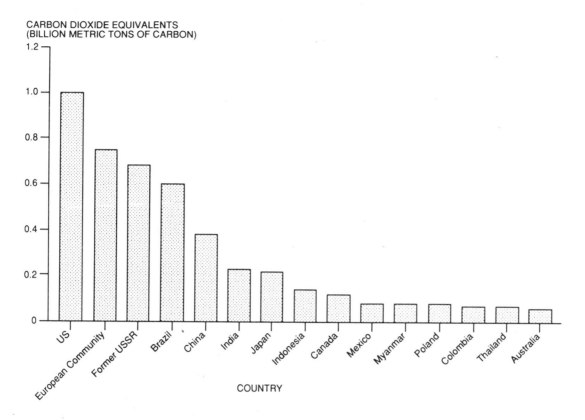

CARBON DIOXIDE EQUIVALENTS
(BILLION METRIC TONS OF CARBON)

COUNTRY

Figure 3.11 *Countries with the highest net greenhouse gas emissions for 1987. Redrawn after World Resources Institute (1990).*

by 26 per cent. Between 1950 and 1980, CO_2 emission increased by an estimated 586 per cent in the developing countries (Plate 3.1), 337 per cent in the former Soviet Union and Eastern Europe, 91 per cent in North America, and 125 per cent in Western Europe, the rest being made up by other developed countries (Pain 1989a): Figure 3.11 shows the main countries responsible for the emissions of CO_2. It is interesting to note that the principal responsibility for producing these emissions rests with the developed and industrialised countries. The USA is the largest emitter of CO_2, accounting for 24 per cent of global emissions; the UK accounts for 3 per cent of emissions, 96 per cent of which comes from the burning of fossil fuels for energy use, mostly from electricity generation (*Climate Change: Our National Programme for CO_2 Emissions* 1992, UK Department of the Environment). Also, in the UK, just over 50 per cent of the 1990 CO_2 emissions were accounted for by the use of private cars, and more than 25 per cent by the industrial, commercial and public sectors' use of road transport.

The EU's total CO_2 emissions represent approximately 13 per cent of global CO_2 emissions, compared with 24 per cent for the USA, 5 per cent

for Japan, and 25 per cent for Eastern Europe and the former Soviet Union (CONCAWE motor vehicle emission regulations and fuel specifications – 1992 update; CONCAWE is the oil companies' European organisation for environmental and health protection, established in 1963). In Britain, the peak emission of 190 Mt in 1979 was followed by a drop, but from 1984 to 1987 there was an 18 Mt increase to 171 Mt. In 1987, power stations accounted for 37 per cent of the total emissions of CO_2, with 20 per cent from industry, 16 per cent from transport, 14 per cent from domestic combustion, and 13 per cent from offices and other sources.

Transport accounts for nearly one-third of the total global energy consumption and contributes around 25 per cent of the world CO_2 output, as well as CFCs, methane and nitrous oxide (Greenpeace 1990). North America and Europe each possesses more than one-third of the world's vehicles, which was 400 million in 1985. In the EU, 26 per cent of the total anthropogenic CO_2 emissions come from transport. Despite these figures, there is still relatively little concerted effort aimed at energy efficiency and reducing the harmful emissions of greenhouse gases from car exhausts.

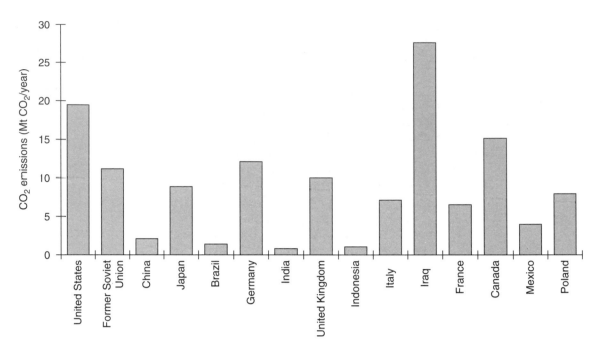

Figure 3.12   *Per capita CO_2 emissions from industrial processes 1991 for the 15 countries with the highest total green-house gas emissions 1991. Data from World Resources Institute (1994–95).*

The generation of CO_2 as a result of the combustion of fossil fuels suggests that if present trends continue, then its concentration will double every 50 years. In 1989 and 1990, the combustion of fossil fuels added an estimated annual 6.0 ± 0.5 gigatonnes of carbon (GtC) to the atmosphere (of which the main contributory nations are shown in Figure 3.12), compared with 5.7 ± 0.5 GtC in 1987; the estimated total release of carbon in the form of CO_2 from the oil wells of Kuwait in 1991 was 0.065 Gt, equivalent to about 1 per cent of the total annual anthropogenic emissions (IPCC 1990, 1992). Monitoring of the increase in CO_2, however, shows an increase that is only 50 per cent of the predicted level and this difference is ascribed to the ameliorating or buffering effect of reactions in the world's oceans and atmosphere. Some researchers have estimated that a doubling of the present CO_2 level to 600 ppmv will cause an average rise in global temperature of about 8°C (Maddox 1989). These figures do not take account of any feedback mechanisms that might serve to reduce the predicted temperature rise, for example an increased abundance of clouds with their cooling effects.

To understand the details of these feedback systems, it is necessary to identify the major global storage and transfer of carbon. The carbon cycle is illustrated in Figure 1.15A.

Dying forests may contribute to the greenhouse effect. This is because decaying vegetation releases CO_2 and H_2O, and also as part of the respiration process, trees convert CO_2 to O_2. Deforestation could be sending an annual 4 billion tonnes of CO_2 into the atmosphere that would otherwise be taken up by plants during their metabolic processes, double the most commonly quoted estimates (Pearce 1989a). The IPCC reports (1990, 1992) quote an annual average net flux to the atmosphere of 1.6 ± 1.0 GtC from land use during the 1980s.

Methane

Methane (CH_4), an atmospheric trace gas involved in many chemical reactions in the troposphere and stratosphere, initially received relatively little attention as a greenhouse gas, but in the past few years this has been rectified. The 1992 IPCC report stated that a methane molecule has 11 times more global warming potential than a molecule of carbon dioxide. Total annual anthropogenic and natural emissions of CH_4 are 500 Tg (1 Tg = 10^{12} g) (IPCC 1992 update). Since 1950, concentration levels of methane have been rising at 1 per cent per annum (10–16 ppbv), which is four times the rate of increase of carbon dioxide and could lead to methane becoming the principal greenhouse gas within 50 years (Pearce 1989b). The IPCC report (1990), *Climate Change*, puts the annual increase of CH_4 at 14–17 ppbv, giving a present atmospheric concentration of 1,700 ppbv. Atmospheric CH_4 concentrations have more than doubled since the Industrial Revolution as a result of increased rice production, cattle rearing, biomass burning, coal mining and the ventilation of natural gas (*ibid.*). Prior to the Industrial Revolution, atmospheric concentration of CH_4 was almost constant at about 0.65 ppmv, whereas in 1990 it reached 1.72 ppmv (Badr *et al.* 1992a, b).

The main sources and sinks of CH_4 input into the atmosphere, expressed in global average Tg per annum, are given in Table 3.6 (IPCC 1992). The net annual average input of CH_4 to the atmosphere is estimated at about 32 Tg. A further source of CH_4 emissions is from natural gas leakage in distribution systems, and from livestock – including cows' burps and farts! It has been estimated that the annual emission of methane from the world's cattle is close to 100 Mt (Pearce 1989b). If only we could harness this!

Perhaps one of the most surprising aspects of the source of CH_4 production is the role played by termites. There are an estimated 250,000 billion termites in the world, which inhabit approximately two thirds of the land area and consume something in the region of a third of the global vegetation. In 1982, scientists from the then West Germany, USA and Kenya discussed the importance of termites as contributors of CH_4 to the atmosphere and concluded that termites could account for about one-third of the annual global emission of methane or 150 Mt, although more recent evaluations of the contribution from termites have considerably revised this figure downwards to an annual 5 Mt.

Estimates of CH_4 emissions from various sources, however, remain uncertain, particularly compared with those for CO_2. Indeed, these IPCC figures exclude perhaps the largest source of CH_4, which may be degassing of the mantle at mid-ocean ridges and from the bacterial breakdown of organic matter buried in sediments at continental margins and within lakes. Studies of carbon isotopes in CH_4 molecules suggest that approximately 100 Tg, or 20 per cent, of atmospheric methane was produced long ago and, as already stated, is currently escaping from melting permafrost, present as gas hydrates, coal seams, oil reservoirs, and rocks beneath the oceans and natural gas sources, i.e. it is of fossil origin.

Nitrous oxide

Nitrous oxide (N_2O) is an important trace gas in the atmosphere. The main anthropogenic sources of nitrous oxide are fertilisers, fossil fuel combustion and

Table 3.6 *Estimated sources and sinks of methane (Tg CH_4 per annum).*

	Annual release	Range
Sources		
Natural:		
Wetlands (bogs, swamps, tundra, etc.)	115	(100–200)
Termites	20	(10–50)
Ocean	10	(5–20)
Fresh water	5	(1–25)
CH_4 hydrate	5	(0–5)
Anthropogenic:		
Coal mining, natural gas and petroleum industries	100	(70–120)
Rice paddies	60	(20–150)
Enteric fermentation	80	(65–100)
Animal wastes	25	(20–30)
Domestic sewage treatment	25	?
Landfills	30	(20–70)
Biomass burning	40	(20–80)
Sinks		
Atmospheric (tropospheric + stratospheric) removal	470	(420–520)
Removal by soils	30	(15–45)
Atmospheric increase	32	(28–37)

Source: IPCC *Climate Change* 1992.

various synthetic chemical manufacturing processes, e.g. nylon production. The global N_2O concentration has been rising at a rate of 0.2–0.3 per cent per annum, reaching about 310 ppbv in 1990 (Badr and Probert 1992a, b). The increased atmospheric concentrations of N_2O are of concern because of its role in destroying the ozone layer as a result of producing nitric oxide in the stratosphere, and because N_2O contributes to the atmospheric greenhouse effect. Currently, estimates of individual N_2O sources and their emission rates are poorly constrained, with the IPCC estimated sources and sinks given in Table 3.7. Data from Antarctic ice cores show that atmospheric N_2O concentrations were about 30 per cent lower during the Last Glacial Maximum compared with the Holocene epoch (also, see Chapter 2), and with present-day N_2O concentrations unprecedented in the past 45 ka, suggesting that the recent increases in atmospheric N_2O are due to human activities (Leuenberger and Siegenthaler 1992).

Tropospheric (ground-level) ozone

Tropospheric, or **low-level**, **ozone** (not to be confused with stratospheric ozone) is a greenhouse

Table 3.7 *Estimated sources and sinks of nitrous oxide (Tg N per annum).*

Sources	
Natural:	
Oceans	1.4–2.6
Tropical soils	
Wet forests	2.2–3.7
Dry savannahs	0.5–2.0
Temperate soils	
Forests	0.05–2.0
Grasslands	?
Anthropogenic:	
Cultivated soils	0.03–3.0
Biomass burning	0.2–1.0
Stationary combustion	0.1–0.3
Mobile sources	0.2–0.6
Adipic acid production	0.4–0.6
Nitric acid production	0.1–0.3
Sinks	
Removal by soil	?
Photolysis in the stratosphere	7–13
Atmospheric increase	3–4.5

Source: IPCC *Climate Change* 1992.

gas that is toxic to plants, humans and other organisms. In the Northern Hemisphere, the growth in surface emissions of nitrogen dioxide and hydrocarbons leads to increased concentrations of ozone in the troposphere. A recent study by Johnson *et al.* (1992) has shown that the radiative forcing of surface temperatures is most sensitive to changes in tropospheric ozone at a height of about 12 km, where aircraft emissions of nitrogen oxides are at a maximum, and where the model sensitivity of ozone to NO_x emissions is enhanced. The model of Johnson *et al.* (1992) shows that the radiative forcing of surface temperatures is approximately 30 times

more sensitive to the emissions of NO_x from aircraft than to surface emissions: their study also found that the impact on global warming of increases in tropospheric ozone due to increases in the surface emissions of NO_x have been overestimated by a factor of up to 5 (including IPCC report, *Climate Change*, 1990), because of errors in the calculations of the ozone budget. Compared with the Northern Hemisphere, the Southern Hemisphere is 60 per cent more sensitive to changes in the emissions of NO_x, since it receives only 18 per cent of the total emissions (Johnson *et al.* 1992). In the atmosphere, hydroxyl ions are capable of ameliorating much of the harmful effects of gases such as NO_x by oxidising them to less harmful substances (see Box 3.5).

A study by Oltmans and Levy (1992) suggests that it is the natural processes, not the anthropogenic sources of pollution, that control the seasonal cycle of tropospheric ozone over the western North Atlantic; even though springtime daily average O_3 concentrations at Bermuda exceed 70 ppbv, and in 1989 hourly readings surpassed the Canadian air quality limit of 80 ppbv. Continuous measurements of tropospheric ozone from Bermuda (32°N, 65°W) and Barbados (13°N, 60°W) indicate that the high levels of O_3 are transported from the unpolluted upper troposphere at altitudes greater than 5 km above the northern USA and Canada (*ibid.*). In support of their conclusions, Oltmans and Levy pointed out that in Barbados the seasonal and diurnal variations in surface O_3 are virtually identical to those measured at Samoa in the tropical South Pacific, far removed from anthropogenic sources of pollution, and where the low levels of NO_x ensure that natural processes control surface ozone levels. They also note that during the summer, when surface O_3 concentrations over the eastern USA can exceed 70 ppbv due to pollution, in Bermuda typical measurements are 15–25 ppbv.

BOX 3.5 ATMOSPHERIC CLEANSERS: HYDROXYL RADICALS

The **hydroxyl radical** is the main cleansing agent in the atmosphere. It removes chemical compounds that are considered pollutants by oxidising them to less harmful substances. Amongst the gases that hydroxyl radicals deal with by oxidation processes are CH_4, CO, and formaldehyde (HCHO), the latter being converted to CO_2. Nitrogen oxides (NO_x) are oxidised to nitric acid, and SO_2 to SO_3, which dissolves in clouds to form sulphuric acid (H_2SO_4). After most of these reactions, hydroxyl radicals are returned to the atmosphere, and therefore are able to react again for further cleansing of atmospheric pollutants.

Hydroxyl radicals are produced by the action of sunlight with ozone in the troposphere, with the greatest production in the equatorial regions. They are also produced by some reactions related to urban pollution. Levels of hydroxyl radicals in the atmosphere, however, are not well known and there is a fear that anthropogenically produced CH_4, CO, and NO_x will greatly reduce the levels and effectiveness of hydroxyl radicals as efficient atmospheric cleansing agents. Table 3.8 shows the estimated sources and sinks of carbon monoxide (IPCC report 1992).

Table 3.8 *Estimated sources and sinks of carbon monoxide (Tg CO per annum).*

	WMO (1985)	Seiler and Conrad (1987)	Khalil and Rasmussen (1990)	Crutzen and Zimmerman (1991)
Primary sources				
Fossil fuel	440	640 ± 200	400–1,000	500
Biomass burning	640	1,000 ± 600	335–1,400	600
Plants	–	75 ± 25	50–200	–
Oceans	20	100 ± 90	20–80	–
Secondary sources				
NMHC oxidation	660	900 ± 500	300–1,400	600
Methane oxidation	600	600 ± 300	400–1,000	630
Sinks				
OH reaction	900 ± 700	2,000 ± 600	2200	2,050
Soil uptake	256	390 ± 140	250	280
Stratospheric oxidation	–	110 ± 30	100	–

NMHC = non-methane hydrocarbons
Source: IPCC *Climate Change* 1992.

Climate and the greenhouse effect: a bleak future?

As the greenhouse effect takes a strong hold on the planet, perhaps by the middle of the next century, then the world climates and climatic belts will look very different to today. Temperatures near the poles have been estimated by some studies as getting up to 12°C higher. Of course, not only will the temperature patterns look very different to the present, but rainfall or precipitation patterns will change so that parts of the Earth become drier and others wetter. The altered temperature and rainfall patterns will cause a dramatic shift in the position and extent of some vegetation belts, while others show little or no change.

Worldwide, climate change will bring about a shift in the more local climatic vegetation belts, with some narrowing and others widening. Ecological niches will be affected to differing extents and in varying ways (see Box 3.6). Predictions of a mediterranean-type climate for Britain by the middle of the next century mean that the types of crops grown at present and the natural vegetation will change. Many of the grain crops, such as wheat, could be replaced by olives and grapes. This vision of a more equable climate for Britain might, at first, appear rather pleasant, but will these predicted changes have deleterious knock-on effects on the food chains and animal life that rely upon the present balance? Again, what price must be paid to slow down this global warming?

There will be other important implications that result from the global warming. Large volumes of water now locked up in the Antarctic and Arctic as ice sheets and glaciers may be released into the hydrosphere. Sea level may rise by an amount that will be significant, although experts differ in their estimates of this figure from over 10 cm to nearly 1.5 metres. The lower estimates may seem insignificant to many people, but in fact even these relatively small sea level rises will cause the flooding of extensive areas of dry land. Table 3.9 gives the IPCC (1990) estimated contributions over the past 100 years to global sea level rise from the thermal expansion of the oceans, and the melting of glaciers, small ice caps, the Greenland ice sheet, and the Antarctic ice sheet. The result of any significant rise in sea level will be the marginalisation and destruction of large areas of coastal lowlands as agricultural land and habitats for various flora and fauna diminish. IPCC (1990) estimates of future global sea level rise are given in Table 3.10.

There is debate on the feedback mechanisms associated with global temperature changes and atmospheric moisture content. Most current GCMs assume that global warming will be associated with an increase in atmospheric water vapour content or

Table 3.9 *Estimated contributions to sea level rise over the past 100 years (cm).*

	Low	Best estimate	High
Thermal expansion	2	4	6
Glaciers/small ice caps	1.5	4	7
Greenland ice sheet	1	2.5	4
Antarctic ice sheet	–5	0	5
Total	–0.5	10.5	22
Observed	10	15	20

After IPCC 1990.

BOX 3.6 VULNERABILITY OF ECOSYSTEMS TO CLIMATE CHANGE

Human-induced climate change adds an important additional stress to most environments, but particularly those that are already affected by pollution, increasing resource demands and non-sustainable management practices. The impacts, however, are difficult to quantify, and existing studies are limited in scope. Predicting the environmental changes is difficult because the systems are subject to multiple climatic and non-climatic stresses, the interactions of which are not always linear or additive. The Second Assessment Report (SAR) of the Intergovernmental Panel on Climate Change (IPCC) emphasises that successful adaptation depends upon technological advances, institutional arrangements, availability of financing and information exchange. The vulnerability of human health and socio-economic systems increases as the economic circumstances and institutional infrastructure decreases. The detection of climate-induced changes in most ecological and social systems will prove extremely difficult to detect in the coming decades. Furthermore, unexpected changes cannot be ruled out. Further research and monitoring is, therefore, essential to improve regional-scale climate projections; to understand the responses of human health, and ecological and socio-economic systems to changes in climate and other stresses; and to improve the efficiency and cost-effectiveness of adaptation strategies.

The sensitivity and adaptation of some selected sensitive ecosystems to global warming are briefly summarised below as outlined in the SAR of the IPCC (1995).

Forests

Approximately one-third of the world's forests will undergo major changes in broad vegetation types. The greatest effects will occur in high latitudes and the least in the tropics. Climate change will occur at a more rapid rate than the speed at which forest species grow, reproduce and re-establish themselves. Therefore, the species composition of forests will change; some forests may disappear and may be replaced by new ecosystems.

Rangelands

No major alterations are likely to occur in tropical regions, but in temperate rangelands growing seasons will change and the boundaries between grassland, forest and shrubland will change.

Deserts and desertification

Deserts are likely to become more extreme, becoming hotter and not significantly wetter. Desertification is likely to increase and may become irreversible in some areas.

Cryosphere

One-half of the existing glacier mass could disappear over the next 100 years, but little change in the extent of the Greenland and Antarctic ice sheets is expected over the next 50–100 years. Changes in glacier extent and snow cover may also affect the river flow and water supplies.

Mountain regions

The altitudinal distribution of vegetation is projected to shift to higher elevations and some species may become extinct as habitats are lost as the belts rise above the mountain tops.

Lakes, streams and wetlands

Water temperatures, flow regimes and water levels will change, altering bioproductivity and species distribution. Increases in flow variability, particularly the frequency and duration of large floods and droughts, are likely to occur. The geographic distribution of wetlands is likely to shift with changes in temperature and precipitation.

Coastal systems

A rise in sea level and changes in storms may result in increased coastal flooding and erosion, increased salinity in estuaries and fresh-water aquifers, alterations in tidal ranges, changes in sediment and nutrient transport, and chemical and microbiological contamination in coastal areas. Coastal ecosystems particularly at risk include salt-water marshes, mangrove ecosystems, coastal wetlands, coral reefs and atolls, and river deltas.

Oceans

Changes in sea level, oceanic circulation and vertical mixing, and reductions in sea ice cover are likely to occur. This will result in changes in nutrient availability, biological productivity, the structure and function of marine ecosystems, and the heat and storage capacity of the oceans, which control important feedback systems.

Source: IPCC World Wide Web Site

Table 3.10 *Estimates of future global sea level rise (cm).*

	Thermal expansion	Alpine	Greenland	Antarctica	Best estimate	Range[f]	To (year)
Gornitz (1982)	20	20 (combined)			40		2050
Revelle (1983)	30	12	13		71[b]		2080
Hoffman et al. (1983)	28–115	28–230 (combined)				56–345	2100
						26–39	2025
PRB (1985)	c	10–30	10–30	–10–100		10–160	2100
Hoffman et al. (1986)	28–83	12–37	6–27	12–220		58–367	2100
						10–21	2025
Robin (1986)[d]	30–60[d]	20 ± 10[d]	to +10[d]	to –10[d]	80[i]	25–1,659	2080
Thomas (1986)	28–83	14–35	9–45	13–80	100	60–230	2100
Villach (1987) (Jaeger, 1988)[d]					30	–2–51	2025
Raper et al. (1990)	4–18	2–19	1–4	–2–3	21[g]	5–44[g]	2030
Oerlemans (1989)					20	0–40	2025
Van der Veen (1988)[h]	8–16	10–25	0–10	–5–0		28–66	2085

[a] from the 1980s
[b] total includes additional 17 cm for trend extrapolation
[c] not considered
[d] for global warming of 3.5°C
[f] extreme ranges, not always directly comparable
[g] internally consistent synthesis of components
[h] for a global warming of 2–4°C
[i] estimated from global sea level and temperature change from 1880–1980 and global warming of 3.5 ± 2.0°C for 1980–2080
After IPCC 1990.

moisture. This assumption has been challenged by some scientists, who contend that global warming would increase air convection, leading to a drying of the middle atmosphere, thereby providing a negative feedback to counteract any greenhouse effect. The consensus of scientific opinion, however, is that most current GCMs make appropriate allowances for the amplifying effect of water vapour – a view that appears to be supported by recent satellite observations.

A likely scenario is that increased rises in global temperature may lead to increased precipitation in the currently arid polar regions as the Arctic Ocean becomes more free of sea ice. The effect will be a reduction in aridity, which will lead to the growth of glaciers and ice sheets rather than their gross melting (Miller and de Vernal 1992). Figure 3.13 illustrates some of the possible changes in thickness of the Greenland ice sheet that might be expected to occur over the next 200 years assuming a global warming of 6°C every fifty years. Note that the ice sheet thickens in the high centre and thins at the low elevations along its fringe. The geological data over the past 130,000 years support the idea that

greenhouse warming, which is expected to be most pronounced in the Arctic, coupled with decreasing summer insolation, may lead to more snow deposition than melting at high northern latitudes and thus to ice sheet growth (*ibid.*) (Plate 3.2).

Global climate change may increase the range of infectious diseases, especially those spread by insects and water. These include cholera, malaria, and yellow and dengue fever. These problems could be more immediate than the consequences of rising sea levels. If the average global increase of 4°C occurs by the year 2100, the number of deaths caused by mosquito-borne malaria would be likely to increase by two to three million each year (Stone 1995: Table 3.11). The outbreak of dengue fever in 1990 in Texas, where fourteen cases were reported and more than 900 cases across the Rio Grande, in Reynosa, Mexico, is thought to have been the result of recent changes in weather patterns in that region. Dengue fever is a mosquito-borne tropical disease that results in fevers and respiratory problems, and in some cases it can be fatal. It is thought that the mild winter allowed the dengue-bearing mosquitoes to multiply and create more opportunities for transmission of the

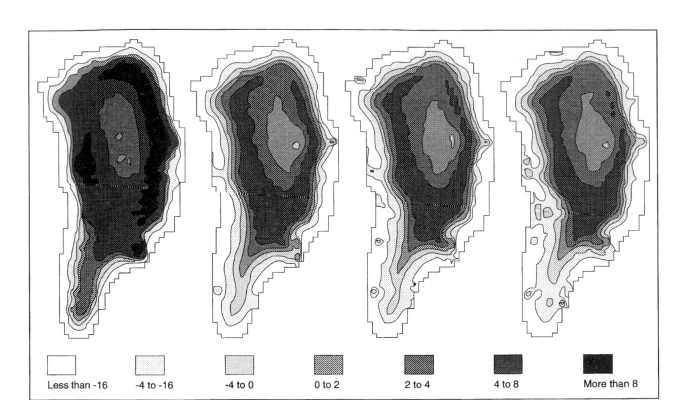

Figure 3.13 *Numerical modelling of the changes in ice thickness (shown in metres) of the Greenland ice sheet at 50-year intervals for the next 200 years, assuming a total stepped global warming of 6°C. Note the thickening of the high centre and the thinning of the lower altitude fringe, especially in the south. Redrawn after Sugden and Hulton (1994).*

disease, while the hotter summer raised water temperatures, which increased the breeding of mosquitoes carrying the disease, as well as their tendency to bite more frequently in hot weather. Many believe that such outbreaks are a sombre warning of things to come.

A greenhouse conspiracy? Myth or reality?

Will global warming lead to melting of the polar ice, to release large quantities of sea water so that sea level rises and countries such as the Maldives will be drowned? This section looks at some of the

Table 3.11 *Major tropical diseases likely to spread with global warming.*

Disease	Vector	Population at risk	Prevalence of infection	Present distribution	Likelihood of altered distribution with warming
Malaria	mosquito	2,100 million	270 million	(sub)tropics	+++
Schistosomiasis	water snail	600 million	200 million	(sub)tropics	++
Filariasis	mosquito	900 million	90 million	(sub)tropics	+
Onchocerciasis (river blindness)	blackfly	90 million	18 million	Africa/Latin America	+
African trypan-osomiasis (sleeping sickness)	tsetse fly	50 million	25,000 new cases/year	tropical Africa	+
Dengue fever	mosquito	estimates unavailable		tropics	++
Yellow fever	mosquito	estimates unavailable		tropical South America and Africa	+

+, likely; ++, very likely; +++, highly likely
Source: Stone 1995.

Plate 3.2 View looking north from Axel Heiberg Island at 78°N across the frozen seas of Greely Fjord towards the Arctic Ocean. This sea only partially melts in summer as large 'leads' open (centre of the frame) and the surface of the ice begins to melt and form ponds. This ice plays an important role in the hydrosphere–atmosphere interactions at high latitudes. An understanding of its dynamics is critical for accurate modelling of global climatic change.

arguments and debate surrounding any global warming. There are four main pillars that are most frequently used to support the view that the Earth is currently experiencing an anthropogenically created greenhouse effect or global warming: first, that the Earth's climate record shows that global temperature has increased and sea levels have risen; second, that carbon dioxide has been the primary cause of these changes; third, based on predictions of climate models that a doubling of atmospheric carbon dioxide will result in increased mean global temperature of 2–5°C; and finally, that the underlying physics is widely assumed to prove that carbon dioxide is a greenhouse gas and that further increases will result in increases in global temperature. Expert opinion remains divided on these issues.

One of the central issues focuses on the validity of the actual temperature measurements made over the last one hundred years or so and the way in which they can be interpreted. The thermometer record, so it would seem, cannot be taken at face value. Worldwide, more than 60,000 measurements are taken every day, amounting to a staggering 22 million annual measurements! But weather stations are not evenly distributed around the world; most are in the heavily populated, developed, regions of the Northern Hemisphere, with far fewer in the Southern Hemisphere. Also, the oceans are almost unrepresented in this data set, yet they cover more than three-quarters of the globe. Even more significantly, most weather stations are sited in urban areas, where temperatures are invariably warmer than the surrounding countryside. For example, Phoenix, Arizona, is frequently 10°C warmer than its suburbs. This temperature difference, known as the **heat-island effect**, is not due to global warming, but because urban areas release additional heat into the atmosphere. As urban areas have grown, so too have average urban temperatures risen through human activities. Some studies have even suggested that villages with as few as 300 inhabitants can cause urban warming of up to 0.3°C per decade, the amount proffered for global warming this century!

Even allowing for the heat-island effect, many critics argue that this effect is underestimated in the climate models that are used to support global warming. Historical measurements of sea-surface temperatures have also been unreliable. In the past, most measurements were made on water samples collected in canvas buckets lifted out of the sea onto a boat. During this process, some of the sea water evaporates and cools the sea water in the uninsulated bucket. The result is an underestimate of the actual temperature. Later, more reliable measurements were gathered from sea water in the intake of ships' engines. These results were roughly 0.5°C higher than the measurements using the earlier technique. Today, satellite measurement of sea-surface temperature is routinely used, and provides both rapid and more consistent data. Unfortunately, due to the poor sampling techniques in the past, there is not a reliable historical record of the long-term changes in sea-surface temperatures. Graphs of global temperature change over the past decade have been produced from satellite data, with a precision of about one hundredth of a degree per month, and they do not appear to support the global warming hypothesis. Spencer, a physicist at the NASA Marshall Space Flight Centre, University of Alabama, Huntsville, USA, has concluded from his analysis of this satellite data that 'over the entire ten-year period there was no net warming or cooling'. So, while over the last ten years the thermometer record shows an underlying upward trend in temperature, the satellite data appear to show that the Earth was warmer in the first half of the 1980s and cooler in the latter part. A judicious choice of time frame in the last century can be used to suggest global warming or cooling. The temperature data, so the critics claim, is at best ambiguous.

Another area of debate centres around the predicted rise in sea level due to global warming. The popular press has carried figures of up to 20 m of sea level rise during the next century, but a consensus of sensible scientific estimates gives figures closer to a 0.5 m rise or less. Evidence for changes in sea level come from tidal gauges, generally located in harbours and estuaries. Though thousands of measurements are available annually, controversy surrounds the interpretation of the data. An underlying cause for concern is whether the measurements chart the vertical movement of land relative to a fixed sea level, or the converse. Obviously, individual cases can be interpreted with varying degrees of certainty. After large earthquakes, scientists are generally able to estimate the vertical movement of land. In the British Isles, for purely geological reasons, sea level is falling in the north of Scotland and rising in the southeast. The fall in the northeast of Scotland is due to the vertical rebound, or **isostasy**, after the weight of ice was removed from this region, along with Scandinavia and other northern land masses, after the last glaciation.

And what of the reports that the extent of sea ice is diminishing because of global warming and the melting of the polar ice? Submarines passing under the polar ice have reported that the ice at specific locations is now thinner than it was a decade ago. But satellite data gathered daily over the past fifteen years do not appear to corroborate the notion of melting ice, since they suggest no change. So, the sceptics of global warming argue that there is no evidence of an imminent greenhouse world with higher sea level. In the early 1970s, the media even talked of global cooling and the dawn of a new ice age! More recently, Vaughan and Doake (1996) examined the meteorological records for the past fifty years from the Antarctic Peninsula and they showed that there is a measurable retreat of the Antarctic ice shelves on a millennial time scale. They caution, however, that the retreats may not be unique or even unusual but rather an expression of natural oscillatory advances and retreats of ice shelf fronts and, furthermore, they do not claim that these changes cannot necessarily be 'ascribed to a global warming magnified by regional temperature/sea-ice feedback,

or if this is a natural oscillation as a result of the same feedback' (*ibid.*).

Debate also surrounds the reliability of climatic models that are used to make predictions about the future global climate. Sceptics argue that the uncertainties in these computer models, together with their lack of sophistication for simulating actual climatic conditions, render them at best inaccurate and at worst misleading. These arguments are not, in themselves, a case against global warming but rather an attempt to exploit the uncertainties that arise from modelling global climatic change.

The sceptics of global warming stress that the climate models tend to underemphasise the importance of negative feedback mechanisms, which may stabilise any potential runaway greenhouse effect. Also, the term 'greenhouse gas' has misleading connotations when associated only with CO_2 and CH_4, because water vapour is actually the most common greenhouse gas, yet it is ignored in most articles. All these gases absorb and radiate heat energy in varying ways that depend upon many complex, interlinked factors such as their position in the atmosphere and the relative concentration of the cocktail of gases in the atmosphere. For example, convection currents complicate the heat budget of the atmosphere.

There are even experts who claim that an increase in atmospheric CO_2 could have beneficial effects on plant growth. It is interesting to note that plants evolved at a time when atmospheric CO_2 levels were probably 5–10 times greater than present levels. But what is good for plants may not be good for the human species and the continuation of civilisation.

Perhaps the critical argument centres around the link between atmospheric CO_2 levels (and other greenhouse gases) and global climate. Detailed studies of atmospheric CO_2 levels and palaeotemperatures following the most recent deglaciation show that the rise in CO_2 levels significantly preceded the rise in local sea-surface temperatures (Shackleton 1990). These data were gathered from the ice core record and deep-sea sediment cores by techniques such as:

● the UK-37 method, in which temperature is estimated from the ratio of various organic molecules (di-unsaturated to tri-unsaturated C_{37} alka-dienones), which are specifically associated with a type of algae known as prymnestophyte algae or coccoliths, or

● the identification of the influx of warmer water marine planktonic organisms such as the foraminifera *Globorotalia menardii*, which is a marine microfossil.

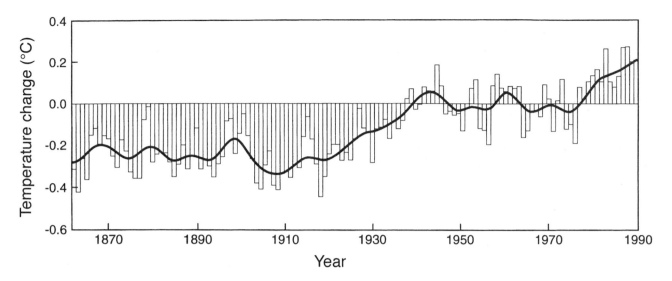

Figure 3.14 *Global mean combined land–air and sea-surface temperatures from 1861 to 1989, plotted relative to the average (0.0) for the years 1951 to 1980. Note that the rise in temperature has not taken place at a consistent rate: noticeable increases occurred between 1910 and 1940, and since the early 1970s (after Houghton et al. 1995).*

So, changes in atmospheric CO_2 levels appear to drive changes in sea-surface and linked atmospheric temperatures as suggested by the proponents of global warming and not, as the sceptics would have it, the other way around. The link between atmospheric CO_2 levels and global temperature change appears robust.

In September 1990, the report of Working Group 1 of the Intergovernmental Panel on Climate Change (IPCC), set up jointly by the World Meteorological Organisation and the United Nations Environment Programme, was published. After examining the scientific evidence, this document led the international experts to the conclusion that:

- 'there is a natural greenhouse effect which already keeps the Earth warmer than it would otherwise be',
- 'emissions resulting from human activities are substantially increasing the atmospheric concentrations of the greenhouse gases: carbon dioxide, methane, chlorofluorocarbons (CFCs) and nitrous oxide. These increases will enhance the greenhouse effect, resulting on average in an additional warming of the Earth's surface. The main greenhouse gas, water vapour, will increase in response to global warming and further enhance it.'

Following the previous IPCC reports (1990, 1992 and 1994), the 1995 Second Assessment Report (SAR) of the IPCC, highlighted the following certainties (Source: IPCC World Wide Web Site):

- Since the late nineteenth century, there has been an increase in global mean surface temperature of about 0.3–0.6°C. This change is unlikely to be entirely natural in origin (Figure 3.14).
- Global sea level has risen by 10–25 cm over the past 100 years. Much of the rise may be related to the increase in global mean temperature.
- Since the inception of instrumental climate records in 1860, recent years have been amongst the warmest on record. This is despite the global cooling effect of the 1991 Mount Pinatubo eruption.
- Night-time temperatures over land have generally increased more than daytime temperatures.
- Regional climate changes are also evident. For example, the recent warming has been greatest over the mid-latitude continents in winter and spring, with a few areas of cooling, such as the North Atlantic Ocean. Precipitation has increased over land in high latitudes of the Northern Hemisphere, especially during the cold season.
- The 1990 to mid-1995 persistent warm phase of the El Niño Southern Oscillation was unusual in the context of the past 120 years.

The SAR also emphasised the main uncertainties in the ability to project and detect future climate change. These include (Source: IPCC World Wide Web Site):

- The estimation of future emissions and biogeochemical cycling (including sources and sinks) of greenhouse gases, aerosols and aerosol precursors

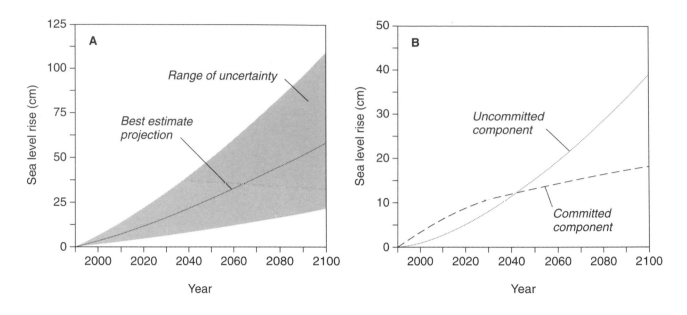

Figure 3.15 *1994 IPCC report predictions for the amount by which global sea level will rise between the years 1990 and 2100 under a mid-range rate of increased greenhouse gas emissions, the business-as-usual scenario (BAU), which produces the best estimate (solid line) of a 58 cm rise in global (eustatic) sea level by 2100, with a range of uncertainty (grey area) from a 21 to a 105 cm rise by 2100. Redrawn after Houghton et al. (1995).*

and projections of future concentrations and radiative properties.

- The representation of climate processes in models, particularly feedbacks associated with clouds, oceans, sea ice and vegetation, in order to improve projections of rates and regional patterns of climate change.
- The systematic collection of long-term instrumental and proxy observations of variables in the climate system (e.g. solar output, atmospheric energy balance components, hydrological cycles, ocean characteristics and ecosystem changes) for the purpose of model testing, assessment of temporal and regional variability, and detection and attribution studies.

Of course, the exact consequences of global warming remain uncertain, but one thing is certain: dismissing global warming or inaction can serve only to put an unacceptable risk on the survival of life on Earth, certainly for human civilisation.

The IPCC document on climate change models a number of scenarios for predicted levels of change in the atmospheric concentrations of greenhouse gases, and the resulting changes in climate that might reasonably be expected to occur under the various 'options'. One of these predictions has been termed

the business-as-usual scenario, under which the emissions of greenhouse gases continues at current rates. In this case, the IPCC estimates that during the next century: (a) global mean temperature will increase by 0.3°C per decade (with an uncertainty range of 0.2–0.5°C per decade), which is greater than that seen over the last 10,000 years, and (b) global mean sea level will rise by about 6 cm per decade (with an uncertainty range of 3–10 cm per decade), mainly because of the thermal expansion of the oceans and the melting of some land ice. These predictions suggest that global mean temperatures will be about 1°C above the present value by 2025, and global mean sea level will have risen by about 20 cm by 2030 (Figure 3.15).

More recently, revised projections of future global greenhouse gas warming suggest that by 2100, with a rise of about 0.5°C by 2010, the increase relative to 1990 will vary between 0.62–2.31°C and 1.61–5.15°C, depending upon whether CO_2 levels are 2 or 5.5 times the pre-industrial CO_2 concentrations, respectively (Schlesinger and Jiang 1991). Evidence for global warming is coming from places as remote as northwest Tasmania, at 1,040 m above sea level on the slopes of Mount Read and around Lake Johnston, where the width of growth rings from Huon pine trees (*Lagarostrobes franklinii*) well above

---

BOX 3.7 CALCULATING GREENHOUSE GAS EMISSIONS

Gas emissions are calculated by multiplying fuel consumption by a **carbon emission factor**. The carbon emission factor is the amount of CO_2 released through the combustion of a specified quantity of fuel, e.g. one litre or 1 tonne. Since fuels contain varying amounts of carbon, they are associated with different carbon emission factors.

Predictions about future emissions are calculated by multiplying projected fuel consumption values (taking account of likely economic indicators, such as trends in fuel prices, etc.) with the appropriate carbon emission factors. Clearly, errors are associated with such calculations, probably in the range 5–10 per cent. Figure 3.19 shows various scenarios for the future estimates of CO_2 emissions that were considered in the 1992 IPCC report. Note the variability between estimates, the only similarity being that almost all increase with time.

CO_2 budgets are calculated as 44/12 of that for carbon, because the CO_2 molecule contains one carbon atom (atomic weight relative to a hydrogen atom = 12) and two oxygen atoms (atomic weight = 16), giving a molecular weight of $12 + (2 \times 16) = 44$.

---

their normal altitude range suggests that the temperature rise during the last twenty-five years has been much greater than at any time since AD 900 (Cook *et al.* 1991). The tree ring index, obtained by subtracting the growth from natural maturation from the thickness of the ring, can be used to interpret past climatic conditions. The tree ring index from the Tasmanian pines suggests a mean temperature rise of just over 1°C since 1965.

In 1992, the IPCC revised figures for the effects of greenhouse gas emissions (see Box 3.7 on calculating greenhouse gas emissions; IPCC report 1992, see Wigley and Raper 1992). These new figures result from taking into account new policies already implemented or proposed for controlling CO_2 emissions and halocarbon production, and allow for recent political changes. In addition, they are based on a wide range of socio-economic factors that influence the development of emissions in the absence of unilateral or multilateral efforts to reduce them. The various scenarios that are presented differ from each other because they make different assumptions about, for example, population growth, economic growth, technological developments, resource limitations, fuel mixes and agricultural development. The new climate models also include results for global mean thermal expansion of the oceans, a principal component of future rises in sea level. From the range of possibilities, the IPCC shows low, middle and high estimates of global mean temperature change, global mean sea level change and radiative forcing. The results are less severe than previous estimates but remain greater than the limits of natural variability. For example, middle estimates suggest that by 2100, global mean temperature will rise by about 3.5°C, and global mean sea level will have risen by around 50–60 cm (*ibid.*). The range of solutions for low-, middle- and high-temperature and sea level projections, based solely on the anthropogenic component

of future change, show that over the period 1990–2100, warming will be between 1.7 and 3.8°C, with corresponding sea level rises of between 22 and 115 cm (*ibid.*). These revised, reduced rates of projected future change are still four to five times those that occurred over the past century. The 1992 IPCC update, however, has revised the global warming potential (GWP) of CFCs downwards from being thousands of times more potent, molecule for molecule compared with carbon dioxide, to zero: this is because CFCs produce two opposing effects, destroying stratospheric ozone as well as being greenhouse gases.

In 1994, the IPCC again revised the GWPs of the main greenhouse gases and calculated GWPs for a number of new species, particularly HCFCs, HFCs and perfluorocarbons (PFCs).

In the 1992 IPCC report, six gas emission scenarios (known as IS92 scenarios) were described based on assumptions regarding economic, demographic and policy factors (Figure 3.16a). Figure 3.16b shows the likely resultant atmospheric CO_2 concentrations based on several carbon cycle models. None shows a stabilisation before 2100. Following this study, the 1994 IPCC report investigated the greenhouse gas emission profiles that would lead to stabilisation of concentrations of greenhouse gases in the atmosphere. Using the same carbon cycle models to calculate future concentrations of CO_2, concentration profiles were derived that stabilise at CO_2 concentrations from 350 to 750 ppmbv (Figure 3.17). Figure 3.18 shows the model-derived profiles of total anthropogenic CO_2 emissions that lead to stabilisation following the concentration profiles in Figure 3.17. The implication of these models is that stabilisation will occur only if emissions are reduced below 1990 levels, but the different measures of reduction will result in stabilisation occurring at different times.

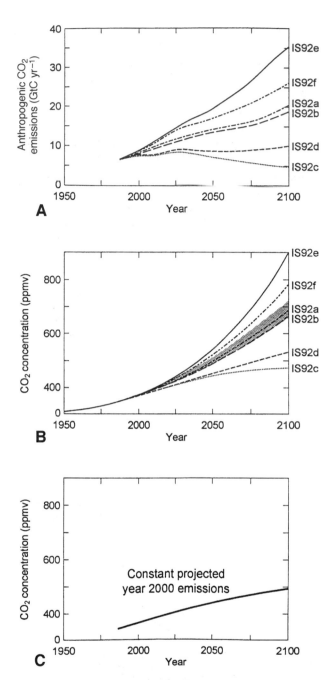

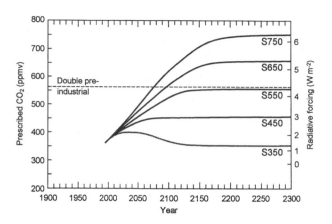

Figure 3.17 *Profiles of atmospheric CO_2 concentration leading to stabilisation at 350, 450, 550, 650 and 750 ppmbv. Doubled pre-industrial CO_2 concentration is 560 ppmbv. The radiative forcing resulting from the increase in CO_2 relative to pre-industrial levels is marked on the right-hand axis. Note the non-linear nature of the relationship between CO_2 concentration change and radiative forcing. Redrawn after Houghton et al. (1995).*

Figure 3.16 *(A) Prescribed anthropogenic CO_2 emissions from fossil fuel use, deforestation and cement production for the IS92 scenarios. (B) Atmospheric CO_2 concentrations calculated from the scenarios IS92a–f emissions scenarios (Leggett et al. 1992) using the Bern model, a mid-range carbon cycle model (Siegenthaler and Joos 1992). The typical range of results from different carbon cycle models is indicated by the shaded area. (C) CO_2 concentrations resulting from constant projected year 2000 emissions (using the model of Wigley 1993). Redrawn after Houghton et al. (1995).*

Natural phenomena and atmospheric change

Volcanoes

It is not only human activities that contribute to the gases that may cause changes in atmospheric turbidity or optical clarity, global warming, or cooling (by depleting the ozone layer). Natural causes may be very important, for example volcanic eruptions. Volcanoes can emit huge quantities of greenhouse gases, including CO_2. Mount Etna in Sicily, for example, is amongst the world's most actively degassing volcanoes. Data from the eruptions of Etna between 1975 and 1987 led Allard *et al.* (1991) to a conservative estimate of approximately 25 Mt of CO_2 per year, equivalent to the output from four 1,000-megawatt conventional coal-fired power stations, but still insignificant compared with the annual global emission of 5 GtC from the combustion of fossil fuels. Its SO_2 emission rate is also very high, at about 10 per cent of the global total for volcanic degassing. This emission from Etna is roughly an order of magnitude greater than that of Kilauea in the Hawaiian Islands, another well-studied volcano. While this figure is indeed large, the CO_2 emissions from Etna are only 0.07 per cent of the annual anthropogenic CO_2 contribution to the

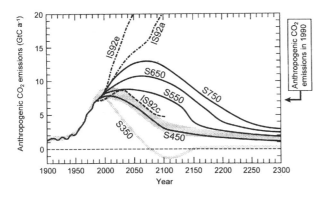

Figure 3.18 *Illustrative anthropogenic emissions of CO$_2$ leading to stabilisation at concentrations of 350, 450, 550, 650 and 750 ppmbv following the profiles shown in Figure 3.17, using a mid-range carbon cycle model. The range of results from different models is indicated on the 450 ppmbv profile. The emissions for the IS92a, c and e scenarios are also shown. The negative emissions for stabilisation at 350 ppmbv are an artefact of the particular concentration profile imposed. Redrawn after Houghton* et al. *(1995).*

atmosphere (*ibid.*). The global rate of CO$_2$ emissions from all the subaerial, and submarine, volcanoes is not precisely known, but it is probably somewhere in the region of 130–175 Mt per year (Gerlach 1991).

Volcanic eruptions, particularly the more explosive types, release chlorine (Cl) and fluorine (F) compounds into the stratosphere to produce 'halogen pollution'. Hydrogen chloride (HCl) and hydrogen fluoride (HF) are the main halogen compounds released during volcanic eruptions, with estimated annual yields of 0.4 x 10^6 to 11 x 10^6 tonnes of HCl and between 0.06 x 10^6 and 6 x 10^6 tonnes of HF. Approximately 10 per cent of these

Figure 3.19 *Energy-related global CO$_2$ emissions for various scenarios. Shaded areas indicate coverage of IS92 scenarios. Numbers compared with various scenarios. For reference to various energy-related global CO$_2$ emissions scenarios, refer to supplementary table on pp. 299–300 in Houghton* et al. *(1995). Redrawn after Houghton* et al. *(1995).*

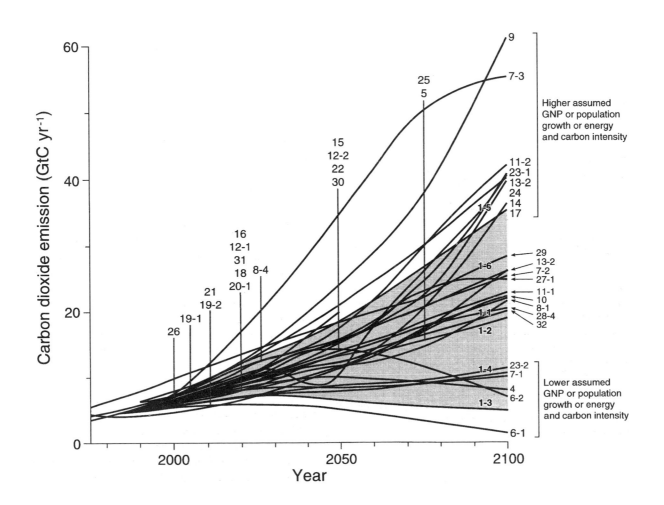

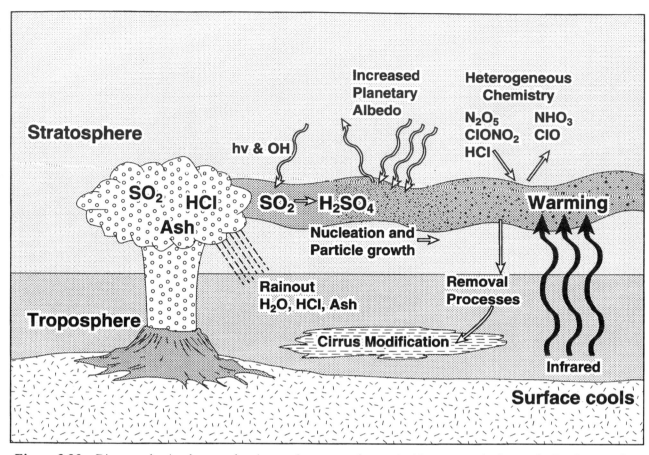

Figure 3.20 *Diagram showing how a volcanic eruption can produce a significant perturbation to the Earth–atmosphere system by injecting material into the stratosphere. Redrawn after McCormick et al. (1995).*

gases are produced in explosive types of eruption alone, where the exhalative gases are injected into the stratosphere. Of course, volcanic eruptions do not occur at regular frequencies, or time intervals, nor are they equally spaced around the Earth. Thus, if the atmospheric levels of anthropogenically created CFCs are at sub-critical concentrations, then it might conceivably take only one or two particularly large explosive volcanic eruptions to cause stratospheric-ozone-destroying chemicals to exceed a critical threshold level, and cause an accelerated depletion of ozone. Such a scenario could lead to global cooling. Figure 3.20 illustrates the volcanic and resultant atmosphere processes that may lead to climate change.

Earth scientists need to understand more about the role of volcanic eruptions in contributing to the overall levels of greenhouse gases and/or destroyers of the stratospheric ozone layer. This is because of the notion of a critical threshold level beyond which the consequences may be very grave for life on Earth. A very small additional amount of CFCs released by

human activities could cause a very large change in climatic conditions. On a less alarmist level, scientists need to increase their understanding because further research allows them to gauge the 'natural' concentrations of chlorine and fluorine compounds in the stratosphere that result from volcanic eruptions and use these figures as a benchmark or yardstick against which to calibrate the effects of human activities in destabilising the atmosphere.

Volcanoes also eject sulphate particles into the lower stratosphere, which could form surfaces on which heterogeneous reactions occur. Such crystal surfaces are therefore catalysts, just like the ice crystals in very cold clouds, and require further research to assess their role and potency in the reactions that deplete the ozone shield.

Not only are volcanic eruptions capable of emitting gases that can lead to global warming or cooling, but they may precipitate a 'volcanic winter'. The eruption of Toba in Sumatra, 73,500 years ago, created the largest known volcanic event in the Quaternary; the eruption is estimated to have lofted

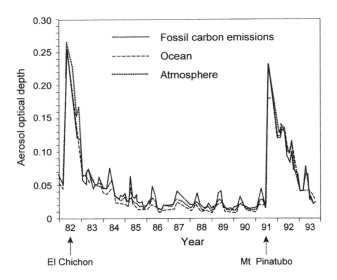

Figure 3.21 *Variation of aerosol optical depth following the Mount Pinatubo and El Chichon volcanic eruptions, and the subsequent removal of aerosols over several years following the eruptions. Redrawn after Dutton and Christy (1992).*

Table 3.12 *Major twentieth-century volcanic eruptions.*

Volcano	Date	Estimated aerosol loading (Tg)
Stratospheric background	possible 1979	< 1
Katmai	June 1912	20
Agung	March 1963	16–30
Fuego	October 1974	3–6
El Chinchon	April 1982	12
Mount Pinatubo	June 1991	30
Cerro Hudson	August 1991	3

After McCormick *et al.* 1995.

approximately 10^{15} grams each of fine ash and sulphur gases to heights of 27–37 km (Rampino and Self 1992; also see Chapter 2). The injection of all this volcanic material into the atmosphere may have caused a decrease in the amount of solar radiation reaching the Earth's surface and, therefore, led to a global cooling, estimated by Rampino and Self as a decrease of between 3–5°C lasting up to a few years. Stable oxygen isotope data suggest that the eruption of Toba occurred during a period of rapid ice growth and falling global sea level, and Rampino and Self proposed that the eruption could have accelerated the deterioration in global climate. The cool weather in 1992–93 may be a consequence of the eruption of Mount Pinatubo in the Philippines on 11 June 1991, which ejected very large amounts of volcanic dust into the upper atmosphere to reduce the solar flux to the Earth's surface (Table 3.12).

The ash falls from Mount Pinatubo caused the deaths of several hundred people, and the evacuation of tens of thousands. The aerosol cloud was the largest since Krakatau in 1883, which was estimated to be 25–30 Mt. Substantially greater than El Chichon (1982, 12 Mt) and Mount St Helens (1980, 0.5 Mt), the Pinatubo eruption injected the aerosols, comprising liquid droplets of approximately 25 per cent water and 75 per cent sulphuric acid, to heights of 15–25 km, i.e. to the same level as the ozone layer. Figure 3.21 shows the variation of aerosol optical depth in Hawaii following the eruptions of El Chichon and Mount Pinatubo. The aerosols remain for only a few years, but may be sufficient to reduce global temperatures by about 0.5°C. Furthermore, Figure 3.22 shows the global mean ozone concentrations for the period prior to the eruption of Mount Pinatubo and the reduced concentrations after the eruption. The aerosols absorbed most strongly at the infrared end of the spectrum and scattered solar radiation back into Space. The stratosphere, therefore, warmed while the troposphere cooled. Observations by Vogelmann *et al.* (1992) on the biologically effective ultraviolet light (UV-B$_E$) at the Earth's surface showed increased surface sunlight intensity immediately following the eruption, from which they calculated that the effect of ozone depletion outweighed that of increased scattering. Also, satellite imaging of stratospheric ozone revealed a depletion of as much as 15–25 per cent at high latitudes, and in November 1991, a depletion of 20 per cent was observed over Boulder, Colorado (Westrich and Gerlach 1992; Brasseur 1992). This marked ozone depletion was originally thought to be caused by volcanogenic chlorine. This cause, however, was probably minor because of the rapid down-flushing of chlorine as hydrochloric acid. The depletion is now considered to be the result of frozen sulphuric acid droplets, which act in a similar way to the stratospheric ice particles in polar regions, providing surfaces for chlorine-releasing chemical reactions involving anthropogenic CFCs, which caused ozone loss. Figure 3.23 shows the negative radiative forcing of Mount Pinatubo compared with the radiative forcing due to anthropogenic gases and aerosols. The effects from Pinatubo were short-lived, but they were sufficient to offset the global warming due to positive radiative forcing during 1992 to 1994.

Volcanic eruptions appear capable of creating both positive and negative feedback in the global climate system, depending upon their timing in relation to overall global cooling or warming due to other causes

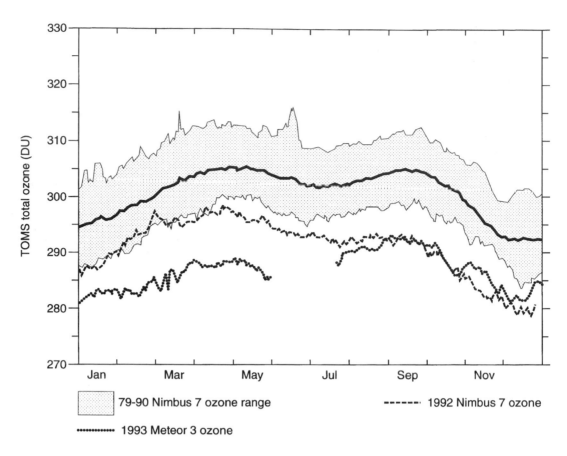

Figure 3.22   *Global mean ozone from the total ozone mapping spectrometer (TOMS) measured as a function of time. Line 1 (the dashed line): TOMS/Nimbus 7 measurements for 1992; Line 2 (the dotted line): TOMS/Meteor 3 measurements for 1993. These two lines are superimposed on the range of TOMS-observed global (65°N–65°S) mean ozone between 1979 and 1990 (Line 3 – the shaded area). Line 4 (the bold line) is the average total ozone for this period. TOMS data show a notable decrease in column ozone amounts beginning in early 1992 and persisting to the end of 1993. The loss reached as much as 6 per cent of the column in April 1992. Redrawn after McCormick et al. (1995).*

(see Chapter 2). There is a useful review of climate change and volcanic activity over the past 500 years by Bradley and Jones (1992), also discussed in detail in a UGU Special Report by the American Geophysical Union (1992a). In summary, the average surface cooling resulting from major volcanic eruptions is of the order of 0.2–0.5°C for a short duration, up to perhaps about five years after a major eruption (Self *et al.* 1981, Rampino and Self 1992), although local temperature reductions over similar time periods may be as high as 1.5°C (Porter 1986). Proxy data, such as acidity levels in acid cores, by measuring electrical conduction, provide a temporal comparison for possible volcanic events, and their possible relationship with any global cooling (Hammer *et al.* 1980, Rampino and Self 1992).

International action on global atmospheric pollution

Atmospheric ozone depletion

In September 1987, the leading industrial nations, including Britain, were signatories to the Montreal Protocol, in which they agreed to specific reductions in the production and use of five of the most harmful CFC ozone-depleting gases known at the time. These CFCs are used in many aerosol sprays, refrigerants, solvents and plastic foams. The 1987 agreement followed on from an international agreement that was much less stringent and which was made under the Vienna Convention in 1985.

The leading nations agreed to attempt to find substitute chemicals to replace the most dangerous CFCs, one of which is CFC-11. A possible substitute for CFC-11 is CFC-22, which, although less toxic,

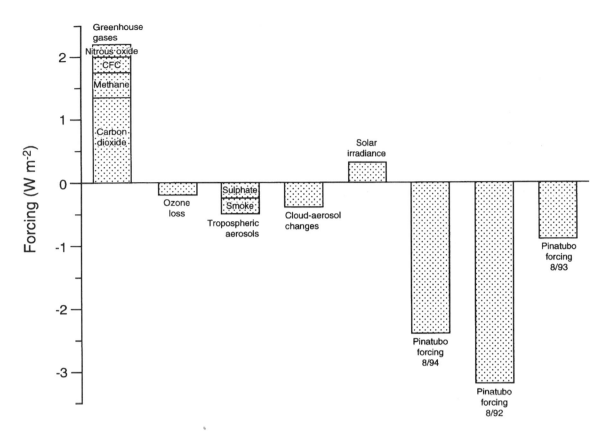

Figure 3.23 *Radiative forcing resulting from the Mount Pinatubo volcanic eruption in 1991 compared with the radiative forcing due to anthropogenic greenhouse gas and particulate emissions. Redrawn after McCormick et al. (1995).*

still represents a considerable threat. For example, it has been calculated that CFC-22 is about 20 times less harmful than CFC-11. CFC-11 has a half-life of about 75 years, compared with 20 years for CFC-22. Over the next 20–30 years, however, when the ozone layer may be most rapidly depleted, it is estimated that CFC-22 is likely to be 20 per cent as damaging as CFC-11. In fact, Peter Fabian of the Max Planck Institute for Aeronomy, Göttingen, believes that if CFC-22 is released into the atmosphere at the present rates, then by the end of this century it will have destroyed as much ozone as the two most common CFCs, CFC-11 and CFC-12.

CFC-22 is not in the Montreal Protocol. It may indeed be safer than CFC-11, but it is still harmful. Replacing one dangerous CFC with another that is perhaps cumulatively just as harmful should be seriously questioned.

In Helsinki, in early May 1989, 81 nations agreed in principle to ban eight industrial chemicals that damage the stratospheric ozone layer by the year 2000. These chemicals include five CFCs and three halon gases. At the Helsinki meeting, Eileen Claussen, of the USA's Environmental Protection Agency, predicted that even if CFCs are totally phased out and banned by the turn of the twenty-first century, chlorine concentrations will still rise from existing levels of 2.7 to 8.9 parts per billion (ppbbv) by 2010: mainly because of the long time taken for CFCs to break down in the atmosphere. A spokesperson for Greenpeace claimed that another greenhouse gas, methyl chloroform (a solvent and cleaning fluid) is increasing in the atmosphere at the rate of 7 per cent per annum. Many of the nations that met in Helsinki and agreed in principle to ban eight ozone-depleting chemicals hoped to include Greenpeace's proposals in the Montreal Protocol in 1990.

In early March 1989, the British government organised an international conference on atmospheric ozone depletion at which representatives attended from 124 other nations. The conference was in essence a stage-managed affair. Even before the conference opened, the European Community (EC), the United States and Canada had agreed to a complete phasing out of the five CFCs and three halons covered by the Montreal Protocol. The protocol had stipulated a 50 per cent reduction in the production and consumption of these materials by the year 2000 and so a total ban by the EC, USA and Canada was justifiably seen as a resounding success story by the media and general public.

One of the outstanding problems was succinctly put by Prime Minister Margaret Thatcher at the conference when she stated that the actions of the developed, industrialised world would not be sufficient by themselves to reverse the deterioration of the ozone layer. Many developing nations are only just beginning to industrialise and wish to import and manufacture refrigerants, refrigerators, plastics, air-conditioning plants and electronic goods, all of which involve the use of ozone-depleting CFCs and halons. These nations cannot, and should not, have to await the development and widespread installation of safer chemicals to replace the harmful CFCs and halons in the products they believe they need for development. This point was made at the conference by President Daniel Arap Moi of Kenya, who went on to ask that the industrialised world take moral and scientific responsibility for providing the technology and economic circumstances that would allow the developing nations effectively to leap-frog the CFC and halon technology.

At the London conference, 1990, another optimistic sign for the future was the announcement that 20 additional countries were to sign the Montreal Protocol, taking the total number of signatories to 59, although, at the time of this book's going to press, the protocol has been ratified in only 32 of the signatory countries. Representatives from India and China said that their nations would be amenable to signing the protocol if the industrialised countries made a clear and attractive financial commitment to setting up a fund for adequate research and development to produce alternative chemical substances, and then transfer this technology free of charge to those nations. A multilateral fund of US$160–240 million over three years was established to support the phasing out in developing countries that consume less than 0.3 kg of CFCs per capita.

At an international meeting of the parties to the Montreal Protocol, in November 1992 in Copenhagen, revised controls on ozone-depleting substances were agreed. The deadlines for the global phase-out for most chemicals was brought forward, and the multilateral fund was re-authorised on a permanent basis. However, a controversial part of the Copenhagen package was the allowance of a sharp increase in the use of HCFCs, as a transitional measure to substitute for CFCs as they are phased out. The Copenhagen package consisted of the following agreements:

- For *CFCs*, the phase-out date was brought forward from January 2000 to 1 January 1996, with a 75 per cent reduction, based on 1986 levels, by 1 January 1994. The EC proposed an interim 85 per cent reduction by 1 January 1994.

- *Carbon tetrachloride* should be phased out by January 1996 rather than 2000 as originally proposed, with an 85 per cent reduction, based on 1989 levels, by 1 January 1995. The EC proposed an interim 85 per cent reduction by 1 January 1994.

- *Halons* should be phased out by January 1996, again brought forward from the January 2000 target date. The EC proposed a phase-out by January 1994.

- *Methyl chloroform* should be phased out by January 1996, brought forward from January 2005, with a 50 per cent reduction on 1989 levels by January 1994. However, the 'essential use' of halons may continue beyond the new proposed phase-out dates, and a UNEP panel was to have prepared an assessment of such essential usage, with the announcement of a decision scheduled for 1994.

- *Methyl bromide* consumption was to be pegged at 1991 levels by 1995. More stringent controls were to await scientific evaluation by two UNEP panels, which were scheduled to report in 1995.

- *HBFCs*, although not in general use, should be phased out by January 1996. This was the first time that HBFCs had come under any control.

- *HCFC* use was to be capped in January 1996 at a level amounting to the sum of their consumption in 1989 and 3.1 per cent of the level of consumption of CFCs in 1989. This formula arose in order to take into account existing consumption of HCFCs, which were already high in some countries in 1989, and also in recognition of their role as transitional substitutes for CFCs. The Copenhagen amendments incorporated controls for the first time on HCFC use, which is to be cut to 65 per cent of the 1996 consumption level by 2004 and to 35 per cent by 2010, with a total ban by 2030.

Consumption levels of CFCs are calculated in ODP-weighted tonnes. The Copenhagen amendments included the wording that party nations 'shall endeavour to' achieve the various targets and, therefore, they cannot be considered as absolutely binding commitments, more statements of intent.

The amounts of CFCs produced and consumed have decreased by 40 per cent since 1986 and, under the 1990 London and 1992 Copenhagen amendments to the Montreal Protocol on protecting the ozone layer, will continue to fall, but the release of CFCs already in use means that atmospheric concentrations of CFCs will continue to rise for some years. At least the problem has been tackled at an international level with considerable effects. Unfortunately, an illegal market is developing in smuggling CFCs into countries such as the former Soviet Union and parts of the USA, but governments are attempting to police this. The developing countries, however, must be offered incentives to discourage them from using CFCs in the amounts that have been used by the industrialised countries.

In November 1993, the fifth meeting of the parties was held in Bangkok. No changes to the core commitments were made, but US$510 million was committed for three years to the multilateral fund and recommendations were made that there be no essential-use exemptions to the 1994 halon phase-out in industrial countries.

The sixth meeting of the parties was in Nairobi in October 1994, and although there were no major changes to the core commitments, they recommended that 11,000 tonnes of essential-use exemptions be granted in the case of the 1996 CFC phase-out target for industrial countries. In November 1995, the seventh meeting of the parties was held in Vienna, commemorating the tenth anniversary of the Vienna Convention.

The Ozone Secretariat is responsible for overseeing controls of ozone depletion. It acts as the secretariat for the Vienna Convention for the Protection of the Ozone Layer and for the Montreal Protocol on Substances that Deplete the Ozone Layer. The secretariat is based at the UNEP in Nairobi, Kenya. Its duties include arranging and servicing the conferences of the parties, meetings of the parties, their committees, bureaux, working groups and assessment panels. It is responsible for arranging to implement the decisions of the meetings, as well as monitoring the implementation of the convention and protocol. The secretariat is also responsible for representing the convention and the protocol in the relevant international bodies and for providing information.

Greenhouse gas emissions

In February 1991, the US government proposed a strategy to limit global greenhouse warming by suggesting that nations should seek a comprehensive framework for the emission of greenhouse gases in preference to focusing on a single gas. In many respects, although this is a self-evident truth, one has to ask whether this proposal by the USA was designed to frustrate efforts to obtain a phased reduction in the emissions of CFCs and CO_2, or a genuine attempt to encourage a more balanced perspective on the task at hand.

A thorny issue concerning the production of greenhouse gases is the ways in which industry can substitute one form of energy generation with another. Many environmentalists would not regard nuclear energy as an alternative to conventional fossil fuel power stations and the solution to reducing emissions of CO_2. In fact, earlier in 1989 at the public inquiry into plans for a nuclear power station at Hinkley, Somerset, David Fisk, the chief scientist at the UK Department of the Environment, gave written evidence which stated that

> nuclear power alone will not provide the solution to the greenhouse problem but the continuation of a nuclear contribution to the electricity supply will provide a diversity of fuel sources which will play a part, together with energy efficiency and renewable resources.
>
> (Pain 1989b)

Much more research is needed to assess the impact on the global climate of human activities. There is a need to quantify cause and effect, and what courses of action will maximise the chances of ameliorating or reducing the destructive potential of human activities on the world climate. Predictions about economic growth and trends in fuel prices, in the context of any overall energy strategy or policy, will have a major impact on determining the acceptability to individual countries of any proposed greenhouse gas emissions targets and the means to achieve these. Many people, including environmental groups, support energy conservation and reducing energy demand as a means of controlling the emission of greenhouse gases, with cost-effective energy efficiency measures. There is also considerable debate as to the relative merits of using regulations versus economic instruments to control greenhouse gas emissions.

The issues raised by the greenhouse effect are inherently very political. There are scientists interested in exploiting the current level of media interest in order to give their 'climatic' research greater status and to persuade funding bodies to direct more research grants their way. There are environmentalists who are only too willing to believe every gloomy prediction that an anthropogenically enhanced greenhouse effect may have, and who use the information to scare people into donating money to their cause. They are all too often unable to judge the scientific content of any argument about global warming, but

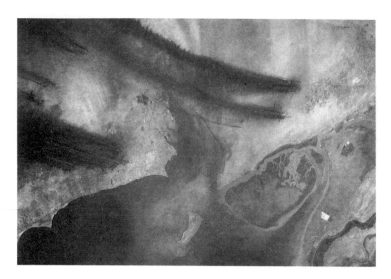

Plate 3.3 *Satellite image of oil fires in Kuwait.*
Courtesy of NASA.

simply react in a visceral manner. In essence, anyone who gets involved in the issue of global warming due to an anthropogenically enhanced greenhouse effect reacts in a *political* way. It is impossible to be objective, but we owe it to others, as well as to ourselves, to try to become as informed as is possible about the ramifications of the greenhouse issue. Opinions expressed with some measure of knowledge allow for constructive debate and thoughtful action, but, naturally, there are vested economic interests that can so quickly come to dominate the issues raised by the greenhouse effect.

Even the United Kingdom Atomic Energy Authority (UKAEA) has seen an 'angle' in the greenhouse debate. In January 1990, it published a glossy brochure entitled *Nuclear Power and the Greenhouse Effect*. Its 30 pages set out to persuade people that nuclear power, with the tarnished image it seems to have acquired, is the answer to human energy needs and offers an environmentally friendly alternative to the dirty conventional fossil fuel power stations that emit so much greenhouse gas. In fact, the last paragraph states:

> The use of nuclear power could be readily expanded. By 2020 it could be reducing energy CO_2 emissions by 30 per cent of what they would otherwise have been, effectively reducing global warming by 15 per cent. In the longer term it could offer much deeper cuts in greenhouse gas emissions from energy.

The Gulf War of 1991 was probably the first war in history that really made people consider the impact on the environment that mass destruction can wreak along with the human injury and death toll. The Kuwait oil fires released approximately 240 Mt of CO_2, equivalent to about 1 per cent of the annual global emissions of CO_2 (Plate 3.3). The last oil well-head fire, of about 750, was finally extinguished in Kuwait on Wednesday 6 November 1991, bringing the total cost to an estimated US$1 billion. But the effect of these fires was enormous regional pollution in the area of the Persian Gulf. Many environmentalists predicted that the Iraqi fire-raising in Kuwait would precipitate a catastrophic effect on global climate, including a substantial cooling of the Earth's surface as solar radiation was blocked by the soot clouds, akin to a nuclear winter scenario, with the massive failure of the Asian monsoons, crop failure and other catastrophic climate changes, none of which appear to have actually come about. Whilst it is true that the deliberate burning of Kuwait's oil wells by the Iraqis was an act of gross environmental vandalism, the global climatic impact was exaggerated. Studies now show that the smoke from the oil wells and refineries was not injected high enough into the atmosphere to cover large parts of the Northern Hemisphere, nor was it produced in sufficient amounts to cause a measurable temperature change, or a failure of the monsoons, and only a small increase in the global CO_2 budget resulted (Small 1991, Johnson *et al.* 1991). However, by late 1992 and early 1993, the local effects of this eco-terrorism appeared to have been severe. A study commissioned by the Saudi government, and published in an EC report in 1993, showed that earlier estimates of 1–2 million barrels of oil being dumped in the Gulf have been revised upwards to about 11 million barrels, and the Saudi Meteorology and Environment Protection Administration claims that 2.7 million barrels are still clogging the beaches of Saudi Arabia, Iran and Qatar, with an additional 2.6 million barrels possibly trapped beneath the sea surface. The smoke clouds that blotted out the sunlight caused a 10°C fall in sea-surface temperatures, and with it the destruction of plankton and fish larvae in the Gulf, which are at the bottom of the food chain. This led to a fish famine, and also cannibalism amongst the tens of thousands of starving seabirds. Despite the initial hyperbole by environmentalists, the media and other concerned people over the potential effects of the fires on global climate, the local and regional effects appear very serious. The Gulf War should have taught people to respect the planet more, even at times when human self-respect is at an all-time low. It is possible to stop wars, but their effects could be far harder to control in the ensuing peace.

In fact, the eruption of Mount Pinatubo in the Philippines more than offset any warming effects that may have been caused by the Kuwait oil fires, when it ejected 3–5 km^3 of magma and 20 Mt of sulphur into the atmosphere. Although most of the ash and dust settled rapidly, small sulphate particles, or aerosols, remain in the atmosphere, where they reflect incoming solar radiation back into Space. Using simple climate models, it has been predicted that the effects of the Mount Pinatubo eruption will be a global cooling of up to 0.5°C over the next few years.

An international treaty on pollution in the atmosphere is overdue. For a number of years throughout the 1980s, negotiations along this road were bogged down. A round of negotiations in Nairobi in late September 1991 ended in deadlock because of the continued US refusal to set targets and schedules for the emission of greenhouse gases. The argument put forward by the delegates from the USA was that setting targets is too costly and would discourage other countries from signing a climate convention. Japan also did not want targets, but rather was in favour of a 'pledge-and-review' treaty, whereby nations would agree to cut their emissions of greenhouse gases and then monitor the results. The Japanese proposed that nations stabilise their CO_2 emissions to 1990 levels by the year 2000 (something enshrined in the Rio Earth Summit – see next paragraph), also supported by Britain, Canada and the USA. The European Commission, however, insisted on more stringent measures. So an impasse developed without any agreement on mutually acceptable emission targets and schedules. Once the storm clouds had passed on the negotiations over a climate treaty, a much more optimistic note was struck in 1992 in Rio.

In June 1992, at the United Nations Earth Summit in Rio de Janeiro (for details, see Chapter 10), many countries supported the UN Framework Convention on Climate Change, committing them to reducing emissions of the greenhouse gases CO_2, CH_4 and N_2O to their 1990 levels by the year 2000. CFCs, which are covered by the Montreal Protocol, are not part of the Earth Summit Convention. The United Nations Framework Convention on Climate Change was established in response to the growing concern over the threat of climate change. It was presented at the Rio Summit. The convention is designed as a launching pad for potential further action and to allow countries to discuss the issue of climate change before they all agree that it is a problem. The framework is designed to allow countries to weaken and strengthen the treaty in response to new scientific developments as the dynamics of climate change begin to be understood. The treaty took effect on 21 March 1994 and as of 18 April 1996, the United Nations had received 158 ratifications. Countries ratifying the convention are called 'Parties to the Convention'. The Conference of Parties acts as the supreme body of the convention. In essence, it acts to review the implementation of the convention, to examine the obligations of the parties, and to facilitate the gathering of data and the exchange of information. The Conference of Parties met for the first time in March 1995 and will meet yearly thereafter. Two subsidiary bodies assist the Conference of Parties, one for scientific and technological advice, and the other for implementation.

The Objective (Article 2) of the convention is to achieve:

> stabilization of greenhouse gas concentrations in the atmosphere at a level that would prevent dangerous anthropogenic interference with the climate system. Such a level must be achieved within a time-frame sufficient to allow ecosystems to adapt naturally to climate change, to ensure that food production is not threatened and to enable economic development to proceed in a sustainable manner.
>
> UNEP/WMO Information Unit on Climate Change and web site for the Secretariat for the UN FCCC

The convention acknowledges that a change in the Earth's climate and its adverse effects are a common concern of humankind. It is concerned that human activities have been substantially increasing the atmospheric concentrations of greenhouse gases, which will increase the Earth's temperature and may adversely affect natural ecosystems and humankind. It acknowledges that the largest share of historical and current global emissions of greenhouse gases has originated in developed countries, whereas developing countries' per capita emissions are relatively low, but that the share of global emissions originating in developing countries will grow to meet their social and development needs. In this respect, the convention places the major responsibility, in relation to financial and technological transfer, on the 24 countries belonging to the Organisation for Economic Co-operation and Development (OECD).

The convention is aware of the role and importance in terrestrial and marine ecosystems of sinks and reservoirs of greenhouse gases. It acknowledges the many uncertainties in predicting climate change, particularly with regard to the timing, magnitude and regional patterns of change. Furthermore, the convention is conscious of the analytical work on climate change and the need for the exchange and coordination of scientific research to understand and address the environmental, social and economic problems associated with climate change.

The convention recalls earlier conventions, declarations and General Assembly resolutions that are relevant to climate change including, for example, the Vienna Convention for the Protection of the Ozone Layer, 1985; the Montreal Protocol on Substances that Deplete the Ozone Layer, 1987, and its amendment on 29 June 1990; and the Ministerial Declaration of the Second World Climate Conference, 1990. It reaffirms the principle of sovereignty of states in international cooperation to address climate change and the principle of sovereign rights of states to exploit their own resources and pursue their own environmental and developmental policies, and their responsibility to ensure that their activities do not cause damage to the environment of other states. It recognises the need for developed countries to take immediate action in a flexible manner on the basis of clear priorities as a first step towards a comprehensive response strategy on a variety of scales.

Furthermore, the convention recognises the special problems of sensitive and vulnerable regions such as low-lying and small island countries, mountain regions, and arid and semi-arid regions. It also appreciates the difficulties of developing countries, especially those whose economies are particularly dependent on fossil fuel production, use and exportation. In this respect it is also concerned to help coordinate social and economic development to achieve sustainable growth and eradicate poverty.

In summary, Article 3.3 of the convention provides guidance on decision-making where there is a lack of scientific certainty, so that parties should:

> take precautionary measures to anticipate, present or minimize the causes of climate change and mitigate its adverse effects. Where there are threats of serious or irreversible damage, lack of full scientific certainty should not be used as a reason for postponing such measures, taking into account that policies and measures to deal with climate change should be cost effective so as to ensure global benefits at the lowest possible cost. To achieve this, such policies and measures should take into account different socio-economic contexts, be comprehensive, cover all relevant sources, sinks and reservoirs of greenhouse gases and adaptation and comprise all economic sectors. Efforts to address climate change may be carried out cooperately by interested Parties.

The UK's commitment to reducing its CO_2 emissions is set out in a Department of the Environment (DoE) consultative document, *Climate Change: Our National Programme for CO_2 Emissions*. Also, in the UK's 1992 March Finance Bill (the 'Budget'), higher excise duties were added to petrol and diesel prices, and a value-added tax (VAT) was imposed on domestic fuel and power, which came into effect in 1994 at 17.5 per cent. The Government claimed this was part of a broader policy on controlling environmental pollution although, in this particular case, cynics interpreted this move as a straightforward revenue-raising tax.

On 7 April 1995, representatives from 118 countries attending the two-week Berlin conference on climate change agreed to the Berlin Mandate, which committed them to an agreement within the next two years on future reduction targets and timetables for implementing them when the present agreements to stabilise greenhouse gas emissions run out in 2000, i.e. for the first time there will be emissions targets and timetables set for beyond 2000. This meeting was the first of the biannual meetings agreed to by the 118 signatories to the United Nations Climate Change Convention drawn up at the Earth Summit in Rio in 1992. While most developed nations were satisfied with the outcome, the 32 members of AOSIS, the Alliance of Small Island States, together with pressure groups such as Greenpeace and the World-Wide Fund for Nature, were disappointed with the lack of any firm agreement at this stage, as they fear that low-lying islands and coastal lowlands will suffer severe flooding and even be submerged beneath the rising sea level if action is not taken. They wanted a reduction of 20 per cent in greenhouse gas emissions by 2005. As in Rio, poor countries attending the Berlin conference made no commitments. Acting upon a request by the Russian Federation, the Berlin Mandate will require nations to promote 'carbon sinks' – natural areas of forest and wetlands that can absorb and store greenhouse gases. As a result of the agreement, a permanent secretariat of 43 people was set up in Bonn with a budget equivalent to £12 million over the next two years with the job of getting negotiations under way on reducing any anthropogenically created climate change. Agreement at the 1997 meeting will mean that the protocol for targets should come into force as international law by the year 2000.

At the request of governments, the Intergovernmental Panel on Climate Change prepared a Second Assessment Report (SAR), which provides a comprehensive assessment of new and recent literature

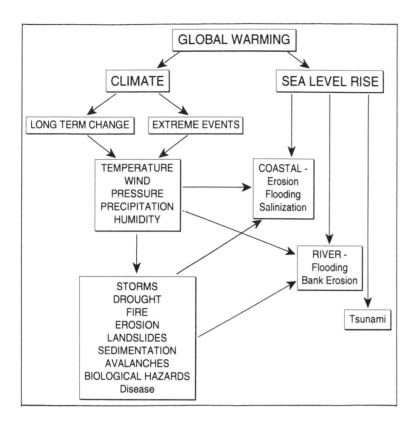

Figure 3.24 *Global warming may trigger a cascade of hazard effects – both directly through the mechanism of climate change and indirectly via sea level rise. Redrawn after Campbell and Ericksen (1990).*

on climate change. Three working groups were established:

- IPCC Working Group I: The Science of Climate Change.
- IPCC Working Group II: Scientific–Technical Analysis of Impacts, Adaptations, and Mitigation of Climate Change.
- IPCC Working Group III: The Economic and Social Dimensions of Climate Change.

The SAR is presented in three volumes and is approximately 2,000 pages long. In addition, a Synthesis Report brings together material contained in the three volumes of the full report that is relevant to governments' interpretation of the UN Framework Convention on Climate Change (UNFCCC) objective. The Synthesis Report is the consequence of an IPCC decision, following a resolution of the Executive Council of the WMO (July 1992), to include an examination of approaches to Article 2, the Objective of the UNFCCC, in its work programme. In October 1994, the IPCC organised a workshop in Fortaleza, Brazil, to address this subject and assembled a team of workers, who produced a draft synthesis report, which was submitted for expert and government review and comment. The final draft synthesis was approved by the IPCC at its eleventh session in Rome on 11–15

December 1995. The synthesis presents information on the scientific and technical issues related to Article 2, drawing on the underlying IPCC Second Assessment Report. The synthesis, however, is not a simple summary of the IPCC Second Assessment Report, and the Summaries for Policymakers of the three IPCC working groups should also be consulted for a summary of the SAR. Together these three volumes of the IPCC Working Groups and the Synthesis Report are known under the common title, *Climate Change 1995* (Source: IPCC World Wide Web Site).

Finally, global warming could trigger a cascade of natural hazard effects, both directly through the meteorological processes associated with climate change, and indirectly because of rising sea level. Figure 3.24 is a flow chart to summarise the potential natural hazards, all of which have socio-economic and environmental consequences which need to be considered at a variety of scales.

Conclusions

Governments are beginning to take action to control the amount of global warming caused by anthropogenically created greenhouse gases. In the USA, the US Office of Technology Assessment (OTA) has

BOX 3.8 CASE STUDY: THE UK CLIMATE CHANGE PROGRAMME

The UK Climate Change Programme was published in January 1994 in response to the UN Framework Convention on Climate Change (UNFCCC). It set out policies and measures aimed at returning emissions of each of the main greenhouse gases to 1990 levels by the year 2000. In March 1995, the Department of Trade and Industry published *Energy Paper 65*. This reported on the policy developments and progress of the measures undertaken on the UK Climate Change Programme. It stated that new energy projections indicated that carbon dioxide emissions will be 6–13 MtC (4–8 per cent) below 1990 levels by the year 2000, and they are confident of meeting the UNFCCC emission targets for this date. This is a result of the reduction in carbon intensity of electricity supply from an increased investment in combined-cycle gas turbines (CCGT) and the improved performance of nuclear generation. Additional measures included the Home Energy Conservation Act 1995, which introduced a minimum standard of energy per-

formance expected from new buildings and existing buildings whenever they are substantially altered; and revised expectations of savings from individual parts of the programme, including the Energy Saving Trust, valued-added tax on domestic fuel and power, and increased road fuel duties (UK Department of the Environment, 1995). A broad aim of working towards 1,500 MW of new renewable generated capacity by the year 2000 is a particularly notable commitment of the UK Climate Change Programme. The Non-Fossil Fuel Obligation (NFFO) in England, Wales and Northern Ireland, and the Scottish Renewables Obligation (SRO) in Scotland provide the stimulus for a move towards renewable energy. Under the NFFO, electricity distribution companies are required to secure specific amounts of electricity from renewable sources of energy. The UK is also participating in several European Union programmes that aim to aid renewable energy. These include ALTENER, which promotes greater use of renewables for energy production, and JOULE-THERMIE, which aims to reduce emissions through improved conversion and use of energy and the introduction of renewable energy into Europe's energy systems.

developed two scenarios assuming specified action levels for energy supply, energy conservation and forest management, which would limit CO_2 increases to 15 per cent by 2015 ('moderate'), or a 20 per cent reduction in CO_2 emissions over 1987 levels ('tough'). The OTA evaluation of these contrasting scenarios is that the 'moderate' option is achievable with net savings to the economy, whereas the 'tough' option would range from a small overall saving to a cost equivalent to 1.8 per cent of GNP. As a comparison, less detailed UK studies suggest that an 88 per cent reduction in CO_2 emissions relative to their 1987 levels, mainly through energy efficiency measures, is achievable at no net cost.

As for future control of the emission of greenhouse gases, international agencies are beginning to debate the need for severe energy-carbon taxes. On 25 September 1991, the European Commission proposed an energy and carbon tax that would help reduce the emission of greenhouse gases, but raise some fuel costs by up to 60 per cent. The proposal would involve the imposition of a tax of $10 on a barrel of oil over seven years. By the year 2000 this would raise industrial coal prices by 60.6 per cent, natural gas by 31 per cent, heating oil by 40 per cent and industrial electricity by 16 per cent. Renewable energy resources, such as wind and tidal power, would be exempt from the tax – the status of nuclear power remaining undecided. Inefficient

household appliances, such as washing machines and cookers, could also attract a new tax.

The debate on these proposed new taxes is just heating up, in an attempt to cool the planet down. The verdict of the jury will affect all of humanity. It is a debate that runs to the very heart of the need to reduce the emission of greenhouse gases. Is the only real way to curb these emissions to impose severe taxes on the production and consumption of energy? How will individual governments in Europe react to the imposition of energy taxes by the EU in terms of their perceived sovereignty? But, worldwide, proposals for an energy-carbon tax appear to be one of the most practical ways forward for controlling the emissions of greenhouse gases. Finally, the UN Convention on Climate Change contains no specific commitments from the signatory nations to control the emissions of greenhouse gases after the year 2000. And without any post-2000 commitments, in countries such as the UK, greenhouse gas emissions are projected to rise steeply, because the savings in emissions gained from changing to gas-fired (from coal- and oil-fired) power stations will have been exhausted. Also, global transport-related emissions will continue to rise. The issues surrounding the emissions of greenhouse gases are going to be at the fore of both national and international discussions, certainly well into the next century.

Chapter 3: Key points

1   Stratospheric ozone forms part of an important natural shield around the Earth that absorbs large amounts of UV-B radiation from the Sun, thereby reducing the harmful effects of too much radiation reaching the Earth's surface. Reducing stratospheric O_3 levels results in a cooling of the stratosphere and warming of the troposphere; the net result is a cooling of the Earth's surface, but the balance is strongly influenced by latitude, altitude and seasons. Significant depletions in stratospheric ozone have been recorded since 1977 and are, at least in part, the result of anthropogenic emissions of CFCs and other chemicals, as well as from natural processes such as volcanic eruptions, which produce aerosols and acids. Depletion of stratospheric ozone is now common over the poles following the Antarctic spring, when supercooled clouds act as a catalyst for the heterogeneous reactions that destroy the ozone. An increase in atmospheric CO_2 could lead to tropospheric warming and is likely to result in increased cloud cover, which may activate O_3-depleting species and thereby increase the rate of O_3 depletion, but this sequence of events remains controversial. Tropospheric ozone concentrations have been increasing in polluted regions, where it is an important greenhouse gas, and contributes to photochemical smogs.

2   The Earth's atmosphere is warmed by the so-called 'greenhouse effect', where short-wavelength (ultraviolet) solar radiation reaches the Earth's surface and is re-radiated as long-wavelength (infrared) radiation back up into the atmosphere, where it is absorbed by greenhouse gases, which warm the Earth. The most important greenhouse gases are water vapour, carbon dioxide, methane, tropospheric ozone, nitrous oxide, ammonia, CFCs and the halons. These have different global warming potentials (GWPs) and varying residence times in the atmosphere. An increase in these atmospheric gases may lead to global warming. However, there are many negative feedbacks that counteract the effects of any potential global warming caused by these gases, and the real nature of increased concentrations of the greenhouse gases remains debatable. Sophisticated 'general circulation models' for the interaction between the ocean–atmosphere–land systems (GCMs) have been developed by various organisations, including the Intergovernmental Panel on Climatic Change (IPCC), as a means of assessing the climatic sensitivity to changing atmospheric concentrations of greenhouse gases. The IPCC suggests that a doubling of atmospheric CO_2 is likely to increase the global mean surface temperature by 1.5–4.5°C.

Evidence for human-induced global warming is fiercely debated in some quarters. It is difficult to compare historical and modern meteorological data, and although snow and Arctic sea ice cover have decreased over the last decade, it is not known if this is due to natural perturbations in the Earth's atmosphere and hydrosphere or anthropogenic causes. Arctic geotherms, however, provide evidence in the form of thermal anomalies to suggest a warming in the order of 2–4°C over the last century (IPCC). If greenhouse gas emissions continue, global warming may cause changes in global weather patterns, amongst which there could be an increased severity of storms and droughts, and an overall rise in sea level of the order of 50–60 cm by 2100, caused both by thermal expansion of the oceans and melting of ice caps and glaciers.

3   Natural phenomena also cause global climate change, e.g. volcanic eruptions and El Niño events. Volcanic eruptions emit huge quantities of aerosols, which reduce incoming solar radiation and lead to global cooling, usually of the order of between 0.2–0.5°C for short durations (< 5 years). Greenhouse gases (CO_2), ozone-depleting gases (Cl_2, F_2) and gases that may act as catalysts for ozone-depleting reactions (SO_2, which forms H_2SO_4) are also emitted by volcanic activity, and cause either negative or positive feedback mechanisms for global climate change.

4   International action on global atmospheric pollution has resulted in attempts to reduce ozone depletion (Montreal Protocol, 1987; with amendments in Helsinki, 1989, and Copenhagen, 1992) and greenhouse gas emissions (UN Framework Convention on Climate Change, 1992). It remains to be seen whether the developed and developing nations can meet the targets set by the various international agreements on atmospheric pollution.

Chapter 3: Further reading

Boyle, S. and Ardill, J. 1989. *The Greenhouse Effect.* London: Hodder and Stoughton, 298 pp.
A highly readable, if slightly dated, introductory book explaining the nature of the greenhouse effect. Although written for the lay person with little or no scientific background, it may also be useful for social science students wishing to gain a reasonable understanding of the causes and effects of any global warming caused by a greenhouse effect.

Graedel, T.E. and Crutzen, P.J. 1993. *Atmospheric Change: An Earth System Perspective.* New York: W.H. Freeman & Co., 446 pp.
This book provides the basis of an excellent introduction to atmospheric chemical processes, probably at a level somewhat advanced for a generalist course, but it serves as a useful textbook for science-based students.

Houghton, J.T., Jenkins, G.J. and Ephraums, J.J. (eds) 1990. *Climate Change: The IPCC Scientific Assessment.* Cambridge: Cambridge University Press, 365 pp.

Houghton, J.T., Callander, B.A. and Varney, S.K. (eds) 1992. *Climate Change 1992: The Supplementary Report to the IPCC Scientific Assessment.* Cambridge: Cambridge University Press, 200 pp.

Houghton, J.T., Meira Filho, L.G., Lee, H., Callander, B.A., Haites, E., Harris, N. and Maskell, K. (eds) (Intergovernmental Panel on Climate Change) 1995. *Climate Change 1994. Radiative Forcing of Climate Change and an Evaluation of the IPCC IS92 Emission Scenario.* Cambridge: Cambridge University Press, 339 pp.
These three reports are the result of Working Group I of the Intergovernmental Panel on Climatic Change, set up by the World Meteorological Organization and the United Nations Environment Programme. They are essential reading and reference material for anyone interested in global climate change. The reports assess the potential effects that human activity may have on the Earth's climate, and include sections on changes in the concentrations of atmospheric greenhouse gases; modelling of the global climate system; computer prediction of climate change; observed climate change over the last century; the detection of climate change due to human activities; changes in global sea levels due to global warming; the response of ecosystems to global climate change; and the research required to narrow the uncertainties in future predictions of global climate change. The reports consider worst-case, business-as-usual and best-case scenarios for global climate change using various assumptions about the levels of anthropogenic emissions of greenhouse gases.

Kemp, D.D. 1994 (second edition). *Global Environmental Issues: A Climatological Approach.* London: Routledge, 220 pp.
A climatological approach to global problems that includes good sections on the greenhouse effect, acid rain, ozone depletion, drought and the possible effects of a nuclear winter. Appropriate for students in environmental studies and geography wishing to analyse the role of climatology in environmental change.

Nilsson, S. and Pitt, D. 1991. *Mountain World in Danger: Climate Change in the Forests and Mountains of Europe.* London: Earthscan, 196 pp.
An account of the possible changes in the forests and mountain environments of Europe as a result of global warming and acid rain. Possible strategies in response to these changes are discussed in the hope that steps may be undertaken by policy-makers to retard the potentially harmful impact. Good summaries of the IPCC's conclusions and international conventions and declarations are provided in the text and in a series of appendices.

O'Riordan, T. and Jager, J. (eds) 1995. *Politics of Climate Change.* London: Routledge.
This volume provides a critical analysis of the political, moral and legal responses to climate change in the midst of significant socio-economic policy shifts. The book examines how climate change was put on the policy agenda of the EU, and how the United Nations Framework Convention and the subsequent Conference of Parties evolved. The contributors analyse the climate change policies of different nations and reductions of greenhouse gas emissions, and legal aspects of external competence and moral obligations, and they assess the political significance of the European experience within the wider, global perspective of America and Asia.

Parry, M. 1990. *Climatic Change and World Agriculture.* London: Earthscan.
Written by the chief scientist on the Intergovernmental Panel on Climate Change (IPCC), and concerned with the potential impacts of climatic change on agriculture. It provides a good account of the likely patterns of change in climate and world agriculture as a consequence of global warming. Emphasis is placed on the uncertainties associated with this issue, the sensitivity of the world food system, vegetational patterns and animal life to global climate change, the geographical limits to different types of farming, and the range of possible ways to adapt agriculture to mitigate any potential hazards caused by global warming.

Paterson, M. 1996. *Global Warming and Global Politics.* London: Routledge.
The book looks at the major theories within the discipline of international relations, and considers the emergence of global warming as a political issue.

Whyte, I. 1995. *Climatic Change and Human Society.* London: Edward Arnold.
This text examines the various ways that climatic change can interact with society. This is a particularly useful text for students of geography and environmental studies.

I'm a goin' back out 'fore the rain starts a fallin',
I'll walk to the depth of the deepest black forest,
Where the people are many and their hands are all empty,
Where the pellets of poison are flooding their waters,
Where the home in the valley meets the damp dirty prison,
Where the executioner's face is always well hidden,
Where hunger is ugly, where souls are forgotten,
Where black is the colour, where none is the number,
And I'll tell it and think it and speak it and breathe it,
And reflect it from the mountains so all souls can see it,
Then I'll stand on the ocean until I start sinkin'.
But I'll know my song well before I start singin',

And it's a hard, it's a hard, it's a hard, it's a hard,
It's a hard rain's a gonna fall.

Bob Dylan, 'A Hard Rain's a Gonna Fall'

The Earth's atmosphere supports life, yet the skies over much of the urban areas in Europe and North America are affected by toxins from power stations, factories and vehicles. 'Choking Greeks look to the gods for help' was one of the front-page leaders in Britain in *The Times* on Wednesday 2 October 1991. As temperatures rose to an unusual 36°C in Athens on 1 October, levels of the main pollutant gas, nitrogen dioxide, reached an average of 561 mg m^{-3} in the city centre, well above the 500 mg limit above which emergency measures have to be introduced. Indeed, in parts of Athens, the levels of nitrogen dioxide reached 696 mg m^{-3}, the previous record being 683 mg recorded on 10 June 1991. As a result of the cocktail of toxic gases over Athens on 1 October, locally named *nephos*, more than 200 people went to hospital with respiratory and cardiac problems, and the government banned the use of private cars in the city centre between 6 a.m. and 5 p.m. Scenarios such as this are becoming more common in major cities throughout the world because of atmospheric pollution, especially because of the use of private motor vehicles pumping out atmospheric pollutants, some of which, such as nitrogen dioxide, contribute to acidic deposition.

The natural balance of gases and reactions in the atmosphere sustains not only human beings but all the life on Earth. Organisms, including humans, use and produce many gases, including oxygen, carbon dioxide, nitrogen and sulphur compounds. Human activities have drastically altered the natural balance of such gases and have contributed new and extremely harmful toxic gases and other substances to the atmosphere. Figure 4.1 illustrates the various processes that lead to the formation of acidic deposition. While rain is naturally slightly acidic, human activities can increase the acidity and push natural systems over a critical threshold level and bring about atmospheric–climatic changes that have adverse environmental effects.

Two end-member philosophies to pollution may be applied. One is to *concentrate and contain*, while the other is to *dilute and disperse*. Both options have associated pros and cons. Traditionally, industry is the main contributor to acidic deposition, which is exacerbated by its common concentration in industrialised areas resulting in maximum pollution tending to be geographically contained. Arguably, by sharing the misery of pollution through geographically more distributed factories and energy plants, the effects of atmospheric pollution could actually be reduced with the more efficient dispersal of pollutants before they reached threshold concentrations. In reality, commercial and demographic considerations mean that belts or regions of industrial activity are typical of any country. One of the main causes of atmospheric pollution is that the polluter tries to dispose of gaseous pollutants into the atmosphere, in an attempt to dilute and disperse.

The long- and intermediate-term consequences of artificially changing the composition of the atmosphere are not fully understood. The principal chemicals that produce acidic deposition are sulphur dioxide (SO_2), nitrogen oxides (NO_x) and hydrocarbons. Collectively, nitric oxide (NO) and nitrogen dioxide (NO_2) are referred to as NO_x.

Human awareness tends only to surface as necessity dictates, as the effects of pollution are experienced, and not through some sort of spontaneous environmental consciousness that is separated from human exploitation of the Earth's resources. Acidic deposition, however, was appreciated in the middle of the 1850s, for example the 35-year-old Scottish chemist, Robert Angus Smith, in 1852 published a paper in the *Memoirs and Proceedings of the Manchester Literary and Philosophical Society* 'On the air and rain of Manchester'. The increasing acidity of the precipitation is referred to as **acid rain**.

The noxious effects of pollution caused by the burning of coal have been known about for centuries, though they were not identified then as acidic deposition. In England, during the reign of Edward I (1272–1307), the use of sea coal (with its high levels of sulphur and trace elements), washed ashore from

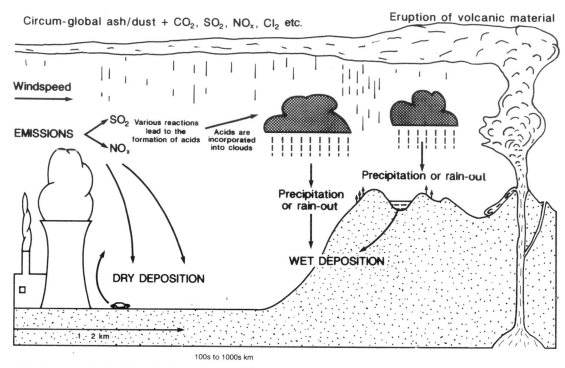

Circum-global ash/dust + CO_2, SO_2, NO_x, Cl_2 etc.

Eruption of volcanic material

Windspeed

EMISSIONS

SO_2 Various reactions lead to the formation of acids

NO_x

Acids are incorporated into clouds

Precipitation or rain-out

Precipitation or rain-out

DRY DEPOSITION

WET DEPOSITION

1 - 2 km

100s to 1000s km

Figure 4.1 *Processes involved in the formation and deposition of acid rain.*

exposed deposits of coal, was prohibited by royal proclamation. A third offence was punishable by execution!

The phrase **acidic deposition** covers a multitude of atmospheric conditions, not only as rain but acid mists and acid fog that have to be contended with, together with photochemical smogs, which may be acidic. Perhaps one of the most infamous smogs of recent times was the 'pea-souper' smog of December 1952 in London, when an excess of about 4,000 deaths were attributed to bronchial causes arising from this smog. The smog lasted five days from 5 December and was caused by warm air holding down a vast acidic cloud, with no wind to disperse it. By Tuesday 9 December, the acid smog extended to a radius of 30 km from the centre of London. The acid in the smog was probably mainly sulphuric acid (H_2SO_4). Current estimates of the pH (a measure of acidity or alkalinity of a substance – see Box 4.1) of the smog are that it was between 1.4 and 1.9 – more acidic than lemon juice. Even today, London is not immune from acidic smogs. Over the weekend 13–15 December 1991, severe smog conditions returned to London, when on Friday 13 December concentrations of nitrogen dioxide at street level reached 293 parts per billion, well above the recommended EC maximum values, and the highest since 1976. The conditions arose because of

a high-pressure system that brought settled weather and created warm dry air at higher altitudes. The air at ground level cooled and cold moist air started to rise, but was trapped below a layer of warmer air. The cold air stagnated because of a lack of wind to mix the air masses. A density inversion was therefore created over London and the pollutants remained trapped near street level to cause a smog. The heavy use of vehicles in the city served only to exacerbate the situation. On 9 September 1989, an acidic mist rolled over the east coast of England, affecting 1,000 square miles from the Humber estuary to Great Yarmouth in Norfolk, with a pH as low as 2.0. This incident was blamed on heavy industry and vehicle fumes from Germany and Poland (*The Sunday Times*, 22 October 1989). The droplets of sulphuric and nitric acid were sufficiently strong to kill the leaves on thousands of trees overnight, turning them brown or black, and corrode aluminium instruments. Previously, Britain's worst recorded acidic deposition was in 1974 at Pitlochry in Scotland, when the mist had a pH level of about 2.4, stronger than vinegar.

Other examples of serious incidents such as the 1952 London smog include the two days in August 1984 when Athens suffered an acid smog that led to the hospitalisation of 500 people and the Greek government declaring an emergency situation. Other cities around the globe, such as Mexico City, Tokyo

---

BOX 4.1 pH

Acidity and alkalinity are measured on a logarithmic 'pH scale' (defined as the negative logarithm (reciprocal) of the hydrogen ion [H⁺] concentration):

$$pH = -\log_{10} [H^+]$$

Pure water is neutral and has a pH of 7, but natural rain water is slightly acidic with a pH of about 6.5, because of the carbonic acid (H_2CO_3) resulting from the dissolved atmospheric carbon dioxide within the rain water. Lower values occur in acids, while pH values greater than 7 indicate alkaline substances. The further a pH value is away from 7 (towards 0 or 14 at both extremes), then the more acidic or alkaline that substance.

In the eastern USA, during the summer, water sampled near the base of clouds (where acidity is greatest) reveals typical pH values of 3.6, although pH 2.6 has been recorded. Also, in the Los Angeles area, fog with a pH of 2 has been measured, equivalent to the acidity of lemon juice. Vegetated areas on mountains and hillsides are easily exposed to these very acidic situations during low cloud conditions. During rain, the very acid levels are reduced through dilution, typically by 0.5 to 1 pH unit. Thus, the average pH value of rain during the summer months in the northeast USA is about 4.2. In Britain, in Yorkshire and the East Midlands, the average pH of rain water is below 4.3 (Simpson 1990).

---

and Los Angeles, suffer severe air pollution and acid precipitation on a fairly regular basis.

The effects of atmospheric pollution are now widespread and are of increasing concern among environmental health workers who are involved in the study of lung disease. Figure 4.2, for example, shows the risk to human health resulting from SO_2 emissions. In Britain, for example, asthma now affects one in seven children, and chronic lung diseases such as bronchitis and emphysema have become more widespread in recent years. After heart disease, lung disease is now the major cause of illness and disability in Britain. Lung cancer causes a third of all cancer deaths in the male population, and the number of people dying from lung cancer has increased drastically in recent years. In the USA, for example, 18,000 people died from lung cancer in 1950; in 1994 it killed 153,000 (Lee and Manning 1995). It is the only disease in the Western world that has shown a steady increase in recent years. Lung disease may be caused by a variety of means, ranging from viral or bacterial infections to airborne particles (e.g. silicon and coal dust) and gaseous pollutants. There is growing evidence that a large proportion of lung diseases result from human effects on the atmosphere. Where lung disease is linked to atmospheric pollution it is usually referred to as environmental lung disease.

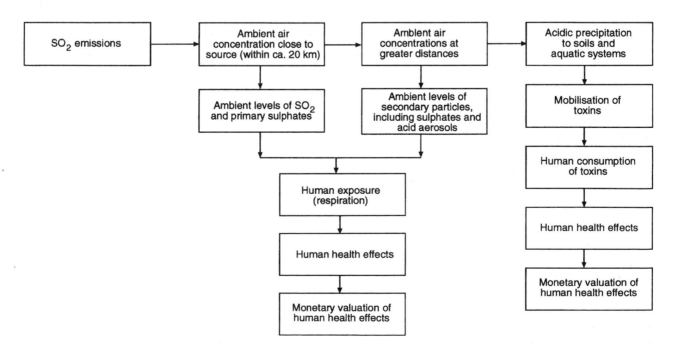

Figure 4.2 *Overview of human health effects resulting from SO_2 emissions. After the [US] EPA (1995).*

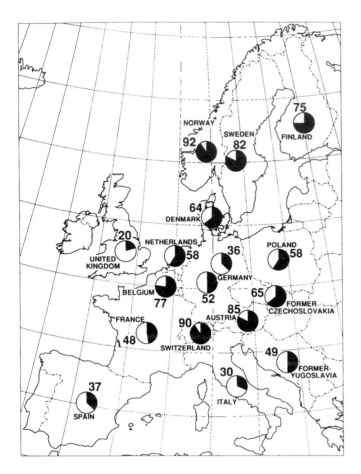

Figure 4.3 *The contribution, shown as a percentage, of external sources (solid portions in pie diagrams) to the amount of sulphur deposited in some European countries. Percentages are based on EMEP model calculations for 1978–1982. Redrawn after the National Environment Protection Board (1987).*

Most atmospheric pollutants are linked to industrial processes, for example smoke from factories, pollen from agricultural activities, dust from building construction, asbestos from insulating material, and SO_2 and NO_x from power stations. People involved in such activities are clearly more vulnerable and may, as a result, suffer from occupational lung diseases. Diseases of the lungs may become apparent within minutes, over months or very slowly over decades. Particularly worrying is the long-term effects of the rapid increase in urban pollution, especially from traffic exhaust emissions. Many of these emissions comprise very small particles known as PM10 (i.e. < 10 μm in diameter), which are produced mostly by diesel vehicles. Since these particles are very small they are able to enter deep into the lungs, into the alveoli and capillaries. They may inflame the lungs and cause the blood to become more viscous, thereby increas-

ing the risk of clotting, leading to both heart and lung disease. Hamer (1994) estimates that as many as 10,000 deaths in Britain each year may be the result of the inhalation of PM10s. Similarly, gaseous pollutants are able to penetrate deep into the lung. A reduction in atmospheric pollutants is, therefore, of prime importance in helping to increase public health.

Measuring the effects of acidic deposition

Acidic deposition has already produced many visible and measurable effects. This century, the first areas to be recognised as showing serious damage as a result of acidic deposition were found downwind of the major industrial centres of Britain, mainland Europe and North America. Scandinavia and Canada were identified early in the 1980s as having suffered severe damage from acidic deposition. Much larger areas of Europe are now known to be affected as well as less industrialised nations such as China, India and South Africa. Although sulphur – which in atmospheric chemical reactions forms sulphuric acid – is transported from eastern North America to Europe, the amounts are small compared with the European sources. Sulphur pollution is the result of sources both within and outside individual countries. A study by the National Environment Protection Board (1987) has revealed the contribution, expressed as percentages, of external and national sources of sulphur throughout Western Europe (Figure 4.3). From this study, covering the years 1978–1982, it can be seen that Norway received 92 per cent of its sulphur deposition from external sources, compared with only 20 per cent from outside sources for the UK. Maps such as these reveal the case for stringent international control on sulphur emissions because of the large amounts of pollution that are exported from heavily industrialised regions to other parts of the world.

In eastern North America, anthropogenic sources of sulphur overwhelm natural sources (Galloway and Rodhe 1991). Figure 4.4 shows the estimated amounts of sulphur (Tg S yr^{-1}) transported eastwards from eastern North America (USA and Canada), and Figure 4.5 shows a schematic sulphur budget for the western North Atlantic Ocean atmosphere. Sulphur is removed from the atmosphere by both **wet deposition** and **dry deposition**. Any uncertainties in the fluxes are due, at least in part, to a lack of direct measurements and because of remaining ignorance about the processes leading to dry deposition.

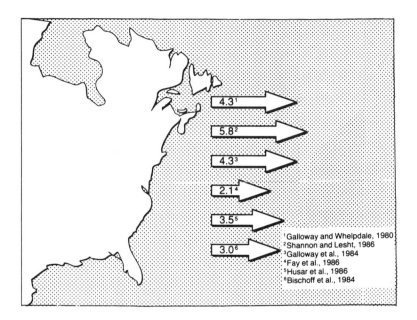

Figure 4.4 *Estimates of the absolute amounts (Tg yr⁻¹) of sulphur advected eastward from eastern North America (USA and Canada) to the western North Atlantic Ocean. Redrawn after Galloway (1990) in Last and Watling (1991).*

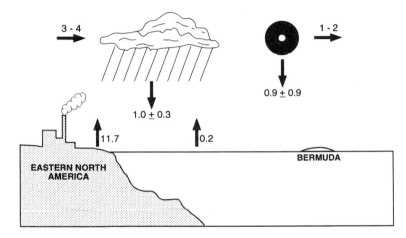

Figure 4.5 *Sulphur budget (Tg yr⁻¹) for the western North Atlantic Ocean atmosphere. Redrawn after Galloway and Whelpdale (1987) in Last and Watling (1991).*

Throughout many parts of the Northern Hemisphere, trees are being poisoned and killed, and soils are becoming too acidic to support plants (Plate 4.1). It is estimated that more than 65 per cent of the trees in the UK, and more than 50 per cent of the forests in the pre-1990 West Germany, the Netherlands and Switzerland, are damaged. Acidic deposition affects the processes of respiration and photosynthesis in plants by damaging the breathing pores, or stomata (through cell damage), and also because it partly removes and dissolves the wax coatings that allow plants to regulate their growing conditions within the cells (Wolfenden and Mansfield 1991). The main way in which plants are damaged by acidic deposition, however, is by the acidification of ground water, impairing nutrient fixing by a plant's roots. Crops are also being affected and producing lower yields. Others, however, would

argue that vegetation degradation may be due to the increase of poisonous anthropogenically created tropospheric O_3.

Lakes have been acidified in North America and Scandinavia. An estimated 20,000 lakes are acidified in Sweden, 20 per cent of which support no fish life. Stone buildings are being eaten away and badly discoloured.

The actual effects of acidic deposition on any environment vary according to whether the ground on which it falls is itself wet or dry, local climatic conditions, the way and rate at which ground-water run-off occurs, together with the type of vegetation. In urbanised areas, the degree of acidic corrosion of structures will depend upon the types of materials; for example, limestone buildings will suffer much more than sandstone buildings in a given area, because limestone unlike most sandstones is readily

Plate 4.1 *(A) Nickel smelting plants at Nikel in north-western Russia, near Murmansk, emitting large amounts of SO₂ and other environmentally harmful gases. (B) Dying forests and dead silver birch trees on the hillsides surrounding Nikel – the result of acidic deposition. The tall chimney stacks actually carry the noxious emissions away from the town of Nikel in the valley around the smelters, with the result that trees within the town remain relatively unaffected unlike those of the surrounding forests. Photographs taken in August 1993.*

soluble in weak acidic solutions. This point is well illustrated by the weathering of many historic monuments, such as St Paul's Cathedral in London, which has suffered considerable damage over the past few hundred years. This is constructed out of Portland limestone, a brilliant white limestone which, prior to recent industrial cleaning, was discoloured to a sooty black, and preferentially hollowed out where the flow of acid rain waters is channelled across the surface of the facing stones; along these areas of stone, bulbous precipitates of gypsum ($CaSO_4$) form beneath anvils and gargoyles as a result of the reaction between acidic deposition (particularly sulphuric acid, H_2SO_4) and the limestone ($CaCO_3$).

Scientists monitoring the degree of weathering suggest that annually as much as 0.62 μm of the surface of the limestone is lost, which over the period since the construction of St Paul's amounts to about 15 mm (Sharp *et al.* 1982). Acidic deposition can also cause damage to property such as the paintwork on vehicles. In the USA, some vehicle manufacturers use acid-resistant paints to reduce the potential damage. At an average cost of about US$5 for each new vehicle, the total cost of implementing this scheme is approximately US$61 million per year (EPA World Wide Web Site). Thus, acidic deposition has so far affected the aquatic life in fresh-water lakes and rivers, forest ecosystems and other flora in the countryside and towns, and contaminated ground water and coastal waters.

What is acidic deposition?

Acidic deposition occurs because of the atmosphere's continual effort to cleanse itself of various pollutants that are introduced into the air. The water droplets in clouds absorb and adsorb particulate matter and gas molecules out of the air. Not all such substances are removed by rain or precipitation, but instead remain suspended in clouds and moisture.

There are essentially two kinds of pollutant, gaseous and particulate, each of which can be defined on the basis of its mode of formation as primary or secondary pollutants. **Primary pollutants** are produced directly from industrial and domestic activities, whereas **secondary pollutants** are created in the atmosphere by chemical processes acting on primary pollutants. On this basis, Last (in Last and Watling 1991) defines the four groups of pollutants as shown in Table 4.1.

The industrially produced sulphur dioxide, together with nitrogen oxides, are easily converted into sulphuric and nitric acids, respectively, and they accumulate as acidic deposition (see Tables 4.2 and 4.3 for 1992 IPCC-estimated sources and sinks of short-lived sulphur gases, and estimated sources of nitrogen oxides, respectively). Sunlight, free oxygen and water are the abundant ingredients that allow these reactions to take place. The reactions occur mainly in the lowest 10–12 km of the atmosphere (the troposphere). The acids, although dilute, have considerable corrosive capability. They are changing and polluting the natural environment over years rather than decades or centuries.

Plate 14 *Oil wells in Kuwait, October 1991, still burning ten months after the Gulf War.*
Courtesy of Michael McKinnon/The Environmental Picture Library.

Plate 15 *American workers capping an oil well head after it was sabotaged during the Gulf War in August 1991.*
Courtesy of Michael McKinnon/ The Environmental Picture Library.

Plate 16 *Smog over Mexico City.*
Courtesy of Liba Taylor/ Panos Pictures.

Plate 17 *Greening arid land. Artificial rotational irrigation systems – by centre pivot irrigators – west of the Rocky Mountains, USA.*

Plate 18 *Wheal Jane tin mine, the source of the Fal Estuary pollution, UK, in 1991.*
Courtesy of Adrian Evans/Panos Pictures.

Plate 19 *The grounded* Braer *oil tanker off Sumburgh Head in the Shetland Isles, Scotland, leaking oil into the sea in January 1993.*
Courtesy of G. Glendell/ The Environmental Picture Library.

Table 4.1 *Principal chemical pollutants (not all involved in acid rain).*

Gaseous, primary
Sulphur dioxide (SO_2)
Nitric oxide (NO) ⎤ collectively referred to as
Nitrogen dioxide (NO_2)⎦ oxides of nitrogen (NO_x)
Hydrocarbons (HC)
Ammonia (NH_3)
Carbon dioxide (CO_2)

Gaseous, secondary
NO_2 from oxidation of NO
Ozone (O_3) and other photochemical oxidants
  formed in the lower atmosphere by the action of
  sunlight on mixtures of NO_x and hydrocarbons
Nitric acid from the oxidation of NO_x

Particulate, primary
Fuel ash
Metallic particles

Particulate, secondary
The reaction products of sulphuric acid and nitric
  acid with other atmospheric constituents, notably
  ammonia (NH_4HSO_4, NH_4NO_3, etc.)
Sulphuric and nitric acids formed by the oxidation of
  SO_2 and NO_x respectively

Source: Last and Watling 1991.

Table 4.2 *Estimated sources and sinks of short-lived sulphur gases (Tg S per annum).*

Anthropogenic emissions (mainly SO_2)	70–80
Biomass burning (SO_2)	0.8–2.5
Oceans (dimethyl sulphide)	10–50
Soils and plants (DMS and SO_2)	0.2–4
Volcanic emissions (mainly SO_2)	7–10

Source: IPCC *Climate Change* 1992.

Table 4.3 *Estimated sources of nitrogen oxides (Tg N per annum).*

Natural	
Soils	5–20[1]
Lightning	2–20[2]
Transport and stratosphere	1
Anthropogenic	
Fossil fuel combustion	24[3]
Biomass burning	2.5–13[4]
Tropospheric aircraft	0.6

[1] Dognon *et al.* 1991; [2] Atherton *et al.* 1991; [3] Hameed and Dignon 1992; [4] Dignon and Penner 1991.
Source: IPCC *Climate Change* 1992.

The chemical reactions that lead to acidic deposition begin as a quantum of sunlight energy (a photon) hits an ozone molecule (O_3) to form free oxygen (O_2) and a single atom of oxygen, which is very reactive. The single oxygen atom reacts with a water molecule (H_2O) to produce two electrically charged, negative, **hydroxyl radicals** (HO^-).

Although these hydroxyl ions constitute less than one part per trillion in the atmosphere, they are effectively inexhaustible and the oxidation reactions they trigger actually generate more hydroxyl ions as a by-product of the reactions! For example, the oxidation of sulphur dioxide produces **hydroperoxyl radicals** (HO_2^{3-}), which react with nitric oxide (NO) to produce nitrogen dioxide (NO_2) and a new hydroxyl radical.

Individual hydroxyl radicals are therefore capable of oxidising millions of sulphur-containing molecules; it is only the amount of pollutant in the atmosphere that determines the amount of acid ultimately generated. The relatively scarce nitrogen dioxide in the atmosphere reacts with the hydroxyl ions to form nitric acid, HNO_3. The nitric acid then acts as a trigger for the reactions that produce sulphuric acid (H_2SO_4) from sulphur dioxide. Nitrous oxide and nitric acid are produced naturally during lightning storms, and anthropogenically from the exhaust fumes of super-

sonic aircraft; they were also formed in atmospheric nuclear explosions during the 1950s and 1960s.

The usual way the nitric and sulphuric acids reach the ground is in water droplets from clouds, as rain. This is not the only way though, because, for example, sulphuric acid generated in gas-phase reactions can condense to form microscopic droplets (0.1 to 2 µm in diameter) that constitute part of the haze seen over the eastern United States of America during summer months. Some of these particles settle to the ground, and vegetation can also absorb SO_2 gas directly from the atmosphere. This process is known as dry deposition.

In the stratosphere, the rate-limiting step for the oxidation of SO_2 to H_2SO_4 is believed to be the reaction of SO_2 with the hydroxyl ion, OH^-, leading through a sequence of reactions to the production of new sulphate particles, or condensation on pre-existing particles. Sulphur dioxide also acts as a catalyst in the production of ozone, but it can inhibit ozone production by absorbing incoming solar radiation, since it absorbs radiation strongly in the range 235–180 nm, weakly in the range 340–260 nm, and very weakly in the range 390–340 nm.

When oxidised in the atmosphere, sulphur compounds produce non-sea-salt sulphate aerosols, which may act as cloud condensation nuclei. In this way, non-sea-salt sulphate aerosols may affect local and/or global climate change, since the amount, density and size distribution of droplets in clouds can alter cloud albedo. As most acidic aerosols are produced in the Northern Hemisphere, any increase in SO_2 emissions could lead to the possibility of the Northern Hemisphere cooling relative to the Southern Hemisphere. Observations of differences in hemispheric mean temperatures combined with results from a simple climate model suggest that the upper limit is sufficiently large that any effects caused by SO_2-derived forcing of global climate may have significantly offset the changes in temperature resulting from the greenhouse effect (Wigley 1989). Of course, if, as seems likely, global SO_2 emissions are successfully halted or reversed then the net effect could be to accelerate the rate of greenhouse-gas-induced warming (*ibid.*).

The chemical reactions that produce acidified ground can occur in two ways, through what are called dry- and wet-phase reactions. Dry-, or gas-phase reactions typically occur near the pollution source, such as a power station, whereas wet-, or aqueous-phase reactions are characteristic of more distant areas and involve chemical reactions within water droplets, as in clouds, smogs, mists, sleet and snow. Dry deposition generally occurs within 300 km, as opposed to wet deposition, which takes place up to 1,000 km away from the pollution source.

Acid ground waters are caused by a combination of high precipitation (rainfall) from polluted air and clouds, rapid surface-water run-off (for example via streams), low temperatures, thin soil cover low in metals such as calcium (as in limestones), which may reduce or neutralise the acidity and, generally, steep slopes. In temperate climates, such as in Britain, the greatest stream run-off occurs during the spring, as snow melts from upland areas, and in the autumn storms. Typically, more than 70 per cent of the precipitation ends up as stream run-off, although this figure is extremely variable from year to year, and within a single catchment area.

The polluters

Natural processes introduce 'pollutants' into the atmosphere that, amongst other results, may also produce acidic deposition. These surface processes operate over time scales much longer than human life spans and include volcanic eruptions, while bacterial action in soil may also contribute to the atmospheric substances that produce acidic deposition.

Human activities, however, have accelerated the release of such harmful chemicals into the atmosphere and hydrosphere and at levels that make it very hard for the Earth to cope. Over the past century, human activities have increased global emissions of sulphur gases by a factor of about three, leading to increased sulphate aerosol concentrations, mainly in the Northern Hemisphere (Langner *et al.* 1992). Figure 4.6 schematically shows the fluxes of atmospheric sulphur species (excluding sea salts and soil dust) in different parts of the sulphur cycle. Much of the sulphate aerosol results from SO_2 emissions that have been oxidised by in-cloud processes.

The combustion of fossil fuels releases toxins into the atmosphere. Coal, for example, may contain up to about 5 per cent sulphur, compared with oil with typical values of up to roughly 3 per cent. Approximately 120 Mt of SO_2 is emitted annually across the world from fossil fuel power stations, metal-smelting factories and other industries utilising fossil fuels such as oil and coal (Plate 4.2).

The main countries responsible for this pollution are the former Soviet Union, USA, China, Poland, Germany (especially what was until 1990 East Germany), Canada, the UK, Spain, Italy and Czechoslovakia, in decreasing order of amounts of SO_2 emitted. The largest single anthropogenic source of SO_2 emission in the world is the copper- and nickel-smelting plants in Sudbury, Ontario, which produced an annual discharge of about 630,000 tonnes of SO_2 in the early 1980s.

Conventional power stations burning fossil fuels are amongst the worst culprits for causing acidic deposition by the emission of sulphur dioxide and nitrogen oxides. Oil refineries are another significant source of SO_2 emissions. In Britain, along roughly 20 km of the Aire valley, Yorkshire, there is one of the largest concentrations of power stations in the UK belching out gaseous toxins. Three of these conventional power stations, Ferrybridge, Drax and Eggborough, burn coal from the nearby new Selby coalfield and generate 20 per cent of Britain's electricity and an almost equivalent percentage of the country's total sulphur emissions. Their annual emission of SO_2 is about 600,000 tonnes, excluding other gaseous pollutants. The power stations in valleys such as the Aire have created the acidic deposition that has poisoned lakes in Scandinavia.

The former Soviet Union is the world's main producer of SO_2; 60 per cent of Soviet electricity is generated from conventional fossil-fuel-burning power stations, which contribute 40 per cent of the SO_2 emissions. In addition, the countries to its west

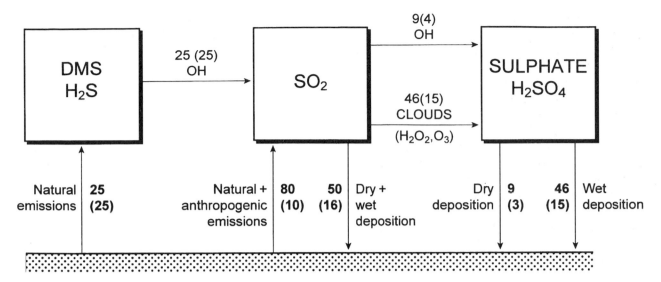

Figure 4.6 *Schematic representation of fluxes of atmospheric sulphur species (excluding sea salts and soil dust) in different parts of the sulphur cycle. Numbers represent fluxes in Tg S yr^{-1}, with pre-industrial fluxes shown in parentheses (from Langner* et al. *1992). Much of the sulphate aerosols result from sulphur dioxide emissions that have been oxidised by in-cloud processes. Redrawn after IPCC report 1994 (Houghton* et al. *1995).*

contributed an estimated 5 Mt of sulphur compounds annually from 1979 to 1981. The secrecy in the former Soviet Union made it difficult to assess the real amount and extent of acid pollution, but in 1984 at the Munich Conference the Soviet delegation admitted damage was occurring. In 1984, the newspaper *Pravda* reported that forests were dying from atmospheric pollution in the neighbourhood of the automobile factory city of Togliatti, near Kuibyshev in the Ural Mountains, and that lakes in the Kola Peninsula and Karelia were suffering acidification.

The USA is second only to the former Soviet Union in its total emissions of SO_2. The principal offending states, in decreasing order, are Ohio, Pennsylvania, Indiana, Illinois, Missouri, Wisconsin, Kentucky, Florida, West Virginia and Tennessee. Clearly, the most polluted states are in the heavily industrialised sectors of the USA. As in Europe, where the polluters are polluting neighbouring countries, the same is true of the USA, which is polluting Mexico with acids from copper smelters in New Mexico and southern Arizona. Other parts of the USA are also guilty. In California, in the town of Corona del Mar south of Los Angeles, a winter fog was recorded in 1984 with a pH of 1.69.

The eastern parts of Germany are the most heavily industrialised, with the former East Germany emitting an estimated annual per capita 235 kg of SO_2 – the total annual emissions being in the region of 4 Mt. Between 90 and 95 per cent of the SO_2 comes

from the combustion of lignite. In fact, this region is responsible for extracting more than 25 per cent of the world total of lignite and, not surprisingly, therefore, it constitutes its principal fossil fuel. In the former East Germany, acid damage is reported to have affected more than 10 per cent of its forests, with soil acidification being severe on the Czechoslovakian border in the Erzebirge Mountains. As is the case for the former Soviet Union, figures are not widely publicised and it is hard to substantiate any available data.

The former Czechoslovakia and East Germany are the biggest per capita producers of SO_2 in the world. Czechoslovakia produces an estimated quantity of more than 3 Mt of SO_2 annually.

Poland is probably the country most affected by acid pollution, due to the combination of its internally generated industrial emissions and because of its proximity to other major industrialised nations whose gaseous toxins contribute to the atmospheric pollution. At the 1984 Munich Conference, the Poles admitted that their annual SO_2 emissions were running at 4.3 Mt, with a projected figure of 4.9 Mt in 1990. Their problem is that Poland tends to extract and burn coals and lignite with high sulphur values; it has, however, planned an emission control programme from 1991, which includes the use of fluidised-bed combustion furnaces to reduce pollution levels.

The industrialised developing countries, such as those in Latin America, Southeast Asia, India

Plate 4.2 *(A) The processing plant and smelter at Chuquicamata in the Atacama Desert of northern Chile. Chuquicamata is the world's largest copper mine, and produces 150,000 tonnes per day of copper sulphide ore. (B) Sulphur dioxide and other fumes from the smelter are often carried by prevailing winds into the giant (4 × 2.4 km wide, 0.7 km deep) open-pit copper mine, where they combine with diesel fumes from heavy machinery to produce a thick, acrid smog, even on the brightest sunny days. The desert environment means that the pollution effects of the acid deposition are not monitored and, therefore, the environmental damage remains unassessed.*
Courtesy of Jeremy P. Richards, University of Leicester.

and southern Africa, all have growing problems of acidic deposition and acidified soils due to industrial pollution. China, for example, has an annual SO_2 emission of more than 18 Mt, with only the USA and the old Soviet Union having greater levels. And 85 per cent of this SO_2 comes from coal-burning power stations and industries.

Britain produces annual SO_2 emissions of about 3.5 Mt, with approximately 1.7 Mt of NO_x, at 1984 figures. Rain in Britain is typically 100–150 per cent more acidic than unpolluted rain, with figures from Bush in Scotland showing the rain to be more than 600 times as acidic! Acid pollution in Western Europe is also a major problem, with the 19 constituent countries being in close proximity and, generally, heavily industrialised.

In Western Europe, another source of acidic deposition is nitrogen oxides pollution, with 30–50 per cent attributed to emissions from motor vehicles, and 30–40 per cent from fossil-fuel-burning power stations. Evaporation from nitrate-based fertilisers used in agriculture can contribute significant proportions. Where fossil fuels are burned, the greater the combustion temperature, the larger the amount of nitrogen oxides given off. Where light oils and gas are burned, almost 100 per cent of the nitrogen is oxidised, 40–50 per cent in heavy oils and 5–40 per cent in coal.

While the role of SO_4^{2-} ions in acidification of soils and waters, and the reversibility of SO_4^{2-}-induced acidification is now measurable with reasonable accuracy, the effects of increased NO_3^- leaching from terrestrial ecosystems and their influence on acidification is less well understood (Wright and Hauhs 1991). Figure 4.7 shows, schematically, the causal chain linking SO_2 and NO_x emissions to soil acidification, and forest and aquatic effects. In temperate coniferous forests, nitrogen acts as a growth-limiting nutrient, therefore nitrogen inputs associated with acid deposition will probably be sequestered by the biomass and thus kept in terrestrial ecosystems. Evidence appears to suggest that areas subject to prolonged heavy loads of SO_2 and nitrogen compounds become nitrogen-saturated, resulting in greater amounts of the nitrogen input being leached into surface and ground-water run-off rather than being retained in terrestrial ecosystems (*ibid.*). Nitrogen saturation can have either a natural (e.g. from fires) or an anthropogenic cause (e.g. clearing vegetation cover). A study from the Hartz Mountains, Germany, measuring nitrate concentrations in run-off at two catchments, Dicke Bramke and Lange Bramke, shows the steadily increasing levels since the late 1970s and mid-1980s, respectively, despite constant nitrogen deposition (Figure 4.8), the increased leaching of NO_3^- apparently being associated with forest decline (*ibid.*). Similar results have been reported from other regions, for example Norway.

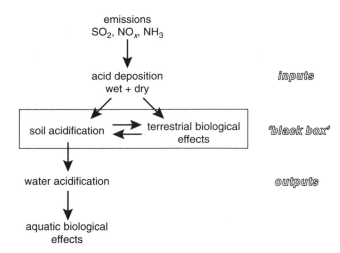

Figure 4.7 *Causal chain linking emissions of SO₂ and NOₓ to soil acidification, forest effects and aquatic effects. Reversibility of acidification may be delayed at one or more links. Some of the effects caused by long-term acidic deposition may be irreversible. A traditional approach to the study of cause–effect relationships has been to treat the terrestrial ecosystem as a 'black box'. Inputs and outputs are measured and used to deduce processes occurring within the box. Redrawn after Hauhs and Wright (1987) in Last and Watling (1991).*

Volcanoes as natural contributors to acidic deposition

Volcanoes are the major natural contributory source to acidic deposition, often outweighing present anthropogenic emissions of sulphur dioxide. Indeed, in the geological past, major phases of volcanic activity on Earth have injected enormous quantities of SO₂ into the atmosphere to affect global climatic and oceanographic conditions.

SO₂ from volcanic eruptions can significantly affect the Earth's atmospheric chemistry and climate. Non-explosive volcanoes out-gas at relatively constant rates (e.g. Mount Etna) and contribute an estimated annual SO₂ flux of about 9 Mt (Stoiber *et al.* 1987). In contrast, explosive volcanic eruptions, such as Mount Pinatubo, are more difficult to quantify, with estimates ranging between 1.5 and 50 Mt, i.e. 15–30 Mt of SO₂, with oxidation of the SO₂ to sulphate aerosols within 1–2 months (McCormick and Velga 1992) – an explosive eruption of 3–5 km³ of total dense-rock equivalent, with sufficient sulphur to form 20 Mt of SO₂ cloud in the stratosphere (Westrich and Gerlach 1992), up to approximately 25 per cent of the current total anthropogenic SO₂ emissions (Bluth *et al.* 1993). Using data from the Total Ozone Mapping Spectrometer (TOMS) aboard the NASA

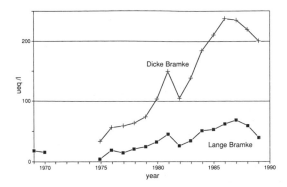

Figure 4.8 *Nitrate concentrations in run-off at two catchments: Lange Bramke and Dicke Bramke in the Hartz Mountains, Germany. At Lange Bramke, NO₃⁻ levels have increased steadily since the mid-1980s despite constant nitrogen deposition. At the adjacent catchment, Dicke Bramke, the increase began several years earlier and concentrations are now about twice those measured at Lange Bramke. Increased leaching of NO₃⁻ is apparently associated with forest decline. Redrawn after Hauhs (1990) in Last and Watling (1991).*

Nimbus 7 satellite over the period November 1978 to May 1993, Bluth *et al.* propose an annual flux of about 4 Mt SO₂ from explosive volcanism, less than 50 per cent of that derived from non-explosive volcanism. The total mean annual flux of SO₂ into the atmosphere, therefore, is about 13 Mt, or approximately 5–10 per cent of the current anthropogenic emissions (*ibid.*).

Acidified ground

Acidic deposition may fall hundreds of miles away from the source of the pollution. In extreme cases, the acidity can be reduced or neutralised by falling on alkaline soils, for example those associated with limestone areas. It is also possible to reduce the acidity by various immobilising and buffering chemical reactions, even in slightly acidic soils such as are typical of the evergreen forests of northern Europe, the USA and Canada.

One such buffering reaction is referred to as cation exchange, in which positively charged ions, or cations, substitute for the hydrogen ions (H⁺) in the acids. Such substitution reactions involve the replacement of hydrogen ions typically by calcium, magnesium and various other metals. These 'substitute' metal ions are released from rocks and minerals during chemical weathering, and so are available to replace the hydrogen ions in the acids. All these reactions occur in an aqueous or water phase, i.e. rain water or ground water. Clay minerals in the soil can release electrically charged particles or ions, such as the acid-producing hydrogen (H^+) and alkali-producing hydroxyl (OH^-) ions. These ions can then react with other minerals and ions in the soil or bedrock to release particles that can cause either a positive or negative feedback. As an example, if clay minerals release calcium ions (Ca^{2+}), these can reduce the acidity of the pore waters in chemical reactions that tend to remove the excess H^+ ions.

In order to evaluate the response of forest ecosystems to changing chemical environments, it is important that the total cation export from a watershed can be resolved into the component produced by chemical weathering reactions of the minerals and that resulting from the removal of exchangeable (plant-available) cations in the soil. One potential method for discriminating between these components is by using $^{87}Sr/^{86}Sr$ ratios to fingerprint the cation sources. Using this technique in high-elevation forest ecosystems in the Adirondack Mountains, New York state, Miller *et al.* (1993) showed that mineral chemical weathering reactions release about 70 per cent and soil cation-exchange reactions about 30 per cent of the annual strontium export. Additionally, by applying these results and the ratios of the principal cations to strontium in the local glacial till, Miller *et al.* estimated the release of Ca^{2+}, Mg^{2+}, K^+ and Na^+ from chemical weathering, and they concluded that the annual loss of cations from the total soil cation-exchangeable reservoir appears to be replaced by present rates of chemical weathering, meaning that the watershed is not suffering an immediate threat from acidification caused by atmospheric acidic deposition. Since the strontium data suggest that 50–60 per cent of the strontium in the organic soil horizon exchangeable and vegetation cation reservoirs has an atmospheric provenance, any reduction in the atmospheric cation inputs, associated with the continued strong acid anion inputs, could lead to a significant depletion of this cation reservoir (*ibid.*).

Carbon dioxide (CO_2) is dissolved in ground water and rain water, which upon reaction with rocks and minerals during chemical weathering releases metal ions. During chemical weathering of rocks, negatively charged radicals, called anions, are also released, such as bicarbonate ions (HCO_3^-). The addition of sulphuric acid to soil allows the sulphate radicals (SO_4^{2-}) to displace and substitute for the calcium and magnesium ions in rocks and minerals, thereby freeing them for other chemical reactions.

The rate of these substitution (or cation-exchange) reactions depends upon many interrelated factors, the principal ones being bedrock geology, soil, vegetation and the chemistry and characteristics of the precipitation. In certain climatic belts and soil types, the ground-water run-off will not lose its acidity, because either the appropriate chemical reactions are inhibited or the environment is already too acidic. Tundra landscapes, where the ground is frozen for much of the year, are just such regions. Earth scientists are beginning to develop quantitative methods for assessing the susceptibility of a soil to acidification, called its **acid susceptibility**, which is based on its chemistry. Clearly, this approach will allow much better modelling of cause and effect for predicting the consequences of acidic deposition upon different soils or bedrock.

There are two main sources of the hydrogen ions (H^+) that cause acidity. One is external, that is from the atmosphere, and it deposits sulphuric acid (H_2SO_4), hydrochloric acid (HCl), carbonic acid (H_2CO_3) and nitric acid (HNO_3). The other is from within the soil itself, resulting from microbial activity releasing CO_2, which then reacts with water to produce carbonic acid and organic acids. Much more research is needed to evaluate the relative importance of both sources of acidity in ground water.

The bedrock geology may be very important in inhibiting cation-exchange reactions. Areas of thin and immature soil cover above quartz-rich sandstones and granites are good examples. In such soils and bedrock, the quartz (SiO_2) is not only very resistant to weathering, but it also does not contain the all-important metal ions that are necessary for cation exchange. Thus, ground-water run-off through such sandy soil or bedrock is unable to buffer the acidity of the water. Ground-water run-off in these regions will, therefore, reach rivers and lakes essentially in its original acidic state, without the benefit of buffering reactions to reduce the acidity.

Many biogeochemical cycles for nitrogen ignore the importance of nitrogen sources from rocks as an important component in the acidification process. In the Klamath Mountains of northern California, where soils have a pH of less than 4.5, the oxidation of ammonium ions released from mica schist bedrock produces significant amounts of nitric acid, leading to the leaching of nutrient cations and the mobilisation of potentially toxic levels of aqueous aluminium (Dahlgren 1994). In neighbouring healthy forest soils, nitrate sequestration by plants ameliorates this effect, but in impoverished soils erosion and leaching have caused a serious depletion in nutrients, impeding plant recolonisation – possibly triggered by a natural perturbation such as a small forest fire (*ibid.*).

Countries such as the UK can be divided up into various areas with different degrees of susceptibility to developing acidic ground and ground waters, the so-called hard- and soft-water areas. The large parts of the UK underlain by limestone are unlikely ever to become seriously affected by acidic ground waters, because the limestone minerals should be able to buffer any percolating acidic waters. Thus, the large areas where ground waters are abstracted for public use from chalk (e.g. southern England), Jurassic limestones and other carbonate aquifers (throughout much of central and eastern England), and the Carboniferous limestones (mainly of central and northern England), are unlikely to develop serious problems with acidic waters. By the same token, these areas have ground waters with a lot of dissolved mineral salts such as calcium carbonate, and it is these that end up being precipitated out of solution in water pipes, kettles and other domestic appliances to form 'scum' or 'scale' from so-called hard water. Soft water comes from areas where there are relatively few dissolved mineral salts, where the water tends to be more acidic, such as upland regions dominated by such rocks as slate and granite.

Where thick, well-developed, or mature, soil profiles exist, as in many temperate latitudes, the soils may contain large amounts of electrically charged cations, which can react with the nitrate or sulphate ions to buffer the acidity of the precipitation and percolating ground water. As long as the buffering capacity of these soils is not exhausted, then there is a good chance that the ground-water run-off reaching rivers and lakes will have lost at least some of its acidity.

Areas in Britain where the bedrock geology is mainly granite or sandstone, however, do contain 'low-alkalinity' or acidic waters. Examples include the ground water associated with Quaternary glacial and fluvial sands and gravels, Tertiary sands, Carboniferous Coal Measures sandstones and Millstone Grit, Cretaceous Greensand strata, Devonian sandstones (especially in southwest England), the Middle Jurassic sands of North Yorkshire, Lower Palaeozoic and Precambrian strata (apart from much of the Silurian), and granites. Such areas tend to be characterised by acid surface waters.

An important aspect of areas with acid ground water is that corrosion is more pronounced in metal well linings and pipes because these regions are associated with the enhanced mobilisation of toxic metal or base elements from any bedrock that the acidic waters come into contact with. Lead (Pb), copper (Cu), cadmium (Cd) and aluminium (Al) are some of the better known of these toxic metals.

Government reports in Britain suggest that, in addition to industrial pollution of the atmosphere, the planting of large areas of coniferous forest in Scotland, Wales and western England has caused abnormally high levels of acidity in nearby lakes (reported in *The Guardian,* 15 April 1989): it appears that forests with a closed canopy are capable of reducing acid decomposition in water by as much as 30 per cent in sensitive moorland areas. A closed canopy reduces the available sunlight reaching the ground and, therefore, causes a decrease in surface temperatures, slowing the rate of acidic deposition.

Many forest trees accelerate the rate of release of acids and potentially harmful aluminium into the soils, which are then washed into rivers and lakes. Also, the increased acidity of lake waters encourages the release of poisonous heavy metals into the waters. This is because the acidic waters allow more effective leaching of soils such that the exchangeable cations (such as Ca^{2+} and Mg^{2+}), which may help to reduce the acidity or neutralise the waters, are quickly released into ground-water and stream run-off. In short, these exchangeable cations are not allowed to stay around long enough to do the job of reacting with the acids and reducing their potency. Currently, about 8 per cent of Britain is aforested, with most new or proposed plantations in upland areas. As aforestation continues, so the problem of acid soils in such areas can only worsen.

Acid rivers and lakes

Ground water may lose its acidity upon entering lakes with the right sort of chemistry. If the water body contains bicarbonate and other negatively charged basic ions from the chemical weathering of rocks and minerals, then they can neutralise the incoming acidic water. Earth scientists refer to this as water's **acid-neutralising capacity**, or **ANC**. The ANC value of a lake is therefore a measure of its susceptibility to becoming acidic. Rivers and lakes with a high ANC may be immune to acidification and neutralise the impact of acidic deposition. If the ANC value is zero, then such water bodies are very susceptible to becoming acidified as a result of acidic deposition, especially if they are sited near to the pollution sources.

Acidified rivers and lakes have tell-tale chemical signatures: the water tends to have a lower pH or alkalinity, and it contains greater amounts of sulphates and aluminium for any level of base cations; the indigenous organisms may well show signs of alteration in terms of physical characteristics, population density and types of communities, and there may even be little or no aquatic life in the case of very polluted water systems. In one of the earliest conclusive pieces of research aimed at investigating acidification and fish populations, a study of 26 streams and 22 lochs in Galloway, southwest Scotland, Harriman *et al.* (1987, but see also Harriman and Morrison 1982) showed that the absence of brown trout was linked to acidic deposition, in some instances exacerbated by aforestation and not nutrient availability, over-fishing, problems of access or suitable spawning grounds. Some seasonality may be present in a lake's chemistry in that at certain times of the year it has a high ANC value, which is reduced by buffering reactions during other months. These lakes are clearly potentially on the brink of becoming acidified permanently and need careful monitoring and precautionary measures to stop their deterioration.

High levels of aluminium in lakes may not always be due to acidic deposition releasing this element. Fluorine released into water from bedrock such as granite, for example, can increase the ability of aluminium to be dissolved into the water.

Acidification of fresh water has deleterious effects on the activities of microbes, plants and animals in poorly buffered ecosystems (Muniz 1991). The biota can be influenced directly, or in more subtle ways, e.g. causing a change in the balance between acid-sensitive and acid-tolerant species at different trophic levels. Figure 4.9 shows just one example of how acidity affects the number of species of zooplankton in lakes in southern and western Norway.

One approach to evaluating lake water acidification utilises palaeolimnology, for example diatom microfossils in the lake sediments, which provide an excellent means of measuring past water chemistry. Diatom species and abundance levels are affected by pH. For example, studies of diatom fossils from lake sediment cores from the Round Loch of Glenhead, Scotland, with dates interpolated using ^{206}Pb dating, show a relatively stable pH of 5.6 until about 1850, after which there was a reduction to a pH of approximately 4.8 (Figure 4.10) (Jones *et al.* 1989, Birks *et al.* 1990). This technique has been extended to many lakes in the UK, Europe and North America and has led to a better understanding of the dose–response relationship (see Battarbee and Allott 1993, and references therein).

In order to reconstruct past levels of acidity, Quaternary lake sediments have been accurately dated and analysed for many different components using diatoms and fly-ash particles (e.g. work on Holocene lake sediments in Scotland by Battarbee (1984, 1986, 1992), and from North America and Europe by Charles *et al.* (1989)).

In the United States, a committee of the National Academy of Sciences (NAS) reported in 1986 on the pH and alkalinity of the water, which roughly equates with the ANC and is a measure of the buffering ability of the water, in hundreds of lakes in New York, New Hampshire and Wisconsin. The committee compared data from the period of the 1920s to 1940s to that of the present day (see report, *Acid Deposition:*

BOX 4.2 NITROGEN DEPOSITION AND FOREST DECLINE

One of the main problems in evaluating the effects of nitrogen deposition on forest ecosystems has been the difficulties associated with using a direct method of tracing the ultimate fate of the deposited nitrogen. The fate of nitrate deposited from the atmosphere into an ecosystem, referred to as $NO_3^-_{atm}$, is essentially controlled by the interaction between nitrification processes that dilute atmospheric nitrate, NO_3^- consumption via plant uptake, denitrification, and fixation by uptake in microbial biomass. Nitrate originating in the atmosphere possesses isotopic abundances of nitrogen and oxygen that are measurably different to those of soil nitrate. While nitrification is commonly believed to be a precursor to NO_3^- leaching in soils (e.g. Gundersen 1992), some studies suggest that NO_3^- output in spring water appears to be linked to nitrogen deposition from air pollution (e.g. Driscoll *et al.* 1989). Typical atmospheric nitrate values, $NO_3^-_{atm}$, of $\delta^{18}O$ range between 52.5 and 60.9‰, compared with nitrate values from microbial nitrification in soil, which vary

between 0.8 and 5.8‰ – as for every three oxygen atoms two originate from soil water and one from the atmosphere. Utilising such isotopic fingerprinting techniques in a study of the isotope ratios of nitrogen, $\delta^{15}N_{nitrate}$, and oxygen, $\delta^{18}O_{nitrate}$, in spring water from eight forest watersheds in northeast Bavaria, Germany, in healthy, slightly declining and limed sites, it was found that only 16–30 per cent of the nitrate in spring water came directly from the atmosphere without being processed in the soil, but in the severely damaged sites, virtually all the atmospheric nitrate ended up in the spring water (Durka *et al.* 1994). The results of the Bavarian study suggest that:

- acid-induced forest decline significantly inhibits the consumption of nitrate by micro-organisms in the soil and trees;
- liming soils to ameliorate the effects of acidification can restore the consumption of atmospheric nitrate; and
- in limed ecosystems, total nitrate output remains high due to internal nitrate production within that system.

(*ibid.*)

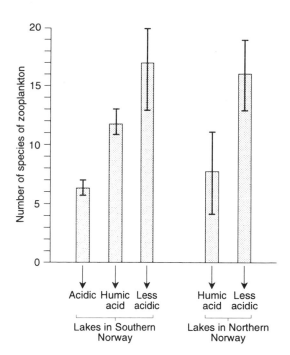

Figure 4.9 *Association between acidity and numbers of species of zooplankton, including Crustacea, Rotatoria and Chaoborus larvae, in lakes in southern and western Norway. Redrawn after Raddum* et al. *(1980) in Last and Watling (1991).*

Long-Term Trends, by the Committee on Monitoring and Assessment of Trends in Acid Deposition, 1986; Mohnen 1988). The committee found that, on average, pH and alkalinity has increased in the Wisconsin lakes while remaining essentially the same in New York and New Hampshire. In New York state, however (including the Adirondack Mountains), a trend towards acidification of many lakes was recorded. Acidified lakes were also found in eastern Pennsylvania and Michigan.

The National Academy of Sciences committee was also able to make some further disturbing observations concerning the aquatic life in these lakes. In particular, in eleven of the sampled lakes in the Adirondack Mountains, a change in pH was accompanied by a change in the assemblages of brown algae and diatoms. Six of these eleven lakes had demonstrably become more acidic since the 1930s, with present pH values of less than 5.2. The most rapid acidification occurred in the period immediately prior to the 1980s. Acidic deposition was the only plausible explanation that the committee could find for this acidification.

In the Adirondack Mountains, the lakes have low pH values, and therefore are very acidic, both because

of the extremely acidic deposition in western New York state (as low as pH = 4.1), and also as a result of the quartz-rich, granite-floored lakes and soils with their poor buffering ability. This is a good example of how the bedrock geology conspires to make the ground water maintain its acidity after the acidic deposition has occurred.

Many lakes and streams, especially those at high altitudes, are particularly sensitive to *episodic acidification*, which occurs during brief periods of low pH levels associated with snow-melt or heavy rainfall. In the central Appalachians, the Mid-Atlantic coastal plain and the Adirondack Mountains, many additional lakes and streams become temporarily acidic during storms and snow-melts, leading to large-scale mortality in fish populations. In the Adirondack Mountains, approximately 70 per cent of sensitive lakes are at risk of episodic acidification. This is more than three times the number of chronically acidic lakes. Furthermore, during an acidic episode in the central Appalachians, approximately 30 per cent of the sensitive streams arc likely to become acidic (EPA World Wide Web Site).

There are, however, large numbers of acidic lakes, such as in Florida and southern New England, which are believed to owe their acidity not to acidic deposition but to organic acids produced as vegetation decays. In part, the acidity of these lakes is also as a result of pollution from agricultural fertilisers finding their way into the water.

In the eastern USA, geochemical investigations of selected watersheds, hydrological flow paths (e.g. rivers and subsurface seepage) and lakes suggests that the bedrock geology is probably the most important factor in controlling the chemistry of the surface waters and, therefore, water acidity. Rainfall, or wet precipitation, can be very acidic. Individual storms with a pH as low as 3.4 and up to 5.6 have been recorded in this area.

Threatened forests and woodlands

Acidic deposition is responsible not only for adversely affecting many of the world's rivers, lakes and ground water, but it also appears to be threatening large parts of the flora, particularly trees in Europe and North America. In the former West Germany, where the effect of acidic deposition on trees has been so dramatic in the past few decades, they have coined the term *waldsterben,* or 'forest death'.

False-colour satellite imagery is used to detect damaged crops and forests in various parts of the world. This technique is useful because in the infrared

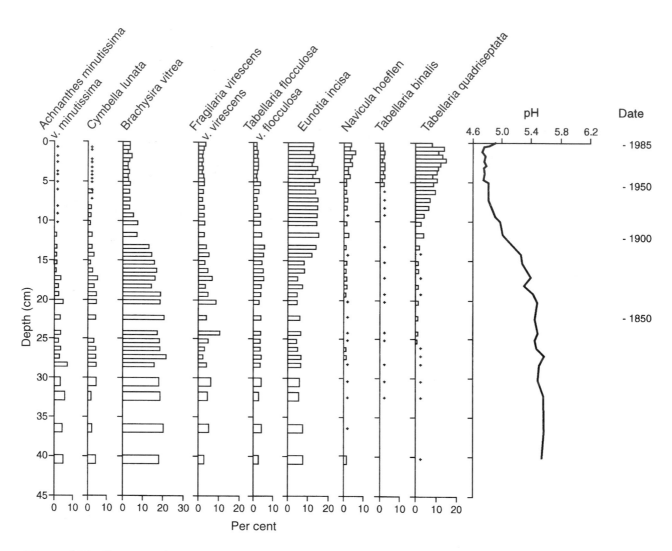

Figure 4.10 *Changes in diatom assemblages and reconstructed pH for a sediment core from the Round Loch of Glenhead, Scotland. Dates are interpolated from ^{210}Pb ages. Redrawn after Jones* et al. *(1989) and Birks* et al. *(1990) in Battarbee (1983).*

part of the spectrum, it is most sensitive to varying amounts of chlorophyll, which is produced during photosynthesis in plants. The satellite pictures have revealed large areas of dying and diseased woodland and forest, probably caused by acidic deposition.

In North America, statistics on trees in the Adirondack Mountains, the White Mountains of New Hampshire and the Green Mountains of Vermont have shown a death rate of the red spruce – which grows above 850 m altitude, where it is most susceptible to the effects of acidic deposition – exceeding 50 per cent in the last 25 years. Below this altitude, there has also been a deterioration in the health of both softwood and hardwood trees.

Earth scientists do not really understand the chain of chemical reactions leading to the death or loss of vitality in many parts of the world's woodlands and forests. Possible reasons include the leaching of potassium, calcium, magnesium and other important chemical elements from the soil by percolating acidic ground water derived from acidic deposition. Without such chemical nutrients, the health of trees suffers and they die. Acidic deposition can also upset the ecosystems in soil by causing the impoverishment and death of micro-organisms that help to release the all-important nutrients. Percolating acidic ground water may also release large quantities of aluminium, which then competes successfully with other elements such as calcium for binding sites in the roots of trees. Tree growth then slows down. High nitrate levels in soils, for example from nitric acid, can affect the symbiotic fungi that colonise the roots of conifer trees, providing both nutrients to the trees and helping to protect them against disease. Acidic depo-

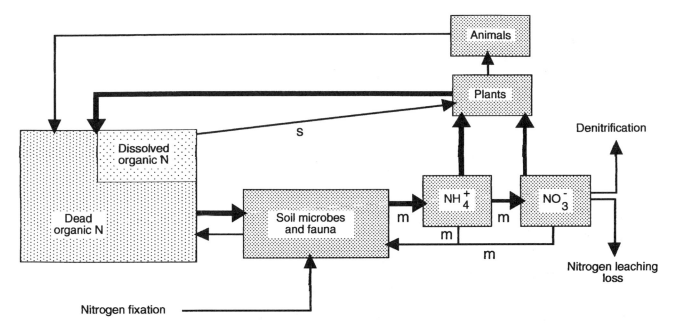

Figure 4.11 *Main fluxes (arrows) and sinks (boxes) for nitrogen in terrestrial ecosystems. Arrow thickness is approximately proportional to the magnitude of the net flux. 'm' denotes microbial controls, conventionally believed to limit the rate of nitrogen cycling. The path marked 's' short-circuits these conventional microbial controls and may be important as an alternative pathway for the transfer of nitrogen within infertile ecosystems. Redrawn after Northup et al. (1995).*

sition also affects the leaves of trees and other plants by damaging the plant cells. Acidic rain or acidic clouds and fog will leach out nutrients such as calcium, potassium and magnesium from conifer needles at a faster rate than they can be replenished through the tree's root system.

Not all acidified soils are hostile to plant life. An interesting aspect of plant communities that have become adapted to infertile and strongly acidic soils is their sustainability despite low nitrogen availability and propensity for the loss of nitrogen from the ecosystem. In a study of the pygmy forest in the Ecological Staircase, northern California, a series of coastal terraces, Northup *et al.* (1995) found that a pine (*Pinus muricata*) can strongly influence the release of dissolved organic nitrogen (DON) in soils through the production of polyphenols in leaf litter. Plants adapted to relatively infertile and acidic soils can utilise phenol-bound organic matter to give them a competitive edge over plants that use only mineralised nitrogen – produced by the action of soil microbes and other fauna on dead organic nitrogen converted to ammonium (NH_4^+) and then nitrate (NO_3^-). This study, for the first time, showed that plants inhabiting infertile and acidic soils can obtain their nitrogen directly from organic sources, thereby effectively by-passing the mineralisation step which has traditionally been regarded as the rate-

determining step in terrestrial nitrogen cycling (Figure 4.11).

Recovery and sensitivity of ecosystems

Studies, particularly from North American (Canadian) and Norwegian (e.g. the Reversing Acidification in Norway – RAIN – project) lakes, are now showing that recovery or reversal of acidified ground and ground-water/run-off can occur following reductions in acidic deposition, the rate depending on the sensitivity and degree to which ecosystems have been affected (Wright and Hauhs 1991). Table 4.4 shows examples of surface water acidification and its reversal, the results suggesting continued acidification and slightly decreased acidic deposition, but reversal and recovery in response to large decreases in deposition.

It is possible to construct maps to indicate the sensitivity of different areas to acidic deposition, for example Figure 4.12 shows the relative sensitivity of ecosystems in Europe to acidic deposition (after Chadwick and Kuylenstierna 1990), based on a scale of 1 to 5, the least and most sensitive, respectively.

The five sensitivity classes of ecosystems to acidic deposition proposed by Hornung *et al.* (1986) are shown in Table 4.5. More sophisticated attributes of these five sensitivity classes have been added by Chadwick and Kuylenstierna (1990), by focusing on four groups of factors, each of which is subdivided into two or four categories:

1 bedrock geology;
2 soil type;
3 land use; and
4 amounts (not quality) of rain.

Arguably, these refinements actually add more subjectivity to the sensitivity classification (Last 1991).

Another facet of sensitivity studies is in the use of models to predict the likely future trends in acidic deposition. For example, Figure 4.13 shows current estimated annual deposition of sulphur from natural and anthropogenic sources. Using predictions that by the year 2020, rates of sulphur emissions due solely to population increases will be about 30 Tg S yr^{-1} (Galloway 1989), and assuming a modest increase in per capita energy consumption, Galloway and Rodhe (1991) forecast that global sulphur emissions will increase to about 100 Tg S yr^{-1}, with an environmental impact covering a much greater area than the

regions that will have most of the increased population, namely Asia, South America and Africa. A realistic increase in sulphur emissions of 15 Tg S yr^{-1} in Southeast Asia, for example, could alter global sulphur deposition as shown in Figure 4.13. The value of 15 Tg S yr^{-1} was chosen because of the intense convective vertical mixing of air currents over Southeast Asia, resulting in the upward transport to the free troposphere of substantial amounts of sulphur, which then, due to the stronger and more prolonged winds, can ensure the horizontal dispersal over considerable distances across the low-latitude belts compared with emissions in more temperate, higher latitude regions. Figure 4.14 summarises the global problem of acidic emissions and precipitation in the late 1980s, revealing the regions of acid emissions, and regions where soils are likely to suffer damage by acidic deposition, together with present and future problem areas (after Rodhe *et al.* 1988). Figure 4.15 shows detailed variations in the acidity of rain in Europe and North America.

Recoverability and *reversibility* within ecosystems depends not only on reduced acidic deposition but

Table 4.4 *Summary of examples of surface water acidification and its reversal. The examples are grouped by recent trends in the amounts of acidic deposition. The data suggest continued acidification with constant and slightly decreased acidic deposition, but reversal and recovery in response to large decreases in deposition.*

Acid deposition trend	Start year	End year	No. years	SO$_4$* deposition (meq m^{-2} yr^{-1}) start	end	change (%)	Surface water ueq l^{-1} (start year) Ca*+ Mg*	SO$_4$*	alk	(end year) Ca*+ Mg*	SO$_4$*	alk	Fobs.	Trend, surface water acidification
Constant														
White Oak Run	1980	1987	8	100	115	15	63	77	24	67	94	12	0.24	Acidifying
Slight decrease														
Plastic Lake	1979	1989	11	77	62	−19	146	137	21	141	135	7	2.50	Acidifying
Large decrease														
Clearwater Lake	1973	1989	17	140	100	−29	460	575	−49	365	313	−22	0.36	Recovering
Hubbard Brook	1965	1986	22	135	90	−33	76	124	−25	63	102	−23	0.59	No change
Loch Enoch	1978	1988	11	75	50	−33	19	62	−79	16	27	−25	0.09	Recovering
RAIN KIM	1984	1989	6	58	16	−72	22	125	−108	1	35	−33	0.23	Recovering

Source: Wright and Hauhs (1991).

Table 4.5 *Sensitivity classes for ecosystems.*

	Soil/rock combination	Occurrence of acidic waters
Class 1	Acid soils over rocks with little or no buffering capacity	Acid waters will occur at all flow levels
Class 2	Acid soils over rocks with low buffering capacity	Acid waters likely at all flow levels
Class 3	Acid soils over rocks with moderate buffering capacity	Acid waters may occur at high flow levels
Class 4	Acid soils over rocks with infinite buffering capacity	Acid waters could occur at very high flows
Class 5	Non-acid soils over any rock type	Acid waters will not occur

Source: Hornung *et al.* 1986.

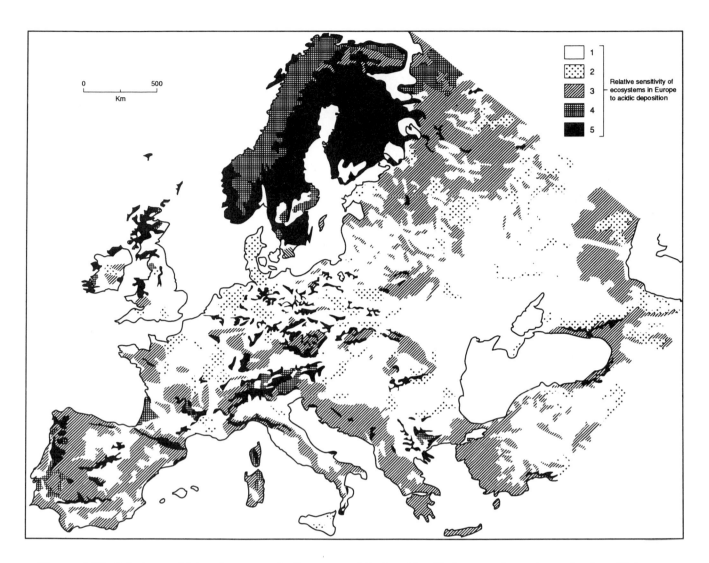

Figure 4.12 *Relative sensitivity of ecosystems in Europe to acidic deposition (1–5, on an increasing scale of sensitivity – see Table 4.5). A preliminary assessment of the sensitivities of aquatic and terrestrial ecosystems. Redrawn after Chadwick and Kuylenstierna (1990).*

is also a function of the sensitivity of catchment soils. In most cases, the most acidified lakes will show the most rapid initial improvement in response to decreased loads, although complete recovery may take much longer. Methods of measuring water acidification include the **Henriksen empirical chemical approach**, involving a comparison of the original pre-acidification alkalinity and present alkalinity, since acidification equates with the loss of alkalinity, such that the **acidification index, AI**, is given by:

$$AI = A_o - A_t$$

where A_o is the original pre-acidification alkalinity, and A_t is the present alkalinity.

Another method of measuring water acidification is by using a palaeolimnological approach. Computer

models are available to simulate the physio-chemical conditions resulting from sulphur loading of catchments, e.g. the Model of Acidification of Groundwaters in Catchments (MAGIC) (Cosby *et al.* 1985). For data input, these models require the amount and characteristics of the rainfall, run-off, soil and the history of acidic (sulphur) deposition. Once calibrated, such models have the potential to predict future responses to specified acidic loads, and to reconstruct past histories.

Perhaps one of the most natural means of reversing the acidification of lakes is stimulating primary biological productivity by the addition of phosphate fertiliser. Acidified lake waters can be neutralised or have their acidity reduced by adding a base (expressed as alkalinity), but the resulting calcium-rich water

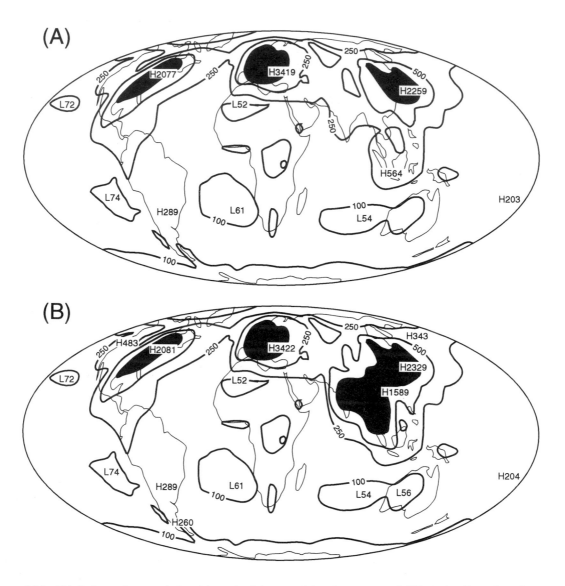

Figure 4.13 (A) Estimated annual deposition of sulphur resulting from natural (58 Tg S yr⁻¹) and anthropogenic sources (10 Tg S yr⁻¹). Units are in mg S m⁻² yr⁻¹. (B) Estimated annual deposition of sulphur following an assumed increased emission of 15 Tg S yr⁻¹ in Southeast Asia. Other emissions are in (A). 'H' and 'L' deposition refer to maximum and minimum depositional amounts, respectively. Redrawn after Rodhe et al. (1991) in Last and Watling (1991).

supports a flora and fauna that are different to that in natural soft-water lacustrine ecosystems. In an experiment in Seathwaite Tarn in the English Lake District, Davison *et al.* (1995) added phosphate fertiliser to the lake to stimulate primary production, which generated sufficient base to permit the assimilation of nitrate to increase the pH of the acidified lake waters without radically changing the natural ecosystem. Phytoplankton growth was not excessive and increased biological productivity was observed at all trophic levels, with the predicted additional benefit that further base should be generated

by the anoxic decomposition of organic matter on the lake bed (*ibid.*).

Mopping up the mess: clean technologies

The philosophy behind national approaches to reducing acid deposition, together with other forms of environmental damage, varies considerably. In Germany, for example, the clean-up philosophy is to use the 'best available technology', whereas in the

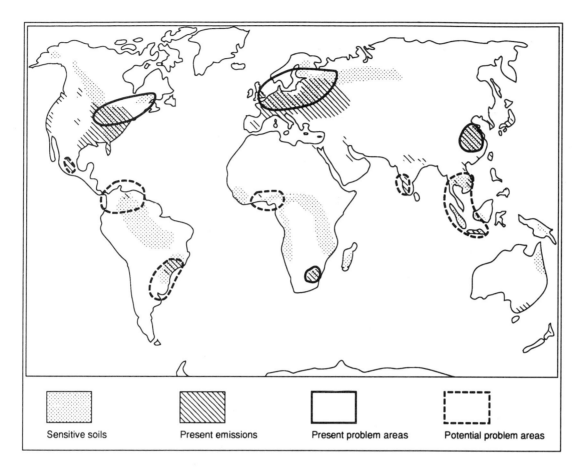

Figure 4.14 *The global problem of acidic emissions and precipitation during the late 1980s. The map shows regions of acidic emissions, areas where soils are likely to suffer damage from acid rain, and present and future problem zones. Redrawn after Rodhe et al. (1988).*

UK it is the 'best available techniques not entailing excessive cost' (known by the acronym BATNEEC).

So what can be done, and what is being done, to reduce the effects of acidic deposition? First, ways must be found to reduce the harmful chemical emissions from our traditional power stations, which run on fossil fuels such as coal. Second, industry and domestic consumers must burn 'cleaner' coals such as those with low levels of sulphur and other very harmful chemicals. These coals may be naturally low in sulphur or may have been treated by various washing processes. Such approaches have already produced lower levels of sulphur emission from conventional fossil fuel power stations in West Germany before 1990, and in North America and Japan.

In Germany, for example, wet limestone is introduced into the hot gaseous emissions of some power stations, where it can scavenge up to 90 per cent of the sulphur dioxide. This process, unfortunately, does not reduce the amount of nitrogen oxides given off

and it leads to a slightly reduced efficiency for the power station; the sulphur-rich waste also presents a disposal problem. The present solution is a classic case of swings and roundabouts. Nevertheless, society must weigh up the benefits and disadvantages of cleaning up the environment. Furthermore, research and development should continue to seek better, more environmentally sound means of cleaning up the atmosphere.

Other new technologies for reducing the noxious emissions from coal-fired power stations include the process called 'atmospheric fluidised-bed combustion'. In this process, a turbulent bed of coal and limestone particles is suspended (or fluidised) by high-velocity, upward-streaming air, with combustion occurring at a steady and lower temperature than in more conventional burners. As a result of this process, the amount of harmful nitrogen oxides is reduced, and much of the sulphur dioxide reacts with the limestone.

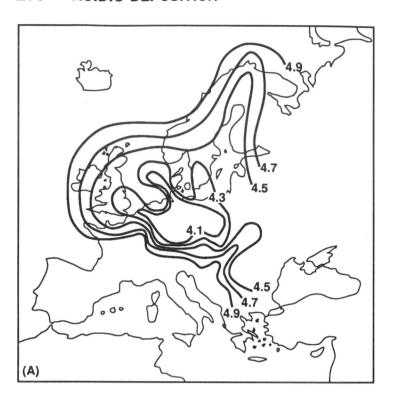

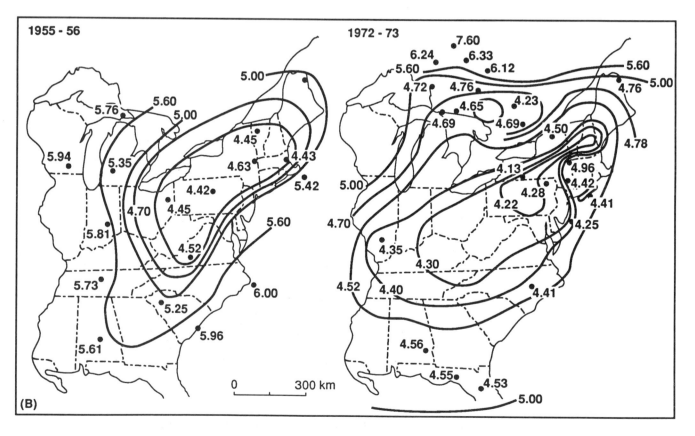

Figure 4.15 *Variation in the acidity of rain in (A) Europe, and (B) eastern North America. Numbers refer to pH. Redrawn after Blunden and Reddish (1991).*

In Germany, a new filtering process is being developed to remove both the SO_2 and NO_x from coal in conventional power stations. The technology appears capable of removing these pollutants from the cheapest and dirtiest lignites. The cleaning process uses very porous pellets of **active coke** about 5 mm in diameter, which catalyse reactions to remove the pollutants. The most impressive performance of this new technology is at a power station near Arzberg in northeastern Bavaria, which burns brown coal (lignate) from the former Czechoslovakia with 4,000 mg of sulphur m^{-3} of coal. The filtering process left only 10 mg m^{-3} of sulphur (Coghlan 1991).

In Britain, the electricity generating companies are currently evaluating the use of sea water to remove the sulphurous gases emitted by conventional fossil-fuel-burning power stations. Basically, the polluted flue gas goes through a pre-scrubber process to remove any residual ash, along with chlorides and fluorides. It then enters the main scrubber, or absorber, where the counterflow of sea water removes about 90 per cent of the sulphur dioxide from the gas. The acidified water is then funnelled into an aeration tank, where it is aerated to convert the sulphite to sulphate, neutralise the pH and drive off the carbon dioxide and oxygenated water. The cleaned effluent is then pumped into the sea, and the clean gas discharged. Although sea water could be used successfully in any desulphurisation process, the water would also pick up any heavy metals such as mercury and selenium from the coal, which could then find their way into the effluent discharged back into the sea. The disadvantages could outweigh the benefits. If these potential problems can be overcome, then the sea water method for desulphurisation could provide an alternative to the use of limestone, which generates mountains of gypsum.

On the domestic front, the noxious emissions from cars can be reduced by conservation, for example through limiting private vehicle use to only those journeys where public transport is not available. Naturally, this means some sacrifices in terms of convenience, but unless a longer-term perspective on atmospheric pollution and acidic deposition is taken, then the counter-measures that eventually have to be taken will prove all the more austere. Motor car manufacturers should continually strive, via research and development, to produce 'cleaner' internal combustion engines.

Governments should enact restrictive legislation that sets tough acceptable standards of emissions and offers financial incentives to encourage vehicle manufacturers to market 'cleaner' transport. It is also the responsibility of governments to make public transport convenient and economically attractive to the populace. While governments undeniably must take a considerable share of the responsibility for a less polluted atmosphere, individual members of society should not shirk the responsibility to conserve energy and, thereby, reduce the causes of atmospheric pollution and acidic deposition.

Major oil companies are now becoming much more environmentally aware. In April 1991, BP Chemicals Nitriles and Nitrogen division, for example, announced in its company newsletter, *BP Chemicals World*, a US$17 million plan to build a new nitric acid plant to replace the 35-year-old unit at the Lima Complex, Cleveland, in the USA. It is claimed that the new plant, which will boost Lima's nitric acid production capacity by 27 per cent to 90,000 tonnes a year, will result in a 95 per cent reduction in nitrogen oxide emissions. Clearly, if industrialists are really capable of tying increased production to significantly reduced environmental risk, then these are the sorts of developments that merit a cautious welcome.

In 1989, Britain committed itself to spending £1.8 billion in a programme to reduce acidic deposition by using a cleaning-up process in its conventional, fossil fuel power stations. The problem is that this new process may actually pose a different threat to some of the country's most beautiful countryside. Millions of tonnes of limestone will be required in order to extract the sulphur dioxide from the combustion furnaces. This limestone will have to come from areas of natural beauty, including the national parks of the Yorkshire Dales and the Peak District, where the geology is most suitable for obtaining the relatively pure limestone needed in the power stations.

Environmental groups are understandably concerned that a substantial expansion in the quarrying operations in these beauty spots will both desecrate such areas and lead to a drastic reduction in the ecological niches available for certain wildlife, particularly birds. Another major problem is that the chemical reaction of limestone and sulphur dioxide produces calcium sulphate ($CaSO_4$), more commonly known as **gypsum**. Although gypsum has industrial uses, such as in the production of plasterboard and as a road-fill material, large amounts have to be dumped as industrial waste.

Funding research into acidic deposition

The effects of acidic deposition are now widely appreciated, and some of the problems of acidic deposition are being tackled. In recent years, despite the move away from research into acidic deposition, with its attendant media interest on a very low back-burner, studies have shown that acid pollution in the coal-mining provinces of southwest China is now approaching levels found only in the most polluted parts of the United States. David Schindler, from the Canadian government's Freshwater Laboratory, believes that the acidification of large areas of the temperate lands, with far less diversity of species than in the tropical regions, makes them much more vulnerable to catastrophic disruption with the loss of a few species. Schindler also believes that acidic deposition may have made ecosystems such as Europe's forests more vulnerable to global climatic change associated with the greenhouse effect:

> Greenhouse warming may enhance the effects of acid precipitation on boreal lakes and streams. If greenhouse droughts cause water tables to rise, then a sudden rainstorm could release many years of accumulated sulphuric acid in a single deadly acid pulse.

The argument appears to be that droughts would lead to enhanced vertical movement (by capillary action) of water through the soil column because of increased evaporation at the surface. This would lead to increased salinisation. If soil ground water is reduced during droughts, then there would be accelerated oxidation of sulphides (e.g. iron sulphide or pyrite) and organic material previously below the water table. Consequently, any sudden rainfall could release the oxidised sulphides, as sulphates, into surface run-off as acidified waters.

The fallout of nitrogen in acidic deposition might also lead to the emission of greenhouse gases such as nitrous oxide from acidified soils. Scientists such as Melillo, from the Woods Hole Marine Biological Laboratory, Massachusetts, USA, has estimated that air pollution deposits about 18 Mt of nitrogen yr^{-1} in the temperate Northern Hemisphere, which has increased the natural emission of nitrous oxide. Excess nitrogen in acidified soils might reduce their ability to absorb one of the other important greenhouse gases, methane (CH_4). Such a reduction in the soil 'sink' for CH_4 could have been an important, yet little appreciated, factor in the doubling of atmospheric concentrations of methane since pre-industrial times (Pearce 1990).

A countervailing argument is that acidic deposition could act as a negative feedback mechanism to reduce the potency of the greenhouse effect. Nitrogen fallout in the oceans might allow increased biological activity so that the oceans could absorb greater amounts of atmospheric CO_2, the main greenhouse gas.

Wigley, Head of the Climatic Research Unit, at the University of East Anglia, UK, has suggested that sulphate in acidic deposition might act as a 'seed' for the formation of clouds that protect the planet from the Sun's rays, and so reduce any greenhouse effect. As Lovelock highlighted in his book *The Ages of Gaia*, marine plankton give off sulphurous particles, called dimethylsulphide, which also act as seeds in the formation of clouds. So, acidic deposition may actually exert some negative feedback mechanisms that reduce the amount of global warming (Charlson *et al.* 1992).

Acidic deposition needs continued study and monitoring if the prospects for the future are to be accurately predicted. Its harmful, polluting, effects need to be minimised. As the study of the greenhouse effect and global warming gathers pace, so too should research into any positive or negative feedback mechanisms that might enhance or reduce acidic deposition. From a brief look at research into acidic deposition, the following section deals with international action on acidic deposition.

The US Acid Rain Program

This section is based on data obtained from the US Environmental Protection Agency's (EPA) World Wide Web Site.

In an attempt to continue to reduce the problem of acidic deposition in the USA, Title IV of the Clean Air Act Amendments 1990 was implemented and overseen by the government-funded EPA. Title IV is referred to as the US Acid Rain Program and its primary goal is the reduction of annual SO_2 emissions to 10 Mt below 1980 levels by 2010. The Act also calls for a 2 Mt reduction in NO_x emissions by the year 2000.

The US Acid Rain Program provides an innovative model and a prototype that illustrates the effectiveness of legislation in the reduction and management of emissions that cause acidic deposition and associated pollution. The EPA gained broad input into the development of the Acid Rain Program by consulting representatives from the various stake-holders. Throughout its development and implementation it maintained an open-door policy to optimise its effectiveness. The ultimate goal of the Acid Rain Program is to achieve significant environmental and public health benefits through reductions in the emissions of SO_2 and NO_x, the primary causes of acidic deposition. The Acid Rain Program has three primary objectives:

- Achieve environmental benefits through reductions in SO_2 and NO_x emissions.
- Facilitate active trading of allowances and use of other compliance options to minimise compliance costs, maximise economic efficiency, and permit strong economic growth.
- Promote pollution prevention and energy-efficient strategies and technologies.

To achieve the reductions in SO_2 emissions, the law requires a two-phase tightening of the restrictions placed on fossil-fuel-fired power plants. Phase I began in 1995 and affects the largest power plants in the USA, associated with the greatest levels of atmospheric pollution. A total of 263 boilers at 110 mostly coal-burning electric utility plants, located in 21 eastern and mid-western states, and an additional 162 boilers, joined Phase I of the programme as substitution or compensation units, bringing the total of Phase I affected units to 425. Preliminary emissions data indicate that SO_2 emissions at these units nationwide were reduced by almost 40 per cent below their required level, that is, from 8.7 to 5.3 Mt of SO_2. Phase II will begin in the year 2000 and aims to tighten the annual emission limits on the larger, higher-emitting plants. It will also set restrictions on smaller, cleaner plants that are fired by coal, oil and gas.

The programme introduces an allowance trading system that harnesses the incentives of the free market to reduce pollution. Under this system the utility units are allocated an allowance based on their historic fuel consumption and a specific emissions rate. Each allowance permits a unit to emit 1 tonne of SO_2 during or after a specific year. For each tonne of SO_2 discharged in a given year, one allowance is retired, that is, it can no longer be used. These allowances may be bought, sold or banked. Regardless of the number of allowances a source holds, it may not emit at levels that would violate federal or state limits under Title I of the Clean Air Act to protect public health. If a source exceeds the SO_2 emission allowance it must pay a penalty of US$2,000 per tonne of SO_2 and offset the excess SO_2 emissions with allowances in an amount equivalent to the excess. The total allowances during Phase II of the programme will be 8.95 million, setting a permanent ceiling on emissions to ensure environmental benefits will be achieved and maintained.

The US Congress also established an Opt-in Program to include additional SO_2-emitting sources, allowing sources to enter the program on a voluntary basis, reducing their emissions and receiving their own acidic deposition allowances. The allowance trading system, therefore, provides an incentive for utilities to prevent pollution, since for each tonne of SO_2 that a utility avoids emitting, one fewer allowance must be retired. Utilities that reduce emissions through energy efficiency and renewable energy are able to sell, use or bank their surplus allowances. The Acid Rain Program also has a reserve of 300,000 SO_2 allowances, which may be awarded to utilities that employ efficient and/or renewable energy measures to produce electricity.

The NO_x programme embodies many of the same principles embodied in the SO_2 trading programme, but it does not 'cap' NO_x emissions as the SO_2 programme does, nor does it utilise an allowance trading system. Emission limitations for NO_x boilers focus on the emission rate to be achieved. Two options for compliance with the emission limitation are provided:

- Compliance with an individual emission rate for a boiler.
- Averaging of emission rates over two or more units to meet an overall emissions rate limitation.

These options give utilities flexibility to meet the emission limitations in the most cost-effective way and allow for the further development of technologies to reduce the cost of compliance. If the utility properly installs and maintains the appropriate control equipment designed to meet the emissions limitations established in the regulations, but is still unable to meet the limitation, the NO_x programme allows the utility to apply for an alternative emission limitation (AEL), which corresponds to the level that the utility believes to be achievable.

Like the SO_2 emission reduction requirements, the NO_x programme is implemented in two phases. Phase I of the programme began on 1 January 1996 and affects dry-bottom wall-fired boilers and tangentially fired boilers, with limitations of 0.50 and 0.45 lb of NO_x per MBTU (million British thermal units of heat input) averaged over the year, respectively. A significant proportion of the reduction in NO_x will be achieved by the installation of low-NO_x burner (LNB) technologies in coal-fired utility boilers. Phase II will begin in the year 2000 and will set lower emissions on boilers that were subject to regulation in Phase I, and it will establish new limitations to include cell-burner technology, cyclone boilers, wet-bottom boilers and other types of coal-fired boiler. The EPA continues to modify the targets under the Acid Rain Program. On 19 January 1996, for example, it issued a press release proposing a rule under the Clean Air Act to reduce NO_x emissions from electric power plants further, by 820,000 tonnes annually by the year 2000.

Under the Acid Rain Program, each boiler plant must continuously measure and record its emissions of SO_2, NO_x and CO_2, together with the volumetric flow and opacity of the exhaust. In most cases, a Continuous Emissions Monitoring System (CEMS) must be used to provide the EPA with unit reports of hourly emissions data at quarterly intervals. In April 1995, selected utilities and the EPA began to test direct electronic transmission of data via dial-up modem to the EPA computer system. In the future, this is expected to become the common method of data submission. The use of modems allows data to be recorded in the Emissions Tracking System (ETS), which serves as a repository of emissions data for the utility industry. This method should ensure confidence in the allowance transactions by certifying the existence and quantity of the commodity being traded, and it also ensures that NO_x averaging plans are working.

Throughout the USA, the US Geological Survey plays an instrumental role in operating and maintaining a nationwide network of precipitation monitoring stations. This programme, the National Atmospheric Deposition Program/National Trends Network (NADP/NTN), has operated since 1978 and is currently responsible for collecting weekly precipitation samples from approximately 200 sites nationwide. This programme is essential for evaluating the effectiveness of emission control efforts and in determining the trend in precipitation data, as well as assessing water quality at watershed, regional and national levels. It is essential that such programmes continue to provide a further check on the effects of emissions.

The environmental benefits of the Acid Rain Program are predicted to be quite considerable. The EPA has developed a Regional Acid Deposition Model (RADM) to assess the effectiveness of the benefits of the sulphate reductions under Title IV of the 1990 Clean Air Act Amendments (EPA 1995). It is designed to model the ambient sulphate aerosol concentrations for grid cells of 80 by 80 km over the entire eastern USA mathematically. Figure 4.16A shows the actual 1985 emissions, and Figure 4.16B shows the estimated 1997 emissions with Title IV implemented. The model then compares the probable concentrations and distributions of sulphate aerosols without Title IV (Figure 4.16C) and with Title IV (Figure 4.16D).

The RADM is a useful model to examine the effectiveness of Title IV and it allows the EPA to evaluate the likely human health benefits for the eastern USA. The vulnerability of sensitive ecosystems will be reduced, and the quality of surface water will be improved. The attraction of a region will be improved, and there should be a decrease in smog and increased visibility, particularly in the eastern USA. The corrosion of buildings and other property is expected to decrease, with a substantial saving in the costs of protective and remedial measures. Regional health is expected to improve – currently 25 per cent of particles inhaled comprise anthropogenic sulphate aerosols. A reduction in these is expected to reduce mortality and morbidity from lung disorders such as asthma and bronchitis. Although there is a large body of epidemiological literature on the relationship between atmospheric particulate matter and health effects, there is much debate over the extent to which sulphate aerosols may cause illness. The effects of 'piggyback' toxins, that is metals and other toxins that attach themselves to the surface of sulphate aerosol molecules, is not known and has been little researched. An assessment of the human health benefits from sulphate reduction under Title IV of the 1990 Clean Air Act Amendments by the EPA, however, provides a likely scenario on the effects of sulphate aerosols on human health. Considering the economic perspective, the EPA estimates that a reduction in SO_2 to the target levels is likely to save between US$12 billion and US$78 billion, with an estimated mean value of US$40 billion in medical bills throughout the eastern USA by 2010.

International action on acidic deposition

Although acidic deposition has been known about for decades, its impact on the environment was consistently denied by politicians until the 1980s. Ronald Reagan spent much of his first term in the White House denying Canadian claims that acidic deposition caused by emissions from American power stations was killing fish north of the 49th Parallel in Canada and destroying maple trees. It was not until 19 March 1986 that the British government, under Prime Minister Margaret Thatcher, freely accepted the link between acidic deposition, water pollution and the death of fish in Norwegian and Swedish lakes. About 17 per cent of the acid that falls over Norway comes from Britain.

In the USA, the main legislation against air pollution is the 1963 Clean Air Act, considerably amended in 1970. In 1970, a ten-year plan was conceived to set standards for national air quality, including SO_2, NO_x, O_3 and hydrocarbons – all of which can contribute to acidic deposition. The required standards were never met, and in 1977 there were

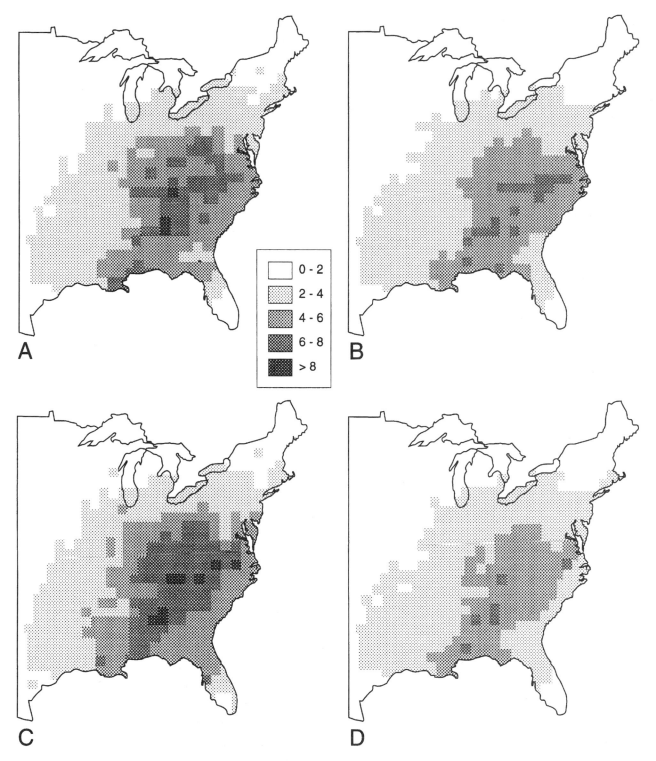

Figure 4.16 *RADM 50th percentile annual sulphate concentrations (μg m⁻³) for (A) 1985 base case; (B) 1997 with Title IV; (C) 2010 without Title IV; and (D) 2010 with Title IV. After [US] EPA (1995).*

amendments to the act in order to set less stringent targets. Despite verbal pledges to meet the issue of acid deposition head-on, President Reagan stalled; by the time George Bush assumed the presidency, very little had changed regarding controls on the polluters and levels of acceptable air quality.

Europe is the most heavily populated and industrialised continent; it is also the most polluted. More than 50 per cent of the worldwide global sulphur- and nitrogen-related pollution comes from Europe. Until recently, it seemed reasonable to assume that European politicians did not show an appropriate level of concern by encouraging sufficient investment in research into the problem of acidic deposition and stimulating debate over the best ways of tackling the problem. What has been done and what is being done to counter the effects of acidic deposition?

There have been various international meetings to address the problems associated with air pollution and acidic deposition. Following is a brief summary of just some of the most significant.

In the early 1970s, the Scandinavian countries seemed to be a lone voice in the wilderness crying out for greater environmental awareness. However, the wind of change blew with the 1975 Helsinki Conference on Security and Co-operation, at which the Soviet leader, Leonid Brezhnev, proposed that attempts be made to tackle the three pan-European issues, energy, the environment and transport. The Convention on Long-Range Transboundary Air Pollution arose from this and was signed by 35 countries, including the USA and Britain, in November 1979 in Geneva. This was the first environmental agreement to involve North America, Western Europe and the Eastern Bloc. The treaty, since ratified by most of the participating nations (35 by March 1985 and also the EC), dealt not only with voluntary controls on levels of SO_2 emissions, but also with other atmospheric pollutants including various sulphur, nitrogen and chlorine compounds, polycyclic aromatic hydrocarbons, and heavy metals.

The Conference on the Acidification of the Environment was held in Stockholm in 1982 and resulted in international agreement to endeavour to reduce acid pollution to the environment. This conference came about at a time when the first reports of serious acid rain damage to trees in the then West Germany were becoming common knowledge – the West German Green Party won its first Landtag seats in 1981, and in March 1983 its first seats in the Bundestag.

Amongst the agreed measures, a threshold value of $0.5 \text{ g S m}^{-2} \text{ yr}^{-1}$ was accepted as the level above which sensitive aquatic ecosystems would begin to suffer adverse effects. The conference was divided into two separate sessions, an initial scientific and technical session, followed by a ministerial meeting. At the latter meeting, ministers from many of the major industrialised nations accepted the seriousness of the acidification of the environment and agreed to concerted efforts to reduce pollutants such as SO_2 and NO_x emissions using available technologies.

The 30% Club came about following a meeting of the 1982 Stockholm Convention in 1983, with a Scandinavian proposal for a mutual 30 per cent reduction in emissions of SO_2 within a decade, from 1983 to 1993, with the 1980 levels of emissions taken as the baseline. The proposal was rejected by the executive body of the convention in Geneva during the 1983 meeting. Amongst the members of the executive body who voted against the proposal were the USA, Britain, France and the Eastern Bloc countries.

Frustrated by the lack of progress on this issue, the nations supporting the Scandinavian proposal met in Ottawa in March 1984 to sign their own agreement. In 1984, the members of the 30% Club who pledged themselves to a 30 per cent reduction in SO_2 emissions over a decade were Sweden, Finland, Switzerland, the Soviet Union, Luxembourg, Belgium, Bulgaria, East Germany, Ukraine, Belorussia, Liechtenstein, Italy and Czechoslovakia; Hungary joined in 1985. The Netherlands and Denmark agreed to a 40 per cent reduction by 1995, while a 50 per cent reduction in SO_2 emissions was pledged by France by 1990, West Germany and Austria by 1993, Norway and Canada by 1994, and Denmark by 1995.

The Multilateral Conference on the Environment held in Munich from 24–27 June 1984 was attended by the environment ministers from the 1979 Convention countries, with the twofold aim of encouraging additional nations to join the 30% Club and exploring further ways of reducing both SO_2 and NO_x emission levels. Environment ministers from the countries within the Stockholm Convention of 1982 also attended.

At the conference, it was requested as a matter of some urgency that, by 1993, the executive body of the United Nations Economic Commission for Europe adopt a specific agreement for a reduction in annual sulphur emissions, or their transboundary levels (fluxes). It was also recommended that total annual emissions, or transboundary fluxes, of NO_x be reduced by 1995. It was at this meeting that Patrick Jenkin, the British Environment Minister, declared the 'indivisibility of environmental and economic policies'.

The Helsinki Meeting, in July 1985, was the third meeting of the executive body (UN Economic Commission for Europe) of the 1982 Stockholm Convention, which drew up a protocol for reducing sulphur emissions. The protocol bound its signatories to a 30 per cent reduction in SO_2 emissions at source by 1993 from the 1980 levels – in other words, on the same terms as those embodied in the 30% Club. This was a protocol to the 1979 Convention on Long-Range Transboundary Air Pollution.

By September 1987, 16 nations had become signatories to the agreement, which embodied most of the sentiments of the original Scandinavian proposal that led to the creation of the 30% Club. Countries such as Britain, Poland and the USA did not become signatories to the agreement. Poland, whose sulphur emissions had actually been increasing since 1980, made the excuse that it would not have the technology to control its emissions until the 1990s.

The International Conference on Acidification and its Policy Implications in Amsterdam in May 1986 focused on a number of issues, perhaps the most notable of which concerned NO_x emissions. It was not until 1988 that a draft agreement was reached for the fixing of emission levels for NO_x at 1987 levels, to be reached by 1994.

At the Sofia meeting in 1988, 25 nations signed the protocol that was formulated as a result of the Amsterdam conference in 1986. Twelve EC countries agreed to even more stringent targets to reduce NO_x emission levels by 30 per cent by 1998. The USA did not agree to the 30 per cent reduction but signed the protocol. This was also a protocol to the 1979 Convention.

On 1 March 1993, negotiations opened in Geneva under the auspices of the UN Economic Commission for Europe for a new 'sulphur protocol' aimed at reducing SO_2 emissions. It was intended that this protocol might be signed at the end of 1993, and that it would replace the 30% Club, which, with notable exceptions such as the UK, had been signed by most Europeans. The new protocol will establish *different* clean-up targets for individual countries, and the date by which these targets should be met is under negotiation. The targets or thresholds are based on the concept of perceived 'critical loads' of acidic deposition that any sensitive ecosystem (e.g. lakes, forests, etc.) can tolerate before being harmed. For many countries in the EC, it appears that on economic grounds it will prove too costly to achieve the critical load targets, so it has been proposed that states should reduce the gap between the critical load and existing acidic deposition fallout by 50 per cent – the so-called '50 per cent gap reduction'. Since

1980, European SO_2 emissions from fossil-fuel-burning (coal and oil) power stations have been reduced considerably, partly because of the installation of 'flue-gas desulphurisation' equipment in power stations, but also as a result of the changeover to other energy sources such as natural gas and nuclear power. Examples of reduced SO_2 emissions since 1980 are 50 per cent in the Netherlands and Belgium, 70 per cent in the former West Germany and Austria, and 60 per cent in France – but only 25 per cent in Britain, virtually all due to the closure of small coal-fired power stations (Pearce 1993). Along with other members of the EC, Britain was a signatory to the 1987 Large Combustion Plant Directive requiring countries to reduce SO_2 emissions, e.g. for the UK to 60 per cent of its 1980 levels by 2003. Scientists at the UK government's Institute of Terrestrial Ecology in Cambridgeshire, however, argue that even these reductions are insufficient to get many of Europe's most threatened environments below the critical loads. The directive sets different levels of reduction of NO_x and SO_2 for each EC country.

On 14 June 1994, the UK signed a treaty in Oslo, supported by the EU and 29 other countries, including for the first time many in Eastern Europe, to reduce acidic deposition pollution by 2010 to 80 per cent of the 1980 levels. This is the first treaty to be based on the concept of 'critical loads', i.e. levels of SO_2 that scientifically are deemed tolerable by the natural environment. In order to meet the 979,800 tonnes of sulphur target by 2010, the UK must cut its 1994 levels by about 68 per cent. The UK government hopes to achieve this target at least partly through the switch from coal-fired to gas-fired power stations, which emit less sulphur, and by introducing acidic deposition 'scrubbers' on coal-fired power stations. Also, on 13 June 1994, the UK ratified a treaty on air pollution in which the UK is committed to reduce levels of volatile organic compounds (VOCs), which contribute to ground-level ozone, which causes respiratory ailments, by 30 per cent of their 1994 levels by 1999. VOCs are found mainly in the emissions from motor vehicles, petrol stations and food manufacture.

Endpiece

In preparation for the UN negotiations, European countries have been compiling their critical load maps for soil and fresh water, but there are clear signs of some countries attempting to 'adjust' the figures. Two years ago Britain, for example, drew up a critical load map for soil, which the British government claimed showed that only 8 per cent of its soil would experience greater levels of acid deposition than the critical load. A 1992 report prepared by the environmental pressure group Friends of the Earth, however, claims that had the UK adopted the mapping methods employed by other European countries, then this 8 per cent figure would actually have been 47 per cent. In the first critical load maps, the UK was divided into 1-km squares and assigned a critical load for its soils. Later maps by government scientists lumped these small areas into 100 km² boxes and, as in previous maps, assigned a critical load based on the dominant, rather than most sensitive, soil type. Many Western European countries, however, have assigned critical load values as those which are necessary to protect 95 per cent of the soils in each box (equating to 70 per cent or less for the UK method). Clearly, if countries are going to 'cheat' or simply use statistical methods that massage over any potential problem areas in evaluating the critical load maps, then this is a rather depressing outlook for any concerted international efforts to clean up the natural environment.

Chapter 4: Key points

1   Human activities result in the emission of various pollutants to the atmosphere (principally SO_2, NO_x, NH_3, hydrocarbons and particulate matter), resulting in a number of environmental problems such as poor air quality (e.g. the *nephos* over Athens on 1 October 1991) and acidic deposition with very low pH. Acidic deposition is produced by both natural and anthropogenic activities, and the principal acids produced are carbonic acid (H_2CO_3), formed by the reaction of carbon dioxide and water, sulphuric acid (H_2SO_4), formed by the reaction of SO_2 with H_2O, and nitric acid (HNO_3), which is formed by the reaction of nitrogen oxides with water. The pH of some human-induced acidic deposition has been recorded as low as 2. Volcanoes, which emit NO_x, SO_2 and Cl_2, also produce acidic deposition.

2   Acidic deposition is a particular problem in industrialised regions, where the combustion of fossil fuels releases large quantities of SO_2 into the atmosphere. Some regions and countries are polluted by acidic deposition caused by industrialised parts of other countries up-wind. Acidic deposition causes acidification of ground waters and surface water, damage to life, particularly forests and aquatic life, and building decay. Buffering reactions due to the presence of certain clay minerals, and because of cations in water in some soils and lakes, may reduce the immediate effects of acidic deposition on those environments. The susceptibility of a soil to acidification is quantified as its acid susceptibility. Regions can be classified on this basis. Areas where the soils are calcareous and/or the bedrock is limestone have a low susceptibility compared with areas where clastic and acidic igneous rocks predominate. Acidic ground water may cause corrosion of water pipelines to mobilise toxic metals such as lead (Pb), copper (Cu), cadmium (Cd) and aluminium (Al). The susceptibility of lakes to acidification is measured by the acid-neutralising capacity (ANC). Acidification of lakes has deleterious effects on their ecology. 'Forest death' is also a consequence of acidic deposition, which may be caused by direct cell damage or by a depletion from within the soil of important nutrients that support the plants.

3   Recovery from acidification of the environment can occur and is dependent on the sensitivity of the ecosystem. Reducing acidification involves cutting down on anthropogenic emissions, mainly from fossil-fuel-burning power stations, e.g. by using appropriate clean technologies such as atmospheric fluidised-bed combustion, the use of active coke and cleaning combustion engines in both private and commercial vehicles. Research and development of existing and new clean technologies exists, but without clear incentives for those who use and develop such technologies, and penalties for polluters, it appears that acidic deposition will remain a major regional environmental problem.

4   International conventions and agreements have been signed during the last few decades in order to improve air quality and reduce the emissions of SO_2. These have included the Convention on Long-Range Transboundary Air Pollution in the 1970s; the Conference on the Acidification of the Environment, Stockholm 1982; the 30% Club, formed in Ottawa in 1984; the Multilateral Conference on the Environment, in Munich in 1984; the Helsinki Meeting in 1985; the International Conference on Acidification and its Policy Implications, held in Amsterdam in 1986; and the Sofia Meeting in 1988.

Plate 21 *Craters produced by underground nuclear explosions at the Nevada Test Sites, USA. Courtesy of Rex Features.*

Chapter 4: Further reading

Carter, F.W. and Turnock, D. (eds) 1996. *Environmental Problems in Eastern Europe*. London: Routledge.
This text analyses the major forms of pollution and the resulting decline in the quality of life in each country in Eastern Europe. Air, water, soil and vegetation pollution, dumping of waste, nuclear power and transboundary issues are examined in relation to the role of legislation, political movements, international co-operation, aid and education that strive for solutions to environmental problems.

Journal of the Geological Society of London, 1986. *Geochemical Aspects of Acid Rain*. Thematic set of papers, 619–720 pp.
A useful, if somewhat specialised, thematic set of research-level scientific papers on acid deposition. Recommended reading for courses with an in-depth appreciation of this topic.

Last, F.T. and Watling, R. (eds) 1991. *Acid Deposition: Its Nature and Impacts*. Edinburgh: The Royal Society of Edinburgh, 343 pp.
An excellent volume on acid deposition, produced as edited conference proceedings, and aimed at students and teachers, including researchers, with a good scientific background. This book is a reference work that should be recommended to science-based students undertaking courses in environmental science.

McCormack, J. 1989. *Acid Earth: The Global Threat of Acid Pollution*. London: Earthscan, 225 pp.
This is a highly readable book written by a freelance environmental writer. It should appeal to those wishing to understand the nature of acid deposition and the international context in which nations have sought to reduce the associated problems.

Pearce, F. 1987. *Acid Rain*. Harmondsworth, New York: Penguin Books, 162 pp.
A readable book by the senior editor of *New Scientist* summarising the results of work undertaken by scientists into the problems associated with acidic deposition. The book examines acidic deposition and its effects on the natural environment and people's health, the corrosion of building materials, the acidification of water sources, the damage to parts of the biosphere and environmental degradation, and policy issues relating to national and international efforts aimed at reducing the emissions that cause acidic deposition.

Water, water, everywhere,
And all the boards did shrink;
Water, water, everywhere
Nor any drop to drink.

The very deep did rot; O Christ!
That ever this should be!
Yea, slimy things did crawl with legs
Upon the slimy sea.

About, about, in reel and rout
The death-fires danced all night;
The water, like a witch's oils,
Burnt green, and blue and white.

Samuel Taylor Coleridge, 'The Rime of
the Ancient Mariner'

A blue planet

The Apollo space missions in the 1960s revealed the Earth to countless millions as the 'Blue Planet'. Everyone's imagination was captured by this new vision of the Earth from Space. It appears as a blue planet from Space because of the light-scattering properties, particularly of the water and water vapour in the hydrosphere and atmosphere. The seas and oceans cover about 71 per cent of the Earth's surface (Plate 5.1) and contain more than 3.61 million km^3 of water. Not only did life on Earth begin in water and emerge from water, but animals themselves are made up of more than 70 per cent water, with plants containing over 90 per cent water by weight.

Water as a resource

Water is essential for life on Earth. Within organisms, water provides the medium in which the complex metabolic processes necessary for life take place. Organisms simply cannot function without water and if deprived will rapidly die. Not only do animals and plants need water, but the water must be clean. Human beings are affected by the most subtle variations in water chemistry and supply. According to the World Health Organisation (WHO), an estimated 1,200 million people lack a satisfactory or safe water supply. There are clear disparities between domestic, municipal and industrial water consumption in the developing compared with the developed world (see Tables 5.1 and 5.2). Also, the greatest amount of wasted water occurs in the developed and industrialised nations, for example in Europe and North America.

Water is polluted directly or indirectly by introducing substances that are a hazard to human health (Table 5.3), lead to a reduction in amenities and/or prevent water activities, such as swimming, fishing and other recreational activities. Figure 5.1 shows the world consumption of water expressed as the average annual consumption per person. These figures show that the developed countries use far more water per capita than the non-industrialised developing world. Examples of per capita consumption of water for 1988/1989, in litres per day, are as follows: Austria, 145; Belgium, 108; Denmark, 190; Finland, 151; France, 159;

Plate 5.1 Part of the hydrological cycle – the coupled ocean–atmosphere system. Calm seas (top) and rough seas (bottom); Whitesands Bay, South Wales.

Table 5.1 *Domestic and municipal water consumption.*

Region	1980s				Year 2000 projection			
	Population (millions)	Water withdrawal (km³)	Consumptive use (km³)	Waste water (km³)	Population (millions)	Water withdrawal (km³)	Consumptive use (km³)	Waste water (km³)
Europe	496	48	10	38	512	56	8	48
Asia	2,932	88	53	35	3,612	200	100	100
Africa	589	10	7	3	853	30	18	12
North America	411	66	20	46	489	90	22	68
South America	279	24	14	10	367	40	20	20
Australia and Oceania	26	4.1	1.2	2.9	30	5.5	1.5	4
Former USSR	282	23	5	18	310	35	5	30
World total	5,015	263.1	110.2	152.9	6,173	456.5	174.5	282

Source: *World Resources 1990–91*, a report by the World Resources Institute in collaboration with the UN Environment Programme and UN Development Programme 1990.

Table 5.2 *Water use in industry.*

Region	1980s				Year 2000 projection			
	Population (millions)	Water withdrawal (km³)	Consumptive use (km³)	Waste water (km³)	Population (millions)	Water withdrawal (km³)	Consumptive use (km³)	Waste water (km³)
Europe	496	193	19	174	512	200–300	30–35	170–265
Asia	2,932	118	30	88	3,612	320–340	65–70	255–270
Africa	589	6.5	2	4.5	853	30–35	5–10	25
North America	411	294	29	265	489	360–370	50–60	310
South America	279	30	6	24	367	100–110	20–25	80–85
Australia and Oceania	26	1.4	0.1	1.3	30	3.0–3.5	0.5	2.5–3.0
Former USSR	282	117	12	105	310	140–150	20–25	120–125
World total	5,015	759.9	98.1	661.8	6,173	1,153–1,308.5	190–225.5	962.5–1,083

Source: *World Resources 1990–91*, a report by the World Resources Institute in collaboration with the UN Environment Programme and UN Development Programme 1990.

Hungary, 205; Italy, (1984), 220; Luxembourg, 176; the Netherlands, 167; Spain, 126; Switzerland, 264; Sweden, 194; the UK, 136; and former West Germany, 145 (Water Services Association 1992, Table 9A).

Water supply is not only a problem for the developing world. In California, for example, there is a looming water supply crisis. The Sacramento River index, which measures the water flow in five northern Californian rivers that feed the state's largest reservoirs, is at one of its lowest levels in history. The water-parched city of Santa Barbara on the Pacific coastline is leading the way after being afflicted by a five-year drought, by turning to the ocean. In May 1991, the city council approved the immediate construction of a US$37.4 million desalination plant intended to provide 9.1 billion litres of water per year for Santa Barbara and two neighbouring communities; when complete, it will be the largest municipal desalination plant. Until this scheme was approved, some Californian communities, such as the agricultural community of Goleta, had considered importing fresh water by ocean-going tankers from Canada. Other communities in California, such as San Luis Obispo, are also on the verge of approving desalination plants.

In developed countries, droughts (also see Chapter 8) are usually manifested in the issuing of drought orders and the imposition of temporary bans on the excessive use of water, for example, hose-pipe bans, etc. In some countries water meters are being introduced to charge by volume, and to influence demand, although most European countries already have them. It appears that metering water can reduce demand by between 7 and 55 per cent (OECD 1987), but typically averaging between 10 and 15 per cent in regions that have similar climates and consumption patterns to Britain (OFWAT 1990). In a detailed 1993 report, POST (the UK Parliamentary Office of Science and Technology) has indicated that leakage control programmes appear to give higher savings than a universal metering programme, even where external meters with an ensuing reduction in supply pipe leakage are used. In England and Wales, about 22 per cent of the water input to the supply system is lost from the water companies' distribution systems, with a further 8 per cent lost from customers' supply pipes (*ibid*.). Interestingly, it is possible to regard water supply leakages as partially beneficial since they are a means of ground-water recharge in urban areas.

Rivers

Human communities have been polluting water since civilisation began. The necessity for water led early civilisations, such as those in Mesopotamia (the land between the Tigris and Euphrates), in Egypt along the Nile, along the Indus in Pakistan, and in China along the Yellow and Yangtze Rivers, to develop along the courses of these rivers.

Rivers are polluted by adding organic waste, including human and animal excreta, and fibrous agricultural waste from plants that are harvested. Rivers are frequently used as sewers to discard waste. Accounts of Elizabethan London describe people taking their dirty sewage down to the Thames, where the bucket was emptied and refilled with river water for household use (Cook 1989). Much of this waste is degraded by microbes in the water through a natural process called self-purification. When populations are large, and excessive amounts of waste are produced that end up in the water supplies, the natural process of self-purification cannot keep pace with the input of pollutants, so the water quality rapidly deteriorates. Such organic waste is responsible for the transmission of infectious diseases such as cholera, typhoid, dysentery and diarrhoea. Inadequate nutrition and poor medical care in many developing countries exacerbate

Table 5.3 *Classification of water-related infections.*

Category	Infection	Pathogenic agent
Faecal–oral (waterborne or water-washed):	Diarrhoeas and dysentries	
	Amoebiasis	P
	Campylobacter enteritis	B
	Cholera	B
	E. coli diarrhoea	B
	Giardiasis	P
	Rotavirus diarrhoea	V
	Salmonellosis	B
	Shigellosis (bacillary dysentry)	B
	Enteric fevers	
	Typhoid	B
	Paratyphoid	B
	Poliomyelitis	V
	Ascariasis (giant roundworm)	H
	Trichuriasis (whipworm)	H
	Taenia solium taeniasis (pork tapeworm)	H
Water-washed:		
Skin and eye infections	Infectious skin diseases	M
	Infectious eye diseases	M
	Louse-borne typhus	R
Other	Louse-borne relapsing fever	S
Water-based:		
Penetrating skin	Schistosomiasis	H
Ingested	Dracunculiasis (guinea worm)	H
	Clonochiasis	H
	Others	H
Water-related insect vector:		
	Trypanosomiasis (sleeping sickness)	P
Biting near water	Filariasis	H
Breeding in water	Malaria	P
	Onchocerciasis (river blindness)	H
	Mosquito-borne viruses	
	Yellow fever	V
	Dengue	V
	Others	V

B, bacterium; P, protozoan; S, spirochaete; M, miscellaneous; H, helminth; R, rickettsia; V, virus.
Adapted from Feachem 1984 and Nash 1993.

the health problems associated with polluted waters (Plate 5.2). As long ago as 1389 in England, Richard I outlawed the dumping of dung, filth and garbage into streams and rivers near cities, but humans continued to discharge their waste into water supplies (*ibid*.). The results were frequent cholera epidemics such as those experienced in Victorian London, which killed many people.

It was not until the turn of the century that people such as Dr John Snow in London made the

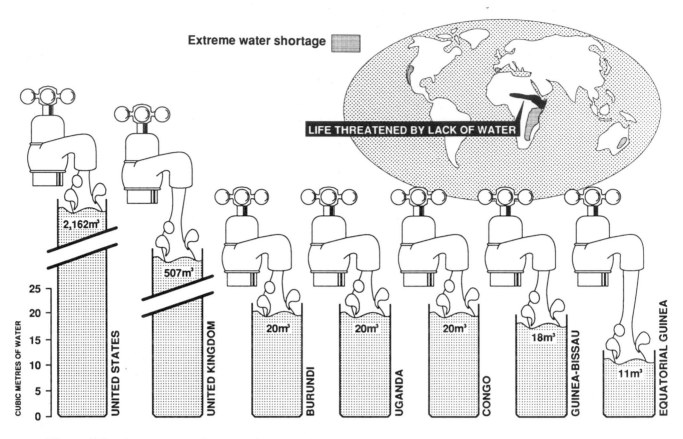

Figure 5.1   *Average per capita annual consumption of water in selected countries. Redrawn after World Resources Institute (1990).*

Plate 5.2 *Polluted river in Kathmandu, Nepal. Rubbish and human sewage are dumped into the river and the local people collect water for domestic use.*

connection between bad water quality and cholera epidemics. Laws were passed in the British Parliament, including the Public Health Acts of 1848, 1872 and 1875, the Rivers Pollution Prevention Act, 1876, and the Land Drainage Act, 1930. These restricted the type and quality of pollutant that could be disposed of within sewerage systems, and made pollution a prosecutable offence. More recently, the EC has introduced regulations, directives and new standards for water quality, e.g. the 1973 Community's First Action Programme on the Environment. Stronger legislation was enforced in the USA, especially during Richard Nixon's presidency in the early 1970s, which paid particular regard to the dumping of waste at sea. Similarly, the Canadians introduced the Arctic Water Pollution Bill in 1970, restricting waste disposal and shipping activities in Arctic waters.

Despite such legislation, pollution by organic waste is still prevalent in Europe and North America, and throughout the rest of the world (Nash 1993). For example, in many countries raw sewage is flushed directly into the sea via outfalls. Today, around the coastlines of the Mediterranean and on many British beaches, it is a common sight to see human excrement, toilet paper, sanitary towels, disposable nappies and condoms, something that the new EC legislation was designed to stop.

Additionally, rivers are still being polluted, not only by traditional organic wastes but also by industrial wastes, which now form the most common and hazardous pollutants. Industrial waste includes radioactive chemicals, dangerous organic chemicals, nitrates, heavy metals and oil. Table 5.4 shows the US Environmental Protection Agency's estimates for the amount of impaired water in the United States. These data on the extent of pollution are clearly a cause for some concern. In 1989–90, there were more than 25,000 recorded pollution incidents in England and Wales, about 1,000 more than in the preceding year and approximately twice as many as in 1982. Of course, the anti-alarmist, business-as-usual lobby put this increase in reported pollution incidents down to enhanced public awareness of environmental issues. It seems reasonable to assume that at least some of these statistics are indeed due to increased public concern, but it would be foolish to dismiss the figures on this basis. On 11 December 1991, the National Rivers Authority (NRA) in England and Wales released a report which showed that overall water standards have been dropping for the past decade. At the same time as these results were released, the NRA announced a five-year plan to force industrialists, water companies and farmers to improve the quality of water in rivers, canals and estuaries. Legally binding standards would be introduced for water quality.

The increase in pollution is disturbing. Even when an incident is reported in the UK little action is taken against the polluter. For instance, in 1986 only 254 prosecutions were made out of a total of more than 20,000 incidents. The maximum penalty that a magistrate can impose is a mere £2,000, a very small slice of an industrialist's profit when compared with the expense of treating waste before it is discharged. In the USA, the situation is better because individuals can initiate law suits against polluters: penalties may be up to US$10,000 per day for violations, with criminal penalties of between US$25,000 and US$42,500 for a first offence, and US$50,000 for a second offence. The more serious environmental damage can lead to imprisonment.

More disturbing is the fact that the less developed countries may have even greater problems than the developed countries, since government legislation and implementation is crude due to poor resources.

Table 5.4 *Impaired waters in the USA, by causes and sources of pollutants.*

Causes	Per cent of rivers, lakes and estuaries surveyed determined to be impaired		
	River length	Lake area	Estuary area
Siltation	42.4	25.4	6.7
Nutrients	26.6	48.8	49.6
Pathogens	18.6	8.6	48.1
Organic enrichment	14.6	25.3	29.0
Metals	10.8	7.4	9.5
Pesticides	10.3	5.3	1.0
Suspended soils	6.2	7.5	
Salinity	6.1	14.3	
Flow alteration	5.8	3.3	
Habitat modification	5.7	11.3	
pH	5.1	5.1	0.4
Priority organics		8.2	4.1
Oil and grease			23.4
Unknown toxins			5.1
Other inorganics			0.4
Ammonia			0.2

Sources	Per cent of rivers, lakes and estuaries surveyed determined to be impaired		
	River length	Lake area	Estuary area
Agriculture	55.2	58.2	18.6
Municipal	16.3	15.1	53.1
Resource extraction	13.0	4.2	34.2
Hydrological/habitat modification	12.9	33.1	4.8
Storm sewers/run-off	8.8	27.7	28.5
Silviculture	8.6	0.9	1.6
Industrial	8.5	7.7	12.1
Construction	6.3	3.3	12.5
Land disposal	4.4	26.5	27.4
Combined sewers	3.7	0.3	10.3

Source: US Environmental Protection Agency 1990, Gleick 1993.

There are still areas, such as in the Indian subcontinent and Latin America, where streams are used as open sewers as well as a source of water for household use (Plates 5.3 and 5.4). Bedding (1989) graphically described such a situation in Cairo and drew attention to the high infant mortality rate (131 in 1,000 children die before the age of 5). This is mostly a result of gastro-enteritis, cholera and typhoid related to poor sanitation and water quality. A new network of sewers, pumping stations and treatment works is scheduled for completion in Cairo, the Greater Cairo Wastewater Project. Raising finance for such schemes, estimated at well over £200 million for the Cairo project, is not easy in a developing country.

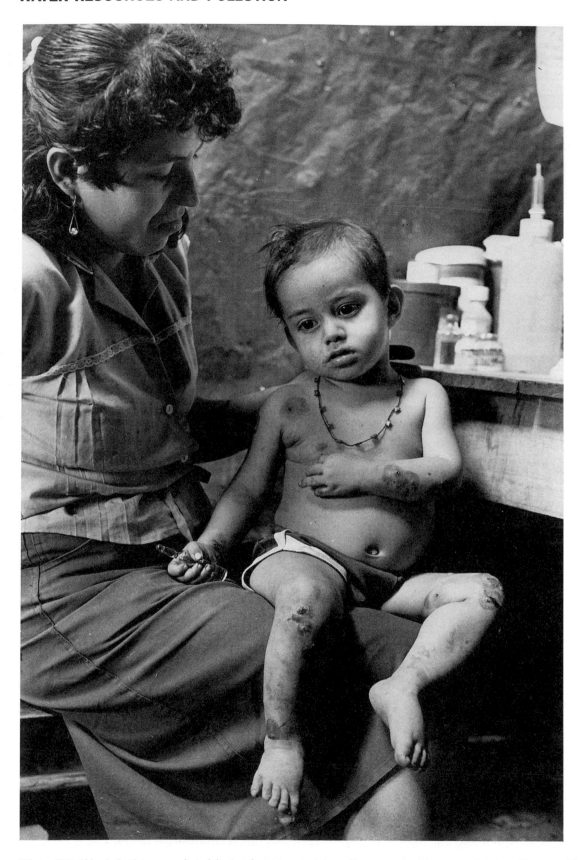

Plate 5.3 *Skin infection exacerbated by inadequate nutrition, dirty water and poor sanitation, Communidad Santa Martha, El Salvador. These problems are being rectified by local health programmes.*
Courtesy of Rhodri Jones/Oxfam.

Plate 5.4 *Washing clothes in a stream – a feature of everyday life, Communidad Santa Martha, El Salvador.*
Courtesy of Rhodri Jones/Oxfam.

Oceans and seas

The oceans and seas have acted as the dumping ground for waste for many centuries, but the problem has only become very serious in the industrialised and highly militarised world of the twentieth century. The waste ranges from domestic sewage, oil, detergents, pesticides, toxic metals and **polychlorinated biphenyls** (**PCBs**), to radioactive waste. The persistence of such pollutants in the oceans varies considerably (Figure 5.2). Much of this chapter is devoted to some of these pollutants.

Every year since 1986, eight million tonnes of raw sewage has been dumped into the ocean at the Mid-Atlantic Bight, an area about 100 miles off the coast of New York and New Jersey (American Geophysical Union 1992b). When sewage was first dumped here, it was believed to be a safe and ideal site because of the strong ocean currents and deep water. More recently, scientific studies co-ordinated by the US

NOAA's National Undersea Research Program have revealed the extent of sewage sludge accumulation at the site.

Sediment traps were suspended between 5 and 100 m above the ocean bottom and the trapped material tested for silver, the trace metal that is one of the most sensitive inorganic tracers of increasing amounts of sewage. Background levels in the oceans are believed to be approximately 0.1 ppm, and elevated amounts would be regarded as around 5 ppm. The sediment traps contained material with silver levels up to 16 ppm, an order of magnitude greater than the background level. At the site, sea urchins were found to contain as much as 25 per cent sewage-derived organic material in their body tissue, showing that they are actively feeding on the sewage. The implications of this type of study have yet to be fully evaluated but, whatever the official government views on the dumping of raw sewage at sea, it is clear that the sludge is not removed as fast as it

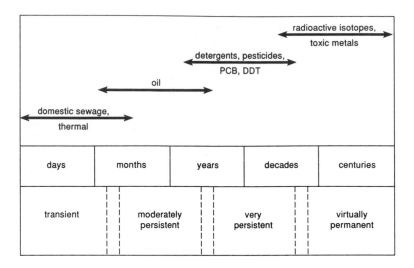

Figure 5.2 *The persistence of pollutants in the oceans. Redrawn after Smith and Warr (1991).*

is discharged. The sea water is becoming polluted by potentially toxic waste. At present, the UK dumps sewage sludge in the North Sea, both from coastal outflow pipes and from ships, but under an EC directive this will cease in 1998.

The water cycle

In order to understand what happens to pollutants in water, it is necessary to examine the global water cycle, or what is known as the **hydrological cycle**. This was discussed in detail in Chapter 1 (see Figure 1.14), but it has direct implications for understanding the dynamics of water pollution.

The precipitation of water may introduce pollutants such as sulphur dioxide (SO_2), which is partly responsible for the acidification of water in acid lakes (Chapter 4), airborne lead and many radioactive substances (Chapter 6). Precipitation, however, may have the positive effect of diluting polluted surface waters, and it can then be regarded as essentially pure water. The precipitation may then be stored for long periods as ice within mountain glaciers or ice caps, or it may flow over the surface of the Earth, transporting any pollutants, perhaps picked up on its journey towards the sea. The water that percolates into the ground may also carry pollutants with it to cause polluted ground water.

Ground water in many parts of the world is a vital water resource. In deserts and semi-arid regions it may be the only source of water. The flow of surface and ground water to the sea carries the pollutants, which become concentrated as the sea water evaporates. Evaporation from the soil may also lead to a form of water pollution known as **salinisation**. This involves the drawing of salts and pollutants towards the surface as soil water evaporates or is taken up by plants and is replaced by water that comes from depth. This concentrates salts as the water evaporates, causing great problems in semi-arid regions, especially where irrigation is practised. Postel (1993) emphasises the problems of salinisation. She presented data showing that approximately 15 million ha, mainly in India, China, Pakistan, Iran and Iraq, is experiencing serious reductions in crop yields because of salinisation. In Egypt and Pakistan, waterlogging and salinity account for a reduction in crop yields by 30 per cent, and in Mexico, salinisation has reduced crop yields by an amount equivalent to *c*. 1 Mt of grain per year, enough to feed five million people.

In addition to understanding the path of pollutants in the hydrological cycle, it is important to consider the fate of pollutants when organisms eat, drink or inhale polluted water or water vapour. Animals form part of a hierarchical system, which in ecological parlance is referred to as a **food chain**. Put simply, organisms at the bottom of this chain obtain their food from plants, while predatory creatures further up the chain consume those below. Humans constitute the top of many of these chains. For example, plants may be eaten by small fish, which in turn are preyed on by larger fish, and fish are then consumed by humans. In reality the different chains interlink in a more complex way to form what ecologists call **food webs** (Figure 5.3).

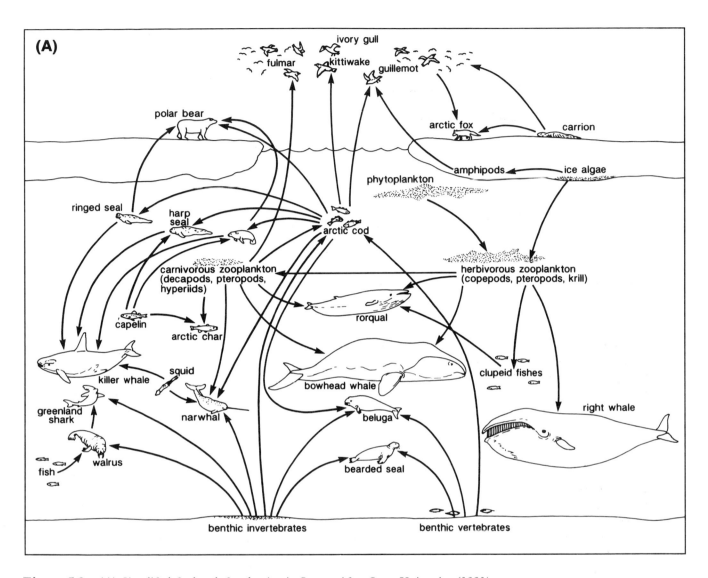

Figure 5.3 *(A) Simplified food web for the Arctic Ocean. After Open University (1991).*

There is a relative accumulation of pollutants up the food chain in more complex organisms, at least for toxins that are stored in some way, i.e. by **bioaccumulation** (Figure 5.4). Plants that are affected, for example, may be eaten by small animals with little adverse effect, but as the relative pollution levels rise toxins become concentrated higher up the food chain, where the effects can be magnified. It is not quite this simple, because higher up a food chain, larger creatures may have a greater resistance or ability to deal with poisoning.

This knock-on effect in a polluted food chain was illustrated vividly in Rachel Carson's famous book, *Silent Spring,* published in 1962. The book begins by describing a picturesque town in the heart of America renowned for its bird and fish life. It then goes on to describe a strange blight that swept the area. The birds and fish began to die until, one spring, there was no more dawn chorus and no fish in the rivers. What Rachel Carson was describing was the effects of pesticide poisoning on the area and how the poisoned insects, which formed part of the food chain, transferred this poison further up the food chain to the higher organisms, such as birds and fish. She paid particular attention to the pesticide **dichloro-diphenyl-trichloro-ethane (DDT)**, which was widely used throughout the world at that time. DDT was commonly washed off the crops into streams and into the ground. The plants absorbed this water containing DDT, and small creatures could

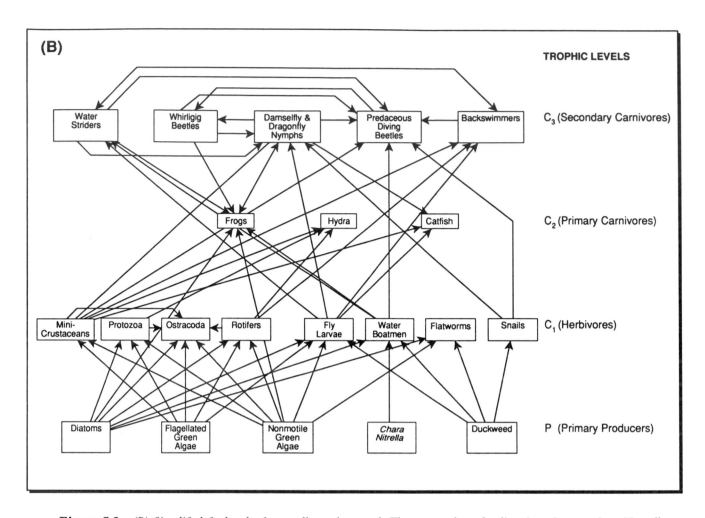

Figure 5.3 (B) *Simplified food web of a small meadow pond. The arrows show the direction of energy flow. Not all species are confined to one level, especially carnivores, which may occupy all trophic levels. Some animals, particularly insects, may occupy different levels at different times of their life cycle. Redrawn after Clapman (1973).*

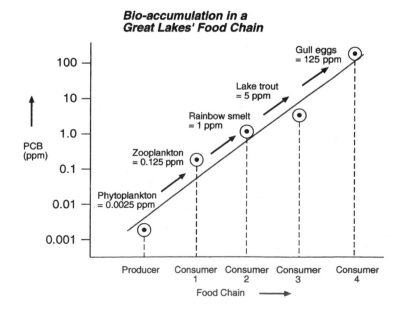

assimilate the poisons into their bodies by drinking the polluted water. The affected organisms then passed the pollutant up the food chain. Later, this pesticide was banned in the USA, but it is still used today in some less developed countries, such as India, where it is used in cotton production and malaria suppression (Nash 1993). Other pollutants are widely used today that can have similar effects and that can just as easily be transmitted up the food chain, yet because their effects are either less publicised or less well understood, they continue to be used. Table 5.5 lists the major pesticides that are still being used, and Table 5.6 illustrates the extent of their impact in developing countries. Unfortunately, multinational

Figure 5.4 *Increased concentrations of PCBs up trophic levels within the food chain of organisms in the North American Great Lakes. Redrawn after Marsh and Grossa (1996).*

Table 5.5 *Major pesticides.*

	Insecticides	
Type	*Examples*	*Persistence*
Chlorinated hydrocarbons	DDT, aldrin, dieldrin, endrin, heptachlor, toxaphene, lindane, chlordane, kepone, mirex	High (2–15 years)
Organophosphates	Malathion, parathion, monocrotophos, methamidophos, methyl parathion, DDVP	Low to moderate (normally 1–12 weeks, but some can last several years)
Carbamates	Carbaryl, maneb, priopoxor, mexicabate, aldicarb, aminocarb	Usually low (days to weeks)
Pyrethroids	Pemethrin, decamethrin	Usually low (days to weeks)
	Herbicides	
Type	*Examples*	*Effects*
Contact	Triazines such as atrazine and paraquat	Kills foliage by interfering with photosynthesis
Systemic	Phenoxy compounds such as 2,4-D, 2,4,5-T and silvex; substitute ureas such as diuron, norea, fenuron and other nitrogen-containing compounds such as daminozide (Alar), glyphsate	Absorption creates excess growth hormones; plants die because they cannot obtain enough nutrients to sustain their greatly accelerated growth
Soil sterilants	Trifluralin, diphenamid, dalaphon, butylate	Kills soil micro-organisms essential to plant growth; most also act as systemic herbicides

Source: G. Tyler Miller Jr., *Living in the Environment* (Wadsworth, Belmont, California, 1990), p. 551.

corporations continue to market products that are banned in the developed countries because they are deleterious to natural ecosystems. Profit is the only motive.

To understand the causes and effects of pollutants in the main groups of toxic chemicals, they are considered in the following sections.

Sewage and sludge disposal

Sewage is a cocktail of dissolved and suspended materials, and its quality is defined in terms of the **suspended solids (SS)**, **biochemical oxygen demand (BOD)** and ammonia content. Raw, untreated sewage typically contains 1–7 per cent solids. A general formula for the dry solids of sewage sludge is $C_5H_7NO_2$. The SS is determined by weighing after filtration of a known volume of sample through a standard glass-fibre filter paper, expressed as mg l^{-1}. The BOD is determined by sample incubation at 20°C for five days, with the amount of oxygen consumed also being given in mg l^{-1}. Domestic sewage, with about 1,000 mg l^{-1} of impurities, comprises about two-thirds organic material,

Table 5.6 *Estimated annual impact of pesticides in developing countries.*

Population groups at risk	*Character of exposure*	*Estimated overall annual impact*
Pesticide formulators, mixers, applicators, pickers, and suicides and mass poisoning	Single and short-term, very high-level exposure	3,000,000 exposed, 220,000 deaths
Pesticide manufacturers, formulators, mixers, applicators, and pickers	Long-term, high-level exposure	735,000 suffer from specific chronic effects of long-term exposure
All population groups	Long-term low-level exposure	37,000 suffer from chronic effects of long-term exposure (cancer)

Source: World Health Organisation (WHO), *Our Planet, Our Health* (WHO, Geneva, 1992), p. 81.

BOX 5.1 SEWAGE TREATMENT

Sewage treatment usually involves a preliminary screening stage to remove larger suspended and floating material, followed by primary sedimentation to separate out solids, and then a secondary, biological, treatment of the sludge, commonly associated with an anaerobic stage in a digester and an aerobic stage in settlement tanks and/or lagoons, the latter processes being referred to as digestion. Sludge treatment and disposal can account for 40 per cent of the operating costs of a waste water treatment plant (Lester 1990). An important part of the treatment of sewage sludge is the reduction of the water content, or dewatering, thereby allowing easier handling and reduced transport costs since the volume of sludge is much less: 30 Mt of wet solid corresponds to 1.25 Mt of dry solid. Dewatering processes vary considerably in both duration and effect, e.g. it may take more than two months using drying beds (giving about 25 per cent solid content of the sludge), or 2–18 hours under pressure filtration at 700 kPa in cloth-lined chambers to form cakes typically of 25–50 per cent solids. Alternatively, vacuum filtration, operating with a pressure difference of about 70 kPa, produces sludge with 15–20 per cent solids.

the main constituents being nitrogen compounds. Between countries, standards vary considerably, and within the EU minimum standards and controls for municipal waste water and the disposal of sewage sludge are being introduced through the Urban Waste Water Treatment Directive (91/271/EEC). This requires secondary treatment of municipal waste water as a minimum, except in areas designated as less sensitive, where primary treatment alone may be permitted. Also, the directive requires the cessation of sludge dumping at sea by 1998.

Disposal of sewage sludge, either to land or sea, is frequently based on economic rather than environmental considerations, although in many countries this is changing. Both land and sea disposal have pros and cons. Application of sludge to agricultural land, either in liquid or dried form, commonly as soil/subsoil injection, has the advantage of resource recovery but may cause more immediate environmental damage than disposal at sea. Land disposal can lead to contamination of ground water and surface streams by viruses, bacteria, protozoans and other pathogens, persistent toxic organic compounds, and toxic heavy metals, and therefore requires careful monitoring. On the credit side, sludge, with its high nitrogen and phosphorus content, provides a valuable fertiliser.

The balance between various sewage sludge disposal routes varies between countries, being dependent on factors such as appropriate technology, perceived acceptable standards, and purely economic considerations. In England and Wales, sewage sludge disposal for 1991/92 was as follows: 51 per cent farmland, 22 per cent sea, 11 per cent landfill, 7 per cent incineration, and 9 per cent other (Water Services Association 1992). As the dumping of sludge at sea is reduced and finally ceases, other forms of disposal are becoming more important, e.g. incineration. There are plans to build Europe's largest waste incinerator in southeast London, projected to generate 100 MW of electricity from 1.5 Mt yr^{-1} of waste. Thames Water plans to build two incinerators capable of burning 0.5 Mt yr^{-1} of sewage sludge. Throughout the UK, as in other parts of Europe, other schemes are under way for the construction of incinerators to burn domestic waste and sewage sludge. These changes have implications for local air quality and broader environmental change. Incineration produces more immediate air pollution and the release of CO_2 into the atmosphere, whereas sea dumping leads to the release of more CH_4. Incinerating dry sludge releases about 53 per cent of its weight in carbon, equivalent to approximately 195 per cent CO_2. In calculating CO_2 budgets, however, there is little to no net contribution to atmospheric CO_2 levels, because life-cycle analysis shows that it represents a short-cycle sequestration and release of CO_2.

In countries such as England and Wales, **sewerage** systems carry the sewage to the treatment site or discharge point via **foul sewers** (carrying only domestic and industrial effluent) and **storm sewers** (discharged directly to natural water courses). In older towns and cities, however, combined foul and storm water systems are used, leading to large differences in the flow of sewage during storms. In Britain, water pollution was a recognised problem even in 1850, and in 1876 Parliament passed the first act to control water pollution; between 1898 and 1915, the Royal Commission on Sewage Disposal outlined the objectives of, and requirements for, sewage treatment. The main purpose of sewage treatment is to destroy disease-carrying organisms and to remove, or at least dilute, toxic chemicals; other aspects include the reduction of unpleasant odours (see Box 5.1). The indirect reuse of water is common in some countries and tends to be on the increase, e.g. the UK, where it is predicted that by the year 2000 the amount of reuse will have approximately doubled from 1990 levels. Where reuse is practised, nitrate concentrations in river water are an area of concern and subject to monitoring; high levels of nitrate in potable (drinking) water can cause methaemoglobinaemia or 'blue-baby syndrome'.

Traditional organic waste, microbes, bacteria and viruses

Human excrement discharged in an untreated form into rivers and seas has already been mentioned and constitutes a major pollutant. Such discharges are frequent in developed countries as well as in less developed countries. In England and Wales, many of the 6,500 sewage treatment plants discharge sewage effluent into the rivers, rather than the open sea or estuaries, and are old and cannot handle the increased amounts of sewage. About 20 per cent of the UK sewers were laid before 1914, and many were designed to carry rain water as well as sewage. A major programme to improve and upgrade these plants is currently under way.

The result of using manure from animal husbandry can be considered equally detrimental to the environment, a point emphasised in an article on the production of manure by piggeries in Denmark (Armstrong 1988). These piggeries produce around 94 million tonnes of manure each year, much of which cannot be treated or disposed of adequately. Not only does this manure produce ammonia gas (NH_3) to pollute the atmosphere and damage vegetation, but it also produces nitrates, which are washed into streams (considered in more detail in a later section). The manure concentrates heavy metals in soils to toxic levels. These heavy metals, such as copper, cadmium and zinc, which in high enough concentrations become poisonous, were originally added to the pigs' feed to increase their growth, but they are discharged by the pigs in their excreta. When the manure is used as a fertiliser or stockpiled, the metals may then leach out of the manure to end up in streams and ground water. Large amounts of these heavy metals may be retained in the soil and can lead to the death of earthworms, which are essential for the breakdown and aeration of the soil. The knock-on effect of decreased soil breakdown is soil erosion, which then accelerates the input of the heavy metals to stream water. A vicious circle results. The heavy metals are then washed into the drainage systems to find their way into the water used for human consumption and can ultimately cause serious illnesses (discussed in a later section). If manure is properly composted, however, rather than being applied in a raw state, then its nutrient value can be reclaimed, in which case many of the problems outlined above should not occur.

Excrement is often associated with particular microbes, bacteria and viruses, among which cholera and typhoid are numbered as particularly rampant and dangerous. In recent years, there have been several outbreaks of animal parasites, such as cryptosporidia. For example, affected water supplies in the Swindon area of the UK during February 1989 threatened 20,000 people and caused many to suffer diarrhoea and vomiting. Though these effects were mild, there should be concern for those people with suppressed immune systems, such as babies, people on particular immuno-suppressive drugs and other vulnerable people. They may be more adversely affected by such disorders and could die.

Other cases of cryptosporidia have been reported, including a similar outbreak in Ayrshire in 1988, while a case of giardiasis in Leeds, in 1980, infected 3,000 people. Giardia (*Giardia lamblia*) is a particularly unpleasant parasite that causes dysentery and possibly death in weak individuals. Giardia affixes itself to the gut lining, multiplies extremely rapidly and is very difficult to eradicate without strong antibiotics. It is the most abundant animal parasite, and the most common cause of water-borne disease in tropical developing countries, but the number of cases of giardiasis in Europe, even amongst people who have not travelled abroad, has been increasing in recent years, particularly because of its resistance to chlorine (Hibler and Hancock 1990). Another disease that is on the increase is leptospirosis, commonly known as 'sewerman's disease', which can be contracted from contact with rats' urine. Fifteen people died in the UK from this disease in 1989. Some experts believe that modern farm slurry discharges are responsible for the rising infection rates.

The Scottish Home and Health Department has been carrying out intensive investigations with universities and laboratories in Scotland and is particularly concerned about the parasitic organisms in water supplies. The results indicate that in many of the large water bodies, especially recreational lakes, river water, sewage and even treated drinking water, there are viable parasitic cysts. This poses a particularly daunting threat to water supplies.

Nitrates

Pollution of rivers, seas and drinking water by nitrates from fertilisers and traditional organic waste has become a serious problem. Nitrates, however, are essential chemicals for the metabolic functions of plants. They are produced naturally by nitrofixing bacteria, which fix nitrogen from the atmosphere in the decomposing organic matter and form nitrates. Nitrates are also formed by the weathering of rocks during the formation of soil.

In areas where intensive agriculture is practised, the soil may become depleted in nitrates, with poor crop yields. To remedy this, artificial fertilisers are added to the soil, or muck naturally rich in nitrates is spread over agricultural land. This effectively makes the soil more fertile. If there are excessive quantities of nitrates, however, they percolate into the ground-water supplies, flow into streams and rivers, and eventually reach the sea or lakes, where they become concentrated. Nitrates may also be concentrated in ground water, which tends to be as a very long-term effect due to the slow 'turnover' of ground-water reservoirs. This may lead to several detrimental effects on aquatic and marine ecosystems, and may lead to public health problems when drinking water is contaminated.

The quantity of artificial nitrates used on arable land in attempts to improve agricultural yields has increased considerably over the last twenty years. In the UK, for example, 1.3 Mt of nitrates was used in 1986, twice that of 1966. Natural fertilisers or manure also produce large quantities of nitrates. Animal husbandry also produces large amounts of excrement, both manure and urine, which requires proper composting prior to its use as fertilisers: this represents a huge potential resource of organic fertiliser, and in combination with carbon-rich material (such as waste paper) can transform this waste into a useful nitrate-rich fertiliser. Unfortunately, this excrement is often leaked or intentionally released into water supplies in order to dispose of the large quantities produced by animal husbandry. Soil itself will naturally produce nitrates by the microbial activity of nitrofixing bacteria by supplying nitrogen as nitrates, directly to the plant. When the plant decays nitrates are deposited in the soil.

The EC has set an acceptable limit of 50 mg per litre (mg l^{-1}) of nitrate concentration in public water supplies. The World Health Organisation (WHO) recommends a limit of 100 mg of nitrates per litre. It is still not known if these levels are safe, but they are commonly exceeded. In 1984, the WHO produced evidence to show that nitrate pollution was responsible for 'blue-baby' births (methaemoglobinaemia). In this disease, nitrates in the human stomach are converted to nitrites, and infants are particularly susceptible because of the relatively low acidity in their stomachs, which causes almost all the ingested nitrates to be converted to nitrites, whereas in adults only about 5 per cent is converted. The nitrites react with bacteria in the gut and deplete the blood oxygen levels, which then affects the brain and heart muscles (Horte et al. 1991). It has even been suggested that some forms of stomach cancer result from nitrate poisoning. The EC has identified fifty-two areas in Britain where concentrations exceed permitted nitrate levels and it is estimated that some 1.3 million people are affected. Mothers have been told to use bottled water for their babies because boiling water does not remove nitrates.

At Rothamsted Experimental Station in Hertfordshire, UK, research into the effects of nitrates in soils has identified the main conditions under which nitrates leach into water supplies. Suggestions have been made for improved use of nitrate-rich fertilisers (Addiscott 1988). Little, however, is being done to reduce the levels of nitrates in contaminated water.

It is possible to remove nitrates from water chemically. Various techniques that are technically available to transfer nitrates from one body of water to another but not used on any significant commercial basis, include selective ion exchange, reverse osmosis, electrodialysis and distillation. Biological processes are also available. One of the new nitrate-removal processes involves using aluminium powder to reduce the nitrate to less harmful ammonia, nitrogen and nitrite (Murphy 1991). With such a range of nitrate-removal processes available, there really is little reason for the current complacency.

A blooming problem

A superabundance of nitrates in lakes and seas, introduced by ground water and streams rich in nitrates, causes additional problems. Often, other compounds such as phosphates will also be present with the nitrates derived from detergents washed or flushed into the water sources. In lakes, the great increase of these chemicals, either naturally or accelerated by pollution, many of which are nutrients for some organism, will initially produce prolific growth of algae. As the algae grow, they reduce, and may eventually cut off, the light to other plants that are necessary in the processes that oxygenate bottom waters. The larger amounts of decaying plant material being rotted by bacteria on the bottom stimulate the production of greater numbers of bacteria, which in turn further deplete the water of oxygen. The low free-oxygen levels rapidly cause the fish to die, and before long, a diverse living ecosystem may become lifeless apart from bacteria. Scientists call this process **eutrophication** and describe lakes as being **eutrophic** when this process has occurred.

Many of the Great Lakes in the USA suffer to varying extents from this condition. Similar conditions may occur in sea water, where green and red algae become very prolific when nitrates and phosphates are introduced into the water above threshold levels. Such phenomena are called **algal blooms**. Red blooms are produced by single-celled algae that turn the water red, green or brown, although some species do not colour the water. Some types may be toxic and have poisoned fish, sea mammals and humans. They are a common sight in tropical waters but may occur naturally along coasts where cold nutrient-rich bottom waters rise to the surface by a process known as **upwelling**. These blooms are associated with cycles of increased algal growth and large fish populations living on the algae and the new supply of nutrients. The consequent over-productivity can result in a depletion of free oxygen in the water and eventually lead to large-scale suffocation and death of marine life. Similar cycles are experienced in areas where pollutants such as nitrates and phosphates are introduced into the sea by waste disposal. Up to 50 per cent of global 'new' production of biomass in the oceans caused by upwelling occurs in the eastern equatorial Pacific.

In 1972, a massive red tide stretched from Maine to Massachusetts following a September hurricane. Since then shellfish toxicity has been detected virtually every year in the region. In 1987, fourteen humpback whales died in Cape Cod Bay, and fishermen and tourists who had eaten mussels from Prince Edward Island started complaining of respiratory problems and eye irritation, and some developed symptoms suggesting neurotoxic poisons. This poisoning was attributed to toxins from red blooms that had accumulated high up the food chain (Anderson 1994). Recently, blooms of algae have begun to threaten the waters of the North Sea and the Mediterranean. Beaches that are now greatly infected include those between Trieste and Rimini, and around the coasts of Malta, Marseilles, Barcelona, Algiers and Alexandria (Pearce 1995a). Milne (1989) described especially alarming blooms of green algae in the seas adjacent to the major estuaries of the North Sea, and illustrated the widespread extent of elevated nutrient levels and how they combine to create oxygen-deficient waters. Pearce (1995a) identified four main estuaries in the UK that are experiencing toxic algal blooms: the Thames, Humber, Tyne and Forth. The biggest problem, however, is in areas along the eastern regions of the North Sea, such as the Rhine, Rhem, Weser and some of the Norwegian fjords. Fortunately, he reports, at the 1987 North Sea ministerial conference held in London, most countries agreed to try to cut the input of nutrients by 50 per cent by 1995 to help reduce the problem.

Dangerous organic chemicals

Amongst the class of dangerous organic chemicals, one of the principal chemical elements is chlorine. Organic chemicals containing chlorine are commonly referred to as 'chlorinated hydrocarbons'. The most notorious are the polychlorinated biphenyls (PCBs) and, to a lesser extent now, dichloro-diphenyl-trichloro-ethane (DDT).

PCBs are used mainly in the manufacture of paints, plastics, adhesives, hydraulic fluids, and electrical components such as generators. Most of the PCBs that eventually enter the world's rivers and seas do so from the atmosphere, having been transported as aerosols. The pesticide DDT is air-sprayed onto crops, and can be transported considerable distances in the atmosphere as well as through rivers and lakes to the sea.

DDT is essentially non-biodegradable because, when it does eventually break down, it forms equally harmful chemicals with a life span measured in decades. PCBs are biodegradable by the action of micro-organisms, but the process takes a long time. PCBs are non-inflammable, unless incinerated at very high temperatures (up to 1,200°C) to stop the formation of very toxic chemicals called furans and dioxins. They are, therefore, very difficult to break down. PCBs are actually more persistent than DDT. Both DDT and PCBs concentrate in the fatty tissues of living organisms and may become concentrated along the food chain as one animal eats another. In many developed nations, legislation has made the use of DDT illegal, but it is still widely used in the agricultural economy of developing countries.

The adverse effects of PCBs and DDT are most pronounced in animals with a backbone, the vertebrates. For example, higher than expected levels of PCBs were found in the 12,000 fish-eating seabirds washed up on the shores of Britain in 1969. These chemicals inhibit normal growth in animal populations and cause the bones of seabirds to be abnormally thin. Thinner bones are weaker and so the affected birds are unable to sustain their flight in the normal way, which makes them more vulnerable to exhausting injury and predation. Ironically, DDT has brought many environmental problems, including the development of DDT-resistant strains of malarial mosquitoes.

PCBs can pose a threat to ocean mammals such as seals, polar bears and whales, which in an extreme case could face extinction without action to stop the release of PCBs into the atmosphere and seas. Recent analyses of the blubber of dead carnivorous marine mammals show a substantial increase in the quantity of absorbed PCBs. High PCB levels in these animals cause infertility. At a level of 50 ppm of PCBs in male blubber (e.g. in whales), there is a sharp cut-off in the production of sperm. In the summer of 1991, the Ministry of Agriculture in the UK discovered PCB levels of up to 320 ppm in a dead baby bottlenose dolphin found off Dyfed, Wales, and 93 ppm in a dead porpoise in Cardigan Bay, Wales. Twenty-two bottlenose dolphins are known to have died in Cardigan Bay in 1989. Deep-ocean killer whales have been shown to contain 410 ppm of PCBs, with values of 833 ppm in dolphins off the coast of Europe, considerable distances away from sources of PCB pollution. Even polar bears in the Arctic have been found to have levels of 10 ppm of PCBs. Cummins, a geneticist at the University of Western Ontario, Canada, has suggested that 'should only 15 per cent of the world's PCBs at present in use, storage, or simply dumped in developing countries, ever enter the oceans they would be sufficient to cause the extinction of most, if not all, marine mammals and the chemical fouling of ocean fisheries, rendering them unsuitable for use by humans'. This view, although at the most alarmist end of the spectrum, is shared to varying degrees by many other scientists. The toxicity of PCBs is unquestionable. The issue is whether or not the environmental consequences of dumping PCBs in the sea will be appreciated at an international level in time to safeguard the survival of marine and other life.

Tributyltin (TBT), a type of **organotin** used as an anti-fouling paint on boats, has also been shown to cause serious damage to marine life such as shellfish. France banned its use in 1982, whereas the British government set limits on the amount of TBT in anti-fouling paints. In 1986, scientists showed that these limits were still too high and shellfish were still being seriously affected. Estuaries where pleasure craft are common still show levels of poisonous chemicals, including TBT, that are unacceptable. As an example, during the summer of 1986, the Crouch River estuary in Essex, UK, showed concentrations of organotins varying from 50–130 nanograms per litre. The target set by the British government for water-quality is 20 nanograms per litre. In that same year, five out of the eight estuaries investigated by the

Ministry of Agriculture, Fisheries and Food (MAFF) did not meet the target.

In Los Angeles, California, the 15,000-strong Surfrider Foundation brought a successful lawsuit against the Louisiana-Pacific Corporation and the Simpson Paper Company for polluting the ocean with effluent from their paper mills near Eureka, including dioxins and other toxic chemicals. These companies had to pay US$5.8 million in fines and invest more than US$50 million in ways of reducing the effects of the dangerous effluent. The financial package, negotiated with the help of the US Environment Protection Agency (EPA), also included provision that the pulp companies do not harm the abalone mollusc, two species of echinoid (sea urchins), and the seaweed, kelp. Settlements like this are unusual at the moment but pave the way for the possible pattern of future lawsuits and ensuing agreements.

There has been much speculation about the global spread of organochlorine compounds. A process called *global distillation* has been proposed, in which organic pollutants move through the atmosphere from relatively warm source regions and condense at colder, higher latitudes onto bodies of water, vegetation and soil. Simonich and Hites (1995) were able to examine this process by considering the global distribution of twenty-two potentially harmful organochlorine compounds present in tree bark samples from ninety sites around the world. Their results showed high concentrations in both developing countries and industrial countries, which continue to be highly contaminated, even though the use of many of the compounds is now restricted. Relatively volatile compounds such as HCBs and HCHs are readily distilled to colder higher latitudes, whereas less volatile compounds such as endosulfan and DDT are not. The global distribution of less volatile compounds was controlled by the regulations and usage of the particular organochlorine compound within the country of origin. An understanding of both the dynamics of transportation and the effects of dangerous organic chemicals is critical because of their potentially harmful effects.

Heavy metals

Mercury, lead, arsenic, tin, cadmium, cobalt, selenium, manganese and copper are examples of what are called **heavy metals**. The natural weathering of rocks on land and volcanic activity introduces these heavy metals into rivers and the sea, but in many instances they are found at unnaturally high concentrations because of mining activities and other industrial processes. Rivers may transport large amounts of these metals to the seas and oceans, where they become even more concentrated. In large enough doses, these metals can prove lethal to organisms, including humans. Heavy metals, therefore, constitute another major category of pollutants in rivers, coastal waters and seas.

Figure 5.5 shows the increase in toxic metals since the beginning of this century in the River Rhine as a result of increased industrialisation. Notice the steady increase of contaminants up until the late 1960s and early 1970s, when levels began to decrease with the increase in environmental awareness and legislation.

The most common toxic pollutants are lead salts (e.g. methyl lead), mercury and arsenic. If these accumulate in the body they can cause brain damage and even death. A particularly serious case that has been well documented was the poisoning of people from the fishing community of Minamata in Japan in the 1950s. The symptoms were numbness of the limbs, speech impairment and loss of co-ordination. The cause was eventually traced to methyl mercury introduced into the area from a nearby factory, but it was some time before the factory admitted liability (Waldbott 1978).

Pure mercury is not as toxic as mercury vapour and soluble mercury compounds. It was originally thought that mercury deposited into a lake or pond would sink to the bottom and remain harmless. But it is now known that bacteria can convert mercury into methylmercury ion (CH_3Hg^+), which is highly soluble and poisonous. This can get into the food chain and ultimately poison humans. Mercury pollution is particularly common in regions where noble metals have been or are being mined, because mercury is important in the extraction process. Its use was particularly widespread between the sixteenth and twentieth centuries in South and Central America, where it was used in a process called amalgamation to extract silver. Nriagu (1993) calculated the amounts of mercury that were lost during this process between 1580 and 1820 in South and Central America to have been 292–1,085 tonnes yr^{-1}, averaging 527 tonnes yr^{-1}, with a cumulative loss of 126,000 tonnes. From 1820 to 1900, the extraction of silver accelerated, but new processing techniques reduced the amount of mercury used in the extraction of silver. Nevertheless, the total losses of mercury during this period were about 70,000 tonnes. Similar discharges of mercury are occurring today in the Amazon basin as a result of the extraction of gold. Lacerda and Salomons (1991) estimated that 1.3 to

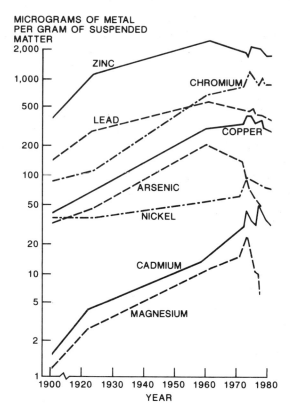

Figure 5.5 *Changes in metal contamination associated with suspended matter in the River Rhine. Redrawn after World Resources Institute (1990).*

1.7 kg of mercury is lost for every kg of gold recovered, and 90–120 tonnes yr^{-1} of gold is produced. Nriagu (1993) estimated that about 10 per cent of the mercury used in the abstraction of silver was lost during transport and storage, 25–30 per cent was left behind in residue or removed by streams, and 60–65 per cent was released into the atmosphere during the processing of the silver.

Modern abstraction of gold produces similar percentage releases into the atmosphere. Atmospheric fluxes of mercury produced by modern industries are in the order of 910 to 6,200 tonnes yr^{-1}, compared to the atmospheric fluxes of mercury in South America between 1587 and 1820 of 180–705 tonnes yr^{-1}. Mercury abandoned in waste tips and deposited in aquatic sediments has the potential to methylate and be released into the atmosphere. Thus silver production was responsible for much of the atmospheric mercury pollution in the past, and the legacy of mercury deposits produced by silver mining and modern gold mining is probably one of the greatest sources of global mercury pollution. The local and regional effects of mercury poisoning are not really known, and there is little literature and few reports on the problem, but given the past and present discharges the effects are probably severe and concentrated in the mining areas of South and Central America, as well as adding to the global distribution of mercury by atmospheric circulation.

Mercury poisoning has also occurred in areas of the Amazon where no mining is undertaken. Mercury has an affinity for organics in soil and it can be methylated by biotic and abiotic processes, to be channelled into the food chain. Veiga *et al.* (1994) attributed the poisoning of people in some areas of the Amazon to high levels of mercury in the fish they eat and the release of mercury into the atmosphere by burning during deforestation. They emphasise that the high rates of deforestation in the Amazon may contribute 25 per cent more mercury to the atmosphere than mining activities, and the mercury is methylated, i.e. in a form more dangerous than is emitted by the gold extraction.

Lead in old water piping can cause serious health problems. In March 1989, the EC published a report showing lead levels in drinking water exceeded those it set in 1985 in every Scottish region. The principal culprit was lead piping. The maximum acceptable lead level set by the EC is 50 mg l^{-1}. The British government has set its own standard at double this figure. Excessive lead levels, even at low dosages, are known to cause hyperactivity in children and to affect their performance in attainment and ability tests. Lead causes kidney damage, metabolic interference, and central and peripheral nervous system toxicity, and it suppresses **biosynthesis** of protein, and nerve and red blood cell formation (Nash 1993). Some experts believe that there is no safe level of lead in drinking water and that the government should provide grants to all householders to remove any existing lead fittings in their water systems.

The single biggest source of lead poisoning, however, is probably car fumes. This lead is introduced into the air by engine exhausts and is blown or washed by rain into water sources. In 1969, Murozumi *et al.* showed that since ancient times the concentration of lead in Greenland snow has increased by a factor of 200, and they concluded that this was probably a result of the use of alkyl-leaded petrol. Partly because of these findings, the USA and other Western countries limited the use of additives in petrol, and concentrations of lead in Greenland snow decreased from 1970 onwards. Using $^{206}Pb/^{207}Pb$ ratios, Rosman *et al.* (1993) were able to trace the sources of the lead pollution and showed that the USA was the major contributor up until 1976, since when, because of regulation, the amounts have decreased. European sources of pollution continued at a lower but steady rate (Box 5.2). As long as people still choose to use leaded rather than unleaded petrol, there is little that we can do to reduce this form of pollution of the environment. Taxation may be used to discourage the use of leaded petrol. Fortunately, with the lower price of unleaded petrol and people's increased awareness of environmental issues the amount of leaded petrol used is slowly being reduced.

Arsenic poisoning is beginning to be considered a major threat to people who use contaminated ground waters (Table 5.7). Arsenic poisoning is cumulative, building up in the body over many years as a result of drinking contaminated water. It causes inflamed eyes (conjunctivitis) and skin lesions, in severe cases melanomas and other cancers, and a form of gangrene, which ultimately leads to death. In West Bengal, as many as 200,000 people are affected by arsenic poisoning (Pearce 1995c). Arsenic levels from wells are between 20 and 70 times the WHO limits. Arsenic poisoning of ground water in this region is the result of increased well-drilling for irrigation in the 1960s. The use of the wells lowered the water table beneath arsenic-bearing sulphides in the rock. The iron sulphides that contained the arsenic were exposed to air and oxidised, releasing the arsenic into the ground water. Arsenic poisoning also occurs in mining areas where rocks that contain iron sulphides are exhumed and oxidised to release the arsenic into water supplies. Such conditions are found around spoil tips associated with the old tin mines in Cornwall.

Aluminium is also a health hazard and is a relatively new addition to the list of chemical elements that are frequently cited in the media as being at

Table 5.7 *Arsenic poisoning throughout the world. The WHO suggests that the safe level of arsenic in drinking water is 10 mg l^{-1}.*

Region	Concentrations of arsenic in water (mg l^{-1})	Numbers of people poisoned
West Bengal	≤ 2,000	200,000
Taiwan	≤ 600	20,000
Inner Mongolia		50,000
Lagunera region of Mexico		20,000
Antofagasta, Chile	≤ 800	20,000
Cordoba, Argentina		10,000
Obuasi, Ghana	175	effects unknown

Source: Pearce 1995.

BOX 5.2 IDENTIFYING SOURCES OF METAL POLLUTANTS

Sceptics argue that there is little direct proof of sea pollution by heavy metals. Sophisticated analytical techniques available today, however, can show the source of pollutants. By measuring the ratio of lead $^{204}Pb/^{206}Pb$ and $^{207}Pb/^{208}Pb$ isotopes (an element whose atom may have different numbers of neutrons in its nucleus to give several atomic varieties of different masses) in near-surface waters of the Pacific Ocean off San Francisco, scientists have been able to trace local pollution to ^{204}Pb, ^{206}Pb, ^{207}Pb and ^{208}Pb industrial lead and polluting waters thousands of kilometres west across the other side of the ocean in Asia! These contaminated ocean currents come to the surface, or upwell, as a cold-water current off California (Fanning 1989). The lead was introduced to the water column from industrial lead aerosols and advected across the Pacific under the influence of the prevailing westerly wind system. Analysing lead isotopes in Antarctic waters, Flegal et al. (1993) were also able to show that anthropogenic inputs of lead extend to the most remote regions.

The analysis of heavy metals that accumulate in the shells of clams, cockles, limpets and other shellfish provides another technique to trace and identify marine pollution. The shells of these organisms are normally crushed and the concentration of metals absorbed over the creatures' lifetimes are measured. New techniques, such as those being developed by Bill Perkins and Nick Pearce at the University of Wales in Aberystwyth, are now able to measure the concentrations of pollutants in seasonal growth layers, thus providing a history of pollution over a creature's lifetime (Coghlan 1994). Using the clam, Arctica islandica, which lives for 80 to 100 years, they have been able to trace a history of pollution in Cardigan Bay, Wales, over the last sixty years, identifying periods of increased pollution that can be attributed to waste discharges from disused mines. Such techniques promise to make polluters more accountable for their actions, and provide more information on the mechanics and nature of pollution.

---

unacceptably high levels in some sources of drinking water. In January 1989, an article appeared in the medical journal *Lancet* that linked high levels of aluminium in tap water to Alzheimer's disease, the most common cause of senile dementia. It is estimated that perhaps one in twenty people over sixty years of age may be affected by this disease. High levels of aluminium tend to occur in naturally acidified water, not only in areas of pollution but also where certain geological rock and soil types exist; for example, above-average levels have been measured in northern England, especially around the River Tyne.

The only known cause of Alzheimer's disease is a genetic defect. A genetic defect in chromosome 21 has been identified by John Hardy of St Mary's Hospital Medical School in London in approximately 5 per cent of the people diagnosed as having Alzheimer's disease (Ferry 1989). Chris Martyn from the Epidemiology Unit of the US Medical Research Council has found that in districts where drinking water contains more than 0.01 mg l^{-1} of aluminium, the incidence of Alzheimer's disease was 1.3–1.5 times higher than in areas with lower aluminium levels.

Using a laser microprobe mass analysis technique, Dan Perl of Mount Sinai Medical School, New York, has identified significantly higher levels of aluminium in the brain cells of people who suffer from 'Guam disease' (named after the Pacific island of Guam, where it is endemic), allied to Parkinson's disease, which causes senile dementia. Perl believes that there are probably close links between these various diseases that cause dementia and elements such as aluminium. The present EU standard for aluminium in drinking water is 0.2 mg^{-1}. There is little available data to say whether even this permitted amount is too much. Statistics from the Ministry of Agriculture, Fisheries and Food in the UK show that wheat contains 2 mg of aluminium kg^{-1}, and dry tea leaves have as much as 1,000 mg kg^{-1}. The Ministry also calculates that the average person has a daily intake of about 6 mg, 10 per cent coming from drinking water and the remaining 90 per cent from food.

The effects of increased aluminium levels in drinking water were dramatically reported in the UK on the BBC television programme *Panorama* on 20 March 1989. The programme revealed what has become known as the 'Camelford incident'. This occurred because a lorry driver delivering aluminium sulphate to the Lowermoor water treatment works in north Cornwall for use as a water purifier dumped 20 tonnes of the hazardous chemical into a purified water tank by mistake. This water was then released to the main supply and delivered to 7,000 households and 20,000 people in the Camelford area. The result was a discoloration of household water and an increase in its acidity, which helped dissolve copper, lead and zinc, further adding to the metal pollutants in the water.

The people of Camelford developed sickness, diarrhoea, mouth and nose ulcers, blistered tongues, bloody urine, and aching muscles. Those suffering from arthritis found themselves in excruciating pain, and some people began to show signs of Alzheimer's disease with a loss of memory. Later, farmers reported that calves were born at only half their normal weight, breeding rabbits became sterile and piglets were born with both male and female parts. The water was found to contain 6,000 times the maximum levels considered safe by the EC and the WHO. The situation was exacerbated by the denial that the water had been polluted by the South West Water Authority (SSWA), which was responsible for the treatment plant. The denials continued despite the complaints by local people and an independent report, on 11 July, by a local biologist and expert on water pollution, Dr Douglas Cross. In fact, the *Panorama* report revealed that the SSWA had known about the pollutant since 8 January but continued to be economical with the truth about the incident, even putting an advertisement in the local paper addressed to the residents of the Camelford area assuring them that the water was fit to drink. The long-term effects have yet to be evaluated.

A new generation of phosphate-eliminating sewage treatment plants, and the increasing prohibition of phosphates in detergents, has led to considerable reductions in phosphate concentrations in surface waters, but this has created new concerns. Replacing phosphates in detergents by complexing agents could cause the increased mobilisation of various toxic heavy metals, which would result in serious pollution to surface waters. However, as a result of a twelve-year analysis of phosphate, manganese and cadmium levels in the River Glatt, Switzerland, and the adjacent aquifer infiltrated by the river water, von Gunten and Lienert (1993) postulate that the lower phosphate levels have decreased the amount of oxidisable organic carbon, thereby creating less reducing conditions in the infiltrating water. Consequently, there is decreased reductive dissolution of Mn and Cd in the ground water and, therefore, an unexpected improvement in drinking water quality with respect to some toxic metals.

Radioactive waste

Radioactive waste forms one of the most frightening pollutants. Much of the waste has a whole range of disastrous short- to long-term effects, from the mutation of genes and the birth of deformed organisms to the development of cancer and death in humans. Worst of all is that radioactive waste is a legacy to future generations. The nature and effects of radioactive waste are dealt with in Chapter 6 on nuclear issues.

Oil pollution

Worldwide, oil pollution has grown steadily with the increased transport and use of oil. It is estimated that over 3.6 Mt of oil is spilled into the sea every year, mainly as a result of shipping accidents involving oil tankers (Figure 5.6). There are many other sources of oil spills, including natural seeps from the sea bed, for example from oil fields via fissures in rocks. Other seeps include offshore oil seeps from drilling and oil exploration, airborne oil droplets from unburnt fuel and from combustion engines on land, and discharges from shipping, either because of washing out engines or tanks, discharges of water used as ballast from oil tankers, or deliberate discharges into the sea as ecological terrorism.

In 1967, the *Torrey Canyon* ran aground off south-west England, leaking 80,000 tonnes of crude oil, which affected 32 km of coastline and killed 40,000–100,000 sea birds. One-third of this oil was burned using bombs, napalm and rockets fired from fighter-bombers and thousands of gallons of oil were removed using oil-dispersing detergents. On 4 February 1970, the supertanker *Arrow* ran aground in Chedabucto Bay, Nova Scotia, releasing 10,000 tonnes of oil to contaminate 300 km of coastline. On 6 August 1983 off the Cape coast of South Africa, the Spanish supertanker, *Castillo de Belver* caught fire and split in two while fully laden with 200,000 tonnes of crude oil. The slick that developed was about 30 km long and up to 5 km wide. On 16 March 1987, the *Amoco Cadiz* ran aground on rocks near Portsall, France, releasing 227,000 tonnes of crude oil and contaminating a 300 km stretch of the Brittany coast. In April 1991, a blazing oil tanker still laden with 100,000 tonnes of oil sank to the sea bed off the Italian Riviera. Of the initial total 140,000 tonnes, 40,000 tonnes had been burned in the fire. Initial oil spills are often exacerbated by the rapid spread of oil across the surface of water aided by waves, wind, tides and ocean currents. This can make even a relatively small spill a widespread problem, but in certain circumstances it actually helps to disperse the thick oil and break it up into less serious patches.

The five largest recorded oil spills, excluding the 1991 Persian Gulf oil slick, were all the result of oil tanker accidents. Recent major oil spills include that of the 17-year-old single-hulled oil tanker *Braer*, operating under a Liberian flag of convenience for the American oil company Ultramar, which ran aground on the southern tip of the Shetland Isles, Scotland, on Tuesday 5 January 1993, after losing all engine power in storm conditions. By the following

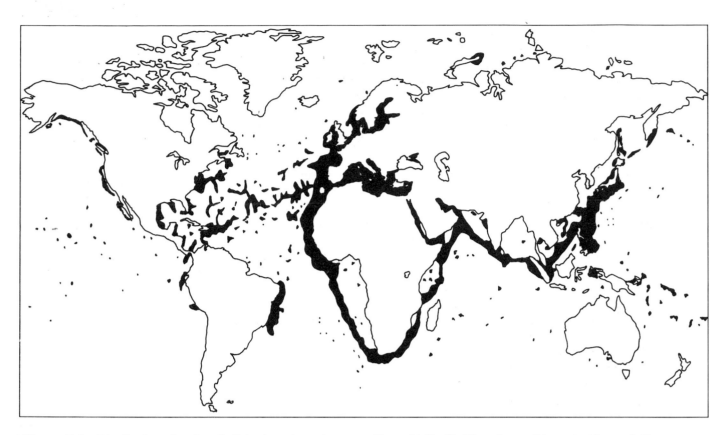

Figure 5.6 *Distribution of major oil slicks throughout the oceans (shown in black). There is a striking coincidence of oil spillage with major shipping routes, for example off the west coast of Africa. Redrawn after Mysak and Lin (1990).*

Wednesday, 12 January, all the 84,500 tonnes of light crude oil that the 89,730-tonne *Braer* had been carrying had escaped into the stormy seas, which had not abated since the disaster first happened. Hurricane-force winds prevented a full-scale clean-up operation from getting under way (although detergents were sprayed from six DC-3 aircraft fitted with infrared tracking equipment in an attempt to disperse the oil slick) and also stopped any attempts at salvaging the stricken tanker before it broke up on 11 January. This oil spill affected important, internationally recognised bird and otter communities around the coast of the Shetlands. In mid-February 1996, the supertanker *Sea Empress* ran aground at St Ann's Head, Dyfed in Wales, to leak about 60,000 tonnes of its 147,000-tonne oil cargo (Plate 5.5). A 13-km-long oil slick formed off Carmarthen Bay to threaten internationally important bird sanctuaries (Box 5.3).

The second-largest oil spill occurred on 24 March 1989, when the *Exxon Valdez* ran aground on Bligh Island, Alaska, leaking 10 million gallons of crude oil over more than 2,300 km² of water and affecting hundreds of miles of shoreline. The spilt oil was often over 10 cm thick along the shore and it is estimated

that less than 4 per cent of it has been removed so far. Amongst the dead were an estimated 3,500–5,500 otters, 200 harbour seals and almost 400,000 birds. More than one year after the catastrophe, scientists studying the effects of the pollution found themselves

Plate 5.5 *Oil slick on beach in South Wales from the* Sea Empress, *1996.*

BOX 5.3 CASE STUDY OF A SUPER-TANKER OIL SPILL: THE *SEA EMPRESS*, 1996

On 15 February 1996, the 147,000-tonne *Sea Empress* went aground on the rocks at St Ann's Head near the entrance to Milford Haven harbour in South Wales. The ship was on its way to discharge its cargo of North Sea oil at the Texaco refinery in Milford Haven. The pilots of the supertanker three times recommended that it be towed back out to sea, but this was overruled by those running the salvage operation, who wanted to tow the ship into the estuary to berth her at an old Esso jetty. On the evening of 19 February, the tugs that were holding it in place, however, lost control in fierce winds and the *Sea Empress* drifted a quarter of a mile down the coast, spilling more oil than before. It took six more days before salvors eventually brought her to berth at Milford Haven. The operation was slowed down by the poor weather, a small outbreak of fire and the danger of the ship exploding, as well as procrastination among the salvors, poor implementation of remedial programmes to cope with the spillage and an inadequate contingency programme.

As a result, 60,000 tonnes of oil were spilled, spreading over 250 km along the south and west coasts of Wales, washing up on beaches and reaching as far as Lundy, an island in the Bristol Channel, about 65 km to the south. Many regions it affected were designated Sites of Special Scientific Interest (SSSIs) and bird sanctuaries. More than 20,000 birds were affected in the first weeks, and the eventual death toll is estimated to have reached 50,000. The effect on the marine mammals in this area has still not been assessed. Environmentalists argue that this oil spill was the biggest environmental disaster since the *Torrey Canyon* in 1967 (Midgley and Nuttall 1996). It is believed to have been worse than the *Braer* disaster, because the oil was denser and more like an emulsion, and the concentration of marine life is larger than in the

Shetland Isles, with sanctuaries on islands such as Skomer, Lundy and Grassholm. Also, the time of the year exacerbated the problems, because spring birds had begun to return to nest and breed, and these will therefore have been affected in greater numbers (*ibid.*).

Immediately after the disaster, 25,000 tonnes of oil was successfully transferred from the tanker, and several marine pollution control planes sprayed detergent onto a five-mile slick that was drifting eastwards, while relief teams started to collect affected wildlife, and clean-up teams began to collect the oil from the beaches. The clean-up operation, however, is estimated to have cost £10 million, and compensation claims are likely to be in the region of £18 million to £20 million, particularly from fishermen and workers in the tourist industry.

This was the second incident in four months, with the Swedish tanker *Borga* grounding off St Ann's in October 1995. Fortunately, the *Borga* was double-hulled and there was no spillage, unlike the *Sea Empress*, which was three years old and did not have a protective outer hull, as is now required. The entrance to Milford Haven is quite treacherous because of the narrow channels of deep water, high tidal range and fast currents within the estuary. The *Sea Empress* was managed by a Glasgow company, Acomarit (UK) Ltd, and like the *Braer* it was Liberian-registered. It was manned by a Russian crew. The British National Union of Marine, Aviation and Shipping Transport argued that such ships are not subject to the same kind of inspection as British ships and are therefore at greater risk of a disaster occurring (Bowcott 1996).

Millions of tonnes of oil are brought into Milford Haven each year and there are now plans to transport Orimulsion, a fuel that contains particularly noxious chemicals, some of which mimic oestrogen, to fuel Pembrokeshire's power stations (Lean 1996). The consequences of an Orimulsion spillage would be catastrophic, and the history of Milford Haven does not leave much room for optimism.

---

increasingly embroiled in litigation as lawsuits were filed against Exxon. Researchers were hired by Exxon, the state of Alaska and the US government to assess the extent of the damage. Besides the two lawsuits with the highest profile, those of the state of Alaska and the US Department of Justice, which each sought millions of US dollars in compensation, there are about another 150 lawsuits by environmental groups, local organisations and individuals. So, with all this litigation under way, many of the professional assessments of the environmental damage remained unpublished and sensitive. And with the information remaining secret, the wider scientific community could not consider the actual damage and begin to use the data to develop strategies to reduce any environmental impact that a future major oil spill might have.

Major oil spills in coastal waters raise a number of important issues about the transport of oil and the

risk of oil pollution. Amongst these issues, the principal concerns are:

- ageing oil tankers, commonly only single-hulled, sailing under flags of convenience with inexperienced international crews who may have problems communicating with each other in an emergency, all done in order to save money at the expense of the environmental and safety factors;
- the routes taken by oil tankers, travelling close to ecologically sensitive and important coastal habitats for birds and other marine life;
- the procedures to be adopted by the crew when a tanker appears to be out of control, and the responsibility of captains for taking the decision to radio a mayday call for assistance without awaiting a response from the owners; and
- clean-up procedures and technology following a major oil pollution event, for example the desira-

Plate 22 *(previous page) The damaged reactor number 4, the* Sarcophagus, *at Chernobyl following the accident in 1986.* Courtesy of Rex Features.

Plate 23 *Reactor number 4 at Chernobyl, which was damaged during the accident in 1986. The reactor is now contained within a new construction, the* Sarcophagus, *but radiation is still 40 times the normal background radiation levels. Note the unprotected workers in the foreground.* Courtesy of Anders Gunnartz/Panos Pictures.

Plate 24 *The Hoover Dam in Nevada, famed for the earthquakes it initiated.* Courtesy of Comstock.

bility of using detergents, which also cause marine pollution, in order to break up the oil.

In January and February 1991, during the Gulf War between Iraq and the American-led, United Nations-backed allied forces, Iraq's President, Saddam Hussein, resorted to ecological terrorism by ordering oil to be discharged into the Gulf of Arabia off the coast of Kuwait by the Iraqi invasion forces. Additionally, so bad was the pollution caused by the huge plumes of smoke from more than 600 oil-well heads deliberately set ablaze by the Iraqis during the Gulf War that thick layers of soot and sludge were found on the mountains of Kashmir more than 2,000 miles away. Saddam Hussein's scorched-earth policy in Kuwait may have far-reaching consequences for the environment and it will be many years before these can be fully assessed as part of the cost of the Gulf War.

The deliberate sabotage and discharge of vast quantities of oil from oil refineries by the Iraqi army created one of the biggest oil slicks ever known, though the exact quantity of oil released remains uncertain. Initial estimates of 10–11 million barrels, used by environmental groups and politicians, were subsequently revised down to about two million barrels. In fact, 50 per cent of the oil spills evaporated within the first 24 hours. The oil slick spread over 150 km along the coasts of Kuwait and Saudi Arabia, and had catastrophic effects on the ecosystems of Abu Ali Bay, but it did not affect the large oil refinery port of Jubail farther south. The effects of pollution were enhanced over the extensive inter-tidal zones or flats, commonly 2 km wide, by the low tidal range in the Gulf. It is in these zones, which support vast quantities of life, that much of the early devastation occurred.

Abu Ali Bay became the focal point for conservationists and environmentalists to wade in oily waters and to present the aftermath of the Gulf War as an ecosystem in crisis. The sea bed was smothered with tar that rained down through the water column as balls, and it was sterilised of life below the slick, apart from the proliferation of oil-consuming bacteria. The darkest nightmares of pollution were indeed realised here, but further offshore the coral reefs and fish life, together with the rest of the marine life and fishing catches, seemed to survive unscathed. Was this really the environmental catastrophe that the media and environmentalists offered to the world? The question was discussed in many political and scientific arenas across the world (e.g. Bakan *et al.* 1991, Browning *et al.* 1991). After assessing all the evidence, the verdict was that the Gulf War was more of a human than an ecological disaster. Yes, environmental damage has been done, but the amount is, thankfully, much smaller than had originally been feared. In fact, the Gulf is being polluted not so much by Saddam Hussein's environmental terrorism as by the slow and inexorable long-term spillage of oil through accidents and deliberate acts. In the Gulf, about one million barrels of oil per year finds its way into the sea water, and three-quarters of Saudi Arabia's beaches were polluted before the Gulf War.

The effect of oil pollution can be disastrous for many ecosystems. The behaviour of oil when spilled into the sea, and its dispersal and degradation, are shown in Figures 5.7A and 5.7B. Oil, especially gasoline, affects fish because it is toxic, coating the gills and causing suffocation, and also because it covers the surface of the sea, thus inhibiting the diffusion of oxygen into the water. Oil sinks to the bottom of the sea, where it may immobilise sperm and reduce fertilisation rates. Similarly, oil affects birds by coating their feathers, which reduces buoyancy and insulation. Lipid pneumonia and intestinal irritation can result. It may also affect birds' eggs, reducing the permeability of the egg, and it may cut down the food supply as the fish life dies.

Oil pollution also affects many other organisms, such as plants and invertebrates, all an essential part of food webs and ecosystems. In the Gulf, oil represents a major threat with the destruction of vast quantities of sea grass in the subtidal zone (only a few metres deep) and algae in the inter-tidal zone. These plants form the base of important food webs, providing the food for many invertebrates, which in turn provide food for higher organisms. Destruction of these food webs means the death of vast numbers of marine animals, many of which are already under threat, such as the manatee (sea cow), which grazes on sea grass.

The obvious immediate effects of oil spillage on humans is to poison the marine life, thus reducing the consumption of fish and fish yields. Oil is inflammable and explosive. Beaches contaminated with oil are unattractive for recreation, to such an extent that tourism in polluted areas may decline along with investment from tourist revenue. Inland, oil can pollute local water supplies and cause the build-up of various toxins. As an example, the extraction of oil from the oil shales of Estonia is causing serious pollution problems, with local water wells being polluted and large concentrations of heavy metals in the Baltic Sea. The technology for this oil production is decades old and needs updating to clean up the area.

Various methods are available to try and constrain the spread of an oil spill and to disperse the oil. These

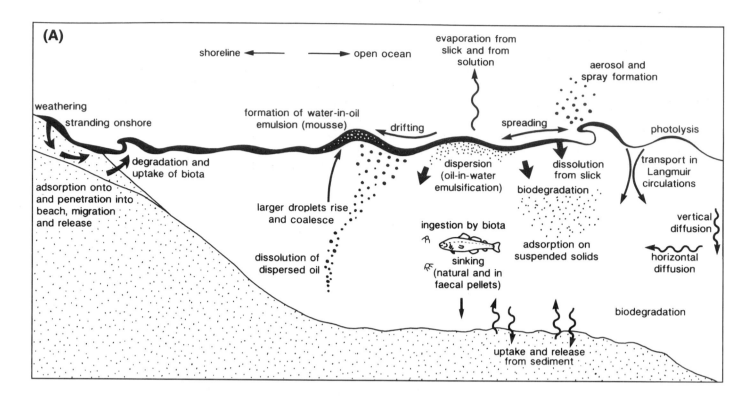

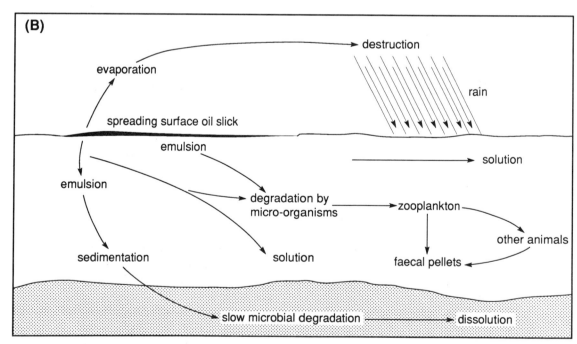

Figure 5.7 (A) The behaviour of oil released into the sea. After Open University (1991). (B) The persistence of pollution in the oceans, its dispersal and degradation. Redrawn after Smith and Warr (1991).

include burning the oil, using high-capacity pumps to draw off the surface oil, adhesion collectors (which are mechanical devices that scrape the oil up from the surface of the water), the use of sinking agents (e.g. carbosand to settle the oil onto the sea bed), and using emulsifiers and dispersants to break the oil into droplets to increase its surface area and so aid bacterial degradation of the oil. Biological agents such as desulfovibrio and desulfomaculum are also employed, and act by using the oil as a food source. Bragg *et al.* (1994) reported that the effectiveness of fertiliser applications, which stimulate the growth of hydrocarbon-degrading micro-organisms within the intertidal zone, had significantly increased the rates of bioremediation for the *Exxon Valdez* oil spill. The rate of **biodegradation** depends mainly on the concentration of nitrogen near the shoreline, the oil loading, and the extent to which biodegradation has already taken place. Swannell (1993) points out that although remediation using microbes was successful on the cobbly, rocky shorelines of Alaska, the technique might be more limited if oil spills occurred on mudflats, fine sand and salt marshes, where oxygen is in short supply and biodegradation would be reduced.

International treaties on oil pollution were initially conceived in 1926 and 1935, but it was not until 1954 that the first treaty was signed. This was the International Convention for Prevention of Pollution of the Sea by Oil (OILPOL), requiring all tankers to keep the oil content of discharges below 100 ppm when within 50 miles of land, but it was unregulated on the high seas. Governments were required to impose fines large enough to deter violation and to provide facilities to receive waste oil. Minor amendments to the treaty were made in 1962 and 1969, so that tankers would consolidate ballast and improve their cleaning operations. In 1983, the International Convention for Prevention of Pollution from Ships (MARPOL) took effect, retaining the 1969 discharge standards but requiring that all new tankers should be fitted with segregated ballast tanks (SBTs), which remove a major source of oil pollution from ships. MARPOL allowed governments to detain foreign-flagged ships for violating oil pollution laws. To date, seven countries have exercised that right. Mitchell (1995) emphasises that for these treaties to be effective there must be sufficient economic incentive to comply and the authority for monitoring and enforcement. Any future amendments to the treaties must, therefore, take these factors into consideration.

A recent issue involving the oil industry arose over the proposal by Shell Oil Company to dump the redundant *Brent Spar* oil platform at sea in the North Atlantic (Box 5.4). Since Shell was thwarted in its plans to dump the *Brent Spar* because of a highly publicised and concerted effort by the international environmental pressure group Greenpeace to force a change of plans, the future of dumping oil platforms at sea remains unresolved.

The short-term effects of oil pollution are easy to observe. The long-term effects, however, on ecosystems are still not known. Particularly disturbing may be the effect on the benthic fauna (sea-floor dwelling animals) as oil settles on the ocean floor. These latter risks have not really been assessed.

It is not only the oceans that are polluted by oil spills. In mid-1994, major leaks from the Khararyaga–Usinsk regional oil pipeline in northern Russia became a major environmental concern as it polluted large areas of forest and wetland. The pipeline stretches across 160 km, carrying as much as 220,000 barrels of oil each day from the Pechora oil fields to a larger pipeline, which carries the oil into the Russian heartland. Oil has been pumped from this area since the mid-1970s. During this time, small leaks were ignored or simply bulldozed over or burned, but since 1994, however, the pipeline has leaked some 730,000 barrels of oil (> 100,000 tonnes: Menon 1996). This is nearly three times the amount spilt by the *Exxon Valdez*. The oil has seeped from rusted pipes into forested, boggy land and streams. By late 1994, the oil had begun to enter the Kolva River, the main water source for local people. In March 1995, the World Bank and a European redevelopment fund provided a US$124 million loan to help with the clean-up. By October 1995, approximately 90 per cent of the oil had been cleaned up by the Alaska based oil clean-up company, Hartec Management Consultants. To aid the clean-up, however, many large dams and roads were constructed, and there is much concern over the environmental degradation associated with their construction.

This oil spill is a sombre warning of possible future catastrophes, since most of Siberia's pipelines are old and corroded, and are forced to carry ever more oil to the cash-starved Russian economy.

Beaches

Many of the world's beaches, both in well-known coastal resorts and those that are more remote, are suffering from serious water pollution. Untreated sewage, oil, industrial chemicals and radioactive waste are all contributors to the pollution. Plastics discarded from ships and land form another major type of pollution on many of the world's beaches. Most of these plastics are not biodegradable and therefore represent a cumulative form of pollution.

Untreated sewage is the most common culprit. Many of the beaches in the western Mediterranean are unsightly and represent a health hazard because of sewage. This problem is not confined to Europe. Industrial waste and raw sewage threaten the famous Guanabara Bay beaches beneath the statue of Christ

BOX 5.4 DECOMMISSIONING MARINE OIL RIGS: LESSONS FROM THE *BRENT SPAR* PLATFORM INCIDENT

In any extractive industry there is always the eventual need to decommission industrial plant and machinery but, perhaps besides the nuclear industry, this is nowhere more sensitive an issue than in the oil industry. The case of the proposed dumping at sea of the *Brent Spar* platform in the northern North Sea, owned and operated by the Shell Oil Company, became a *cause célèbre* in that its decommissioning raised a whole spectrum of environmental issues and a polarisation of attitudes between the Shell Oil Company and the environmental pressure group Greenpeace. From the public's perspective, the conflicting scientific claims made by each side, and the lack of agreement between so-called independent experts, not only made any balanced news reporting difficult but also tended to act against the Shell Oil Company and in favour of Greenpeace and the anti-dumping lobby.

Retrospectively, it seems fair to conclude that the emotive reaction to the *Brent Spar* incident stemmed from inaccurate estimates by Greenpeace of the actual quantities of potentially toxic chemicals being left on the platform. To worldwide media attention, Greenpeace activists had occupied the platform and refused to leave, while Shell had tried various tactics to evacuate them, finally being thwarted by the British government's *volte face* from explicit support for Shell's dumping at sea policy to its bringing pressure to bear on the company not to dump the *Brent Spar* at sea, but rather to tow it to shallower waters in a Norwegian fjord for manual dismantling. Although at the time the action by Greenpeace was widely seen as a victory in terms of stopping the dumping of this redundant oil platform at sea, the wildly inaccurate figures used by Greenpeace later allowed Shell to win the intellectual high ground. The issues raised by this incident, however, as to the best course of action, and as a precedent for future dumping of oil platforms at sea, remain unresolved and open to diametrically opposed views as to what should happen in the future.

Some Earth scientists have argued that if oil platforms like the *Brent Spar* were dumped on the flanks of mid-ocean ridges – the sites of the creation of new ocean floor – such as along the Mid-Atlantic Ridge, where natural hydrothermal systems can vent 5×10^5 to 5×10^6 tonnes yr^{-1} of metals, then the amount of heavy metals in a discarded oil platform would be orders of magnitude less and, therefore, any environmental damage would probably be minimal (Nisbet and Fowler 1995). Indeed, it is even possible that the dumping of extra metals would act as a nutrient source for the local ecosystem (*ibid.*). Even if these arguments were to be accepted both by governments and the companies involved, and the flanks of oceanic spreading centres were selected for the disposal of redundant oil platforms, there are still moral-ethical and environmental arguments against this. Hydrothermal vent systems along mid-ocean ridges are the sites of very spectacular and delicate, slow-growing sulphide chimneys, called **black smokers** because of the black, metal sulphide-rich gases that emanate from them. These areas support a diverse and unique biota that includes grey limpets, polychaete worm species and giant clams, which live at depths of up to thousands of metres around the vents, where fluids emerge at temperatures $> 350°C$. These hydrothermal vent fields require much more research before there is a complete inventory of their biota. The dumping of oil platforms over such sites, which arguably should be designated as international submarine wilderness parks for the benefit of future generations and without exploitation (cf. the Antarctica Treaty) – as sites of special scientific interest – could cause environmental damage, both physical and chemical. Also, there are strong environmental arguments against using the ocean floors as a vast dumping ground for human waste simply because they are deep and out of sight, instead of taking full responsibility for dismantling, making safe and/or recycling, even at some considerable economic cost, the waste and by-products of this industrial age.

that towers over Rio de Janeiro. The whales that once inhabited the bay have gone and the dolphin population has dwindled.

In Australia, Sydney's Bondi Beach, the third-largest tourist attraction in the country, is unsafe for bathing two days out of five because of the deposits of raw sewage and industrial waste (Beder 1990). Sydney has three main outfalls and four minor ones for the discharge of sewage. The largest discharges 640 million litres (Ml) of sewage per day in dry weather, while the outfall on the northern headland of Bondi provides 165 Ml of sewage a day. Immediately north of Sydney harbour, the third outfall contributes 325 Ml of sewage a day. All of this sewage undergoes only a primary treatment process to remove most of the solids. In the USA, secondary treatment is mandatory. A 1987 study of the levels of heavy metals and pesticides in fish caught near one of the sewage outfalls (Malabar), leaked to the *Sydney Morning Herald* in January 1989, found dangerously high levels of organochlorines (such as benzene hexachloride, DDT and heptachlor epoxide). In eight red morwong that were caught 300 m from the outfall, the average level of benzene hexachloride was 122 times greater than the recommended limit set by the Australian National Health and Medical Research Council.

In Britain, around the nuclear power station of Sellafield, the once popular beaches on the Irish Sea coast west of the Lake District are contaminated by sea water with high levels of radioactive materials.

Ground water

After the oceans, ground water is the Earth's largest storage of water, comprising approximately 50×10^6 km^3, of which 4×10^6 km^3 is fresh water (Price 1991). It provides 98 per cent of the available global fresh water (Hiscock 1994). In the UK, for example, ground water makes up 35 per cent of the total public fresh water supply. Ground water is also important because it helps sustain the flow in rivers and forms an important component within the hydrological cycle (see Chapter 1). The study of ground water, hydrogcology, is one of the fastest-growing branches of geology, particularly because more money is spent on obtaining water than petroleum, and because of the growing concern over the pollution of this valuable resource by seepage of pollutants from contaminated ground.

Water is stored in the ground mainly within small pores between grains in rocks. The **porosity** of a rock is dependent upon the grain size distribution, the grain shape, the degree to which the grains are compacted and cement within the rock. It is strictly defined as the ratio of voids to total rock volume and may vary from about 26–53 per cent for unlithified sand to 5–30 per cent for a sandstone, and 0–10 per cent for dense crystalline rock to as much as 45 per cent when the crystalline rock is weathered. The ability of a rock to store water, therefore, relies heavily on its ability to hold water – i.e. porosity. But water must also be able to flow through the rock, to supply rivers as well as providing a usable source for humans. The **permeability** of a rock describes its ability to allow water to move through it. Permeability depends on the connectivity of pores and cavities, and discontinuities such as bedding planes, joints, fissures and geological faults. Rocks that are both porous and permeable enough to store water constitute **aquifers**. As an example, Figure 5.8 shows the major aquifers in the United States: one of the largest continuous aquifers in the world is the Ogallala Aquifer, which stretches under the Great Plains, from South Dakota to Texas.

Water enters an aquifer by percolating through the ground or from other ground-water sources, a process known as recharging. The level to which all the pores are filled with water is called the **water table**; below this is the **saturated zone**, where all the pores are filled with water. Above the water table, the **unsaturated zone** may contain water drawn up by capillary action or water that is percolating downwards towards the water table. Water supplies may be trapped within aquifers by **confining layers** of impermeable rock. The trapped water, however, can be tapped by drilling wells. This water may be under **artesian pressure**, resulting from the height of the column of water overlying it, referred to as the head. This **hydraulic head** (h_L) controls the height at which water stands in a well. The level at which water stands in a well defines an imaginary surface, which is known as the *potentiometric surface*. When this surface extends above the ground, the well will overflow or springs will develop.

Darcy's Law describes the rate of flow of water through rock (Q). It essentially describes the flow of water through a cylinder of rock, such that $Q = A(h_L/l)$ where h_L/l is the hydraulic gradient (l is the distance along the cylinder), and A is the cross-sectional area of the cylinder. The law is commonly cited as:

$$Q = KA(h_L/l)$$

where K is a constant of proportionality, known as the *hydraulic conductivity*. K is measured in ms^{-1} and really describes the ability of a rock to allow water to flow through it. This can be extremely slow, ranging from 3×10^{-14} to 2×10^{-10} ms^{-1} for unfractured crystalline rocks to 3×10^{-4} to 3×10^{-2} ms^{-1} for unlithified gravels (Domenico and Schwartz 1990).

Darcy's Law forms the basis for much of ground-water studies, allowing hydrogeologists to determine how quickly ground-water sources can be recharged after being tapped and how fast pollutants can be dispersed through the ground.

Although the slow movement of ground water allows much of the bacterial contamination in the original surface or near-surface waters to die and decay by the time the ground water reaches the well source, it also means that pollutants, which do not degrade over time, may be present within ground-water sources many years after they have percolated into the ground. Dispersion is a phenomenon that is important in understanding the pollution of ground water. It describes how a pollutant is spread out over time to form a plume along and perpendicular to the flow direction from the point where it was released. This phenomenon means that the flow will become diluted, and the period of time the pollutant will travel past a point will be increased considerably. In this respect, pesticides, many of which are now banned, and nitrates from farming activities that have percolated into ground waters, will remain in ground-water sources for long periods and will have widespread effects. Hydrogeologists are also becoming increasingly concerned about pollutants that were buried in landfill sites many decades ago, before regulation and accurate documentation. Table 5.8 lists the twenty most common pollutants

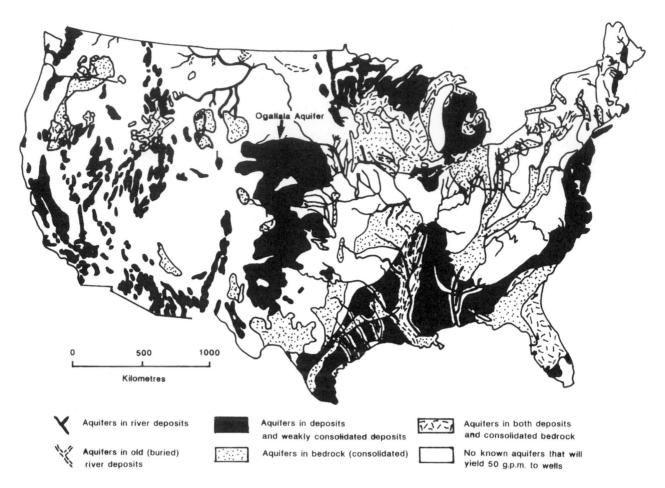

Figure 5.8   *The major aquifers in the United States. Only those yielding ≥ 250 litres min⁻¹ of water are shown. Redrawn after Marsh and Grossa (1996).*

associated with landfill sites in the USA. The contamination of ground water may occur as a result of many other activities, as shown in Table 5.9. The main dangerous pollutants associated with each type of contamination are listed in Table 5.10.

Once ground water has been polluted the remedial measures are expensive. There are three basic ways of dealing with the pollutant: containment; contaminant withdrawal; and *in situ* treatment of contaminants. Containment involves stopping the spread by using such measures as constructing slurry walls of low-permeability material; driving interlocking steel sheet piles into the ground; grouting; and the use of geomembranes. Contaminant withdrawal involves pumping to remove the contaminant, interception systems such as drains and trenches to collect the contaminant close to the water table; soil venting to remove volatile organic compounds, such as soil gases, from the unsaturated zone; and excavation to remove the contaminant disposal elsewhere. Disposing of halogenated organic compounds, including organochlorine pesticides such as DDT,

PCBs, and dioxins is difficult because of their high toxicity and their persistence, and most techniques involve high-temperature chemical reactions that produce toxic emissions or by-products. Rowland *et al.* (1994), however, have recently developed a technique that involves the mechanical milling of toxic waste at low temperatures which breaks down the toxic molecules producing non-toxic by-products. This promises to provide a safe and efficient way of dealing with toxic waste. The third method of remediation involves the *in situ* treatment of waste by either biological or chemical degradation techniques. Chemical or biological sources, such as bacteria, are introduced into the ground to degrade, detoxify or immobilise the contaminant (Domenico and Schwartz 1990). Alternatively, management options can be taken to avoid health problems or the shutdown of a source of contamination without specifically dealing with the contamination in the ground. These options are shown and assessed in Table 5.11.

In the UK, there has been great concern over the risk of contamination of the chalk aquifer that under-

Table 5.8 *The twenty most abundant organic compounds found at 183 waste-disposal sites in the USA.*

Rank	Ground water contaminant	Sites
1	Trichloroethane	63
2	Methylene chloride	57
3	Tetrachloroethane	57
4	Toluene	57
5	1,1-Dichloroethane	52
6	bis-2-ethylhexylphthalate	52
7	Benzene	50
8	1,2-trans-Dichloroethane	50
9	1,1,1-Trichloroethane	49
10	Chloroform	46
11	Ethyl benzene	46
12	1,2-Dichloroethane	39
13	1,1-Dichloroethylene	37
14	Phenol	35
15	Vinyl chloride	30
16	Chlorobenzene	30
17	Di-*n*-butyl phthalate	28
18	Naphthalene	23
19	Chloroethane	23
20	Acetone	22

Source: After Plumb and Pitchford 1985 in Domenico and Schwatz 1990.

lies London. This provides much of the water for millions of people who live in London. Of particular concern is the likelihood that a major spillage of hazardous chemicals could occur as a result of a traffic accident at one of the major motorway junctions around the M25, which circles London. The British Geological Survey has undertaken detailed studies at the major junctions to assess the effects and possible mitigation procedures should an accident occur. As human activity accelerates the potential dangers to ground-water sources increase.

Another concern with ground water is its continued depletion as a result of population growth and increased urbanisation, which places greater demand on water resources. In Chicago, the abstraction of water has lowered the water table by 200 m (Goudie 1993b), and Figure 5.9 shows the declining levels in ground water in the London area prior to major development and up to 1985. Lowering of ground-water levels creates a variety of problems, ranging from subsidence as rock compresses as the water that once filled pores and helped support the bulk of the rock is lost, to a reduction in industrial activities as water sources are depleted. In coastal regions, a thin fresh-water layer overlies saline ground waters. When ground water is abstracted, the lowering of the water table induces a corresponding rise in the saline–fresh water interface. Saline water

migrates inland and may eventually reach the well. This intrusion of salt water is one of the major problems for people living in small island communities and coastal regions. Careful management of water sources is therefore very necessary in such environments.

Hydropolitics

Water is the most precious resource on Earth, more valuable than gold and diamonds, oil or land. Humans have controlled the course of rivers for thousands of years in order to irrigate land and make it more fertile (Plates 5.6 and 5.7). Its superabundance, however,

Plate 5.6 *Providing clean water at Communidad Santa Martha, El Salvador.*
Courtesy of Rhodri Jones/Oxfam.

Table 5.9 *Sources of ground water contamination.*

CATEGORY I – Sources designed to discharge substances

Subsurface percolation (e.g. septic tanks and cesspools)
Injection well
Hazardous waste
Non-hazardous waste (e.g. brine disposal and drainage)
Non-waste (e.g. enhanced recovery, artificial recycling, solution mining, *in situ* mining)
Land application
Waste water (e.g. spray irrigation)
Waste water by-products (e.g. sludge)
Hazardous waste
Non-hazardous waste

CATEGORY II – Sources designed to store, treat, and/or dispose of substances; discharge through unplanned release

Landfills
Hazardous industrial waste
Non-hazardous industrial waste
Municipal sanitary
Open dumps including illegal dumping (waste)
Residential (or local) disposal (waste)
Surface impoundments
Hazardous waste
Non-hazardous waste
Waste tailings
Waste piles
Hazardous waste
Non-hazardous waste
Material stockpiles (non-waste)
Graveyards
Animal burial
Above-ground storage tanks
Hazardous waste
Non-hazardous waste
Non-waste
Underground storage tanks
Hazardous waste
Non-hazardous waste
Non-waste
Containers
Hazardous waste
Non-hazardous waste
Non-waste
Open burning and detonation sites
Radioactive disposal sites

CATEGORY III – Sources designed to retain substances during transport or transmission

Pipelines
Hazardous waste
Non-hazardous waste
Non-waste
Materials transport and transfer operations
Hazardous waste
Non-hazardous waste
Non-waste

CATEGORY IV – Sources discharging substances as consequence of other planned activities

Irrigation practices (e.g. return flow)
Pesticide applications
Fertiliser applications
Animal feeding operations
De-icing salts applications
Urban run-off
Percolation of atmospheric pollutants
Mining and mine drainage
Surface mine related
Underground mine related

CATEGORY V – Sources providing conduit or induced discharge through altered flow patterns

Production wells
Oil (and gas) well
Geothermal and heat recovery wells
Water supply wells
Other wells (non-waste)
Monitoring wells
Exploration wells
Construction excavation

CATEGORY VI – Naturally occurring sources whose discharge is created and/or exacerbated by human activity

Ground water–surface water interactions
Natural leaching
Salt-water intrusion/brackish water upconing (or intrusion of other poor quality natural water)

Source: Office of Technology Assessment 1984.

commonly makes it an under-valued resource, although this is not the case in every country. In countries and regions where there is a scarcity of clean water, water management is a potential source of co-operation or conflict. Water is a commodity to be bargained with, bought and sold, and used to influence policy in neighbouring countries. Figure 5.10 shows the total water and global fresh-water reserves.

In the Middle East, where the climate is arid, rivers are few, population growth is rapid (only exceeded by Africa), nations are already hostile towards one another, and most of the significant rivers flow through more than one country, the conditions are ripe for political tension and military conflagrations over water (see Box 5.5). Examples of tension over water resources are many. Israel currently taps an

Table 5.10 *Occurrence of organic contaminants in relation to potential sources.*

	Organic chemicals			
	Aromatic hydrocarbons	Oxygenated hydrocarbons	Hydrocarbons with specific elements	Other hydrocarbons
CATEGORY I				
Subsurface percolation	■□	■□	■□	■□
Injection well	□	□	■□	□
Land application				
Waste water				■□
Waste water by-products				■□
Hazardous waste	■□		■□	
CATEGORY II				
Landfills	■□	■□	■□	■□
Open dumps	■□	■□	■□	■□
Residential disposal	■□	■□	■□	■□
Surface impoundments	■□	■□	■□	■□
Waste tailings				
Waste piles				
Material stockpiles				
Graveyards				
Animal burial				
Above-ground storage tanks	□	□	□	□
Underground storage tanks	■□	■□	■□	■□
Containers	□	□	□	□
Open burning and detonation sites				
Radioactive disposal sites				
CATEGORY III				
Pipelines	□	□	□	■□
Materials transport				
and transfer operation	□	■□	■□	■□
CATEGORY IV				
Irrigation practice				
Pesticide applications			■□	
Fertiliser applications			■□	
Animal feeding operations				
De-icing salts applications				
Urban run-off	□	□	■□	□
Percolation of atmospheric pollutants				
Mining and mine drainage				
Surface mine related				
Underground mine related				

Key: ■ Contaminant in class has been found in ground water associated with source.
   □ Potential exists for contaminant in class to be found in ground water associated with source.
Source: Office of Technology Assessment 1984, after Domenico and Schwartz 1990.

aquifer in the West Bank, and has done so since the Six-Day War in 1967. The Palestinians claim that much of this water is transferred to the rest of Israel or used by Israeli settlers in the occupied territory.

In the 1950s, Israel diverted waters from the Sea of Galilee (Lake Tiberias) to the densely populated coastal strip and farther south into the Negev Desert. The result was to divert saline springs into Jordan and so degrade the quality and quantity of useful water flowing into that country. In the 1960s, Jordan planned to dam a tributary of the River Jordan, the Yarmuk, which rises in Syria and flows along part of the border between the two countries, but also forms a border between Israel and Jordan. Israel would not entertain the project without Jordan's guarantee of its right to Yarmuk waters. Shortly afterwards, the Six-Day War ensued, Israel occupied the Golan Heights and the site of the proposed dam, and the

Table 5.11 *Assessment of management alternatives to the remediation of contaminated ground waters.*

Technique	Principal components of uncertainty affecting performance	Development status	
		Summary*	Remarks
Limit/ terminate aquifer use	Ability to shut down domestic due to possible public resistance The ability to enforce usage patterns in cases of environmental exposure (e.g. to sport fish)	b	Historically this is a common response to aquifer contamination
Develop alternative water supply	Availability of water supply alternatives, especially in water-short areas, which may, in turn, limit the long-term growth of an area	b	In conjunction with limiting/terminating aquifer use, alternative water supply development is a frequently implemented response
Purchase alternative water supply	Reliance on imports, especially in water-short areas, which may be terminated or depleted Potential opposition to inter-basin transfers	b	This is a frequently implemented response although generally considered a short-term solution
Source removal	Increased contaminant migration (e.g. via breakage of drums or additional infiltration during precipitation) Availability of secure disposal options Extent of contamination and resulting costs	a	Conventional construction techniques are used for source removal although substantial increases in health and safety precautions are required for ground-water contamination applications. Current activity already involves significant health and safety measures
Monitoring	Undetected plume migration because of improper placement or sampling of wells Mistakes are difficult to detect until a problem occurs or back-up wells around key exposure points are installed	a,b	Conventional technology is used for monitoring ground-water contamination problems and conducting hydrological investigations. If methods are used properly, reliable plume delineation and migration data can be generated
Health advisories	The ability to enforce usage patterns in cases of environmental exposure The ability to shut down domestic wells due to possible public resistance	b	This option is a conventional practice of state and local health departments
Accept increased risk	The ability to predict contaminant migration Corrective action alternatives can be more expensive as the contaminant spreads out (i.e. a larger plume)	b	Historically this option is the response to many contamination incidents. Impacts on population are unclear

*Key: a – Technology is proven; performance data are available from applications to ground-water contamination problems.
    b – Technology has been applied historically, for example before the development of regulatory programmes and consideration of potential long-term impact.
Source: Office of Technology Assessment 1984, after Domenico and Schwartz 1990.

Jordanians shelved the scheme. Jordan and Syria now have a new site for a dam and reservoir, but to date international funding has not been forthcoming without the agreement of all the parties involved. Israel has so far declined to support this scheme. Actually, Syria already uses so much of the Yarmuk waters that the Jordanians have a considerably reduced potential water resource.

Yarmuk-type sagas do not end with Israel, Jordan and Syria. Syria relies upon the waters of the River Euphrates, which passes upstream into Turkey. In order to irrigate southeastern Anatolia, Turkey is currently constructing dams along the course of the Euphrates, such as the Ataturk Dam, and the Tigris. The Euphrates and the Tigris flow through Iraq to the Persian Gulf, so like other countries in the Middle

East, it is also concerned about any upstream activities to tap, divert and store water. Gleick (1994) stresses the need for a comprehensive framework for planning and managing shared water resources in the Middle East to help reduce political tensions and aid long-term sustainable economic development. The goals of such a framework might include identifying mechanisms for implementing joint projects; identifying minimum water requirements; developing and practising water efficiency; developing means of shifting water use within regions such as through water banks; and providing new sources.

In wars, enormous damage to water supplies is a common occurrence. For example, in January 1993 in the Balkan conflict, retreating Serbian soldiers blew up the reservoir behind the Peruca Dam in Krajina; fortunately, the reservoir drained safely, removing the risk to about 20,000 people inhabiting the valley below. In mid-1994, more than fourteen water supply lines were cut in Bosnia; some were in the UN 'safe areas' of Bihac and Srebrenica, and in central Bosnia more than 200,000 people were left without water when supplies were cut off to the towns of Vares, Vitez and Zenica (Pearce 1994b). Similarly, during the 1994 civil war in Yemen, soldiers destroyed a water plant at Bir Nasser, cutting off supplies to 350,000 people in Aden. Destroying the enemy's water supplies has a notorious pedigree – about 4,500 years ago in Mesopotamia, along the Euphrates, the king of Umma destroyed the banks along the irrigation canals to unleash torrents of water on his downstream neighbours at Girsu; in the sixteenth century, William of Orange deliberately flooded large parts of the southern Netherlands to impede the Spanish; and in the seventeenth century, the French under Louis XIV were defeated by the Dutch using the same tactics. In 1938, during the Sino-Japanese War, Chinese generals ordered dykes to be broken on the Yellow River to halt the enemy advance, inadvertently killing between 300,000 and a million of their own people (Pearce 1995e).

In November 1994, the International Committee of the Red Cross (ICRC) launched a campaign to improve the protection of water supplies and to increase the supply of water engineers during war because of the devastating consequences to civilians (Pearce 1994b). Article 54 of the Geneva Convention of 1949 prohibits attacks on objects that are indispensable to the survival of civilian populations unless they are in direct support of military action. In the Gulf War, Allied forces assiduously avoided waterworks while bombing Baghdad, but nearly every water installation was put out of operation. The ICRC believes that the 1949 Geneva Con-

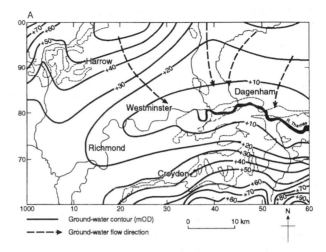

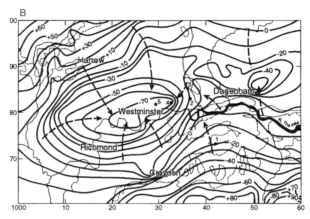

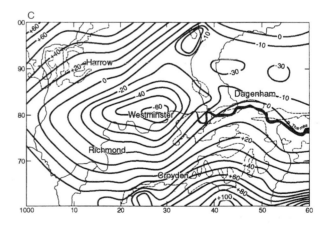

Figure 5.9 *Levels of ground water in the London area: (A) prior to major development; (B) in 1965; and (C) in 1985. Redrawn after Goudie (1993b), Wilkinson and Brassington (1991).*

vention does not really protect people who live in modern cities and that there should be specific outlawing of destruction to waterworks. Today, the World Bank and other funding agencies are

Plate 5.7 *Artificial irrigation of the upper reaches of the Indus River in Ladakh, north of the high Himalayas, has created a fertile valley capable of supporting many villages and small towns.*

increasingly refusing to provide investment for dams in regions of the world where there are extreme political tensions, such as war zones. The Danube is another example of an international conflict over water resources in what was until very recently at least in part a war zone (see Box 5.6).

On a more optimistic note, some governments are becoming more aware of the need to balance the human demand for water resources with a concern for the effects on the environment. In France, the demand for more water from the Loire is posing a threat to the fragile marshes and wetlands along the course of the river and its tributaries. Along the Mediterranean coastline, the Camargue has lost roughly half of its wetlands since 1945 because of drainage and development associated with tourism. In 1980, the French wetlands, covering an area exceeding 1.2 million ha, were designated as being of international importance for wildlife, yet by 1990 more than half this area had been drained or was at some risk (Purseglove 1991). There are now more than twenty hydroelectric dams and other barriers along the Rhône between Geneva and Arles – something that has radically altered the vegetation and

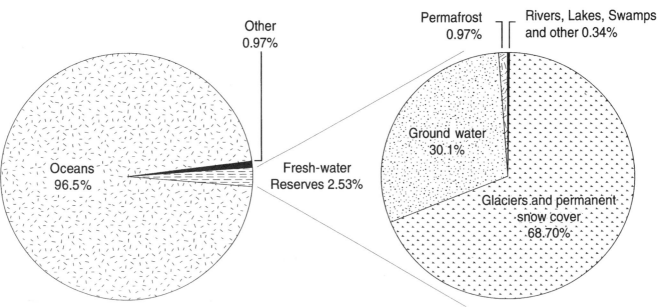

Figure 5.10 *Global total water and fresh-water reserves. Redrawn after World Resources Institute (1994–1995).*

BOX 5.5 THE DEAD SEA

Today, the Dead Sea is shrinking and the River Jordan, which has provided water for civilisations for thousands of years, has a much reduced flow. Jordan, a land-locked desert nation and one of the driest countries in the world, is suffering extreme water shortages. These problems have arisen because of the extreme diversion of water for irrigation and domestic use, particularly by the Israelis, who in 1964 began diverting most of the River Jordan's annual 1,200 million m³ of water.

In response the Jordanians under King Hussein have devised a plan to connect the Dead Sea with the Red Sea by a 240-km-long canal. Following the signing of the peace treaty in October 1994 between Jordan and Israel, Hussein is seeking international financial support for this scheme as part of a 'peace dividend'.

ecosystems along this mighty river. In France, growing environmental awareness, and the lessons that have been learned from the over-zealous development of the major water courses, have actually led to the scrapping of various plans for new dams; for example, pressure from conservationists helped to reverse a decision to construct two new dams in the upper reaches of the River Loire. To date, four major projects for the construction of dams along the Loire and its tributaries have been abandoned, partly as a result of lobbying by environmentalists (Villerest, Grangent, Naussac I and Naussac II – see map in Purseglove 1991).

In England and Wales, the privatisation of water for party political reasons, through the Water Act 1989, consolidated into the Water Resources Act 1991 and the Water Industry Act 1991, led to the establishment of ten water and sewage companies having responsibility for providing clean water and sewage treatment, together with its disposal. Additionally, another twenty-two companies provide water only. Also, the Office of Water Services (OFWAT) and the National Rivers Authority (NRA) were created to regulate and monitor the economic and environmental aspects of water, respectively. This privatisation has raised many issues, including the

BOX 5.6 THE DANUBE

The Danube is Central Europe's largest river, rising in the Black Forest, Germany, and flowing through the Austrian Alps, the Central European Plain (Slovakia, Hungary, Croatia, Serbia and Romania) to the Black Sea. Stretches of the flood plain are associated with ecologically important wetlands and flooded forests. Also, the delta is Europe's largest underground water reservoir with a capacity of *c.* 20 km³ of water.

In the 1960s, the governments of Hungary and Yugoslavia planned a raised ship canal to link the Rivers Rhine and Danube, allowing ships to travel from the North Sea to the Black Sea, bypassing the wetlands, which were difficult to navigate. A treaty to undertake this and construct two dams (the Gabcikovo Dam in Slovakia and the Hagymaros Dam in Hungary) was signed in 1977. In 1992, Slovak engineers diverted the River Danube along a new 35-km-long ship canal to a hydroelectric dam. Environmentalists claimed that this would lead to an ecological disaster as one of Europe's last flooded forests would dry out, and wells would empty to become polluted and fetid. The Hungarians asserted that the diversion was illegal as they were being deprived of water that was also rightfully theirs, and that millions of people would lose their domestic water supply.

In Hungary, particularly, concerned scientists and large sectors of the general public, who were sympathetic to the environmental lobby, perceived the project as very much a part of the unpopular Communist regime, and they also believed that their water resources were being threatened. In 1988, tens of thousands of Hungarians had taken to the streets in a popular uprising that included the issue of the diversion of the River Danube and this revolution brought down the Communist regime. Since then, construction work on the Nagymaros Dam has been abandoned with only the foundations. In September 1992, the Hungarians formally abandoned the 1977 treaty on environmental grounds and requested the Slovaks to do likewise.

The Slovaks came to see the Gabcikovo Dam as inextricably tied up in their nationalist struggle against the former Czechoslovakia and refused to accede to the Hungarian demands. Slovak engineers then diverted the River Danube along a raised canal, which had been constructed between Cunovo, in the west, and Gabcikovo. Both nations entered into an official and frequently acrimonious dispute, taking the case to the International Court of Justice in the Hague in 1995, where a ruling is awaited.

Three years on, the ecological disaster that was claimed has not occurred: the forest remains flooded, wells are mainly clean and full, and the trees healthy. Slovak wetlands are the healthiest in thirty years. Since May 1993, the Slovaks have routinely diverted up to 10 per cent of the new canal water flow onto the wetlands near Dobrohost, together with the construction of raised banks and culverts to irrigate the area artificially. After recharging the wetlands, the water table near Bratislava has risen by 3 m, and 50 km farther downstream it is 0.5 m higher than before. In July 1994, the World-Wild Fund for Nature, originally one of the most outspoken groups against the Slovak dam, abandoned its campaign.

fundamental question of whether the responsibility for managing and distributing something as necessary and basic to life should be left in the hands of companies whose first concern is profit.

These examples of the politics of water management serve to show just how precious and potentially explosive are the issues concerning water. Co-operation and sensible planning between countries can enhance the living standards and quality of life for a whole region. Selfish and thoughtless water management, driven by greed and power politics, can create regional tensions and even wars.

International controls on marine pollution

The first major international control on marine pollution was the International Convention for Prevention of Pollution of the Sea by Oil (OILPOL), adopted in 1954 and coming into effect in 1958. Oilpol followed earlier unsuccessful attempts in 1926 and 1934, and was aimed at controlling operational pollution from ships at sea. The main requirements embodied in Oilpol included proscription of the deliberate discharge of oily substances at specified distances from coastlines, encouraging the loading of oil from above the waterline (LOT, or load-on-top method), retention of residue in slop tanks, and that tankers should maintain and carry an oil record book.

The Safety of Life at Sea (SOLAS) Convention and the Collision Regulations were instigated to reduce oil spills. The *Torrey Canyon* supertanker oil spill of 1967 led to the Tanker Owners' Voluntary Agreement concerning Liability for Oil Pollution (TOVALOP). The *Torrey Canyon* incident helped pave the way for the adoption at the Brussels Conference in 1969 of the International Convention relating to Intervention on the High Seas in case of Oil Pollution Casualties, which came into force in 1973. Another outcome of the Brussels Conference was the adoption of the International Convention on Civil Liability for Oil Pollution Damage, which did not come into force until 1975. In 1971, the International Oil Pollution Compensation Fund was introduced, but it did not come into force until 1978. Also agreed in 1973 was the International Convention for the Prevention of Pollution from Ships (MARPOL), while 1972 saw the introduction of the Oslo Convention for Prevention of Marine Pollution by Dumping from Ships and Aircraft, followed in 1974 by the Paris Convention for Prevention of Marine Pollution from Land-Based Sources. Subsequent amendments to MARPOL were

introduced for oil pollution in 1978 as Annex I, coming into effect in 1983; noxious liquid substances in bulk as Annex II, coming into effect in 1987; packaged dangerous goods as Annex III in 1989; and sewage from ships as Annex IV, adopted on a voluntary basis by many countries.

In 1978, in order to increase the professionalism of mariners in charge of potentially hazardous cargoes, the United Nations International Maritime Organisation (UN IMO; established in 1958 as the Intergovernmental Maritime Consultative Organisation, IMCO) introduced the Convention on the Standards of Training Certification and Watchkeeping for Seafarers. In 1982, the UN Law of the Sea Convention was introduced as a comprehensive framework of rules to cover all forms of marine pollution. Part of this convention involved the setting of revised maritime jurisdictional zones. Under the 1982 convention, the sovereign jurisdiction of a coastal state, the *territorial sea*, was defined to 12 nautical miles, and its *contiguous zone* was extended to 24 nautical miles compared with the 12 nautical miles set in the 1958 Geneva Convention. A state's exclusive economic zone (EEZ) extends out to 200 nautical miles from the territorial sea baseline.

In 1991, the International Conference on Water and the Environment (ICWE) was convened in Dublin to prepare recommendations for the United Nations Conference on the Environment and Development (UNCED) in Rio de Janeiro in 1992. This became known as the Dublin Statement, which provided a sound basis for the management of water resources. Most of the recommendations were incorporated into the UNCED Agenda 21 document, the blueprint for action into the twenty-first century (see Box 5.7).

Conclusions

In conclusion, it can be seen that there are many different ways in which water becomes polluted and many types of pollutant. Polluted waters not only affect delicately balanced or fragile ecosystems, but they can also have detrimental or catastrophic effects on human health. Today, environmental pollution is becoming more prevalent as populations rise and humans put greater and greater stress on the environment through intensive agricultural and industrial processes that produce large amounts of waste as by-products. Table 5.12 shows the US Environmental Protection Agency's (EPA) list of priority pollutants. Although it does not include nitrates, sewage and oil, it illustrates the vast risk of water contamination

BOX 5.7 THE ICWE DUBLIN STATEMENT

The main principles from the Dublin Statement are summarised from Young *et al.* (1994) and include:

Principle No. 1: Fresh water is a finite and vulnerable resource, essential to sustain life, development and the environment.

Since water sustains life, effective management of water resources demands a holistic approach, linking social and economic development with protection of natural ecosystems. Effective management links land and water uses across the whole of a catchment area or ground-water aquifer.

Principle No. 2: Water development and management should be based on a participatory approach, involving users, planners and policy-makers at all levels.

The participatory approach involves raising awareness of the importance of water among policy-makers and the general public. It means that decisions are taken at the lowest appropriate level, with full public consultation and involvement of users in the planning and implementation of water projects.

Principle No. 3: Women play a central part in the provision, management and safeguarding of water.

The pivotal role of women as providers and users of water and guardians of the living environment has seldom been reflected in institutional arrangements for the development and management of water resources. Acceptance and implementation of this principle requires positive policies to address women's specific needs and to equip and empower women to participate at all levels in water resources programmes, including decision-making and implementation, in ways defined by them.

Principle No. 4: Water has an economic value in all its competing uses and should be recognised as an economic good.

Within this principle, it is vital to recognise the basic right of all human beings to have access to clean water and sanitation at an affordable price. Past failure to recognise the economic value of water has led to wasteful and environmentally damaging uses of the resource. Managing water as an economic good is an important way of achieving efficient and equitable use, and of encouraging conservation and protection of water resources.

---

from human activities. Implementation of legislation and monitoring of water bodies by organisations such as the EPA will help to reduce the likely risks of one or more of these pollutants entering our water sources. Unless such activities are carefully controlled and pollutants more effectively dealt with, the consequences may be disastrous for the whole planet, let alone specific geographical areas.

National and international legislation and public awareness are paramount in the prevention and control of water pollution. In the USA, the Water Quality Act of 1965 required states to set standards for water and effluents. This was superseded in 1972, when Congress took control of water pollution with the Federal Water Pollution Control Act Amendments, permitting the EPA to set standards for water and air emissions. This was followed by the Safe Drinking Water Act of 1974, intended to protect human health and guarantee safe public drinking water supplies. In addition, the Comprehensive Response, Compensation, and Liability Act (see the *Contaminated land* section in Chapter 9) was established in 1980 to allow a swift response to health hazards caused by pollutants. Howe (1991) evaluated these acts and emphasised that they may be strengthened and made more cost-effective by integrating planning and management across the states; greater use of economic evaluation in setting the standards and in the search for least-cost methods of achieving these standards by increasing economic incentives; and better monitoring of pollution so that

polluters will know that the authorities are working and offenders are being identified and punished.

Pearce (1995) reported on the present problems facing the Mediterranean Sea, notably the effects during the summer of 1994, which included the death of more than a thousand dolphins, killed by a morbillivirus between Morocco and Greece, and the plagues of jellyfish and red tides in the Aegean. All these were possibly connected in some way to the huge pollution pressures in the Mediterranean and the sewage they create. This is not surprising, because more than 130 million people live along the coasts of the Mediterranean and 100 million tourists visit the region each year, pouring more than 500 Mt yr^{-1} of raw sewage into the sea. Pollution problems of this type were predicted twenty years ago, when all the nations bordering the sea, except Albania, agreed to be part of the Mediterranean Action Plan (MAP). Unfortunately, these countries have failed to reduce the problems and many have not even paid their dues to the secretariat (Table 5.13).

It is also discouraging to see so many governments neglecting their responsibilities, for example the privatisation of the Water Boards in Great Britain principally for commercial gain, without properly policeable water standards. Cook (1989) discussed the ethics and politics of privatising the Water Boards in the UK. She came to the same conclusion that the Mayor of Birmingham, Joseph Chamberlain, reached almost 100 years ago when trying to obtain clean water and sanitation for his city:

Table 5.12   *US Environmental Protection Agency list of priority pollutants. Organic compounds are subdivided into four categories according to the method of analysis.*

Base-neutral extractables

Acenaphthene	Diethyl phthalate
Acenaphthylene	Dimethyl phthalate
Anthracene	2,4-Dinitrotoluene
Benzidine	2,6-Dinitrotoluene
Benzo[*a*]anthracene	Di-*n*-octyl phthalate
Benzo[*b*]fluoranthene	1,2-Diphenylhydrazine
Benzo[*k*]fluoranthene	Fluoranthene
Benzo[*ghi*]perylene	Hexachlorobenzene
Benzo[*a*]pyrene	Hexachlorobutadene
Bis(2-chloroethoxy) methane	Hexachlorocyclopentadene
Bis(2-chloroethyl) ether	Hexachloroethane
Bis(2-chloroisopropyl) ether	Indeno[1,2,3-*cd*] pyrene
Bis(2-ethylhexyl) phthalate	Isophorone
4-Bromnophenyl phenyl ether	Naphthalene
Butyl benzyl phthalate	Nitrobenzene
2-Chlorophenyl phenyl ether	*N*-Nitrosodimethylamine
Chrysene	*N*-Nitrosodiphenylamine
Dibenzo[*a,b*]anthracene	*N*-Nitrosodi-*n*-propylamine
Di-*n*-butyl phthalate	Phenanthrene
1,2-Dichlorobenzene	Pyrene
1,3-Dichlorobenzene	1,2,4-Trichlorobenzene
1,4-Dichlorobenzene	2,3,7,8-Tetrachloro-
3,3¹-Dichlorobenzidine	dibenzo-*p*-dioxin

Acid extractables

p-Chloro-*m*-cresol	2-Nitrophenol
2-Chlorophenol	4-Nitrophenol
2,4-Dichlorophenol	Pentachlorophenol
2,4-Dimethylphenol	Phenol
4,6-Dinitro-*o*-cresol	2,4,6-Trichlorophenol
2,4-Dinitrophenol	Total phenols

Volatiles

Acrolein	1,2-Dichloroethane
Acrylonitrile	1,1-Dichloroethane
Benzene	*trans*-1,2-Dichloroethylene
Bis(chloromethyl) ether	1,2-Dichloropropane
Bromodichloromethane	*cis*-1,3-Dichloropropane
Bromoform	*trans*-1,3-Dichloropropane
Bromomethane	Ethylbenzene
Carbon tetrachloride	Methyl chloride
Chlorobenzene	1,1,2,2-Tetrachloroethane
Chloroethane	Toluene
2-Chloroethyl vinyl ether	1,1,1-Trichloroethane
Chloroform	1,1,2-Trichloroethane
Chloromethane	Trichloroethylene
Dibromochloromethane	Trichlorofluoromethane
Dichlorodifluoromethane	Vinyl chloride
1,1-Dichloroethane	

Pesticides

Aldrin	Dieldrin	PCB-1016*
α-BHC	α-Endosulfan	PCB-1221*
β-BHC	β-Endosulfan	PCB-1232*
γ-BHC	Endosulfan sulphate	PCB-1242*
δ-BHC	Endrin	PCB-1248*
Chlordane	Endrin aldehyde	PCB-1254*
4,4'-DDD	Heptachlor	PCB-1260*
4,4'-DDE	Heptachlor expocide	Toxaphene
4,4'-DDT		
		* not pesticides

Inorganics

Antimony	Chromium	Nickel
Arsenic	Copper	Selenium
Asbestos	Cyanides	Silver
Beryllium	Lead	Thallium
Cadmium	Mercury	Zinc

Source: Domenico and Schwartz 1990.

It is difficult and indeed almost impossible to reconcile the rights and interests of the public with the claims of an individual company seeking as its natural and legitimate object, the largest private gain.

On an international scale, it is even more disconcerting to witness just how little effort is being made to combat water pollution. Planet Earth may change from a beautiful life-giving blue to a barren polluted brown before its natural time unless collective responsibility for water resources and their cleanliness is taken now.

Table 5.13 *Ten objectives set by the Mediterranean Action Plan in 1985 for 1995. None has yet been achieved.*

- Measures to reduce the amount of contaminated water discharged by ships into the sea, including facilities at ports to offload dirty ballast waters and oily residues.
- Sewage treatment works for all cities with over 100,000 people, plus appropriate sewage outfall pipes into the sea or treatment plants for towns with over 10,000 people.
- Environmental impact assessments for development projects such as new ports, marinas and holiday resorts.
- Improved navigation safety to minimise the risk of collision, such as better shipping lanes in and out of ports, especially for ships carrying toxic cargoes.
- Protection of endangered marine species such as monk seals and sea turtles.
- Reduced industrial pollution.
- Identification and protection of 100 historic coastal sites.
- Identification and protection of 50 marine sites and coastal conservation sites.
- More protection against forest fires and soil loss to minimise run-off into rivers.
- Reduction of acid rain.

After Pearce 1995.

Chapter 5: Key points

1 Clean and abundant water is an essential resource for life. There are water shortages in many countries and human activities have polluted water resources throughout history. Many of the world's beaches are suffering severe pollution, some with dangerously high levels of toxic chemicals. An understanding of the hydrological cycle is important in understanding the routes that pollutants can travel through the ecosphere, particularly through food chains and webs. There are many types of pollutant, both natural and created by human activities.

2 Sewage sludge is one of the most common pollutants, often being disposed of in an untreated form, or raw state, into streams and seas. Treatment involves filtering, biological digestion and drying to produce sewage sludge, which is easier to handle and dispose of. Sewage treatment also aims to destroy disease-carrying organisms, e.g. cholera, typhoid, cryptosporidia, leptospirosis and giardia. Sewage sludge properly treated can provide a valuable fertiliser, or energy source through incineration to generate electricity. The decay of organic matter produces NH_3 and nitrates, which, if in sufficient concentrations, may contaminate the atmosphere and water sources.

3 Nitrates are produced by traditional and artificial fertilisers, and may be washed into water sources. Nitrates are believed to be responsible for diseases such as methaemoglobinaemia and stomach cancer, and are responsible for the eutrophication of lakes and algal blooms, which cause the depletion of free oxygen in standing bodies of water such as lakes, and the consequent suffocation of aquatic animals. Nitrates may be chemically removed from water by cation exchange, reverse osmosis, electrodialysis, distillation and biological processes, but these are expensive to implement.

4 Dangerous organic chemicals (chlorinated hydrocarbons, notably PCBs and DDT, and organotins) are manufactured for paints, plastics, adhesives, hydraulic fluids, electrical components, defoliants, pesticides and anti-fouling paints, and the waste from such industrial processes is commonly dumped into water courses. Such chemicals take a long time to biodegrade or break down; therefore they can concentrate in food chains to poison animals and plants.

5 Heavy metals (principally mercury, lead, arsenic, selenium, cobalt, copper and magnesium) are produced naturally by weathering, but industrial processes discharge large quantities into water courses, which may be at toxic levels. The metals become concentrated at high trophic levels within food webs, and may have serious effects on life, e.g. brain damage and even death in animals. Lead piping for water and the use of aluminium in cooking utensils may further concentrate these metals in the human body. Aluminium may be a contributory factor in Alzheimer's disease and Guam disease. The sources of heavy metal pollutants can be traced using isotopic methods, and this may provide an effective means of tracing metal pollution to specific polluters.

6 Radioactive waste has the potential to pose a very serious threat to health if it gets into ground water. It is briefly discussed in this chapter, but dealt with more fully in Chapter 6.

7 Oil pollution is a major environmental problem and results mainly from shipping accidents, offshore oil exploration, oil droplets from unburnt fuel, combustion engines on land, deliberate discharges from ships (particularly during tank-cleaning operations), natural seepage from the sea bed, and ecological terrorism. Oil pollution affects many ecosystems: poisoning organisms by coating fish gills, leading to suffocation; immobilising fish sperm; coating birds' feathers to impede flight and reduce insulation, leading to death; reducing the permeability of birds' eggs; causing lipid pneumonia in aquatic mammals; and seriously disrupting food webs. Clean-up technologies include containing the spread of oil by using booms, dispersal, and break-up of oil slicks using detergents, combustion, collecting the oil and bacterial degradation.

8   Ground water provides the largest available supply of fresh water and is important in sustaining the flow of rivers and a vital component of the hydro-logical cycle. The dynamics of ground-water flow are important for resource management, and the mitigation of ground-water pollution and its remediation.

9   The importance of water as a valuable resource causes international and regional conflicts, and governments are becoming increasingly aware of the need for long-term agreements between nations that must share common water resources.

Chapter 5: Further reading

Clark, R.B. 1989. *Marine Pollution*. Oxford: Clarendon Press.

Domenico, P.A. and Schwartz, F.W. 1990. *Physical and Chemical Hydrogeology*. Chichester: John Wiley & Sons, 824 pp.
An excellent comprehensive text on the physical and chemical aspects of ground water. The text is rather mathematical and may be too specialised for many readers, but there are excellent chapters on contaminant hydrology and remediation.

Hinrichsen, D. 1990. *Our Common Seas: Coasts in Crisis*. London: Earthscan, 192 pp.
This book is based on UNEP data and describes the growing pressures on coastal ecosystems throughout the world. Case studies are provided to illustrate the local successes in protecting marine and coastal environments.

Kliot, N. 1993. *Water Resources and Conflict in the Middle East*. London: Routledge.
This book examines the hydrological, social, economic, political and legal issues in the Middle East. It shows how water shortages threaten the renewal of conflict and disruption in the Euphrates, Tigris, Nile and Jordan basins.

Mason, C.F. 1996. *Biology of Freshwater Pollution* (third edition). Harlow: Longman Scientific & Technical.
This text provides an excellent and comprehensive overview of aspects of fresh-water pollution. Useful sections include new developments in European and UK water resources management, including the Environmental Protection Bill; a mathematical model of eutrophication; and case studies.

Open University 1991. *Case Studies in Oceanography and Marine Affairs*. Oxford: Pergamon Press, 248 pp.
A well-illustrated textbook produced to support an Open University course on oceanography. It examines marine resources and activities, and the development and nature of international conventions on the sea, and it provides case studies on the Arctic Ocean and the Galapagos Islands. This book emphasises the complex interactions between the political, economic and environmental aspects of development within the marine realm.

Price, M. 1985. *Introducing Ground-water*. London: Chapman & Hall, 195 pp.
An excellent introductory text to hydrogeological/hydrological aspects of ground water. The book is written for non-specialists and uses minimal technical and mathematical formulae. There are thirteen chapters including: water underground; water circulation; caverns and capillaries; soil water; ground water in motion; more about aquifers; springs and rivers; deserts and droughts; water wells; measurements and models; water quality; ground water, friend or foe?; some current problems.

Be advis'd;
Heat not a furnace for your foe so hot
That it do singe yourself. We may outrun
By violent swiftness that which we run at,
And lose by overrunning.

Spoken by Duke of Norfolk,
William Shakespeare,
All is True (*Henry VIII*), Act I, Scene i

Death or salvation?

The nuclear age has promised both death and salvation. Today, in the Nuclear and Space Age, humankind has come a long way, technologically, from the heady early days when the atom was split. Many of the physicists and other scientists involved in the early atomic experiments wanted to believe in 'atoms for peace'. At its inception, nuclear power offered an endless supply of clean and cheap energy, but it also offered destruction and fear.

Military strategists, however, saw things differently. Nuclear weapons would be the ultimate deterrent in a brave new world of mutually assured destruction. Nuclear weapons would keep the peace and if conflagrations occurred, then they would ensure the speedy restoration of the military superiority of the superpowers. Thus, through the late 1940s and 1950s, post-war reconstructionist, 'we've never had it so good', USA and Europe, the early days of atomic research seemed to offer something for everyone.

Despite the mood of optimism about the dawning of a Nuclear Age, the horrors of Hiroshima and Nagasaki had made their mark. A vocal minority of public opinion foresaw the problems that a nuclear future would bring to humankind. They believed that nuclear weapons should never be used, that the consequences of using nuclear weapons could never be justified, and that we should 'Ban the Bomb'. So, by the start of the 1960s, there was open public debate over the morality of nuclear weapons. And, of course, there is the ongoing debate over the sheer financial issues of a nuclear arms programme. Table 6.1 gives the military and education expenditures for selected countries in 1987, and Figure 6.1 shows the money spent per capita on defence in 1990 in NATO countries (in thousands of US$). There are many people who would argue that in the face of a world recession and the need for greater social provision and life chances for a country's citizens, these levels of expenditure are far too high. The debate is complicated and many vested interests are involved in the arguments for and against high levels of expenditure on nuclear arms.

Nuclear waste disposal and accidents at nuclear power stations were perceived as much less of an issue. They only became common currency in the media in the 1970s with the dawning of the 'environmentally conscious era'. Now the two big nuclear issues are whether or not nuclear energy should be used, allied to concern over the production of nuclear waste and its disposal, and concern over the morality of nuclear weapons, the arms race and the verification of treaties to police nuclear arms. These issues are explored in this chapter, although nuclear energy as a fuel is dealt with in more depth in Chapter 7.

Historical perspective

The Nuclear Age can be considered to have dawned in 1919 when Ernest Rutherford and his team at Cambridge University in the UK smashed the atom, converting a nucleus of nitrogen into a nucleus of hydrogen (eight years after Rutherford had discovered the atomic nucleus). On 19 December 1938 in Berlin, at the Kaiser Wilhelm Institute, two nuclear chemists, Otto Hahn and Fritz Strassman, split the uranium atom into roughly two equal parts and recognised that **nuclear fission** had occurred. Actually, four years earlier, Enrico Fermi, working with Frederic and Irene Joilet-Curie, had created artificial radioactivity by bombarding a target with alpha particles, but they were not aware of the full significance of their discovery (Box 6.1).

Research into nuclear physics and chemistry was given its real momentum on 9 October 1941, when President Roosevelt sanctioned the US construction of an atomic bomb. In 1945, President Harry S. Truman made the fateful decision to use the atomic bomb against Japan. *Little Boy* was detonated at an altitude of just under 2,000 feet over Hiroshima on 6 August and over 70,000 people died, whilst *Fat Man* was dropped on Nagasaki on 9 August 1945,

Table 6.1   *Military and education expenditures in selected countries.*

	Military expenditure (1987)		Educational expenditure (latest year obtainable; % of GNP or GDP)	Ratio of military to educational expenditures	Literacy rate (%)
	$ billion	% GNP or GDP			
Australia	4.99	2.5	5.6	0.45	99.5
Bangladesh	0.32	1.8	2.1	0.86	33.1
Botswana	0.02	2.2	6.0	0.37	70.8
Bulgaria	6.66	10.3	7.1	1.45	95.5
Burkina Faso	0.05	3.1	2.5	1.24	13.2
Canada	8.84	2.2	7.4	0.30	95.6
China	20.66	4.4	2.7	1.63	72.6
Costa Rica	0.02	0.6	5.2	0.11	92.6
Cuba	1.60	5.4	6.3	0.86	96.0
Egypt	6.53	9.2	5.5	1.67	44.9
Ethiopia	0.44	8.5	3.9	2.18	3.7
France	34.83	4.0	6.1	0.66	98.8
Guyana	0.34	8.9	10.1	0.88	95.9
Iran	21.12	7.9	3.8	2.08	61.8
Iraq	16.70	30.7	3.8	8.08	45.9
Japan	24.32	1.0	5.1	0.20	100.0
Lesotho	0.01	2.3	3.5	0.66	73.6
Mauritania	0.04	4.2	7.9	0.53	28.0
Mozambique	0.10	8.4	1.2	7.00	16.6
Nigeria	0.18	0.8	1.8	0.44	42.4
Pakistan	2.23	6.5	2.1	3.10	25.6
Peru	2.20	4.9	2.9	1.69	87.0
Senegal	0.10	2.2	4.7	0.47	22.5
South Africa	3.40	4.4	3.8	1.15	79.3
Tanzania	0.08	3.3	4.3	0.77	85.0
USSR	303.00	12.3	7.0	1.76	99.0
UK	31.58	4.7	5.2	0.90	100.0
United States	396.20	6.5	7.5	0.87	95.5
Venezuela	1.38	3.6	6.8	0.53	89.6
Zambia	0.17	6.6	5.4	1.22	68.6
Zimbabwe	0.28	5.0	7.9	0.63	76.0

Note: literacy is differently defined in various countries. Thus the UK and Japanese figures reflect all considered capable of reading, Canadian figure excludes functionally illiterate.
GNP/GDP = gross national/domestic product.
Source: *Encyclopedia Britannica Yearbook 1990.*

causing around 35,000 deaths. Only weeks earlier, on 16 July 1945 in the New Mexico desert near Alamogordo, the first American test of a nuclear bomb, code-named *Trinity*, took place under the direction of Robert Oppenheimer. This story is dramatically told by John Newhouse in his excellent book, *The Nuclear Age: from Hiroshima to Star Wars*.

Before considering various other aspects of nuclear issues, it is worth briefly considering the history of nuclear arms treaties between the superpowers. Obviously, this book cannot explore the subtle ramifications of these treaties, but a précis of the main agreements gives some insight into the pace and nature of such talks and treaties. Perhaps the pro-

genitor of the succession of treaties was the 'Atoms for Peace' speech by US President Dwight D. Eisenhower on 8 December 1953, and the forming in 1957 of the International Atomic Energy Agency (IAEA), committed to promoting the peaceful use of nuclear energy and preventing its destructive use. Box 6.2 summarises the principal international treaties that chart the progress of nuclear arms negotiations and agreements to date.

To date, six nations have openly tested nuclear weapons – the USA, the Soviet Union, Britain, China, France and India – but they are currently observing a moratorium on testing, with the exception of China. The USA, Britain, Russia, China, France and the former Soviet republics of Belorussia, Kazakhstan and the Ukraine are declared nuclear weapons states with the capability of delivering nuclear weapons, whereas of the undeclared nuclear weapons states (Israel, India and Pakistan) only Israel is known to possess deliverable weapons. Algeria, Iran, Iraq, Libya, North Korea and Syria are known to be working on obtaining nuclear weapons, while Argentina, Brazil, South Africa, South Korea and Taiwan are believed to have ceased developing a nuclear capability. States which probably possess or will soon have a nuclear capability are often referred to as 'threshold nations'. While the nuclear nations promote non-proliferation of nuclear weapons, many of the threshold nations are still moving towards joining the nuclear club, mainly because of the perceived threat to national security from neighbouring states. The disparity in nuclear capability between nations has created two diametrically opposed philosophies. One is that peace will be achieved through the containment of nuclear weapons and non-proliferation. The other is that peace will come first through nuclear proliferation in the developing countries and later 'denuclearisation' by every nation.

Threshold nations complain that the IAEA treats them differently to the nuclear nations by imposing stricter verification procedures such as on-site inspections, seismic and satellite monitoring, perimeter monitoring, and checking for the diversification of nuclear fuels. Most inspectors come from the nuclear nations, and it is alleged that they may connive to cheat by turning a blind eye to the movement of prohibited nuclear materials between nuclear states. More than 4 tonnes of plutonium appears to have gone missing from the British nuclear industry, apparently to the US weapons programme (Hassard 1992). Issues such as these generate a distrust by the threshold nations of the nuclear states such as India, which has refused to sign the NPT because it perceives it as merely maintaining the nuclear *status quo*. There is little indication that the world is addressing these issues head on, and without clear international agreements between nuclear and non-nuclear nations, the problems are being shelved for the near and intermediate-term future. In its attempts to contain the threat to world peace caused by a spread in the number of states with a nuclear weapons capability, the USA and its allies cannot guarantee the current national integrity of Iran against Iraq, India against China, the Arab states against Israel, North Korea against South Korea, or Pakistan against India – and it is tensions such as these that fuel the scramble to acquire nuclear weapons. Proliferation cannot be stopped, but international treaties can slow the pace of growth and

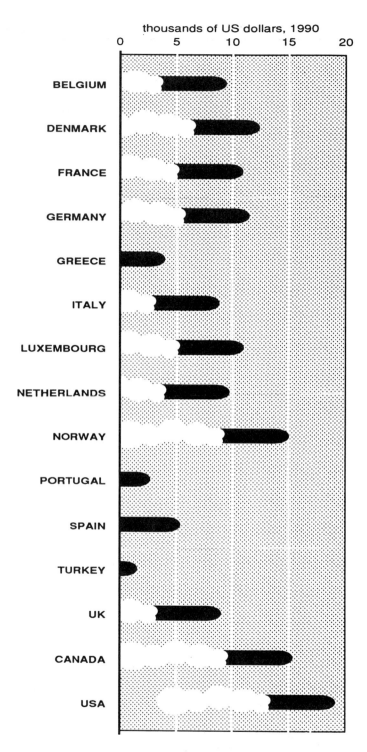

Figure 6.1 *Comparisons between the military expenditure per capita of the top fifteen developed countries. From* The Environmental Guardian *(11 February 1992).*

BOX 6.1 RADIOACTIVITY

An atom is made up of a **nucleus** of neutrons and protons, and one or more electrons, which revolve around the nucleus. The total number of protons in the nucleus constitutes its **atomic number**, while the total number of protons and neutrons is referred to as the **atomic mass number** of the element. Isotopes are elements that have the same atomic number but a different atomic mass number. A **radioisotope** is an isotope that spontaneously undergoes radioactive decay, a process known as **radioactivity**. During the decay process the radioisotope changes from one isotope to another and emits one or more forms of radiation. Three types of nuclear radiation are produced during radioactive decay, involving the emission of **alpha (α) particles, beta (β) particles** or **gamma (γ) rays** (Figure 6.2). Alpha particles comprise two protons and two neutrons (essentially a helium nucleus). Radioactive decay involving alpha particles results in a decrease of both the atomic number and the atomic mass number of the radioisotope. The original radioisotope is known as the **parent** and after decay the product is known as the **daughter**. Alpha particles have the greatest mass of all types of nuclear radiation and because of this they travel only approximately 5 to 8 cm in air. They travel between 50 and 80 μm in human tissue, which is more dense and, therefore, to cause cell damage, alpha radiation must originate in close proximity to the cell. Beta decay occurs when a neutron in the nucleus of an isotope spontaneously changes into a proton, which remains in the nucleus while an electron (beta particle) is emitted. A proton may change into a neutron, or a neutron may be transformed into a proton, and as a result of this process another particle, a neutrino, is also ejected. A neutrino is a particle with no rest mass. In this case the atomic mass number is reduced. Beta particles can travel farther through air than alpha particles, but they are blocked by a thin sheet of metal or block of wood. Strontium-90 is an example of a beta emitter with a half-life of 28.1 years. Figure 6.2 illustrates alpha and beta decay for Radon 222 (^{222}Rn) and Lead 214 (^{214}Pb), respectively. Gamma rays are a type of electromagnetic radiation similar to x-rays, but they are more energetic and can penetrate thick shielding.

Different radioisotopes behave differently; some decay only by alpha radiation, while others include two or more decay processes. Some radioisotopes, particularly the very heavy elements (those with high atomic mass numbers), will undergo a series of radioactive decay steps before they reach a non-radioactive isotope (**stable isotope**). This is known as a **decay series** and Figure 6.3 illustrates the uranium 238 (^{238}U) decay chain and the types of radiation associated with each of the decay steps. The rate at which the decays occur will vary between different steps within this decay series. The time required for one-half of a given amount of the radioisotope to decay to another form is known as its **half-life**. Each radioactive isotope has its own unique and unchanging half-life. For example, ^{238}U has a half-life of 4.5 billion years, while ^{222}Rn has a relatively short half-life of 3.8 days, and the half-life of protactinium 234 (^{234}Pr) is only 1.2 minutes.

Nuclear energy involves the changes that take place within the nucleus of the atom. Two main processes can be used to release energy: **nuclear fission** and **nuclear fusion**. Nuclear fission involves splitting the atom's nucleus into smaller fragments, while fusion involves combining the atomic nuclei to form heavier nuclei. Uranium 235 (^{235}U) is the only naturally occurring fissionable material, and it is essential for the production of nuclear energy. Fission reactors split ^{235}U by bombarding it with neutrons. When a neutron strikes the uranium 235 nucleus, it produces fission fragments, free neutrons and heat energy (Figure 6.4). The released neutrons may then strike another ^{235}U atom, releasing more neutrons, fission fragments and heat energy. The process continues and a chain reaction develops. The fast-moving neutrons must be slowed down (moderated) to increase the probability of fission. Water is normally used as the **moderator**. The heat energy that is created is used to produce steam to run turbines to generate electricity. Fusion involves combining light elements such as hydrogen isotopes (deuterium and tritium) to form heavier elements such as helium. As fusion occurs, heat energy is released. For fusion to occur there must be extremely high temperatures (of the order of 100 million °C for the isotopes of hydrogen) and a high density of the fuel elements. At these high temperatures the atoms are stripped of their electrons and an electrically neutral material comprising positively charged nuclei and negatively charged electrons is formed. This is known as a **plasma**. The fuels for fusion, namely deuterium and tritium, are very cheap to extract and the potential economic benefits of fusion are therefore considerable. There are many developmental problems, however, to be solved before controlled nuclear fusion will become a viable energy source (Furth 1995).

contain the time when many more nations will have joined the nuclear club.

Radioactive fallout from nuclear tests

Atmospheric nuclear tests between 1945 and 1980 were responsible for putting a large, but still unquantified, amount of nuclear fallout into the atmosphere.

Atolls, oceanic islands representing the tops of volcanoes and commonly fringed by coral reefs, were amongst the main test sites for nuclear explosions, e.g. Bikini and Eniwetok Atolls in the American territory of the Marshall Islands – where more than 59 atomic and hydrogen bombs were detonated between 1946 and 1958 (Stoddart 1968) – and more recently in 1995 the underground testing on Mururoa Atoll (Anderson 1995, Patel 1995). The

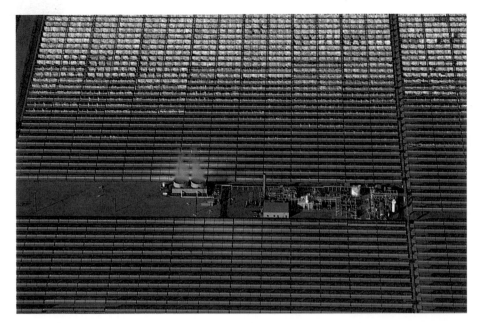

Plate 25 An economically viable solar power plant at Kramer Junction, California.
Courtesy of Georg Gerste/ Comstock.

Plate 26 Tidal power station at La Rance, France.
Courtesy of Mathew Boysons/Panos Pictures.

Plate 27 Wind turbines southwest of Los Angeles, California.

Plate 28 *Surface expression of geothermal energy as 'Old Faithful' geyser, Yellowstone National Park, Wyoming. Old Faithful erupts every 30–120 minutes, for between 1.5–5 minutes, ejecting super-heated steam to a height of 35–60 m.*

Plate 29 *Geothermal energy plant south of Turangi, Lake Taupo area, New Zealand.*

Limited Test Ban Treaty of 1963 led to the cessation of most testing, but some non-signatory countries, such as France, continued testing until 1980 and more recently in 1995. These continued tests deposited large quantities of additional **radionucleides** such as strontium-90 (^{90}Sr) and caesium-137 (^{137}Cs), which found their way, via precipitation, into the hydrosphere. These issues are dealt with at greater length in Chapter 5, where their impact on hydrological systems is considered, that is within rivers, lakes and seas.

Radioactive waste is very much a legacy the twentieth century has bequeathed to future generations. Table 6.2 (A, B) lists the typical yields of the so-called 'actinide group' of chemical elements, in curies, together with their typical yields of fission products. The main radioactive elements involved in polluting the environment and causing ill-health and death are the isotopes of iodine, strontium, caesium and ruthenium. Iodine accumulates in the thyroid gland. Strontium, which is chemically similar to calcium, is absorbed through the walls of the intestine and collects in the bones. Caesium behaves chemically in a similar manner to potassium and, therefore, can be distributed throughout the body in much the same way. Ruthenium has no chemical

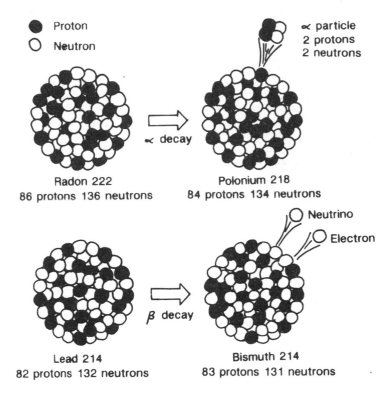

Figure 6.2 *Idealised diagrams showing (A) α-decay and (B) β-decay processes. Redrawn after Botkin and Keller (1995).*

BOX 6.2 INTERNATIONAL NUCLEAR ARMS AGREEMENTS

In 1963, the Limited Test Ban Treaty (LTBT) was signed by President John F. Kennedy for the USA and Premier Nikita Khrushchev for the Soviet Union; it banned nuclear tests in the atmosphere, underwater or in space. All tests have been underground since 1972. Kennedy had ideally wanted a Comprehensive Test Ban Treaty (CTBT) to stop the testing of all nuclear weapons, but the LTBT represented the best compromise. Indeed, Kennedy believed it was the greatest achievement of his presidency.

In 1968, the Non-Proliferation Treaty (NPT) was proposed on 1 July in Geneva and signed by more than 100 states (notably, unsigned at the time by the French). Those states with nuclear weapons undertook to make every effort to rid themselves of their nuclear weapons and to provide access to peaceful nuclear energy for the non-nuclear signatory states, the latter agreeing not to attempt to acquire nuclear weapons. The treaty grew from an 18-nation Committee on Disarmament, established in 1962 under the auspices of the United Nations General Assembly. The NPT forbade the superpowers helping others to acquire a nuclear capability – a course of action that would have been extremely unlikely in any event. This was a somewhat asymmetric agreement between the nuclear and non-nuclear states. China refused to sign the NPT and, subsequently, is believed

to have traded nuclear secrets with South Africa, Pakistan, Iraq, North Korea and Algeria. The NPT was ratified by Britain in 1970, and by the USA, the Soviet Union and 40 non-nuclear states. Today, more than 140 nations have signed the NPT, recent signatories being France, China and South Africa. Significant non-signers include Argentina, Brazil, India, Iran and Pakistan.

In 1972, the Strategic Arms Limitation Treaty I (SALT I) was signed by US President Richard Nixon and Soviet leader Leonid Brezhnev in Moscow on 26 May. This treaty, in two parts, comprised an anti-ballistic missile (ABM) treaty limiting both superpowers to just two ABM sites, further reduced to a single site each under an agreement in 1974, together with an interim offensive arms agreement to set ceilings on the deployment of certain nuclear weapons. Amongst these weapons, it was agreed that the Soviet Union could deploy 740 submarine-launched ballistic missiles (SLBMs) on 62 submarines, with the USA being restricted to 710 SLBMs on 44 submarines. It was also agreed that no additional inter-continental ballistic missile (ICBM) launchers could be added to those that were operational by 1 July 1972, while SLBMs were limited to those operational in May 1972, new 'heavy' ICBMs could not be deployed, and future modernisation and replacement was restricted to a one-for-one basis. According to the memoirs of Dr Henry Kissinger, US Secretary of State under Nixon, it was the land-based ICBMs, known as the SS-7s, that the USA was most eager to get rid of. The SS-7s had a range of over 6,000 miles and carried warheads of 6 megatons, and 70 were reasonably invulnerable in hardened underground silos.

Also, in 1972, a Biological Weapons Convention was signed by the superpowers to ban biological weapons and order the destruction of existing stockpiles.

In 1974, the Threshold Test Ban Treaty (TTBT) was signed as an agreement between the USA and the Soviet Union not to carry out nuclear tests of more than 150 kilotons explosive yield, equivalent to 150 thousand tonnes of TNT, or 10 times the Hiroshima bomb. Despite the treaty, both sides continued to deploy warheads with greater kilotonnage, such as the US Minuteman 3 and Peacekeeper missiles with 350-kiloton warheads and Soviet ICBMs such as the 550-kiloton SS-19 and the 3.6-megaton SS-17.

In 1976, the Peaceful Nuclear Explosions Treaty (PNET) was formulated between the Soviet Union and the USA. In essence, this treaty came about because the superpowers did not want a loophole in any existing international nuclear arms control agreements whereby it would be possible for either side to make a military device as a peaceful explosive, e.g. the theoretical use of nuclear devices in the construction of canals or to divert waterways (Kissinger 1982). Detailed verification procedures formed an important part of this treaty; for example, under this treaty the Soviets allowed, for the first time, on-site inspection. Because the treaty concerned more technical details compared with earlier treaties, it generated relatively little debate. Instead of attracting the bitter animosity and discussion that was associated with the SALT agreement, it attracted media indifference. In fact, the treaty was never ratified by the US Senate.

In 1979, the Strategic Arms Limitation Treaty II (SALT II) was signed by US President Jimmy Carter and Soviet President Leonid Brezhnev to limit the strategic missiles and bombers deployed by both parties to a maximum of 2,400. Certain restrictions were to apply within this broad framework. Weapons equipped with multiple independently targeted re-entry vehicle (MIRV) warheads and air-launched cruise missiles were limited to a maximum of 1,320. Either side could deploy no more than one new ICBM before December 1985. ICBMs could not carry more than 10 warheads per missile and a ceiling of 1,200 ballistic missiles carrying MIRV warheads was imposed.

On 8 December 1987, the Intermediate Nuclear Forces (INF) Treaty was signed by US President Ronald Reagan and Soviet leader Mikhail Gorbachev to eliminate an entire class of intermediate-range nuclear weapons (Plate 6.1). It was agreed to remove and destroy 1,286 nuclear missiles, containing more than 2,000 warheads, from Europe and Asia. The treaty involved the most stringent and comprehensive verification procedures to date in any arms control agreement.

In 1991, the Strategic Arms Reduction Treaty (START) was signed on 31 July in Moscow by Presidents George Bush for the USA and Mikhail Gorbachev for the Soviet Union. This treaty was the result of nine years of diplomatic groundwork and is designed to reduce Soviet and US strategic nuclear weapons by around 35 per cent on each side over a seven-year period. It will still leave the Soviets with 7,000 warheads and the Americans with 9,000 (reported in the *Financial Times*, 1 August 1991), and it does not address ground-based multiple-warhead missiles or sea-launched cruise missiles. So, while undeniably releasing some of the pressure for over-arming on the part of the superpowers, the treaty may well be viewed in the fullness of time as a major PR exercise rather than as a significant and irreversible reduction in the nuclear capability of either side. On 27 September 1991, President George Bush further announced that the USA would destroy 3,050 nuclear weapons, many of which are currently based in Europe or at sea.

A week later, on 5 October, President Mikhail Gorbachev offered to match the sweeping cuts in nuclear weapons announced by George Bush. The Soviet offer amounted to the most far-reaching reductions in the nuclear arsenal since the start of the Cold War, and went much further than originally embodied in the START signed in June 1991.

The proposal envisaged a 1998 target figure of 5,000 long-range strategic nuclear warheads instead of the 6,000 proposed in the June treaty. Gorbachev also proposed that the USA and the Soviet Union enter immediate negotiations on additional radical cuts of around 50 per cent in strategic offensive weapons. As with the US proposal, the Soviets would effect a complete elimination of all nuclear artillery shells and warheads for tactical missiles, and remove all tactical nuclear weapons from submarines and ships. In addition, they would remove all heavy bombers armed with nuclear weapons from alert status, order a one-year moratorium on nuclear tests, and cut 700,000 jobs in their four million-strong army. Gorbachev pledged that all nuclear artillery munitions removed from tactical missiles would be destroyed, together with at least some from the submarines and ships. He also announced that there would be moves to stop work on modified short-range missiles for heavy bombers and a small mobile ICBM; six nuclear missile submarines, with a total of 92 launchers, would be removed from active service; more than 500 ICBMs, including 134 with MIRVs, would be stepped down from day-to-day alert, and plans to construct new launchers for ICBMs on railcars would be scrapped, with the cold-storage of the missiles. The offer is currently under discussion by the superpowers.

START 2 was signed on Sunday 3 January 1992, in Moscow by Presidents George Bush for the USA and Boris Yeltsin for Russia. This treaty will in effect deny both sides a first-strike capability by eliminating about 15,000 warheads by 2003, leaving the USA with 3,500 and Russia with 3,000 warheads. Amongst the terms of the treaty, Russian SS-18 missiles are to be scrapped, and American B-1 and B-52 bombers converted to conventional use. The goals to be reached by 2003 include the following: bomber warheads, 750 Russian and 1,250 American; submarine-launched warheads, 1,750 Russian and 1,750 American; land-based missile warheads, 500 Russian and 500 American; and the elimination of all land-based multiple-warhead ICBMs. The implementation of START 2 is to be achieved in two stages over ten years.

START 2 cannot take effect until START 1 has been ratified by the three former Soviet republics of Ukraine, Belorussia and Kazahkstan: Ukraine has also not yet signed the non-proliferation treaty, because it claims that the Western funds to help to destroy its 176 long-range missiles are inadequate

In January 1994, the Ukraine entered into negotiations with the USA to destroy its nuclear arsenal in exchange for US$12 billion. The deal, signed in Moscow by US President Bill Clinton and Ukrainian President Leonid Kravchuk, means that the Ukraine will dismantle the 1,200 nuclear warheads on its SS-24 and SS-19 missiles, which were left in its territory when the Soviet Union collapsed. The missiles will be shipped to Russia for destruction, and the enriched uranium reprocessed in the USA. The deal requires ratification by the Ukrainian parliament, which has yet to be done.

On 24 September 1996, the five declared nuclear powers, the United States, Russia, China, Britain and France, and many other nations, signed the Comprehensive Test Ban Treaty to outlaw all explosive nuclear tests. After signing this treaty, President Clinton told the United States General Assembly, 'This is the longest-sought, hardest-fought prize in arms control history.' The ratification of this treaty requires the signature of 44 nations with a nuclear power industry. The long-term prospects for the ratification of the treaty are jeopardised by the opposition to the treaty from the nuclear 'threshold states' of India and Pakistan, both of which refused to sign.

---

analogue with a biological function. Figure 6.5 shows the various ways in which radioactive substances can reach individuals.

The effects of radioactive iodine can be ameliorated by saturating the thyroid gland with safe iodine through iodine tablets. Radioactive iodine has a short **half-life** and will decay quickly to safe levels soon after the event. Isotopes of strontium and caesium, however, are widely distributed throughout the body and their accumulation cannot be controlled by the aid of tablets. Also their half-lifes are long (40 days to 30 years), which results in prolonged effects.

Several sources are responsible for generating radioactive pollutants. One natural source of radiation which leaks into the atmosphere is from rocks and sediments rich in radioactive elements. The following section deals in particular with one of the most common naturally occurring radioactive gases, radon.

Radon gas

Radon gas (see Box 6.3) is a major component of the **background radiation** dose received by populations in certain geographic areas; it is, for example, the biggest contributor to radiation exposure in Britain, where, of an average total annual radiation dose per person of 2.5 millisieverts (mSv), radon contributes almost 50 per cent (approximately 1.2 mSv), compared with 12 per cent from medical sources and only 0.1 per cent due to anthropogenic nuclear discharges. Since the 1960s, studies such as that undertaken by the National Radiological Protection Board (NRPB) in the UK have drawn tentative links between the high incidence of cancers (e.g. lung cancer and leukaemia) and high concentrations of

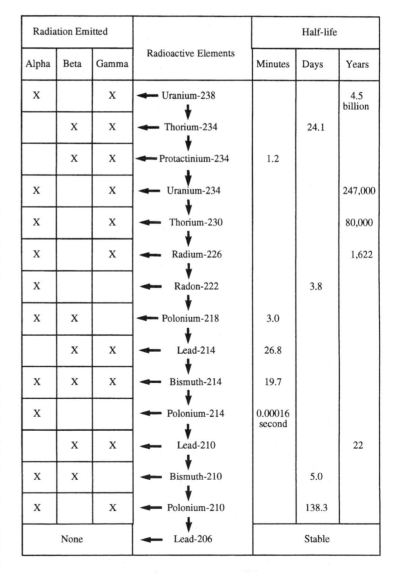

Radiation Emitted			Radioactive Elements	Half-life		
Alpha	Beta	Gamma		Minutes	Days	Years
X		X	Uranium-238			4.5 billion
	X	X	Thorium-234		24.1	
	X	X	Protactinium-234	1.2		
X		X	Uranium-234			247,000
X		X	Thorium-230			80,000
X		X	Radium-226			1,622
X			Radon-222		3.8	
X	X		Polonium-218	3.0		
	X	X	Lead-214	26.8		
X	X	X	Bismuth-214	19.7		
X			Polonium-214	0.00016 second		
	X	X	Lead-210			22
X	X		Bismuth-210		5.0	
X		X	Polonium-210		138.3	
None			Lead-206	Stable		

Figure 6.3   *Uranium-238 (^{238}U) decay series. Redrawn after Botkin and Keller (1995).*

Table 6.2 *Typical yields of (A) actinides*, and (B) fission products†.*

Element	Half-life (years)	Decay interval**									
		0.3 yr (g)	(Ci)	10 yr (g)	(Ci)	500 yr (g)	(Ci)	10,000 yr (g)	(Ci)	100,000 yr (g)	(Ci)
Neptunium-237 (^{237}Np)	2.1×10^6	760	0.59	760	0.59	786	0.61	810	0.63	790	0.61
Plutonium-238 (^{238}Pu)	86	5.8	105	5.5	100	0.1	1.8	–	–	–	–
Plutonium-239 (^{239}Pu)	24,400	27.5	1.7	27.5	1.7	32	2.0	59	3.6	4.5	0.3
Plutonium-240 (^{240}Pu)	6,580	8.5	2.0	19.2	4.5	38.4	8.8	13.9	3.2	–	–
Plutonium-241 (^{241}Pu)	13.2	4	464	2.4	273	–	–	–	–	–	–
Plutonium-242 (^{242}Pu)	379,000	2	0.009	2	0.009	2	0.009	2	0.009	1.7	0.007
Americium-241 (^{241}Am)	462	54	189	55.6	198	29.5	103	–	–	–	–
Americium-243 (^{242}Am)	7,370	82	17.0	8.2	17.0	77	16.0	31	6.5	–	0.001
Curium-244 (^{244}Cm)	17.6	30	2,570	19.7	1,700	–	–	–	–	–	–
Total											
grams/tonne fuel		974		974		965		916	100	796	50
Actinides, including daughter products (approx.)			200,000		10,000		900				

* Actinides = series of chemical elements beginning with actinium-89 and continuing to lawrencium-103. Includes intermediate and long half-lives in the waste stream from processing 1 tonne of light-water reactor fuel irradiated to 33×10^9 watt-days (thermal) per tonne of fuel.

** g = grams per tonne of fuel; Ci = Curies.

Element	Half-life	Curies remaining			
		10 yr	100 yr	500 yr	1,000 yr
^{144}Ce/^{144}Pr	285 days	300	–	–	–
^{106}Ru/^{106}Rh	367 days	1,100	–	–	–
^{155}Eu	1.8 yr	160	–	–	–
^{134}Cs	2.1 yr	8,300	–	–	–
^{125}Sb/^{125}Te	2.7 yr	980	–	–	–
^{90}Sr/^{90}Y	28.1 yr	1.2×10^5	1.32×10^4	0.6	–
^{137}Cs/^{137}Ba	30 yr	1.6×10^5	2.1×10^4	2	–
^{151}Sm	90 yr	1,100	520	30	0.4
^{99}Tc	2.1×10^5 yr	15	15	15	15
^{93}Zr	9×10^5 yr	3.7	3.7	3.7	3.7
^{135}Cs	2×10^6 yr	1.7	1.7	1.7	1.7
^{107}Pd	7×10^6 yr	0.013	0.013	0.013	0.013
^{128}I	17×10^6 yr	0.025	0.025	0.025	0.025
Total curies (approx.)		300,000	35,000	53	22

† Includes intermediate and long half-lives from processing 1 tonne of light-water reactor fuel irradiated to 33×10^9 watt-days (thermal) per tonne of fuel.
Chemical element symbols: Ce = cerium; Pr = protactinium; Ru = ruthenium; Rh = rhodium; Eu = europium; Cs = caesium; Sb = antimony; Te = tellurium; Sr = strontium; Y = yttrium; Ba = barium; Sm = samarium; Tc = technetium; Zr = zirconium; Pd = palladium; I = iodine.
Source: Cassedy and Grossman 1990 (reproduced from: *The Nuclear Fuel Cycle*, The Union of Concerned Scientists, The MIT Press, Cambridge, Mass., 1975).

household radon levels. The NRPB has calculated that as many as 2,000 deaths per year in the UK may be the result of the radiation produced by radon, and under certain circumstances the increased risk of contracting lung cancer may be comparable with those risks faced by heavy smokers. Epidemiologists, however, argue that a direct link between radon and an increased risk of contracting cancer cannot be proved statistically.

Many factors control the rate of migration and the concentration of radon in the human environment. Concentrations of atmospheric radon are generally highest in regions comprising rocks that contain the parent sources, namely uranium and thorium, and

have effective pathways for the migration of the gas, for example granites and some sedimentary rocks such as certain types of shales and ironstones, particularly where there are **phosphate nodules**. The concentrations, however, cannot be directly correlated with the concentrations of uranium and thorium in rock, because the uranium is present in different host minerals, some of which are more easily weathered, and so release higher concentrations of radon more effectively. The rate at which radon passes through the ground is important in controlling its concentration at the surface or within the soil. Rocks with higher permeability (inter-connectedness of cavities), e.g. highly fissured and faulted rocks, allow radon, in the form of gas or dissolved in ground water, to migrate at a faster rate into the

human environment. Some of the highest concentrations of radon are often associated with springs rather than known uranium-rich rocks. This is because ground-water pressure decreases when it comes to the surface allowing the release of dissolved gases such as radon from the contaminated water.

During rain storms, pore spaces within soils and rocks fill with water and prohibit the migration of radon and allow its build-up within the soil, but when the ground dries radon is released in high concentrations. Diurnal variations may occur because dew inhibits the release of radon from the soil at night and in the early morning, while radon is released as the soil dries out during the day. Also, reduced atmospheric pressure during a cyclonic depression increases the soil–atmosphere pressure

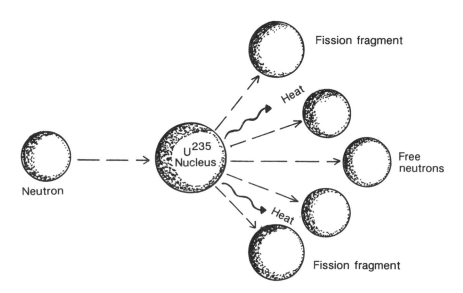

Figure 6.4 *Fission of ^{235}U. A neutron strikes the ^{235}U nucleus to produce fission fragments and three neutrons, together with the release of thermal energy. The released neutrons may then each strike another ^{235}U atom, releasing more neutrons, fission fragments and energy. As the process continues a chain reaction develops. Redrawn after Botkin and Keller (1995).*

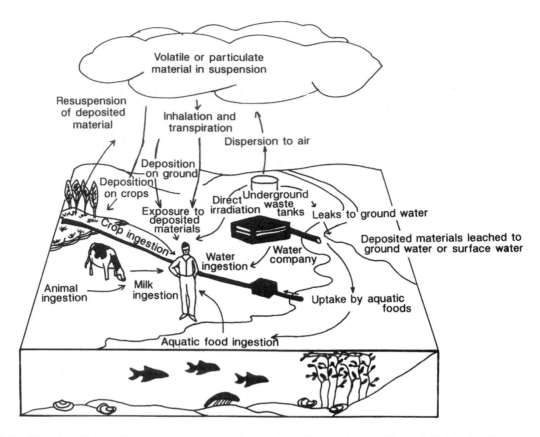

Figure 6.5 *Ways in which radioactive substances reach people. Redrawn after Botkin and Keller (1995).*

gradient, which may result in an increase in the rate of release of radon from the soil and help concentrate it in the atmosphere (Miller and Ball 1969, Miller and Ostle 1973). These factors contribute to complex variations in atmospheric radon concentrations.

Atmospheric radon, however, is easily dispersed by the wind; it is only when its dispersal is inhibited that it becomes a hazard. The most common way in which it is concentrated is by becoming trapped within a building, particularly houses. Figure 6.6 shows the main ways that radon can enter and be concentrated in the home. The majority of the radon enters through the floor from the ground and accumulates in cavities where there are gaps between the floor and the ground. The concentration of radon in a building is aided by the lower indoor pressure compared with the ambient atmospheric pressure, creating a pressure gradient, which drags air into the building. This is greatest when buildings have chimneys. Radon may also enter a house by the release of the dissolved gas in shower and bath water, and from walls that have been constructed from rocks such as granite or gypsum board, which have high uranium concentrations.

Insulated houses (i.e. double-glazed, draught-proofed) impede the escape of radon and hence may have relatively high concentrations of the gas. The variation in radon concentrations depends on human activity. Levels of radon vary annually, the highest concentrations being between November and March, when the home is kept warm and ventilation is reduced. Daily variations are very common, with the highest concentrations at night when people are in bed and radon can accumulate, and with lower concentrations in the morning and evening, when people open and close doors and windows, thereby releasing the radon into the atmosphere.

The average amount of radon in household air in Britain is 20.5 Bq m^{-3} (1 becquerel (Bq) is equivalent to 1 atomic disintegration per second), but the NRPB has found levels of more than 200 Bq m^{-3}. In the UK, the Radiation Regulations 1985 dictate that annual radon doses should not exceed 50 mSv, equivalent to a radon level of 1,000 Bq m^{-3}. Areas where there is a probability of more than 1 per cent of the households having levels above 200 Bq m^{-3} have been designated as 'affected areas'. Radiation levels in some 'affected areas' may exceed 8,000 Bq m^{-3}. The safe concentration of radon is difficult to assess because the

BOX 6.3 RADON

Radon (Rn) is a naturally occurring colourless, odourless and tasteless radioactive gas produced by the decay of uranium (U) and thorium (Th) in rocks and soils. There are three main isotopes, radon-219, radon-220 and the most important in terms of its effects on humans, radon-222. Exposure to radon increases the risk of lung cancer, because once inhaled the gas sits in the respiratory tract and lungs. Radon-222 has a half-life of 3.82 days and is produced in the decay chain starting with uranium-238 (^{238}U); ^{220}Rn (thoron), which has a half-life of 55.3 seconds, is produced by a decay series of ^{232}Th; and ^{219}Rn (actinon), with a half-life of 3.92 seconds, is formed by the decay of ^{235}U. With a longer half-life, ^{222}Rn has more opportunity to migrate from rocks and soils into the human environment, in homes and places of work. Alpha particles are formed and released into the environment during many of the disintegrations involved in the decay series of ^{238}U through radium-226 and finally to the stable daughter isotope lead-206 (^{206}Pb). Alpha particles can cause tissue damage, but as these particles are relatively large and possess a high electrical charge they do not travel easily through clothing or skin. They may, however, enter the body in drinking water and can be inhaled during respiration. Since ^{222}Rn has a half-life that is appreciably longer than the time required for clearance of gas from the respiratory tract, few alpha particles are produced while the radon is in the lungs and thus few alpha particles enter the body. The daughter products, however, which include polonium-219 (half-life 3.05 minutes), lead-214 (half-life 26.8 minutes) and bismuth-214 (half-life 19.7 minutes) have shorter half-lives and are more important in irradiating the lungs and contaminating drinking water, thereby increasing the risk of lung cancer.

relationship between radon exposure and the incidence of lung cancer is based on studies of workers in uranium mines, who received their doses in a totally different environmental setting. Also, it is difficult to assess and measure the concentrations of household radon because of the annual and diurnal variations, discussed above. In Britain, the NRPB has measured radon levels using a track–etch method on a plastic such as cellulose nitrate, which is sensitive to the passage of alpha particles. With the aid of such measurements and good public co-operation, the NRPB published its first assessment of a potentially affected area which included the counties of Devon and Cornwall, in 1990. No areas in Cornwall and Devon had concentrations below the 1 per cent probability of homes being above the recommended level (also referred to as the 'action level') of 200 Bq m^{-3}, giving a 3 per cent lifetime risk factor of a fatal cancer. This action level represented a compromise by the British government between setting very low risk levels for cancer versus the cost of the remedial work that would be necessary to modify existing houses and buildings if an even safer radiation dose level were fixed: at the set level, the NRPB estimates that up to 100,000 homes in Britain (60,000 in southwest England alone where there are large areas of granite basement) may exceed the safety limits. As a result of this study, the British government launched a campaign to

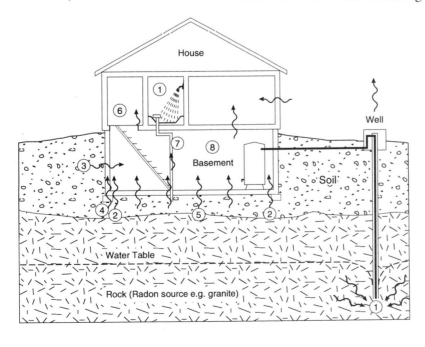

Figure 6.6 *Ways in which radon can enter homes via 1 = ground water and domestic water for drinking, cooking and bathing; 2 = construction joints; 3 = cracks in walls; 4 = cavities in walls; 5 = cracks in solid floors; 6 = cracks in suspended floors; 7 = service pipes; 8 = radon concentrating in basements.*

encourage householders to reduce radon levels in their homes by altering their construction and design, part of which included a document produced by the Department of the Environment (DoE) entitled *The Householder's Guide to Radon*. Recommendations to help reduce radon levels include regularly ventilating the house, sealing the floor and depressurising spaces beneath ground or basement floors. Governments are becoming more concerned by the likelihood of a build-up of radon in houses, and surveys are under way in many developed countries to identify areas of high radon concentrations and to assess the acceptable levels of radon for houses.

Nuclear reactors

Nuclear energy is currently obtained from splitting atoms, termed nuclear fission. Typical yields of fission products are shown in Tables 6.1A and 6.1B, giving some indication of the longevity of many of these radioactive substances. Intensive research is currently under way to harness this energy by combining light atomic nuclei in the process of nuclear fusion. Fusion is seen as having an enormous advantage over fission because it should produce much less radioactive waste as by-products, and the principal raw material is heavy water, containing deuterium, something that is effectively in limitless supply.

In order to develop and successfully install nuclear fusion power stations, it is going to take astronomical amounts of investment and time. Nuclear fusion is still not technically, let alone commercially, viable.

The production of nuclear energy is dealt with in this chapter rather than Chapter 7 (Energy resources) because reactors are used not only for commercial energy production but also to enrich fuels for nuclear weapons. Figure 6.7 shows several different types of nuclear reactor. These use different mixtures of fuel, moderator and coolant.

Pressurised-water reactors (PWRs) are small, with cores 2 to 3 m across. They use ordinary liquid water, which is kept under pressure (*c.* 150 bars) and at a temperature of 295–330°C, as the moderator and coolant. The fuel is slightly enriched in ^{235}U ($\leq$ 3 per cent). These reactors are used in submarines and provide an output of about 1.3 GW.

Boiling-light-water reactors are also small reactors. The water in these, however, is at a lower pressure than in the PWRs and the water is allowed to boil. These reactors are designed so that if the reactor power increases and more boiling occurs, the reactivity of the reactor drops. This is a good safety factor.

Advanced gas-cooled reactors (AGRs) use uranium oxide fuels clad in steel, with a graphite moderator and carbon dioxide coolant gas, which is at a pressure of about 40 bars. AGRs operate at high temperatures, so they need a back-up cooling system in case of an emergency. These types of reactor are characteristic of the British nuclear programme.

Canadian CANDU reactors use natural, unenriched uranium fuel and heavy water as both a moderator and coolant. The moderator is contained within a cylindrical steel vessel called a calandria, which is about 6 m long and 7 m in diameter, with 380 tubes passing horizontally through it. Each tube contains an inner tube that carries fuel, with helium between the two tubes. This allows the moderating heavy water in the calandria to be maintained at relatively low temperatures (65°C) and pressures, and the coolant is under a pressure of 87 bars at a temperature of between 250 and 293°C. The pressure tubes act as a safety factor so that if depressurisation occurs the whole reactor will not be affected.

Graphite-moderated, water-cooled reactors of the RBMK type are of Soviet design and are now being discontinued. Unfortunately, they had inadequate control rods, no secondary shut-down system, a graphite moderator at 700°C, which could burn spontaneously in air, and inadequate containment structures. One of the major design problems with RBMKs is that as the reactor power increases, the reactivity of the reactor also increases. This positive feedback can lead to 'runaway', with catastrophic effects. In addition, it is not known if safety systems can be overridden during an emergency, so that the control rods can be withdrawn to reduce the danger.

Radioactive waste

Radioactive or nuclear waste is created from the production of nuclear weapons, in electricity generation using nuclear power, and in smaller quantities from medical practice, some industrial activities and certain types of scientific research (obviously, mainly but not exclusively from nuclear research).

The range of activities associated with the operation of civil nuclear reactors, from uranium extraction, enrichment, fuel fabrication, reactor operation, spent-fuel storage, spent-fuel reprocessing, and other aspects of waste management, is referred to as the **nuclear-fuel cycle** (NFC: Figure 6.8). The decommissioning of commercial nuclear reactors is not usually assumed to be part of the NFC (Berkhout 1991).

The disposal of radioactive waste is one of the most sensitive of environmental issues. Nobody,

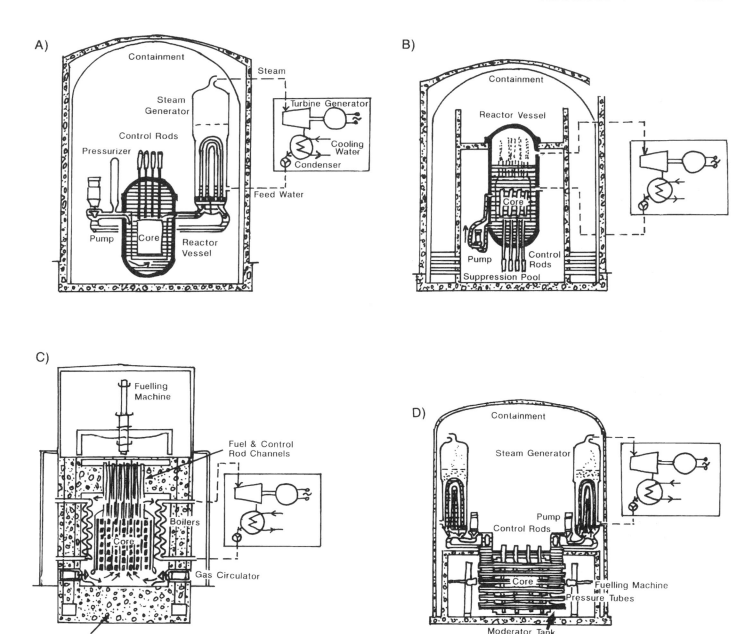

Figure 6.7 *Types of nuclear reactor. (A) Pressurised water reactor; (B) Boiling water reactor; (C) Advanced gas-cooled reactor; (D) CANDU heavy-water reactor. Redrawn after Rippon (1994).*

whether pro or anti the creation of radioactive waste through nuclear energy and other means, wants to see this waste dumped in their proverbial backyard ('not in my back yard', or **NIMBY**), so issues involving the disposal of radioactive waste include a very large measure of selfish motives mixed with the altruistic.

Radioactive substances dumped at sea can pollute the water as low-level radioactive waste. In the USA, dumping was terminated in 1970, and in European countries in 1982, but many other countries continue

to dump their waste in the oceans. Since 1975, all radioactive waste disposal at sea has been conducted in accordance with the London Dumping Convention.

Many radioactive wastes are extremely toxic, with their radioactivity decreasing exponentially through time – the half-life being the time taken for half the original amount of material to decay. Half-lives and toxicity vary greatly amongst the radioactive elements. Tables 6.2A and 6.2B show the fission products and actinides (a group of fourteen intensely

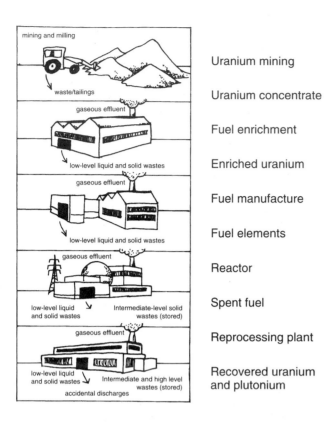

mining and milling

Uranium mining

waste/tailings

Uranium concentrate

gaseous effluent

Fuel enrichment

low-level liquid and solid wastes

Enriched uranium

gaseous effluent

Fuel manufacture

low-level liquid and solid wastes

Fuel elements

gaseous effluent

Reactor

low-level liquid and solid wastes   Intermediate-level solid wastes (stored)

Spent fuel

gaseous effluent

Reprocessing plant

low-level liquid and solid wastes   Intermediate and high level wastes (stored)

Recovered uranium and plutonium

accidental discharges

Figure 6.8 *The nuclear fuel cycle. Redrawn after Blunden and Reddish (1991).*

radioactive elements, with increasingly short half-lives for those with the highest atomic numbers) that would be typical for 1 tonne of spent liquid reactor fuel. From these tables, it can be seen that half-lives vary up to millions of years, and typically range over at least tens to thousands of years, far longer than human recorded history. The main point of presenting these tables is to stress that any nuclear accidents can contaminate the natural environment into the foreseeable future, and beyond!

Radioactive waste is introduced into water as liquid effluent discharged from two commercial nuclear-fuel reprocessing plants (one in the UK and the other in France). Long-lived nuclides such as strontium-90 (^{90}Sr), caesium-137 (^{137}Cs) and ruthenium-106 (^{106}Ru) are frequently discharged. An example of such an incident occurred on 11 November 1983, when Greenpeace informed the public that the beach below Sellafield nuclear plant in Cumbria, UK, had become highly contaminated by radioactive water from the Irish Sea that had been discharged from Sellafield. On 7 December the government issued a warning to the public that a 25-mile stretch of beach had become contaminated to a level of over 1,000 times the normal background radiation. Ironically,

independent studies of the same beach showed that it had begun to be contaminated as far back as 1979, but little had been done to counteract the problem.

In addition to the beach pollution at Sellafield, the environmental pressure group, Friends of the Earth, revealed new high levels of an isotope of caesium (^{137}Cs) and also americium (^{241}Am), a particularly dangerous isotope associated with plutonium, in the River Esk, which flows into the sea at Ravenglass in Cumbria. Friends of the Earth were particularly concerned about the health of anyone spending more than twenty hours a week along the river bank, because they would be at risk from inhaling radioactive dust blown up from the drying sediment along the river banks in quantities that could exceed the recommended safety levels.

The transport of radioactive substances is a hazardous process. In the North Sea, on 28 August 1984, for example, the French cargo ship, *Mont Louis*, carrying 450 tonnes of uranium-rich material, collided with a ferry and capsized off Belgium. Disasters of this type that involve radioactive material in transit have happened and, undoubtedly, will occur again. They give cause for concern and demand more stringent international legislation to minimise the dangers of handling and transporting such waste.

Scientists have also shown concern about the risks of dumping radioactive waste in salt mines. In 1991, however, in a US$800 million project, the American government planned to store 4,250 barrels of radioactive waste in underground caverns in New Mexico. The salt could collapse to make access impossible, or at least extremely difficult, should the containers leak.

Accidents at nuclear power stations can cause serious contamination of the environment. Particularly memorable examples of this type of pollution occurred at Chernobyl in 1986, Windscale (now called Sellafield) in 1957 and Three Mile Island in 1979. The effects of polluted rain water from the Chernobyl accident in the old Soviet Union in April 1984 have still to be fully assessed. The Soviets stated that there was little effect on the waters around Chernobyl, with water supplies to the main city of the province, Cave, remaining uninterrupted after the accident. There has been no real assessment of how the fallout polluted water supplies, let alone how waters might be treated if contaminated. Some authorities, such as Heinz Hansen (Riso National Laboratory, Denmark), believe the concentration of radioactive nuclei in contaminated water supplied to humans is small compared with other sources, which concentrate the nuclei, such as dairy products, fish, fruits and vegetables. Many of the food sources, however, that concentrate these nuclei may derive them from water sources (O'Neill 1989).

Underground and sub-sea repositories for radio-active waste storage are being adopted by many governments and/or waste-disposal companies as the most acceptable option. This is exemplified by the US and British decisions. For example, in Britain, on 11 December 1991, the company responsible for the disposal of radioactive waste from reprocessing spent reactor fuel rods, NIREX, announced that the chosen site for the disposal of low-level and inter-mediate-level radioactive waste (LLW and ILW) is to be an inland underground site near Sellafield in Cumbria. There are, however, still proponents of an alternative undersea site below the Irish Sea off the coast near Sellafield. Both options have serious draw-backs, but a sub-sea site would have been the worst choice, principally because it would have posed the greatest difficulties in monitoring the radioactive waste. Furthermore, the retrieval of such waste from below the sea bed would be logistically much more difficult than if it were stored underground.

Following exploratory drilling, NIREX decided to dispose of the ILW and LLW within the deeply buried volcanic 'basement' to this area, known as the Borrowdale Volcanic Series. These rocks are exten-sively exposed at the surface throughout much of the Lake District and are therefore well known to geologists. Where the Borrowdale Volcanic Series is exposed in outcrops, the rocks are invariably highly folded, fractured, jointed and dislocated by geolog-ical faults. In the absence of readily available detailed data on the equivalent deeply buried Borrowdale Volcanic Series rocks offshore from Sellafield there is no reason to assume that they are not similarly fractured, jointed and faulted. Many of the geolog-ical faults are vertical to near-vertical and, therefore, not easy to detect in vertical boreholes. Such defor-mation of the rocks has created fluid-filled and gas-filled cavities at all scales, from the microscopic to the macroscopic, which is known as **porosity**. The inter-connectedness of these spaces is known as **permeability**. Should radioactive waste escape following some unforeseen accident it is through this porosity and permeability that it will move, possibly towards the surface in ground-water systems. The highly fractured rocks associated with geological faults are probably the principal pathways for fluid movement. This could occur in both the under-ground and sub-sea sites.

Sealing the fracture porosity and destruction of the associated permeability to acceptable minimum values in the Borrowdale Volcanic Series will present major scientific and technical problems.

Prior to making a firm decision as to where to bury the radioactive waste, it may well have been much better to await a detailed quantitative under-standing of the fracture porosity in the proposed host rocks designated to contain the LLW and ILW, together with an open and public debate, with all the relevant information freely available for inspec-tion by independent bodies. Supporters of storage of ILW and LLW in a site below the sea floor must accept that such an option has all the potential prob-lems associated with inland underground storage, but with the additional complications imposed by oper-ating in the sea, where there are also possibilities for shipping accidents and bad weather conditions hampering storage and/or retrieval.

To minimise the long-term risk to the environ-ment caused by the disposal of radioactive waste, and until the under-sea repository option has been fully evaluated, the storage of LLW and ILW should perhaps be in surface sites, where the waste would be most readily retrievable and accessible for ease of monitoring in preference to its disposal in the poorly understood deep repository in the Irish Sea. Inland underground storage would be preferable to sub-sea storage, but neither option is satisfactory given the sites from which NIREX must choose. In order to gain more information, additional boreholes are needed to supplement the very few drilled to date.

Box 6.4 exams the US Nuclear Regulatory Commission to provide an example of how nuclear waste is regulated on a national scale as a means of ensuring adequate protection of public health and safety.

Nuclear accidents and fallout

One of the ways in which past thermonuclear explo-sions can be detected is from suitably located high-latitude ice cores. Atmospheric thermonuclear weapons testing (ATWT) generates large amounts of beta activity, a radionucleide, and associated nitrous oxide (N_2O) approximately 10^{10} g of NO per megaton, equivalent to about 2×10^{10} g of nitric acid (HNO_3), commonly measured as nitrate (NO_3^-). The mechanism by which this occurs is the fixation of nitrogen by nuclear fireballs, which generate atmos-pheric nitrogen oxides (NO_x), and ultimately nitric acid (HNO_3), which falls as rain or snow. A study of ice and snow cores from Mount Logan ($60°36'N$, $140°30'W$), in the Yukon Territory, Canada, collected in 1980 at an altitude of 5,340 m and cali-brated to an accuracy of less than half a year, revealed that during the era of intense ATWT, between 1952

BOX 6.4 US NUCLEAR WASTE

The US Nuclear Regulatory Commission (NRC) was established to ensure adequate protection of public health and safety, common defence and security, and the environment in the use of nuclear materials in the USA. Its responsibilities include the regulation of commercial nuclear power reactors: non-power research, testing and training reactors; fuel-cycle facilities; medical, academic, and industrial uses of nuclear materials; and the transport, storage and disposal of nuclear materials and waste.

The NRC has developed a classification system for low-level waste based on its potential hazards and has specified disposal and waste form requirements for each of the three general classes of waste – A, B and C. Class A waste has lower concentrations of radioactive material than Class C. The volume of radioactive waste varies from year to year. Approximately 800,000 ft^3 of low-level radioactive waste was disposed of in 1993. This was 45 per cent less than in 1992, and was probably due to an effort by the nuclear industry to minimise waste generation and reduce the volume of waste by compaction and incineration.

The NRC enforces the Low-Level Radioactive Waste Policy Amendments Act (LLRWPAA) of 1985, which provides for a system of milestones, incentives and penalties to ensure that states and disposal facilities will be responsible for their own waste. Active, licensed disposal facilities include Barnwell in South Carolina, Hanford in Washington and Clive in Utah. Other disposal facilities include Ward Valley, California; Boyd County, Nebraska; Hudspeth County, Texas; Wake County, North Carolina; and Beatty, Nevada, which ceased disposal operations on 1 January 1993.

In 1993, approximately 28,000 tonnes of spent nuclear fuel was stored at commercial nuclear power reactors. By 2003, the amount is expected to be approximately 48,000 tonnes. This spent fuel is regarded as high-level radioactive waste. In 1990, the NRC amended its regulations to authorise licensees to store spent fuel at reactor sites in NRC-approved casks. The Nuclear Waste Policy Act of 1982 and the Nuclear Waste Policy Amendments Act of 1987 specify a detailed approach for high-level radioactive waste disposal, with the US Department of Energy (DOE) having operational responsibility. The NRC is responsible for the transportation, storage and geological disposal of the waste. The Amendments Act designated a candidate site for a high-level deep repository at Yucca Mountain, Nevada. The DOE is responsible for the determination of the site's suitability.

Sources: http://www.nrc.gov/org.html; http://www.nrc.gov/radwaste.html; http://www.ymp.gov/ref_shlf/yms/ymstoc.htm.

---

and 1980, the observed all-time peak pulse in gross beta activity coincides with the largest nitrate kick (Holdsworth 1986). After taking account of other factors that might have led to these peaks, including volcanic eruptions, Holdsworth concluded that ATWT was indeed responsible for a significant part of the nitrate content.

The following sections explore some of the consequences of the most well-documented accidents at nuclear power stations.

Three Mile Island

On 26 March 1979, the Three Mile Island nuclear power plant near Harrisburg, Pennsylvania in the USA, underwent an accident sequence that almost led to a **meltdown** of the reactor core. The accident was part human error and part machine failure. The initial action in the sequence of events on that day was human error, because a major feedwater valve was erroneously left closed at the end of a routine maintenance exercise the day before the accident. The plant operators were unaware of the closed valve for several critical minutes. They also received false information on the correct setting of another valve. Both of these faults led to the operators taking counterproductive corrective action in shutting down the cooling water pumps, which resulted in a drop in the coolant level in the reactor core. The reactor underwent a near meltdown, in which it overheated towards a critical state, before these errors were observed and remedial action taken.

The consequences of this accident were that reactor number 2 was seriously damaged and within the containment building, the entire area became prohibitively radioactive. Fortunately, nobody was killed.

The ensuing clean-up operation, still unfinished more than two decades after the event, nearly bankrupted the operating company, General Public Utilities, which was left with an estimated bill for US$1 billion to make the area safe and repair the damage. The significance of this accident at Three Mile Island is that it represented a turning point in public awareness about the costs of nuclear power and nuclear technology. Since the incident, no more US-commissioned nuclear power stations have been constructed. Even those people in favour of nuclear power stations tend to favour the option of 'not in my back yard'.

Chernobyl

The Chernobyl accident must rank as the worst in the history of nuclear power. On 26 April 1986, reactor number 4, now christened 'the Sarcophagus', at the nuclear power plant of Chernobyl in the Ukraine suffered a meltdown to release large amounts of radioactive material, which spread across much of northern Europe and the Soviet Union. The accident was part human error and part design flaw. In the explosion, the 2,000-tonne lid of the reactor was dislodged and fell back into the reactor to become precariously balanced at an angle. The floor of the reactor dropped by about 4 m in the explosion. Radioactive elements released into the atmosphere included iodine-131, strontium-90 and caesium-137. The nuclear fuel in the reactor overheated and mixed with the sand used to line the reactor, then flowed like lava into the rooms below the reactor to solidify as a highly radioactive glass. It was only when the Soviet clean-up team, known as the 'liquidators' or 'bio-robots', came to examine the reactor core that, to their consternation, they found only the twisted remains of the cooling rods.

The clean-up operations at Chernobyl have been a catalogue of disasters, mismanagement and under-funding. Even today, after the Soviets had requested international aid, little help and equipment has been forthcoming. Western governments seem to see Chernobyl as a Soviet disaster and not a global catastrophe. Of the estimated 600,000 people who have been involved in the clean-up, about 250,000 have already received their lifetime's safe dosage of radiation. Many have died from **acute radiation sickness** (**ARS**) and suffered radiation burns from the intense fallout at Chernobyl, but the exact numbers are unknown. Vladimir Chernousenko, the scientific director of the 30-km radius exclusion zone around Chernobyl, has estimated that 7,000–10,000 people could have died in the clean-up, but the Soviet authorities officially gave a figure of 250–350 dead.

In this and the next century, the Sarcophagus will never be safe. Amongst the plans that have been proposed to contain the further release of radiation at Chernobyl have been suggestions to bury the reactor in sand (but the reactor fuel might overheat and cause a nuclear explosion), embed it in concrete (making it difficult to monitor the state of the solidified nuclear fuel), or erect a second sarcophagus over the damaged reactor to seal it hermetically for at least a few hundred years.

It is obvious that the results of this single nuclear accident at Chernobyl cannot easily be contained, yet governments are busy planning and constructing more nuclear power stations, which will increase the risks of still further unforeseen disasters. No reactor design, however well engineered, can be totally safe. It is a question of acceptable levels of risk measured against the perceived benefits of nuclear energy that have to be weighed in the balance. Over and above these arguments, there is the often unstated military requirement for nuclear power stations to process and reprocess radioactive material for weapons. Governments certainly find it easier to hide these military motives behind the smokescreen of a debate about nuclear energy.

Tomsk-7

In early April 1993, an explosion and fire at a plutonium plant east of Moscow, the Tomsk-7 military complex in Siberia, caused by an explosion in a tank of uranium, led to radioactivity covering the entire complex 14 miles outside the town of Tomsk. The explosion occurred underground when nitric acid was being added to uranium in a stainless steel tank, which caused the temperature to rise rapidly. At reprocessing plants such as Tomsk-7, the spent nuclear fuel is separated by dissolving it in acid to allow the recovery of the uranium and plutonium. In the case of the Tomsk-7 plant, waste gases in the uranium tank were ignited, blowing the concrete shield off the tank in a large explosion.

This type of accident is amongst the most serious that can occur at a reprocessing plant. On the Russian scale of nuclear accidents, from 1 to 7, the latter being the most serious, the Tomsk-7 incident was initially rated by the Russians as level 3 to 4, or possibly higher. Later analysis revised the seriousness downwards to a relatively minor incident. However, radiation levels in the contaminated area were measured at between several milliröentgens and several röentgens, substantially more than the maximum permitted annual dosage for nuclear workers. Although prevailing winds caused the radioactive cloud to move away from populated areas and drift northeastwards, the radioactivity poses an environmental hazard, and to date has resulted in a large-scale clean-up operation with extensive topsoil removal and raised concerns about the melting snow finding its way into water courses to affect human health. Box 6.5 lists some of the major nuclear accidents that have taken place in the USA, or elsewhere as a result of US activities, during recent years. These data illustrate the vulnerability of the environment to nuclear activity, and the ease with which nuclear accidents can occur.

BOX 6.5 US NUCLEAR ACCIDENTS

Nuclear power and nuclear devices have not enjoyed a safe history in the USA. A large number of incidents mar the safety record of nuclear plants, facilities, bombers and ships, resulting in numerous deaths and injuries. The following list is a selection of some of the most recent events involving nuclear devices and facilities. The data have been taken from a comprehensive list of US nuclear accidents that was compiled by Allen Lutins and is available on the Internet: http://www. nitehawk.com/alleycat/welcome.html. The list makes fascinating reading and it highlights how easily an accident can occur, stressing the risk from nuclear industry (Plate 6.2).

Power plants

22 March 1975
A technician checking for air leaks with a lighted candle caused $100 million in damage when insulation caught fire at the Browns Ferry reactor in Decatur, Alabama. The fire burned out electrical controls, lowering the cooling water to dangerous levels, and requiring a manual shut-down of the plant.

28 March 1979
A major accident at the Three Mile Island nuclear plant near Middletown, Pennsylvania. At 4.00 a.m., a series of human and mechanical failures nearly triggered a nuclear disaster. By 8.00 a.m., after cooling water was lost and temperatures soared above 5,000°C, the top half of the reactor's 150-tonne core collapsed and melted. Contaminated coolant water escaped into a nearby building, releasing radioactive gases, leading to as many as 200,000 people fleeing the region. Despite claims by the nuclear industry that 'no one died at Three Mile Island', a major study by Dr Ernest J. Sternglass, professor of radiation physics at the University of Pittsburgh, showed that the accident led to a minimum of 430 infant deaths.

1981
The Critical Mass Energy Project of Public Citizen Inc. reported that there were 4,060 mishaps and 140 serious events at nuclear power plants in 1981, up from 3,804 mishaps and 104 serious events the previous year.

11 February 1981
An auxiliary unit operator, working his first day on the new job without proper training, inadvertently opened a valve, which led to the contamination of eight men by 110,000 gallons of radioactive coolant sprayed into the containment building of the Tennessee Valley Authority's Sequoyah I plant in Tennessee.

1982
The Critical Mass Energy Project of Public Citizen Inc. reported that 84,322 power plant workers were exposed to radiation in 1982, up from 82,183 the previous year.

25 January 1982
A steam generator pipe broke at the Rochester Gas and Electric Company's Ginna plant near Rochester, New York. Fifteen thousand gallons of radioactive coolant spilled onto the plant floor, and small amounts of radioactive steam escaped into the air.

15–16 January 1983
Nearly 208,000 gallons of water with low-level radioactive contamination was accidentally dumped into the Tennessee River at the Browns Ferry power plant.

1988
The Critical Mass Energy Project of Public Citizen Inc. reported that there were 2,810 accidents in US commercial nuclear power plants in 1987, down slightly from 2,836 accidents in 1986.

25 February 1993
A catastrophe at the Salem 1 reactor in New Jersey was averted by just 90 seconds when the plant was shut down manually, following the failure of automatic shutdown systems to act properly. The same automatic systems had failed to respond in an incident three days before, and other problems plagued this plant as well, such as a 3,000-gallon leak of radioactive water in June 1981 at the Salem 2 reactor, a 23,000-gallon leak of 'mildly' radioactive water (which splashed onto 16 workers) in February 1982, and radioactive gas leaks in March 1981 and September 1982 from Salem 1.

28 May 1993
The Nuclear Regulatory Commission released a warning to the operators of 34 nuclear reactors around the USA that the instruments used to measure levels of water in the reactor could give false readings during routine shutdowns and fail to detect important leaks. A failure to detect falling water levels could have resulted, potentially leading to a meltdown.

Bombs and bombers

In addition to the list below, at least 50 nuclear weapons lie on the sea floor due to US and Soviet accidents.

24 January 1961
A B-52 bomber with nuclear bombs fell apart in mid-air over North Carolina, killing three of the eight crewmen and releasing two 24-megaton nuclear bombs. One bomb parachuted to the ground and was recovered; the other fell free and landed in waterlogged farmland, never to be found. When the recovered bomb was studied, it was found that five of its six safety devices had failed.

14 March 1961
A B-52 bomber with nuclear bombs crashed in California while on a training mission.

13 January 1964
A B-52 bomber with two nuclear weapons crashed near Cumberland, Maryland.

17 January 1966
A B-52 bomber collided with an Air Force KC-135 jet tanker while refuelling over the coast of Spain, killing eight of the eleven crew members and igniting the KC-135's 40,000 gallons of jet fuel. Two hydrogen bombs ruptured, scattering radioactive particles over the fields of Palomares; a third landed intact near the village of Palomares; the fourth was lost at sea 12 miles off the coast of Palomares and required a search by thousands of men working for three months to recover it. Approximately 1,500 tonnes of radioactive soil and tomato plants were removed to the USA for burial at a nuclear waste dump in Aiken, South Carolina. The USA eventually settled claims by 522 Palomares residents at a cost of $600,000 and gave the town a gift of a $200,000 desalination plant.

22 January 1968
A B-52 bomber crashed seven miles south of Thule Air Force Base in Greenland, scattering the radioactive fragments of four hydrogen bombs over the terrain after a fire broke out in the navigator's compartment. The contaminated ice and airplane debris were sent back to the USA, with the bomb fragments going back to the manufacturer in Amarillo, Texas. The incident outraged the people of Denmark (which owns Greenland and which prohibits nuclear weapons over its territory) and led to massive anti-USA demonstrations.

21 May 1968
The *USS Scorpion*, a nuclear-powered attack submarine assumed to have been carrying four to six nuclear weapons sank mysteriously on this day. It was eventually photographed lying on the bottom of the ocean, where all 99 of its crew were lost. Details of the accident remained classified until November 1993, when the Navy admitted that it had suspected all along that the *Scorpion* had accidentally been torpedoed by an American vessel.

14 January 1969
A series of explosions aboard the nuclear aircraft carrier *Enterprise* left 17 dead and 85 injured.

16 May 1969
The *USS Guitarro*, a $50 million nuclear submarine undergoing final fitting in San Francisco Bay, sank to the bottom as water poured into a forward compartment. A subcommittee of the House Armed Services Committee later found the Navy guilty of 'inexcusable carelessness' in connection with the event.

2 November 1981
A fully armed Poseidon missile was accidentally dropped 17 feet from a crane in Scotland during a transfer operation between a US submarine and its mother ship.

Nuclear bomb tests and testing facilities

9 December 1968
Clouds of radioactive steam from a nuclear test in Nevada broke through the ground, releasing fallout and violating the Limited Nuclear Test Ban Treaty signed five years earlier.

18 December 1970
An underground nuclear test in Nevada resulted in a cloud of radioactive steam being thrust 8,000 feet in the air over Wyoming.

Processing, storage, shipping and disposal

The list below is just a few of many incidents. In addition, from 1946 to 1970 approximately 90,000 canisters of radioactive waste were jettisoned in fifty ocean dumps around the US coast, including prime fishing areas. The waste included contaminated tools, chemicals and laboratory glassware from weapons laboratories and commercial/medical facilities.

July 1979
A dam holding radioactive uranium mill tailings broke, sending an estimated 100 million gallons of radioactive liquid and 1,100 tonnes of solid waste downstream at Church Rock, New Mexico.

August 1979
Highly enriched uranium was released from a top-secret nuclear fuel plant near Erwin, Tennessee. About 1,000 people were contaminated with up to five times as much radiation as would normally be received in a year.

19 September 1980
An Air Force repairman doing routine maintenance in a Titan II ICBM silo dropped a wrench socket, which rolled off a work platform and fell to the bottom of the silo. The socket struck the missile, causing a leak from a pressurised fuel tank. The missile complex and surrounding areas were evacuated. Eight and a half hours later, the fuel vapours ignited, causing an explosion that killed an Air Force specialist and injured 21 others. The explosion also blew off the 740-tonne reinforced concrete and steel silo door and catapulted the warhead 600 feet into the air. The silo has since been filled in with gravel, and operations have been transferred to a similar installation at Rock, Kansas.

21 September 1980
Two canisters containing radioactive material fell off a truck on New Jersey's Route 17. The driver, en route from Pennsylvania to Toronto, did not notice the missing cargo until he reached Albany, New York.

24 November 1992
The Sequoyah Fuels Corp. uranium-processing factory in Gore, Oklahoma, closed after repeated citations by the government for violations of nuclear safety and environmental rules. A government investigation revealed that the company had known for years that uranium was leaking into the ground at levels 35,000 times higher than federal law allows.

Nuclear war and a nuclear winter

The view that the consequences of a nuclear war could provoke a 'nuclear winter' was first presented in 1982 by a group of scientists, including Sagan and Turco (1990). In their book, *A Path Where No Man Thought*, Sagan and Turco (1990) portray the possible effects of the detonation of a 1-megaton nuclear explosion a few kilometres above New York city (also, see Turco *et al.* 1984):

> The fireball radiation, traveling at the speed of light, has already ignited flammable structures ten miles and more from the city center. The shock wave, traveling at the speed of sound, has not yet reached the city. . . . As the nuclear shock wave is leaving the city, skyscrapers and most buildings have been blown down. Fires are momentarily extinguished by the blast wave, and smoke is

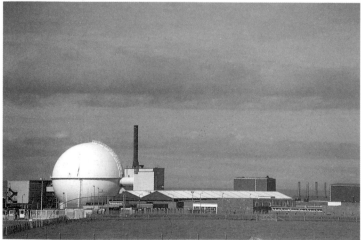

Plate 6.2 *Nuclear power stations at San Clement, California (top) and Dounreay, Scotland (bottom).*

propelled away from the city. Looming over the scene is the mushroom cloud, which sucks debris up to high altitudes – into the lower stratosphere for a groundburst of yield greater than about 200 kilotons. The shock wave has passed. Many fires ignited by the fireball, and others – set, for example, from broken or demolished gas mains – begin to rage. The fires spread and merge over an area of 100 square miles or more. Great clouds of rolling black smoke rise above the fires. The inferno becomes a firestorm. Like a roaring fire in a fireplace with the flue open, but on a vastly larger scale, a huge column of convective air establishes itself, sucking up flames and carrying smoke to high altitudes. Winds in the firestorm can exceed hurricane force. Many days later, hovering over the flattened city is a vast smoke pall extending into the stratosphere. Simultaneous development, and subsequent spreading and merging, of many such soot clouds at many altitudes can lead to a nuclear winter.

The cloud of soot injected into the upper atmosphere (stratosphere) could blanket the Earth, and block the normal passage of sunlight to the Earth's surface. Consequently, there would be a drop in mean air temperature of more than a few degrees Celsius, and because the soot cloud would remain for at least many months, this temperature fall could seriously affect many food chains. The knock-on effect could affect the very survival of many species, including human beings. This bleak scenario of freezing temperatures and decimated food chains does not even take into account the radiation fallout, and the effects of any pyrotoxins (poisons from fires). The idea of a nuclear winter remains controversial, not as a chain of events, but rather the precise effects that would result from a given set of starting conditions, taking account of the magnitude and location of the initial explosion/s, the prevailing climatic conditions (wind direction, etc.), and the amount of explosive matter ejected to various altitudes.

The concept of a nuclear winter has even been applied to the geological column, for the impact of a 10-km diameter meteorite impacting the Earth at the end of the Cretaceous Period, approximately 65 million years ago, causing the extinction of many species, including the dinosaurs (see Chapter 2).

Nuclear arms and verification

Nuclear weapons pose one of the most threatening forces to the future survival of human beings and the entire ecosystems of the planet. We cannot 'uninvent' these ominous weapons, which possess a combined power capable of destroying the world many times over. Indeed, Lord Zuckerman made the incisive comment that '*No one can ban what is not yet invented.*' Treaties will inevitably lag behind scientific and technological advances in any field, including that associated with nuclear weapons. The USA and the former USSR have had this destructive nuclear capability of overkill, or **Mutually Assured Destruction** (MAD), since the 1950s.

The exclusive and highly sought-after membership of the world nuclear club (nations with a nuclear weapons capability) is slowly growing. It appears that secret trading of nuclear information between China and Algeria will lead to Algeria becoming the first Arab nation to possess the nuclear bomb. Within two years, the nuclear plant in the foothills of the Atlas Mountains, 165 miles south of Algiers within a military exclusion zone, is expected to start producing 8 kg of plutonium annually.

Whatever people's personal views about the build-up and ever-increasing sophistication of nuclear weapons, concerned scientists have a duty to seek ways and means by which they can reduce the threat of a nuclear holocaust. This statement instantly begs the question of just what is the concerned scientist's role in such issues? Scientists invented nuclear weapons in the first place because they were given the necessary financial support and environment in which to satisfy the perceived needs of politicians and military strategists. Once nuclear arms had been invented, the decisions about their deployment and development, or reduction, became the responsibility, primarily, of politicians and the armed forces. Nuclear arms issues are essentially moral and political, but the scientist in her or his own right can still make a valuable contribution.

The verification arena is where scientists can contribute to increasing the level of confidence in nuclear treaties that may be signed between nuclear powers. Generally, the confidence that signatories have in their ability to monitor the compliance by other signatories to a nuclear test ban treaty, partial test ban, threshold treaty or any other treaty will tend to determine the levels at which targets or limits are set. It is scientifically possible, for example, to detect and monitor underground explosions, and to discriminate between earthquakes and nuclear explosions.

By carefully analysing the seismic signals that emanate from earthquakes and underground explosions, their characteristic fingerprints can be identified. It has been argued by the British Seismic Verification Research Project (BSVRP) group that about fifteen monitoring stations outside the old Soviet Union could detect an underground explosion of 1 kiloton anywhere within the former USSR, assuming that the nuclear test was not set off during an earthquake, and that there is 'efficient coupling' (BSVRP 1989). Efficient coupling basically means that the explosion is not set off, say, within a large underground cavity where a significant amount of the energy waves from the test are absorbed by the air. Naturally, there are ways of trying to cheat on treaties, for example by attempting to 'decouple' nuclear explosions in underground cavities, or by trying to set them off during a cycle of earthquake activity. As long as nuclear explosions are above a certain very low threshold, or yield, and as long as the superpowers allow a certain amount of monitoring on their territory, then the technology is available to discriminate between earthquakes and explosions, even if both take place over the same time period. In other words, the science of verification is *effective* as long as there is good will on the part of signatories to nuclear treaties – that is, if there is *adequate* verification.

The need for verification

In the arms arena, verification processes are discussed and built into many international agreements in order to provide the means by which partner states to any treaty can attempt to satisfy themselves that the other signatories are indeed abiding by the terms of the agreement. The verification process involves both an initial phase of monitoring and data acquisition, followed by an evaluation of the gathered intelligence data. Most disputes centre on the monitoring processes rather than the evaluation of such information. Agreements and deals can only be struck where parties to a treaty believe that the terms of agreement are fair and equitable, where the signatories feel that what they bargain away at least matches (or is perceived as not exceeding) the accrued benefits and, perhaps most importantly, that the conditions stipulated in the agreement are verifiable. It is in this last arena of verification that scientists can make a valuable contribution.

All sorts of people are involved in the verification arena. Amongst the more scientifically based is the BSVRP, co-ordinated by Aftab Khan (University of Leicester, UK). It was only as recently as the mid-1980s that foreign nationals were allowed to set up seismic monitoring stations in the Soviet Union with the express purpose of monitoring nuclear tests. This agreement came about after the non-governmental

US Natural Resources Defence Council proposed a scientific exchange between the USA and the USSR to demonstrate that verification need not be an obstacle to a comprehensive test ban treaty. Three stations were installed at Bayanul, Karasu and Karkaralinsk, less than 250 km from the test site near Semipalatinsk in east Kazakhstan. In February 1987, however, when the USSR resumed nuclear tests, the American equipment was switched off by the Soviet military. After fresh negotiations, the US scientists agreed to move their equipment from within 250 km to more than 1,000 km from the test sites. It was at this stage that the BSVRP group was invited to join them, with the number of monitoring stations increasing to five.

It has been argued that any strategic arms system that can be monitored for arms control purposes must be vulnerable to a first strike, simply because of the accuracy of modern weapons. There is little point in spending vast sums of money in knowing the exact location of sophisticated weapons if you lack the capacity to neutralise them in the early stages of a conflict. So, first-strike survivability can create problems for arms control negotiations. The development of Midgetman and SS-25 mobile ICBMs, which were more survivable than their predecessors, created a need for even more sophisticated anti-missile missiles. And so the arms race spirals upwards in a vicious circle of ever-increasing costs as the stakes get higher. At what stage can we create a breathing space in the arms race and a chance for more considered judgement? Any development in the nuclear arms race has an associated, limited, 'window of opportunity' for constraint to slow or halt the technological advances.

Ironically, it is the same technological advances which drive the arms race that can improve the verifiability of treaties. Through the 1960s, the improved missile targeting and battle management capabilities facilitated the remote-sensing (satellite) verification of nuclear weapons installations. Technology is very much a double-edged sword in the arena of arms control.

Policing missile deployment

One of the most difficult aspects of verifying nuclear arms agreements is in policing the nature and number of warheads on any missile. Short-range missiles (i.e. with a strike capability up to 300 miles or 482 km) provide part of NATO's flexible response that represents an option intermediate between the long-range missiles and conventional weapons. Of the Allied powers, Britain maintains that this class of nuclear missiles must remain. The so-called third zero option would be the abolition of this entire class of weapons, but Britain argues that such a ban would prove extremely difficult to verify.

One of the most controversial short-range missiles is the supersonic Lance missile, which can carry either nuclear or conventional warheads. Furthermore, both types of warhead are easily interchangeable and it is this aspect that is at the heart of a current controversy between the USA and Britain on one side against what was West Germany before 1990. Germany wishes to negotiate a reduction in the number of short-range missiles that are deployed, whereas the USA and Britain are not in favour of such action. In fact, the USA actually wish to update the Lance missile which was originally deployed in the 1960s, in a programme called the 'follow-on-to-Lance' (FOTL), which is still at an early stage. If FOTL proceeds, then it is intended to increase the range of the Lance missile to 450 km from the present 110 km with a conventional warhead or 135 km with the lighter nuclear warhead. Part of the reason for updating the Lance missile is to increase its accuracy, which, it is argued by Bill Arkin of the Institute of Policy Studies, Washington DC, should allow the fitting of a warhead with a smaller yield, i.e. 1–10 kilotons and an accuracy that is two or three times better than the current accuracy of 300 m at full range, where some of the warheads have a yield of 100 kilotons.

Improving the accuracy of a missile is extremely costly. NATO possesses 692 Lance missiles and 88 launchers, compared with the former Soviet Union's 1,600 launchers for short-range missiles. Both superpowers tend to argue that the number of launchers is more critical than the total count of warheads. The inherent problems in verifying any reductions in short-range missiles that might be negotiated, because of the ease of substituting conventional for nuclear warheads, has led some military analysts to suggest that a possible solution to the problem is to abolish this entire class of weapons. Without the weapons themselves, any launchers detected by either superpower would be a clear violation of a treaty. Other viable solutions include the sealing of the warheads into the missiles and to 'tag' them in some way at the assembly stage or specify the number of allowable launchers without specifying the type of warhead.

The main ways in which military installations and missile launching sites can be monitored is through the use of 'spy' satellites. Such remote sensing of the Earth had generally been jealously guarded by the military until the recent deployment of commercial reconnaissance satellites. The intelligence business became public on 22 February 1986, when an Ariane rocket was launched from French Guiana with a SPOT satellite (SPOT 1) into an orbit 832 km above the Earth. The satellite, jointly owned by French, Swedish and Belgian interests, and managed through the French Space Agency (CNES), is available to the public and media. SPOT, with two cameras and sensors, can deliver a resolution of 10 m in panchromatic or 20 m in the green, red and near-infrared wavebands (Zimmerman 1989). This resolution represents about a threefold improvement on the Landsat 4 and 5 satellites. It was the SPOT satellite that gathered detailed pictures from the USSR of the burning reactor at Chernobyl, the missile early warning radar station at Krasnoyarsk, the Soviet naval base at Severomorsk on the Kola Peninsula, the nuclear testing site at Semipalatinsk, the laser laboratories of Sary Shagan, and the ballistic missile base at Yurya.

Whilst surveillance experts can argue about the quality of the remotely sensed images of military sites, the advent of commercial reconnaissance satellites opens up a proverbial whole new ball game to which non-military (third party) organisations can gain access, at a price, to verifying at least certain aspects of arms control treaties and agreements. The very fact that third party verification of nuclear arms treaties is possible means that a useful deterrence function could be served by independent organisations, including the United Nations, to discourage false allegations being made by any party to an agreement in which they may have a vested interest in non-compliance. Hopefully, in the years ahead, the prevarication over signing nuclear arms treaties, which is based upon arguments over the non-verifiability of agreements, together with the attendant suspicions of cheating, can be largely allayed by designating a non-partisan third party, satellite-based verification. The United Nations may well prove to be the most acceptable organisation to assume this responsibility.

On 20 March 1992 came the first arms control agreement of the post-Cold War period, in which NATO and the former Warsaw Pact countries announced completion of a treaty covering aerial surveillance of all territory from Vancouver to Vladivostok. The treaty sets out conditions under which spy flights can be made, guaranteeing that any information received must be disseminated.to any of the signatory countries to the treaty. It is this kind of open skies treaty that has the potential for removing any climate of fear, and substituting it with mutual trust.

Smuggling radioactive material for nuclear weapons use

Whilst there are strict international rules and careful monitoring of any potential trans-boundary movement of nuclear material that may potentially be used to manufacture nuclear weapons, the possibility of smuggling cannot be completely eliminated. Unlike many other forms of smuggling, the cumulative illegal transfer of even very small quantities of radioactive material may have serious consequences, since such activities have a tendency to destabilise any extant international relationships and create a greater sense of insecurity amongst the signatories to any international nuclear arms treaties.

In 1992, global stocks of plutonium totalled almost 1,100 tonnes, and this is estimated to rise to 1,600–1,700 tonnes by the year 2000 – sufficient to manufacture 200,000 10-kiloton nuclear bombs.

It has been argued by Williams and Woessner (1996) that several of the most dangerous radioactive isotopes, such as ^{235}U and ^{239}U, are only weakly radioactive, thereby making them easier to smuggle across national borders, since they can be shielded from detection by Geiger counters or similar equipment.

The death of the Soviet Union: a new world order

The dramatic overthrow of Soviet President Mikhail Gorbachev in a military coup in the early hours of Monday morning, 19 August 1991, and his replacement by a right-wing, hard-line, so-called Emergency Committee of eight senior communists created international fear. The West anticipated a return to a more isolationist Soviet Union with a rejection of the policies of perestroika and glasnost. The status of the various arms treaties and the future continuation of moves towards significant reductions in the superpowers' nuclear arsenals became unclear.

Fortunately, the abject failure of the military *coup d'état* because of the stance of President Boris Yeltsin and his supporters in the Russian parliament building, the show of popular support from the Soviet citizens themselves, the indecision and disunity amongst the 'gang of eight', with their conflicting military demands that they were never able to meet, the violence through the night of the 20th and into the early hours of 21 August with the aborted military action, the defection of key personnel on the morning after, and the exertion of international pressure and condemnation, allowed the return of Gorbachev on the evening of Wednesday 21 August.

When Gorbachev resigned from his position in the Communist Party on Saturday 24 August, he signed the death knell for the party that had terrorised Soviet citizens for more than 70 years. On Christmas Day 1991, at around 7 p.m. in the Soviet Union (5 p.m. GMT), Gorbachev stepped down as President of the Soviet Union and handed power over to Boris Yeltsin, who became President of the Russian Federation. With this transfer of power went the control of the nuclear arsenals. Shortly before 6 p.m. GMT, the Red Flag was lowered over the Kremlin, symbolising the demise of the Soviet Union and the birth of the Commonwealth of Independent States. In his resignation speech, Gorbachev spoke of the changes that had happened since he assumed power in 1985, and of the new order where free elections, a free press and a multi-party political system had become a reality.

This string of momentous events has served to show the often fragile nature of peace, upon which so much confidence and trust throughout the rest of the world depends. Perhaps, with the new mood of optimism sweeping the dismembered Soviet Union for a more democratic future, there will be greater opportunities for significant arms reductions, particularly in the nuclear sphere. As a cautionary counterbalance, the break-up of the Soviet Union may lead to the new sovereign states claiming control of the nuclear weapons on their territory, as the Ukraine did in October 1991 when it announced its independence. There is, then, the potential for the proliferation of nuclear weapons, not a reduction, as a number of small states overnight become independent nuclear powers. It appears, however, that Russia may be able to assume full control of the nuclear arsenals and in any nuclear arms talks become the superpower that will deal directly with the USA. The full implications of this second revolution in the Soviet Union this century still remain uncertain. A new world order is being born out of the ashes of the fire of Soviet communism, with the potential for more superpower co-operation on issues that affect the global environment.

A major outcome of the decreased international tension between the so-called free world and the former Soviet Union is the decommissioning or scaling down of certain nuclear capabilities. For example, in Britain, the Royal Navy expects to decommission about 10 nuclear submarines by the year 2000, all of which will be contaminated with radioactive waste. If the submarines are dumped at sea, as planned, then, although the nuclear fuel will be removed, there will be marine pollution by at least low-level radioactive waste.

Resumption of nuclear tests by France

In 1995, amidst outrage from much of the rest of the world, the French government resumed nuclear tests at the Muroroa Atoll in the South Pacific. Muroroa Atoll, located about 1,200 km southeast of Tahiti, comprises a coral reef rising up to 3 m above sea level enclosing a lagoon up to 50 m deep. Over the past three decades this fragile ecosystem has been severely damaged by more than 120 nuclear explosions to date.

The French government has justified these nuclear tests on the basis of a need to collect new and essential actualistic data from explosions to improve its nuclear capability, and in order to check that its nuclear weapons will work.

Many concerns have been raised by the resumption of nuclear tests in the Pacific. Broadly, the international political issues raised by the latest round of nuclear test explosions can be summarised as follows:

- Recent hopes for an effective comprehensive test ban treaty (CTBT) have been seriously jeopardised, with the French setting a precedent for other nations to either opt out or simply not comply.
- A cooling of diplomatic relations between France and many of the Pacific-rim nations, particularly those closest to the site of the tests, e.g. New Zealand, Australia and Japan.
- Encouragement of further civil unrest and a more concerted effort by the indigenous Pacific island populations under French control to secede from French control as independent nation states.
- Renewed environmental pollution by radioactivity and physical degradation of the Pacific. These tests can only increase the likelihood of long-term leakage of radioactivity into the Pacific Ocean. The cumulative effect of all these nuclear explosions on the atoll remains unclear, but worst-case scenarios predict that parts of the atoll may collapse because of inherent instabilities caused by the nuclear tests.

Whatever the actual consequences of the renewed French nuclear tests in the South Pacific, they epitomise the actions of a major international nation taking a NIMBY approach to its own perceived requirement to cause radioactive pollution. Also, with many scientists asserting that such explosions can be adequately simulated on computers, there is the unanswered question over the actual real necessity for renewed nuclear testing.

In mid-1996 France signed the Pacific N-test ban in which it will agree to ban nuclear weapons from the Pacific, after having completed its own nuclear tests. Thus, the French are planning to become signatories to the 1985 South Pacific Nuclear Free Zone Treaty, also known as the Treaty of Rarotonga. The decision to sign this treaty was announced in 1995 jointly between France, Britain and the USA. The clear message that France has sent out to the rest of the world is that it is acceptable to sign treaties and agreements that endeavour to limit the rest of the world's military activities but only after satisfying one's own agenda.

Conclusions

The world political arena is entering into a new and uncertain age of American and Russian co-operation. The Cold War of fear is being replaced by an economic and commercial conflict where the technology of the nuclear arms race is being used to earn foreign currency. The vast military industries and infrastructure of the East and West are looking for work. There is now talk of a joint defence programme between America and Russia. Areas of possible co-operation include sharing satellite information about missile attacks and other sensitive military information. The form of any response to an attack on either nation, or a third party, is controversial, because the order to intercept would have to be taken jointly and the mechanism for such a course of action remains unclear. Sharing sensitive military information is by no means an inevitable consequence of the present rapprochement between the USA and the former Soviet Union. The American early warning system deep within Cheyenne Mountain, Colorado, is unlikely ever to route unfiltered military intelligence direct to the Russians, and the converse is true; there will always be a reason to withhold or filter the intelligence. In the Gulf War, intelligence information was passed to the Israelis from Cheyenne Mountain, but even this was filtered. Meanwhile, the second-league nuclear countries, such as Britain, look on with interest as spectators. A consequence of real co-operation between Russia and America could be to render the British nuclear capability obsolete. There are those who caution that greater co-operation between the USA and Russia might accelerate rather than reduce the international nuclear arms race. Countries such as China, Japan, South Korea and India might feel the need to arm more heavily in order to combat any superpower alliance. Perhaps we should be cautiously optimistic about a joint defence programme between America and Russia, because the international consequences have yet to be fully assessed.

Chapter 6: Key points

1 Nuclear technology and nuclear weapons are accepted as a necessary evil by some, but regarded as unacceptable by others. Whatever a nation's viewpoint, the exploitation of nuclear energy and possessing a nuclear arsenal have their associated risks from accidents, together with problems for the disposal of radioactive waste.

2 The Nuclear Age began in Cambridge University, UK, in 1919, with Rutherford's experiments on the structure of the atom. Nuclear fission was first achieved by Hahn and Strassman in 1934. The development of the atomic bomb was sanctioned by Roosevelt in 1941, and the first two bombs were used against Japan in August 1945. Treaties signed to limit arms proliferation and nuclear testing include the 1963 Limited Test Ban Treaty (LTBT); the 1968 Non-Proliferation Treaty (NPT); the 1972 Strategic Arms Limitation Treaty I (SALT I); the 1974 Threshold Test Ban Treaty (TTBT); the 1976 Peaceful Nuclear Explosions Treaty (PNET); the 1979 Strategic Arms Limitation Treaty II (SALT II); the 1987 Intermediate Nuclear Forces (INF) Treaty; the 1991 Strategic Arms Reduction Treaty (START); the 1992 START 2 Treaty; and the 1996 Comprehensive Test Ban Treaty.

3 Policing nuclear missile deployment is difficult, and effective and adequate verification mechanisms are necessary in order to instil confidence in a treaty, and in order to check that no nation is cheating. Mechanisms for verification include seismic verification to monitor underground nuclear testing, remote sensing, scientific exchange programmes, and on-site inspections of nuclear installations.

The collapse of the Soviet Union and the birth of the Commonwealth of Independent States, including the Russian Federation, may result in greater opportunities for significant arms reductions, although real concerns remain over the proliferation of nuclear weapons, and their pre-emptive use by politically unstable regimes in various war zones.

4 Radon is a major component of the natural background radiation dose received by many people, and results from the decay of radioactive minerals in rocks. Tentative links have been drawn between cancer in humans and areas of high radon concentrations. Factors that concentrate radon in the environment include the geological setting and the bedrock type; building design and the materials used; water sources; and atmospheric conditions. Governments have commissioned surveys to identify areas of high risk and have provided recommendations to help reduce the effects within the home.

5   Nuclear waste is created during the production of nuclear weapons and energy, medical products and scientific research. Waste management in the nuclear power industry includes the nuclear-fuel cycle. Prior to 1970 for the USA and 1982 for Europe, waste could be dumped at sea, but since 1975 all dumping has had to comply with the London Dumping Convention. The half-life and toxicity of radioactive chemicals vary greatly, a factor that determines the risk and magnitude of any potential contamination. The transport of nuclear waste is hazardous, and accidents may affect very large areas for extremely long time periods. Acceptable underground nuclear waste disposal relies on many factors, including the geological setting and bedrock geology, ground-water movement, the nature of the containers in which the waste is sealed, and its monitorability and retrievability.

6   Nuclear accidents at power stations have occurred for a variety of reasons, but all involved some degree of human error; they include Three Mile Island in 1979, Chernobyl in 1986, and Tomsk-7 in 1993. These accidents serve to highlight the potential dangers, clean-up problems and costs, and the environmental damage.

Chapter 6: Further reading

Berkhout, F. 1991. *Radioactive Waste: Politics and Technology*. London: Routledge, 256 pp.
This is a companion book for students in environmental studies, geography and public administration. The book focuses on radioactive waste management and disposal policies in three European countries – the UK, Germany and Sweden. A detailed historical account of the policy processes in these three countries is presented, and the theoretical and public policy implications are evaluated. A particular strength of this book is in its comparative approach, and the way in which Berkhout sets the issue of radwaste management at the centre of the current debate about nuclear power, the environment and society.

Newhouse, J. 1989. *The Nuclear Age: From Hiroshima to Star Wars*. London: Michael Joseph, 486 pp.
An excellent history of post-World War II events and personalities behind one of the single most important issues of the past fifty years – nuclear weapons and the nuclear arms race. This book is well researched and highly readable. It tells a dramatic story of confrontation and rapprochement, of scientific and technological advance, of diplomatic wrangling and blatant shows of military strength, through the Cold War and the Cuban Missile Crisis, to the Star Wars programme, and the more recent US-Soviet Union arms agreements. It is an invaluable account of the Nuclear Age.

Park, C.C. 1989 *Chernobyl: the Long Shadow*. London: Routledge, 224 pp.

Pasqualetti, M.J. (ed.) 1990 *Nuclear Decommissioning and Society: Public Links to a New Technology*. London: Routledge.

Sheehan, M.J. 1988. *Arms Control: Theory and Practice*. Oxford: Blackwell, 188 pp.
A useful analysis of the origins and development of arms control, and the issues that underpin arms control, the problems of verifying treaties, and the political context in which arms control negotiations, both domestic and international, are considered. This is a useful supplementary book for any student interested in understanding arms control issues.

*Energy is the only life,
and is from the Body;
and Reason is the
bound or outward
circumference of Energy.*

William Blake, 'Marriage of
Heaven and Hell'

This chapter explores the various energy resources from conventional fossil fuels to renewable energy, all of which are considered in the context of projected energy requirements into the twenty-first century. This chapter begins with a look at the first energy crisis to affect much of the world and which caused a global re-evaluation of energy reserves, supply and energy diversity.

A crisis of conscience

Much of the worldwide demand for domestic and commercial energy consumption comes from only a few precious natural resources. They may seem limitless in the context of the short human life-span, something to squander and let future generations replace. Never has the importance of this resource been appreciated as much as during the world energy crisis of October 1973. The crisis was precipitated by the Arab–Israeli war, when the Arab oil-exporting nations cut exports of oil to Israel's ally, the USA. American citizens suddenly became aware of their extreme dependence on oil as a commodity. Following the oil embargo by the Arab nations, the crisis was lessened but did not abate. The 1970s were dogged by successive energy crises. At the end of the 1970s, oil was more than US$30 a barrel, compared with US$3 a barrel at the start of the decade. The Organisation of Petroleum Exporting Countries (OPEC) cartel seemed determined to raise the price of crude oil *ad infinitum*. Domestic bills for oil, gas and electricity rose sharply, fuelling the inflationary spiral of the 1970s throughout the developed world.

A useful outcome of the energy crisis was the realisation that the Earth's **fossil fuels** are a limited, finite resource. There also came the awareness that individuals and countries have a very real responsibility, not only to future generations, but also to themselves as concerned and responsible tenants of this planet to husband these resources in a prudent manner. Whatever is taken from the Earth has a price and a cost for the environment. No energy resource is free, environmentally completely safe or limitless. Energy must be used more wisely in order to minimise the environmental hazards and optimise the efficiency with which it is produced. Energy issues are as much about factual information and scientific arguments as they are about education and changing people's attitudes.

Energy use and energy reserves

Over the next thirty years world energy demand is likely to grow by as much as 70 per cent (Jennings 1996). It has been estimated that humankind annually expends an amount of fossil fuel that it took nature on average about one million years to produce. In 1990, global energy expenditure amounted to an annual 1.3 billion tonnes (Bt) of coal equivalent, four times greater than in 1950, and twenty times more than in 1850.

In 1992, 74.6 per cent of global energy came from fossil fuels, 13.8 per cent from biomass (wood, crop wastes, dung, etc.), 5.9 per cent from hydro-power and 5.6 per cent from nuclear energy (British Petroleum 1993). Approximately 1,200 million people living in developed countries consumed over two-thirds of this total energy supply, while less than one-third went to the 4,100 million people in the developing world (*ibid.*).

Over the past twenty years world energy demand has increased by 35 per cent (World Resources Institute 1994). Together, the USA, the largest producer of commercial energy, and the former USSR, ranking second, account for almost 40 per cent of the world's energy supply. China is third in rank, producing 8.8 per cent, Africa produces 6.4 per cent, and South America 4.3 per cent (*ibid.*).

In terms of energy consumption, the USA is first in rank. In the USA, per capita energy consumption

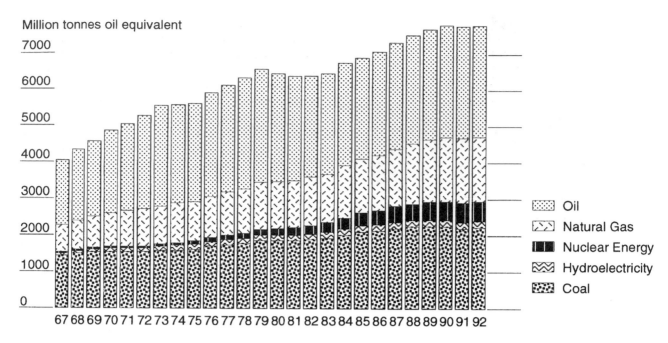

Figure 7.1 *World energy consumption 1967–92. Source: British Petroleum (1993).*

is 320 gigajoules per annum (GJ yr^{-1}), a fall of 4 per cent during the past twenty years, although total consumption has increased by 17 per cent over the same period, but with its energy intensity (energy use per unit of economic output) showing a 17 per cent decrease (*ibid.*). By comparison, per capita per annum energy consumption is 9 GJ in India, 23 GJ in China and less than 1.5 GJ in the nineteen lowest ranking countries, sixteen of which are in Africa: this same group of countries obtain 83–96 per cent of their total energy from traditional fuels (*ibid.*).

For recoverable energy reserves, the following figures are taken from the report by the World Resources Institute in collaboration with the United Nations Environment Programme and the United Nations Development Programme (*ibid.*). Global recoverable energy reserves are dominated by the USA and former USSR, with Southeast Asia and countries around the Persian Gulf controlling 57 per cent of proved recoverable petroleum reserves: Saudi Arabia probably controls about 26 per cent. The former USSR controls 42 per cent of proven recoverable gas reserves and the Persian Gulf 25 per cent. In terms of hydroelectric power generation, the USA (14 per cent) and the former USSR (10 per cent) lead in installed capacity.

The consumption of energy in developing countries is rising rapidly, and by the end of this century will dominate energy markets worldwide. In a report released in April 1994 by the International Energy Agency, energy consumption in East Asia is expected to grow by about 150 per cent by 2010, while in the twenty-two countries that belong to the Organisation for Economic Co-operation and Development (OECD), for the same period the increase is predicted to be 28 per cent. Based on these energy consumption figures, by 2010 carbon dioxide emissions are expected to increase by as much as 160 per cent (to 2.6 Bt yr^{-1}) in East Asia, and by about 29 per cent (to 13.4 Bt yr^{-1}) in the OECD countries. Even allowing for a growth rate in the demand for energy in the developing countries 1–2 per cent lower than the present trend, global demand is likely to exceed 100 million barrels a day of oil equivalent (mbdoe) by 2010, and possibly 200 mbdoe (World Bank 1992, p. 114).

Coal, oil and natural gas account for 74.6 per cent of the global energy used (*British Petroleum Statistical Review* 1993), with nuclear fuel supplying most of the remaining needs. Under-developed and developing nations, however, still tend to rely heavily upon other fuel sources such as wood, crop waste and dung. Oil accounts for roughly 38 per cent of commercial energy consumption, with natural gas contributing about 20 per cent. Figure 7.1 shows world energy consumption. World energy produc-

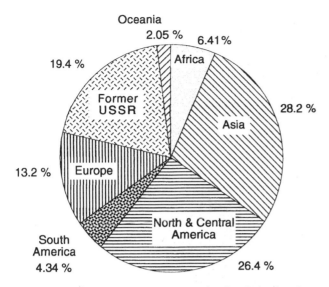

Figure 7.2 *Commercial energy production by region for 1991. Source: World Resources Institute (1994–95).*

United Nations Statistical Office, provide a useful breakdown of the energy production and consumption by region and fuel type. Again, the industrialised countries, especially the OECD countries, consume substantially more liquid fuels than they produce, and have a smaller but nevertheless negative balance of gas and solid fuels, a situation that is less common in developing countries. Figures 7.2 and 7.3 show commercial energy production and consumption, respectively, by region. Figure 7.4 shows the regional energy consumption patterns for 1992 broken down into oil, natural gas, nuclear energy, hydroelectric and coal. Coal remains the prime energy source in Asia and Australasia, while oil and gas account for more than 60 per cent of demand in all other regions. Projected global energy demand is shown in Figure 7.5. It is interesting to note the accelerated growth in commercial energy production in the world's less developed regions compared with the developed countries and the former Soviet Union (Figure 7.6).

Perhaps the biggest challenge for developing countries in relation to energy consumption is to develop and implement technologies that help reduce the emissions of gases and particulate matter (dust and smoke), which have both local and possible global environmental impacts. It is important that societies

tion and consumption for 1991, and reserves and resources of commercial energy for 1990, are shown in Tables 7.1A, B and C, respectively. The growth in world energy demand has stalled since 1990, mainly because of declining energy consumption in non-OECD Europe. Energy data for 1989, from the

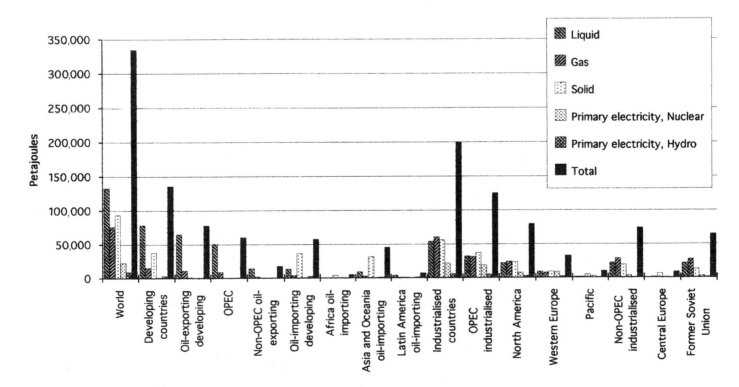

Figure 7.3 *Energy consumption by region and by fuel for 1991. Source: World Resources Institute (1994–95).*

Table 7.1A *Commercial energy production 1991 in petajoules (PJ).*

	Total	Solid	Liquid	Gas	Primary geothermal and wind	Hydro-electricity	Nuclear
World	334,890	93,689	132,992	76,275	1,261	8,049	22,669
Africa	21,335	4,119	14,368	2,598	13	192	46
Asia	94,351	32,336	47,731	9,061	282	1,557	3,391
North and Central America	88,467	23,923	28,800	25,054	784	2,280	7,642
South America	14,541	890	9,810	2,441	0	1,300	101
Europe	44,335	15,285	9,341	8,689	127	1,732	9,177
Former USSR	64,994	12,719	21,708	27,412	1	846	2,313
Oceania	6,867	4,417	1,234	1,021	54	141	0

Table 7.1B *Energy consumption 1991.*

	Commercial energy consumption				Traditional fuels		
	Total	per capita	per constant 1987 US$ of GNP	Imports as percentage of consumption	Total	per capita	Percentage of total consumption
	PJ	GJ	MJ		PJ	MJ	
World	321,430	60	–	–	19,942	3,702	6
Africa	7,871	12	no data	(286)	4,815	7,275	38
Asia	80,374	25	no data	(87)	8,996	2,833	10
North and Central America	96,086	243	no data	(29)	825	2,086	1
South America	9,493	32	no data	(861)	2,748	9,180	22
Europe	68,507	134	no data	115	598	1,171	1
Former USSR	54,730	193	no data	(13)	792	2,797	1
Oceania	4,367	161	no data	7	185	6,837	4

Table 7.1C *Reserves and resources of commercial energy 1990.*

	Anthracite and bituminous coals Proved reserves in place	Sub-bituminous coals and lignite Proved reserves in place	Crude oil Proved recoverable reserves	Natural gas Proved recoverable reserves	Uranium Recoverable at < US$80 per kg	Hydroelectric Known exploitable potential
	Mt	Mt	Mt	billion m^{-3}	tonnes	MW
World	1,212,852	743,193	134,792	128,852	1,410,040	no data
Africa	131,841	1,506	9,005	8,178	547,020	no data
Asia	335,204	135,304	95,137	45,148	4,320	no data
North and Central America	231,947	219,051	10,478	10,129	247,900	no data
South America	26,225	15,312	9,872	4,695	83,530	no data
Europe	311,366	162,587	2,145	5,180	58,270	no data
Former USSR	130,000	157,000	8,000	54,530	no data	3,831,000
Oceania	66,269	52,433	204	991	469,000	no data

Source: World Resources Institute (1994–95).

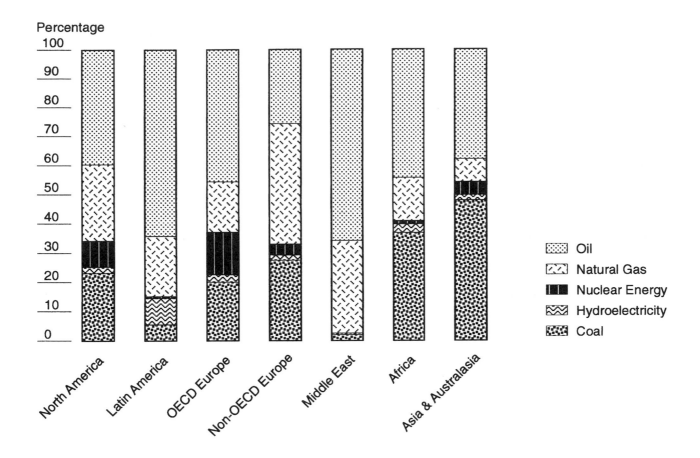

Figure 7.4 *Regional energy consumption pattern for 1992. Source: British Petroleum (1993).*

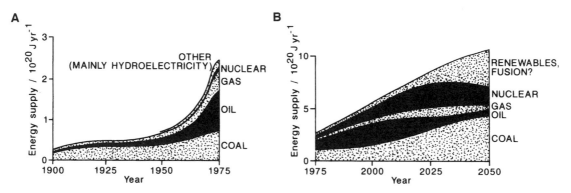

Figure 7.5 *(A) World energy supplies 1900–1975. (B) Projected demand in energy requirements to 2050. Redrawn after Blunden and Reddish (1991).*

endeavour to use preferentially those energy resources that create the least pollutants as by-products. Natural gas, for example, is 'cleaner' than oil and other fossil fuels. The combustion of natural gas releases 14 kg of CO_2 for every billion joules of energy produced, compared with oil at 20 kg and coal at 24 kg.

The way in which developed countries provide their energy services to the developing world is important for the following reasons (outlined by

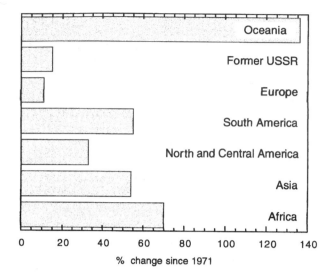

Figure 7.6 *Percentage change in commercial energy production between 1971 and 1991. Source: World Resources Institute (1994–95).*

the US Office of Technology Assessment (OTA) 1992b):

- international political stability, through steady, broad-based economic growth;
- humanitarian concerns, helping developing countries to meet their energy requirements;
- trade and competitiveness, facilitating the internationalisation of economic growth;
- global environmental issues will only be tackled on an international basis if there is co-operation between the developed and developing countries;
- global oil markets, if World Energy Conference predictions are accurate, will change such that developing countries will account for 90 per cent of the increased world oil consumption between 1985 and 2020, and this may cause both higher prices and greater price instability, with impacts on all countries in terms of inflation, balance of trade and overall economic performance; and
- global financial markets are affected by the indebtedness of the developing countries, a large part of which was incurred through building their energy sectors, and unless debt repayments are restructured then they will contribute to international instability in the world's financial institutions.

One of the most significant outcomes of the global energy crisis in the 1970s has been the realisation that a nation should not place too much energy *dependence* on a very limited basket of fuels. As a consequence, energy *diversification* has become more important to many countries. In February 1996, these points were eloquently put by the chairman of Shell Transport and Trading Company plc and group managing director of the Royal Dutch/Shell Group of companies when he stated that:

> It seems likely that we have found all the really large petroleum provinces ... sources of energy supply are likely to have become highly diversified by around 2060.

The issues raised by statements such as these are addressed later in this chapter.

Units of energy

Before going on to look at an historical perspective of energy, and dealing with energy resources, it is important to mention ways in which power output and energy consumption are measured. Power output and consumption are measured in megawatts (MW), electrical output as megawatts electrical (MWe): 1 MW is 1,000 kilowatts (kW). Alternatively, energy is measured in petajoules (PJ), where 1 PJ is equivalent to 10^{15} Joules or 947.8×10^9 BTUs (British Thermal Units): 1 Watt = 1 J s^{-1}, 1 kWh = 3.6×10^6 J and 1 Calory = 4.186 J.

There are various ways of measuring energy consumption, for example as the amount of energy used to produce a given amount of economic output (the energy:GDP ratio). As an illustration, in the UK the shift from coal to gas has helped to improve the energy:GDP ratio because the use of gas involves cleaner technologies and cleaner fuels, e.g. differences in the balance of fuel used to generate electricity; per unit of electricity generated from gas, there is an approximately 50 per cent reduction in the CO_2 emissions per unit of electricity compared with that derived from coal.

An historical perspective

Until the Industrial Revolution began in England in the eighteenth century, demand for energy resources was relatively modest. Water power, and fire power from wood and peat were the principal means of obtaining energy. Archaeological evidence in caves in the Peking area of China shows that early humans were utilising fire power by at least 400,000 BC.

The natural energy resources of wood, fossil fuels such as peat and coal, water power from streams and rivers, and wind energy harnessed in windmills, seemed adequate and did not pose any long-term environmental problems. Any atmospheric pollution seemed to be merely a local phenomenon associated with certain areas such as the 'Potteries' in the north-west English Midlands. Nobody really thought that there was any serious threat to the environment, much less to planet Earth. How could they? Science and research had not advanced to the sophistication of this age, which is capable of establishing complex cause-and-effect relationships about the global balance of atmospheric gases.

By the early 1970s, natural gas was considered a fossil fuel for premium use such as in domestic heating and cooking, with oil as the major intermediate and, perhaps, long-term energy resource. The Gulf of Mexico and the North Sea were fast becoming the largest areas of high-tech exploration and discovery. Coal seemed like yesterday's fuel, expensive to mine and dirty to use.

Then, in 1974, OPEC abruptly quadrupled the world oil price. It is probably fair to say that this dramatic increase precipitated the world economic recession that became evident in the late 1970s and early 1980s. Politicians and industrialists suddenly became obsessed with a need for alternative, viable energy resources and nuclear energy appeared to provide the answer.

The 1980s saw the groundswell of environmentalists lobbying for safer, 'greener', energy resources, and a less profligate use of these resources, together with an increasingly shaken faith in nuclear energy, especially after Chernobyl. It is within this new-found green political climate that energy resources and resource planning must be placed. While issues such as the greenhouse effect, acidic deposition and water pollution tend to dominate the headlines (with the exception in 1992 of the debate over energy policy associated with the British government's proposal to close more than thirty coal mines – something that has now been done), energy issues tend to take second place – even though acidic deposition, for example, is caused by the emissions particularly from conventional coal-fuelled power stations, and greenhouse gas emissions are heavily linked to fossil fuel use.

Conventional fossil fuels

The three main conventional fossil fuels are coal, crude oil and natural gas. Other fossil fuels include peat and oil shales. These fossil fuels constitute the largest source of greenhouse gases.

In order to discuss fossil fuels, it is necessary to define what is meant by the term mineral/fossil fuel

proven reserves. Proven reserves do not represent the total amount of reserves that are estimated to be ultimately recoverable, but rather that part of the reserves judged to be recoverable under the extant economic and operating conditions. Proven reserves, therefore, are dependent upon world commodity prices, the state of exploration and recovery technology. As an example, due to changes in economic conditions and because of exploration, over the past twenty years proven reserves of oil and gas have increased. Reserve lifetime is a measure of the duration of the reserves at existing rates of extraction and demand. At current consumption rates, world coal reserves have a predicted life span of 200–400 years, whereas for oil the projected longevity of supply is very dependent on region, e.g. more than 100 years for some Middle Eastern states, but with a global lifetime of about 56 years, averaging thirteen years for the USA and UK (Blunden and Reddish 1991). Any figures such as these must be treated with circumspection as their derivation involves considerable uncertainties, and political-commercial calculations.

Estimating the reserves of fossil fuels is difficult. In an attempt to do this, scientists and economists use a variety of techniques, which include using the physical evidence for known locations and predictive or statistical methods that make assumptions about the size and type of reserves that may be discovered. The estimated resources available to humankind will always outstrip the actual recoverable reserves. Fossil fuels are being depleted by human activities at a rate 100,000 times faster than they are being formed (Davis 1990). Extractive technology can change to such an extent, however, that reserves that seemed inaccessible become within human reach. So, the figures that follow should not be taken as gospel but rather as a reasonable estimate of what is available.

World coal reserves have been estimated at 3,160 billion barrels of oil equivalent (BBOE), compared with a world natural gas figure of 425 BBOE, and an estimate for oil of 700×10^9 barrels (Table 5.5 in Cassedy and Grossman 1990). Of course, reserves are not the same as recoverable resources. The estimated recoverable resources are 25,600 BBOE for coal, 1,200 BBOE for natural gas and 1,863 billion barrels for oil (*ibid.*). In 1950, the world's proven reserves of oil and gas stood at 30 billion tons (Imperial) of oil equivalent (btoe), whereas today they exceed 250 btoe, despite a total world consumption over the past forty years of 100 btoe (World Bank 1992).

At present, most electricity-generating processes use fossil fuel power stations, which emit greenhouse gases, in particular CO_2. In a Friends of the

***Table* 7.2** *Emissions associated with production/saving of 1,000 MW of electricity.*

Method	CO$_2$ tonnes per annum
Coal	5,912,000
Nuclear	230,000
Hydroelectric	78,000
Wind	54,000
Tidal	52,000
Loft insulation	24,000
Low-energy lighting	12,000

Source: Mortimer 1989.

Earth submission to the UK Parliamentary Select Committee (Mortimer 1989), figures were presented for the CO$_2$ emissions associated with the production, or saving, of 1,000 MW of electricity 'taking into account related mining and fuel-producing processes', expressed as CO$_2$ tonnes yr^{-1}; these are given in Table 7.2.

Coal

Between 1950 and the present, proven coal reserves have risen from 450 to 570 btoe (World Bank 1992), updated to 694 btoe (Bowler 1993). Global coal resources (potential) are at least 10,000 btoe. Coal remains the largest fossil fuel resource, but coal extraction and combustion in power stations has a relatively high economic and social cost – acidic deposition, for example, is caused (but not solely) by coal-burning power plants emitting sulphur dioxide. Underground mining can be dangerous because of accidents caused by gas explosions and roof collapse. Fatalities and long-term illnesses such as **pneumoconiosis** (lung disease) are even more commonplace, and are caused by miners inhaling fine coal dust over prolonged periods.

Coal is formed from the remains of vegetation that grew in wetlands and swamps (collectively called **mires**). Under favourable climatic and burial conditions, the decaying plants form peat, which is then converted into various ranks of coal. About 10 m of peat typically forms in between 4,000 and 100,000 years, with 10 m of peat compacting and transforming into approximately 1 m of coal. The ideal environments in which peat, and subsequently coal, can form are associated with major river deltas, where the rate of subsidence and burial occurs at about the same rate as the vegetation is being established. There are different types, or ranks, of coal – **anthracite**, **bituminous**, sub-bituminous and **lignite** – all of which have different calorific (energy) values

and produce varying amounts of polluting gases. Anthracite has the highest calorific value, with lignite having the least.

In the nineteenth century, the ascendancy of coal as the major fossil fuel was closely tied to the Industrial Revolution, which began at Ironbridge in the English Midlands. Coal has been used mainly to generate steam to drive turbine engines and make electricity. The importance of coal as a global energy resource peaked in the 1920s, when it accounted for roughly 70 per cent of fuel use. Today, coal supplies only 26 per cent of the world's energy needs. Oil and gas have replaced coal as the main fossil fuel, a trend that began after World War II, but that really accelerated in the 1960s and 1970s. For comparative purposes, quantities of coal are expressed as billions of barrels of oil equivalent (BBOE).

In the West, many of the world's most accessible and easily extractable coal reserves are now nearing depletion or are worked out. The remaining coal seams are often thinner and broken by geological faults, which makes the prediction of their extent and extraction more difficult.

The old mining communities that were so commonplace in the US Appalachian Mountains (and epitomised in films such as *Coalminer's Daughter*), in central and northern Britain, and throughout large parts of Belgium, France and Germany, are in decline. The decline of coal mining and the associated redundancies are an emotive issue. Countries such as Germany provide heavy subsidies for their coal production because of the social aspects of providing employment and for reasons of energy security. In the late 1980s, nations such as the UK actively encouraged the development of competition in power generation following the privatisation of the Central Electricity Generating Board (split into two generator companies, PowerGen and National Power) by promoting the so-called 'dash-for-gas', involving the construction of new combined-cycle gas turbine stations (CCGTs). In Britain, open-cast mines are common, but their output is falling, not because of a lack of recoverable reserves, but because of local opposition to their development on the basis of environmental considerations ('not in my back yard' – commonly referred to as NIMBY) (Plate 7.1). There are total proven reserves of 300 Mt, 25 Mt more are found each year that would require deep mining to extract, and British Coal estimates coal reserves at 130 Bt.

In Britain, on 13 October 1992, the announcement by the government of the closure of thirty-one of the country's remaining fifty deep coal mines, and 30,000 job losses from the industry, caused a public outcry. The result of protests from many trade union-ists, economists and the general public resulted in the President of the Board of Trade giving a stay of execution for twenty-one of the thirty-one pits, with the government promising a wide-ranging review. Since then it has transpired that even more coal mines have been closed. The response of the mining communities to these closures illustrates the deep-seated concern of ordinary working people at the prospect of large-scale unemployment as a traditional energy industry undergoes rapid social upheaval. The effects of such closures will have far-reaching socio-economic impacts in a society where the high rate of unemployment makes finding alternative employ-ment very unlikely. The issues surrounding the pit closures in the UK are discussed in some depth in a special issue of the journal *New Scientist* (Bowler 1993, Charles 1993, Cross 1993, Maitlis and Rourke 1993, Ridley 1993).

While countries such as the UK may be able to meet their environmental obligations through turning to natural gas, other nations such as India and China have a rapidly rising energy demand that they will need to meet, and very large indigenous coal reserves. If such fossil fuel reserves are to be used in an environmentally acceptable way, clean coal technologies (see Box 7.1) are therefore very impor-tant to these nations. Unfortunately, the widespread

Plate 7.1 *Open-cast coal mine in the East Midlands, UK.*

adoption of clean coal technologies is unlikely to happen unless developed countries implement the technology first, and offer technology transfer capa-bilities. Commercially available clean coal technolo-gies, which are new and different to earlier processes, include **fluidised bed combustion** and **gasification**, together with **hybrid systems** such as British Coal's Topping Cycle, at a developmental stage (Box 7.1).

BOX 7.1 CLEAN COAL TECHNOLOGIES

Coal combustion is associated with relatively high emis-sions of greenhouse gases (e.g. CO_2), gases that lead to acidic deposition (e.g. SO_2), and particulate matter, which causes human health problems (e.g. hydrocarbon particulates). For these reasons there has been consid-erable research to develop 'clean coal' technologies (CCTs) which might provide both increased energy efficiency and reduced atmospheric pollution (Table 7.3). In the USA, there is growing pressure on the electricity-generating industry to use the cleanest coals, i.e. the low-ash, low-sulphur coals. The choice is between the black coals with their higher calorific value, but also higher sulphur levels, which occur in relatively thin seams in the Appalachian coal fields, and the younger coals with lower calorific values and less sulphur, which occur in thicker seams farther west. In order to compensate for the shift towards increased coal mining in the more westerly coalfields, the coal producers of the eastern USA are looking to greater exports to European markets.

CCTs are seen by many as providing improved energy efficiencies and lower emissions (see POST 1992a), but because of coal's high carbon content, CO_2 emissions from coal will always be higher than from gas – with the carbon/energy ratio of gas being half that of coal.

In Europe, SO_2 and NO_x emissions limits are given in the EC Large Combustion Plant Directive. For the UK, this directive stipulates a 20 per cent reduction in SO_2 emissions from the levels in 1980 by 1993, 40 per cent by 1998, and 60 per cent by 2003. Also, by 1993, UK NO_x emissions should be reduced by 15 per cent, and 30 per cent by 1998. As for CO_2 emissions, the UK is committed to no more than its 1990 levels by the year 2000.

In many industrialised countries, there are ongoing changes in the balance of primary energy sources because of both economic and political considerations. In the UK, for example, in 1990, about 40,000 MW of coal-fired generating capacity supplied 68 per cent of the nation's electricity, with nuclear energy supplying about 21 per cent and oil approximately 9 per cent. The proposed introduction of 10,000 MW and 20,000 MW gas-fired combined cycle gas turbines (CCGTs), seen as having the advantage of contributing less CO_2, NO_x and SO_2 than coal-burning power stations, will change this balance.

In conventional pulverised fuel power stations, coal is ground or powdered, and then undergoes combustion to give high-pressure steam, which drives turbines. Flue gas desulphurisation removes up to 90 per cent of the SO_2 produced. A by-product of this process is large amounts of the calcium sulphate mineral, gypsum, with industrial uses such as in the manufacture of plaster-board. In some industrialised countries such as Germany, flue gas desulphurisation has been routinely fitted to coal-fired power stations. As part of the UK's package of measures to reduce SO_2 emissions, flue gas desulphurisation is being fitted to some of the existing coal-fired power stations, such as the Drax power station, and possibly Ferrybridge.

NO_x emissions can be reduced by 30–50 per cent, depending on the type and age of the power station, by installing 'low NO_x' burners, and 50–80 per cent reductions are possible through flue gas cleaning techniques such as selective non-catalytic reduction or the more costly selective catalytic reduction.

In fluidised bed combustion, either under pressure (pressurised fluidised bed combustion), or at normal atmospheric pressure (circulating fluidised bed combustion), a mixture of coal and crushed limestone is partially suspended by an upward moving stream of air or oxygen from the bottom of the combustion chamber. The limestone can absorb around 90 per cent of the SO_2, and NO_x production is inhibited by the relatively low operating temperatures. Worldwide, there are approximately 200 operational circulating fluidised bed combustion plants, with four in the UK. The technology is commercially proven for plants having up to 150 MWe capacity. Larger-scale demonstration plants, however, are under way, such as in France, where the 250 MWe Electricité de France plant is under construction. In pressurised flu-idised bed combustion plants, coal is burned at pressures of 10–20 atmospheres, with additional energy efficiency being achieved by allowing the hot pressurised combustion gas into the gas turbine along with the steam. Countries where pressurised fluidised bed combustion plants are operational or planned include the USA, Japan, Britain, Sweden, Spain and Eastern Europe. Integrated gasification combined-cycle plants produce gas from coal by reacting the coal with steam and air/oxygen in a gasifier; about 99 per cent of the SO_2 is removed from the gas, along with other impurities, and the gas is used to drive a gas turbine. Worldwide, gasifier plants are operational or planned in the USA (with strong government support as part of its clean coal strategy), Britain, Germany, the Netherlands and Spain.

The most desirable technical aspects of combustion and gasification are synthesised in hybrid combined cycle systems. In such power stations, some coal is used to produce steam and drive steam turbines by combustion in a fluidised bed, while the rest is converted into hot gas to drive gas turbines. An example of a hybrid system is in the UK, at Grimethorpe, where British Coal has pioneered the Topping Cycle, and other pilot plants have been constructed in the USA and Germany.

For future energy production, other clean coal technologies are being developed that may provide more than 50 per cent energy efficiency, e.g. those that combine coal gasifiers and fuel cells, and also magneto-hydrodynamics, but these are not expected to become commercial plants for at least twenty years.

Table 7.3 *Comparisons of energy efficiencies and emissions.*

Technology	Generating efficiency*	SO_2 removal %	NO_x emissions mg m^{-3}
Pulverised fuel + flue gas desulphurisation + low NO_x burners	38–39	90	500–650
Circulating fluidised bed combustion	39–40	90	100–300
Pressurised fluidised bed combustion	41–43	90	150–300
Integrated gasification combined cycle	43–44	99	120–300
Hybrid combined cycle (British Coal Topping Cycle	46–47	90	150–300
EC Large Combustion Plant Directive (for high-sulphur fuel)		90	650

* Based on lower heating value
Source: UK Parliamentary Office of Science and Technology, Briefing Note 38, 1992.

Oil

The past few decades may well be seen retrospectively as the 'Age of Oil', to the extent that Daniel Yergin (1991) refers to the present generation as the 'Hydrocarbon Society'.

Oil is the world's largest and most pervasive business, with commercial and political influence transcending national frontiers. Indeed, the American company Standard Oil became one of the first truly multinational enterprises (MNEs). The first oilwell, which was drilled by Edwin Drake near Titusville in Pennsylvania, struck oil on 27 August 1859 at about 69 feet below the ground. The advent of the internal combustion engine ensured the supremacy of oil as a fuel. Many of the international conflicts this century have involved a struggle for military and political control over oil fields or oil supplies, for example the Suez Crisis of 1956 and the invasion of Kuwait by Iraq in 1991. Many household names are associated with the history of oil, such as the entrepreneurs Rockefeller, J. Paul Getty and Armand Hammer, but civilisations have been aware of oil for thousands of years – bitumen seepages were tapped around 3000 BC in Mesopotamia and, in the first century AD, bitumen's cauterising capacity in medicine was commented upon by the Roman naturalist Pliny. Oil, in the form of tar or bitumen, found an early use in warfare. Homer, in the *Iliad*, describes the Trojans using simple firebombs, probably of tar set alight. Bitumen also had a more constructive use, e.g. as a building mortar throughout the ancient Middle East. Today, up to thousands of years later, civilisations' uses of oil and its derivatives remain essentially unaltered, merely more sophisticated.

Most oil and natural gas is formed from the remains of marine micro-organisms that died and accumulated on the sea floor. Under favourable burial temperature (**geothermal gradient**) conditions, the organic chemical compounds are transformed into relatively short-chain hydrocarbon molecules. If the temperature is too high, the oil is broken down into volatiles or natural gas, of which methane (CH_4) is the principal constituent. Over time, the oil and gas migrate through the microscopic cavities, or pores, of a sediment or rock and become trapped in certain geological formations to form hydrocarbon reservoirs. In order for the oil to be trapped in a reservoir there needs to be a seal, typically formed by impermeable mudstone or shales. Many of the major oil and gas fields occur in ancient sedimentary environments associated with continental margins that were created by the break-up and rifting apart of land masses in what were equatorial to temperate latitudes. It is in such sites that the lake and marine conditions are favourable to high organic productivity, fast burial of organic carbon, and appropriate geothermal gradients and sediment burial histories.

In the early 1970s, about 40 per cent of global fossil-fuel use was of oil, whereas today the figure has dropped to around 38 per cent. In 1988, global oil consumption rose by 3.1 per cent, with oil production averaging 8.8 Mt per day (World Resources Institute 1990). Of this quantity, US consumption is about 25 per cent of the total global figure and constitutes approximately 50 per cent of the oil used by all the OECD countries, which together account for more than half the world's oil demand. In 1988, the global production of oil was 5,313,303 Mt, with consumption being slightly greater at 5,326,785 Mt.

The developed nations use far more oil than they produce (Plate 7.2). For example, in 1988 North America and Western Europe produced 22,857 PJ and 8,290 PJ, respectively, but consumed 36,171 PJ and 24,875 PJ, respectively (1 Mt of oil equivalent = 41.87 PJ). In contrast, in the same year Africa produced 10,991 PJ and consumed 3,609 PJ of oil,

Plate 7.2 *Oil platform offshore from New Orleans, Gulf of Mexico.*

and the Middle East produced 30,954 PJ and consumed 5,669 PJ.

In 1984, in an analysis of the likely effects of an oil import crisis, the US Congress Office of Technology Assessment (OTA) concluded that the USA then had the technical capability to replace 3.6 MB/D (million barrels per day) of oil imports, equivalent to a curtailment of 70 per cent of US net imports and a loss of 20 per cent of US oil supplies, which was regarded as a comfortable margin in any realistically projected oil crisis (OTA E-503 report 1991). The Gulf War in 1991 refocused American concern about the security of energy supplies. By 1990, US petroleum consumption had risen from 15 to 17 MB/D, domestic production had decreased from 10.3 to 9.2 MB/D, oil imports had increased from 5 to 8 MB/D, and the share of US oil needs supplied by imports had risen from 33 per cent to more than 40 per cent (*ibid.*). The OTA concluded that if a 70 per cent curtailment in oil supplies were to affect the USA – believed by some experts to be an unlikely but not impossible situation – and assuming a similar scenario to that in the 1984 report, then using all currently available oil

Plate 7.3 *Oil pump tapping oil shales in the Eocene Green River Formation, Green River basin, Wyoming, USA. Oil pumps such as these are commonly referred to as 'nodding donkies'.*

replacement technologies it could displace only 2.9 MB/D of 1989 oil use within five years. This replacement potential has to be offset against an ongoing decline in domestic US oil production, which gives an effective replacement capability of between 1.7 and 2.8 MB/D (Plate 7.3). All this adds up to the fact that between 1984 and 1991 (the dates of the two OTA reports) the USA has reduced its ability to respond effectively to a serious and prolonged oil shortage, and in such an event would fall some 5 MB/D short of its 1984 capability. Perhaps authoritative reports such as these are providing at least some of the impetus for nations investing in alternative renewable energy resources.

Following the Gulf War in 1991, the return of Kuwait oil to the world market had a minor effect on crude oil prices, with a very modest fall in the price of Brent crude – a reliable global indicator of international crude oil prices. In recent years, the highest growth rates in oil consumption have been recorded in Asia, in particular South Korea with a 21.2 per cent increase in 1992 (British Petroleum 1993).

In many developed countries, oil consumption is falling because of improved energy efficiency and reduced oil intensity (oil used per dollar of GNP), associated with energy diversification and the shift to other fuels such as natural gas, nuclear energy, etc., increased strategic petroleum reserves, and in some regions environmental considerations. Finally, it is important to distinguish oil import *dependence* from oil import *vulnerability*. Economic dependence on imported oil, measured as a percentage of domestic consumption met by imported oil, can contribute to import vulnerability but does not in itself necessarily cause import dependence.

> Import vulnerability arises out of the degree and nature of import dependence, the potential harm to the economic and social welfare from a severe disruption in physical supplies or prices, or its duration, and the likelihood of such a disruption occurring.
>
> OTA-E-503 1991

The magnitude of a nation's dependence on oil imports, however, remains a cause for concern and a central part of any considerations that help to formulate energy policy, and it provides one of the main reasons for energy diversification programmes.

Natural gas

From 1965 to the present, proven reserves of natural gas have increased fivefold to 100 btoe (World Bank 1992). Natural gas will probably be the fastest-growing energy resource. The present global use of natural gas stands at roughly 19 per cent and this figure is predicted to rise. Natural gas provides an energy alternative to oil and coal, and it is a cleaner fuel in terms of atmospheric pollution. In recent years, overall demand for natural gas has slowed, particularly in the developed countries. World demand for gas decreased in 1992 by around 0.4 per cent (with a reduction of more than 5 per cent on 1991 demand) because non-OECD Europe cancelled out increases elsewhere (British Petroleum 1993). In 1992, most growth in demand was in the **LDCs**, with South Korea showing the largest increase in gas consumption with a 3.5 per cent rise from 1991 to 1992 (*ibid.*).

Although the exploitation of natural gas as an efficient energy source is well advanced, current predictions are that global reserves will not last beyond about 120 years, whereas economically recoverable coal reserves might last for another 1,500 years (Fulkerson *et al.* 1990). Natural gas could last up to three times as long as the 120-year figure if unconventional sources were to be tapped, or if it becomes economically viable to recover the less accessible reserves.

Leakage and losses of CH_4 from distribution systems, including pipelines, oil/gas wells and domestic use, provide a contribution to anthropogenic emissions of greenhouse gases (Chapter 3). Estimates for annual leakage rates range between 1 and 5 per cent of the total amount used each year in Europe and North America, and may be as high as 10 per cent in countries such as the former Soviet Union. The total global quantity could reach as much as 50 Mt per year, but more accurate estimates need to be made.

Nuclear energy

In 1953, 'Atoms for Peace' was the slogan and programme launched by President Dwight D. Eisenhower at the United Nations. Commercial nuclear energy was portrayed as the panacea – a limitless, inexpensive and clean energy resource. Many people perceive the underlying motive behind the development of nuclear energy as a military need for still greater resources, encapsulated in phrases of the day such as 'a bigger bang for a buck' and 'massive retaliation'.

Conventional nuclear power uses mainly uranium in the process of **nuclear fission**, in which atomic particles are split and large amounts of energy released (see Box 6.1). The nuclear particles released from a fission event cause further fission of neighbouring uranium atoms in **nuclear chain reactions**. The heat that is generated by these chain reactions is harnessed to raise steam to drive electricity-generating turbines.

There are a number of different types of nuclear reactor, such as Magnox stations, pressurised water reactors (PWRs) and advanced gas-cooled reactors (AGRs) (see Chapter 6). Nuclear power has been a priority for many post-Second World War governments.

The British civil nuclear energy programme provides a good example of one country's approach to the issue of a changing energy policy. The Magnox programme was initiated in the 1950s, even though the Conservative government of the day did not expect it to be economic (House of Commons Energy Committee 1990). In 1964, the Labour government authorised a second reactor programme, the AGRs, involving a twenty-fold scaling-up from a small prototype. By the early 1970s, the AGR programme had run into serious difficulties, both economic and technological. Another reactor type, the steam-generating heavy water reactor, was also briefly considered but shelved in favour of continuing with the AGR programme. In 1979, the newly elected Conservative government announced a third programme, the PWRs, with an intended electricity-generating capacity of 15 GW and costing (at 1979 prices) £15 billion over ten years from 1982. In Britain, the first PWR, Sizewell B, was formally authorised by the government in 1987, and was

scheduled for completion in 1994. The other three identical PWRs that had been approved were effectively abandoned until the 1994 review of Britain's nuclear programme. The common themes running throughout the nuclear programme have been the escalating cost of reprocessing spent fuel and the estimated cost of decommissioning reactors.

Nuclear energy was seen as the sensible, modern answer to future energy requirements until nuclear accidents such as Three Mile Island and Chernobyl (see Chapter 6). Environmentalists then began to emphasise the negative factors such as safety and the issue of the dumping of nuclear waste. In the USA, no new nuclear power station has been commissioned since 1978, at least partly as a consequence of the nuclear accident at Three Mile Island in 1979. The other factor cited by environmentalists is that it is simply uneconomic to order any new nuclear power plants. There are currently moves by the nuclear industry to persuade Congress to simplify the way in which operating licences are obtained, something that has already been begun by the US Nuclear Regulatory Commission (NRC). The principal obstacle is that two hearings are required to make operational any nuclear power station in the USA – one prior to construction, and the second following construction but before the NRC grants an operating licence. Investors are disinclined to put their money behind such a lengthy process from commissioning to operation.

Critics of nuclear power argue that, in the final analysis, the lower limit of accident probability may be taken to an irreducible minimum level of human error, and that how ever technically safe a nuclear reactor may be, the level of risk is still unacceptably high. This is, of course, a value judgement. Many environmentalists accept a level of accident risk that is generally considerably lower than that which is acceptable to industrialists. Politicians, depending upon their allegiance, will express a complete spectrum of opinions.

The governments of some countries have decided to reduce drastically, or eliminate, the use of nuclear energy. In January 1991, the Swedish government committed itself to phasing out the use of nuclear power by 2010. Sweden actually generates more nuclear power per capita than any other country – 50 per cent of its energy comes from twelve nuclear power plants. At the same time, Sweden has allocated 3.8 billion krona (£352 million) over the next five years to developing alternative energy resources and to improving energy efficiency. These resolutions by Sweden also include a pledge to maintain high levels of employment and to an annual economic growth rate of 1.9 per cent. By the same token, it is committed to not building any more new hydro-electric dams – the source of its remaining energy. The final sting in the tail for Sweden is that it has also agreed that by 2010, its emissions of CO_2 will be reduced to its 1986 levels. Can these lofty goals be realised? The balance in this tricky equation seems to depend on improved energy efficiency (MacKenzie 1991).

All of the foregoing discussion has concerned nuclear fission. The goal of energy from **nuclear fusion**, released by the bringing together of light atoms such as the isotopes of hydrogen known as deuterium and tritium, is still a long way off. On the evening of Saturday 9 November 1991, there was a false alarm when at the Culham laboratories in Oxfordshire, England, in a £75 million per year project jointly funded by fourteen European countries, including Britain, nuclear scientists believed that they had made the breakthrough in an experiment lasting only two minutes, in which their furnace, named Torus, achieved temperatures of 300 million °C – twenty times hotter than the core of the Sun. They thought that, for the first time, they had succeeded in creating a controlled fusion experiment that released nuclear power, other than in a nuclear bomb. Later attempts at repeating this experiment failed, and the scientific consensus is that the experiment never generated power by nuclear fusion.

Nuclear energy issues such as the reprocessing of spent fuel rods, the enrichment of nuclear fuel for military purposes and accidents at nuclear power stations are all dealt with in Chapter 6.

Hydrogen energy

One of the exciting new energy resources is the use of energy from sunlight to produce **chemical fuels**, for example hydrogen energy (see Box 7.2), supported by some as an easily transported and readily storable fuel, two attributes that have advantages over the limitations associated with converting solar energy directly into electricity. Since most of the regions of the Earth where there is ample sunlight for efficient solar energy plants to be constructed tend to be in remote areas away from large population centres, transportable chemical fuels may well provide an attractive energy proposition. Many early attempts to generate chemical fuels involved producing the light and combustible gas hydrogen by the **electrolysis** of water. Upon combustion of hydrogen, water is produced. If the electricity used to burn the hydrogen is generated from a non-fossil fuel such as wind, hydro- or solar power, then this process can be seen as environmentally benign.

Renewable energy

Over the next decades, energy based on fossil fuels will continue to provide most of the world's requirements, and it is unlikely that overall present energy systems will change radically. It is likely, however, that within the second quarter of the next century the extent and depletion of oil and gas reserves will have begun to be reflected in energy prices (World Energy Commission 1992). Global CO_2 emissions

---

BOX 7.2 PRODUCING HYDROGEN ENERGY

One of the ways to produce hydrogen is by the high-temperature decomposition of sulphuric acid (H_2SO_4), which releases water and sulphur dioxide (SO_2). The SO_2 then reacts with iodine and water to produce hydrogen iodide (HI) and H_2SO_4, followed by the thermal decomposition of the HI molecules to yield free hydrogen (H_2) and iodine (I_2) gas. More recently, methods for producing hydrogen have included the electrolytic process, where SO_2 and H_2O react to release H_2SO_4 and H_2, the advantage being that only 0.29 volts is required, compared with about 2 volts for the high-temperature decomposition of H_2SO_4. Another electrolytic process involves the reaction of bromine (Br_2), SO_2 and H_2O to produce hydrogen bromide (HBr) and H_2SO_4, followed by the application of 0.62 volts to the HBr molecules to release free H_2 and Br_2. Today, the efficiency of producing electricity commercially from solar sources is around 12 per cent, but 70 per cent for electrolysing water, with an overall efficiency of about 8 per cent (Dostrovsky 1991).

Although it is theoretically possible to generate hydrogen from water by heating it to more than 2,000°C (obtainable from concentrated solar energy), the technology is not yet available to stop the hydrogen and oxygen recombining to form water vapour as the gases cool. Other sources of hydrogen include fossil fuels and plant material, the latter being known as biomass. The biomass option requires temperatures of only 700–900°C and steam in the absence of air to produce a gaseous cocktail of H_2 and carbon monoxide (CO). It is feasible to ensure a mix of H_2 and CO in the ratios of 2:1 and 3:1 and to use this so-called syngas to synthesise liquid fuels such as pure hydrogen, methanol and gasoline. This biomass technology is still in its infancy but deserves sustained industrial and government funding because it provides the potential for the production of energy using renewable resources that are environmentally more friendly.

will rise throughout the 1990s, and Western Europe will not achieve the target 20 per cent reduction by 2005 as set out in Toronto (see Chapter 3). The World Energy Commission believes that the target is achievable by 2020, provided that an appropriate mix of energy is available and that effective incentives are given to promote energy saving. With these factors in mind, there is a need to consider new technologies for transport fuels, and for government/EU assistance in achieving their technological and commercial viability.

Renewable energy resources (Table 7.4) are becoming more attractive because, unlike conventional fossil fuels, they offer 'cleaner' technologies associated with lower emissions of greenhouse gases, and they do not contribute to acidic deposition. Furthermore, they offer energy diversification and, therefore, greater energy security. Many developed countries are encouraging programmes of utilising renewables, such as the Non-Fossil Fuel Obligation (NFFO) in the UK, which requires regional electricity-generating companies to provide customers with a certain proportion of their electricity supply from nuclear and renewable sources. The NFFO and its associated Fossil Fuel Levy were set up by the British government in 1989 under the Act that privatised the UK electricity industry, mainly as a subsidy for nuclear power, and to some extent to offer support for renewables, the latter being added to give it an acceptable environmental gloss.

Some countries obtain extra revenue for the development of alternative energy sources from the conventional fossil fuel sector. For example, in the UK, the Fossil Fuel Levy is used to impose an 11 per cent surcharge on all electricity generated from fossil fuels, and the revenue raised (£1.25 billion in 1990/91) subsidises electricity produced from non-fossil fuel sources – mainly nuclear energy and renewables, such as tidal and wind power. Clearly, this type of taxation carries many implications for any energy policy, not least of which is the positive discrimination against fossil fuels in favour of alternatives and renewables.

The following sections examine the principal renewables.

Hydroelectric energy

Water power is amongst the oldest of energy resources. Before the Industrial Revolution in the nineteenth century, water mills were a common sight. **Hydroelectric power** (hydro-power) is widely utilised around the world as a clean, renewable

Table 7.4 *Estimates of selected global renewable energy resources at the surface of the Earth.*

Resource	Estimated as recoverable	Resource base
Solar radiation	1,000 TW	90,000 TW
Wind	10 TW	1,200 TW
Waves	0.5 TW	3 TW
Tides	0.1 TW	30 TW
Geothermal flow	–	30 TW
Biomass standing crop	–	450 TW years
Geothermal heat stored	>50 TW years	10^{11} TW years
Kinetic energy stored in atmospheric and oceanic circulation		32 TW years

NB: As these energy resources are renewable, they are described in terms of energy flow (i.e. power) in TW, except for the standing crops of biomass, and geothermal and kinetic energy stored, which are described in terrawatt years (1 TW year = 3.16×10^{19} joules).
Source: CEC (1992); Harrison (1992).

energy resource, although capital investment costs are normally very high for the construction of dams and reservoirs. Nevertheless, hydroelectric power may prove to be one of the most acceptable 'compromise' energy sources.

Hydro-power is currently the world's largest renewable energy resource. Worldwide, there are about 15,000 dams (Boyle 1996). Since so much information is available on regional watersheds, hydro-resources tend to be the best audited of all the renewable energy options. These resources are defined as potential hydroelectric capacity, measured for example in kilowatts or megawatts. It has been estimated that global potential hydroelectric capacity is enormous – at 2.2 million MW, 'double the present installed world generating capacity for power plants of all types and sizes' (Cassedy and Grossman 1990).

Hydroelectric schemes range from a capacity of a few hundred to more than 10,000 MW, and are distinguished by the location and type of dam/reservoir, the rated power output (capacity), the effective head of water (i.e. low, medium or high), and the type of installed turbines.

There are environmental problems associated with diverting water courses and the construction of hydroelectric dams, including the loss of vegetation, ecological niches in upland and mountainous areas, and agricultural land, and displacement of the population. Changes in river courses may affect the animal, plant and fish life.

Apart from the obvious problems of dam failure and the resultant catastrophe of flooding causing loss

Plate 7.4 *Wind turbines southwest of Los Angeles, California. Panorama of valley filled with turbines (top) and close up view of turbines (bottom).*

Table 7.5 *Dams and seismicity.*

Dam, country	Height of dam (m)	Reservoir capacity (billion m³)	Earthquakes since construction
Glen Canyon, USA	220	33	nil
Hoover, USA	220	38	medium
Kariba, Zambia/ Zimbabwe	130	160	strong
Oroville, USA	240	4	slight
Warrangamba, Australia	140	2	medium

Source: Boyle (1996).

of life, many additional problems are associated with dam construction and reservoirs. These include the flooding of ecologically important areas; for example, the Kariba Dam in Zimbabwe flooded 5,100 km² of land and destroyed one of the most natural habitats of the rhinoceros and elephant. A further example is the Tucurui Dam in the Amazon, which caused the destruction of virgin rainforest and displaced thousands of people. The Volta Dam in Ghana displaced 78,000 people from 700 towns and villages, and was associated with a string of resettlement problems and increased pressures on neighbouring areas. Such flooding causes the decay of drowned vegetation, leading to acidification of the lake waters and the anaerobic conversion of organic matter to greenhouse gases such as methane.

Other problems include modification to surface- and ground-water hydrology. In northern Quebec, for example, recently constructed hydroelectric power stations have altered the hydrology of an area comparable in size with Switzerland, much of this region originally having been forest. Since the construction of the Hoover Dam, which impounded Lake Mead on the Colorado River in 1935, earthquake activity appears to have increased, and this has been linked with the increased height of ground waters and loading of the underlying rocks due to reservoir construction. Prior to construction of the Hoover Dam, no earthquakes had been noted in that area: since 1935, more than 1,000 earthquakes strong enough to be felt by the local population have been recorded. Worldwide since then, more than fifteen other reservoir constructions are known to have generated large earthquakes, including the magnitude 5–6.5 earthquake associated with the Koyna Dam in India, which caused the loss of 177 lives (Table 7.5).

Despite some of the negative aspects of using water power to generate electricity, hydroelectric schemes are seen by many people as one the most acceptable, least environmentally damaging, energy resources. Clearly, the careful and sensitive selection of sites for hydroelectric power plants and dams can ameliorate many of the potential problems. Very small-scale hydroelectric ('micro-hydro' and 'mini-hydro') power plants, however, can be much more appropriate in many rural situations, and without many of the side-effects associated with large-scale dam construction.

Wind power

Wind power (Plate 7.4) is one of the most under-utilised renewable energy resources. Wind farms are capital-intensive but cheap to run, with greater wind speeds providing more power, and hence more cost-efficient plants. While many countries still seem uninterested in a serious commitment to wind power, countries such as Denmark appear to be blowing full-steam ahead. Table 7.6 shows the national targets for wind power development for selected countries.

The largest wind farm in Denmark is sited in the west of Jutland at Velling, where there are 100 wind turbines with a capacity of about 13 MW. Denmark also has the world's first offshore wind farm, which will help to supply the targeted 10 per cent of its energy requirements by the year 2000. The farm, which is expected to generate 12 million kWh of electricity, has eleven wind turbines, sited in up to 5 m of water 1.5–3 km offshore from Vindeby on Lolland Island in the Baltic Sea. The co-operative company, Elkraft, that runs the wind farm estimates that if the turbines have a 20-year life span, then the electricity will cost about 0.63 Danish krone per kWh (£0.058). The turbines are controlled and monitored on land through optical-fibre cables built into the 12-kilovolt power cable on the sea floor that transfers the wind energy from the offshore turbines to land.

In the Netherlands, the government has committed itself to the largest programme of windmill construction since the seventeenth century. It wishes to encourage the construction of 2,000 new windmills and increase wind power by an equivalent 2,000 per cent by the end of this century. Its intention is to generate more than 1,000 MW of wind power. Currently, there are about 1,000 windmills in the country. In order to encourage this growth, the government will provide subsidies of up to 40 per cent to build the windmills, which will make the cost of wind power about equal to that of nuclear or conventional fossil-fuel-generated power. This energy resource, however, is far cleaner. By developing more energy-efficient windmills, the government hopes to increase the power output and so reduce the unit costs.

Those who are sympathetic to the technology of renewable energy believe that Britain is squandering the best wind in Europe. In the UK in 1990, only six wind projects were included in the list of renewable energy projects covered by the first NFFO, which was set by the Department of the Environment in Britain to allow them to become established: Subsequently, in the second NFFO, this was increased to 220 MW at fifty-eight sites. In a recent UK government white paper (Department of the Environment), the contribution from wind power has been set at 1,500 MW by the turn of the century.

Table 7.6 *National targets for wind power development in selected countries.*

Country	Target	at 25% load factor output would be (TWh yr^{-1})
Denmark	c. 1,200 MW of wind power by 2000	2.6
Germany	250 MW of wind power by 1995	0.55
Greece	400 MW of wind power by 2000	0.88
Italy	600 MW of wind power by 2000	1.3
Netherlands	400 MW of wind power by 1995	0.88
	1,000 MW of wind power by 2000	2.2
California, USA	10% of electricity from wind by 2000	c. 30.0
India	5,000 Mw of wind power by 2000	11.0

1 TW = 1 million kW.
Source: House of Commons Energy Committee Fourth Report on Renewable Energy 1992, vol., 1, p. xvi. London: HMSO.

As a comparison, countries such as Denmark have set their sights on much higher targets, with a projected 1.5 per cent of its electricity coming solely from wind power, and 10 per cent by the year 2000. Even India has set a target of 5,000 MW by the beginning of the next century. Environmentalists in Britain believe that the low target (<2 per cent) in the NFFO is because the Department of the Environment is offering only very short-term contracts of up to eight years. This time limit arose because the EU would not allow a subsidy on nuclear power to go beyond 1998, but the UK government is seeking agreement from the EU to extend support for renewables. Yet the House of Commons Energy Committee (1992) has stated that 'the wind energy resource in the United Kingdom is particularly rich', with the technical potential of onshore wind energy as 45 TWh yr^{-1}, and offshore as high as 140 TWh yr^{-1}. The British Wind Association estimates that the UK has 40 per cent of Europe's total realisable wind energy potential. During the past decade or more, there have been considerable technical and economic improvements in wind energy technology (Figure 7.7), all of which have served to make this renewable more viable.

Hopefully, governments will encourage the harnessing of wind energy through attractive contracts to prospective companies, and by allocating a specified proportion of their energy requirements to wind resources. The environmental impacts of wind farms are that they can cause local noise pollution and electromagnetic interference. Perhaps the major problem is that they are seen by many as unsightly. This has been a particular problem in the UK, where many of the windiest sites are in areas

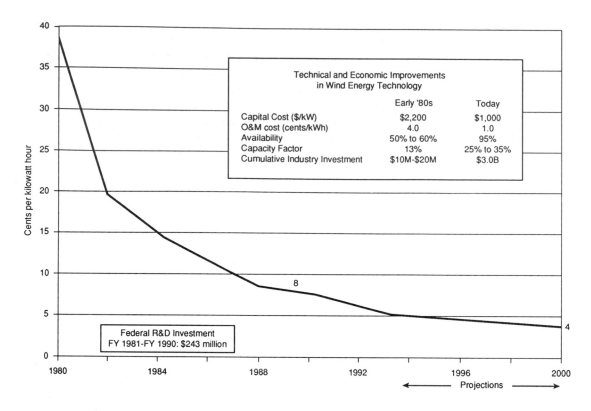

Figure 7.7 *Cost of electricity from wind in the USA. Redrawn after [UK] House of Commons Energy Committee (1992).*

Tidal energy

Tidal power, which is inexhaustible and intermittent (although absolutely predictable and reliable), utilises the generally twice-daily rise and fall of the tides to generate electricity. Coastlines around the world are divided into micro-tidal, where the tidal range is less than 2 m; meso-tidal, with 2–4 m tidal ranges, and macro-tidal, where the range is greater than 4 m. Macro-tidal coasts with tidal ranges greater than 15 m (as in the Bay of Fundy, Nova Scotia) provide the best conditions for harnessing tidal energy.

In the UK, where the EU estimate the technical potential of tidal power as 53 TWh, with about 90 per cent of this figure in eight estuaries, or about 20 per cent of current electricity demand in England and Wales (House of Commons Energy Committee 1992), there are major tidal energy schemes for the Severn, Mersey, Wye, Conway and Humber estuaries, along with other relatively small sites. The tech-

nology is largely proven, but for such major engineering projects, the capital costs are high, although offset against this is the lifetime of the tidal barrages, expected to be around 120 years. Cost estimates are in the range of £10.2 billion for the Severn barrage (8,640 MW, or 17 TWh annual output), £966 million for the Mersey barrage (700 MW, or 1.4 TWh annual output), and £72.5 million for the Conway barrage, with annual running costs of £86 million, £17.6 million and £600,000, respectively. Estimated electricity prices for the Severn barrage are £0.06 per kWh for a 16.5-year payback period, and £0.05 per kWh for 20 years (House of Commons Energy Committee 1992). Broadly comparable prices are estimated for the other tidal barrages.

A £4.4 million preliminary feasibility study for erecting a tidal power barrage across the Severn estuary was completed a few years ago, with the conclusion that there are no overriding environmental or technical barriers to the project. This study was financed by the Department of the Environment (DoE), the then Central Electricity Generating Board (CEGB) and a consortium of construction companies named the Severn Tidal Power Group. The 10-mile barrage would be erected from Bream Head in

Somerset to Lavernock in South Wales, with a projected cost of about £9 billion.

The Severn tidal barrage would have a power output equivalent to five nuclear reactors but would cost six times as much as a pressurised water reactor (PWR) such as that constructed at Sizewell, Suffolk (£1.5 billion). The tidal power station will house 200 turbines and generate more than 7,000 MW of electricity at a unit cost of £0.0379 per kWh (PWRs produce electricity at £0.0224 per kWh). Although the superficial economic comparisons between the tidal power station and a PWR suggest that the latter is better value, the working life of the PWR at Sizewell will be 35 years, whilst it will be 120 years for the Severn barrage. The barrage scheme would also employ an estimated 30,000 people over its nine-year construction period. Some would argue that, on balance, the economic and environmental case seems to favour tidal power stations over nuclear schemes such as the proposed Severn estuary scheme.

Tidal barrages can have a significant environmental impact, for example by impairing fish reproduction, because spawning takes place in areas where fresh water and saline marine waters mix, which in the case of the Severn could move the zone up to 30 km seaward. Also, barrage construction would reduce the area of mud flats washed by the tides, leading to more plant growth upstream of a barrage, and corresponding reductions in the amount and diversity of invertebrate animal life. The ability of estuaries to support wintering and migrating bird populations could be affected, although this is an area of dispute with conflicting views and, therefore, in need of further research. Tidal barrages can also adversely affect migratory and spawning fish populations.

Wave power

The energy associated with waves can be used to generate electricity. Wave energy is very much in the research and development (R&D) stage, with few operational devices, and considerable uncertainties over costs. The UK and Norway have led R&D into wave energy.

Wave energy may involve small-scale shoreline (or onshore) and large-scale offshore devices. Offshore devices have potential energy levels of three to four times those of the inshore devices. One example of the technology, known as the 'Edinburgh Duck', or 'Salter's Duck', utilises a large hinged plate that moves up and down in response to passing waves. The problem with this, as is the case for other wave-energy devices, is the reliability of the technology,

but an attraction is that any environmental impacts are likely to be minimal.

The world's first commercial wave power station sank off the Scottish coast during a storm on 2 August 1995, less than one month after its launch. It foundered because of holes in its ballast tanks caused by the severe wave battering. It was developed by a consortium of private investors and companies together with a US$700,000 grant from the EU's Joule programme for R&D into non-nuclear energy, under the umbrella name OSPREY (Ocean Swell Powered Renewable Energy). OSPREY 1 was designed to generate electricity for 2,000 homes for twenty-five years. Despite this setback, the consortium is to install a replacement, OSPREY 2, a very similar design to OSPREY 1.

Solar energy

The Sun is the main source of energy – it allows photosynthesis, the process whereby plants convert solar energy into chemical energy, which is then changed into fossil fuels. Every year, energy from the Sun, or solar radiation, bathes the Earth's surface in roughly 15,000 times the current global energy supply – equivalent to an estimated annual energy supply of approximately 178,000 TW or 5.6 million exajoules (1 TWh = 1,000 million kWh; 1 exajoule = 10^{18} joules). Dostrovsky (1991) gives the amount of solar radiation reaching the Earth's surface annually as 3.9 million exajoules (equivalent to the amount of heat released by the combustion of 22 Mt of oil); 30 per cent of this total solar budget, however, is reflected back into Space and 50 per cent is absorbed, converted into heat and re-radiated. The remaining 20 per cent powers the hydrological cycle, but only a mere 0.06 per cent drives photosynthesis. The Sun is directly responsible for solar power, wind power and hydro-power.

Table 7.7 shows the solar radiation in selected countries. It is clear that many developing countries have considerable potential for developing solar energy plants, and Box 7.3 summarises the process of converting solar energy into electricity.

Solar power has come a long way since 1876, when two British researchers, W.G. Adams and A.E. Day, were the first to convert sunlight into electricity using a selenium cell. Energy from the Sun, or solar energy, is now being used in many countries worldwide. Not only is electricity generated from solar energy for domestic and commercial use, but many everyday gadgets can run on solar power such as calculators, watches, personal cassette players,

Table 7.7 Solar radiation in selected countries.

Region	Solar radiation ($kWh\ m^{-2}yr^{-1}$)
Mali	2,490
Niger	2,450
Mexico	2,080
Sierra Leone	2,000
Venezuela	2,000
India	1,950
Brazil	1,880
Chile	1,630

Source: OTA-E-516 1992 (compiled from: California Energy Commission, *Renewable Energy Resources Market Analysis of the World*, P500–87–015, pp. 14–16).

telephones and radios. Satellites also rely on solar energy.

In the USA, electricity produced by solar power costs about 30 cents per kWh, but this cost is dropping rapidly. In February 1989, a Californian company opened a plant to convert sunlight into heat energy. This solar–thermal plant expects to produce power at less than 3 cents per kWh. As a comparison, in the USA, where conventional power stations use fossil fuels to drive electricity generators, running costs are about 3 cents per kWh.

Southern California plugs into the largest power socket ever built. Here, the sun-drenched Mojave Desert is home to 600,000 computer-driven parabolic mirrors pointed skywards to collect the energy in the Sun's rays. This power farm reaps the almost daily harvest of solar energy and testifies to the practicality and utility of natural alternative energy resources. The Mojave Desert power farm at Kramer Junction is the largest of three solar energy complexes operated by Luz International. The farm covers an area of about 1,000 acres and generates 90 per cent of the world's grid-connected solar energy – 275 MW. The parabolic mirrors track the Sun to focus its rays onto pipes filled with synthetic oil, which is heated to about 390°C. The super-heated oil is then used to boil water to drive steam turbines. It is as simple as that. A cocktail of sunlight, air and water demonstrates the viability of this energy resource.

In Europe, the largest solar energy plant is situated near Koblenz in Germany, at Kobern-Gondorf on the slopes of a former moselle vineyard. It can generate 340 kWh at peak capacity. Just outside Chur in eastern Switzerland, the motorway is lined with solar panels for many hundreds of metres. This kind of approach to supplementing other energy resources provides an unobtrusive and sensible location for such panels. Other countries should follow suit.

As to investment in R&D for solar energy, amongst the market leaders is Japan, which spends about US$50 million annually, compared with US$43 million in 1990 by the United States. In fact, in 1989, in the absence of rapid results, the Reagan administration cut R&D to US$35 million from its 1981 high of US$150 million (Spinks 1990).

Countries where solar energy is used tend to supplement their electricity grid by solar energy rather than see it as a mainstay of their energy budget. The environmental argument, however, is that even if solar power is marginally more expensive, it should still be used since the long-term cost to the planet through continuing to burn fossil fuels at present rates will be greater.

BOX 7.3 CONVERTING SUNLIGHT ENERGY

The energy from the Sun strikes the Earth every day as a bombardment of packets of light energy, termed **photons**, the sum of which is commonly referred to as the **solar flux**. It is stored in photoelectric cells, either for immediate or later use. Naturally, countries with the most potential for using solar energy are those that have the sunniest weather, thus this energy source is more appropriate in some countries than others.

There are two main methods of converting sunlight to electricity. One is to convert the sunlight to heat or thermal energy, and then use the heat to raise steam to generate electricity. This method has now been eclipsed by the **photo-voltaic cell**, a device for changing solar energy directly into electricity using semiconductors (Figure 7.8). Efficiency, although improving, remains a problem with these cells. Their efficiency varies between 15–29 per cent (monocrystalline silicon cell) and 12–18 per cent (polycrystalline silicon cell). The US Department of Energy aims to improve this to 35 per cent, with 40 per cent efficiency being regarded as the theoretical maximum. Many types of crystals are now being used to make solar cells, such as copper indium diselenide, gallium arsenide and gallium antimonide. Gallium arsenide appears to be one of the most efficient semiconductors, but unfortunately it is also one of the most costly. There is also much interest in amorphous silicon cells, which have a lower efficiency than crystalline cells but are cheaper to produce. Despite all these difficulties, the increased efficiency and reduced costs of solar panels have conspired to reduce the cost of solar power to a current average of about US$0.50 per kWh from a 1980 figure of around US$1.50 per kWh.

In the future, it may be viable to use photo-voltaic panels as cladding on buildings to supply electricity during the summer months to run air-conditioning systems.

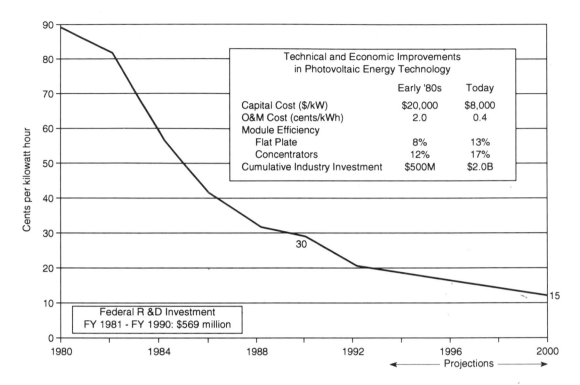

Figure 7.8 *Cost of electricity from photo-voltaics in the USA. Redrawn after [UK] House of Commons Energy Committee (1992).*

Geothermal energy

The Earth continuously emanates heat or thermal energy from rocks and molten magmas at depth, referred to as **geothermal energy**. It is an essentially limitless resource that is currently under-utilised, and is estimated worldwide at almost 5 GW (US Congress Office of Technology Assessment 1992b). The two main sources of geothermal energy are heat extraction from rocks and the tapping of hot water, including oceanic thermal energy by heat exchange processes (Box 7.4). Table 7.8 shows the geothermal electricity generated in the USA and selected developing countries in 1990.

There are geothermal energy programmes in a number of countries, for example in the USA in northern California, where more than 2,000 MW of power is produced. Some other very suitable countries where the geothermal gradients are high include Iceland, the Philippines, Mexico and Italy (Plate 7.5). In Britain, technical problems have bedevilled the Hot Dry Rock Project, which aimed to tap geothermal energy from the Cornish granites, and its future remains uncertain. In the port of Southampton, UK, geothermal energy is tapped from hot salt waters (brines) below the city, which are used to heat public and private buildings. The brine is pumped out of the well at a rate of about 12 litres per second and

---

BOX 7.4 GEOTHERMAL TECHNOLOGIES

Technologies for geothermal heat extraction include direct steam, single-flash, double-flash and binary systems. In direct steam, the simplest of the technologies, steam is piped directly from the subsurface reservoirs and used to drive steam turbines to generate electricity. In the single-flash process, the underground hot water is 'flashed' into steam to drive turbines, and in the double-flash process, a second 'flashing' is undertaken to utilise up to 20 per cent of the energy that would otherwise escape. In binary units, an intermediate working fluid is used to transfer energy from the geothermal reservoir to the electricity-generating turbines. An advantage of the binary cycle, which is especially appropriate in developing countries, is that it can use relatively low-temperature resources (e.g. 170–180°C), and this cycle is suitable for relatively small plants, typically 5–10 MWe.

Table 7.8 *Geothermal electricity generation in selected countries in 1990.*

Country	Geothermal capacity (MW)	Technology
USA	2,827	All technologies
Philippines	894	Single flash
Mexico	725	Dry steam, single flash, double flash
Indonesia	142	Dry steam, single flash
El Salvador	95	Single flash, double flash
Nicaragua	70	Single flash
Kenya	45	Single flash
Argentina	<1	Binary
Zambia	<1	Binary

Source: Table 6.11, OTA-E-516 1992.

a temperature of 74°C, giving a basic 1.1 MW of power. Additional energy sources, including a diesel generator and a gas/oil boiler, raise the maximum power output of the scheme to 14 MW. Clearly, although geothermal energy provides only a relatively small energy source, it makes a contribution to providing power and has environmental benefits.

Geothermal energy does have some adverse environmental implications. CO_2 and hydrogen sulphide (H_2S) emissions may be high, the latter being limited in the US to 30 ppbv. Technologies exist to eliminate these emissions, for example so-called binary technology. Water consumption is also high, with the amount required being particularly dependent on the plant design. CO_2 emissions depend on the amount of carbon contained in the particular resource, averaging about 5 per cent of the CO_2 emissions per kWh output from a coal-fired power station. There may be land subsidence above geothermal wells, and water supplies can become contaminated by saline, possibly toxic, water, but these problems can be managed successfully.

Biomass energy

Biomass fuels are produced from plant and animal matter wastes and residues, and account for about 13.8 per cent of current worldwide energy consumption, equivalent to approximately 25 MBO/D, equal to OPEC's current crude oil production (Hall *et al.* 1992, British Petroleum 1993).

Natural degradation of waste under controlled conditions in landfill sites releases energy that can be tapped. It is estimated that the **biodegradation** of 1 tonne of refuse can ideally produce 400 m^3 of gas (landfill gas – see following paragraph), equivalent to 7,500 MJ of heat energy. Generally, less than 25 per cent of the degradable fraction of such waste will break down within the first fifteen years, with most being produced in the early stages. Thus, the efficiency of energy production from decaying material will decrease rapidly with time.

Organic matter such as food products decay quite rapidly, principally through the action of bacteria. This decomposition initially uses free oxygen until the decaying matter becomes oxygen-deficient, when the natural breakdown of organic matter by organisms

Plate 7.5 *Geothermal pumping plant near Grindakiv in Iceland which supplies hot water around the Keflavik Peninsula for domestic use. Cold fresh water is pumped into the ground at the plant and reaches temperatures of 95–125° before being pumped via a system of pipes, which have a total length of 300 km, to the users, who eventually consume the water at about 80°C. (A) The plant and one of the distributary pipes (top). Evidence for the relatively recent volcanicity and high heat flow in this area is provided by the fresh poorly vegetated lava flow in the foreground and the young volcanic cones in the distance. (B) The 'Blue Lagoon' (bottom) at the plant provides hot baths for tourists and Earth scientists.*

occurs without free oxygen as **anaerobic** digestion. This comprises three main stages. First, simple sugars such as glucose ($C_6H_{12}O_6$) are produced. These are then converted into fatty acids, mainly acetic acid, by acetogenic bacteria. Finally, **methanogenic bacteria** under anaerobic conditions break down the organic matter, the organic acids, into **bio-gas**, or landfill gas. All landfill sites release methane (CH_4), a greenhouse gas. By making use of it to produce heat and power, CH_4 is converted to CO_2, a less potent greenhouse gas and its use reduces the release of CO_2 from other fossil fuels that would otherwise have been burned.

Bio-gas is a cocktail of roughly equal amounts of CH_4 and CO_2, a few per cent by volume of hydrogen and nitrogen, traces of sulphur-containing organic molecules and halogens, plus some heavier hydro-carbons. Landfill sites with household waste commonly have temperatures in the region of 35°C, which is ideal for the natural biodegradation stages outlined here. The bio-gas can be burned to generate electricity and/or heat.

The first commercial venture to exploit the production of bio-gas was about fourteen years ago in the USA, at Palos Verdes in California. Countries such as Germany and Britain followed suit a little over a decade ago. About seventy such sites operate in the USA, the largest being at Puente Hills, California, which generates 46 MW, which is sold to the local electricity grid. Approximately fifty production sites operate in former West Germany.

Throughout England and Wales, 300 potential sites for the economic production of landfill gas through biodegradation were identified in 1986 by the former Department of Energy (DoE), now incorporated into the Department of Transport and Industry (DTI). The DoE estimated that the total annual production of bio-gas from these sites alone would be equivalent to 1.3 Mt of coal, at least 1 Mt of which is economically viable to recover. In 1989, just twenty-six of these 300 sites were being exploited. In Bedfordshire, England, for example, London Brick (then part of the Hanson Trust Group) sold off the large open pits created by the extraction of clay for brick making as landfill sites for the dumping of household waste from which commercial bio-gas is marketed. In Britain, estimated savings on conventional fuels from bio-gas production increased from £3.5 million in 1986 to approximately £6 million in 1988. Savings for 1989 were expected to be in the region of £12 million, with a further doubling in 1990.

Although the principal use of bio-gas is in the generation of electricity or supply to nearby boilers and furnaces, another use is as a substitute for high-grade fuels after it has been cleaned. On the debit side, the production of bio-gas from landfill sites raises environmental issues. Biodegradation produces greenhouse gases, including small amounts of CFCs. Until more research has been conducted into the management of landfill sites and the associated production of bio-gas, it is not known if the benefits outweigh the disadvantages. The onus is on the proponents of biodegradable waste and bio-gas to show not only that these processes provide an efficient way of converting unwanted waste into energy, but also that the process need not necessarily harm the environment. The use of this gas provides a way of utilising something that would otherwise escape into the atmosphere as a greenhouse gas.

Not only can energy be produced from the degradation by biological means of organic matter, for instance bacterial decay in the process of biodegradation, but there is also the potential to reclaim or recycle the various metals, glass and plastics that form part of such waste. In a typical year, Britain produces more than 80 Mt of mixed household, commercial and industrial waste. An estimated 28 Mt of this waste is biodegradable. The problems of waste disposal are increasing, both from the point of view of available land for dumping and danger from the by-products of the natural breakdown of the waste, let alone the intrinsic dangers from certain dumped industrial chemicals.

Typical household waste has a calorific value of about one-quarter to one-third that of coal, equivalent to an energy yield of approximately 9 MJ kg^{-1}. A proportion of this energy can be released simply through combustion or incineration to produce electricity. An alternative is to convert the waste into a solid, liquid or gaseous fuel, but this is very much for the future. Combustion is easy, but it releases harmful gases into the environment, including various greenhouse gases.

Road transport and energy

Transport services are a major and integral part of economic development. Many countries have to import large quantities of transport fuels, leading to import dependence. The principal control on traffic growth is GDP and fuel price. The two fastest growing sectors of energy consumption are for electricity supply and transport. Western Europe consumes 18 per cent of world commercial primary energy supplies but contains only 7 per cent of the global proven coal reserves, 2 per cent of proven oil reserves, and 5 per cent of proven natural gas

BOX 7.5 TYPES OF LIQUID BIO-FUEL

Bio-diesel can be produced from most edible oil crops. In Europe, oilseed rape is the principal crop used to produce bio-diesel, from which the oil (called rape seed oil, oilseed rape) is converted into rape methyl ester (RME). In the USA, soya oil is the main source, whereas canola oil is used in Canada. The main opportunity for future energy from edible oils could be a cost reduction through using acid oils and waste oils. As with many industrial monoculture crops, oilseed rape should not be grown continuously on one site because of the risk of disease build-up in the plants, therefore it should be cultivated as a rotational crop.

Oil is extracted from the rape plant by crushing, with 3 tonnes of rape yielding 1 tonne of usable RME. Most diesel engines can run unmodified on unblended rape oil but become clogged after several days, prevented by removing the glycerine from the oil by mixing 1 tonne of rape oil with 110 kg of methanol in the presence of the catalyst nitrogen hydroxide, and heating to temperatures of 40–50°C. The glycerine then settles out to leave the clear liquid RME. Glycerine, with many industrial uses, such as pharmaceuticals, cosmetics and explosives, is a high-value by-product, giving a significant economic credit to RME production. Bio-diesel and ethyl tertiary butyl ether (ETBE) appear to have no serious technical limitations to their use in motor transport. Blends of less than 10 per cent RME with mineral diesel fuel are commonly used since this involves little, if any, modification to existing engines. Engine performance is slightly reduced where RME is used, estimated by the Italian tractor manufacturer, Same, as less than 2 per cent compared with mineral diesel fuel. The high aniline point of RME means that it destroys rubber more easily than mineral diesel oil, a problem that is not manifest within the engine, but requires addressing for distribution systems using rubber hoses, for example, in oil tankers and at petrol stations. The use of stronger rubber in the distribution system, at cost, provides a solution.

The main sources of bio-ethanol (pure ethanol is neat alcohol) are cereals (wheat and barley), sugar beet, maize and surplus wine, and lignocellulose (wood and straw) using either acid or enzyme hydrolysis, the technology for which is not sufficiently advanced as yet and requires further research, for example to improve enzyme efficiency.

The technology for producing bio-ethanol (fuel ethanol) from feedstocks containing sugar and starch is well proven and the processing equipment is commercially available on an industrial scale. Ethanol is produced by fermentation and recovered by a distillation process. Bio-ethanol blended with petrol must have the water content removed (anhydrous ethanol) by a dehydration process after distillation, as even small amounts of water lead to phase separation. For wheat, the yield is about 6.7 tonnes of field-harvested grain ha^{-1}, and about 2.7 kg of grain produces 1 litre of bio-ethanol. Bio-ethanol is used as a blend or as a pure fuel. Vehicle engines require substantial modification to run on pure bio-ethanol, and they cannot then run on conventional petrol. Blended motor fuel containing less than about 10 per cent bio-ethanol does not require such modification.

reserves, and it is therefore a net importer of energy – a situation that is likely to remain into the foreseeable future. Factors such as these lend credence to arguments in favour of developing alternative renewable energy supplies (energy security), and when combined with environmental issues such as clean air requirements conspire to make liquid **bio-fuels** (Box 7.5) worthy of much more attention as possible large-scale substitutes for fossil fuels. An area of present and future concern, and fraught with powerful vested-interest lobbies in preserving the dominance of oil as a fuel in motor transport, is the use of bio-fuels as part of the package in transport policies. This issue is dealt with at some length because of the potential for bio-fuels to substitute for petroleum products.

With the increased use of transport in most countries, the level of atmospheric and other environmental pollution is growing. Transport emissions account for a large part of global emissions of air pollutants that contribute to photochemical smogs and pose health hazards. In India, for example, petrol-fuelled vehicles, mainly two- and three-wheeled, account for 85 per cent of the carbon monoxide, and 35–65 per cent of the hydrocarbons from fossil fuels, with diesel vehicles (buses and trucks) being responsible for 90 per cent of urban nitrogen oxides (NO_x) emissions (US Congress Office of Technology Assessment 1992b).

There are a wide range of alternatives to using fossil fuels for motor transport, and the following section explores the potential of liquid bio-fuels.

Liquid bio-fuels for motor transport

Bio-fuels for motor transport are produced from plant and animal matter, wastes and residues, as (1) a group of organic chemicals called esters, which substitute for diesel, and (2) alcohols, such as bio-ethanol, bio-methanol, and a derivative from ethanol as a high-octane blending fuel, called ethyl tertiary-butyl ether (ETBE). Bio-fuels can be pure fuels or blended with conventional petroleum products, typically less than 10 per cent since most engines can run on such mixtures without modification because they act as oxygenates (organic chemicals containing oxygen and having properties as a fuel that are

Plate 30 Crops of oilseed rape in southern England. The oil from these plants can be converted into rape methyl ester (RME) for use as biodiesel, either as a substitute for conventional diesel or as a blend.

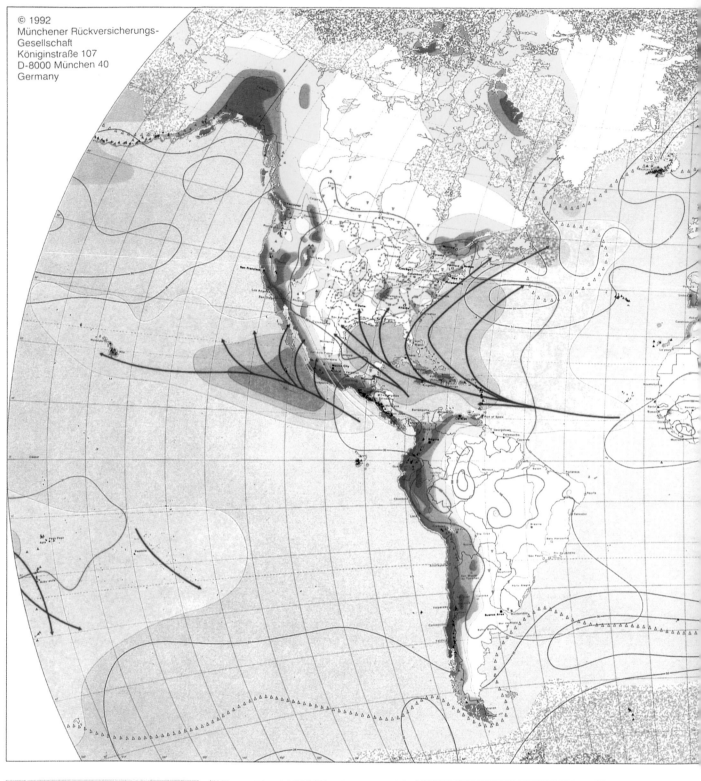

© 1992
Münchener Rückversicherungs-
Gesellschaft
Königinstraße 107
D-8000 München 40
Germany

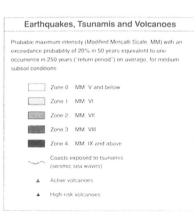

Earthquakes, Tsunamis and Volcanoes

Probable maximum intensity (Modified Mercalli Scale, MM) with an
exceedance probability of 20% in 50 years equivalent to one
occurrence in 250 years ("return period") on average, for medium
subsoil conditions

- Zone 0   MM V and below
- Zone 1   MM VI
- Zone 2   MM VII
- Zone 3   MM VIII
- Zone 4   MM IX and above

〰 Coasts exposed to tsunamis
(seismic sea waves)

▲ Active volcanoes

▲ High-risk volcanoes

Further Natural Hazards, Other

△ △ △ Limit of iceberg drift

▢ Temporary pack ice

▨ Permanent pack ice

▨ Sea fog frequency above 30% (July)

—100— Isoline of thunderstorm days per year

▫	Bombay	more than 1 million inhabitants
○	Chimbote	100 000 to 1 million inhabitants
○	Townsville	less than 100,000 inhabitants
●	Bonn	capital city
▢	**Sydney**	MR office abroad

〰 State borders

(These should not be regarded as official.)

〰 Rivers

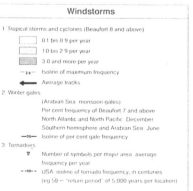

Windstorms

1 Tropical storms and cyclones (Beaufort 8 and above)

- ☐ 0.1 bis 0.9 per year
- ☐ 1.0 bis 2.9 per year
- ▨ 3.0 and more per year

—25— Isoline of maximum frequency

➤ Average tracks

2. Winter gales

(Arabian Sea: monsoon gales)

Per cent frequency of Beaufort 7 and above
North Atlantic and North Pacific: December
Southern hemisphere and Arabian Sea: June

—20— Isoline of per cent gale frequency

3. Tornadoes

▽ Number of symbols per major area: average
frequency per year

—50— USA isoline of tornado frequency, in centuries
(eg 50 = "return period" of 5,000 years per location)

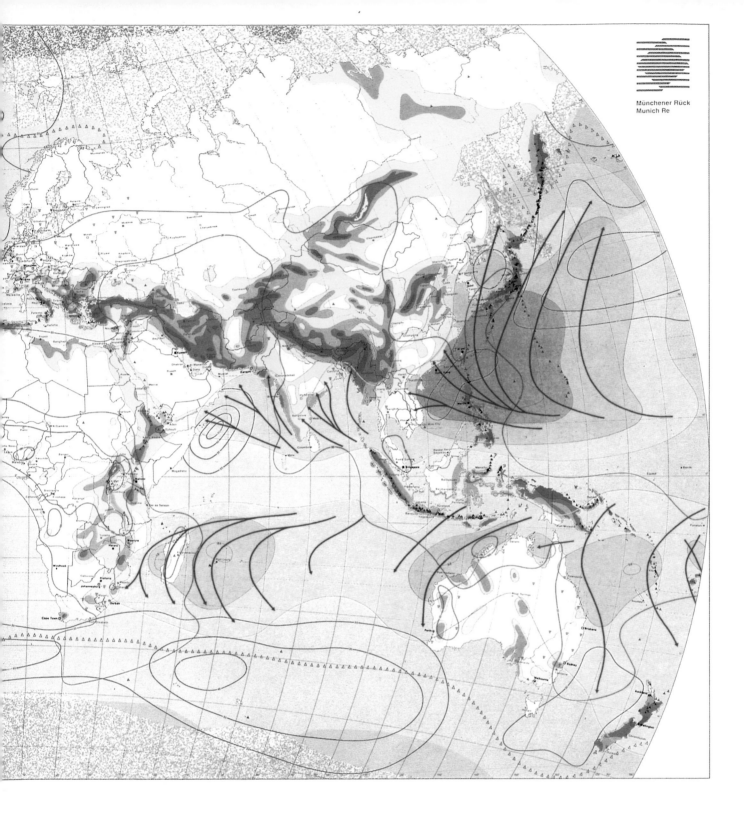

Plate 31 *World map of natural hazards.* Courtesy of Münchener Rück (Munich Re, Germany).

Plate 32 *A pyroclastic cloud produced by the eruption of Mount Pinatubo in the Philippines in June 1991.*
Courtesy of Alberto Garcia/ SABA Katz Pictures.

Plate 33 *Extensive flood damage in Charles County, Missouri, produced when the Mississippi flooded in 1993.*
Courtesy of Comstock.

BOX 7.6 ENERGY BALANCE AND CO$_2$ BALANCE OF BIO-FUELS

Most studies of bio-fuels indicate a positive energy balance (i.e. the energy value of the fuel obtained is greater than the total amount of energy input at all stages in its production), although the magnitude of the energy credits are a common source of dispute. An Energy Technology Support Unit (ETSU) study (Culshaw and Butler 1993) suggests that the energy ratio for bio-diesel production (output:input) ranges between 1.3 and 3.8, depending on the use made of the by-products. In 1991, the British consultancy, Environmental Resources Ltd (ERL), on behalf of the European Fuels Oxygenates Association (EFOA), examined the energy inputs and associated CO$_2$ emissions arising from the production of fertilisers and pesticides, farming operations, crop harvesting, transport, processing and drying, and onward transport of bio-ethanol and by-products. The calculations were undertaken for a modern industrial plant producing 80,000 tonnes of fuel annually from four principal crops that are the main candidates for large-scale bio-ethanol production in Europe, that is, wheat, sugar beet, sweet sorghum and Jerusalem artichoke. The fate of the by-products from bio-ethanol refineries is crucial to the overall energy and CO$_2$ balance. In its report, ERL showed that only the currently uneconomic option of using the by-products from bio-ethanol production as fuels would yield a reasonable energy credit and a significant reduction in CO$_2$ emissions.

---

compatible with petrol, for example alcohols and esters). Energy forestry, including the production of bio-ethanol from woody biomass in coppice crops, is being evaluated, but at current price estimates may be less economic than the arable agricultural biomass. In the longer term, coppiced wood may become more economic as a bio-fuel.

Bio-fuels (Box 7.5) for motor transport are perceived by many people as providing a suite of alternative and renewable motor fuels that cause less environmental pollution (Box 7.6), permit greater energy diversification, decrease the dependence on petroleum products (security of energy supply), and provide a potentially profitable use for fallow agricultural land, for example set-aside land in Europe under the EC Common Agricultural Policy (CAP). The development of bio-fuels in motor transport may depend on favourable tax regimes (just as the launch of lead-free petrol required tax incentives), probably combined with total or partial tax exemptions for pilot studies, something that is permitted under the EC Council Directive on the harmonisation of excise duties, but taken up by few member states. Given that oil prices are in real terms close to their lowest ever (Figure 7.9), the marginal to slightly uneconomic aspects of bio-fuels for use in motor transport may change by the early part of the twenty-first century.

In Europe, Germany, Switzerland, France and Britain are the biggest cultivators of **oilseed rape**. The Italian company Novamont produces and supplies **rape methyl ether** (**RME**) to seventeen Italian cities, the Berlin taxi association, all the taxis in Bologna, buses in Zurich, and the ferry on Lake Como. In Britain, RME-based motor fuel has been used in a pilot study for three public transport buses in Reading. RME also is used in niche markets, for example as a lubricating oil on North Sea oil rigs.

In 1992, Novamont produced about 50,000 tonnes of RME, and in 1993 about 100,000 tonnes using its new 60,000-tonne capacity plant in Livorno, Italy, which is Europe's first and largest plant solely for the purpose of bio-diesel production. Previously, the main customer for Novamont's RME had been the Italian domestic heating market due to environmental considerations, that is the need to cut SO$_2$ emissions drastically in cities such as Milan (where La Scala is heated by bio-diesel). In France, near Compiegne, a 20,000 tonnes yr^{-1} factory has been built. In Austria, where Vogel and Noot are the main manufacturers of equipment for bio-diesel plants, in 1991 almost 20 per cent of the 50,000 ha of rape cultivated was used to produce RME, and a group of agricultural co-operatives ensures its distribution to more than fifty fuel stations.

In Europe, trials to evaluate bio-ethanol as a fuel have been conducted in Germany, Italy and Sweden, with a small public distribution of blended bio-ethanol in France. The USA has major transport bio-fuels programmes using bio-ethanol, such as the '**gasohol**' programme, supported for energy diversification and environmental reasons. Ethanol production in the USA began because of the Cuban embargo on sugar exports to the USA, which led to a market crisis and the decision to expand US domestic production of sugar substantially. This led to over-production, and the conversion of the surplus sugar into energy. In the mid-western USA, bio-ethanol has been used for about the past ten years as a 10 per cent blend in gasoline (gasohol). Brazil is the biggest and most successful user of bio-ethanol for motor fuel. Under the fourteen-year 'proalcohol' initiative in Brazil, set up in response to the world oil crisis in the early to mid-1970s, about one-third of Brazil's twelve million cars used pure ethanol from sugar cane by the early 1990s, with the rest using a

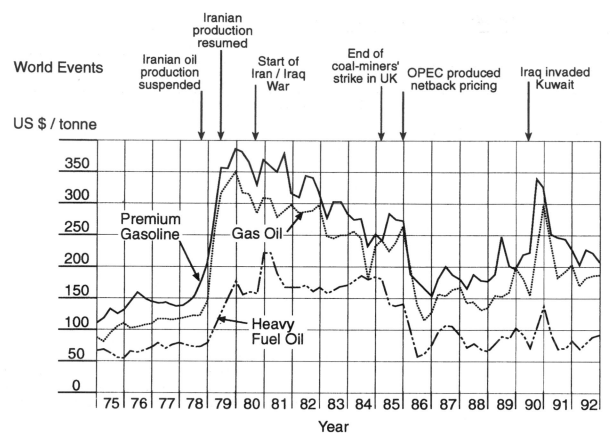

Figure 7.9 *Rotterdam product prices for premium gasoline, gas oil and heavy fuel oil between 1975 and 1992. Source: British Petroleum (1993).*

20 per cent blend. The dry pulp left over from crushing the sugar cane, or 'bagasse', can be used as an animal feed or burned in high-pressure steam turbines to generate electricity, and the potassium-rich liquid waste produced in the distillation of ethanol provides a good fertiliser. Brazil now produces about 10 Mt of bio-ethanol annually, giving major cost reductions through economies of scale, but bio-ethanol still costs about US$40 per barrel to produce.

Despite some commonly held beliefs about the health risks to populations living near areas of extensive oilseed rape cultivation, there is no scientific/medical evidence to support such a contention. A British study by the Department of Environmental and Occupational Medicine, Aberdeen University, Scotland, compared 2,000 people in rural areas, half of whom live in areas extensively cultivated for oilseed rape, but otherwise there were no known differences. The results show no statistical difference in allergy and asthma complaints between the populations, implying that people living in areas of extensive cultivation of oilseed rape are no more prone to asthma

and other allergies than the rural population living outside such areas. There is, however, a 20 per cent greater incidence of complaints associated with allergic reactions to pollen in rural compared with other areas, but this is not specifically related to oilseed rape.

Under the CAP reform agreement reached by the EC Council of Ministers in May 1992, growers of cereals, oilseeds and protein crops (other than those categorised as small producers) were required to set aside 15 per cent of the land on which they were claiming arable support in relation to the 1993 harvest. The change from tonnage- to land-area-based subsidies to farmers for the cultivation of oilseed rape may alter farming practice, because prior to the CAP the higher-yield winter varieties of rape (average 3.2 tonnes ha[-1]) were more popular. Although the spring crop gives lower yields (average 2.2 tonnes ha[-1]), its cultivation requires fewer inputs and, therefore, is cheaper. Rape tends to be grown on heavy soils more suited to winter cereals, which may be unsuitable for spring varieties. If up to 5 per cent of EU gasoline were to come from bio-fuels,

Table 7.9 *Bio-fuel emissions relative to conventional fuels.*

| Pollutant | Main effect | Bio-diesel | Blend with petrol | |
			Bio-ethanol 10%	Bio-ethanol 95%
Sulphur dioxide (SO$_2$)	acid rain	almost 100% less	10% less	60–80% less
Nitrogen oxides (NO$_x$)	acid rain ground-level ozone	up to 20% increase	2% increase	no difference
Unburnt hydrocarbons (HC)	respiratory diseases possible carcinogens	typically, 10–40% reduction	6% increase	up to 15% less
Carbon monoxide (CO)	respiratory diseases	up to 30% less	4% less	up to 8% increase
Particulates	respiratory diseases	up to 40% less	no significant difference	up to 5% increase

Source: [modified from] UK Parliamentary Office of Science and Technology, Briefing Note 41, March 1993.

then at 1992 figures between two and five times the current area of set-aside land, depending on the selected crop, would need to be cultivated for bio-fuel production. In the EU, blended bio-diesel may be economically more viable, at least in the short term, compared with bio-ethanol, principally because its use would involve the least disruption to the present energy production and distribution systems, and best fit present EU farming practices and policies.

More employment in rural areas and associated industries could be created by developing a bio-fuels market, but with job losses in the fossil fuel and allied chemicals industries. Due to the relatively small volume of bio-diesel marketed within the EU, RME producers and suppliers such as Novamont have targeted high-volume users in niche markets, such as public transport bus and taxi operators. Other potential niche markets are where the environmental advantages of the greater biodegradability of bio-diesel can be exploited, as in inland waterways and urban use. The economies of scale are helped by the simplification of the distribution system for bio-diesel in being delivered direct to the high-volume user.

Oil company studies suggest that liquid bio-fuels will only marginally improve EU energy security, with the 5 per cent of gasoline level given over to bio-fuels accounting for a 1.5 per cent elimination of oil imports (equivalent to 0.5 per cent of EU primary energy consumption). The oil companies and closely allied chemicals industries believe that bio-fuels must compete in the present market as pure fuels or as octane enhancers with alternative oxygenates.

A study by the Department of Land Economy at Cambridge University, UK, predicts that within the EU by the year 2000, 1–1.5 million ha of land will be removed from food production, and that this figure will increase to an estimated 5–5.5 million ha by 2010. If the consumption of diesel remains stable and all this land were used for the cultivation of oilseed rape for bio-diesel under a rotational set-aside agricultural system, then by the year 2000 an estimated 10–15 per cent (1–1.5 Mt) of the diesel market could be satisfied by this fuel, growing to 51–56 per cent (5.1–5.6 Mt) by 2010. If such projections were to prove accurate, then it is not surprising that the oil companies and allied chemical industries feel a certain amount of discomfiture at the prospect of the increased use of bio-fuels in motor transport.

Environmentalists, including the producers of bio-fuels, would argue that even though the net volume of bio-fuels in motor transport is small, and is likely to remain so in the near future, the increased use of bio-fuels can help to achieve specified emission standards at stabilised levels and, therefore, warrant government support through encouraging further R&D and additional pilot studies (possibly with tax concessions/exemptions). Ultimately, the use of bio-fuels in motor transport hinges around any perceived environmental benefits (Table 7.9) and increased energy security over and above shorter-term economic factors.

The use of bio-fuels as alternatives to conventional petroleum products requires a substantial shift in public attitudes, probably towards an acceptance of paying a premium for motor bio-fuels, and any new engine technology. Such changes would require a change of the fuel market through subsidies and tax incentives, and possibly a pollution or carbon tax. It could be argued that since energy markets are already regulated through government pricing policies, a shift towards energy being sold at prices that fully cover all costs – including environmental costs (often referred to as internalising the externalities) – may lead to bio-fuels becoming more competitively priced compared with conventional petroleum products, at least as niche-market fuels. In the USA, tax concessions/exemptions give a subsidy of about 60 cents per gallon of bio-fuel gasoline. In California, exhaust emission control standards are more stringent than in any other state, and it is this factor rather than purely economic considerations that is stimulating the expansion of bio-fuels in motor transport.

Energy efficiency

The fundamental problem with the consumption of non-renewable energy resources is that individuals and countries simply use too much. Human activities are invariably wasteful of valuable energy. Present attitudes tend to accept wastefulness and encourage a business-as-usual approach. Energy conservation is popular as a buzz phrase but not in action. The growing public concern over poor air quality and the emissions of greenhouse gases associated with the production of electricity (see Table 7.2), however, have led to governments, particularly in the developed countries, taking a more serious view of these issues, including proposals for a carbon or carbon/energy tax in the EU (Box 7.7).

Carbon taxes provide a means of reducing greenhouse gas emissions, while also raising state revenue. Such taxes help to ensure that environmental goals can be met efficiently and equitably, while maintaining an environmentally acceptable level of economic activity. A tax on the polluter commensurate with the cost of removing the pollution encourages cleaner fuel technologies and acts as a brake on the profligate use of fossil fuels. The revenue from such taxes can be redirected to reduce other forms of pollution and/or invested in developing new technologies. In the USA, since 1991, proposals for a carbon tax have been promoted in the Maryland, Minnesota and California legislatures. The

---

BOX 7.7 EU CARBON OR ENERGY/CARBON TAX

A measure under active consideration by the European Commission to reduce the emissions of the main greenhouse gas, carbon dioxide, is the proposed carbon or carbon/energy tax. This proposed tax would be calculated on the basis of 50 per cent related to the carbon content of the fuels, and 50 per cent to their energy content, increasing gradually over seven years from US$3 on each barrel of oil equivalent to US$10 per barrel of oil equivalent by the year 2000; these figures assumed that the tax came into effect in 1993/4. The tax would reduce energy consumption in general, but particularly with regard to the use of coal (House of Commons Trade and Industry Committee 1993).

From its inception, the proposal was consistently supported by Germany, the Netherlands, Belgium, Denmark, Italy and Luxembourg. The other member states were more equivocal and/or opposed the imposition of this tax. The poorer EU countries, Spain, Portugal, Greece and Ireland, claimed it would hinder their economic growth. France claimed the tax would be unfair because much of its electricity generation is from nuclear power and, therefore, its CO_2 emissions are low anyway. The UK claims that its domestic taxation policy (e.g. increased value-added tax, VAT, on domestic fuel and a 10 per cent increase in the duty on fuels, imposed in the 1993 March Finance Bill (the Budget) represents an adequate response to seeking to control future CO_2 emissions.

It has been argued that even the possibility of a carbon/energy tax may already have had an impact, because electricity-generating companies now perceive coal-fired power stations as having a much higher risk attached to them. Clearly, such increased risk encourages the use of other fuels, for example natural gas. There are those who argue that a carbon/energy tax will help promote cost-effective energy investment, although energy-intensive industries may well be damaged. Countries such as the UK would probably find that a carbon/energy tax would have only a small impact on its CO_2 emissions, with EU estimates of UK gross energy consumption in the year 2000 being only 2.5 per cent lower and fossil fuel consumption 3.5 per cent lower compared with a situation without the proposed tax (*ibid.*).

A strong argument in favour of a carbon/energy tax is that it would curb CO_2 emissions from the transport sector, which is the fastest growing area of fossil fuel consumption. The transport sector, however, is notoriously unreactive to small changes in price, therefore there is perhaps little reason to be optimistic that modest taxation changes would have much of an impact. Although the impact of a carbon/energy tax would probably be small in reducing CO_2 emissions, because gross energy demand appears to react in only a modest way to fiscal measures, it may well be that nations can achieve any set emissions targets only through a mixture of both large- and small-impact measures.

Center for Global Change, University of Maryland, has developed a state carbon tax model that provides an assessment of the role of a carbon tax for individual states in the USA. For a given level of carbon tax rate the computer model evaluates the strategy, which will depend upon a particular state's legislation and requirements.

International action to reduce greenhouse gas emissions and generally improve air quality is dealt with in more detail in Chapters 3 and 4.

The 1973 oil price rises forced energy planners to rethink their strategies. Unfortunately, from the standpoint of energy resource management, the 1980s saw a return to relatively cheap fossil fuels and so discouraged, on purely economic considerations, further improvements in energy efficiency. The International Energy Agency (IEA) has estimated that if available, economically viable ways of conserving energy were implemented, by the turn of the century energy efficiency could be 30 per cent greater than the present level.

The need to conserve both existing reserves of fossil fuels and the growing awareness of environmental impact have led to increased energy-efficient technologies (see Table 7.10 for comparison of energy efficiencies of fossil-fuel power plant systems). Necessity has been the mother of invention. Energy-efficient buildings that are better insulated, optimise available natural light and are constructed from better materials, new types of combustion furnaces (e.g. condensing furnaces that require 28 per cent less fuel to produce an equivalent amount of energy compared with conventional furnaces), the cogeneration of power (e.g. heat and electricity from the same process) and a greater reliance on public transport are but a few of the ways in which people can use energy resources more sensibly.

Cars and light commercial vehicles consume one out of every three barrels of oil produced and, in the USA, contribute 15 per cent of the total CO_2 emissions into the atmosphere. Attempts both to reduce the polluting capability of vehicles and conserve energy (e.g. through the use of unleaded petrol and better engine design) are becoming increasingly popular throughout the developed world. Electric cars and trams provide cleaner, and less noisy, forms of public transport. In many European cities, such as Amsterdam, Oslo, Geneva and Basel, the trams are a pleasant feature of the urban landscape.

Most existing housing has not been designed with energy efficiency in mind. Energy-efficient homes are less harmful to the environment, save money and can be much more pleasant to live in, since if they are

Table 7.10 *Energy efficiencies of fossil fuel power plant systems.*

Power plant system	Current efficiency %
Current US power plants (overall average)	33
Oil and natural gas combustion	
Aircraft derivative turbines	40
Combined-cycle systems	47
Coal combustion	
Atmospheric fluidised bed combustion	38
Pressurised fluidised bed combustion (using a combined cycle)	42–44
Integrated gasification combined cycle	43
Fuel cells	40–60
Cogeneration	up to 85

Source: World Resources 1990–91, A Report by the World Resources Institute in collaboration with the UN Environment Programme and UN Development Programme (1990).

well designed they will conserve heat in winter and be cooler in summer.

There are many ways of making homes more energy efficient (see Oliver 1991). Using energy-efficient low-wattage light bulbs, turning lights off when leaving rooms empty, using better insulation, double-glazing, and draught-proofing around doors and windows are but a few of the ways. New homes should be constructed to minimum standards. Windows facing south and east are the most efficient at receiving the Sun's rays – and are also brighter. Depending upon the latitude and climate of a country, its housing should be designed so that windows preferentially face the appropriate directions and are of a suitable size.

In countries with a high amount of sunshine hours, solar panels can supplement, or provide all, domestic energy requirements. **Superinsulation**, where buildings are sealed against draughts with a ventilation system that is controllable to allow fresh air in winter, saves significant amounts of energy. In Sweden, legislation now decrees that all new houses must have less than three changes of air per hour. Superinsulated homes may have fewer than one air change per hour, compared with many homes where there are typically far more. In another attempt at being more environmentally conscious, CFC-free plastic foams are available for cavity filling.

Some countries, such as the UK, have appeared to pursue a policy of increasing fuel and power prices, imposing value-added tax (VAT) in 1993, thereby leaving the consumer to decide priorities in terms of both expenditure and improving energy efficiency. An alternative means of encouraging energy efficiency and energy conservation is by using **demand-side** management (DSM), first developed in California in 1975 as a response to the steep price rises following the first oil crisis in the early 1970s (see Cragg 1993).

DSM is advocated by environmental groups and energy economists in many countries because it makes electricity-supply companies into suppliers of energy services. In DSM, the supplier tries to reduce demand for energy rather than building a new power plant to meet the increased demand. DSM is a cheaper option than constructing new power stations at considerable cost, because the electricity supplier distributes energy-efficient consumer goods, such as thermal insulation and energy-efficient light bulbs, and could offer, for example, free energy audits to large energy consumers. At least part, if not the entire cost of energy-efficiency and energy-conservation measures can be passed on to the consumer, but this should be much less than that incurred by building new power plants. This has the added advantage that the output of pollutants from power stations is reduced. Examples of DSM in action include South California Edison, in 1988, providing 450,000 energy-efficient light bulbs to low-income customers (thereby saving an estimated 8 MW of generating capacity); in 1987, fifty-nine US power companies offered rebates to customers if they purchased more energy-efficient refrigerators, and the Tennessee Valley Authority provided US$250 million as interest-free loans for improving domestic insulation. In Washington DC, the environmental Worldwatch Institute calculated that by 1988, US measures for energy conservation instigated by the six largest power companies had saved 7.24 GW of electricity-generating capacity. Also, a 1992 report by the Boston-based Goodman Group, commissioned by Greenpeace, estimated that, besides creating 80,000 new jobs to date, for every US$1 spent on DSM, US$1.50–1.75 is saved on constructing new power plants. The Goodman report also emphasises the enormous potential of DSM, since currently only about 1 per cent of the turnover of power companies is invested in energy efficiency. DSM should be an integral part of both developed and developing countries' energy policy.

Finally, an exciting and challenging future means of energy efficiency and conservation will be through the use of so-called **smart materials**, made of polymers (very long chain molecules) that can actually sense and respond to external stimuli by changing density, colour and other physical properties. Smart materials behave like organisms, which is in many respects the antithesis of conventional building materials such as concrete. Smart material technologies are currently being developed, with plans on the drawing table for a smart building in Tokyo. Such smart buildings will use much less energy than those constructed from conventional materials, because the outer skin will be designed to respond to changes in ambient atmospheric and interior temperature and pressure, and to be able to respond, as required, to computer-input commands to increase insulation, warm or cool parts of, or the entire, building, etc. In other words, smart materials will be alive and aware of their environment and, therefore, able to capitalise on changes in the environment.

Vehicle emissions

Motor vehicles have a major impact on the environment (Box 7.8). Exhaust emissions, emissions control technology and legislation are major environmental issues associated with bio-fuels in motor transport.

BOX 7.8 BIO-FUELS AND MOTOR TRANSPORT EMISSIONS

Bio-fuels may provide a contribution to reducing the emission of greenhouse gases, and gases which contribute to photochemical smogs and acid rain. Amongst these, bio-ethanol and bio-diesel are considered here. The high volatility of ethanol produces evaporative emissions 5–220 per cent above those of conventional petrol, the effect being moderated by the presence of a co-solvent (e.g. tertiary butyl ether) used to increase the solubility of ethanol and water in petrol, which reduces the possibility of phase separation. Compared with conventional diesel, bio-ethanol combustion gives reduced HC and CO emissions, variable higher or lower NO_x, and increased emissions of aldehydes (which are toxic and contribute to the development of smog and low-level ozone) and esters. Particulates (mainly carbon particles) are not significantly different, but other emissions, such as the carcinogenic organic chemicals called alkanes, alkenes, ketones and aldehydes, are worse, but these increases are offset against the greater emissions of an even more carcinogenic group of organic chemicals from mineral diesel called polycyclicaromatic carbon compounds (PACs). Additionally RME can cause problems with lubricants, exhaust catalyst durability, and poor fuel consumption. NO_x levels can be reduced slightly by altering engine configuration (e.g. timing) and ignition temperature, at a trade-off of increased particulates (unburnt and partially burnt HC) but still well below those from mineral diesel. RME combustion also produces exhaust fumes that many regard as having an unpleasant odour, but this can be rectified with additives in the bio-diesel. More bio-diesel is retained in the engine lubricating oil because it is less volatile than mineral diesel, a factor that can lead to greater emissions of particulates from the bio-diesel when it is ignited.

Catalysts are being used to help meet the increasingly stringent emissions control regulations. It is technically easier, by the addition of oxidative catalysts to the fuel, to remove the increased emissions from the combustion of RME of mainly single carbon-chain aldehydes, alkenes and ketones than the PACs that occur in larger volumes in mineral diesel or petrol. Adding such catalysts (e.g. platinum–rhodium) to bio-fuels entails a relatively small increase in cost.

Many of the EU countries are committed to stabilising emissions of CO_2 at 1990 levels by the year 2000. In Britain, an Energy Technology Support Unit (ETSU) study (Culshaw and Butler 1993) calculated that for each litre of bio-diesel used, 1.5 kg of CO_2 is saved (i.e. CO_2 emitted from the substituted fuel minus CO_2 emitted during bio-diesel production). A 1993 report by Germany's Federal Environment Office (UBA) shows that if RME displaced 400,000 tonnes yr^{-1} of fossil fuels, equivalent to 640,000 tonnes of CO_2, it would account for 0.5–0.7 per cent of Germany's total emissions, a small part of the 25 per cent target reduction. Also, the UBA report shows the consumption of 1 kg of mineral diesel fuel, including the preliminary chain, releases 3.40–3.49 kg CO_2, together with 0.9 kg of emissions from other climate-relevant trace gases expressed in CO_2 equivalents, giving a total of 3.5–3.6 kg CO_2 equivalent per kg for mineral diesel, compared with 1.9–2.3 kg CO_2 kg^{-1} for substituted bio-diesel, which takes into account the production (agriculture), processing (rape seed oil extraction and esterification) and transport; if by-products are given CO_2 emission credits, e.g. glycerine and oilseed rape groats, then the CO_2 balance is 1.2 kg CO_2 kg^{-1} for rape seed oil and 0.8–1.4 kg CO_2 kg^{-1} for RME, depending upon the recoverability of the glycerine, giving 3.5–4.0 kg CO_2 kg^{-1} for rape seed oil and 2.7–4.4 kg CO_2 for RME. For many countries, at current and realistically projected levels of oilseed rape cultivation, RME will make only a small contribution to the overall target reduction levels for CO_2.

Emissions include evaporative hydrocarbons, occurring throughout the fuel system (dependent upon the volatility characteristics of the fuel), and combustion emissions, i.e. unburnt hydrocarbons (HC), carbon monoxide (CO) and mixed nitrogen oxides (NO_x), the latter contributing to acid rain. Exhaust combustion emissions include particulates such as carbon particles, which can be coated with various hydrocarbons. There are other emissions, such as formaldehyde, which generally occurs as a trace gas and can be particularly toxic and carcinogenic.

Exhaust emissions standards between countries are very variable, with the Californian legislation and planned standards being, as has always been the case in the USA, the most stringent (Table 7.11A). California has introduced a plan for the progressive reduction of vehicular emissions to help the state achieve national air quality standards by 2010. The plan includes the progressive introduction of transitional low-emission vehicles (TLEV), low-emission vehicles (LEV), and eventually ultra low-emission vehicles (ULEV) and zero-emission vehicles (ZEV). ZEVs are vehicles that produce zero exhaust and evaporative emissions of any pollutant, with California committed to ZEVs accounting for a minimum of 2 per cent of sales by 1998 and 10 per cent by 2003 (Table 7.11B). In the USA, Title 2 (relating to motor vehicles, fuels and their emissions) of the revised Clean Air Act, signed by President Bush in November 1989, following protracted negotiations between the House of Representatives and the Senate, included:

- the imposition of tighter exhaust emissions standards;
- the establishment of compliance testing and maintenance programmes;
- the establishment of a reformulated gasoline programme;
- legislation relating to clean fuel vehicles, which could lead to the introduction of alternative fuels;
- legislation covering operators of vehicle fleets in areas where there are specific air quality problems; and
- reaffirmation of the rights of individual states with particular air quality problems to set more severe emissions standards, but which must not exceed the Californian standards, a restriction imposed so that vehicle manufacturers would not have to produce customised vehicles for each state, but rather two models, one to meet federal standards and the other to comply with the laws of California.

The inconclusive, and often apparently contradictory results of the exhaust emissions tests are due to different engine performance characteristics and test cycles used, causing greater differences in rates of emissions than may be attributed to the use of the bio-fuel itself. Without similar test data sets from different laboratories, available emissions results suffer from limitations due to non-comparability. In Europe, a standard emissions test involves sampling emissions at various steady-state engine speeds, whereas a more representative measure of the inservice situation may

Table 7.11 (A) Vehicle exhaust emissions legislation in California; (B) Planned emissions control standards in California.

(A)

		1991/92	1993	TLEV	LEV	ULEV	ZEV
HC	g/mi	0.39	0.25	0.125	0.075	0.04	0
	g/km	0.24	0.15	0.078	0.047	0.025	
CO	g/mi	7.0	3.4	3.4	3.4	1.7	0
	g/km	4.375	2.13	2.13	2.13	1.06	
NO$_x$	g/mi	0.4	0.4	0.4	0.2	0.2	0
	g/km	0.25	0.25	0.25	0.125	0.125	
HCHO	g/mi		0.015	0.015	0.015	0.008	0
	g/km		0.0093	0.0093	0.0093	0.0093	

(B)

	1991/92 (%)	1993 (%)	TLEV (%)	LEV (%)	ULEV (%)	ZEV (%)	NMOG (g/mi) for fleet average
MY 91/92	100						0.390
MY 1993	60	40					0.334
MY 1994	10	80	10				0.250
MY 1995		85	15				0.231
MY 1996		80	20				0.225
MY 1997		73		25	2		0.202
MY 1998		48		48	2	2	0.157
MY 1999		23		73	2	2	0.113
MY 2000				96	2	2	0.073
MY 2001				90	5	5	0.070
MY 2002				85	10	5	0.068
MY 2003				75	15	10	0.062

TLEV = Transitional low-emission vehicle
ULEV = Ultra low-emission vehicle
NMOG = Non-methane organic gases
g/mi = grams per mile
LEV = Low-emission vehicle
ZEV = Zero-emission vehicle
HCHO = Formaldehyde
g/km = grams per kilometre
Source: US Office of Technology Assessment

be the US transient testing, since the latter includes conditions that better approximate to acceleration and deceleration under varying engine conditions. It is important that emissions tests are undertaken on in-service vehicle engines, which requires more pilot studies.

Bio-fuel combustion produces insignificant amounts of sulphur compounds such as SO_2 (because most bio-fuels contain much lower amounts of sulphur compounds than petroleum products), which contributes to acid rain.

The combustion of bio-fuels is often referred to as 'CO$_2$ neutral' because potentially as much CO_2 is sequestered from the atmosphere during plant growth as is released during combustion. The CO_2 budget, however, must account for the energy expended in activities associated with planting, cultivating, harvesting, transportation and processing, which, if generated from fossil fuels, will have a net contributory effect on atmospheric CO_2 levels. Even

allowing for the use of substituted bio-fuel energy at all production stages, the use of pesticides and fertilisers manufactured from petroleum products, and other energy inputs that use fossil fuels, may result in a net input of CO_2 into the atmosphere. Increased use of pesticides and fertilisers could further reduce any energy and CO_2 credits. To achieve an overall reduction in atmospheric CO_2 levels relative to mineral (fossil) fuels, the waste vegetable matter from bio-fuel cultivation must be burned as an energy source rather than used as an animal feedstuff, something that is less economically attractive. An argument expressed most strongly by the oil companies is that CO_2 emission targets are better met through technical improvements to the diesel engine aimed at increasing engine efficiency and reducing fuel consumption, and that the use of bio-fuels as alternatives to fossil fuels provides, at best, a very small reduction in atmospheric CO_2 levels.

Table 7.12 *Comparison of environmental impact of electric power generation technologies.*

Emissions of pollutants from electric power generation: the total fuel cycle[a] *(tons per gigawatt hour)*

Energy source	CO_2	NO_2	SO_2	TSP	CO	HC	Nuclear waste	Total
Conventional coal	1058.2	2.986	2.971	1.626	0.267	0.102	NA	1066.1
Natural gas IGCC	824.0	0.251	0.336	1.176	TR	TR	NA	825.8
Nuclear	8.6	0.034	0.029	0.003	0.018	0.001	3.641	12.3
Photo-voltaic	5.9	0.008	0.023	0.017	0.003	0.002	NA	5.9
Biomass[b]	0*	0.614	0.154	0.512	11.361	0.768	NA	13.4
Geothermal	56.8	TR	TR	TR	TR	TR	NA	56.8
Wind	7.4	TR	TR	TR	TR	TR	NA	7.4
Solar thermal	3.6	TR	TR	TR	TR	TR	NA	3.6
Hydro-power	6.6	TR	TR	TR	TR	TR	NA	6.6

*With biomass fuel regrowth programme.
TSP: Total suspended particles.
NA: Not applicable.
TR: Trace elements.
HC: Hydrocarbons.
IGCC: Integrated gas turbine combined cycle.
Note: The total fuel cycle includes resource fuel extraction, facility construction and plant operation.
[a] Meridian Corporation, *Energy System Emissions and Material Requirements*, pp. 25–29.
[b] Carbon dioxide data adapted by the Council for Renewable Energy Education from Martin, *Environmental Emissions from Energy Technology Systems: The Total Fuel Cycle*, US Department of Energy, Washington DC, spring 1989, p. 5. Other emissions data from *Energy Technologies and the Environment*, *Environmental Information Handbook*. Office of Environmental Analysis, US Department of Energy, October 1988, pp. 333–334.
Source: Council for Renewable Energy Education.

Endpiece

Energy policy, realistically, is not formulated in a political vacuum. Decisions about the energy resources used by a nation depend upon many complex economic, social, political and military factors. A nation's accessible resources play a major role in determining energy policy.

The geographic distribution of energy resources is extremely unequal. Countries such as Japan have virtually no commercially viable large reserves of fossil fuels, and therefore rely upon importing almost all their energy resources. This lack of natural energy resources partly explains the interest in areas such as Antarctica, where the carve-up of territory for economic exploration and exploitation has fortunately been frustrated by the Antarctic Treaty Environmental Protocol 1991, which stopped the destruction of this major natural wilderness (see Chapter 10). Other nations such as the former USSR and South Africa are amongst the world's richest lands in energy resources. Estimates of total versus recoverable reserves are subject to continual re-appraisal, with the result that downwardly revised figures can lead to more aggressive economic and military policies. Examples of downwardly revised estimates include many of the non-OPEC nations such as the former USSR, which currently consumes

roughly 15 per cent of global oil production and, because of its rapidly growing consumption, needs to increase production and oil imports. In 1987, the USA imported US$40 billion of oil, equivalent to about one-third of the nation's trading deficit, and in the same year the Pentagon spent US$15 billion on protecting these supplies of oil.

OPEC controls about 75 per cent of the known crude oil reserves. The dependence on Middle East countries for oil by the countries of the industrialised world means that their economies are directly affected by the pricing of crude oil by OPEC. About 70 per cent of natural gas reserves are held by the former USSR and the Middle East. Of the calculated 950 billion tonnes of global coal reserves, estimated to last another 275 years if present production rates are maintained (Kumar *et al.* 1987), the USA and former USSR each controls 25 per cent. Much of the remaining reserves are in Europe, Africa, Australia and Asia, particularly China. Approximately 20–30 per cent of the energy requirements of industrialised countries come from coal. Almost 75 per cent of China's energy comes from coal.

The environmental lobby against dirty fuels that pollute the atmosphere, produce greenhouse gases and cause acidic deposition has stimulated the search for cleaner alternatives (see Table 7.12). Nuclear fuel has been seen by many people as a viable alternative, though environmentalists express concern about this because of the nuclear waste and ever-present dangers of another, perhaps far more serious, Chernobyl or Three Mile Island disaster. Today, nuclear energy supplies roughly 17 per cent of the world's electricity, with France as the leading country, obtaining about 70 per cent of its electricity by this method, and Japan about 35 per cent.

Although nuclear energy does not produce green-house gases or acidic deposition, the cost of nuclear reactors is very high and public concern over the disposal of nuclear waste and nuclear accidents conspires to make it less attractive as a political option. These perceived disadvantages have meant that the USA has not ordered a new nuclear plant since 1978. Most operational nuclear reactors are of the light-water type, and the pro-nuclear lobby has suggested that advanced reactor designs could restore public confidence in such energy.

The funding of research and development into energy resources can help to predetermine the outcome of decisions to adopt energy policies and to invest further in certain directions. In Britain, for example, there is an enormous difference in the R&D money committed to nuclear energy versus renew-ables, let alone other resources. In the period 1990–1991, the R&D breakdown in millions of pounds was nuclear fast-breeder reactors, 84.6; nuclear fusion, 26.9; other nuclear, 17.5; coal-based energy, 7.6; oil and gas, 10.7; energy-efficient R&D, 11; renewables, 20.3 (Milne 1991). With about two thirds of the total financial support in R&D directed towards nuclear energy, a lot of influ-ential people have made a large commitment to this fuel source, and vested interests would find it corre-spondingly hard to change direction.

As for energy efficiency, many of the measures to utilise renewable energy resources and construct more energy-efficient buildings incur additional costs to society and the individual, both in financial terms and in the way in which people live their lives. There is a price to pay for everything.

It should be the responsibility of governments to encourage a more prudent use of energy. They should provide attractive subsidies to companies that wish to construct power stations or energy farms that rely upon renewable energy, something that the UK non-fossil fuel obligation (NFFO) does. Builders who are prepared to construct houses and other build-ings to improved standards should be given similar incentives. Indeed, in the USA in the late 1970s, relatively generous research budgets were allocated to renewable energy resources, replaced by tax concession schemes in the 1980s. A spin-off from these policies was the 'Californian wind rush', in which about 16,000 privately owned wind turbines were installed, mainly as wind farms, with a gener-ating capacity of 1,500 MW. After the experimental period in which many wind turbines revealed design faults and were poorly sited, the tax concessions were withdrawn in 1985 and the industry contracted. Despite the juddering start to renewable energy resources in many countries, the only way to over-come design flaws and improve energy efficiency in the new breed of power plants, governments must be prepared to invest heavily in these alternative tech-nologies. In Britain, the government's advisory body, the Energy Technology Support Unit (ETSU), has stated that it is feasible for half the nation's energy requirements to be met from renewable energy resources by some time early in the next century. So it is technologically possible, but the big question is whether or not the will is there to achieve this goal. Without real government support for a programme of moving over to a much greater emphasis on renewable energy, societies may be even less prepared for forthcoming energy crises.

The EU is currently undertaking a five-year (1993–1997) R&D plan to promote renewable energy sources within the community, called the ALTENER, and expected to cost 125 million ECU. One of the European Commission proposals is to reduce CO_2 emissions drastically by 2005, as a result of which it will be necessary to increase the contri-bution made by renewable energy resources to meet EU energy demands, triple the amount of electricity produced by renewables, and use more bio-fuels (Table 7.13). The ALTENER plan makes provision to fund four categories of projects concerning:

1 technical studies and evaluations to set forth tech-nical specifications and standards;
2 measures provided for by the member states to extend or create infrastructure for renewable ener-gies, acquire equipment and organise professional training;
3 measures to establish an information network enabling better co-ordination among the activities of the member states; and
4 plan action in the sector of biomass, especially bio-fuels.

It is through the prescient action of individual countries, or groups of nations – as in the case of the EU – that provide substantial funds for renewables that any potential energy crisis in the twenty-first cen-tury can be averted. Naturally, the fossil fuel and allied chemical industries are deploying every commercial and environmental reason to at least slow the wind of change that is blowing through the energy markets. Such delaying tactics will conserve company profits and jobs in those industries over the short term, but as pollution standards and control measures become ever more stringent, and as the price of oil goes up from its present historic low, a greater reliance on renewables will inevitably come about and with it a sharp decline in the use of fossil fuels. At such a stage, nations that have already developed a viable policy of energy diversification and energy security will be best placed to survive (Box 7.9).

On a final but by no means unimportant note, military expenditure and the military policy of the world's leading nations is strongly governed by the need to maintain supplies of energy resources, particularly oil from the Middle East. It is for reasons such as these that any armed conflict in the Middle East, with the associated economic implications, creates such international tension with the potential for local wars to become 'flash points' for the rapid escalation into much larger-scale international conflagrations. Arguably, the Gulf War in 1991 was actually solely about preserving American and European interests in Middle East oil supplies.

Table 7.13 *ALTENER new and renewable energy targets for 2005 in the European Union.*

	Production 1991			Targets 2005		
	GW	TWh	Mtoe	GW	TWh	Mtoe
Electricity						
Small hydro (<10 MW)	5.0	15.0	1.3	10.0	30.0	2.6
Geothermal	0.5	3.0	1.9	1.5	9.0	5.4
Biomass and waste	2.0	6.1	2.7	7.0	20.0	8.6
Wind	0.5	0.9	0.1	8.0	20.0	1.7
Photo-voltaic	0.0	0.0	0.0	0.5	1.0	0.1
Total electricity excluding hydro	8.0	25.2	5.6	27.0	80.0	17.6
Large hydro	74.8	154.5	13.3	88.6	198.5	17.1
Thermal						
Fuelwood			20.0			50.0
Other biomass (bio-gas, waste, etc.)			2.7			2.7
Geothermal			0.4			0.4
Solar			0.2			0.2
Total thermal			23.3			62.2
Bio-fuels			0.0			
Total contribution (excl. large hydro)			29.3			91.6
Total contribution of renewables (incl. large hydro)			42.6			108.7
Total energy consumption			1,160.0			1,400.0
Percentage share of renewables			3.7			7.8

Source: CEC 1992, Harrison 1992.

BOX 7.9 GLOBAL ENERGY SCENARIOS

There are a number of long-term global energy scenarios, of which the following represent selected examples:

World Energy Commission scenarios (World Energy Council 1992, 1993a, b; Figures 7.10A, B and 7.11A).

The World Energy Council, a non-governmental organisation consisting of representatives from the major energy industries in about 100 countries, produced one of the most comprehensive appraisals of possible patterns of world energy demand for the next few decades. The commission prepared three energy cases, each of which incorporates different assumptions (Table 7.14), but all adopt the UN predictions for the growth in world population from 5.3 billion in 1992 to 8.1 billion by 2020.

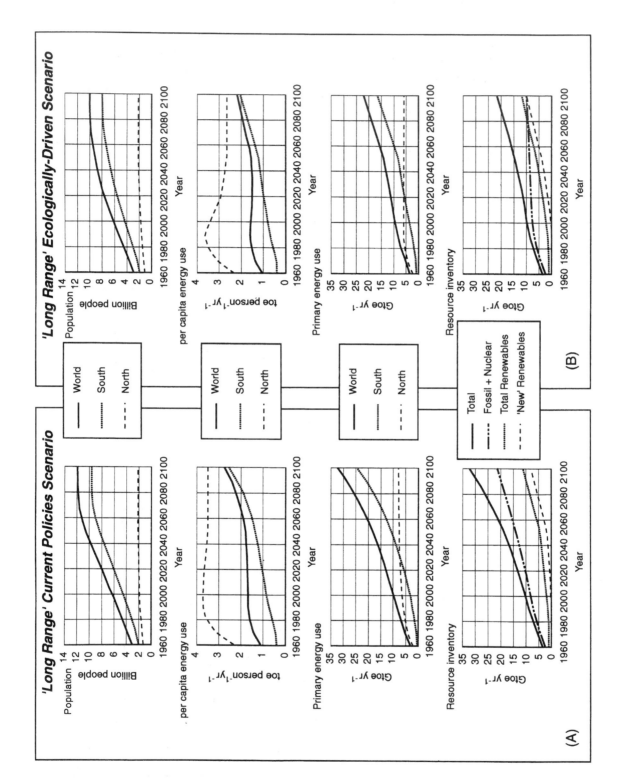

Figure 7.10 World Energy Council (1993b) long-range energy scenarios: (A) 'Long-range current policies' scenario, assuming a continuation of present energy policy trends; (B) 'Long-range ecologically driven' scenario, assuming a higher priority is given to environmental considerations.

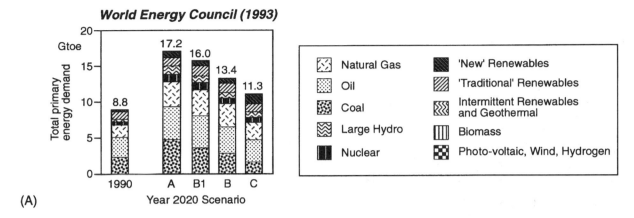

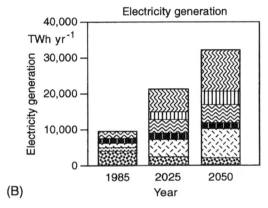

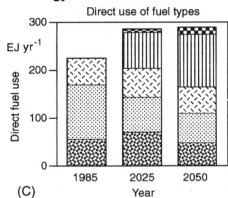

Figure 7.11 *(A) Total primary energy demand and projected energy supply mix in WEC scenario for 2020. Redrawn after WEC (1993a). (B) Electricity generation in the renewables-intensive global energy scenario. Despite a forecast tripling in electricity prices over the period shown, the contribution by renewables is predicted to increase potentially to 60 per cent of the market. Redrawn after Johannson et al. (1993). (C) Redrawn after Johannson et al. (1993).*

Table 7.14 *Energy mix – past and future: global fuel use (Gtoe).*

Energy form	1960	1990	2020		
			Ref.	*EED*	*ED*
Solid	1.4	2.3	3.2	4.8	2.1
Liquid	1.0	2.8	3.7	4.6	2.7
Gas	0.4	1.7	2.8	3.5	2.3
Nuclear	0.0	0.4	0.8	1.0	0.7
Hydro	0.15	0.5	1.0	1.2	0.9
New renewable	0.0	0.2	0.5	0.8	1.5
Traditional	0.5	0.8	1.3	1.2	1.0
Total	3.45	8.7	13.3	17.1	11.2

Ref.,   'Reference' case
EED,   'Enhanced Economic Development' case
ED,   'Ecologically Driven' case
Source: World Energy Council 1993.

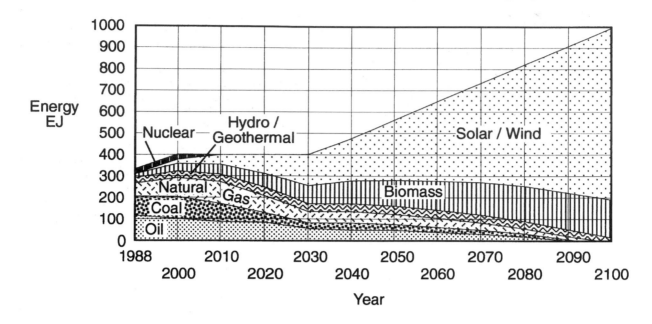

Figure 7.12 *The Greenpeace fossil-free energy scenario (FFES), in which world energy demand increases by a factor of approximately 2.5 by 2100. By this date, all fossil and nuclear fuels are envisaged as phased out and replaced by a mix of solar, hydro-, wind, biomass and geothermal energy sources. Redrawn after Lazarus* et al. *(1993).*

The WEC 'reference' case is for a 'moderate' annual growth of 3.3 per cent in the world economy, giving a total world GDP of about US$56 trillion (million million) by 2020, and a total world primary energy demand of 13.3 Gtoe. In contrast, the WEC 'enhanced economic development' case assumes an annual growth rate of 3.8 per cent for the world economy, giving a world GDP of about US$65 trillion by 2020, and a total world energy demand of about 17.2 Gtoe. The WEC 'ecologically driven' scenario assumes a greater reliance upon renewables and energy conservation but maintains an annual growth rate of 3.3 per cent for the world economy, resulting in a total world energy demand of 11.3 Gtoe.

The renewables-intensive global energy strategy (RIGES) (Johansson *et al.* 1993; Figures 7.11B, C).
RIGES employs similar assumptions to those adopted by the WEC, but assumes a greater degree of successful energy efficiency and energy conservation than the WEC ecologically driven case. Under the RIGES scenario, it is suggested that renewable energy sources could increase their share of electricity generation from the current level of about 20 per cent to more than 60 per cent by 2050. Also, under the RIGES scenario, renewable energy other than in the direct production of electricity could reach a figure of about 40 per cent of total world energy demand.

The Shell Scenarios (Booth and Elliott 1995, Shell 1995, both cited in Boyle 1996).
In 1995 the Shell International Petroleum Company published its long-term future prospects for world energy demand and supply as two scenarios. The 'sustained growth' scenario predicts that world energy demand will increase from the current 400 EJ to about 1,500 EJ by 2060. Amongst its assumptions are that world population will reach 11 billion by the end of the twenty-first century, a 3 per cent per annum growth in world GDP, and a growth in world energy demand of approximately 2 per cent per annum. In its 'dematerialisation' scenario, world energy demand increases at about 1 per cent per annum to about 1,000 EJ by 2060. The latter case assumes an extreme scenario where world consumption of energy and materials is at a perceived realistic minimum.

The Greenpeace fossil-free energy scenario (FFES) (Lazarus *et al.* 1993; Figure 7.12).
Amongst the most optimistic of energy scenarios is that developed for the environmental pressure group Greenpeace International by the Stockholm Environment Institute, Boston Center, USA. The FFES adopts the UN prediction of a world population of 11 billion by 2100, with world GDP by this date showing a fourteen-fold increase from 1988. In the FFES, world primary energy demand increases to 1,000 EJ or about 24 Gtoe by 2100, and all fossil and nuclear fuels are assumed to have been phased out, to be replaced by a mixture of solar, hydroelectric, biomass, wind and geothermal energy resources. In the FFES, greenhouse gas emissions specifically from fossil fuels are assumed to have been eliminated, resulting in a rise in global mean surface temperature of about 1.5°C, compared with the predicted 4°C if the use of fossil fuels continues at present rates (IPCC 1994).

Chapter 7: Key points

1 Industrialisation, population growth, increased living standards, concerns about pollution and serious depletion of many traditional energy resources, such as fossil fuels, have increased the need to develop renewable and cleaner energy resources. Most of the global energy supply comes from finite resources (77 per cent essentially from fossil fuels), renewables (18 per cent hydro-power, wood, crop waste, dung and wind), and nuclear energy (*c.* 5 per cent).

2 Conventional fossil fuels (coal, oil and gas) have relatively low reserve lifetimes, at most measured in hundreds of years. Extraction and use of fossil fuels causes many environmental problems, including chemical pollution, disfigurement of the landscape, oil and gas spillages from transporters, mainly tankers, pipelines and installations. Despite the environmental costs, current and projected future energy demand make it very unlikely that fossil fuels will not continue to make a major contribution to world energy markets.

3 In the conventional fossil fuel sector, environmental awareness has led to improved, cleaner technologies. Clean coal technologies, for example, are being developed, which include fluidised bed combustion, flue gas desulphurisation, gas-fired combined-cycle gas turbines, gasification, and British Coal's Topping Cycle. To reduce atmospheric pollution, many governments in the developed countries have agreed to reduce their emissions by target dates, but such agreements have yet to make any significant headway in some of countries that represent the future major polluters, such as China and the former Soviet Union, two of the largest users of coal.

4 Nuclear energy, produced by nuclear fission, has been developed worldwide since the Second World War. The industry has encountered many problems during this time, including several nuclear accidents and the problems associated with the safe disposal of nuclear waste. The problems and poor public image of the nuclear industry have resulted in some countries deciding to reduce their dependence on nuclear power, and even mothball nuclear power plants. Decommissioning nuclear power stations is costly and may contribute to long-term environmental pollution.

5 In the future, hydrogen energy, a form of chemical fuel, which could be easily transported and readily stored, may provide a major part of the world's energy demand, but the remaining technological problems are immense.

6 Renewable energy resources include hydroelectric, wind power, tidal energy, wave power, solar energy, geothermal energy and biomass energy. Technologies are being developed to increase the energy efficiency of these resources. Environmental problems, however, are associated with renewables, including the disruption of ecosystems, visual pollution and the current high production costs. Also, established energy businesses and cartels operate to minimise the potential competition for as long as possible.

7 Cumulatively, the use of private vehicles results in the use of excessively large amounts of energy and contributes to poor air quality and other forms of environmental pollution. Improved public transportation systems and better urban planning could lead to a considerable reduction in pollution. Liquid bio-fuels, including bio diesel and bio-ethanol, as substitutes for conventional petrol and diesel, might provide environmentally more friendly alternatives, although the technology to use these substitutes is insufficiently developed, and the hydrocarbon and allied chemical industries have strong vested interests in slowing the introduction of bio-fuels. Without a distortion of the traditional fuel markets through favourable tax incentives and subsidies it seems unlikely that bio-fuels will become important until at least the middle of the twenty-first century.

8 Attempts to reduce fuel consumption and improve air quality, including a reduction in the emissions of greenhouse gases caused by the combustion of fossil fuels, include proposals for the introduction of an energy/carbon tax, the increased use of energy-efficient technologies, energy conservation, and a greater reliance on renewable energy resources.

Chapter 7: Further reading

Boyle, G. (ed.) 1996. *Renewable Energy: Power for a Sustainable Future.* Oxford: Oxford University Press/The Open University, 479 pp.
An excellent review of renewable energy resources. This book will tend to appeal to students undertaking courses with a strong emphasis on the technology and science of renewable energy.

Blunden, J. and Reddish, A. (eds) 1991. *Energy, Resources and Environment.* London: Hodder & Stoughton, 339 pp.
A well-written and well-illustrated textbook dealing with the nature of energy and mineral resources, their extraction, refining and disposal, and the environmental problems associated with obtaining these resources. The book explores the possible alternatives to conventional energy resources, including substitution and recycling, energy conservation, solar energy, wind power, and water energy. It also considers the political implications of using such alternative energy resources. There is a whole chapter devoted to an examination of the politics associated with the disposal of radioactive waste.

Cassedy, E.S. and Grossman, P.Z. 1990. *Introduction to Energy Resources, Technology, and Society.* Cambridge: Cambridge University Press, 338 pp.

In this textbook, the authors explore energy issues and examine the benefits and problems associated with energy technology. The book is in three parts: Part I, dealing with energy resources and technology; Part II, examining power generation, the technology and its effects; and Part III, evaluating energy technology in the future. This book will prove useful as supplementary reading for students with a general interest in energy issues, particularly energy technology.

Flood, M. 1991. *Energy Without End*. Friends of the Earth. 76 pp.

Johansson, T.D., Kelly, H., Reddy, A.K.N. and Williams, R.H. (eds) 1993. *Renewable Energy for Fuels and Electricity*. London: Earthscan, in association with the United Nations, 1,200 pp.

A comprehensive and authoritative assessment of the technical and commercial potential for renewable forms of energy.

Ramage, J. 1988 *Energy: A Guidebook*. Oxford: Oxford University Press, 337 pp.

Scientific American 1990. Special Issue: *Energy for Planet Earth*. 263, 20–115.

Stockholm Environment Institute 1993. *Energy Without Oil: The Technical and Economic Feasibility of Phasing out Global Oil Use*.

Yergin, D. 1991. *The Prize: the Quest for Oil, Money and Power*. London: Simon & Schuster, 885 pp.

The river went on raising and raising for ten or twelve days, till at last it was over the banks. The water was three or four foot deep on the island in the low places and on the Illinois bottom. On that side it was a good many miles wide; but on the Missouri side it was the same old distance across – a half a mile – because the Missouri shore was just a wall of high bluffs.

Mark Twain
The Adventures of Huckleberry Finn (1884)

The landscape is fashioned by a wide variety of natural processes, which may occur in such a manner as to precipitate a catastrophic effect on biotic systems. Natural processes can be considered to pose a natural hazard when they are capable of posing a threat to life, actually causing mortalities (Table 8.1), or causing damage to buildings and agricultural land (Table 8.2). A natural hazard becomes a catastrophe for humans if a situation develops in which the damage to people, property or society is sufficient to require a long recovery or rehabilitation process. Floods, hurricanes, tornadoes, **tsunamis** (earthquake-generated sea waves), volcanoes, earthquakes and large fires are the most common natural hazards that produce catastrophes. Particular natural hazards are concentrated in distinct geographic regions. The reasons for this distribution are discussed in this chapter. Although natural hazards are normally discussed in terms of their human impact, wherever there is severe damage to any part of an ecosystem the causal event should be regarded as a natural hazard.

During the past few decades, if statistical databases are to be taken at face value, there has been an apparent worldwide increase in the damage done by natural hazards. In reality this apparent increase may simply be a function of an increase in the reportage and media coverage, with the enhanced ability to transfer news and events more efficiently and effectively around the world. It may, however, be a function of the increased global population and its concentration in marginal areas of the world, which are less suited to habitation because of the inherent associated risks from natural disasters. Alternatively, any increase, if real, may be due, in part, to human-induced environmental change, leading to the crossing of critical thresholds, above which natural disasters are much more likely.

If climatic hazards during the past few decades are considered, for example, it seems that the most fierce storm this century was experienced in September 1988, when Hurricane Gilbert devastated the Caribbean with reported wind speeds of up to 349 km hr^{-1}. In the South Pacific between 1941 and 1980, five severe typhoons were recorded in the Fiji Islands, compared with six between 1981 and 1985. The North Atlantic has also experienced severe storms, with three major hurricane depressions between October 1987 and January 1990. In southern England, the storm of October 1987 was the most violent experienced in this area in 300 years (Thompson 1989).

There have also been severe droughts since the 1960s in the Sahel, with 1984 being the driest year this century (Hulme and Kelly 1993). Recent droughts in the Mid-west states of the USA were reminiscent of the '**dust bowl**' conditions of the 1930s. Extreme drought conditions have been more frequent in southern Britain in the last fifteen years, with four severe drought years – 1976, 1983, 1989 and 1995 (Hamer 1992, Thompson 1992). Whether such conditions are produced by natural fluctuations

Table 8.1 *Risk of death from involuntary hazards.*

Involuntary risk	Risk of death/ person/year
Struck by automobile (USA)	1 in 20,000
Struck by automobile (UK)	1 in 16,600
Floods (USA)	1 in 455,000
Earthquake (California)	1 in 588,000
Tornadoes (Mid-west)	1 in 455,000
Lightning (UK)	1 in 10 million
Falling aircraft (USA)	1 in 10 million
Falling aircraft (UK)	1 in 50 million
Explosion, pressure vessel (USA)	1 in 20 million
Release from atomic power station	
At site boundary (USA)	1 in 10 million
At 1 km (UK)	1 in 10 million
Flooding of dike (Netherlands)	1 in 10 million
Bites of venomous creature (UK)	1 in 5 million
Leukaemia	1 in 12,500
Influenza	1 in 5,000
Meteorite	1 in 100 billion

After Dinman 1980, in Smith 1992.

Table 8.2 *Federally declared disasters in the USA, 1965–85.*

Type of disaster	Number	Federal outlay (thousands of current $)	Federal outlay (thousands of 1982 $)
Ice and snow events	19	151,427	205,511
Hurricanes/tropical storms	39	1,173,141	1,947,939
Earthquakes	7	203,881	405,706
Dam and levee failures	7	55,764	80,806
Rains, storms and flooding*	337	1,684,702	2,439,852
High winds and waves	2	125,313	120,536
Coastal storms and flooding	7	158,261	205,357
Tornadoes	109	441,685	648,352
Drought/water shortage	4	1,134	5,344
Totals	531	3,995,803	6,059,403

*includes land, mud and debris flows and slides.
Source: Federal Emergency Management Agency, DMIS Reports, 1965–1985; quoted in Rubin *et al.* 1986. After Smith 1992.

in the Earth's climate or are stimulated by human activities has yet to be resolved. This chapter examines the causes and consequences of the main natural hazards.

The effect on populations of a disaster resulting from a natural hazard may be direct, involving injury, death and damage to property, or indirect through a knock-on effect from reduced economic resources caused by the catastrophe. Sometimes, disasters occur through everyday natural processes, for example as occurred on 26 July 1987, when the Greek government declared a state of emergency because a national heat wave had led to the death of more than 700 people.

Natural hazards are generally considered in terms of their magnitude (the intensity of the energy released) and their frequency, that is their recurrence time. There is an inverse relationship between magnitude and frequency – the larger the magnitude, the less frequent the event (*cf.* Wolman and Miller 1960). The impact of a natural hazard on humans and other organisms is not directly related to the magnitude of the natural event. An earthquake in mountainous and unpopulated regions clearly will not have the same effect as if the earthquake had occurred in the vicinity of a crowded city. The devastation is dependent upon human factors, such as the concentration of population within the area, and the remedial or preventive measures that may be taken to mitigate the likelihood of the natural hazard occurring or its impact. A society that is prepared for a natural hazard is more likely to suffer less than one where totally unexpected and unpredicted events occur.

There are many examples where the magnitude of a natural hazard does not necessarily relate to the scale of the ensuing disaster. The largest recorded earthquake of recent time occurred in southern Chile in 1960, and caused the death of approximately 5,700 people. The energy released by this earthquake was more than ten times that of the 1976 Tangshan earthquake in northeast China, which killed over 650,000 people, with some estimates putting the fatalities in the latter incident at more than one million. Northeast China is one of the most densely populated regions of the world, while southern Chile has a considerably smaller population density. So, the difference in fatalities between these two examples is a function of the size and density of the populations in an earthquake-prone region.

Urbanised areas are clearly more at risk than the rural open countryside because potentially large numbers of fatalities can result during any single disaster. As the population increases and more demands are placed on the Earth's limited resources, the potential increases for natural processes to become hazards. With increasing awareness and understanding of the Earth's natural processes, however, many remedial measures can be taken to safeguard against a likely catastrophic event.

Total deaths due to natural hazards, however, are on the increase, simply because disasters affect areas with large and growing populations. In monetary terms, the total loss of property due to damage has also been increasing as urban areas increase in size and rural areas become more developed.

In an attempt to reduce the likelihood of a major catastrophe, the inexact science of **risk** assessment and risk management has evolved. The risk of a particular hazard is related to the probability of its occurrence multiplied by the predicted consequences.

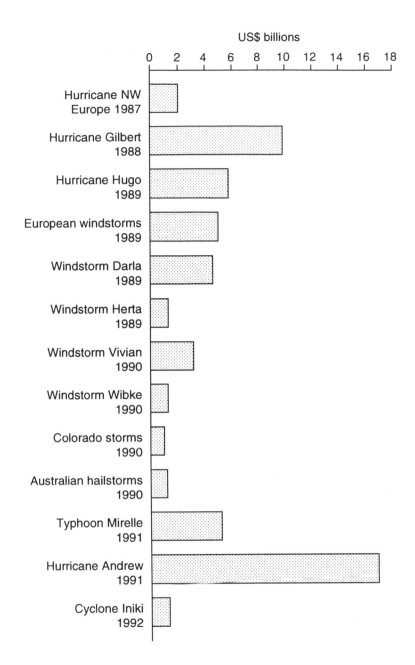

US$ billions

| Hurricane NW Europe 1987 |
| Hurricane Gilbert 1988 |
| Hurricane Hugo 1989 |
| European windstorms 1989 |
| Windstorm Darla 1989 |
| Windstorm Herta 1989 |
| Windstorm Vivian 1990 |
| Windstorm Wibke 1990 |
| Colorado storms 1990 |
| Australian hailstorms 1990 |
| Typhoon Mirelle 1991 |
| Hurricane Andrew 1991 |
| Cyclone Iniki 1992 |

Figure 8.1 *Catastrophic insurance losses (in US$ billions) for selected major global natural disasters from 1987 to the present. Redrawn after* The Times, *5 April 1993.*

It is very difficult to assess the consequences of a likely event and to estimate the probability of its occurrence. As a result, the awareness of individuals and policy-making bodies such as governments varies considerably. Although much is understood about the different types of natural hazard, frequently all too little is done to reduce the risks. Additional problems arise because of the poor communication between scientists working on aspects of particular natural hazards, the policy-makers and the media – and, naturally, those who are actually at risk.

From a financial perspective, the insurance industry has a vested interest in risk assessment related to natural hazards. Within the past few years, there has been a string of natural disasters associated with severe weather, and insurance companies have had to pay out large sums of money to meet the claims (Figure 8.1). There are people who believe that world weather patterns are changing rapidly because of global warming influenced by human activities, while others argue that global climate changes are both natural and cyclical over a time frame of hundreds of years.

Minimising the risks from natural hazards takes many forms. Good land planning, 'hazard-proof' construction, insurance, evacuation programmes and disaster preparedness are among the most common. In less developed countries, the populace often has no choice but to bear the brunt of a disaster, with all its attendant loss of life and damage. The degree to which 'adjustments' are made depends on the type of hazard, how the population perceives the hazard and, not least, the cost of taking preventive action. Severe blizzards, for example in northern Canada, are not considered a hazard by Canadians as they are a common occurrence during the winter, whereas, in the UK, they represent a dangerous natural hazard because they occur infrequently and the population is generally unprepared for such freak conditions. When severe blizzards occur in the UK, they often result in some fatalities, particularly of the infirm and elderly, and consequently they attract considerable media coverage because they are perceived as a natural hazard. Heat waves can be just as hazardous as cold spells where people are not prepared for the temperature conditions. Bridges and Helfand (1968), and Oechsli and Buechly (1970), for example, have described the significant increase in fatalities related to particular hot spells in the USA.

In order to cope better with natural hazards, it is necessary to understand the causes and effects of such hazards, and something of the dynamics of natural processes, that is, the way they come about and just how they occur. This chapter reviews the nature of the various types of natural hazard, which can be considered in three groups of processes: geological, meteorological and biological.

Geological hazards are of two main types. First, there are those hazards that are driven by the Earth's internal energy. These include earthquakes, volcanoes and tsunamis. The second type may be described as hazards resulting from land-surface processes or Earth surface processes. It is these processes that are primarily responsible for shaping the landscape. Earth surface processes are dependent on the atmospheric, climatic and weather conditions, the vegetation type and cover, topography, drainage patterns, bedrock type and geological structure, the tidal regime in coastal regions, and the way in which the land is used. The ability of objects to fall and do damage by virtue of gravity, also called gravitational potential energy, is very important as it is responsible for major processes such as the movement of rock, ice and snow masses down slopes, the flow and form of rivers, and tides. Solar energy is also important because it controls the climatic conditions that influence these processes (see Chapters 1, 2 and 3). Hazards directly or indirectly caused by gravity and climatic conditions include landslides, rock and snow avalanches, river flooding, collapsing soils, icebergs and **jökulhlaups**.

Meteorological hazards are driven primarily by the Sun's energy. These include tornadoes, hurricanes, floods, lightning strikes and resulting fires, droughts, cold fronts, fogs, hailstorms, blizzards and snow, windstorms, sandstorms, and frosts.

Biological hazards include epidemics, proliferation of pests, endemic parasites, and the invasion of areas by insects or plants. The spread of these hazards is influenced by the local and regional conditions, which include climate and the available food sources for pests. These conditions are essentially ultimately controlled by the Sun's energy.

Some natural hazards fall into more than one of the above categories as they may have two or more causes. A corollary of this is that similar effects may be produced by different causes. Floods, for example, may be produced by meteorological factors such as heavy rainfall, sea surges resulting from typhoons, and spring snow melts. They may also be the result of geological factors, such as the bursting of landslide-dammed lakes (jökulhlaups), or they may be caused by processes originating within the Earth, such as earthquake-generated tsunamis. In coastal areas, submergence due to land-surface movements associated with an earthquake can effectively be instantaneous. Examples include the famous Good Friday Alaskan earthquake, which caused widespread subsidence of several metres along the coastal regions around Anchorage (Eckel 1970).

Some natural hazards are induced or initiated by previous hazards. A good example of these cases is where landslides and fires commonly occur after many large earthquakes. In 1920, approximately 200,000 people are believed to have died as a result of an earthquake in Gansu province, central China. This was not the result of collapsing buildings during the earthquake but of major landslides produced by the mobilisation of wind-blown sediment (loess) as the ground shook (Derbyshire *et al.* 1991). Similarly, the 1906 earthquake in San Francisco was followed by extensive fires, which could not be extinguished due to the depleted water supply resulting from water mains broken by the earthquake (Lawson *et al.* 1908). These fires are believed to have claimed the majority of the deaths associated with the quake. Another example of the domino effect with natural hazards is where epidemics such as cholera and hepatitis often follow a flood disaster because drinking water becomes contaminated, and medical help and supplies are limited due to poor physical communication, poverty or the destruction of the normal channels of communication and transport. This has been a particularly big and recurrent problem in developing countries, such as in the flood-prone delta regions of Bangladesh.

Hazards that appear initially to be natural may actually have anthropogenic causes, that is human interference with the natural environment. Landslides, for example, may be the result of badly managed land, deforestation or construction works. Such landslides are particularly common in tropical mountain areas, as in the Himalayas and the countries of Southeast Asia. It is here that extensive deforestation is taking place on steep slopes that commonly experience high rainfall. Similarly, some river flooding may owe its origin to badly managed catchment areas, channelisation of rivers and adjacent urban settlements. In these cases the flow of water into the main stream has been accelerated by human activities.

Human-induced earthquakes are another hazard, as experienced in Denver, Colorado, during the period 1962–1965. The quakes were due to the disposal of chemical waste underground at the Rocky Mountain Arsenal. Numerous small quakes in Nevada are believed to have been triggered by underground nuclear testing (Evans 1966). Many Earth scientists believe that human interference with the atmosphere could result in global warming and changes to the atmospheric circulation, which in turn could lead to an increase in natural hazards.

The El Niño event during 1982–1983 is frequently cited to demonstrate the adverse effects of possible global warming. This event occurred because of changes in the direction of the trade winds, which resulted in a warming of equatorial Pacific water. This increased the amount of heat energy supplied by the ocean to the atmosphere and consequently caused droughts in eastern Australia, Central America and East Africa. It also increased precipitation, resulting in the flooding of large areas of Bolivia, Peru, Ecuador and the western coastal regions of the USA, and, in addition, the warm oceans may well have caused more tropical cyclones, thus further adding to the occurrence of natural hazards during 1982–1983 (Golnaraghi and Kaul 1995). El Niño events appear to have a frequency of about every seven years (Wuethrich 1995), although their frequency appears to have increased during the past two decades.

The range of types of natural hazard are considerable. To help understand and appreciate these in more detail and to evaluate better the preventive measures that can be taken to predict a likely catastrophe and prevent a disaster, the main natural hazards will be considered.

Earthquakes

Since the beginning of this century, more than 1,500,000 people have died because of earthquakes. The loss of life, terror and general destruction associated with earthquakes make them front-page news. On 19 April 1906, an earthquake shook San Francisco, with up to 1,000 people being killed through falling debris and the associated fires (Lawson *et al.* 1908). On 25 December 1972, about 10,000 people are believed to have lost their lives when tremors over a two-hour period shook the Nicaraguan capital of Managua. In 1988, in Armenia in the then Soviet Union, an earthquake rocked the region around Spitak and Leninakan and killed 25,000 people. On 17 January 1994, the Northridge earthquake resulted in the death of 61 people and was rated the second most expensive natural disaster in US history (after Hurricane Andrew). A year later, on 17 January 1995, more than 4,800 people were killed and more than 25,000 people were injured when a major earthquake shook Kobe in Japan. The list goes on, and a quick glance at Table 8.3 illustrates even more clearly the catastrophic effects of recent major earthquakes and tsunamis, exceptionally large waves created during an earthquake (*tsunami* is the Japanese word for 'harbour wave').

Earthquakes pose a particularly serious natural hazard, with all their related phenomena such as landslides, fires, **liquefaction** of the ground (a temporary change to a liquid-like state), and virtually instantaneous changes in land-surface elevation, which can cause flooding of coastal areas and the disruption of ground-water supplies and communication links (Plate 8.1), as well as tsunamis. The 1960 Chilean earthquake created a tsunami that travelled right across the Pacific to cause damage in Japan (Figure 8.2). The energy released during an earthquake, its magnitude, is measured on the **Richter scale** and is recorded on a **seismometer**. The amount of energy released by the largest earthquake recorded (M8.5–8.7) was equivalent to 60,000 1-megaton hydrogen bombs. The effects of earthquakes are also measured semi-quantitatively using the modified **Mercalli scale** of intensity (Table 8.4). The intensity of a particular earthquake at a location is dependent upon the ground conditions, the distance from the source of the earthquake (or focus) and its magnitude.

Earth scientists explain the cause of earthquakes by the theory of **plate tectonics** (see Chapter 1). The relative motion of lithospheric plates produces **stress** (= force per unit area) along their boundaries.

Table 8.3 *Toll of recent large earthquakes and tsunami.*

Earthquakes

Date	Location	Magnitude	Deaths	Damage (millions of US$)
1995	Kobe, Japan	7.2	4,800	30,000
1994	Northridge, California	6.8	61	20,000
1993	Maharashtra, India	6.4	30,000	280
1990	Luzon, Philippines	7.7	1,621	2,000
1990	Northwest Iran	7.7	40,000	7,000
1989	San Francisco	6.9	62	6,000
1988	Northwest Armenia	6.8	55,000	14,000
1985	Tangshan, China	8.2	242,000	5,600
1970	Northern Peru	7.7	66,794	500

Tsunami

Date	Location of source	Location of deaths	Height	Deaths
1992	Nicaragua	Nicaragua	>10 m	
1983	Sea of Japan	Japan, Korea	15 m	107
1979	Indonesia	Indonesia	10 m	187
1976	Celebes Sea	Philippines	30 m	5,000
1964	Alaska	Alaska, Aleutian Is, California	32 m	122
1960	Chile	Chile, Hawaii, Japan	25 m	1,260

After Begley 1995 and Crystal 1993.

The strain or movement produced by these stresses may be reduced slowly if the plates are able to slide past each other with ease. If there is no easy and smooth slip, stress builds up because of the frictional forces between plate boundaries. This stress build-up may be released suddenly and violently. The result is a burst of energy, which is referred to as an earth-quake. The build-up of stress is not only along the actual boundary between the plates, but it also occurs along zones of high localised **strain** associated with the boundaries. These zones comprise highly deformed, folded, faulted and jointed rocks.

Large cracks across which the Earth's surface has been displaced are called **geological faults**. It is along such faults that the sudden release of accu-mulated stress occurs to allow some relative move-ment (strain) between the opposing blocks. Typically, in any single earthquake event, movement may be measured in metres, and in very large-magnitude earthquakes up to tens of metres. This is supported

Plate 8.1 *A view looking across the remains of a small town in the Garhwal Himalaya, northern India. This was totally destroyed when an earthquake (magnitude 7.1 on the Richter scale) shook the area in October 1991. Seventy people and numerous cattle died as their poorly constructed houses and barns collapsed around them. Until this event, no large earthquakes had been recorded in the region during historical times, and the area was considered a seismic gap within an otherwise highly seismic mountain belt. At present, a major dam is being constructed at Tehri, less than 100 km down the valley from the epicentre of this earthquake, causing concern for the future safety of those people living there.*

Plate 34 *The Mississippi River breaking through a levee at Valmere, Illinois, during the Great Flood of 1993.*
Courtesy of Cameron Davidson/Comstock.

Plate 35 *Human-induced landslide on the steep slopes of the Mid-Levels in Hong Kong, caused by the construction of a large tower block, the remains of which can be seen scattered in the landslide debris.*
Courtesy of the Geotechnical Control Office, Hong Kong.

Plate 36 Dry rock and powder avalanche in July 1995 in the Swiss Alps, a few kilometres south of the drainage divide in the Simplon Pass. Many of the tall pine trees are about 30 m high.

Plate 37 Flooding is a common sight in delta regions of Bangladesh. It is due to a combination of heavy rains and sea surges produced by tropical storms and cyclones during the monsoon season.

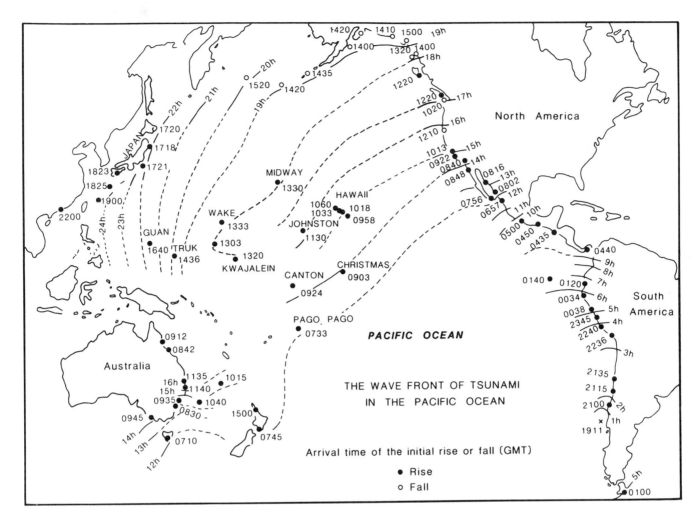

Figure 8.2 *Tsunami wave front in the Pacific Ocean following the May 1960 Chilean earthquake. After Pickering* et al. *(1991).*

by examining the distribution of past earthquakes throughout the world. Since the Earth's surface comprises seven main plates, the plate boundaries are limited to several well-defined zones. Currently, the three most important seismic zones are the Alpine–Himalayan Belt, where the continental Eurasian plate is colliding with the continental Africa and Indian plates; the Circum-Pacific Belt, where the oceanic Pacific plates are colliding with the continental plates of Eurasia, South America, Australia and North America; and the Mid-Atlantic Ridge, where the oceanic crust of the Atlantic is spreading apart to generate new ocean crust.

Earthquakes can also occur within plates, away from plate boundaries, and geologists are still trying to find appropriate explanations for their occurrence. It seems likely that such within-plate seismic activity owes its origin to the accommodation of forces that

build up as a result of the differential movements along the main plate boundaries on a sub-spherical (actually an oblate spheroid), and because of mantle processes operating directly under the various plates. Examples of within-plate earthquakes include the Charleston (South Carolina, USA) earthquake of 1886, which killed 2,000 people (Dutton 1889); the New Madrid earthquake (Missouri, USA) of 1811 (Fuller 1912); and the 1993 earthquake in Maharashtra, India (Jayaraman 1993a).

An understanding of the mechanism of faulting is essential in earthquake hazard assessment. Clearly, the earthquake hazard is a result of the accumulated build-up of stress and its sudden release. Seismologists (geologists or physicists who study earthquakes) can study the hazards from earthquakes in a number of ways. First, zones susceptible to earthquakes can be identified. Figure 8.3, for example,

Table 8.4 *The Mercalli scale of earthquake intensity.*

Scale	Intensity	Description of effect	Maximum acceleration (mm sec⁻¹)	Corresponding Richter scale	Approximate energy released in explosive equivalent
I	Instrumental	detected only on seismographs	< 10		1 lb TNT
II	Feeble	some people feel it	< 25		
III	Slight	felt by people resting; like a large truck rumbling by	< 50	< 4.2	
IV	Moderate	felt by people walking; loose objects rattle on shelves	< 100		
V	Slightly strong	sleepers awake; church bells ring	< 250	< 4.8	
VI	Strong	trees sway; suspended objects swing; objects fall off shelves	< 500	< 5.4	Small atom bomb, 20,000 tons TNT (20 kilotons)
VII	Very strong	mild alarm; walls crack; plaster falls	$< 1,000$	< 6.1	Hydrogen bomb, 1 megaton
VIII	Destructive	moving cars uncontrollable; chimney falls and masonry fractures; poorly constructed buildings damaged	$< 2,500$		
IX	Ruinous	some houses collapse; ground cracks; pipes break open	$< 5,000$	< 6.9	
X	Disastrous	ground cracks profusely; many buildings destroyed; liquefaction and landslides widespread	$< 7,500$	< 7.3	
XI	Very disastrous	most buildings and bridges collapse; roads, railways, pipes and cables destroyed; general triggering of other hazards	$< 9,800$	< 8.1	60,000 1-megaton bombs
XII	Catastrophic	total destruction; trees driven from ground; ground rises and falls in waves	$> 9,800$	> 8.1	

Source: Bryant 1990. *Natural Hazards*, Cambridge University Press.

shows a seismic risk map of the United States (Costa and Baker 1981). This sort of map is constructed by assembling data from past earthquakes and studying their distribution. The focus of an earthquake, the position at depth within the Earth from where the earthquake may have originated, can be located using an array of seismometers (instruments used for detecting earthquakes), which are distributed around the world and within earthquake-prone areas (Plate 8.1). Active geological faults may be identified when the earthquake focus has been located. The fault can be traced across a region using geological mapping and subsurface seismic techniques and, to a certain degree, since movement along the fault may disrupt drainage and create scarps or small lakes parallel to its length, topographic mapping using the surface

expression of the fault. Once an active fault has been identified, movement along the fault can be studied using precise surveying techniques; and the accumulation of strain may be measured using **strain gauges** – instruments that deform under stress, the amount of strain being converted electronically into absolute measurements.

Strain may be localised along a fault so that stress is released and movement occurs just within one zone, while stress may continue to accumulate elsewhere along the fault. Once stress has been released as strain along one fault segment, it is unlikely that a major earthquake will occur along this stretch for some time, probably not before the stress has been released from other lengths of the fault that are still being stressed. Using this principle of **seismic gaps**,

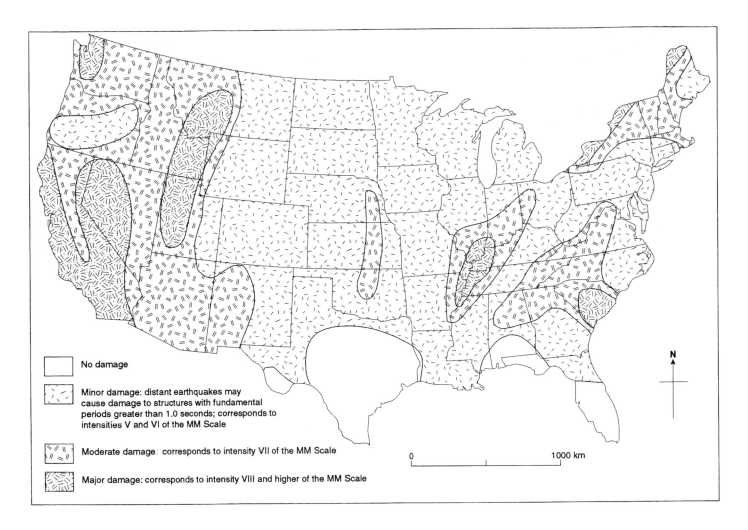

Figure 8.3 *Seismic risk map of the United States. Redrawn after Cooke and Doornkamp (1990), Costa and Baker (1981).*

as well as other studies, seismologists are able to estimate the earthquake risk along particular stretches of faults. This has been done for the San Andreas Fault in California, where seismologists have identified two major seismic gaps where stress has been accumulating for a long period compared with the other stretches of the fault, which have experienced continued movement and release of stress. Dolan *et al.* (1995) believe that numbers and sizes of earthquakes recorded in the Los Angeles metropolitan area of the San Andreas Fault system are too small when compared with the strain that has accumulated over the period of measurement. This might suggest that there is a lull between clusters of moderate earthquakes or that it represents part of a centuries-long inter-seismic period between much larger quakes. They suggest that the latter case is more plausible, and they have estimated that earthquakes of magnitude 7.2 to 7.6 have a return period of about 140 years for that region. Furthermore, they suggest that the accumulated strain may be released along any one or more of the tangle of small fault systems associated with the main San Andreas Fault in the Los Angeles metropolitan area. These small fault systems are similar to the fault that ruptured at Northridge in 1994. Unfortunately, many of these smaller faults have not been mapped in detail, or they have not been identified, as they are concealed beneath folded strata.

Seismologists also seek evidence regarding the frequency or **recurrence intervals** of earthquakes. Archaeological evidence and historical records are used to compare the time intervals between successive earthquakes. This has been particularly successful in China, where accurate and detailed records have been kept for more than 2,000 years (Plate 8.2).

Plate 8.2 *Reconstruction of the earliest seismometer, built by Chang Heng in* AD *132 and now displayed at the Institute of Seismology, Academica Sinica, in Lanzhou, China. When an earthquake occurred the urn would rock and one of the dragons would drop a small ball from its mouth into the mouth of the frog below, and therefore the frog would indicate the direction from which the earthquake wave had travelled.*

Geological evidence is also used such as the detailed study of the displacement of sediment layers that lie astride fault lines. One of the most famous studies was made in California at Pallet Creek, which traverses the San Andreas Fault. Here sediments that were deposited by streams record a history of the last 1,700 years. The displacements of the sedimentary layers were measured and dated using radiocarbon dating techniques. The sediments showed twelve earthquake events, giving a recurrence interval of about 145 years. The last movement of the San Andreas Fault at Pallet Creek occurred about 120 years ago. This suggests that an earthquake is likely in this area within the very near future. Such detailed studies provide predictions for the San Andreas Fault and related faults, which suggests that there is a 25 per cent probability that an M8 earthquake will occur along the San Andreas Fault east of San Diego within the next few decades, and a 75 per cent probability of an M6–7 earthquake along a stretch south of San Jose over the next few decades (Wesson and Wallace 1985). Either event will cause large-scale devastation to parts of California. Such research is referred to as **palaeoseismology**, and many forms of geological evidence can be used to reconstruct past seismic events, which is essential for determining recurrence intervals over long periods. One of the most recent studies, for example, has shown that a magnitude 8 to 9 earthquake probably occurred in the Pacific northwest of North America about 300 years ago. This is based on evidence from coastal marshes between northern California and Vancouver Island, which shows that the coastline in this region subsided almost instantaneously about 300 years ago. Modelling of the likely size of the tsunami that would have swept across the Pacific Ocean to the coast of

Japan gives a 2 m high wave. Historical data in Japan confirm the existence of a 2 m high tsunami on 27 January 1700 (Kerr 1995b).

Seismologists believe that earthquakes will follow a fractal pattern distribution, the larger the event the less frequently it will occur (Panel on Seismic Hazard Analysis 1988). Using a model in which earthquake distribution is fractal and knowing the slip rate in the Los Angeles metropolitan area, Hough (1995) predicted that a magnitude 7.4 to 7.5 earthquake will occur every 245 to 325 years, and that there should be approximately six events of magnitude 6.6 in the same span of time. Although the various studies along the San Andreas Fault have calculated slightly different recurrence intervals for a large earthquake, there is strong agreement that a large quake is likely to occur soon.

Some studies of recurrence intervals are very convincing, such as the regular frequency of earthquakes associated with the Parkfield Fault in California. So convincing was the regularity of earthquakes in this region that the US Geological Survey (USGS) forecast a magnitude 6 earthquake in 1993 and issued a warning. Fortunately or unfortunately, the USGS is still waiting for this quake. This illustrates the unreliability of forecasting, and when predictions of this nature fail, confidence in seismological studies is sadly destroyed.

The short-term prediction of earthquakes makes use of phenomena that have been observed before an earthquake. These are called precursors and include changes in ground level; emission of radon gas; a change in the electrical resistivity (the ability to carry an electric current) of the ground; a decrease in the velocity of small earthquake waves recorded in the area; a decrease in the number of small earth-

quakes, although often there may be many fore-shocks; and changes in water levels within wells. Not all these precursors are observed before every earthquake. For example, before the 1995 Kobe earthquake there were few fore-shocks and no notable ground deformation. There were, however, notable hydrogeochemical anomalies and changes in the chemistry, temperature and water table in ground water, as well as a ground-water radon anomaly before the quake (Igarashi *et al.* 1995, Tsunogai and Wakita 1995, King *et al.* 1995). Observations and studies like these have the potential to aid future earthquake predicting.

These phenomena have been explained using what the seismologists call the '**dilatancy–diffusion model**'. This model suggests that dilation of the ground accompanies the increased build-up of stress before a major earthquake event. As the ground dilates, cracks develop, allowing gases such as radon to escape from the Earth's interior, and water to enter the ground and decrease its electrical resistivity and/or produce changes in water levels and chemistry within wells. A decrease in earthquake wave velocity is explained because earthquake waves have to travel though air and water, which slows them down. After a major earthquake event, several small after-shocks may occur before the region stabilises and stress once more begins to build up.

On the basis of such evidence, scientists may be able to make accurate earthquake predictions in the future. The first successful prediction was made in 1975 in Haicheng, China, where 90,000 people were evacuated before an M7.5 earthquake totally destroyed 90 per cent of the city (Raleigh *et al.* 1977). In the summer of 1989, the USGS warned that there was a high probability of a quake occurring within a few days along the Loma Prieta stretch of the San Andreas Fault. It was wrong about the date, but on 17 October 1989, a violent quake destroyed much of Loma Prieta and parts of San Francisco and Oakland. However, research into earthquake prediction still has a long way to go before accurate medium- and long-term predictions and their effects can be made. For example, the 1992 Nicaragua earthquake generated a tsunami that was disproportionately large for the earthquake surface-wave magnitude of 7 (Kanamori and Kikuchi 1993). The probable reason for this discrepancy was that slip on the plate interface filled with soft, water-rich sediments of the **accretionary prism** caused the rupture process to be slower than in most subduction zones where earthquakes are associated with thrusting (*ibid.*), i.e. the strain rate was slower due to the intrinsic weakness of the sediments. An impor-

tant practical corollary of this scientific observation is that to reduce the hazards from this type of earthquake, tsunami warning systems using long-period (≥100 seconds) waves are necessary.

The prevention of earthquakes is also a major area of study. Scientists suggest that the stress accumulated in particular faults may be relieved slowly by underground nuclear explosions or by the pumping of water into the fault zone (Pakiser *et al.* 1969). However, legislation allowing such experimental methods to be used is a long way off – it is simply not known if such action would prevent the likely occurrence or inadvertently increase the likelihood and magnitude of an earthquake. In the USA, scientists from the Earthquake Engineering Research Center at the University of California at Berkeley released the results of laboratory experiments showing that buildings supported on special steel ball-bearings have a significantly improved chance of withstanding the impact of severe earthquakes. The ball-bearings rest at the centre of concave steel 'bowls', and during an earthquake they absorb much of the earthquake energy by deforming and moving in response to the cyclic stresses of the quake. These results were presented in May 1993 at the National Earthquake Conference in Memphis, Tennessee, and suggest that the ball-bearings can reduce the stresses on buildings by up to 80 per cent. The results have proved so impressive that the US Court of Appeals Building in San Francisco, which suffered damage during the 1989 earthquake, is to have 256 ball-bearings installed underneath. In the future, buildings constructed from **smart materials** will be earthquake-compensating in relation to passing earthquake shock waves and, therefore, much more able to withstand severe earthquakes.

Each additional earthquake provides new lessons for disaster preparedness and new insights into the science of seismology. Although Japan is one of the leading countries in the science of seismology, the earthquake that devastated the container port of Kobe in southern central Japan (Honshu) on 17 January 1995 showed how unprepared Japan actually was and has provided many important lessons for future earthquake preparedness, particularly concerning construction design and searching for trapped and injured people. The earthquake, measuring 7.2 on the Richter scale, shook the ground for just 20 seconds after 5.48 a.m. and resulted in the collapse of extensive areas of old wood-frame and stucco houses with tiled roofs. Many multi-storey buildings collapsed. Commonly, the weakest floor, usually the ground floor, compressed and collapsed under the weight of the building. Elevated sections of highways collapsed as the concrete in the supports crumbled and disintegrated. Fires spread rapidly as the small stoves on which people were preparing breakfast toppled over. In total, more than 50,000 buildings where destroyed. Most were not built to resist earthquakes; many of the stronger buildings collapsed because they were built on soft sediments, which liquefied during the shaking. In addition, the catastrophe was exacerbated by the slow response of relief workers and the lack of rescue equipment such as fibre optics, which are used to search for buried people, and helicopters, which are essential for soaking burning buildings (Begley 1995). This lack of preparedness was somewhat surprising, because every first of September, on the anniversary of the 1923 Tokyo quake, which killed 140,000 people, the Japanese practise earthquake drills and rescues. The disaster highlighted how unprepared Japan is for a major earthquake. Politicians are now rethinking disaster preparedness policies and are revising building codes.

The main reason for the severity and type of damage during the Kobe earthquake was the location of the earthquake and the importance of the shock waves that propagated along the surface. The ground acceleration, i.e. the horizontal acceleration, was actually greater than the force of gravity, so that some buildings, including bridges, were severely shaken such that the lower floors or a particular higher floor in a multi-storey building collapsed, while the remaining floors were left relatively intact. Bridge supports moved momentarily to allow the bridge spans that they supported to collapse but leave the supports intact.

The construction of earthquake hazard maps is often a useful start in helping to mitigate the hazard. One of the first completed was for the San Francisco Bay area (Wesson *et al.* 1975). Among some of the hazards mapped were areas of flooding, subsidence and landsliding. Today, geographical information systems are being developed for southern California collating a variety of data and incorporating early warning systems, emergency rescue plans, and systems to cut off gas and electricity supplies to areas that may have been affected by any quake. Japan is also developing a similar system for Tokyo. In most regions of the world, however, sophisticated warning systems are not available and hazard mapping is more realistic and cost-efficient. Figure 8.4 provides an example of a hazard map for part of the Himalayas, northern India, where landsliding has resulted from earthquakes and heavy rain. This was produced by combining settlement, geological, geomorphological, hydrological, vegetation and soil mapping. Unfortunately, few of these maps are available for engineers and managers to utilise in safe planning.

Many Earth scientists are beginning to believe that earthquake prediction is too complex, particularly when many earthquakes occur without identifiable precursors. Earthquakes often occur on little-known faults, for example the 1994 Northridge earthquake which was triggered by a rupture on an unmapped fault, or they may occur away from plate boundaries, such as the intra-plate quake in Maharashtra in 1993. It may, therefore, be more productive to invest time and money into research that will help to minimise the effects of earthquakes, such as building design and disaster relief programmes.

Building construction can be improved in a variety of ways to withstand earthquakes better. Amongst these designs, safer building includes:

- *Basement isolation* with foundations mounted on rubber or ball-bearings to permit the ground to move below so as to help isolate the building from the tremors;
- *Non-eccentric layout* of buildings, since irregular internal and external layout can result in complex torsional and swaying movements, causing greater damage;
- *Flexible buildings* to allow the structure to flex and deform rather than break and collapse;
- *Ground floor strengthening* to help support the overlying structure, rather than the more typical but structurally weaker ground floor design with large open spaces, many doorways, vestibules, and large window areas.

Volcanoes

Most of the world's active volcanoes are situated in the less developed and developing countries, where there are high rates of population growth. Living space, particularly on the fertile volcanic soils, is at a premium. In the more industrialised, developed nations increasing numbers of people are also living on the slopes of active volcanoes, for example Mount Vesuvius and Mount Etna in southern Italy. Volcanic eruptions pose a serious natural hazard. The USGS in Menlo Park, California, has estimated that by the turn of the century, 500 million people will be at risk from active volcanoes (Pendick 1995; see Table 8.5).

Currently, approximately 600 volcanoes are considered active above sea level, and several thousand extinct or dormant volcanoes still retain their volcanic shape. On average, five to fifteen of these volcanoes are active every month, emitting hot fluids, gases and lavas. Most of the active volcanoes are

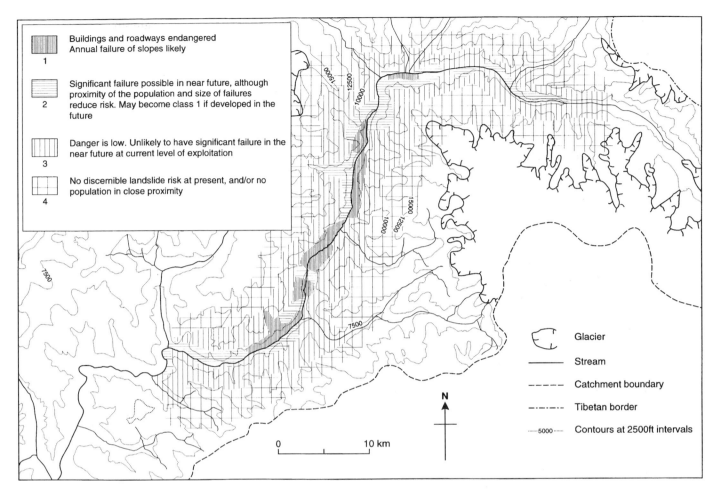

Figure 8.4 *Landslide hazard map for part of the Garhwal Himalaya in northern India. The landslides in this area are initiated by earthquakes and heavy rains during the monsoon season. After Owen* et al. *(1995).*

found in the Pacific region associated with the Circum-Pacific Belt, which is often called the 'Ring of Fire'.

Recent notable eruptions include Mount Pinatubo in the Philippines, about 100 km north of the capital Manila, which erupted after about 600 years of dormancy, with violent outpourings of ash and lava on 12 June 1991 (McCormick *et al.* 1995) (Plate 8.3). A column of ash and vapour was ejected about 25 km into the atmosphere, and 3–5 km³ of magma and 20 Mt of sulphur were ejected, with predictions from simple climate models that this eruption would cause a global cooling of about 0.5°C within the next year. During the two-week eruption, local villages were covered by more than 30 cm of ash. A typhoon exacerbated the situation with high winds and torrential rain. At least 136 people were reported killed, with hundreds injured and more than 100,000 Filipinos fleeing their homes. At the nearby American Clark Air Base, with 14,500 US servicemen and their

dependants, a nuclear alert was issued because cruise missiles, in transit from B-52 bombers in Guam to the USA following the Gulf War, were not in the deepest underground bunkers usually reserved for them: volcanic ash began to block ventilation shafts as personnel were ordered to abandon the base to nature. The thirty-six cruise missile warheads were evacuated from the base by nuclear technicians, who braved the worst of the eruption, and were then shipped out on 17 June aboard the *USS Arkansas*. The missiles themselves, together with 3,300 tonnes of munitions, remained in the bunkers at Clark Base. On 21 June, exploding ammunition dumps in the vicinity of the still-smouldering volcano caused panic in the local population. Although the eruption of Mount Pinatubo was not particularly large or devastating, by historic records, this detailed look at the eruption shows many facets of a volcanic hazard, including the additional inherent dangers in storing nuclear weapons in such areas.

Table 8.5 Major volcanic disasters and caldera crises, 1980–93.

Year	Volcano	Fatalities	Remarks
Volcanic disasters			
1980	Mount St Helens (United States)	57	Worst volcanic disaster in country's recorded history
1982	El Chichón (Mexico)	> 2,000	Worst volcanic disaster in country's recorded history
1982	Galunggung (Indonesia)	27	Relatively few deaths, but tremendous economic loss and human suffering
1985	Nevado del Ruiz (Columbia)	> 25,000	Worst volcanic disaster in country's recorded history
1991	Unzen (Japan)	43	Eruption continuing, 3,000 people still remain evacuated from their homes
1991	Mount Pinatubo (Philippines)	~800*	Largest eruption in country's recorded history; widespread destruction and socio-economic disruption. Given its huge size, fatalities were remarkably few
1993	Mayon (Philippines)	75	Unexpected, relatively small-volume eruptions produced deadly pyroclastic flows; 60,000 people evacuated
Crises at restless calderas†			
1980–84	Long Valley (United States)		Volcanic unrest continuing, but intermittent and less intense than during early 1980s
1982–84	Campi Flegrei (Italy)		Located in heavily populated region near Naples; during height of crisis, volcanic seismicity resulted in collapse of older structures, prompting temporary evacuation of 40,000 people
1983–85	Rabaul (Papua New Guinea)		During height of crisis, officials issued a Stage-2 Alert and were prepared to order evacuations of populace. Crisis abated quickly after issuance of alert

* About 300 deaths during the eruption (mostly from collapse of ash-laden roofs) and 400–500 post-eruption deaths in evacuation camps.
† None of the crises at restless calderas culminated in eruptive activity, but each caused socio-economic impact and serious scientific and public concern.
Source: Tilling and Lipman (1993).

(A)

(B)

Plate 8.3
*Destruction resulting
from the eruption of
Mount Pinatubo in
the Philippines in
June 1991. The burial
and collapse of a
Filipino village (A)
and Clark Airbase
(B), due to thick ash
being deposited during
the eruption.*
Courtesy of US Geological
Survey.

Earlier, on 3 June 1991, a volcano in the Mount Unzen complex near Nagasaki in Japan erupted, resulting in the death of 38 people. In 1792, an eruption of this volcano had killed about 15,000 people. Unlike the volcanic eruption of Mount Unzen, which was successfully predicted, Mount Pinatubo's explosions came as a surprise, basically because unlike Japan, the Philippines lacks a sophisticated and well-funded volcano monitoring system and, in any event, the volcano had not erupted for 600 years. Most of the volcanoes in the Circum-Pacific region are related to **convergent plate boundaries** either oceanic–oceanic plate collision or oceanic–continental collision. In addition, some volcanoes, such as those that form the Hawaiian islands, are believed to be the result of more localised hot magmas rising to the surface from plumes within the mantle. Geologists call the latter 'hot spots'. Other volcanoes are associated with divergent plate boundaries, where new oceanic crust is formed, such as those along the Mid-Atlantic Ridge. Iceland is an example of the surface expression of such volcanoes that have risen from the sea floor. There are two main types of volcano: those associated with divergent oceanic **spreading ridges**, oceanic–oceanic collision and oceanic hot spots, which tend to produce relatively fluid, low-viscosity lavas. The eruptions are usually gentle and the volcanoes have very gentle gradients. On the other hand, volcanoes associated with continental regions tend to have rather viscous lavas and often produce very violent eruptions and steep volcanic cones.

Fortunately, much volcanic activity affects sparsely populated regions, but when major eruptions occur, the destruction may be huge and for nearby large populations it is catastrophic. The effects of volcanic activity are of several main types, including lava flows, ejection of volcanic ash and rock material, poisonous gases, **lahars**, and fires.

Lava flows usually move relatively slowly, and tend not to cause major loss of life. They will, however, destroy any building, construction or farmland that lies in their path. Particularly threatening have been the lava flows that have advanced from Mount Helgafell in Iceland since 1973, which threaten to block the main harbour on the island of Heimaey, Iceland's main fishing port (Williams and Moore 1983). Much effort has been put into containing the lava flows by the construction of walls and also by hydraulically chilling the hot lava to redirect its flow. Such measures have limited and temporary effectiveness. In the early part of 1992, Mount Etna in Sicily began to increase its volcanic activity. Lava continued to pour out of the main vent and fissures, something that had begun in December 1991, and flowed at such a rate that Zafferana village in the path of the advancing lava had to be abandoned. A canal formed that funnelled the lava, allowing it to maintain its high temperature and liquidity and so move forward at up to 16 m hr^{-1}. Attempts were made to impede the path of the lava by channelling it into an artificial pool, but this filled in early April and the lava continued to flow over the top. By mid-April, soldiers were resorting to dropping huge concrete anti-terrorist barriers by helicopter onto the lava and detonating controlled explosions in a further attempt to divert it away from the village. It was possible to slow the advance of the lava temporarily but not, ultimately, to halt its progress.

The eruption of hot volcanic ash and rock creates one of the most severe volcanic hazards. Ash may be ejected over hundreds of miles and may stay in the atmosphere for many years. The immediate effect of the fallout of the ejected material is to destroy animal life and crops, and to cause structural damage to buildings and contaminate water supplies. Probably the most famous example of loss of life due to the ejection of ash occurred in AD 79, when Vesuvius in Italy erupted, burying Pompeii and killing 16,000 people and effectively preserving the city until its rediscovery in 1595.

Ash that stays in the atmosphere for longer periods may affect the amount of solar radiation reaching the Earth's surface and cause darkened days, and possibly may affect global atmospheric circulation. The summers following the great eruption of Krakatau in Indonesia in 1883 were recorded as being darkened, with spectacular sunsets produced by the scattering of light from the increased dust content in the air.

Lahars, mudflows produced by hot fluids, rock and water, often result from melting snow present near the summits of volcanoes. The water or melted snow and other hot fluids help mobilise rock and ash, which, along with associated collapse of slopes during the eruption, results in a melange of material flowing down-slope at fast speeds. One of the most catastrophic lahars occurred in 1985 in Columbia, when a mixture of ash and melted glacial ice flowed from Nevado del Ruiz down the valley, killing over 22,000 people (Voight 1990). The previous year geologists had predicted the event would occur should the volcano erupt, but no preventive action was taken to warn or evacuate the population. Pyroclastic flows and mudflows have been responsible for more than 85 per cent of eruption-related fatalities during this century (Tilling and Lipman 1993).

Predicting volcanic eruptions

Detailed information may be gathered regarding the effects of volcanoes by studying the deposits of previous eruptions. This includes mapping the extent of the erupted materials and dating the deposits to help determine the magnitude and frequency of the eruptions.

Monitoring active volcanoes in order to predict eruptions is common practice in many parts of the world (see *Journal of the Geological Society London* 1991). Large amounts of money are spent each year, particularly in Japan and the USA, especially in the Cascade Range, monitoring active volcanic mountains. The prediction of a volcanic eruption is based on the phenomena associated with the movement of magma towards the surface from depth. Magma moving towards the surface will initiate small earthquakes and may cause a detectable warping of the ground. The upward motion will heat up ground water and release small amounts of various gases through the soil and via small vents, called **fumaroles**. Small eruptions of lavas and ash may also occur before the large eruptive event. Monitoring of these phenomena is useful in predicting large volcanic eruptions. Much was learned by studies of Mount St Helens in the Cascade Range, western USA, by monitoring its eruption in 1980. Promising techniques involve measuring the concentrations of volcanic gases that are released as magma moves towards the surface (Pendick 1995). As the magma advances the confining pressure decreases, and the magma starts to release gas, the least soluble gas first and progressively more soluble gases as the column of magma rises further. For example, a less soluble gas such as CO_2 is released at a lower depth, and as the magma gets higher in the magma chamber more soluble gases such as sulphur dioxide are released. The ratio of gases such as these should provide a measure of the position and rate of movement of magma towards the surface. The collection and measurement of these gases is hazardous, but techniques using instruments such as an infrared telescope and the COSPEC (correlation spectrometer), have aided in the measurement of volcanic gases. The COSPEC was used to measure SO_2 concentration at Mount Pinatubo and was used successfully in helping to predict its eruption. A new device, the VGM (volcanic gas monitor), is currently being tested by Stanley Williams (Arizona State University). This is designed to draw air from the potentially active crater and transmit the data on the concentrations of gases to a remote station (Pendrick 1995). In the case of active volcanoes, the long-period seismicity, typically with periods of 0.2–2 s, is a measure of the pressure fluctuations caused by the unsteady mass transport of magma in the subterranean plumbing system of a volcano. When the seismic activity changes to relatively shallow levels may predict an imminent volcanic eruption due to pressure-induced disruption of the part of a volcano characterised by steam (e.g. Chouet 1996).

A relatively new technique for monitoring active volcanoes is through remote sensing from satellites using infrared radiation. The thermal energy given off by volcanoes can be picked up in the infrared wavelengths using devices such as the AVHRR (advanced very high resolution radiometer) carried by satellites run by the National Oceanic and Atmospheric Administration (NOAA). Plans are also being made to monitor volcanic gas emissions using the Earth Observation Satellite (EOS), which started in 1988.

Lake-water overturn

The sudden release of toxic gases is commonly associated with volcanic eruptions, but other natural processes may be responsible. The build-up of large amounts of dissolved gas at the bottom of lake waters and their catastrophic release can be a potentially serious natural hazard.

The most catastrophic effects of gas release that have been recorded occurred in western Cameroon late in the evening on 26 August 1986 as a result of carbon dioxide and hydrogen being released from the volcanic Lake Nyos, near the town of Wun about 200 miles north of the capital, Yaoundé. A cloud of gas swept along the valleys north of Lake Nyos, leaving in its path about 1,700 people dead, along with numerous livestock. Although carbon dioxide is not poisonous, because most living organisms have evolved to cope with the atmospheric concentrations, it is a very heavy gas and when present in large quantities, will form a thick layer above the ground to displace the lighter oxygen essential for respiration. At the time of the disaster, many people believed that the cause was volcanic, but now scientists think that it was due to the sudden overturn of the stratified water column in Lake Nyos under particular weather conditions. It is believed that gases held under pressure in the bottom layers of Lake Nyos were released by the mixing of lake water due to particularly heavy rainfall, which increased the head of fresh water.

The release of dissolved carbon dioxide from water occurs in an **endothermic reaction**, that is one in

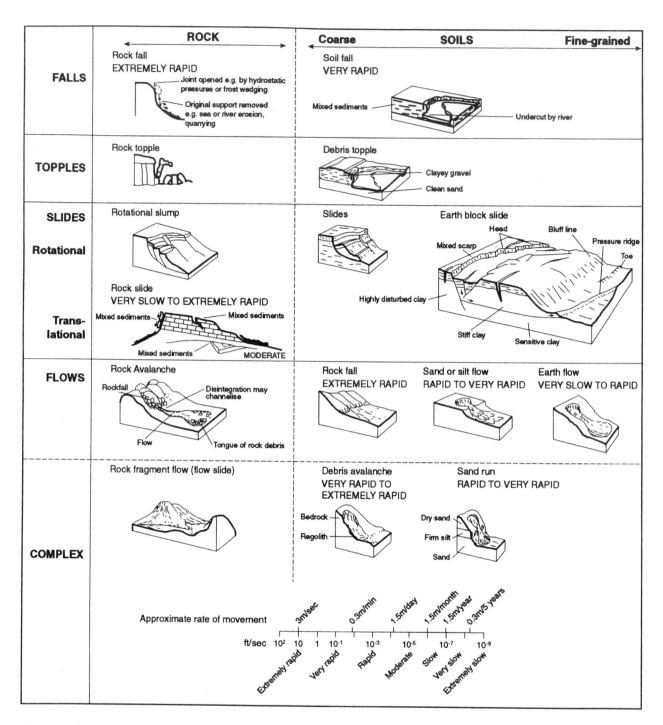

Figure 8.5 *Varnes' classification of landslides (1978).*

which additional energy for the reaction is required, consequently the temperature of both the gas and water involved cool down. The rising gas bubbles expanded and cooled further. It has been estimated that the escaping gas may have been up to 10°C cooler than the deep lake water from where it originated (Freeth 1992). Since carbon dioxide is much denser than air, the gas cloud would have spread across the lake and cooled the surface water, causing it to sink and promote the further rise of deep lake water. A positive feedback loop was probably created, in which cooler surface waters sank to greater depths, resulting in larger quantities of deeper lake water coming to the surface. These deep waters are rich in

Table 8.6 *Major mass movement disasters of the twentieth century.*

Year	Location	Nature	Deaths
1995	Kulu Himalaya, India	Landslides and debris flows	>100
1983	Sala Shan, China	Landslide destroyed four villages	227
1979	Yunjing county, China	Landslides and debris flows	799
1978	Myoko Kogen Machi, Japan	Mudflow, overwhelming ski resort	12
1976	Sau Mau Ping, Hong Kong	Landslide	22
1976	Pahire-Phedi, Nepal	Landslide	150
1975	Zhuanglong county, China	Landslide causing a wave to breach a reservoir dam	500
1974	Mayunmarca, Peru	Rockslide and debris flow	450
1972	West Virginia, USA	Landslide and mudflow in mining tip waste	400
1971	Romania	Mudslide in mining village	45
1971	Quebec, Canada	Landslides in quickclay	31
1970	Mount Huascarán, Peru	Earthquake-triggered ice and rockfall and debris flow	25,000
1966	Aberfan, Wales	Flow slide from colliery waste tip	144
1966	Rio de Janeiro, Brazil	Landslides in shanty towns	279
1963	Vaiont Dam, Italy	Landslide-created flood	c. 3,000
1962	Mount Huascarán, Peru	Ice avalanche and mudflow	c. 4,000
1959	Montana, USA	Earthquake-triggered landslide	26
1956	Santos, Brazil	Landslides	>100
1941	Huaraz, Peru	Avalanche and mudflow	7,000
1936	Leon, Norway	Rockfall into fjord, causing tsunami-like flood wave	73
1920	Gansu, China	Earthquake-triggered flow and slide in loess	200,000

After Goudie 1993a.

ferrous bicarbonate, which, upon coming into contact with the oxygen-rich surface waters, oxidised to produce a characteristic reddish-brown colour due to the formation of hydrated ferric oxide.

Landslides

The movement of material down-slope under the influence of gravity is referred to as **mass movement**. Landslides are one category of mass movement involving falling, sliding and/or gravity sediment flow of rock or weathered rock material, commonly with water, down and out of a slope. Movement is often along well-defined surfaces confined to a limited portion of a slope. There are many different types of landslide: classified on the basis of the type of material involved, plus the speed and mechanism of movement. Landsliding may involve simple fall of debris (rock fall), collapse and fall of rock/snow (avalanche), plastic flow of material (debris flow), sliding of rock (rock slide) or clay and debris (mudslide), the rapid flow and spread of the ground (earth/sensitive clay flow) and rapidly moving mixtures of debris and air and/or water (flow slides). Figure 8.5 illustrates Varnes' (1978) classification of landsliding based on the type of material moved and the mechanism involved. Most slope failure, however, involves a combination of these failure types. Mass movements are responsible for about 600 deaths and millions of dollars of damage each year; in the USA

alone they cause over US$1.5 billion in losses and 25–50 deaths each year (American Institute of Professional Geologists 1993). The most expensive landslide occurred in Utah in 1983, requiring the relocation of a major highway and railway, costing over US$200 million (*ibid.*). Table 8.6 illustrates some of the major mass movement disasters of this century.

Flow slides are the most disastrous type of landslide due to their incredible speeds and their occurrence with little or no warning. Torrential rainfall is probably the main cause of mudslides. In mid-June 1991 in northern Chile, after ten times the annual rainfall fell in under four hours near the city of Antofagasta, a mudslide caused the death of 112 people and injured 750. The mudslide then baked as hard as concrete in the sun, thereby adding to the problems of cleaning up after the catastrophe.

Earth flows are also a big problem, especially in Norway and Canada, where large areas of land consist of **sensitive clays** (Bentley and Smalley 1984). Of all the types of landslide in North America each year, earth flows account for the greatest amount of damage and financial cost. Sensitive clays cause problems because they deform and fail under stress, or due to changes in their moisture content or the chemical composition of the ground water. Failure may occur on very gentle slopes upon which buildings or excavations have been constructed.

Landslide disasters are often composite, i.e. they are triggered by other hazards such as earthquakes, volcanic eruptions or floods. Landsliding costs the United States more than US$2billion each year, with other countries such as Japan (>US$1.5 billion) and Italy (>US$1 billion) following close behind. It is estimated that there are over 600 deaths, not including major catastrophes, each year in the Pacific Basin. For example, in 1970, 18,000 people were killed in Peru as rock debris mixed with snow and ice travelled down the slopes of Mount Huascarán at speeds in excess of 300 km per hour, destroying the town of Ranrahirca (Plafker and Ericksen 1978). This failure, like many other catastrophic landslides in Peru, was initiated by an earthquake. Landsliding tends to be a recurrent problem, especially in the tropics, and the problem is not helped by persistent re-urbanisation soon after the disaster on the still unstable slopes.

Rock falls are a major hazard, particularly in engineering work for bridges and tunnels. On the northern island of Japan, Hokkaido, in early February 1996, a 50,000-tonne slab of rock crashed onto the Toyohama Tunnel, about 50 km northwest of Sapporo, killing twenty people. In a vain attempt to reach any possible survivors, rescue workers attempted to dislodge the enormous block by blasting but were unable to dislodge it until days afterwards.

Snow avalanching is a growing problem in mid- and high-latitude regions as people expand into marginal areas for both recreation and resource exploitation (Armstrong 1984, Butler 1987). In the Khumbu Himalaya in Nepal, more than seventy tourists and guides were killed during a series of snow avalanches, which were the result of particularly heavy snow falls in November 1995. In some areas of the world, for example in Switzerland, snow avalanche warning schemes and hazard mapping have been particularly successful in helping to mitigate the hazard (Frutiger 1980), but most regions, particularly those in developing countries, lack the resources and expertise to implement such measures.

The susceptibility of a slope to failure is dependent on many factors. These include the gradient and length of the slope, the geotechnical properties of the rock (such as the rock strength), cohesion, and if the rock has discontinuities such as joints and bedding planes. The amount of water entering a slope will be a very important factor, which is a function of vegetation, drainage, soil type, the structure of the rock, i.e. if the rock has joints or voids, and the duration and intensity of rainfall. If there are large quantities of loose rock material or soil that are capable of being moved, then erosional processes such as undercutting of the slope by rivers, the

sea or glaciers and the possibility of earthquake or volcanic activity will determine its likelihood of failing. Finally, the use humans make of slopes or adjacent regions can drastically affect slope stability. Particularly susceptible areas are those where there are steep and long slopes with large quantities of weathered rock, high rainfall, the possibility of earthquakes, poor drainage, intense erosion and poor vegetation. Such conditions exist in many tropical Southeast Asian countries, especially where deforestation is taking place rapidly, since the vegetation previously acted as a protective cover. The number of landslides in such regions has increased because of human activities, particularly as a result of the construction of highways (Owen 1996).

Human activities have also been responsible for many landslides in urban areas as slopes are exploited for urbanisation and the creation of artificial slopes. A particularly infamous example occurred on Hong Kong Island, where large tower blocks were built on, and adjacent to, steep tropical slopes. In 1972, major slope failure occurred on one of these slopes in a residential area, the Mid-Levels, completely destroying a newly constructed tower block and severely damaging an adjacent tower. This resulted in 138 deaths (Government of Hong Kong 1972). The most infamous failure of an artificial slope occurred in the small mining village of Aberfan in South Wales. This was caused by the failure of a badly managed coal waste tip. The failure occurred during a morning in October 1966, sending coal waste mixed with water down-slope at speeds of 16–32 km hr^{-1} to engulf a school and adjacent buildings. Most of the 144 people killed were schoolchildren (Welsh Office 1969).

It was through such events as Aberfan that much attention is now given to the potential of a slope to fail. Most large constructions in developed countries will make detailed site investigations to examine the ground conditions to prevent loss of life and preserve property. There is still a reluctance, however, among engineers, especially in less developed countries, to contract detailed site investigations, even though the cost of such studies is small compared with the costs encountered should a catastrophe occur. The hazard from landsliding can be greatly reduced by considering the factors involved in slope failure and attempting to reduce the magnitude of the forces acting on a slope that may lead to failure. Engineering geologists and civil engineers will calculate the factor of safety (F_s) of a slope. This is the ratio of the forces increasing the shear stress (S), i.e. the forces pulling an element of slope downhill, to the forces resisting shearing of the slope (R), such

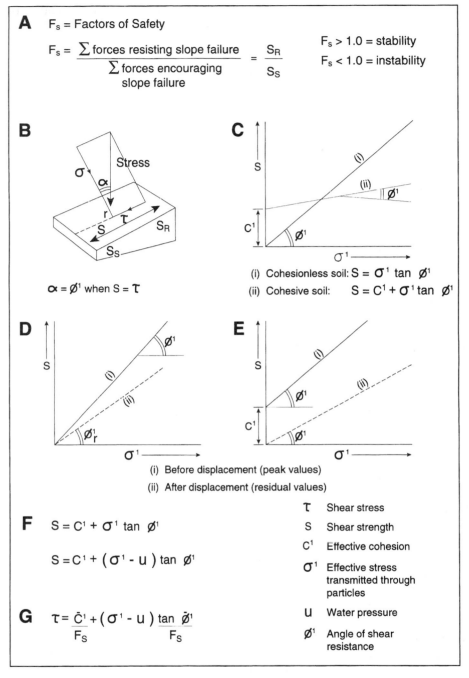

Figure 8.6 *Some fundamental principles of slope stability analysis: (A) Equation for the factor of safety of a slope; (B) stress at a point on a slope divided into two components, the pressure normal to the slope (σ) and the shear stress (τ), which operates in the same plane but in the opposite direction to the shear strength (S). When S = τ, angle a = angle of shearing resistance (φ); (C) shear strength (S) against effective pressure (σ') for cohesionless and cohesive materials; (D) in cohesionless materials, the relationship between peak and residual values of φ' and c'; (E) in cohesive material, σ'r is due to dilation plus a loss of strength resulting from the reorientation of platy particles; (F) Column's equations, which relate shear strength and effective stress, defining the equations of the lines on graphs (C), (D) and (E); (G) relates the shear stress (τ) to the effective stress, cohesion and factor of safety. Redrawn and adapted from Cooke and Doornkamp (1990).*

Table 8.7 *Factors involved in slope failure.*

Factors leading to an increase in shear stress

- Removal of lateral or underlying support
  Undercutting by water (e.g. rivers, waves) or glacier ice
  Weathering of weaker strata at the toe of the slope
  Washing out of granular material by seepage erosion
  Human-made cuts or excavations, draining of lakes or
    reservoirs
- Increased disturbing forces
  Natural accumulations of water, snow, talus
  Human-made pressures (e.g. stockpiles of ore, tips,
    rubbish dumps or buildings)
- Transitory earth stresses
  Earthquakes
  Continual passing of heavy traffic
- Increased internal pressure
  Build-up of pore-water pressures (e.g. in joints and
    cracks)

Factors leading to a decrease in shearing resistance
- Materials
  Beds that decrease in strength if water content increases
    or as a result of stress release following slope formation
  Low internal cohesion
  Bedding planes, joints, foliation, cleavage, faults, etc.
- Weathering changes
  Weathering reduces effective cohesion, and the angle of
    shearing resistance
  Absorption of water leading to changes in the fabric of
    clays
- Pore-water pressure increase
  High water table as a result of increased precipitation or
    human interference such as dam construction

After Goudie 1993a and Cooke and Doornkamp 1990.

that $F_s = R/S$ (Figure 8.6A). When F_s is greater than 1, the slope is stable, whereas for values less than 1 the slope is unstable and has the potential to fail. $F_s < 1$ can occur when S is increasing and/or when R is decreasing. Table 8.7 lists the factors that can lead to values of $F_s < 1$ and subsequent slope failure. A more detailed analysis of a slope's stability requires a geotechnical consideration of the physical properties and mechanics of the slope. Figure 8.6 shows the simplified principles of slope stability analysis.

In an attempt to produce landslide hazard maps that can be used by planners and policy-makers, geologists in some countries are commissioned to produce maps showing the distribution of causal factors that may initiate landslides. Studies are varied, ranging from hazard mapping of debris flow in the Canadian Rocky Mountains (Jackson 1987), and tectonic and monsoon-induced mass movements in the Himalayas (Owen *et al.* 1995; Figure 8.4), to landslides in loess in central China (Derbyshire *et al.* 1991). In the UK, the Department of the Environ-

ment has a computerised database of all known British landslides. This can be used to help planners to identify areas and regions of instability and as an aid to reducing the hazard of building on ancient or active landslides. The most impressive database has been developed by the Hong Kong Geotechnical Control Office. Data include topography, vegetation, hydrology, geology, weathering characteristics, geotechnical characteristics and land use and are available for the whole of Hong Kong. These are published as reports and are computerised in a geographical information system, allowing sound planning. The varied causes of landslides and the increased land-use pressures, however, still make this one of the most common natural hazards.

River flooding

In any season of the year, somewhere in the world, river floods cause the loss of life and the devastation of agricultural land. In mid-September 1992, three days of torrential monsoon rains caused many rivers to swell, such as the Indus, resulting in extensive flooding in northern Pakistan and India, leaving more than 2,000 dead and hundreds of thousands of people homeless. The Ganges–Brahmaputra–Megna drainage basin is just one example of a very densely populated and low-lying coastal region subject to repeated major flooding (Figure 8.7). In the USA throughout the twentieth century, hundreds of people have lost their lives due to flooding (Figure 8.8).

River flooding may occur due to high rainfall, melting snow in spring, or the emptying of lakes when a natural or artificial dam is breached (Figure 8.9). The amount of water overspilling a river bank is a measure of the magnitude of the flood. Generally the larger the flood the less frequent its occurrence. The area of land that is naturally flooded, often annually, is called the 'flood plain' of the active river channel. The hazard of flooding is measured as a function of its destructive capacity in terms of life and property. Hydrologists who study flooding also refer to the magnitude of floods in terms of recurrence intervals. Hydrologists advising engineers and planners in the design of structures and buildings will provide information on the size of a possible flood that may be expected to occur with a recurrence interval of 10, 25, 50 or 100 years (Chow 1964). The magnitude of a natural flood produced by a rainstorm is related to the intensity and duration of that rainstorm, the rate at which the water flows across the land surface toward rivers, and the rate of flow of water through the ground to drainage systems.

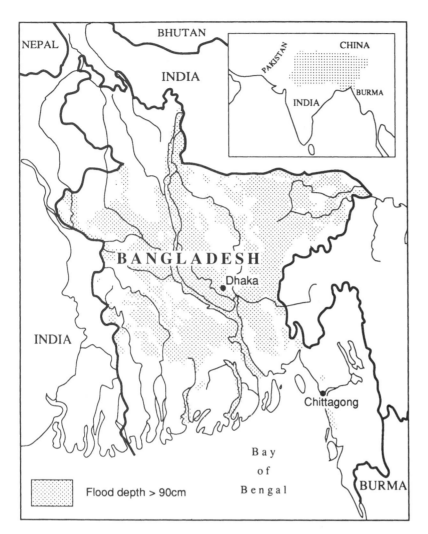

The nature of a rainstorm depends upon the climatic conditions for that region. In desert regions, for example, torrential rainstorms may occur only once or twice a year and when they occur the rainfall is usually very intense and lasts only a short time. In humid high-latitude regions, in contrast, rainstorms may occur intermittently throughout the year, and are usually less intense. In deserts, therefore, river channels tend to fill up quickly, and **flash flooding** is common, whereas in high latitudes flash floods occur less frequently.

The rate at which water can enter river channels during rainstorms depends particularly upon the vegetation cover, which acts to intercept the falling rain and may stop it reaching the ground. The ability of the soil and bedrock to soak up water, the steepness and irregularity of the slope and the number of gullies or rills on that slope will also control the rate of movement of water towards drainage channels. The speed at which run-off water travels through the ground to rivers will determine whether or not flooding is likely.

Large floods occur where there is little vegetation, steep slopes and many gullies, soils that are impervious to water, and rocks that allow water to flow quickly to rivers. This is because the water level rises more quickly in the streams to overspill the banks before the river has time to discharge farther downstream to the sea. Such floods are called 'downstream floods' because the effect of the flooding increases progressively downstream.

Floods produced by the breaching of dams and melting snow tend to be 'upstream floods', which are confined to smaller areas, and their effects are more limited downstream. Flooding may be particularly disastrous along coastal areas if high river discharges coincide with high tides. The high tides prevent the river from releasing the large quantities of water, so that coastal areas are flooded. This was the case in 1953 in the Netherlands, when 1,835 people were drowned as a result of a high sea surge in conjunction with rivers discharging large quantities of flood water into the North Sea.

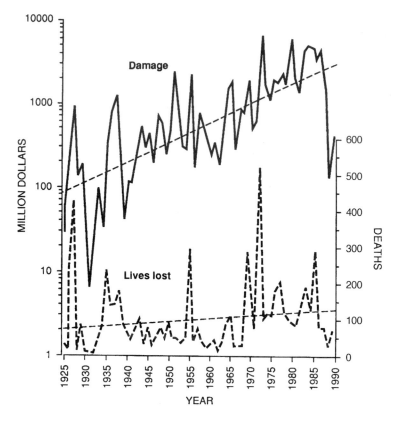

Figure 8.8 *Annual deaths and economic losses caused by flooding in the United States, for the years 1925–1989. Damages are in US$ millions adjusted to 1990 values. Redrawn after Smith (1992).*

In most developed countries, water height in rivers is monitored using **stream gauges**. If the water level rises to a critical height, then flooding is imminent and a flood warning may be issued. In addition, rain gauges are used to monitor the progress of a rainstorm and, from experience and study of a particular river, warnings may be issued if flooding is likely. Rainfall monitoring is more useful than flood gauging because there is a time lag between the peak of the rainstorm and the rise of river level while the rain water flows into the main river. Hence, more time is available to issue an effective warning.

Flood hazard maps are useful in assessing the extent of likely damage and the area of land that needs to be evacuated. Such maps are produced using historical data and evidence from old river strand lines (high-water marks) (Wolman 1971). In addition, they are useful in providing planners and legislators with information for further development.

Unfortunately, today, the risk from flooding is increasing in some areas because of conditions asso-

ciated with human activities. The development of a region usually means that natural vegetation is cleared, and roads, buildings and **storm sewers** constructed with the channelisation of adjacent river courses. The clearing of vegetation reduces the ability of the ground to intercept rain water. The rain may then flow quickly over the impervious surface of concrete and asphalted buildings and roads, and flow rapidly to the rivers along storm sewers and channelised streams. All this means that the rate at which water enters the main drainage channel or river increases, as does the height of water. The likelihood of flooding is correspondingly increased. This can be shown by comparing **flood hydrographs** for a region before and after urbanisation. Figure 8.10A shows a typical flood hydrograph, which is used to describe the characteristics of discharge, the amount of flow passing an area of channel, over a flood period. Note that the discharge reaches its greatest after the maximum rainfall, because the discharge takes time to build up as water collects at the point of measurement from tributary streams and ground-water sources. Figure 8.10B shows the changes in the flood hydrographs for an area before and after urbanisation. The effects of urbanisation can also be illustrated by comparing the changes in mean flood for a drainage area, and the increase in number of flows per year (Figures 8.10C and D).

Urban planners are beginning to consider these factors in their attempts to reduce the likely hazards from floods. One method used to mitigate the effects of floods is to construct **retention ponds** to collect the storm water and to control the release of the water back into the main river.

Some of the most recent devastating flooding in the USA was in the Mississippi River basin in July 1993, the worst in the past twenty years. Following a period of torrential rainfall, when it rained for forty-nine consecutive days, the Missouri River rose to its highest recorded level at St Louis of more than 15 m above normal, causing very extensive flooding of farmland and settlements, and leading to the death of more than thirty people. An estimated 50,000 people were flooded out of their homes, and in Des Moines, Iowa, the flooding left 250,000 inhabitants without clean drinking water or adequate domestic sanitation facilities. As a result, President Clinton promised large amounts of federal aid to mitigate the effects of this flooding, estimated to have cost around US$10 billion.

Opinion is divided, however, as to what measures should be taken to reduce future flooding in the Mississippi basin. The Army Corps of Engineers, which has been channelling and building levees on

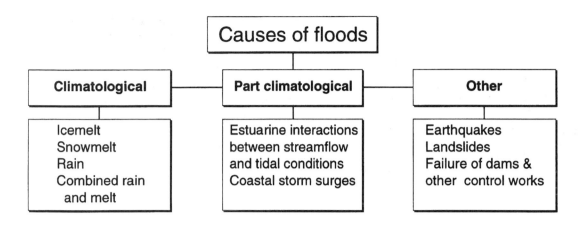

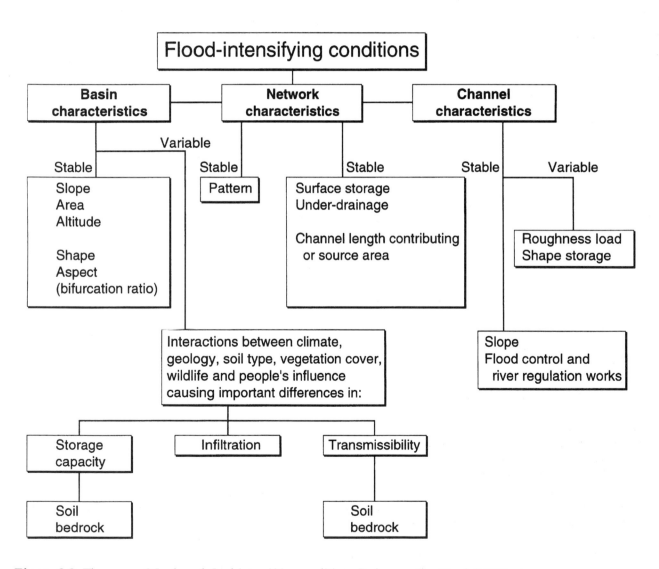

Figure 8.9 *The causes of floods and flood-intensifying conditions. Redrawn after Ward (1978).*

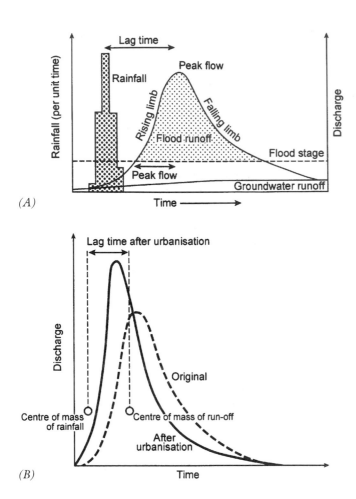

(A)

(B)

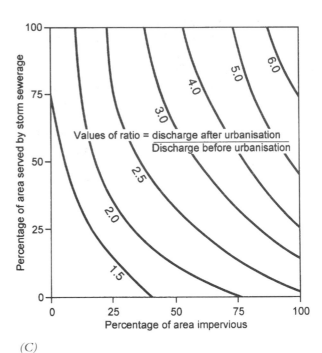

(C)

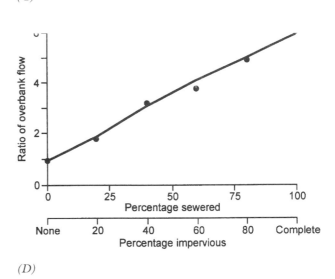

(D)

Figure 8.10 *(A) A typical flood hydrograph, showing the important flood characteristics. (B) Flood hydrographs for an area before and after urbanisation. (C) Urbanisation on mean flood for a 1 mi² drainage area. (D) Increase in the number of flows per annum equal to or exceeding channel capacity for a 1 mi² drainage area, as a ratio to the number of overbank flows before urbanisation. Redrawn after Leopold (1968).*

the Mississippi for the last two hundred years, believe that repairs and improvements to levees are the way to mitigate future threats of flooding. Many environmental groups, however, have blamed the floods on the channelisation schemes themselves. They believe the best remedy is to return the drained flood plains to their original wetland environments (Macilwain 1993 and 1994). The suggestion that channelisation schemes caused the problem and that wetlands would reduce the impact of flooding is hotly disputed by the Army Corps of Engineers. It points out that if the levees had not been present the flooding would have caused an extra US$19 billion of damage. Furthermore, the Army Corps of Engin-

eers believes that a return to wetland environments will only help to contain floods smaller than the hundred-year flood (Macilwain 1994). In examining the details of the 1993 flood and its context in terms of the history of flood management in the Mississippi basin, Myers and White (1993) emphasised the importance of the need for well-planned legislation for disaster preparedness, insurance schemes and plans for remedial measures for the management of river basins of this kind. The 1993 flood illustrates the complex issues involved in the management of large river basins throughout the world. The Mississippi basin, however, has a distinct advantage compared with many other river basins: it is almost completely confined within one nation. Management and planning for flood schemes is not as simple to implement for rivers that flow through many countries with divergent political agendas.

Another recent catastrophe occurred during mid-February 1995, when California experienced some of the worst floods in its history. Westerly Pacific winds brought a succession of violent storms, which over one 24-hour period produced more than 7 inches of rain. Eleven people were killed and 3,000 were made homeless, with estimated damage of over US$200 million (Bogert 1995). Santa Barbara became a virtual island, highways were washed out and its airport was covered in mud and logs. Landsliding, debris flows and stream floods accompanying the flooding exacerbated the problems.

Knox (1993) has shown that there may be a strong correlation between the magnitude and frequency of floods and climate. He examined a 7,000-year record of overbank flood sediments for the tributaries of the upper Mississippi. These geological records showed that during a warmer, drier period about 3,300 to 5,000 years ago, the largest floods were extremely rare and relatively small. After 3,300 years BP, however, the climate became cooler and wetter, and floods that are now considered to have a 500-year recurrence interval were more common. Still larger floods occurred between AD 1250 and 1450, during the transition between the mediaeval warm period and the Little Ice Age. He pointed out that all these changes were associated with annual temperature changes of approximately 1–2°C and changes in mean annual precipitation of 10–20 per cent. Baker (1995) emphasises the implications of this study, stressing the increased threat of flooding that is likely to occur should the world's climate warm up over the next few decades. Such studies help to provide real data for the global climatic modellers to use and to aid policy-makers evaluating the threats that may result from global warming. Questions therefore arise as to whether the recent flooding in the Mississippi basin and California is actually an early warning of global warming.

Glacial hazards

There are four main types of glacial hazard: glacial advances; glacial floods; ice falls; and floating/capsizing icebergs. The degree of hazard depends on the geography of the region and the dynamics of the glacier. Most glaciers are in remote, commonly mountainous areas, where populations are small. However, as some of these areas become more populated due to increasing demands on land space

because of growing populations, and as the areas are utilised for leisure activities, the hazards become more apparent.

Glacial advances constitute a hazard when ice approaches a settlement, threatening to override it. The most famous example of this is the Mer de Glace in the Chamonix Valley, France, which threatened the villages of Les Bois and Le Chatelard in the seventeenth and eighteenth centuries but has since retreated (Tufnell 1984). Glacier movement is usually slow, being of the order of a few metres per year. However, **surging glaciers** occur where rates are measured of the order of several metres per day (Dowdeswell *et al.* 1991). These surging glaciers may pose a threat to settlements, but more commonly they act as dams blocking tributary river valleys and forming lakes, which have the potential to drain catastrophically.

Other glacial floods relate to the release of water from within or beneath the ice (Plate 8.4). The most spectacular examples of these floods have occurred in Iceland, where volcanic activity beneath glaciers melts large quantities of ice with the resulting floods being of extremely high magnitude, having catastrophic effects on the landscape, bridges and roads. These are known as jökulhlaups (Thorarinsson 1974). Jökulhlaups have also caused damage in other parts of the world, notably British Columbia (Mathews and Clague 1993) and Alaska (Strum and Benson 1985).

The advance of valley glaciers may also threaten settlements and people working in glaciated areas, especially where the slopes over which they flow are

Plate 8.4 *Boulder cluster in the Chandra valley. These boulders were transported more than 30 km during a catastrophic flood that drained from an ice-dammed lake in the upper Chandra valley as a glacier retreated at the end of the Last Glacial. Glaciers that are retreating today constitute similar hazards.*

Plate 8.5 *Abandoned village and its dry fields in the Karakoram Mountains, northern Pakistan. This settlement would once have been a fertile area but because glaciers in this area have continued to retreat over the last century, their waters cannot be redirected into this area without the use of expensive pumps.*

convex. Here, the glacier extends and may tumble down-slope as ice falls, colliding with and destroying obstacles in its path.

Where settlements depend on glacial melt waters for irrigation, a retreating glacier may constitute a hazard because the flow of melt water into irrigation systems is reduced. In regions like the Karakoram Mountains, northern Pakistan, a number of glaciers have retreated during this century, resulting in several villages being abandoned, such as the one shown in Plate 8.5.

Icebergs are one of the most common and hazardous glacial phenomena. They are produced where individual glaciers flow into the sea, or at **ice shelves**, where extensive platforms of glacial ice form, such as in Antarctica. The glacier ice floats and may break up by a process called **calving** along its margin to form blocks of ice that may range in size from a few metres to several tens of kilometres. In September 1986, an enormous piece of the Antarctic

Filchner Ice Shelf some 600 km across broke off and formed three separate icebergs, which together covered an area of 13,000 km². These icebergs floated into the Weddell·Sea to be grounded on the shallow sea floor, although one of them, called the A24 iceberg, floated free early in 1991 to travel into the Southern Atlantic Ocean, where it finally melted (Vaughan 1993). In autumn 1991, calving of the Ross Ice Shelf in Antarctica created an iceberg that floated towards the Falkland Islands and measured 50 by 52 km, which is around four times the area of the Isle of Wight in England. The movement of icebergs away from glaciers is governed by ocean currents and to a lesser extent by wind. The routes are often complex and irregular and the iceberg may last a considerable time as melting of ice involves considerable energy and is, therefore, very slow. In addition, the movement of icebergs may be halted for long periods if the sea freezes around them.

They are a persistent problem for shipping in mid and high latitudes, as was dramatically illustrated on 14 April 1912 by the iceberg that collided with the *Titanic*, which subsequently sank. In areas where icebergs are common, continuous information regarding the position of icebergs is issued to shipping on VHS radio. An area particularly prone to iceberg hazard is off the northeast coast of Canada in the Labrador Sea. Icebergs arrive in this region from calving of glaciers on Queen Elizabeth Island (Canadian High Arctic) and in northeast Greenland, having floated down the Davis Strait: they may even travel as far as New York. Iceberg monitoring and management is taken very seriously in the northwest Atlantic off eastern Canada and has involved such schemes as firing jets of water onto the iceberg to redirect its movement. The enormous size of some of these icebergs, such as the recent iceberg floating towards the Falklands, makes the controlled directing of icebergs impossible.

Grove (1987) has discussed these hazards in the light of climate change, and Owen (1995) extends these arguments to consider mountain hazards in the Himalayas by examining the processes associated with glacial fluctuations and possible future global warming. The variability and unpredictability of these hazards, since they are linked to climate, will inevitably increase in the future.

Asteroid and comet impacts

The hazard from the collision of an **asteroid** or **comet** with the Earth is not as unlikely as it may seem. In Chapter 2, the evidence for the impact of an asteroid or comet at the Cretaceous–Tertiary boundary was examined. Its impact probably resulted in the mass extinctions associated with this instant of geological time. The impact probably resulted in the spread of wild fires across the globe producing poisonous gases, and the atmosphere cooled as a result of the smoke and dust associated with the impact. The carbonate compensation depth rose, prohibiting many marine organisms from secreting calcite shells. The implications of a similar impact today are only too obvious.

Asteroids and comets are occasionally perturbed into paths that cross the orbits of planets. Spacecraft exploration has shown from the crater-scarred surfaces of the planets that there has been a slow, but continuous bombardment since the initial deluge of asteroids and comets during **planetary accretion**, which ended approximately 3.8 Ga. Asteroids and comets are remnants of the accretion and formation of the planets before this time. Despite the fact that the surface of the Earth is constantly renewed by erosion and tectonism, over 140 impact craters have been recognised on land. Improved telescopic search techniques continue to identify dozens of new Earth-crossing asteroids each year, and near misses are being reported more frequently.

Most meteoroids that are <50 m in diameter dissipate their energy harmlessly, breaking up as they enter the upper atmosphere and being consumed before reaching the lower atmosphere. A larger projectile, however, could do severe local damage if it hit an urban area. Such impacts occur about once every century or more, but rarely do they hit urban areas. In 1908, an iron meteorite fell over the Tunguska River in Sikhote-Alin, Siberia, resulting in trees being felled over an area of >1,000 km² and a fireball that ignited a large area near the airburst. The fragments created approximately 100 craters greater than 1 m in diameter. This testifies well to the effects of a small impact by such phenomena. Chapman and Morrison (1994) believe that such impacts constitute a much less serious threat than other natural disasters such as floods and earthquakes. However, they suggest that objects greater than 0.5 to 5 km in diameter have the potential to cause global catastrophes. They calculated that the risk from an impact of this size is similar to that of other natural and technological disasters. The difference, however, is that with the exception of a nuclear war, only the global impact catastrophe could lead to the breakdown of civilisation. Their calculations suggest that there is a one in 10,000 chance that a large (approximately 2 km diameter) asteroid or comet will collide with the Earth within the next century.

Our present knowledge of the distribution of such objects is still poor, and there is a need to continue to map their distribution, and to discuss possible means of mitigation. This might include the use of nuclear explosives in deep Space to deflect the orbit of the asteroid or comet away from Earth, obviously an issue that will be controversial among policy-makers. NASA has sponsored several workshops on the hazard, and programmes such as 'Spaceguard Survey' have been proposed that will produce an inventory of all potentially threatening asteroids large enough to cause a catastrophe. The awareness of such a hazard is increasing and over the next few decades it will attract much more attention as our knowledge and understanding of the asteroid threat grows.

Tornadoes

Tornadoes are violent windstorms that take the form of a rotating column of air moving at speeds in excess of 116 km hr⁻¹ in the rotating spiral (Plate 8.6). Tornadoes rarely may reach speeds of 480 km hr⁻¹, but their average speed across land is about 45 km hr⁻¹, and the average path distance is approximately 26 km. The typical diameter of the spiralling air column is 150–600 m, though some have been recorded up to 1.5 km. Tornadoes are compared using the **Fujita intensity scale**.

As they travel across country, tornadoes cause great devastation, with loose objects, including people, being carried skywards. Buildings and crops are severely damaged or destroyed. Houses with closed windows may explode because of the sudden creation

Plate 8.6 *A tornado, seen from a distance of 5 km, in Minnesota, USA, June 1968.*
Courtesy of Associated Press.

of low pressure outside the building induced by the passage of the tornado. In the USA alone, about 750 tornadoes are reported every year. Occasionally, several hundred people may be killed in one day, for example in a region between Canada and Georgia on 3 April 1974, 300 people died due to tornadoes. In the spring of 1995, 484 tornadoes killed 16 people and caused millions of dollars of damage (Davies-Jones 1995).

Reports of loss of life and damage from tornadoes are common. In Kansas and Oklahoma in the Mid-west USA on Friday 26 April 1991, twenty-seven people were killed (mainly in a caravan park near Wichita, where approximately 400 mobile homes were destroyed and twenty-two people left dead) as up to fifty separate tornadoes ripped through a region that is euphemistically called 'Tornado Alley' because of the frequency of its storms. On 27 March 1994, a series of tornadoes and violent storms swept through five southern states in the USA, leaving forty-four people dead, with nineteen of these deaths in Piedmont, Alabama, where a church roof collapsed. These tornadoes developed from a complex depression near the eastern Rockies formed as cool air from the northwest interacted with very warm and moist air masses advected north from the Gulf of Mexico.

Tornadoes are usually associated with severe thunderstorms, when cold air from continental areas converges with warm, wet air from oceanic areas. These storms, and the development of tornadoes, are particularly common phenomena in mid-latitude continental areas during the summer (Fujita 1973). The Mid-west USA, for example, is particularly prone to tornadoes during the months between April and September. These storms, sometimes referred to as a mesocyclone or a supercell, develop along the gust front, the boundary between warm and cold air. They are produced in very unstable environments where winds vary greatly with height and cool, dry air lies atop a mile-deep layer of warm, moist air. The two unstable masses of air are separated by a thin layer of stable air, which confines the underlying air but can be pierced if the Sun warms the lower air or if another weather system invades the region, such as a jetstream, a frontal system or upper-level disturbances in the atmosphere. Air is then drawn upwards, and as pressure decreases with altitude, the rising air expands and cools; eventually water vapour starts to condense and a flat cloud base forms. Latent heat is released as the air condenses. This warms the rising air, causing it to rise to great heights at velocities that exceed 200 km hr^{-1}. The buoyancy of the rising air parcels is partially offset by rain droplets, which form as cloud droplets coalesce. The parcels of rising air lose momentum and descend to heights of about 8 miles above the Earth's surface, flowing out sideways to form anvil-shaped clouds. The falling rain evaporates in the dry mid-level air, causing the air to cool and sink to the Earth, being pulled around the up-draughting air by the storm's rotation. The cool air has a higher relative humidity than the warm air and when forced upwards becomes cloudy at lower heights, forming a lower wall of clouds. These superstorms contain only one or two cells, each with its own coexisting down-draught and broad, rotating up-draught. This allows the storm to maintain a steady and intense state. When the storm is mature or decaying, low-level rotation develops and a tornado may form. The pressure at the centre of a tornado may be as little as 10 per cent of that of the surrounding air. This pressure difference sucks air from all directions towards the low-pressure centre, resulting in winds spiralling inwards and upwards, often forming thick clouds. Eventually the tornado dissipates as the up-draught is cut off near the Earth's surface by cold air flowing out of the storm.

The detailed mechanism by which tornadoes form within such storms is still not fully understood, something that makes their prediction difficult (Davies-Jones 1995). Because exact predictions are impossible at present, and tornadoes are short-lived, they are among the highest-ranking natural hazards. In the USA, a tornado watch is issued in areas where cold fronts develop and tornadoes are likely. When a tornado has been sighted, a warning is issued but, because the path of the tornado is often unpredictable, it is difficult to determine the exact path of damage. Tornadoes, in some instances, can be located using conventional radar. However, the new technology of Doppler radar allows a tornado to be identified up to 20 minutes before touchdown. The USA is currently installing a network of sophisticated Doppler radar across the country to try and improve the efficiency of tornado warnings. Such technology may allow more accurate identification of tornadoes to be made in the near future.

Tropical cyclones

Tropical cyclones are known by various names. They are commonly called hurricanes in the Atlantic–Caribbean, typhoons in the northwest Pacific and cyclones in the Indian Ocean. Cyclones are zones of low pressure, in which the low-pressure region causes the inflow of winds from surrounding regions. In

Plate 38 *Before and after Hurricane Andrew at Biscayne, Florida, in 1992.*
Courtesy of Comstock.

Plate 39 *An electric storm on Grand Calumet Island, Quebec, Canada.*
Courtesy of Comstock.

Plate 40 *Some of the effects of drought in Nigeria.*
Courtesy of Horst Munzig/Comstock.

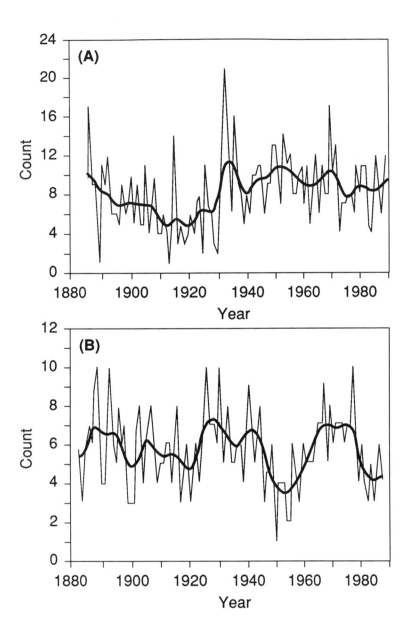

Figure 8.11 *Estimated number of tropical cyclones in (A) the Atlantic and (B) the North Indian Ocean over the past century. Data in (B) are less reliable before 1950. Redrawn after IPCC (1990).*

the tropics, between the latitudes of 5 and 20°, these storms produce winds that can reach 300 km hr⁻¹. They consist of an intense asymmetric vortex of moving air with an average diameter of about 600 km and a height of about 12 km. The vertical development of thick clouds and heavy rain is an integral feature of a tropical cyclone. The 'eye' is relatively tranquil, but away from this centre winds spiral round and upwards at great speeds. Every year, tropical cyclones, which vary considerably in frequency (see Figure 8.11), cause loss of life and damage to property (Figure 8.12 gives US figures over much of this century). In October 1995, for example, Hurricane Opel swept across the Gulf of Mexico, beginning in Mexico, killing ten people and travelling across

Florida, Alabama and Georgia before finally dissipating over Tennessee. In the USA, it left fifteen people dead and caused more than US$2.4 billion of damage, the third costliest storm in US history (Davis and Castaneda 1995).

Tropical cyclones constitute a hazard in three main ways. First, the strong winds can cause considerable damage. In September 1900, in the worst natural disaster recorded in US hurricane history, more than 6,000 people died in the coastal city of Galveston (then the richest in the state of Texas and known as the New York of the South, dealing with more than 1,000 ships a year), and more than 200 people died in smaller towns, as hurricane winds devastated the region. The highest land point in Galveston stood

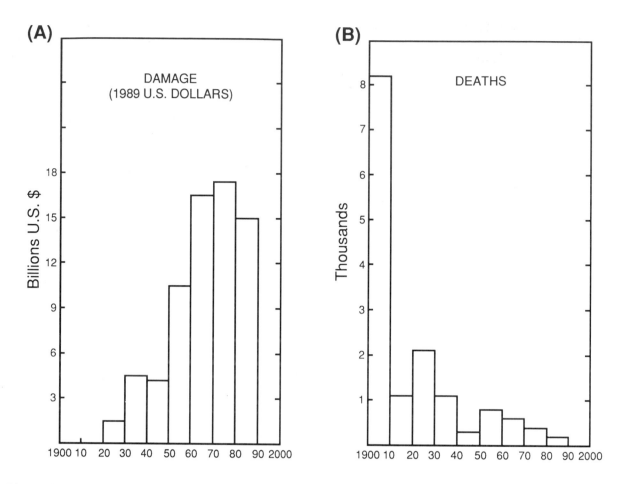

Figure 8.12 *Losses of (A) property and (B) life in the continental United States due to tropical cyclones for the periods 1915–89 and 1900–89, respectively. Redrawn after Gross (1991).*

only 3 m above sea level, but the hurricane brought a storm surge more than 6.5 m high.

The second major problem is the storm surge produced by a tropical cyclone where its eye comes onshore from the open sea. The low barometric pressure caused by the hurricane leads to the development of a dome of water several metres high and 65–80 km long. Such '**coastal set-up**', as it is known, is a particular problem in low-lying delta regions such as Bangladesh or along the coast of the Gulf of Mexico.

In 1970, for example, a surge of just 2 m was responsible for the deaths of as many as 500,000 people in Bangladesh (Figure 8.7). Again, in Bangladesh on 29/30 April 1991, following nine hours of very strong winds gusting up to 140 mph, and high seas with waves up to 6–7 m high, the low-lying coastal areas and offshore islands at about sea level suffered severe flooding and destruction as an estimated 125,000 people lost their lives, thousands of homes were destroyed, about 10 million

were left homeless and hungry, and the harvest was wiped out. Because of the poor sanitary conditions and heat, a cholera epidemic followed the cyclone and many more people lost their lives. The worst flooding and loss of life occurred south of Chittagong, the second city and main port. Poverty drives the Bangladeshi people to eke out a precarious living on these fertile but flat-lying islands and coastal plains covering about 55,000 square miles, generally less than 3 m above sea level and prone to repeated flooding. Following the last really severe tropical cyclone in this region about twenty years ago, 500 shelters were planned, with about fifty completed at the time of the cyclone, but despite warnings to the population and even allowing for the inadequacy of the shelters, it is believed that many people simply did not leave their homes in time. From those who did, there were reports of arriving at shelters to find them full and so being turned away. The scale of these disasters is quite simply awesome.

The third problem is the heavy rainfall commonly associated with tropical cyclones, which causes inland fresh-water flooding. Hurricane Agnes, which was moderate in magnitude, produced extensive floods that were responsible for 122 deaths and US$2 billion of damage to property. On Friday 11 September 1992, on the Hawaiian island of Kauai, Hurricane Iniki severely damaged the homes of more than 50,000 inhabitants, but fortunately without loss of life, when winds gusting to 256 km hr^{-1} and torrential rain swept the island. Waves 7 m high crashed over coastal highways and the island lost all power and telephone services. It was the most powerful hurricane to hit the Hawaiian Islands this century. One month earlier, more than fifty people lost their lives, and thousands of homes were damaged, when Hurricane Andrew roared through the Bahamas, south Florida and Louisiana.

Meteorologists explain the formation of hurricanes using a model they call a '**heat engine**', which is initiated by the evaporation of warm water (>27°C) from tropical seas. Condensation of this water releases latent heat to warm the air and make it buoyant, so that it rises. The result is to reduce the pressure above the sea and cause an inflow of air, that is, wind. As a result of the Earth's rotation, the wind spirals in an anticlockwise direction in the Northern Hemisphere and a clockwise direction in the Southern Hemisphere. The path of a tropical cyclone is also influenced by the Earth's rotation, directing tropical cyclones in a clockwise direction in the Northern Hemisphere, and in an anticlockwise direction in the Southern Hemisphere. If the wind speed does not exceed 61 km hr^{-1}, the cyclone is described as a tropical depression and if less than 119 km hr^{-1}, a tropical storm. It is only when the air speed exceeds 119 km hr^{-1} that the cyclone is described as a cyclone, hurricane or typhoon. A continuous supply of atmospheric moisture is essential to the continued existence of a tropical cyclone, and once it enters a continental area, its energy is quickly dissipated because of the drop in air moisture from sea to land (Anthes 1982, Barry and Chorley 1992).

The prediction of the precise path of tropical cyclones is not an exact science, since they follow unpredictable tracks, although weather satellites can accurately chart their formation and development. Once formed, tropical cyclones are observed continuously and monitored using weather satellites. There is a good correlation between the frequency of hurricanes and El Niños. During El Niño years there is a reduction in the frequency of hurricanes in the tropical Atlantic, up to about 40 per cent fewer than in a normal year. The long-term prediction of tropical cyclones requires continuous historic records going back thousands of years, but the oldest weather station in the world, at Oxford University in the UK, goes back only over two centuries.

In the USA, during the hurricane season from June to November, about 40 million people are at risk. Warnings are issued to populations when a tropical cyclone approaches coastal regions so that preparatory measures can be taken. Unfortunately, many of the areas that are very prone to tropical cyclones are in less developed countries, where resources for preventive measures are inadequate, so that large numbers of people still die as a result of tropical cyclones each year. It is important that the issuing of warnings for an imminent tropical cyclone hitting coastal regions be accurate because the evacuation of people, the closure of businesses and the range of preventive measures can cost very large amounts of money.

In the 1960s, Project Stormfury, a co-operative project between the US Navy and the National Oceanic and Atmospheric Administration in the USA, began to investigate the possibility of modifying or destroying a hurricane. A process was tested called 'cloud seeding', in which silver iodide crystals are dropped into the storm clouds immediately outside the eye of a hurricane so that water vapour condenses on the particles to produce rainfall, which in turn reduces the heat energy produced by the hurricane, causes the eye of the hurricane to expand outwards and hence slow down the maximum wind velocities and hence damage. Project Stormfury actually seeded two hurricanes during the duration of the project, and the scientists involved in the project believed the seeding significantly reduced wind speeds. Many authorities, however, were sceptical about the results from the project and countries such as Cuba protested at the USA altering weather patterns that could result in hurricanes merely being diverted from the USA to neighbouring countries, so it was terminated in 1972. Since then little attention has been given to modifying hurricanes, although their development has been simulated using elaborate numerical and computer modelling.

Tropical cyclones are, by definition, restricted to storms that form in low latitudes. In October 1987, however, in an exceptional storm event that hit southern Britain, hurricane-force winds of more than 120 mph caused more than £100 million of damage, and about 15 million trees fell or were so severely damaged that they had to be felled. This storm had all the characteristics of a hurricane and shows that such storms may rarely affect parts of the world outside the tropical cyclone belt.

Thunderstorms

There are three main hazards associated with thunderstorms: torrential rain, hailstorms and lightning. Thunderstorms are produced as warm air rises rapidly from near the surface of the Earth. As the air rises, more air from the surrounding regions is dragged inwards, which in turn is up-draughted. The rising air cools with height, and water vapour in the air begins to condense, forming thick, dark, billowing clouds that eventually spread out at heights of about 6–12 km to form the characteristic anvil-shaped tops. Torrential rainfall results from the large amount of condensation. Associated with the rapid draught of air, the lower portion of clouds becomes negatively charged. When sufficient electrical potential has built up compared with the positively charged ground, an arc of electrically charged atomic particles, called electrons, passes from the cloud towards the ground, carrying a current of as much as 60,000–100,000 amperes. This arc is the lightning we observe during a storm (Barry and Chorley 1992). Attempts at controlling the effects of lightning using lasers to initiate and direct the strikes are currently being investigated by workers at the Electrical Power Research Institute in California (Muir 1995). The air heats up due to this electrical charge, expands and finally explodes, producing the noise of a thunder clap. The details of this process are still not fully understood.

There are between 1,500 and 2,000 thunderstorms throughout the world at any one moment, with more than 5,000 lightning strikes every minute. On average lightning strikes kill five people in the UK and about ninety-five in the USA every year. Those at greatest risk are people who work outdoors and those who participate in outdoor sports. People out of doors may form positive point sources to which electrical charges are attracted. If lightning is discharged through a person, the large currents are likely to result in the person being severely burned or they may induce heart failure and subsequent death. However, people indoors have also been known to experience electric shocks produced by lightning, usually if the person is earthed while holding a water tap or talking into a telephone.

The most destructive product of a thunderstorm is the hail. It is formed as raindrops within the storm are carried up under the up-draught into the thunderstorm cell, become frozen, and form ice pellets. These pellets increase in size by the attachment of freezing droplets of water. When the size of the hailstones increases such that the up-draught can no longer support their weight, they fall to Earth under the influence of gravity. Some of these hailstones may be larger than hens' eggs, which may cause great damage to crops and property (*cf.* Morgan 1973).

The annual damage from hail each year in the USA exceeds US$300 million. Most of this loss results from the destruction of crops. Hail-suppression programmes have been in progress since the 1960s in the erstwhile Soviet Union, where hail is particularly hazardous to large areas of farmlands. This has involved cloud seeding by firing rockets into the clouds. These rockets explode and release silver iodide crystals. The purpose is to initiate heavy rainfall and to dissipate cloud production, and therefore the heavy up-draughts produced in a thunderstorm that are responsible for elevating rain droplets and their subsequent freezing to form hailstones. However, many scientists are not happy with these experiments, and currently cloud seeding of thunderstorms is neither practised nor permitted by governments.

Snowstorms and cold spells

In the USA, approximately 60 million people live in urban areas that have a high risk of snowstorms (Smith 1992). In mid-latitude regions, heavy snow falls and cold spells are related to the development of mid-latitude cyclonic depressions in winter time. The location of these depressions is controlled by the position of the sub-polar jetstream, whose path meanders and position changes unpredictably, making long-term forecasting and hazard preparation difficult. Snowstorm hazard becomes greater when the snows from earlier storms have not melted away.

Snowstorms and severe cold spells can result in several types of hazard and disruption. Table 8.8 illustrates the variety of disruptions caused by snowstorms as a hierarchial classification. The most noticeable is the disrupting of communications, when highways and rail lines are blocked by thick snow, or when black ice or glaze makes driving hazardous. In the USA and Europe, the effects are enormous, with roads and airports being closed within a few hours after the first heavy snow falls. This disruption is financially expensive, as are safety costs. The road accident rate in the USA, for example, increases by 200 per cent within the first few hours of a snow fall (Bryant 1991). Unprepared stranded motorists caught in a storm may freeze to death, be buried beneath the snow, or die from carbon monoxide poisoning as they try to warm themselves with the vehicle's engine.

The effects of such storms are strongly dependent on the population's preparedness and awareness of

Table 8.8 *Hierarchy of disruptions in urban areas due to snowstorms.*

Activity	*1ST order (paralysing)*	*2ND order (crippling)*	*3RD order (inconvenience)*	*4TH order (nuisance)*	*5TH order (minimal)*
Internal Transportation	Few vehicles moving on city streets City agencies on emergency alert, Police and Fire Departments available for transportation of emergency cases	Accidents at least 200% above average Decline in number of vehicles in CBD Stalled vehicles	Accidents at least 100% above average Traffic movement slowed	Any mention Traffic movement slowed	No press coverage
Retail trade	Extensive closure of retail establishments	Major drop in number of shoppers in CBD Mention of decreased sales	Minor impact		No press coverage
Postponements	Civic events, cultural and athletic	Major and minor events Outdoor activities forced inside	Minor events	Occasional	No press coverage
Manufacturing	Factory shutdowns Major cutbacks in production	Moderate worker absenteeism	Any absenteeism attributable to snowfall		No press coverage
Construction	Major impact on indoor and outdoor operations	Major impact on outdoor activity Moderate indoor cutbacks	Minor effect on outdoor activity	Any mention	No press coverage
Communication	Wire breakage	Overloads	Overloads	Any mention	No press coverage
Power facilities	Widespread failure	Moderate difficulties	Minor difficulties	Any mention	No press coverage
Schools	Official closure of city schools Closure of rural schools	Closure of rural schools Major attendance drops in city schools	Attendance drops in city schools		No press coverage
External[a] Highway	Roads officially closed Vehicles stalled	Extreme driving conditions warning from Highway Partrol Accidents attributed to snow and ice conditions	Hazardous driving conditions warning from Highway Patrol Accidents attributable to snow and ice conditions	Any mention, for example, 'slippery in spots' warning	No press coverage

Continued overleaf

Table 8.8 Continued

(Activity)	1ST order (paralysing)	2ND order (crippling)	3RD order (inconvenience)	4TH order (nuisance)	5TH order (minimal)
Rail	Cancellation or postponement of runs for 12 hours or more Stalled trains	Trains running 4 hours or more behind schedule	Trains behind schedule but less than 4 hours	Any mention	No press coverage
Air	Airport closure	Commercial cancellations	Light plane cancellations Aircraft behind schedule owing to snow and ice conditions	Any mention	No press coverage

CBD = Central Business District.
a Warnings are the key to this classification. They provide excellent indicators because they are so widely publicised.
After Rooney 1967.

the hazard. Cities that experience large annual snow falls are generally more prepared, with effective management programmes for the removal of snow. Many regions expanded programmes for salt- and sand-spreading to reduce the hazard from ice and the problems of snow compaction after a snowstorm. Salting, however, creates its own environmental problems because as the snow/ice melts, salt is washed into water bodies or the ground and can have profound ecological effects. This is becoming a particular problem in Lake Ontario, where increasing lake salinity is threatening fresh-water life (*ibid.*).

Power supplies may be disrupted because power cables may break due to the weight of snow and ice that has accumulated on them. Furthermore, water supplies may be cut off as pipes may be frozen. Many of these may burst, because the water within them expands as it changes phase; when the pipes thaw, substantial amounts of water may be lost, causing flooding and other associated problems. Old people are the most susceptible during these periods, the number of fatalities due to hypothermia increasing. Disaster relief is made difficult because of the difficulty of getting into snow-covered areas. In late December 1995 and early January 1996, these effects were demonstrated dramatically when severe snow-storms and cold spells caused chaos in northern Scotland, especially the Isle of Skye and the Shetland Isles. In these regions, people were without electricity and water for up to two weeks, and relief was hindered by the remoteness of the regions and the lack of communications because of the deep snow that covered highways in the regions. In addition, blizzards hampered the efforts of helicopters bringing relief supplies to these areas.

After a severe snowstorm, flooding may become a hazard as the snow melts. The historic flooding of the Mississippi River in spring 1973 occurred as a result of rapid melting of snow within its catchment during a period of a few days. Snow avalanching is another common consequence of early spring melting of snow packs (Daffern 1989).

Snowstorms and cold spells can have longer-term effects on a community. The infamous 1967 spring snowstorms in the rich agricultural area of southern Alberta resulted in the region being declared a disaster zone (Janz 1968).

Droughts

Droughts are periods of increased dryness due to precipitation falling far short of that expected for a region. Droughts may occur in every type of climate, arid to humid, tropical to tundra, and have profound economic implications for many regions as crops fail and cattle die. Individual droughts vary considerably in nature, effects and duration. Their effects are most evident in semi-arid lands, where increased aridity leads to evaporation of water from the soil and consequently to poor plant growth, increased soil erosion, dust storms, and deficient and polluted brackish water. This is particularly detrimental in developing countries, which are strongly dependent on subsistence agriculture and are capable of doing little to mitigate the problems. The main problem with droughts is not so much the small amounts of precipitation, but its variability. A region can have extremely low precipitation (<100 mm yr^{-1}) and be described as arid, whereas a drought is a temporary phenomenon where rainfall is made less by high temperatures, low humidity and strong winds. There are complex definitions describing the severity of aridity and droughts based on the variability of temperature, precipitation and evaporation, but there is no precise definition in terms of the human impact.

Droughts owe their origin to climatic variations that result in decreased precipitation, although some people believe the effects of droughts are largely caused by bad land management. Specific mechanisms for this variation are unclear, but climatologists consider these variations to be quasi cyclical events. There are natural cycles of weather changes, but the duration and onset of droughts cannot be predicted at present. The effects of a drought also depend on other factors, such as the land's productive capacity, the increased demands on soils and surface waters, and the disaster preparedness and awareness of a society.

Some of the largest droughts have occurred in the Indian subcontinent, e.g. an estimated 3–10 million people died in the 1769–70 drought and, during 1876–78, 3.5 million people died near Madras. The problem of drought in India has been greatly eased during recent decades with the development of water schemes such as the Rajasthan Canal and the Tabela Dam and associated irrigation schemes using the waters from the Indus River. Particularly hard hit in recent years have been the countries of sub-Saharan Africa.

During 1983–85, hundreds of thousands of people died as a result of famine due to poor plant growth and reduced agricultural yields in Ethiopia and the Sudan. The droughts in Africa are related to the movement of the **intertropical convergence zone** (**ITCZ**) north and south of the equator. This produces a distinct dry and wet season. The ITCZ reaches its most southerly extent by January and is between 15 and 20°S in East Africa and about 8°N in West Africa. This allows the development of subtropical high-pressure systems, characterised by dry continental tropical air, to dominate northern Africa. By July or August, the ITCZ has migrated northwards to about 20°N, drawing moist maritime tropical air from the Atlantic Ocean, bringing heavy monsoon weather characterised by southwesterly winds and heavy rains. If the ITCZ fails to move northwards enough, then rains will not fall over northern Africa and droughts will occur. Recently, this has been the case during the period between 1968 and 1973, 1980 to 1987, and 1990. The timing of the movement of the ITCZ is unreliable, further exacerbating the problem. The reasons for this variability are not really understood, but it has been suggested that changes in sea-surface temper-

ature could be responsible for intensifying the strength of atmospheric circulation within the **Hadley cell**. When sea-surface temperatures in the Southern Hemisphere are greater than those in the Northern Hemisphere, the strength of the ITCZ is reduced and there are below average rainfalls in the Sahel (Owen and Ward 1989). Other causal mechanisms may help contribute to this variability, including the ENSO, sunspot activity and changes in albedo related to land degradation.

The additional stresses placed on the land by increasing populations, wars and changes in agricultural practices have contributed to the problems of drought in Africa.

Droughts are also common in developed countries, for example the countries of Western Europe and the USA. Here the associated problems are generally fewer but include decreased water levels in lakes and reservoirs, lower agricultural yields, possible forest fires, and reductions in the production of hydroelectric power and the use of water for recreational purposes.

The early history of the exploration of the western plains of the USA, which became known as the Great American Desert, testify to such droughts. When the first settlers moved into the region in the late 1860s, the weather was wet, but by the 1880s and 1890s, droughts had caused the majority of settlers to move on. Some stayed and experimented with dry farming. Recent droughts on the Great Plains occurred during the mid-1970s, early 1980s, 1987 and 1988 (Kemp 1994). Dendroclimatic evidence shows that particularly severe droughts also occurred in the years around 1752, 1820 and 1862 (Meko 1992). Clearly drought is an integral part of climate on the Great Plains. The droughts are a consequence of aridity due to low rainfall, which is aggravated by falling in the summer months, when evaporation is high, and with a tendency for wet or dry years to run in series.

Famines often accompany droughts, as a result of crop failure and reduced stocks of cattle. Famines, however, are much more complex than simply climatic in origin, involving political and social issues. Sen (1981) describes famines as being the characteristic of some people not having enough to eat, not the characteristic of there being not enough to eat. Africa illustrates this well: during 1983 and 1984, food production in Sudan and Ethiopia declined by 11 and 12.5 per cent, respectively. As a result, there was widespread famine. During the same period, other parts of Africa experienced greater declines in food production: Botswana by 17 per cent and Zimbabwe by 37.5 per cent, but there was no famine in those countries. The difference lies in the tyrannical political regime and conflicts that existed in Ethiopia and Sudan, whereas the governments of Botswana and Zimbabwe were involved in effective famine relief programmes.

Other great famines illustrate this again. The famine in Ireland between 1846 and 1851 was partially a result of potato blight, but during the worst famine years, grain was being exported from Ireland because the English Corn Laws made it too expensive for the Irish to buy. The famine in Bengal in 1943 was the result of the British confiscating rice stores in rural Bengal to reduce the possibility of it being obtained by Japanese invaders. The famine in Bangladesh in 1974 was not the result of rice shortages but the doubling in price because of rumours of shortages. This resulted in a price war and mass starvation. Dando (1980) attributes all the famines in Russia between 971 and 1970 to human factors, and Arnold (1988) attributes the nineteenth-century famines in China to the corrupt Qing Dynasty's failure to maintain the infrastructure of the peasant agricultural systems.

When climatic droughts persist, they may lead to 'desertisation', more commonly termed '**desertification**' to include the human element. The land becomes degraded, with little vegetation and poor soils, which are often deeply eroded. Opinions differ on the relative roles of climate and humans in causing desertification, and if it can be prevented or reversed.

Relieving the effects of drought in vulnerable countries depends on short- and long-term planning, with a consideration of the water demand in terms of land-use change, increased urbanisation, industrial development and other water-related activities. Planning should be supported by monitoring the drought, and developing technologies and projects such as irrigation systems.

Pests

Pests are usually defined as any organisms that are judged to be a threat to human health, comfort and endeavours. Commonly, pests multiply and spread rapidly and, therefore, compete with humans for available food sources such as crops. Pests include micro-organisms such as fungi, bacteria and viruses; invertebrates such as protozoa, flatworms, nematodes, snails, slugs, insects and mites; and vertebrates such as rabbits and many other rodents.

Pests that transmit disease are of particular concern and include mosquitoes transmitting malaria and yellow fever; tsetse fly transmitting sleeping sickness; human lice, fleas, mites and ticks transmitting typhus; blood-sucking bugs carrying the flagellate protozoan *Trypanosoma cruzi*, which is responsible for Chaga's disease (American trypanosomiasis) causing swelling, lymph node enlargement, fever and heart failure; sandflies transmitting leishmaniasis; and the roundworm *Onchocerca volvulus*, which causes river blindness (onchocerciasis). The potentially most serious diseases and disease-carrying pests are discussed below.

Most species that have become pests appear to have adapted to take advantage of human activities associated with increased demands for food production. However, in some cases, the spread of pests has occurred as a result of entirely natural causes, for example locust plagues. Locusts (part of the Orthopteran family Acrididae) have plagued humans for centuries: the Biblical accounts of the plague in Egypt during the time of Moses are well known; in more recent times, the devastation of the prairie farms of Canada and the USA during the 1870s by Rocky Mountain locusts and migratory grasshoppers (*Melanoplus spretus* and *M. sanguinipes*); and even more recently, plagues of locusts in West Africa and the Mediterranean. Locusts can travel vast distances; for example, in 1869 desert locusts (*Schistocerca gregaria*) reached England, probably from West Africa. *S. gregaria* is known to be able to fly in swarms up to 1.5 km high. The size of swarms can be large, for example in 1889 a flight across the Red Sea was estimated to be approximately 2,000 square miles in size. Once formed, a plague is virtually impossible to stop.

Methods of control include destroying the eggs laid by locusts; digging trenches to trap the nymphs (the young stage of the locust, not yet able to fly); using hopperdozers, which are wheeled screens that cause the locusts to fall into troughs containing water and kerosene; using poisoned bait; and spraying swarms and breeding grounds with insecticides. Eradication of the nymphs before they mature and begin to swarm is amongst the most effective methods. A study of the life cycle of the locust is important for the prevention of plagues. In 1945, the Anti-Locust Research Centre was established in London to record and project migrations of swarms, and to study the biology of the locust in order to try and reduce its effects. One of the main problems with control is that locust swarms tend to appear sporadically and unpredictably. This is attributed to the 'phase theory', which suggests that there are two phases of locust: the normal state or solitary phase, in which an individual adjusts its colour to match its immediate surroundings, has a low metabolic and oxygen-intake rate and is sluggish; and the gregarious phase, in which an individual has an affinity to group, is black and yellow in colour, has a high metabolic and oxygen rate, and is very active. If an individual in one or other of these two phases develops within a group dominated by the other phase, it will mature into the other phase. The gregarious phase is considered to be a physiological response to violent fluctuations in the environment. Migratory swarms form in marginal areas, where suitable habitats are scarce. A succession of favourable seasons causes restricted populations to expand and forces them beyond these marginal areas. When less favourable environmental conditions return the enlarged populations in the marginal areas must return to the relatively small permanently habitable areas, and this causes overcrowding leading to swarming.

Today, the transmission of pests is a global problem because of the ease of world travel and global trade. Strict regulations on the importation and quarantine of hazardous food substances and crops, travel of people suffering from diseases, and non-vaccinated people with their associated pests, are enforced by many countries. The risks from pests have increased over recent decades with the change from natural vegetation and traditional small-farming techniques to monoculture farming systems, which are intensive and large. This has led to greater hazards by providing a more uniform food source for plant-eating species to increase rapidly in number. The introduction of new crops to an area may also transfer previously harmless species from their natural habitat to the new abundant food source. Cultural practices such as fertilisation, irrigation and the use of modern harvesting techniques, which leave large amounts of plant litter in the field, also enhance the ability of pests to increase in numbers. The elimination of species that prey on pests and the evolution of insects that become insecticide-resistant exacerbate the problem.

The control of pests has a long history, dating from the ancient Chinese, who used predator ants to control foliage-feeding insects, and the introduction of the Indian mynah bird to Mauritius by Westerners in 1762 to control the red locust, to sophisticated modern physical, chemical and biological controls. Pest control really began to be studied in the eighteenth and nineteenth centuries with increasing populations, industrialisation and agricultural expansion. Methods included physical and chemical controls. Physical methods involved the use of sticky barriers, flooding and burning, but these were effective for only short periods of time. Chemical pesticides began with the use of poisonous plants such as ground tobacco, which was used in France in 1763 to kill aphids. Other natural products that were widely used included nicotine, petroleum, kerosene, creosote and turpentine. Inorganic products such as Paris green, lime sulphur, Bordeaux mixture, hydrogen cyanide and lead arsenate began to be used in the 1800s. Biological means were also used; particularly successful was the introduction of the vedalia beetle (*Rodolia cardinalis*) from Australia to control the spread of cottony-cushion (*Icerya purchasi*) in California in 1888, saving the citrus fruit industry. Plants resistant to pests were also introduced; the control of *Phylloxera*, an aphid-like insect that attacked the European vine, was controlled by grafting a more resistant American strain onto the vine.

The discovery of the insecticide properties of synthetic organic compounds that appeared during the Second World War gave hope for the view of pest-free crops. These included DDT (dichloro-diphenyl-trichloroethane) and BHC (benzene hexachloride), and herbicides such as 2,4-D (2,4-dichlorophenoxyacetic acid). But the serious ecological problems associated with the use of these has led to their ban in many countries.

Current pest controls try to minimise the use of pesticides and combine them with the use of biological methods. These integrated methods include the breeding of pest-resistant crops; crop culture methods that inhibit pest proliferation; the release of predators or parasites; the use of traps that are baited with the pests' **pheromones**; the disruption of reproduction by the release of sterilised pests; and the application of chemical insecticides.

Environmental diseases

Biotic agents are among the largest natural hazards threatening both man and his animals. They include infectious diseases spread by parasites, bacteria and viruses, such as hepatitis B, schistosomiasis (bilharzia), cholera, typhoid, poliomyelitis and malaria. In addition, environmental diseases include **neoplastic diseases** such as environmentally induced cancers, initiated by drinking water supplies or produced in areas where radioactive rocks produce high background radiation. Little information is available on environmentally induced cancers, and the connection between the environment and the prevalence of cancers is uncertain.

Estimates of mortality are difficult to determine for many diseases, commonly because the disease contributes to death but may not appear on any death certificate or statistics, together with poor records being kept in many poor developing nations.

Morbidity statistics are published annually by the World Heath Organisation and provide one of the most reliable estimates. For example, from these tables, it appears that annually there are more than 10 million cases of malaria, over 50,000 cases of polio, some 200 million people infected by schistosomiasis, and approximately 50,000 cases of cholera. Young children, the infirm and the aged are amongst the most susceptible to infection, and there is commonly a clear correlation between poverty and disease, hence the associated high infant mortality rates in developing regions.

The transmission and prevalence of infectious diseases are a function of climate, water supply, sanitation, and socio-economic factors such as nutritional status, hygiene, and population density. Many diseases are transmitted by hosts, such as malaria by the mosquito, schistosomiasis by fresh-water snails,

and sleeping sickness by the tsetse fly. Migration and increased populations of the host help the disease to spread. Eradication of the host or the host's environment helps to contain the spread of disease. Control methods include the spraying of pesticides in lakes and irrigation dykes to kill mosquito larvae and snails carrying schistosomiasis. Most of these infectious diseases can be cured with medical help, especially if they are diagnosed early enough, but they usually have long and debilitating effects. Prevention is considered the best method of control, but unfortunately this can be expensive and difficult to manage in poor, developing countries. Therefore infectious diseases still rank high as a natural hazard. Box 8.1 deals with the most widespread and devastating diseases.

The publication of Richard Preston's book *The Hot Zone* in 1994, the recent movie 'Outbreak' starring

BOX 8.1 COMMON INFECTIOUS DISEASES IN DEVELOPING COUNTRIES

Malaria

Malaria is a serious and chronic relapsing infection in humans, apes, rats, birds and reptiles, and is characterised by periodic paroxysms of chills and fever, anaemia, and enlargement of the spleen, which may lead to fatal complications. Malaria is caused by blood sporozoans of the genus *Plasmodium* and is normally transmitted by *Anopheles* mosquitoes. It can also be transmitted by hypodermic needles and blood transfusions. Hyper-endemic areas include Central and South America, North and Central Africa, the countries bordering the Mediterranean, the Middle East, and East Asia. In many parts of Africa and Southeast Asia, entire populations are infected almost continuously by the disease. It is most common in the tropics because conditions are most favourable to the mosquito.

There are several types of malaria: the most widespread is Vivax (tertian) malaria because it is able to withstand therapy and remains chronic. Falciparum (subtertian or malignant tertian) malaria has the most severe symptoms, tends to cause death, and is confined to the hottest tropical regions, particularly in West Africa. Quartan malaria is present throughout the Mediterranean. Infection from more than one of these forms of malaria can occur in any individual. Effective treatment for malaria has been known since the 1700s, when the bark of the cinchona tree was used, its most active ingredient being quinine; but it was not until much later that the mosquito was recognised to be the transmitter. Anti-malarial drugs may be used to prevent the disease or as a suppressive to reduce or eradicate it. Some individuals have a natural resistance to malaria, e.g. those who carry the sickle-cell trait. An acquired

immunity decreases susceptibility to the disease. Control of mosquito populations is the best prevention, but the use of insecticides such as DDT in the 1950s and 1960s caused other widespread ecological problems. Global eradication programmes have been under way since 1955, when major concern was expressed by the WHO. Unfortunately, many of the countries still threatened by malaria epidemics are poor, and it is difficult to implement control and eradication schemes.

Yellow fever

Yellow fever infects humans, all species of monkeys, and other small mammals. The virus is transmitted by mosquitoes, either between individuals (urban/classical yellow fever) or from a forest mammal to a person (jungle yellow fever). Yellow fever causes headaches, backache, rapidly rising fever, nausea and vomiting, severe haemorrhages in the mucous membranes, and destruction of liver cells resulting in jaundice, and it may result in death (dependent on the virus strain). Typically, the course of the fever is rapid and convalescence is long, but it produces a lifelong immunity in the victim. The disease has plagued the tropics and subtropics over the past two centuries. The USA was subjected to devastating epidemics and outbreaks as far north as Boston, but the last serious outbreak occurred in 1905 in New Orleans and other parts of the southern states. During the nineteenth century, similar plagues paralysed industry and trade throughout the West Indies, Central America, Spain, Italy and England.

The disease is completely preventable by the use of live-virus vaccines or by eradication of the mosquito. One of the most famous eradication campaigns was undertaken to control *Aedes aegypti* mosquitoes during the construction of the Panama Canal, a course of action that saved many thousands of lives.

Typhoid (typhoid fever)

The bacterium *Salmonella typhi* enters the body via contaminated food or water and is absorbed into the bloodstream, causing blood poisoning. Following a 10–14-day incubation period, headaches, lassitude, aching, fever, restlessness resulting in loss of sleep, loss of appetite, nose bleeds, coughs, and diarrhoea or constipation result. Fevers with very high temperatures occur after a further 7–10 days and continue for about 10–14 days, at which time the fever begins to subside, but fluctuates diurnally. Complications include gall bladder inflammation, pneumonia, encephalitis, meningitis, and even heart failure.

Major epidemics have been caused by the pollution of water supplies, by contaminated food and milk, and by flies and persons transmitting the disease. Shellfish grown in polluted waters are a particular cause of the disease. Many cured victims continue to transmit the disease for life, transmitting the bacteria in their faeces, where it may continue to live for months and infect others.

Prevention involves proper sewage treatment, filtration and chlorination of waters, and even the exclusion of carriers from working in the food industry and restaurants. Prophylactic vaccinations are given to people travelling in affected regions, but their effectiveness is limited. Typhoid was prevalent throughout the world up until the beginning of this century, but improved waste disposal management and water treatment have reduced the incidence. It is in the developing countries, where sewage treatment plants are inadequate or non-existent, and where clean water supplies are scarce, that this disease remains a serious problem.

Sleeping sickness (African trypanosomiasis)

Sleeping sickness is transmitted by the tsetse fly, which infects humans and cattle with the flagellate protozoans *Trypanosoma gambiense* and *T. rhodesiense*. Sleeping sickness causes fever and inflammation of the lymph nodes, and it affects the brain and spinal cord, leading to lethargy and frequently death. It is hyper-endemic in central Africa, causing mass mortality in cattle and people. Treatment of the infection by *T. gambiense* includes the use of the synthetic arsenical tryparsamide, but this may result in optic neuritis and loss of vision. Suramin sodium can be used for both infections, but once the infection by *T. rhodesiense* has developed to the toxaemic stage, treatment is ineffective.

Great efforts have been made to control sleeping sickness, including the isolation and treatment of victims; protection from bites; eradicating the tsetse fly by clearing its habitat around villages and using insecticides; the use of prophylactic doses of suramin and damidine compounds for persons entering affected areas; relocation of entire villages from endemic zones to disease-free areas; and in extreme cases, the extermination of the wild game reservoir for the disease. In some areas, these controls have been very effective, particularly in Cameroon, where the locals have persisted with the use of tryparsamide to reduce the problem to very low levels.

Schistosomiasis (bilharzia)

Schistosomiasis is caused by parasitic flatworms of the family Schistosomatidae (commonly called blood flukes), which, as part of their life cycle, live in the blood vessels of humans and other mammals. There are three main types of fluke (*Schistosoma japonicum*, *S. mansoni* and *S. haematotobium*) which are geographically defined and are endemic in Southeast Asia, Africa and South America, and the Middle East and southern Europe. The female fluke is 10–25 mm in length and may release between 300 and 3,500 eggs daily into the blood of a mammal. The release of eggs within the body produces tissue damage, causing allergic reactions, inflammation, coughs, skin eruptions and swelling, and tenderness of the liver, and faeces and urine may contain blood. Chronic stages of the disease affect the body's vital organs, leading to fibrous thickening and loss of elasticity, which may result in serious liver damage, stone formation in the bladder, and bacterial infections of the urinary tract. In extreme cases, eggs may be lodged in the brain and lungs to cause death. The disease is spread by victims releasing eggs from the intestine or bladder into their faeces and urine. The eggs hatch when they reach water, and the larvae swim to a snail host, where they are able to develop within the skin of the snail. Once developed, the fork-tailed larvae (cercariae) leave the snail and swim in water until they find a mammal host, whereupon they penetrate the skin and feed within the blood system, thereby continuing their life cycle.

Early diagnosis and treatment usually ensures recovery from this disease, but since it is common in poor countries, where many people cannot afford medication, prevention provides the best mechanism of control. Preventive measures require a break in the life-cycle of *Schistosoma*, by reducing the possibility of contact between the larvae and potential hosts during bathing or working in standing bodies of water or irrigation canals, places where the host snail abounds and cercariae are released. Attempts have been made to reduce snail populations by molluscicides, which include sodium pentachlorophenate, dinitro-o-cyclo-hexylphenol and copper sulphate. Commonly, irrigation schemes and reservoir construction increase the potential habitats for the host snails and, therefore, have increased the environmental health risks in such regions.

Dustin Hoffman, and the outbreak of Ebola in Kikwit, Zaire, in spring 1995, have highlighted the potential threat of **haemorrhagic fevers**. Patients with these illnesses initially develop a fever, followed by a period in which the patient deteriorates and superficial bleeding develops, where blood seeps from vessels under the skin and bruises appear. Other cardiovascular, digestive, renal and neurological complications may

Table 8.9 Outbreaks of haemorrhagic fever viruses.

1995	More than 190 died from an Ebola outbreak in Kikwit, Zaire
1994	Seven people were infected, six died with Machupo in Bolivia
1994	A researcher at Yale University was accidentally infected with Sabià, but survived
1993	Hantavirus Sin Nombre infected 172 and killed 58 in New Mexico, Colorado and Nevada, after a rodent population grew rapidly
1990	An agricultural engineer died and a laboratory worker fell ill with the arenavirus Sabià in the state of São Paulo, Brazil
1989	More than 100 cases of illness were caused by Guanarito in Venezuela. The epidemic started in a rural community that began to clear the forest
1989	Federal officials were put into panic when monkeys housed in a quarantine facility in Reston, Virginia, USA, started dying from an Ebola-type filovirus
1987	Rift Valley fever followed the damming of the Sengal River in Mauritania
1979	Ebola spread wildly through N'zara and Madiri in Sudan's southern grasslands
1976	Ebola spread wildly through N'zara and Madiri in Sudan's southern grasslands
1976	Ebola killed approximately 300 around a hospital in Yambuka by the Ebola River, Zaire
1970	25 hospital workers and patients suffered from Lassa fever, caused by an arenavirus in Lassa, Nigeria
1970	Following construction of the Aswan Dam, Rift Valley fever infected 200,000, causing 600 deaths
1967	In Germany, several laboratory workers preparing cell cultures from the blood of vervet monkeys died from the Marburg virus
1961	Oropouche, an arbovirus, caused flu-like symptoms in 11,000 residents of Belem, Brazil. Transmitted by the biting midge or sandfly
1951–53	In South Korea, 2,000 United Nations troops were infected with Hantaan
1950s	Machupo caused dozens of deaths in San Joaquin, Bolivia
1940s	Junin killed many agricultural workers in the Argentinian pampas

In addition:

- Dengue fever, caused by a mosquito-borne flavivirus, is spreading from its home territory in Southeast Asia. There were 116,000 infections in Latin America in 1990
- Arenavirus Lassa causes African haemorrhagic fever, infecting 200,000 to 400,000 people annually in West Africa, killing approximately 5,000
- Hantavirus Puumala causes frequent illness in northwest Europe, believed to result from the inhalation of contaminated dust when handling wood
- Hantaviruses have caused illness with renal syndrome for more tham 1,000 years in Central Asia

Source: Le Guenno 1995 and Stone 1995.

follow. In the most serious cases the patient dies from massive haemorrhages or sometimes multiple organ failure. Ebola is the most horrifying haemorrhagic fever. The disease starts with symptoms that may be mistaken for dysentery, but within days blood begins to ooze from every orifice in the patient's body and eventually the internal organs liquefy.

There are four main types of haemorrhagic fever virus: flaviviruses, arenaviruses, bunyaviruses and filoviruses. Table 8.9 summarises the recorded outbreaks of the haemorrhagic fevers caused by these viruses. The flaviviruses include yellow fever, mosquito- and tick-borne viruses and dengue fever. Arenaviruses and bunyaviruses circulate naturally in various populations of animals. Bunyaviruses include hantaviruses, which typically cause an illness known as haemorrhagic fever with renal syndrome. Hantaan and Puumala are hantaviruses carried in the lungs of field mice and bank voles, respectively. Filoviruses are the least understood; they include Ebola and Marburg, which are extremely dangerous. Their incubation period is short, usually less than a week, and several organs are usually attacked, notably the liver. In particular, the viruses attack the blood platelets and the inner surfaces of the blood vessels, leading to uncontrolled bleeding and an accumulation of fluid in the tissues. The Ebola virus is relatively easy to contain, however, because of its short incubation period and because it is transmitted by fluids, but nearly all the outbreaks of Ebola have been exacerbated by impoverished medical facilities and poor hygiene in developing countries. This has helped to increase the transmission of the virus, leading to large fatalities. It is still not known where the Ebola virus hides between epidemics; tests on animals in infected areas have failed to find the host, so the search goes on.

New viruses are continuously being discovered. In 1993, for example, scientists at the Center for Disease Control and Prevention (CDC) in Atlanta identified a new virus from New Mexico. They called it Sin Nombre, Spanish for 'no name'. Haemorrhagic fever viruses are usually called after the place where they first caused outbreaks, such as Ebola after the Ebola River, Zaire, where it was first identified in 1976, and Lassa after Lassa in Nigeria. The most threatening haemorrhagic fever viruses, commonly known as 'emerging pathogens', are mutations or genetic recombinations between existing viruses, or viruses that may have existed for millions of years but have only recently come to light because of changes in environmental conditions. Environmental changes may allow the virus to multiply or mutate and spread in host organisms, becoming more virulent. There is great concern that viruses such as Ebola may mutate into air-borne microbes; their spread and consequences would then be disastrous.

Ecological disruptions appears to be the main cause of most outbreaks of haemorrhagic fever viruses. Increased human population pressures and agricultural activity increase the likelihood of contact with animals carrying the viruses. This is well illustrated with the arenavirus Guanarito in Venezuela, where a rural community clearing a forest region stirred up dust that was contaminated with the urine of a species of cotton rat, the host for the Guanarito virus. Had they not cleared the forest, they would not have come in contact with the host. Dam construction can also cause ecological disruptions, particularly because some bunyaviruses are carried by mosquitoes. Mosquitoes flourish in the new environment, and the increased human populations and cattle now living in the new environment increase the likelihood of contact with the viruses. These factors probably explain the outbreaks of Rift Valley fever in Egypt in 1977 and Mauritania in 1987. Viruses can be contained, however. For example the arenavirus Machupo, carried by vesper mice in Bolivia, which first appeared in 1958, had been prevented since 1974 by extermination campaigns against the rodents that carry it. During the last few years, however, the campaigns were terminated and the disease subsequently reappeared in 1994.

Ecological disruptions also occur naturally. The Sin Nombre virus emerged in the mountains and deserts of New Mexico, Nevada and Colorado after heavier than usual rain and snow falls during the spring of 1993. The exceptional humidity increased the crop of pine kernels, on which the deer mouse, the principal host for the virus, feeds. The deer mouse population increased ten-fold between 1992 and 1993, thus increasing their likely contact with humans.

In order to try and prepare for future outbreaks, the World Health Organisation has established a network for tracking haemorrhagic fevers, and vaccines have been developed against some of the viruses, while active campaigns against the animal hosts are being undertaken in areas of known risk. But we are not sure what nasty surprises await us. Some scientists believe that the mutation of an airborne haemorrhagic fever virus similar to Ebola is the biggest threat to the survival of humankind.

AIDS

Not only is AIDS a global issue of great concern, because of the personal risks to health and as a threat to the health of communities and nations, but it is a global environmental issue since its spread may affect the sustainable development of some developing countries, for example in parts of Africa. It is for reasons such as these that this issue is dealt with in this book.

It was only in 1981 that scientists first recognised AIDS, the acquired immune deficiency syndrome. Already, 10–12 million people have been infected with the human immunodeficiency virus (HIV), which leads to AIDS. The World Health Organisation (WHO) predicts that this figure will treble within the next eight years, with up to 40 million people being infected by the year 2000. Unfortunately for many people in the developed countries, it takes the death of international celebrities such as the film star Rock Hudson in 1985, or Freddie Mercury from the rock band Queen in late 1991, to highlight one of the greatest problems facing humankind today. The spread of the HIV virus still has not been fully assessed, mainly because such studies involve gathering data on people's personal sexual activities and preferences, together with an in-depth understanding of complex behavioural and socio-economic factors.

HIV causes the collapse of the body's immune system, which then makes it fatally susceptible to any infection that a healthy person would fight with ease. AIDS is extremely complex and is not a single virus. In Africa there are at least 100 different strains. The virus attacks T4 helper **lymphocytes** and other cells of the immune system that are present in the bloodstream. These cells play a key role in fighting diseases and are required as a fundamental part of the human immune system. HIV can only replicate itself inside human cells. In order to do this, the virus inserts its genetic material into the healthy cells as a genetic parasite. The virus enters the bloodstream in blood from an infected person. This infection can happen in a number of ways: during intravenous drug use, from infected needles; during sexual activity, including heterosexual acts; or from an infected mother to her baby before the child is born.

Once in the bloodstream, HIV may lay dormant for a latent period of up to 10–15 years. The virus is then activated by secondary co-factors, which are not yet fully understood or even properly identified, but some have suggested that they may include conditions such as gastro-enteritis, tuberculosis, thrush and malnutrition. Once HIV has been activated, the T4 helper cells become depleted in number. When their count falls below a certain threshold, the person is said to have developed full-blown AIDS. At this stage they are susceptible to

Table 8.10 *(A) Global HIV positivity rates; (B) AIDS cases reported and case rate in Africa; (C) Incidence of AIDS in developed countries in 1992.*

A

	Males	*Females*
Sub-Saharan Africa*	1 in 40	1 in 40
North America	1 in 75	1 in 700
Caribbean/Latin America	1 in 125	1 in 500
Western Europe	1 in 200	1 in 1,400
Oceania	1 in 200	1 in 1,400
Asia and Southeast Asia	1 in 2,500	1 in 3,500
Eastern Europe/ former USSR	1 in 4,000	1 in 20,000

B
Namibia	311 (1985–90)	6.5/100,000
Malawi	12,074 (1990)	37.1/100,000
Botswana	216 (1991)	2.3/100,000
South Africa	1,011 (1991)	0.9/100,000

C

Country	AIDS cases per million head of population
United States	1,200
Spain	441
Switzerland	417
France	403
Italy	272
United Kingdom	120
Sweden	89

*Excludes AIDS sufferers. South African data not included.
Source:
(A) WHO, cited in *Science* 19-4-91.
(B) WHO.
(C) Anderson 1993. *Nature*, 360, 65–68.

every sort of infection, and death can occur as a result of the patient contracting any complication, even a cold. HIV is very variable, continuously mutating and developing into many strains in different people within the same city (Nowak and McMichael 1995), so a vaccine against one strain may be ineffective against another.

HIV is known as a **retrovirus**. All organisms, apart from certain viruses, contain their genetic material in the form of **DNA** (deoxyribonucleic acid). Retroviruses are different because their genetic material is in the form of **RNA** (ribonucleic acid), so they have a special enzyme, called **reverse transcriptase**, to convert the RNA into DNA. HIV enters a cell by attaching itself to a so-called receptor molecule on the surface of a cell. Current research suggests that this receptor is something named CD4, present on the surface of T4 helper lymphocytes, **macrophages, microglial cells** in the brain and **dendritic cells**. Inside a cell, the active HIV manufactures more viral proteins to reproduce more viruses. The new viruses, called virions, bud off the cell's outer membrane and go on to infect further healthy cells.

There is also the belief, by people such as Professor Peter Duesburg (University of California at Berkeley), that AIDS may be caused by agents other than HIV, such as drug use. This view is not widely accepted and there is little scientific support for these opinions (Maddox 1993, Griffin 1994, Adam 1994, Shenton and Gildemeister 1994, Dixon 1994).

There are two different HIVs, HIV-1 and HIV-2. HIV-1 has a worldwide distribution, whereas HIV-2 occurs mainly in West Africa. AIDS results from both strains of HIV, but HIV-2 appears to take longer to do so. The origin of HIV is not known. A whole new science has developed, called **seroarchaeology**, to answer this question of provenance. To date, the oldest recorded case is from a man in Zaire in the 1950s. The main spread of the virus was in Africa in the 1970s. Tables 8.10A and 8.10B show the global HIV positivity rates and the number of AIDS cases reported from several African countries. The HIV positivity rates are frightening as the potential number of AIDS cases in the next ten to fifteen years is projected to a staggering level. The WHO estimates that the spread of the virus in Africa is doubling every 34 weeks, and the transmission of the virus to babies is approximately 30 per cent efficient. The WHO worst-case scenario predicts that 40 million people will be HIV-positive by the year 2000, with 6.5 million cases in Africa. Measuring the positivity of the AIDS virus is difficult because of sampling bias, including screening of volunteers, army personnel, prostitutes, blood donors and mothers tested soon after they have given birth. The latter group may prove most reliable for estimating future trends, since they are usually the most representative of the global population.

These frightening statistics predict large mortality rates in the next few decades. This will have profound psychological, social and economic effects. It will be particularly devastating in Africa, where the rates are among the highest and economic and social welfare are least. Controlling the spread of the AIDS virus is also more difficult in these countries. The propagation of knowledge about the AIDS virus and its control is difficult in Africa because of the overall poor level of education and communication. The use of condoms, to reduce the spread of the virus by the transmission of bodily fluids during sexual activity, is still not widely accepted in many African countries. To compound the situation, condoms are often unavailable. Locally, social mores mean that people are often unwilling to use them, as condoms have traditionally been associated with birth control and considered the white person's means of suppressing the Black African. In addition, sexual behaviour patterns and fertility rites differ from those in many Western societies, something that further accelerates the spread of the AIDS virus.

Although there is a large amount of research into HIV and AIDS, it seems unlikely that a comprehensive vaccine will be developed within the next ten years. AIDS, therefore, is one of the biggest threats to humankind in the near future. Fields (1994) strongly maintains that success in controlling the AIDS epidemic is as likely to arise from unrelated areas of research as from AIDS-directed programmes. Broadening the research agenda will help to provide greater insights into the way viruses are transmitted and mechanisms that cause the illness. AIDS may yet prove to be another sinister Malthusian check on population, but to therefore regard AIDS as an acceptable and natural way of controlling population would be tantamount to adopting a neo-Nazi philosophy. Probably the main reason for the lack of action on AIDS in Africa is a lack of money, when many basic health problems compete for a relatively small pot of money, e.g. vaccination programmes and improving water quality for human use.

There has been much discussion recently over whether epidemiologists have greatly exaggerated the potential spread of AIDS into the heterosexual community in developed countries, and even a denial that there is an epidemic in sub-Saharan Africa. In developed countries, there are relatively sophisticated statistics that reveal the temporal evolution of AIDS within different risk groups. Since HIV has an incubation period of up to ten years, current reported incidences of HIV infection have a large margin of error associated with any figures, making it difficult to model the spread of the virus. In developed countries, there has been a steady increase in reported incidence, but there are large differences between countries (Table 8.10C). Anderson (1993) highlights this, and argues that during 1992 there was a signifi-

cant increase in incidence (24.8 per million) in the heterosexual group, suggesting that the epidemic is still in its early stages. Estimates of future incidence are based on mathematical models that attempt to mimic the transmission of the disease, and such models are revised as more data become available. Back-calculating methods such as this show that the rate of transmission has been slowing down in two of the major risk groups, homosexual men and intravenous drug users, whereas the incidence in heterosexuals has been increasing since 1984. Incidence of HIV infection in the developing countries, particularly throughout Africa, is very difficult to calculate because of the poor databases.

Bovine spongiform encephalopathy and Creutzfeldt–Jakob disease

Bovine Spongiform Encephalopathy (BSE), or 'mad cow disease', first diagnosed in cattle in 1986, is a disease that affects the nervous system. It has an incubation period of about four years, after which it leads to lethargy, lack of co-ordination (notably difficulty in moving the hind legs), organ failure and ultimately death. In Britain, there have been more than 158,000 cases of BSE in over 33,000 farms, peaking in 1992 with an average of 700 cattle affected per week. Currently the disease has been reduced, with only seventy cases during the first three months of 1996 (Pearce 1996a).

The outbreak of the disease is attributed to a change in farming methods at the beginning of the 1980s, when farmers started using new cattle feed manufactured from the ground-up remains of other mammals. This was in response to intensification of farming, as the available pastures were not sufficient to feed the growing herds of cattle, and a new feed was needed to maintain the herds. Some of the new cattle feed, however, contained animal protein that was contaminated by the prion disease scrapie. Scrapie is carried by sheep, and their ground-up parts were present in the new cattle feed. Scientists linked BSE to feed containing scrapie and changes in the solvents and temperatures used at the rendering plants where protein was recovered from the animal carcasses to make the cattle feed. BSE was then tentatively linked to Creutzfeldt–Jakob disease (CJD) in humans. It was suggested that infected people had eaten the organs, such as the brain and spinal cord, of cows that were contaminated with BSE. It took more than two years, however, before the Ministry of Agriculture, Fisheries and Food (MAFF) enforced a ban on scrapie-contaminated feed and attempted to ban the inclusion of certain cattle organs in foods for human consumption (Pearce 1996a). The slow response by the MAFF and the long incubation period for the disease compounded the problem and it is now feared that thousands of people in Britain may be infected with CJD.

CJD is a frightening disease. The patient suffers from a gradual loss of memory and personality, co-ordination fails and ultimately death occurs after about six years. Both CJD and BSE are caused by altered forms of the protein PrP, which is normally found in the brain. Once the altered protein begins to appear, the rest of the brain begins to alter into a dangerous form. The disease can be triggered by mutations in the gene for PrP or by an infectious agent. The diseased protein sticks to itself, forming large tangled deposits and killing brain cells. This causes progressive dementia and death. CJD affects about one in a million people each year, but its incubation period is usually several decades, so that almost everybody it affects is elderly, with an average age of about sixty-three years. In late March 1996, however, the CJD Surveillance Unit, which was set up in Edinburgh in 1990 to evaluate the changing patterns of CJD that might be attributed to BSE, identified ten new cases of CJD that had affected younger people. These were aged between eighteen and forty-one, with an average age of twenty-seven years. These victims died unusually quickly, within thirteen months of the onset of symptoms. Examination of their brains showed exceptionally large accumulations of protein fibrils (Pain 1996). These unusual cases prompted the CJD Surveillance Unit to hypothesise that the most likely cause was exposure to BSE. The Spongiform Encephalopathy Advisory Committee was informed and it announced to the public that the most likely explanation for the ten new cases of CJD was exposure to the agent that causes BSE.

The UK government was forced to announce formally that there may be a link between eating BSE-contaminated beef and CJD. It soon became clear that Britain could face an epidemic of CJD along with the collapse of the beef industry. The European Union banned the export of British beef and many shops, particularly fast-food outlets such as McDonalds, stopped selling beef or greatly reduced its price as an incentive to shoppers. If BSE contamination causes CJD, then most of the infection will have taken place in the late 1980s, and the numbers of people affected may be considerable. Recommendations to reduce the hazard have included slaughtering all cattle born before the end of 1990 (approximately 11 million cattle), that is

those that may have been infected with BSE. Such a measure would have considerable environmental implications because of the problems of disposal of such a large number of livestock, and burning and burying the carcasses may not be effective in destroying the disease, which may survive prolonged periods of incubation within soils. The UK government has rejected such a proposal and has recommended that carcasses of all cattle over thirty months old must have their bones removed in specially licensed plants and that trimmings such as vertebrae and nervous and lymphatic tissues should be treated as banned offal (Pearce 1996c).

There is still disagreement on the causes and frequency of CJD, which may have occurred in humans as long ago as 1980, when it was mis-diagnosed as Alzheimer's disease. The incidence of CJD may have been mis-calculated, and its cause may be due to other factors besides BSE contamination. Yet this does not explain the outbreak in the younger age group in 1995 and 1996. The scrapie theory has also been disputed. In Iowa, USA, experimental feeding of scrapie to cattle has shown that infecting cattle with scrapie does result in death within five months, but pathologically the brain does not resemble cattle that have been infected with BSE (Aldhous 1996). This suggests that BSE may not be transferred by scrapie, rather it may have been transferred by contaminated beef that was also present in cattle feed. The UK government's ban on sheep offal, therefore, may not have been appropriate and infection from the cattle feed could be more widespread than initially thought. There is also concern that BSE may be present in pigs and poultry, but this has yet to be considered in detail (Pearce 1996b). The full effects and implications of these diseases are very frightening and have yet to be fully assessed. It may be many years before the full consequences are recognised.

BSE and CJD, therefore, illustrate one of the potential environmental health hazards that may result from changing land use and farm intensification. Such hazards seriously threaten human life and have profound economic implications.

Hazard awareness and response

Attempts to prepare for natural hazards have been made throughout history, for example over 2,000 years ago the early Egyptians constructed flood defences to reduce the effects of the Nile flood water, and in the second century AD, the Chinese made seismometers to study earthquake hazards. Recently, the study of natural hazards has increased, particularly because of increased awareness through better communications and media coverage of natural disasters such as droughts, fires, earthquakes and floods (Plate 8.7). In addition, physical geographers have become increasingly interested in the study of natural hazards because of the need to make their subject relevant and the trend towards the study of neo-catastrophism (Smith 1992). In turn, the study of environmental hazard perception developed as both human and physical geographers saw the need to view their studies in terms of the way managers and decision-makers reacted to hazard mitigation. In the 1980s, emphasis swayed towards the relationship between natural hazards and under-development, and the need to react on an international level to relieve the effects of disasters. In December 1989, the United Nations General Assembly proclaimed the 1990s as the International Decade for Natural Disaster Reduction (Resolution 44/236). As part of this, a special high-level council, a scientific and technical committee and a secretariat were established with the overall aim of reducing through international action, especially in developing countries, loss of life, property damage and social and economic disruption caused by natural disasters. The framework relies on governments formulating their own disaster-mitigation programmes to assess the risks and to reduce the hazard. Today, therefore, hazards are very much a part of public policy in most countries. It is expected that public concern about risks is likely to increase in the future and hazards will be very much part of political agendas (*ibid.*).

Smith (*ibid.*, p.9) emphasises that natural hazards 'result from the conflict of geophysical processes with people and they lie at the interface between what has been called the natural events system and the human use system'. Thus humans are central to the study and consideration of hazards, and it is only when people are present that hazards exist. Natural processes, however, are only hazardous if humans perceive them as being so. The same natural processes that can be hazards to some communities can actually be beneficial to others. Snow, for example, can be considered a resource if it falls on ski slopes and a hazard if it blocks highways. There is a range of variation of the magnitude of processes, beyond which unacceptable damage to humans and their activities will occur, for a process to be considered a hazard. This range may vary due to the economic, social and cultural nature of the people under threat. The risk from a hazard may also vary through time, as a population becomes more vulnerable, or as it adapts to the threats of the hazard, or its perception of the hazard changes. A disaster is also a perceived event, the

Plate 8.7 *Severe fire damage, Yellowstone National Park, Wyoming, USA.*

realisation of the hazard, but its definition is dependent on the perception of the consequences. The UN Disaster Relief Co-ordinator (UNDRO 1984) defines it qualitatively as an event in which a community undergoes severe damage and incurs such losses to its members that the social structure and the essential functions of the society are disrupted or prevented. Much emphasis is now being put on the sociology of hazards as well as the science behind the processes.

Recently populations are beginning to become more responsible for hazards, due to increased awareness regarding land degradation and pollution. In an attempt to reduce environmental hazards, decision-makers may undertake a comprehensive hazard management strategy. The first step in this process is usually a risk assessment. This involves the identification of hazards, an estimate of the probability of the hazard event occurring, and an evaluation of social consequences. Hazard management then involves both assessment of and response to the hazard event. Smith (1992) recognises four chronological stages in this management process:

1 *Pre-disaster planning* which covers a range of activities, including the construction of defensive engineering works, land-use planning and formulation, and dissemination and maintenance of evacuation plans.
2 *Preparedness* reflects the degree of awareness and arrangements for emergency warnings to be issued along with the effectiveness with which public officials can mobilise an evacuation plan.
3 *Response* which deals with the events immediately before and after the event, including reaction to warnings and emergency relief activities.

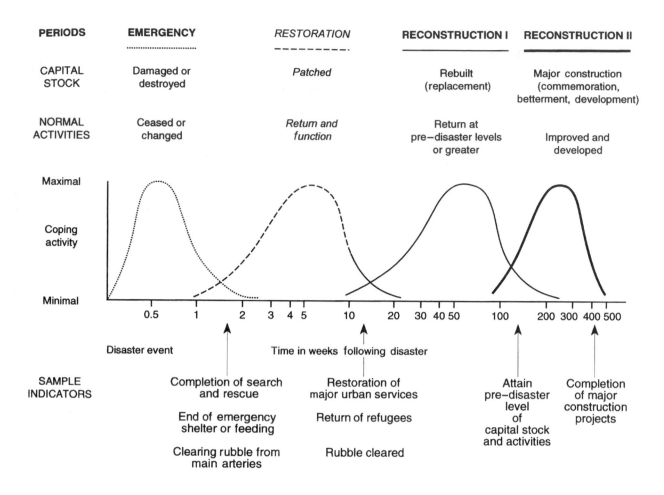

PERIODS	EMERGENCY	RESTORATION	RECONSTRUCTION I	RECONSTRUCTION II
CAPITAL STOCK	Damaged or destroyed	Patched	Rebuilt (replacement)	Major construction (commemoration, betterment, development)
NORMAL ACTIVITIES	Ceased or changed	Return and function	Return at pre–disaster levels or greater	Improved and developed

Maximal

Coping activity

Minimal

0.5   1   2   3 4 5   10   20   30 40 50   100   200 300 400 500

Disaster event

Time in weeks following disaster

SAMPLE INDICATORS	Completion of search and rescue	Restoration of major urban services	Attain pre–disaster level of capital stock and activities	Completion of major construction projects
	End of emergency shelter or feeding	Return of refugees		
	Clearing rubble from main arteries	Rubble cleared		

Figure 8.13 *A sequential model of disaster recovery for urban areas. Redrawn after Haas et al. (1977).*

4 *Recovery and reconstruction* is a much longer-term stage involving rehabilitation taking a few weeks after the event and reconstruction taking several years, with the aim of returning an area to normality.

The magnitude of the disaster is a function of the resilience of a community, i.e. the rate of recovery from a hazardous event, and the community's vulnerability to the disaster, which is a measure of the risk combined with the level of social and economic ability to cope with the hazard event. The process of recovery is complex and may take up to ten years to complete. Figure 8.13 shows the characteristic activities and achievements in relation to the relief, restoration and rehabilitation phases as described by Haas *et al.* (1977) in a four-stage sequential model of disaster recovery for urban areas.

Conclusions

Natural hazards are among the major causes for concern on Earth. Natural processes that become a

hazard have highly variable causes and effects. In order to mitigate the effects, it is necessary to understand the dynamics of the processes involved and to suggest ways that preventive measures can be executed within the socio-economic framework of the area threatened.

Natural hazards are not always detrimental to humankind. Some may provide a service, such as the renewal of mineral nutrients to the soil on flood plains during flooding and adjacent to volcanoes during volcanic eruption, or the creation of new land by Earth movements and volcanism.

Finally, the extent to which humankind is altering the environment is not fully appreciated. If the result is a change in the frequency and magnitude of these natural processes, this may, in turn, affect the occurrence of natural hazards. The following chapter examines the effects on the land, important in helping to understand the consequences of human actions, which may have possible implications in terms of natural hazard assessment.

Chapter 8: Key points

1 Natural hazards are typically unpredictable and may cause death, injury, and destruction of or damage to agricultural land, buildings and communities. The effects of a natural disaster on a community depend upon factors such as the magnitude and extent of the disaster, how prepared the affected population is, and their economic resources to mitigate a potential disaster and/or clean up afterwards. The magnitude and frequency of various natural disasters form part of any risk assessment, and corresponding insurance.

2 Geological hazards result from Earth's internal (tectonic) processes and Earth surface (geomorphic) processes. Earthquakes and volcanic eruptions are a consequence of tectonic processes and are generally located along plate boundaries. Damage may be local and related to building collapse, fires, landsliding and subsidence, tsunamis, flooding, the release of poisonous gases, and associated hazards such as contaminated or depleted water sources, disease, famine, injury and death. Volcanic activity may result in regional and global climate changes. Earth scientists are developing more reliable means of predicting earthquakes and volcanic eruptions.

3 Geomorphological hazards, commonly included with geological hazards, include landsliding, river flooding, glacial hazards, soil erosion, and asteroid and comet impacts. The environmental effects depend upon their magnitude and frequency, which is a function of climate, geology, vegetation and human activity. Mitigating the effects depends upon a knowledge of the dynamics of Earth surface processes, the accurate identification of high-risk zones, improved land-use practices, and the implementation of protective measures to reduce the effects.

4 Meteorological hazards are driven by the Sun's energy and are controlled by atmosphere–ocean systems. The effects are determined by the magnitude and patterns of weather systems, and the disaster preparedness of the threatened population. Included amongst these hazards are tornadoes, tropical cyclones, flooding by heavy rainfall and storm surges, disease associated with contaminated water sources, thunder and lightning damage, hailstorms, and droughts.

5 Biological hazards include pests and environmental diseases. Pests cause great destruction to agriculture and lead to the spread of disease. Such hazards can be controlled by physical, chemical and biological means, but most effectively by integrating all these approaches. Environmental diseases are one of the greatest hazards to health, and can lead to millions of deaths each year. Amongst the most serious diseases to affect humans are malaria, yellow-fever, typhoid, sleeping sickness, and schistosomiasis. Haemorrhagic fevers, such as Ebola and Lassa, are now being considered an increasing threat as human populations expand into marginal areas and cause ecological disruptions. Most diseases can be successfully tackled with good preventive education, early inoculation against disease, improved hygiene, and various artificial controls on the spread of the transmitting agent, e.g. spraying crops, etc. Prompt medical attention can help to cure individuals of many diseases.

Chapter 8: Further reading

Blaikie, P., Cannon, T., Davis, I. and Wisner, B. 1994. *At Risk.* London: Routledge, 320 pp.
This book assesses and defines vulnerability, examining famines and droughts, biological hazards, floods, coastal storms, earthquakes, volcanoes and landslides. It draws together practical and policy conclusions with a view to disaster reduction and the promotion of a safer environment.

Bryant, E.A. 1991. *Natural Hazards.* Cambridge: Cambridge University Press, 294 pp.
An inter-disciplinary treatment of a variety of natural hazards, including oceanographic, climatological, geological and geomorphological hazards. This book is suitable as an introductory undergraduate text on natural hazards.

Chester, D. 1993. *Volcanoes and Society.* London: Edward Arnold, 351 pp.
An informative text on volcanic activity examined from both a geological, and a socio-economic and political perspective. It is well illustrated with many interesting examples of the processes and effects of vulcanicity worldwide. It is an ideal text for university and college students with Earth science and social science backgrounds wishing to pursue the issues associated with environmental risk assessment and natural hazards.

Hewitt, K. 1997. *Regions of Risk.* Harlow: Longman, 389 pp.
This book examines the various aspects of hazards, human vulnerability and disaster. It includes a variety of case studies on natural and technological hazards, as well as an examination of social violence. It emphasises the cultural and social concerns of hazard assessment and response, and provides an examination of the cross-cultural differences and international scope of risk and disaster preparedness and response.

Smith, K. 1992. *Environmental Hazards: Assessing Risk and Reducing Disaster.* London: Routledge, 324 pp.
A comprehensive book covering most of the major environmental hazards, including seismic, mass movement, atmospheric, hydrological and technological hazards. It integrates both the Earth and social sciences and is suitable for undergraduates from a variety of backgrounds, as well as being an important reference source for teachers and researchers.

The parched eviscerate soil
Gapes at the vanity of toil,
Laughs without mirth
   This is the death of earth.
                 T.S. Eliot,
      Four Quartets: 'Little Gidding'

With the rapid growth of the world's population, most societies have been demanding much more from the Earth's resources and, therefore, affecting the land surface at ever-increasing rates. Prehistoric evidence shows that in Palaeolithic times the early hunter-gatherers used fire and, accidentally or intentionally, burned extensive areas of forest (Brain and Sillen 1988). The early agronomists burned large areas of land to create farmland or pasture, they modified the soil by ploughing, altered the drainage by irrigation, introduced or bred new animals and crops, and altered the natural vegetational structure of many regions (Harlan 1986, Davis 1987). In more recent times, humans have destroyed enormous tracts of natural vegetation, excavated large areas of land, greatly modified the landscape, and even created new land.

Many renewable resources are being consumed at rates that far exceed the speed at which they can be regenerated or replenished. Nowhere is this more apparent than in the destruction and deforestation of the rainforests. A hectare of forest can be destroyed within an hour, but it may take several decades for the forest to regenerate itself. A report published in 1991 by the **UN Food and Agricultural Organisation (UN FAO)** estimates that destruction of the tropical rainforests is currently occurring at a rate of 40 million acres per year, mainly as a result of human activities. Secondary effects complicate the problem. For example, rapid degradation of the forest soil accompanies deforestation, the nutrients being washed out by rain. In addition, the organic compounds are no longer replaced in the soil. It may take decades of slow regeneration before the soil can support a forest again. Other effects may lead to changes in slope stability, enhanced soil erosion, increased sediment washed into rivers, changes in regional climate, and increased occurrence of floods.

There are many examples of how uncontrolled or excessive exploitation of the land's natural resources (including vegetation, fossil fuels, minerals, water and land) can have a profound effect on the natural environment, in terms of both ecosystems and the aesthetic beauty of landscapes. This chapter will consider the main effects caused by the exploitation of resources on vegetation, soils, oceans and the landscape.

Human impact on vegetation

Vegetation is important to humans as a primary source of food, as a building material, in manufacturing industries, as a fuel and in medicine. Early in human history, people gathered plants and began to cultivate selected types (Harlan 1986). With agricultural activity came the associated changes in the shape of the landscape.

The first human impact on vegetation, which is still prevalent, is the use of fire (Wertine 1973). Even though over half of the fires that occur are natural, resulting from lightning strikes or spontaneous combustion of decaying organic matter, the rest can be attributed to accidental or deliberate burning by humans. Accidental fires may result from agricultural uses, cigarettes, rubbish tips, children playing with fire, camp fires, trains, and motor vehicles. Deliberate burning is used to clear land, though it can be used to help improve the quality of the soil in arid regions, adding fresh organic material, or as a means of helping to reduce widespread fires. Fires cause a reduction in the natural vegetation; they threaten wildlife, humans and property. Fire produces secondary problems associated with the clearance of vegetation, such as soil erosion, flooding and wind erosion (Imeson 1971). Fires may also provide benefits in an area, adding new mineral matter and nutrients to soils: some plants, such as the jack pine, depend on fire to initiate the dispersal of seeds, their pods bursting upon excessive heating (Goudie 1993b).

Fires are more abundant in hot, arid areas of the world, such as along the coastal regions of the Mediterranean, Australia (particularly near Adelaide),

the northwestern states of the USA and the grasslands of central Africa (Pyne 1982, Powell 1983,
Webster 1986). In these areas, fires devastate
hundreds of square kilometres of land. Smaller fires
are frequent throughout the world, causing only
limited damage. Continued and frequent burning,
however, often reduces the capability of an area to
regenerate itself and to replenish the natural vegetation (Hanes 1971). Particularly problematic is the
frequent misuse of slash-and-burn techniques in
the rainforests, where extensive areas of forest are
cut down and/or burned to provide pasture land for
cattle ranching. In such fire-damaged areas, the land
deteriorates rapidly and the forest is prevented from
regeneration.

The domestication of animals also has a major
impact on the land surface. Heavy grazing of cattle
leads to trampling and compaction of the soil,
reducing its capacity to hold water and altering its
structure. Ultimately this leads to soil erosion, both
by wind and water. Selective grazing of particular
cattle may lead to changes in the nature of the vegetation cover. In the UK, for example, heavily grazed
pasture in Scotland is dominated by bracken, a
successful plant that survives because it is particularly
distasteful and prickly to sheep and cattle (Fenton
1937). Grazing can impede the growth of young
trees, as young saplings are favoured by grazers.
Grazing, however, may have positive effects on the
land, because the animals provide faecal material,
which can act as a natural nitrogenous fertiliser, also
rich in other nutrients. Animals help to propagate
seeds, and grazing may increase species diversity by
opening up new ecological niches. Nibbling of plants
may also encourage vigorous growth (Warren and
Maizels 1976). Many scientists believe that grazing
in grassland areas in fact has little effect on grass,
which is well adapted to withstand grazing.

Major problems are created when humans try to
rear domestic animals in regions not suited to their
lifestyle, especially where human activities alter the
natural vegetation to grassland. This is particularly
true of large areas of Brazil, where the rainforest has
been cut down to produce pasture land (*Economist*
1988, Park 1992). The rainforest soils are relatively
poor in many nutrients and are quickly deprived, as
grass provides little litter to replace the much needed
nutrients. As a result, these areas can be used for
pastures for only a few years, then the area becomes
sterile and soil erosion can quickly devastate the
region, removing the soil and even the deeply weathered bedrock (Sioli 1985).

Deforestation involves the deliberate removal of
forest to create new agricultural or urban land, to
provide wood for building and manufacturing industries, for the exploitation of minerals and fossil fuels,
to create reservoirs for water supplies and hydroelectric energy, to build highways, for fuel, or as a
result of defoliants used to help locate enemies
during wars.

Humans have cleared forests throughout history,
for example there is evidence to suggest that forest
clearance in England goes back 8,900 years (Bush
and Flenley 1987). In Central Europe, a major phase
of deforestation began in the eleventh century. The
forests that once covered most of Central Europe
were almost totally cleared within 200 years (Darby
1956). In North America, before the first colonists
arrived, forest occupied approximately 170 million
hectares between the Mississippi and the Atlantic
seaboard. Today, only about 10 million hectares
remains (Williams 1989). This clearance is illustrated
in Figure 9.1, showing that the major destruction
took place between 1620 and 1920. Currently the
greatest rates of deforestation are occurring in the
humid tropical and equatorial regions of the world,
where the last big forests exist. How long these will
last can only be guessed.

It is not just the rainforests that are substantially
degraded by deforestation. Temperate forests also
suffer considerably. In an experiment involving the
clear-cutting of a small beech–maple–birch forest
ecosystem in New Hampshire, it has been demonstrated that forest clearance can lead to a considerable acceleration in the loss of nutrients (Bormann
et al. 1968). They attributed these losses to a reduction in transpiration, increasing the amount of water
passing through the system (*ibid.*); a reduction in
the surface area of roots meaning that plants were
less able to remove nutrients from the leaching
waters; removal of nutrients in forest products; addi-
tions to the organic substrate available for immediate
mineralisation; and in some instances the production
of a micro-climate move favourable to rapid mineralisation. Clearance in high-latitude forests may cause
other major environmental modifications such as
rapid and substantial permafrost degradation (Linell
1973).

Deforestation not only affects the rainforests and
their ecosystems, but it may also have serious consequences for the adjacent regions. In Jamaica, for
example, deforestation on steep slopes has led to
intense soil erosion and an increase in landslides,
which in turn has increased the suspended sediment
loads in many of the rivers, which drain into the
surrounding coastal seas with fringing coral reefs
(Hughes 1994). Reef-forming corals are very sensitive to sediment concentrations in the water, and

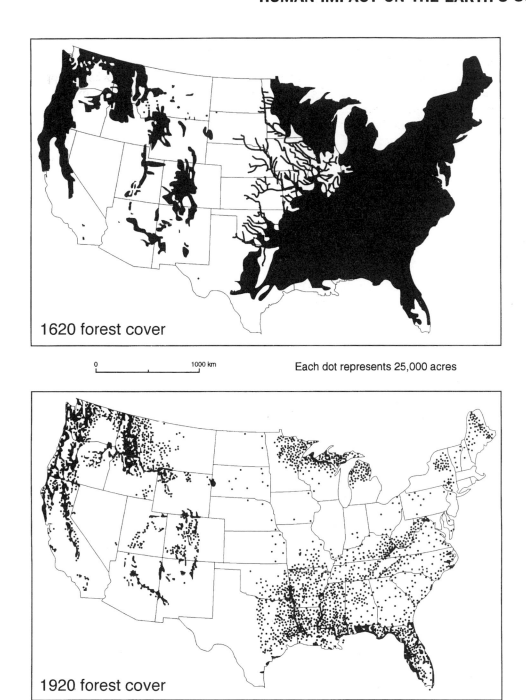

1620 forest cover

0 _____ 1000 km   Each dot represents 25,000 acres

1920 forest cover

Figure 9.1 *The distribution of American natural forest in 1620 and 1920. Redrawn after Williams (1989).*

as increased sediment concentrations reduce the amount of light penetration, coral growth is slowed. More importantly, the increased sediment concentrations poison the coral tissues. As a result, many of the corals around Jamaica have not been able to tolerate the increased land-derived sediment and the reefs have become degraded. Once the coral reefs die, the knock-on effect is the mortality of other animals that depend on the reef for food and are part of the coral reef ecosystem.

The following section focuses on the rainforests, as they represent, arguably, one of the most important natural ecosystems under serious threat from deforestation.

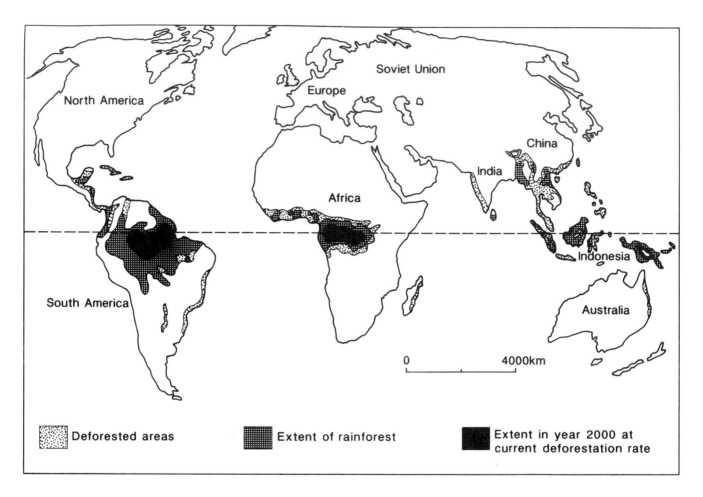

Figure 9.2 *The present extent of tropical and equatorial rainforests, affected and unaffected by deforestation, together with the estimated extent of rainforests by the year 2000 if current levels of deforestation continue. Redrawn after Mannion (1991).*

Deforestation: the rainforests

The tropical and equatorial rainforests are shrinking at an alarming rate because of deforestation (Figures 9.2 and 9.3). There is little sign of a real slow-down in this destruction.

A few thousand years ago, rainforests covered about 14 per cent of the land surface. Today they cover a mere 7 per cent. Much of this has been lost over the last 200 years, mostly following the Second World War. In a study by the UN Food and Agricultural Organisation (UN FAO) published in October 1992, the most thorough to date, and involving satellite and aerial photographic reconnaissance in 88 countries, it was estimated that the rainforests are disappearing at the rate of one acre per second, equivalent to the combined size of England and Wales being lost annually. This annual rate of destruction is 50 per cent faster than a decade ago, and half of this is taking place in Latin America, where 20 million acres of rainforest are devastated annually. As long ago as 1981, the UN estimated that 20 per cent of the rainforests then existing will be completely destroyed by the end of this century. Recent analysis of satellite data for the whole of the Amazon basin shows that deforestation has increased from 78,000 km² in 1978 to 230,000 km² in 1988 within a total forested area of 4,090,000 km²; over the same time period, habitats where biological diversity is seriously threatened have increased from 208,000 km² to 588,000 km² (Skole and Tucker 1993). Figures 9.2 and 9.3 show the present distribution of the rainforests, and the amount, as a percentage, of forest that has been cleared in recent years. These rich, diverse and complex ecosystems will be gone before long, depriving the world of a

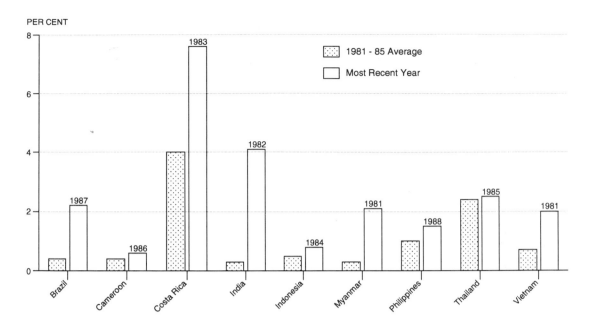

Figure 9.3 *The average percentage of closed forest (i.e. without open spaces) cleared in selected tropical countries in 1981–85, with data also supplied for the most recent year. After World Resources Institute 1990.*

wealth of biodiversity and the potential use of many of its unique biological compounds, often of considerable medical value (Peters *et al.* 1989, Blum 1993).

The reduction of such a vast area of rainforest vegetation may upset global nutrient cycles, especially the O_2 and CO_2 cycles. Figure 9.4 represents the recycling of nutrients in selected ecosystems. Notice the greater rate of flow associated with the tropical rainforests. This in turn may have long-term, but poorly understood, effects on global climate. In the shorter term, water flow may increase over the land surface, as rain will fall directly onto the soil or bedrock, no longer being impeded by the vegetational cover, and will then flow quickly across the surface as stream run-off, or percolate rapidly into the weathered and fractured bedrock as ground water. This increased rate of water flow over the land surface may lead to an increase in the magnitude and frequency of flooding, more rapid and deeper soil erosion, increased suspended sediment loads in rivers, enhanced slope instability, and degradation of adjacent land (Gentry and Lopez-Parodi 1980, Myers 1986, 1988a). Much of the flooding in regions such as Bangladesh has been attributed to the deforestation of the Himalayas, from where the main rivers originate that now drain the mountains more rapidly (Eckholm 1975, Sterling 1976, Myers 1986).

Many interwoven and complex local issues are involved in the deforestation of the tropical rain-

forests. Landowners and others involved in exploiting the rainforests for a quick profit frequently show little concern for the long-term survival of their environment. In South America, particularly in Colombia, large amounts of tropical rainforest are being cleared by drug barons to grow coca to produce cocaine and crack, and to harvest poppies for heroin. The profit motive is simply enormous – with a mark-up at 1992 figures of around 4,000 per cent between Colombia and London (reported on the BBC television programme *Panorama*, 16 November 1992). Amongst this apparent disregard for the rainforests, there are more prescient and brave individuals who have fought and are fighting for the long-term interests of the rainforest. One such individual in Amazonia was Chico Mendes, a Brazilian rubber tapper and self-taught environmentalist (see Revkin 1990). He was murdered in December 1988 by ranchers whose concern was for their short-term profit.

The destruction of the rainforests of Amazonia is at least now well publicised. The rainforests of Madagascar, with their unique plant and animal species, have suffered considerable devastation through deforestation by the land-hungry population, to the extent that the rich biological heritage is in real danger of mass extinction. Satellite images have been used to quantify the amount of destruction of the eastern rainforests in Madagascar,

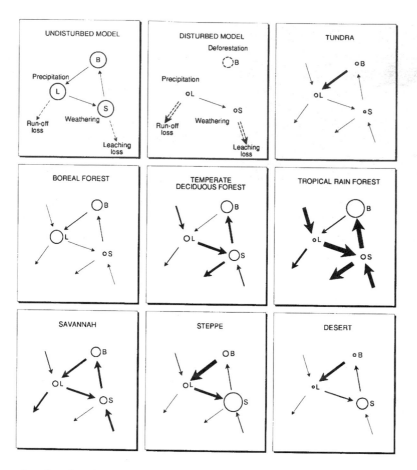

Figure 9.4 *Nutrient cycling in selected ecosystems. In the so-called 'disturbed model' after deforestation, litter decomposes relatively rapidly because the ground is warmer and less water is lost by evaporation and plant transpiration, so more nutrients are lost by surface-water run-off and leaching. The circle size is proportional to the amount in the organic pool. Arrow widths indicate the quantity of nutrient flow, which is expressed as a proportion of that stored in the source's pool. B = biomass; L = litter; S = soil. Redrawn after Tivy (1982).*

revealing that in 1950 there was 7.4 million hectares – an estimated 34 per cent of the original extent – but by 1985 there was only 3.8 million hectares, a reduction of 50 per cent in thirty-five years (Green and Sussman 1990). It is estimated that by the end of the next thirty-five years the only forest that will be left will be restricted to the very steepest slopes (*ibid.*). The natural desire of the Malagasy to better their lives today may be mortgaged at an intolerably high cost for future generations. Fortunately, concerned international agencies are now working in concert with many Malagasy to preserve what remains of the rainforests with their precious inhabitants.

One of the problems in trying to slow the rate of deforestation in the rainforests is their geographical and political location. Most of the rainforests are in less developed countries, many of which are in conflict or at war with adjacent countries. Unfortunately, the economy of most of these countries is depen-

dent, to a large degree, on the exploitation of the rainforests. At present, for example, one of the most destructive projects, La Grande Carajas, will decimate more than 2,000 km² of Brazilian rainforest. It will involve the construction of a reservoir and hydro-electric plant at Tucurui on the Amazon, and the development of a major iron-smelting plant and an associated open-cast mine (Branford and Glock 1985, Tyler 1990). The Brazilian government believes this will help to reduce national expenditure and dependence on imported fossil fuels needed to produce electricity, which constitutes a major part of the country's budget. Brazil plans other larger projects for the future, which will devastate even larger areas of the rainforest (Caufield 1986, Park 1992).

Many additional problems are associated with such construction schemes. Clearance of such large areas of forest usually involves large-scale flooding, with

Internationally, little is being done to control deforestation, although on a national scale many countries are beginning to enforce legislation aimed at controlling the degree to which loggers can exploit the forests. Reserves are being established in many rainforests, for example the Chico Mendes Extractive Reserve in Acre, Brazil, and the Parc Nationale des Volcans in Rwanda, where the famous mountain gorillas and their habitat have been saved from extermination (Park 1992). Unfortunately, many of these restrictions are difficult to enforce or are unrealistic. For example, in the Philippines, the Bureau of Forestry Development has enforced logging standards so that the maximum acceptable damage to a logged-over forest does not exceed 40 per cent. Loggers, however, select the best trees, which clearly degrades the strength of the forest by depleting it of its best gene stock. Second, loggers often return to the forest long before it has been regenerated to its original glory to destroy yet another 40 per cent – leading to even greater degradation. The use of heavy machinery may cause damage to the adjacent vegetation and the soil (Plate 9.1) (Caufield 1986). Experimental studies, such as those being undertaken at the Curua Forestry Station in the Amazon, are providing important information on methods of sustainable exploitation and regeneration of the rainforest (Eden 1989). These studies, however, are only beginning to aid in checking the rate of deforestation in developing regions.

In the insatiable desire for timber, many extravagant and unsubstantiated claims are made by retail outlets as to the source of their solid wood. The timber trade is concerned primarily about profit and has a vested interest in convincing consumers that the timber trade does not damage the rainforests but uses only timber from fast-growing and easily replenished trees. Most logging still destroys forests, whether or not they are tropical. There has to be a real will on the part of traders and consumers alike to use timber only from genuinely sustainable sources. International legislation may prove to be the only satisfactory way forward on this issue.

The World Commission on Forest and Sustainable Development is soon to be established by the UN to help determine the best way that forests can benefit society without being degraded. In order to achieve success, the commission needs to emphasise the global importance of forest ecosystems (see Myers 1995b).

Destruction of the savannahs

Perhaps less well advertised than the destruction of the rainforests is the human threat to Brazil's unique

Plate 9.1 *The use of heavy machinery in deforestation of the Brazilian tropical rainforest.* Courtesy of M. Eden.

trees left to rot and decay in the flood waters. Their decay leads to acidification of the water, which produces poisonous hydrogen sulphide and explosive methane gases. These acidified waters cause faster rates of corrosion of the turbines that produce the electricity in the hydroelectric power plant. Replacement is costly and difficult. The turbines may also become clogged with decaying logs and rafts of weeds, which proliferate soon after flooding. Together with the ensuing mechanical problems and attendant costs, other problems include the spread of diseases such as schistosomiasis and malaria, which are associated with large water bodies (Caufield 1986). Epidemics and fatalities are also associated with many of these developments in developing countries. The indigenous population contract diseases introduced by new migrant workers and can die in quite large numbers. This was a particularly alarming problem and a major issue during the construction of the Trans-Amazon Highway across Brazil. All these problems are difficult and expensive to remedy. Many people believe that as a result such projects as Tucurui will never pay for themselves, and that similar new projects should not be attempted (*ibid.*).

savannahs. About 3 million km² of Brazil – an area larger than the Mediterranean basin – consists of non-forest habitats (Burman 1991). Numbered amongst these are the wonderfully named *cerrado caatinga* and *campo rupestre*, all of which can be grouped under the term savannah. Because these areas do not have the obvious luxuriant vegetation and sheer diversity of species seen in the rainforests, they have tended to be regarded as trash vegetation and so open to thoughtless exploitation. These ecosystems are home to a unique and diverse flora and fauna, including **bromeliads** and orchids.

Open-cast mining occurs on a large scale in these landscapes. There is surface mining for gold in Bahia on the *campo rupestre*, and manganese extraction on the Serra do Cipo. The careless exploitation of the savannahs through the planting of inappropriate crops such as soya bean and sugar cane, and the development of pine plantations or eucalyptus, all serve to decimate habitats where, because of the extremes in climatic conditions, plants are already under natural stress. Plant collectors and the local population take many of the beautiful, rare and valuable flowers and plants at rates much greater than they can be replenished naturally. As the flora diminishes, or is wiped out, the animals and insects lose their habitats and disappear from the savannah.

It is not only Brazil that has large expanses of savannah: 65 per cent of Africa, 60 per cent of Australia, and 10 per cent of the Indian subcontinent, mainland Southeast Asia and Mexico are savannah (Stott 1994). Savannahs are, therefore, a major component of the global biological system, and a knowledge of their dynamics is essential if the potential consequences of human activities in relation to their survival is to be fully appreciated.

Wetlands

Wetlands are ecosystems that constantly contain surface water and are regularly flooded. They cover about 6 per cent of the Earth's land surface, which is just a little less than the tropical rainforests, and contribute about 24 per cent of the world's primary biological productivity. Wetlands are located in every major climatic zone and include mangrove swamps, peat bogs, marshes and fens. The importance of wetlands as a component of biogeochemical cycles has only recently been recognised, particularly their importance in trapping and recycling nutrients because of their very high bio-productivity. Wetlands also provide an important link between aquatic and dry terrestrial environments, helping to mitigate

flooding, protect coastlines from erosion and recharge aquifers.

Unfortunately, wetland ecosystems have been considerably degraded. It is estimated that by 1985, 1.6 million km² of wetlands had been drained for land reclamation (Williams 1990). This was mainly for agriculture. Early drainage schemes included land reclamation in the Netherlands since the sixteenth century; drainage of the English fenlands; and during 1946–1983 approximately 9 per cent of Albania was drained, partly in an attempt to eradicate malaria. In the USA, about 50 per cent of the natural wetlands have been lost, mainly for agricultural use and inland because of dredging canals, creating marinas and port facilities, and in order to allow urban expansion along the coast. In Louisiana, more that 100 km² yr⁻¹ of coastal wetland is being lost by a combination of human and natural factors (Walker *et al.* 1987). The natural factors include sea-level changes, subsidence, compaction and changes in the location of deltaic depocentres, and erosion, particularly during hurricanes. The construction of dams and levees across the Mississippi River has reduced the amount of sediment reaching the coast, which was an essential sedimentary process to keep pace with sea-level rise, while the compaction of deltaic sediments, subsidence, and highway construction altering drainage patterns, have all led to subsidence (Figure 9.5).

Attempts have been made to reverse the loss of wetlands. In the USA, the Wetlands Reserve Program was established in 1990 to try and restore 405,000 ha of privately owned fresh-water wetland that had been drained and farmed (Middleton 1995). The Convention on Wetlands of International Importance, especially as waterfowl habitats, was adopted by eighteen countries at its first conference in Ramsar, Iran in 1971. The convention is now known as the Ramsar Convention, the aim of which is to impose international regulations on land use. Unfortunately, the convention lacks legally enforceable international powers to take action against transgressors, and so the rate of wetland degradation is continuing apace.

Desertification

Deforestation and the degradation of other vegetation, particularly near the margins of deserts, have caused once fertile, vegetated land to become barren, in a process called *desertification*. Binns (1990) discusses the problems of defining desertification and its confusion with the term land degradation. One definition of desertification that we accept here is given by Nelson (1988, p.2) as:

1956   **1978**

Figure 9.5 *Changes from marsh to open water in the Mississippi delta from 1956 to 1978. Artificial controls up-river have reduced the sediment loads and have resulted in decreased replenishment of the wetlands. Redrawn after Goudie (1993b).*

a process of sustained land (soil and vegetation) degradation in arid, semi-arid and dry sub-humid areas, caused at least partially by man [and reducing] productive potential to an extent which can neither be readily reversed by removing the cause nor easily reclaimed without substantial investment.

It has been suggested that desertification should be regarded 'as an extreme form of land degradation occurring where vegetation cover falls below 35 per cent on a long-term basis' (Binns 1990, p.111). In contrast, land degradation can be considered as a significant reduction in the biodiversity and/or usefulness of an area of land due to either natural or anthropogenic influences (*cf.* Johnson and Lewis 1995). It is essential to erect widely acceptable definitions such as these if there is to be a consensus about the extent of desertification, and an analysis of the causal factors leading to desertification and land degradation. Figure 9.6 illustrates the possible causes and processes of desertification. The relative importance of each of the components in this figure needs to be assessed to help in the successful management of desertification.

Desertification and its associated problems are devastating for many parts of the world, especially developing countries. More than 100 countries suffer from the consequences of desertification, and it is estimated that the environmental impacts directly affect 900 million people (Hulme and Kelly 1993). The desert margins of the Sudan, the Sahel region of the southern Sahara Desert, the Gobi Desert in China and the Kalahari Desert in southern Africa are all particularly prone to severe consequences brought about by desertification. The margins of these deserts have advanced by as much as 100 km in the last couple of decades. The United Nations Environment Programme (UNEP) estimates that about 60 per cent of the 3.3 billion hectares of agricultural land outside humid areas is affected to some degree by desertification.

Although the UN claims that desertification caused by human activities is continuing to intensify, many argue that this may not actually be the case (Pearce 1992). The UN world map of desertification, published in 1992, identifies key areas of desertification but states that the margins of error fall within ±10 per cent (UNEP 1992). Any change since its last survey in 1977 falls within this error margin. It

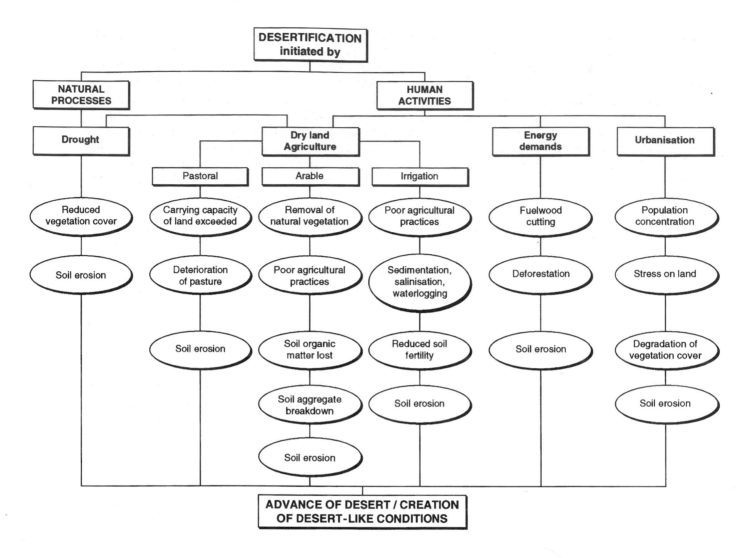

Figure 9.6 *The causes and development of desertification. Redrawn after Kemp (1994).*

is, therefore, not possible from this data to draw any firm conclusions about the degree of desertification. Furthermore, mapping desertification is difficult because remote sensing often gives a false impression of the amount of vegetation cover (*cf.* Tucker *et al.* 1991). Scrublands, for example, are difficult to distinguish from true desert. The dryland areas of Africa, however, have continued to report that desertification has increased since 1977. There are claims that reclamation and irrigation programmes mitigate the problems associated with desertification, which some argue are greatly exaggerated (e.g. Scoging 1993).

Many scientists also argue that desertification has not occurred as a result of human activities (e.g. Hulme 1989), and that land degradation attributed to cattle herding and overgrazing, particularly around watering holes, does not lead to desertification in

the Sahel–Sahara region (Pearce 1992). It has been claimed that cattle herders possess an innate knowledge and sensitivity towards the land, and therefore contribute little to its degradation (*ibid.*). Instead, they argue that the apparent effects of desertification are mainly the result of natural fluctuations in global and regional climate, such as droughts, which are inherent in drylands (Hulme and Kelly 1993). It is also well known that desert margins oscillate by tens of kilometres on a decadal time-frame, depending upon variations in rainfall (Hulme 1989). Droughts, however, may be the result of anthropogenically induced global warming, but this assertion remains unproved. Street-Perrott and Perrott (1990) argue that fluctuations in aridity are due, at least in part, to the development of the North Atlantic Deep Water (NADW), which is an important control on precipitation in North Africa (see Chapter 2). A study of lake levels and sediments from North African lakes reveals longer-term changes in aridity in North Africa (see Chapter 2, work by Street-Perrott and Perrott 1990).

During the Last Glacial Maximum, the Sahara and Sahel regions of North Africa were extremely dry, although there were periods when precipitation in North Africa was much higher than at present, for example 9,000 years BP. Gasse *et al.* (1990) believe that precipitation may have been 125–130 per cent more than today. The evidence for this comes from palaeohydrological indicators, including oxygen and carbon isotope ratios measured in inorganically precipitated carbonates, and also from the mineralogy of sediments from Sebkha Mellala in the Sahara, Algeria and Bougdouma in the Sahel, Niger. These data suggest that during deglaciation the transition from arid to humid conditions occurred synchronously. Comparisons with other palaeoclimatic continental records in Europe, and deep-sea sediments from the North Atlantic Ocean, show the synchronised changes in the coupled ocean–atmosphere system (*ibid.*), controlled by changes in the dynamics of the North Atlantic region and the melt water history of the Laurentide Ice Sheet (Street-Perrott and Perrott 1990, see Chapter 2).

There is much debate as to whether desertification is reversible. Many attempts have been made to reverse desertification, such as irrigating and re-vegetating desert margins, but although many have proven at least partially successful, it is not known how long an area can sustain itself once such maintenance is discontinued.

Mabogunje (1995) discusses the environment challenges that face sub-Saharan Africa and suggests that many of the environmental problems are essentially a consequence of a poor understanding of the nature of desertification and its possible remedies. Furthermore, in many countries where desertification is a problem its effects are exacerbated by unstable political systems linked to a poor economy. Mabogunje is optimistic, however, about the resilience of African nations to cope with desertification and to improve the situation during the next twenty-five years.

There are natural **badlands**, semi-arid landscapes severely eroded and dissected into fantastic geometrical shapes of pinnacles and deep gullies, for example in the USA along the Greybull River in the Bighorn Basin of Wyoming, or the Big Badlands of southwest South Dakota, along the White and Cheyenne Rivers on the eastern and southern edges of the Black Hills. Even in these semi-arid areas, bad management of the land can turn the fertile soil that is available into a 'dust bowl', a story well told in John Steinbeck's novel, *The Grapes of Wrath*, with its images of tattered refugees trudging westwards from Oklahoma to the promised land of California. In the 1920s, wheat farming was very successful in this region. In the early 1930s, however, several years of drought and bad maintenance of the land resulted in millions of acres of wheat shrivelling and thousands of cattle starving to death. In addition, a succession of large wind storms driven by cold air originating from the polar regions in northern Canada blew away the unprotected topsoil, producing devastating dust storms that transported the soil many thousands of miles. This dust also caused many respiratory diseases such as **dust pneumonia**. By the late 1930s the climate had become wetter and government intervention, aided by the advice of soil scientists, returned much of the wheat land to pasture, by introducing new farming practices such as crop rotation, tree planting to act as wind breaks, and the addition of fertilisers to the soil.

One of the Central Asian tragedies of desertification has taken place in the inland Aral Sea, which was once the fourth largest inland water body. However, before reaching the Aral Sea, the water was diverted for irrigation to support a cotton monoculture, to the extent that it has now virtually disappeared, to be replaced by a desert of sand, salt and redundant fishing boats. Since 1960, the water level has fallen by 15 m and the area of the lake has declined from over 66,900 km^2 to less than 33,000 km^2: it is predicted that it will decline to 21,421 km^2 by the year 2000 (Micklin 1992). Recently, a pandemic of anaemia and a host of other diseases such as kidney and thyroid disorders, and stomach and liver cancers has been affecting the people who live around the Aral Sea. The exact causes of these illnesses are not really understood, but they are probably the result of the region's ecological crisis, particularly because most of the water supplies are polluted with very high concentrations of salts and farm chemicals. The remaining fish population, which is a major food source for the region, contains high concentrations of pollutants (Pearce 1995b).

In June 1990, the then Soviet government announced a contest for the best ideas for improving conditions in the Aral Sea region. Some of the suggestions are discussed in Micklin (1992), and Figure 9.7 shows a water redistribution scheme, the Siberian Rivers Diversion Project, which was formulated in the 1970s. Most proposals, including the one shown in Figure 9.7, involve the transfer of waters from Siberian rivers to the Aral Sea, over a distance of some 2,500 km, redirecting water along

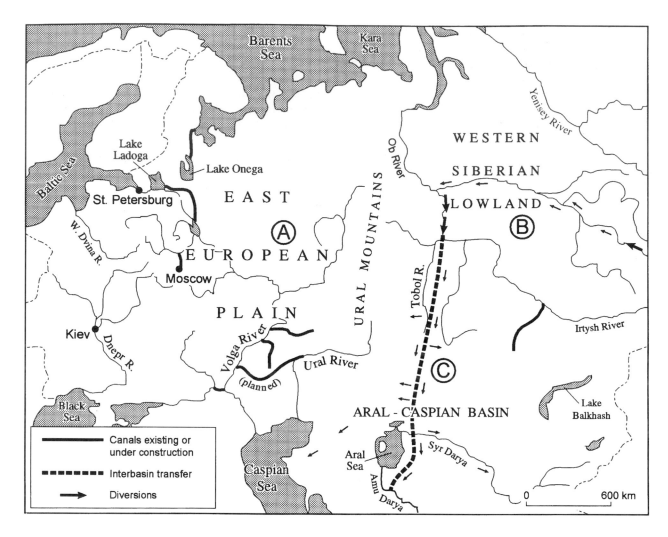

Figure 9.7 *The Siberian Rivers Diversion Project. Redrawn after Petts (1994).*

existing rivers and new canals. Micklin (1988) estimated that 200 km³ of water would need to be transferred to balance the water deficit in the Aral Sea if there is to be adequate domestic, industrial and agricultural water supplies. The negative environmental consequences of such a scheme might include waterlogging, leading to soil degradation along canals, disruption of existing ecosystems, the spread of water-borne parasites and diseases, and a reduction in the size of flood plains downstream from the rivers that have had their waters diverted into the Aral Sea (Petts 1994). The relatively warm waters of the Siberian rivers drain into the Arctic Ocean, and help to warm the sea. A reduction in the amount of warm river water entering the Arctic Ocean, as a consequence of any diversion schemes, could have important climatic responses. In particular, the distribution and the timing of formation of sea ice would be

affected. This could possibly lead to enhanced ocean-water cooling. Various models (e.g. Semtner 1984) have been constructed to attempt to calculate the effects of such river diversions, but the scientific community is still divided over any likely effects of the Siberian diversions. The implementation of such a project has been made more difficult by the dissolution of the Soviet Union. In January 1994, following an agreement and the approval of a development grant from the World Bank and the UNEP, the five new central Asian republics (Uzbekistan, Kazakhstan, Turkmenistan, Kyrghyzstan and Tajikistan), together with Russia, agreed to the establishment of a joint fund aimed at taking ecological steps towards saving the Aral Sea.

Other regions have experienced, and still suffer, similar problems. Drastic changes in farming practices, the trend towards monoculture, increased mechanisation and other potentially harmful farming practices are commonly blamed. Effects are often seen in developing areas, where conservative traditional methods of subsistence farming have been replaced by crops. These crops may not be suited to the region and may result in environmental degradation, including desertification.

Human introduction of exotic plants and animals

The introduction of new plant species, accidentally or intentionally, into an area can have both positive and negative effects. Plants that colonise an area may be more successful that the indigenous flora, and eventually the introduced species may begin to dominate the vegetation. The changes may be advantageous, for example by providing a new food supply, or they may reduce soil erosion. The introduction of the rhododendron shrub from the Himalayas into many parts of Europe is an example of just how successfully an introduced plant can become established. Rhododendron bushes are helpful in stabilising slopes, but they are very difficult to eradicate once established.

The accidental introduction of non-native plants into some areas may lead to devastating ecological consequences, for example **Dutch elm disease** fungus, which was introduced with imported timber to the UK. This caused the death of many thousands of elm trees during the 1970s and was accelerated by the transport of diseased wood. Once the disease was recognised the situation was exacerbated by the slow implementation of measures aimed at restricting its spread (Sarre 1978).

The introduction of non-native animals to a region may have serious effects on its vegetation. This is well illustrated by the introduction of grazing animals such as goats, donkeys, cattle, rabbits and hares to tropical islands. In Laysan Island (Hawaii), for example, much of the vegetation was cleared in a short time and the number of plant species was considerably reduced by the introduction of goats.

The opening of the Suez Canal in 1869 joined two biogeographical provinces – the Mediterranean and the Indian Ocean realms – allowing over 250 Red Sea species to migrate through the canal and start to colonise the Mediterranean. The sudden mixing of formerly isolated communities resulted in greater competition, displacement and depth shifting of some species, often leading to major changes in the various ecosystems. Of particular concern in recent times is the proliferation of the Red Sea jelly-fish, *Rhopilema nomadica*, along the Levant coast in the southeast part of the Mediterranean (Spanier and Galil 1991). The first specimen found in the Mediterranean was in 1977 off the coast of Israel, but by the summer of 1990 an almost continuous belt of jellyfish was present along the northern coast of Israel. The proliferation of jellyfish has had significant impact on local fisheries, causing damage to fish nets; in Haifa Bay, shrimp trawling nearly ceased during July 1990. Intake pipes for the cooling systems of coastal power plants became clogged. The numbers of holiday-makers using beaches dropped dramatically because of concern over the painful stings inflicted by the jellyfish.

There are many examples of the effects of introducing exotic plants and animals into new habitats. There are obvious short-term consequences, but all too often the long-term effects are poorly known and inadequately researched.

Human pollution and vegetation

Human impact on natural vegetation may also result from air and water pollution. Acidic deposition, for example, is considered to be a major culprit in the destruction of millions of trees in high- and mid-latitude regions, and the acidification of lakes. Pollution reduces plant growth and can lead to other forms of environmental degradation, such as soil erosion. This loss of vegetation and a reduction in floral diversity tends to be associated with decreased faunal diversity. Some of these issues are dealt with in the relevant chapters on atmospheric change (Chapter 3), acidic deposition (Chapter 4) and water resources (Chapter 5).

Human impact on soils

Soil is a combination of mineral and organic matter, structurally arranged in layers, and capable of supporting plant and animal life. Soils cannot exist without plants, and plants are dependent on soils for support, air, water and nutrients.

Soils are highly variable in nature. This variation includes their structure, layering, colour, range of particle sizes (particle-size distribution), chemistry, nutrients, acidity, alkalinity, temperature, moisture content, thickness, and organic content and its associated biota (Brady 1990). These properties vary because of differences in the parent material, climate, topography, organic content and the amount of time over which the soil has developed. Jenny (1941) considered soils to be the results of these factors and related them in terms of a factorial model expressed as:

$$S = f(cl, o, r, p, t, \ldots)$$

where S represent the soil type, cl = climate, o = organisms, r = topography, p = parent material and t = time.

Changes in one or more of these factors can lead to a drastic change in the soil properties, altering its physical and chemical nature and ability to support particular plant species. Detrimental effects may result in a reduction in vegetation cover, soil erosion, slope instability, increased flooding and increased suspended sediment loads in rivers and lakes (Morgan 1986). Some of these problems were discussed earlier in this chapter, but the role of human activities in affecting soil properties is explored further here.

The major soil changes resulting from human activities include chemical changes (**salinisation** and **laterisation**), structural changes (compaction), hydrological changes and soil erosion. There are many chemical changes within a soil that may be initiated by human activities. The most widespread and problematic are salinisation and laterisation.

Salinisation involves the accumulation of salts such as sodium chloride (NaCl), potassium chloride (KCl), calcium sulphate ($CaSO_4$) and sodium carbonate (Na_2CO_3) within a soil. Concentrations of these salts make the soil more alkaline and generally restrict or inhibit plant growth. Salinisation may also lead to secondary problems such as soil erosion and poor plant growth. Salinisation can occur naturally in semi-arid and arid areas, where **evapotranspiration** or direct evaporation from the soil exceeds precipitation. It may also occur in coastal regions that have saline ground water (Banks and Richter 1953). In areas where the evaporation of water from the soil is high, water is drawn upwards and evaporated from the soil surface, hence salts are left behind and concentrated near the surface of the soil. This results in a hard salty layer within the soil called a salt pan or concretionary layer.

Salinisation may be induced by irrigation and water abstraction. The abstraction of water leads to a rise in the water table, driving salt precipitation nearer the surface. In coastal regions, withdrawal of underground fresh water, which floats on top of underground saline water (originating from the sea), pulls the saline water beneath it nearer to the surface to contaminate fresh-water supplies. Also, water in wells may become saline and useless without expensive treatment. This is a particularly big problem in coastal regions such as California and Israel, and islands such as Bahrain and Long Island, New York. Salinisation was also a problem in cities such as London and Liverpool during the middle of the last century, until the more effective management of ground-water resources was introduced.

Irrigation also enhances salinisation by increasing the height of the water table in the immediate and adjacent areas over which irrigated water is spread. This leads to the evaporation of water from within the soil, and provides a process by which soil salts can be concentrated and drawn towards the surface. The UN estimates that with the rapid expansion of irrigation schemes in the last twenty years, as much as 25 per cent of irrigated areas have become affected by salinisation, making it a major land management problem. The percentage of saline land with water-logged soils amounts to 50 per cent of the irrigated areas in Iraq, 23 per cent in Pakistan, 30 per cent in Egypt and 15 per cent in Iran. Thomas and Middleton (1993) argue that salinisation may be considered a significant component of desertification processes throughout the world's drylands.

Salinisation may also result from the clearing of native vegetation for agricultural development. In recent years this has been a major problem in Western Australia, where deforestation for agricultural land has reduced evapotranspiration and has led to increased ground-water recharge and rising ground-water levels. Salts stored in the unsaturated soil zone dissolve and are then precipitated at or very near the surface, or the salts are washed into streams to cause the salinisation of both land and streams. Bari and Schofield (1992) have successfully shown that active aforestation, using *Eucalyptus*, substantially lowers the saline water table, thereby reducing the likely problems arising from salinisation. Peck (1978) discusses strategies for reclamation where saline waters occur as seeps at the surface: appropriate management of surface and ground water includes the planting of salt-tolerant vegetation to reduce the draw of saline waters to the surface and to allow the salts to leach from the surface. An understanding of the dynamics of processes such as these, and aforestation strategies, could greatly reduce the loss of land to salinisation and return it to a more fertile state.

Laterisation of the soil is a major problem in the tropics, where soils are enriched in aluminium and iron oxides (see Macfarlane 1976). These metal oxides accumulate due to strong tropical weathering conditions. Minerals and rocks break down to release electrically charged metal ions into the ground water, and these ions are transported, precipitated and concentrated by seasonal wetting and drying of the soil to form layers of metal oxides. Problems start to occur when these lateritic layers become exposed to the air and harden (as **duracrusts**), which inhibits plant growth, which in turn promotes soil erosion and its associated problems. Exposure of these layers may also be due to soil erosion, often as a result of deforestation. Deforestation may lead to increased evaporation of water from the soil, enhancing the process of laterisation.

The extent to which laterisation is a major problem has not been fully assessed, but particular problem areas include northern India, Cameroon and central Africa. Unfortunately, populations in these countries are heavily dependent on the soil for subsistence agriculture and therefore soil degradation is difficult to reduce, as there is increasing pressure on the land caused by population growth.

Changes to a soil structure, its mutual arrangement of grains, can also have a profound effect on soil properties, which include the ability to retain moisture, the transmissibility of water (**permeability**), strength, the degree to which a plant's roots can penetrate, as well as the withdrawal of water from a soil, and its resistance to erosion. The main way by which soil structure can be altered is through compaction – the pushing together of soil particles. Soil compaction can be caused by vehicles driving over its surface, overgrazing, trampling along public footpaths, or by ploughing, which may compress the soil immediately below the ploughed surface (Pidgeon and Soane 1978). Compaction reduces the permeability and **porosity** (amount of air- or water-filled pore spaces) in a soil, with the result that the soil hardness increases. Soil compaction retards or inhibits plant growth and may enhance processes such as soil erosion by wind and water. Soil compaction is a worldwide problem, but it is often greatest in developed countries, where heavy vehicles are more common. Soils are commonly ruined adjacent to building sites, and on old battlefields and military training areas, where heavy vehicles are particularly common. The compaction of soils remains one of the most difficult soil problems to remedy and it may take many decades before a soil can regain its original structure.

Many soils are drained to allow the increased growth of crops, or in order to plant trees and build settlements. In Great Britain, for example, over 60 per cent of the land surface is artificially drained. Draining soil may produce several undesirable effects. Previously wet organic rich soils may begin to decay, producing methane and hydrogen sulphide gases. This decay increases the acidity of a soil and results in reduced biological productivity. As a consequence, the soil volume may decrease as water is extruded,

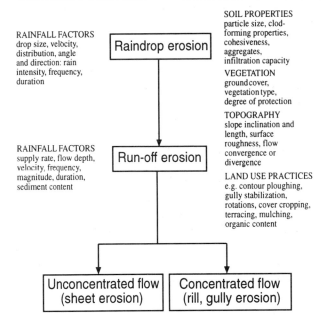

Figure 9.8 Important factors contributing to soil erosion. Redrawn after Cooke and Doornkamp (1990).

leading to subsidence and causing structural problems for buildings. The water table may be lowered, affecting adjacent regions and their soils. Some soils expand and contract due to changes in water content, which will affect engineering structure. Such soils are typically rich in volcanically derived **clay minerals** known as **smectites**. In addition, draining soils may lead to the rapid flow of water over the land surface into rivers and lakes to increase the likelihood of flooding during heavy rainfall and surface-water run-off.

Human activities can also alter the chemistry of soils, by the addition of organic or artificial fertilisers. This action may help to increase agricultural productivity, but it may also be detrimental to the soil, especially if the fertilisers are incorrectly applied – in which case the soil will deteriorate with a consequent reduction in vegetation cover, increased soil erosion and other associated phenomena.

The greatest impact on the soil is caused by soil erosion. Factors leading to increased soil erosion are shown in Figure 9.8. Figure 9.9 shows some of the processes of soil degradation. These factors and processes are complex. Yaalon and Yaron (1966) argue that they should always be considered using a process–response model, incorporating the feedbacks that can lead to one or more changes as a result of

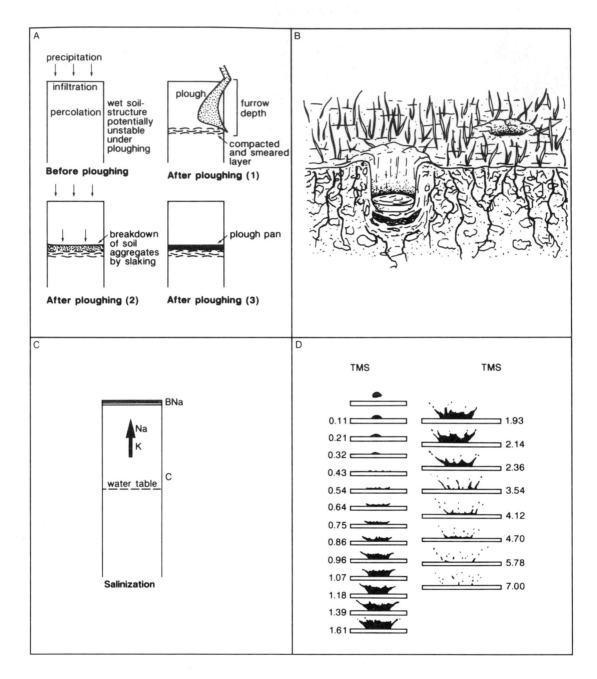

Figure 9.9 *Processes of soil degradation. (A) Formation of a plough pan. After White et al. (1984); (B) Effects of animal hoofprints on soil structure. After Batey (1988); (C) Salinisation of soil. After Knapp (1979); (D) Rain splash. Redrawn after Ghadiri and Payne (1980).*

human interaction with the soil. Increased soil erosion includes the abrasive action of water running over the land surface, the break-up of soil due to the impact of raindrops, and the deflation of soil particles by the wind. The various causal factors that may initiate soil erosion have already been discussed, such as deforestation, grazing, salinisation and later-

isation, compaction and fires. Many of these factors are interlinked and should not be considered in isolation. They may also be accelerated by bad farming techniques, urbanisation, construction schemes, mining, wars and fires. Some of the worst-affected areas are due to a combination of many of the above factors (Plate 9.2).

Plate 9.2 *Deeply weathered granite in Thailand, which has become exposed during the construction of a new road. The slopes are beginning to become gullied and if erosion continues, this will lead to slope stability problems, which will threaten the highway.*

During the last forty years, nearly one-third of the world's arable land has been removed by erosion, and it continues to be lost at a rate of more than 100 million ha yr^{-1} (Pimental *et al.* 1995). Damage from soil erosion is not restricted to the agricultural environment; it also has consequences for a vast range of human activities and facilities. Table 9.1 lists the

types of damage from soil erosion each year for the USA and the estimated costs of preventing these effects.

Although many people might argue that soil erosion is a modern phenomenon accelerated by increasing population pressures, there is much evidence to suggest that humankind has been eroding

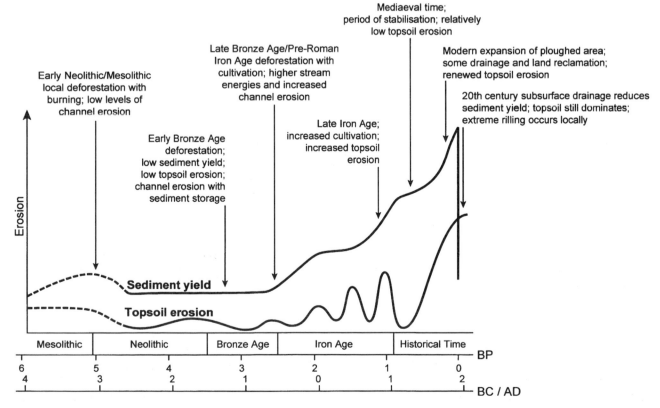

Figure 9.10 *Generalised model of erosional intensity and sources since Mesolithic times in southern Scania, Sweden. Redrawn after Dearing (1991).*

Table 9.1 Damage by wind and water erosion and the cost of erosion prevention each year for the USA.

Type of damage	Costs (millions of dollars)
Wind erosion	
Exterior paint	18.5
Landscaping	2,894.0
Automobiles	134.6
Interior, laundry	986.0
Health	5,371.0
Recreation	223.2
Road maintenance	1.2
Cost to business	3.5
Cost to irrigation and conservation districts	0.1
Total wind erosion costs	9,632.5
Water erosion	
In-stream damage:	
Biological impacts	no estimate
Recreational	2,440.0
Water-storage facilities	841.8
Navigation	683.2
Other in-stream uses	1,098.0
Subtotal in-stream uses	5,063.0
Off-stream effects:	
Flood damage	939.4
Water-conveyance facilities	244.0
Water-treatment facilities	122.0
Other off-stream uses	976.0
Subtotal off-stream	2,318.0
Total water erosion costs	7,381.0
Total costs of wind and water erosion damage	17,013.5
Cost of erosion prevention	8,400.0
Total costs (on and off-site)	44,399.0
Benefit/cost ratio	5.24

Source: Pimental *et al*. 1995.

soil at considerable rates throughout history. Van Andel *et al.* (1990), for example, presented sedimentological evidence for a phase of soil erosion about 500 to 1,000 years after the introduction of farming in Greece – during the Late Neolithic to Early Bronze Age – and several erosional events during the Late Bronze Age and historical times. In the highlands of central Mexico, O'Hara *et al.* (1993) found evidence from lake sediments for three periods of accelerated erosion that were at least as fast as the erosion rates after the Spanish introduction of the plough. Figure 9.10 shows a generalised model of erosional intensity since Mesolithic times in southern Scania, Sweden, based on palaeopedology and sedimentology, and it shows how periods of agricultural change appear to coincide with increased sediment yields, reflecting the erosion of topsoils.

Davis (1976) also showed how changes in land use affect sedimentation. He calculated historical sediment yields at Frain's Lake, Michigan from AD 1800, which purport to show a rapid rise in erosion corresponding to the time of woodland clearance in the 1840s and a subsequent increase caused by farming (Figure 9.11A). Figure 9.11B shows sediment yield data that has been annotated to illustrate that the erosion and the sediment yield respond to these environmental changes. Initially, the response is dramatic, but it is followed by a slow recovery to a steady equilibrium state, but with higher sediment yields. Studies such as these provide an important data set on how the environment can respond to agricultural changes and they provide information on the suitability of various farming methods. These studies also help to answer questions of whether traditional farming methods are really the most beneficial in helping to combat soil erosion.

Increased soil erosion leads to greater suspended sediment loads in rivers and lakes, which in turn can cause the silting up of reservoirs and estuaries. Destruction of vegetation and the exposure of weathered rocks leads to deep gullying and slope instability. Not all gullies, however, are the result of human activities. Wells and Andriamihaja (1993), for example, showed that deep gullies in Madagascar – known as lavaka – are a natural component of the landscape's evolution. These gullies develop on laterite surfaces that have been exposed by tectonic uplift and/or climatic aridification and are initiated by disturbances on hill slopes. Their growth continues by rain-splash erosion, incision by surface-water run-off, earth falls caused by alternate wetting and drying cycles, and seepage erosion. Human activities, however, such as deforestation and the construction of paths and roads can also initiate and accelerate gully formation.

The fastest rates of erosion in the world produce deep gullies in the thick wind-blown sediments (**loess**) of the Loess Plateau, Gansu Province, central China (Plate 9.3). Loess is easily eroded, particularly during heavy rainfall, and the sediment is washed into the Yellow River and redeposited on the flood plains. Plate 9.4 shows a damaged irrigation dyke on the Loess Plateau – a common feature in the fields here and produced by the collapse of a soil pipe due to seepage of water from nearby fields. It is only by careful monitoring and management of erosional and depositional processes that these problems can be tackled successfully in this region, where farming has to support very high population densities. The effects of human-influenced erosion and transport of sediment along the Yellow River can be examined by comparing data from river gauging stations with the

Plate 9.3 *The eroded loess landscape of the Loess Plateau, central China.*

volumes of sediments deposited in the Yellow Sea throughout Holocene time. Milliman *et al.* (1987) showed that prior to extensive agricultural activities on the Loess Plateau, which began about 200 BC, the sediment loads were an order of magnitude lower than today, clearly illustrating the role of human activities in contributing to soil degradation.

Plate 9.4 *A damaged irrigation dyke on the Loess Plateau, central China.*

It is not only agricultural activity that can lead to accelerated erosion. Wolman and Schick (1967) showed that areas cleared for construction may produce more sediment in one year than from many decades of natural or agricultural erosion. They compared the amounts of sediment produced in areas undergoing rapid development near Baltimore, Maryland, and Washington DC with wooded and agricultural areas of the same region. They also examined the effects of increased sediment yield as a result of construction. During construction, sediment loads in streams that previously had little sediment rapidly increased, followed by a steady decline as construction ceased and the landscape became more stable. Increased sedimentation in streams resulted in the blanketing of bottom-dwelling flora and fauna (affecting the ecosystem), reduced light transmission, enhanced abrasive effects caused by the sediment, a greater frequency of bank collapse, and increased flooding.

Much can be done in an attempt to retard soil erosion and conserve soil profiles. Conservation measures include re-vegetation, crop management, slope run-off control, construction of gabions and retaining walls, and the dissemination of information regarding good land-use practices. Pimental *et al.* (1995) discuss the various erosion control technologies, which include ridge planting; no-till cultivation; crop rotation; the use of grass strips; agroforestry; contour planting; the use of cover crops; and windbreaks. They suggest that to determine the appropriate conservation technologies, the soil type, specific crop or pasture, topography, climate and socio-economic conditions of the people living on the site must be considered. Pimental *et al.* also

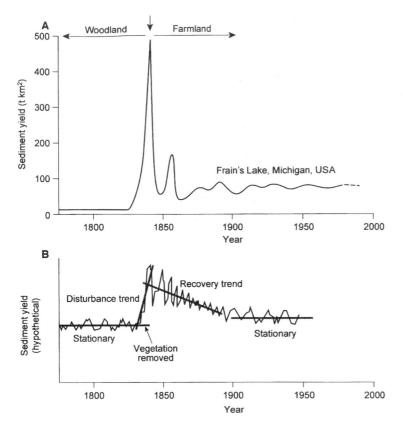

Figure 9.11 *The historical record of sediment yield at Frain's Lake, Michigan since* AD *1800, showing (A) the rapid rise in erosion at the time of woodland clearance and the subsequent erosion level caused by agriculture. (B) The Frain's Lake sediment yield data annotated to show how records can be defined in terms of different states and responses in the sediment system. Redrawn after Davis (1976) and Dearing (1994).*

argue that the implementation of appropriate soil conservation techniques in the USA could reduce soil losses from two- to 1,000-fold and water losses from 1.3- to >20-fold. Some regions have experienced considerable reductions in soil erosion over the past three decades. Meade and Trimble (1974) showed that sediment loads in rivers of the Atlantic drainage of the United States had decreased since 1900 (Figure 9.12). In the main rivers, this was the result of the construction of reservoirs along river courses, while in the tributary streams it was due to the decline in crop farming and the improvement of soil-conservation practices.

The Himalayan dilemma – a lesson in caution

Since the mid-1980s, the media have highlighted the large-scale environmental degradation in the mountains of central Asia, principally the Himalayas

(Plate 9.5). The perspective put forward has been that increased health care, medicine and the reduction of malaria in the post-war years from the 1950s has led to the rapid growth of population in the Indo-Gangetic Plain south of the Himalayas. Within twenty-seven years the population doubled. At the same time, the indigenous populations in the mountains grew, but at slower rates as health care was more difficult to provide in the more remote mountain regions. This population growth led to rapid deforestation of the Himalayan foothills and also of some higher-altitude areas due to the increased demand for timber for fuel and building, together with the need to clear more land for crops to feed both the population and animals. Some environmentalists suggested that the rate of deforestation was so rapid that by the year 2000 no forest would remain in the Himalayan regions. In 1979, the World Bank noted that Nepal had lost half of its forest over the previous thirty-year period, and suggested that by the year 2000 there would be no remaining accessible forest.

Where forest has been cleared on steep slopes, they are cut into terraces forming a stepped topography for the cultivation of wheat, barley, maize and, at lower altitudes, rice. It was argued that this deforestation and change in land use would lead to increased slope instability with more landslides and an accelerated rate of soil erosion once the protective tree cover, which had served to stabilise local soils, was removed. Furthermore, the deforestation caused enhanced surface run-off of rain water, further increasing the rate of soil erosion and leading to flooding. Some of the devastating floods in Bangladesh in the early and mid-1980s have been attributed to high river discharges due to increased run-off in conjunction with the occurrence of tropical cyclones. The growth of islands in the Bay of Bengal was regarded as a consequence of sedimentation from the increased sediment discharge in rivers because of accelerated soil erosion and a greater frequency of landslides: the rapid siltation of reservoirs in hydroelectric power schemes was also attributed to this environmental degradation.

As the forests became smaller, villagers had to travel further for timber. This meant that animal dung was substituted for wood as a major local fuel. The use of dung for fuel drastically reduced its use as a fertiliser and, consequently, soil quality decreased. In turn, the poorer-quality soils became more easily eroded, since the organic matter helped to bind the soil particles together. A vicious circle of enhanced landsliding and flooding ensued. The prospect appeared bleak for the Himalayas.

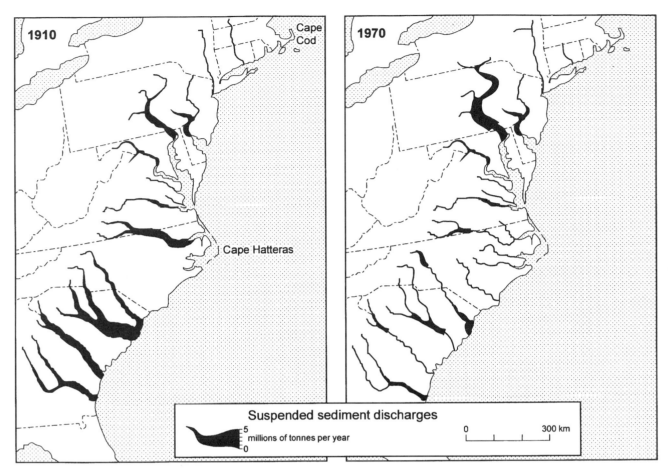

Figure 9.12 *The decline in suspended sediment discharge to the eastern United States between 1910 and 1970 as a result of soil conservation and land-use changes. Redrawn after Meade and Trimble (1974).*

Superficially, these arguments appear very convincing and are still widely used by environmentalists. Based on their work with UNESCO and the World Bank, as well as many other research projects, however, Ives and Messerli (1989) suggested that many

of these arguments were too naive. First, there is no widespread evidence for deforestation on the scale proposed in the Himalayan paradigm. Certainly, some areas have been deforested at alarmingly fast rates, such as the Hengduan Mountains in Sichuan (China)

Plate 9.5 *Mountain people finishing a tree to provide supports for a new house in the Nanga Parbat Himalaya, northern Pakistan. Deforestation for domestic use such as this is essential for these peoples, but it is difficult to reconcile their needs for development with broader environmental issues.*

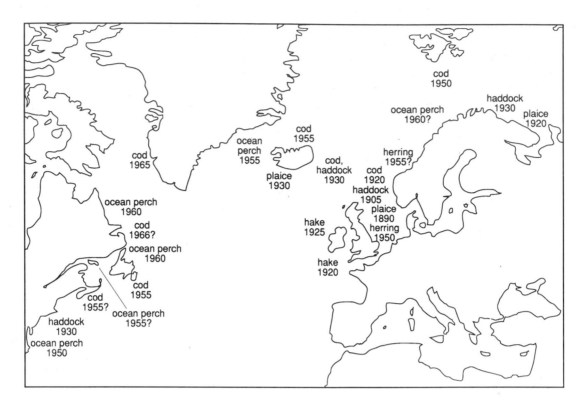

Figure 9.13 *North Atlantic fisheries with dates of the beginning of overfishing for each fish population. Redrawn after Smith and Warr (1991).*

and the lower hills of Nanga Parbat in Pakistan, but in contrast there are areas, such as the Garhwal Himalaya, although degraded earlier this century, that have had a long history of forest management with active aforestation. Today, areas such as the Khumbu Himalaya (Nepal) have a substantial area of land that is designated as a national park – the Sagarmatha National Park (Everest). The future of the forests in the Himalayas seems, after all, not so bleak.

Another major assumption embodied in the arguments for alarming deforestation is that terrace-style farming degrades the valley slopes. In fact, most of the terraces increase slope stability as the slopes become graded: in some cases the best terraced slopes are on ancient landslides because they provide amongst the lowest gradients and therefore are associated with greater slope stability. There have been a few instances of farmers actually initiating landslides in order to produce more stable slopes for farming. The terraces are constructed so that surface-water run-off can be controlled, and the water efficiently utilised in irrigation. Farmers have shown an acute awareness of the need for natural fertilisers and seldom use dung as a fuel.

The assumption that the increased siltation of rivers in the Himalayas and associated drainage basins is linked directly to the degradation of soils brought about by human activities requires closer scrutiny. Most of the rivers receive much of their water from high mountain glaciers, which also supply enormous quantities of silt and mud. Indeed, silt and mud may be supplied from accelerated soil erosion caused by human activities, but to date there have been too few studies specifically aimed at addressing this issue. The silting up of reservoirs is occurring at a disconcerting rate but it is difficult to determine accurately the expected life span of such constructions because of the problems in measuring sediment budgets in large-volume rivers. Most estimates are based on suspended sediment loads, where little consideration is given to the sediment transported along the river bed – the traction load, a quantity that is substantial in the fast-flowing Himalayan rivers. Also, the Himalayan paradigm predicts that landslides supply vast quantities of sediments to the rivers, but studies appear to show that many landslides rarely reach the river, and most of the redeposited sediment is dumped on mid-slopes and benches along the river valleys. Finally, little scientific consideration has been given to the causal factors leading to the growth of islands in the Bay of Bengal; for example, they may have formed in response to entirely natural sedimentary processes rather than because of human activities.

The Himalayan paradigm, hawked as a theory of massive environmental degradation in the Himalayas, has little evidence to substantiate it. This case study epitomises the dangers of environmentalists seeing only what they want to see, and so drawing conclusions in the absence of reliable data.

There is still a paucity of good information on the climatology of the Himalayas – partly because of the difficulty of gathering long-term scientific data in such a politically sensitive region, where climatological data are considered to have military importance and, therefore, are not freely available. Where data are available, they appear to suggest that there are twelve-year cycles in the intensity of the monsoon, similar to the El Niño Southern Oscillations in the South Pacific (see Chapter 2). A coincidence of heavy rainfall during a strong monsoon and a tropical cyclone will lead to increased flooding. Climatic cycles on a longer timescale are even more difficult to reconstruct. Currently, it is not possible to resolve the relative importance of the impact of human activities in the Himalayas from natural processes.

What is a fact in the Himalayan region is that the human population has increased considerably, and this has undoubtedly produced many socio-economic problems and in some areas caused environmental degradation. Ives and Messerli (1989) emphasise that many projected environmental scenarios are not necessarily as simple and straightforward as one might initially believe. There is a need to examine in an integrated manner the physical, social, economic and political aspects of environmental issues. Their book provides a counterbalance to the more alarmist statements and publications on the human impact on the environment.

Human impact on the oceans and seas

The oceans and seas cover more than two-thirds of the Earth's surface. They contain submarine trenches that are deeper than the highest mountains. Life almost certainly evolved from the sea, and there are still more species diversity in the oceans than anywhere else on Earth. Many of the food chains or food webs start with organisms inhabiting the seas and oceans. The ocean–atmosphere system regulates global climate. It is a sensitive thermostat. The seas and oceans are a rich food and mineral resource, but over-exploitation and pollution threaten this vast wilderness. Humans still tend to feel that the vastness of the sea makes it an ideal dumping ground

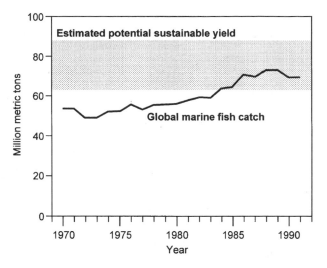

Figure 9.14 *Global marine fish catch and estimated potential sustainable yield, 1970–91. Redrawn after World Resources Institute (1994–95). Sources: Food and Agriculture Organisation of the United Nations (FAO), FAO Fishstat 1992, on diskette (FAO, Rome, 1992); M.A. Robinson, 'Trends and Prospects in World Fisheries', Fisheries Circular No. 772, Food and Agriculture Organisation of the United Nations, FAO, Rome, 1984*

for virtually every type of waste, including toxic chemicals and nuclear waste.

Since the introduction of steam trawlers in the North Sea in the 1880s, vast numbers of fish have been caught more easily and with less cost. The increased fish catches brought a greater demand for fish, both for domestic consumption, and for fish meal and fertilisers. Between the 1950s and 1980s, the global catch of marine fish rose from 20 million tonnes to 70–80 million tonnes and then levelled out (Smith and Warr 1991), leading to a significant depletion of fish stocks, with a major but still poorly understood impact on the marine ecosystems. Figure 9.13 shows the dates of the beginning of over-fishing for different fish populations in the North Atlantic, and Figure 9.14 shows the global marine fish catch and estimated potential sustainable yield.

In the North Atlantic, the Grand Banks off the coast of Newfoundland are one of the richest fishing grounds in the world. In 1992, however, the Canadian government banned fishing on the Banks after the discovery that nearly all the adult cod had disappeared. The reasons for over-fishing are many and include government limits being higher than those advised by many scientists; fishermen cheating on catch quotas; and large errors in the scientific models used for estimating fish catches and the likely consequences (e.g. see Mackenzie 1995). Safina

(1995) attributes the threat to the world's fisheries to political factors, stressing that the fishing lobby is so strong that most politicians have chosen not to heed the obvious signs of the imminent depletion in fish reserves but actually have allowed a massive expansion of the industry.

The main collapse of the traditional fisheries occurred during the 1950s and 1960s, when fishers adopted new technologies. These included radar, electronic navigation aids, satellite positioning systems to improve navigation and relocate fishing grounds, satellite weather maps of water-temperature fronts to indicate where fish will be travelling, and the use of aircraft to spot fish. Ships used enormous trawling nets and drift nets, the largest being up to 40 km long, as well as submerged longlines with thousands of hooks, often up to 130 km long. This sort of fishing was so intense that 80 to 90 per cent of the fish in some populations were removed annually (*ibid.*). During the past two decades, the rate of removal has accelerated, and the reduced stocks have resulted in catches of less desirable fish of lower trophic levels: this has resulted in major disruptions in the food chain, clearly affecting organisms at higher trophic levels.

The ecological problem has been exacerbated by the large quantities of marine life that are inadvertently captured during fishing – known as the **bycatch** or **bykill**. Bycatch can account for one in every four animals that are taken from the sea. Many of these include young fish, marine mammals and diving seabirds. In 1990, for example, high-seas drift nets tangled 42 million animals and, prior to the 1990s, fishing for yellowfin tuna resulted in 400,000 dolphins being killed annually (*ibid.*). Dolphin-safe nets are now used by most fishers, but although they are safe for dolphins, other creature such as turtles, sharks and immature tuna are still part of the bycatch, and the dolphin kill in 1993 was still 4,000. The enormous size of such a bycatch problem has prompted the UN to enact a global ban on drift nets that are longer than 2.5 km: countries such as Italy, France and Ireland still use them.

With the decline in fish stocks in the oceans, **aquaculture** (controlled fish farming) has doubled its output in the last decade to balance the shortfall. Today, more fresh-water fish are provided than from natural, wild fisheries. Aquaculture, however, has not reduced the threat to the world's fisheries since previously 'worthless' wild fish are now used as feedstock, for example for shrimp. Aquaculture results in the construction of pens and/or the removal of vegetation such as mangrove swamps, thereby causing profound environmental degradation. Nutrients used in fish-farming often pollute surrounding environments.

The fishing industry has survived the reduced stocks because of its political strength. Currently, there are far too many fishing boats, and they are so efficient that their numbers could be drastically reduced and still harvest the same catch sizes. Ironically, the cost of fishing exceeds the profits. For example, to catch US$70 billion worth of fish, the fishing industry recently incurred annual costs totalling US$124 billion, the difference being made up by government subsidies such as fuel-tax exemptions, low-interest loans and price controls. Governments continue to subsidise the fishing industry and commonly show a great reluctance to reduce the size of fishing fleets because of the threat of losing votes and becoming unpopular with the unions – since jobs are at stake – while promoting irreversible damage to marine ecosystems.

Fish are the only wildlife still hunted on a large scale. It seems that the threshold conditions that permit fish populations to reproduce at rates sufficient to replenish the fish stocks that are caught has been exceeded in all parts of the Atlantic, Mediterranean and Pacific. The worldwide extraction of fish peaked in 1989 at 82 Mt, after which it seems to have remained about constant or declined a little (Safina 1995). The UN FAO (1995) recently concluded that the operation of the world's fisheries can no longer be sustained, and that substantial damage has already been done to the marine environment. Any solution to the problem of over-fishing must take into account the fact that the economy and major part of the national diet in many countries is dependent upon fish catches. Whilst recognising that over-fishing has caused serious environmental damage to marine ecosystems, nations are unlikely to desist from taking large fish catches from the oceans if the consequence is likely to be large-scale economic impoverishment.

Over-fishing is a global problem necessitating international conventions and treaties, because it involves political disputes over fishing grounds, national diets, and employment for fishermen and allied industries. In order to mitigate the environmental effects caused by over-fishing, catches should be regulated, for example on the basis of the number of fish that can be caught. A more local solution to over-fishing is the development of fish-farming (aquaculture) practices. Intensive fish-farming, however, can create environmental problems, because the introduction of fish food and nutrients into coastal waters to increase fish yields may pollute adjacent waters and seriously disrupt existing aquatic ecosystems (Iwata 1991).

One of the more emotive issues regarding the exploitation of the sea is that of whaling. There are many arguments as well as sound scientific reasons against whaling, particularly because of the threat of extinction, but even this point is disputed. The principal whaling nations, Japan, Norway and Iceland, have used counter-arguments in support of their continued whaling. The arguments for whaling are mostly based on the size of the whale population (Holt 1993). As with the Grand Banks example, modelling and predicting the size of whale populations is difficult, and the uncertainties are large enough to allow those with an economic interest in whaling to dispute the figures for the size of the various whale populations. Despite the uncertainties, surely the risk of losing a whale species is great enough to mean that caution might be the best course of action, that is a truly international ban on whaling (Plate 9.6).

One of the areas of growing concern is the future exploitation of the sea floor. Breuer (1991) makes a good case for the international adoption of a strategy for the sea bed. There are large mineral reserves, for example in the form of manganese nodules. Thirty years ago, these nodules were thought to be worth mining from the sea floor, but the fall in metal prices over the intervening years has made this seem increasingly unlikely in the near future. If metal prices show a significant increase, then renewed interest in mining the sea floor might necessitate international laws and conventions to control the extent and manner in which any exploitation takes place.

The deep ocean floor contains '**black smokers**', or **hydrothermal vent** features, where new oceanic crust is forming, a process that is associated with very large amounts of degassing and fluid escape from the Earth's mantle in what is known as **hydrothermal activity** associated with submarine fissures and volcanoes. Black smokers are tall chimneys built of metal sulphides that have been precipitated from the metal-rich fluids escaping from molten igneous material below the sea floor. Their name derives from the fact that the hot escaping (hydrothermal) fluids produce a very fine suspension of metal sulphide particles in the sea water that looks like black smoke when illuminated by submersibles and camera flash. Together with large amounts of iron, the black smokers also discharge high concentrations of copper, zinc, lead, silver and gold. While the technology to mine these deposits is not yet developed, it must only be a matter of time before a company shows interest in undertaking such a venture. Indeed, the Japanese government has provided US$24 million over five years to evaluate the economic potential of manganese nodules and polymetallic sulphide resources on the sea bed. International

discussions and agreements are needed now, not when the developers move in.

The degradation of coral reefs and atolls has been the focus of much concern in recent years. Table 9.2 shows the various human-induced threats to coral reefs and other tropical marine ecosystems. Hughes (1994), using the Jamaican coral reefs as a particular example, describes how many corals have been degraded by a combination of human and natural disturbances over the past three decades. Here the effects of over-fishing, hurricane damage, and algal blooms, which developed because the main grazer, the echinoid *Diadema antillarum*, was reduced in numbers due to disease, has resulted in widespread destruction of most corals. Bauxite (aluminium oxide) mining in Jamaica has been associated with the construction of loading piers in some of the bays, with the result that parts of the fringing coral reefs were dynamited to clear passageways for ships. The abundance of coral has declined from more than 50 per cent in the late 1970s to <5 per cent today, and the reefs have become dominated by fleshy macroalgae (>90 per cent cover). Hughes emphasises the immediate need for the implementation of scientifically based management programmes and controls on over-fishing in order to avoid further catastrophic damage.

The United Nations Convention on the Law of the Sea (UNCLOS) states that the oceans and seas are a 'common heritage of humankind' that cannot be appropriated by any individual, institution or country; must be managed by and for the benefit of humankind as a whole; must be conserved for future

Plate 9.6 *Greenpeace demonstrators outside the International Whaling Commission's conference in Dublin on 5 May 1995.*
Courtesy of R. Robinson-Owen.

Table 9.2 *Human-induced threats to coral reefs with selected examples (adapted from Middleton 1995) and mangrove and seagrass areas under threat.*

Coral reefs

Threat	Examples
Over-collecting	
Fish	Philippines
Giant clams	Kadavu and islands, Fiji
Pearl oysters	Suwarrow Atoll, Cook Islands
Coral	Vanuatu
Fishing methods	
Dynamiting	Belau, USA
Poison	Uvea Island
Recreational use	
Tourism	Heron Island, Great Barrier Reef
Scuba diving	Puerto Galera, Philippines
Anchor damage	Molokini Island, Hawaii
Siltation due to erosion following land clearance	
Fuelwood collection	Upolu Island, Western Samoa
Deforestation	Ishigaskishima, Yaeyama-retto, Japan
Coastal development	
Causeway construction	Canyon Atoll, Kiribati
Sand mining	Moorea, French Polynesia
Roads and housing	Kenting National Park, Taiwan
Dredging	Johnston Island, Hawaii
Pollution	
Oil spillage	Easter Island (1983)
Pesticide spillage	Nukunonu Atoll, New Zealand (1969)
Urban/industrial	Hong Kong
Thermal	Northwest Guam
Sewage	Micronesia
Military	
Nuclear testing	Mururoa Atoll (1995)
Conventional bombing	Kwajalein Atoll, Marshall Islands (1944)

Habitat	Reason
Mangroves	
Niger River delta, Nigeria	Exploitation for timber, fuel, fodder and human expansion.
Kenya and Tanzania	Cleared for fuelwood, building materials and tourist resorts.
Indus River mouth, Pakistan	Over-exploitation for fuel, fodder and building materials.
Sundarbans, India and Bangladesh	Over-exploitation for fuel, fodder, timber and fishponds.
Malaysia and the Gulf of Thailand	Destruction of fish and shellfish ponds and agricultural land.
Philippines	Destruction for timber, tannin, fuelwood and fish and shellfish ponds.
Indonesia	Massive destruction for logging and woodchip industries, fish and shellfish ponds for building materials and fuelwood.
Queensland, Australia	Town and tourist development.
US south coast, Texas to Florida	Over-development of coastline for urban expansion, resorts, housing estates and rubbish dumps.
Panama	Cleared for fish and shrimp ponds.
Ecuador	Cleared for fish and shrimp ponds.
Caribbean	Main threats: tourism and coastal development.
Seagrasses	
East Africa	Under threat from heavy sedimentation of shallow coastal waters caused by erosion of agricultural lands.
Southeast Asia	Under threat from loss of mangroves, coastal development, urban expansion and bucket dredging for tin.
Caribbean and Gulf of Mexico	Under threat from dredge and fill operations, loss of mangroves, coastal development for tourism, oil production.

After Crystal 1993.

generations; and must be reserved for exclusively peaceful purposes. To date, this convention has been ratified by forty-five nations, but it requires ratification by a further fifteen before it can take effect.

Human impact on the landscape

The land surface is an important resource, permitting amongst other things biodiversity, settlements, agricultural land, mining and recreational use. It is beneath the land surface that mineral resources and fossil fuels are to be found. Human activities are continually modifying the landscape, creating urban areas, roads, airports (Plate 9.7), pits, ponds, spoil heaps, terraces, cuttings, embankments, dykes, canals, reservoirs and areas of subsidence. **Geomorphologists**, scientists who study landforms, consider humans an important land-forming agent, referred to as geomorphological anthropogenic agents. Human activities are an important factor in contributing to the landscape.

A measure of land degradation, known as the potential direct instrumental value (PDIV), represents the capacity for a piece of land to supply direct benefits to humans, such as agriculture, forestry, industry and medicinal production (Daily 1995). It has been estimated that currently 10 per cent of global PDIV of land has already been lost, and that if degradation continues at its present rate the PDIV could be as low as 20 per cent by 2020. Land degradation can be halted and even reversed through the implementation of rehabilitation measures, so that 50 per cent of the loss could be recovered within twenty-five years (*ibid.*). Daily emphasises that rehabilitation of the world's degraded lands is essential in order to meet the need of the growing human population, particularly the need to produce food, feed, biomass energy, fibre and timber.

The rapid increase of population has placed great demands on available living space. The trend towards urbanisation has led to an increase in the size of settlements at an incredible rate and the exploitation of marginal lands. In the latter, natural processes constitute a hazard to people settling in these regions. Advances into these regions are often accompanied by rapid deforestation and de-vegetation, resulting in soil erosion, flooding and other associated problems.

Subsidence is also a problem in areas where the ground is permanently frozen. The upper layers of frozen ground can melt in spring and then refreeze during the winter. Building on these permafrost areas can cause the ground to thaw to a deeper and greater

Plate 9.7 *Clearing the runway of snow at the British Antarctic Survey's main base, Rothera Base, on the Antarctic Peninsula.* Courtesy of Gary Nichols.

extent than normal. The ground may shrink and subside, and water is extruded, resulting in structural damage to buildings. This is a particular problem in the settlements in Arctic regions such as Alaska and Siberia, where special engineering structures have been designed to cope with the extreme conditions of alternate freezing and thawing of the ground. Disturbances to the permafrost may occur in many other ways, which include compaction of vegetation by vehicles, trampling by humans and other animals, and the storage of supplies: all these activities may lead to the demise of any native vegetation. Waste-disposal sites, chemical spills, spoil heaps and the removal of the vegetation cover all contribute to damage to the permafrost zone (Price 1972). Once permafrost is disturbed, its thermal equilibrium is altered, which in itself may inhibit plant growth, and a so-called **thermokarst** develops. The period of recovery in a permafrost layer, the time for thermal equilibrium to be re-established so that plants may once again grow, can be considerable. Using experimental sites in Alaska, Lawson (1986) demonstrated that it may take more than thirty years for recovery. Permafrost environments are one of the most sensitive of ecosystems to environmental change, where plant growth is intrinsically slow and animal life often occurs in relatively small populations, although biodiversity may be high (e.g. with many lichen species, etc.). They represent one of the remaining relatively unexploited natural environments and, therefore, their conservation should be encouraged.

The pressure for land has become so great in many regions of the world that additional living space is created by coastal reclamation, for example in Hong

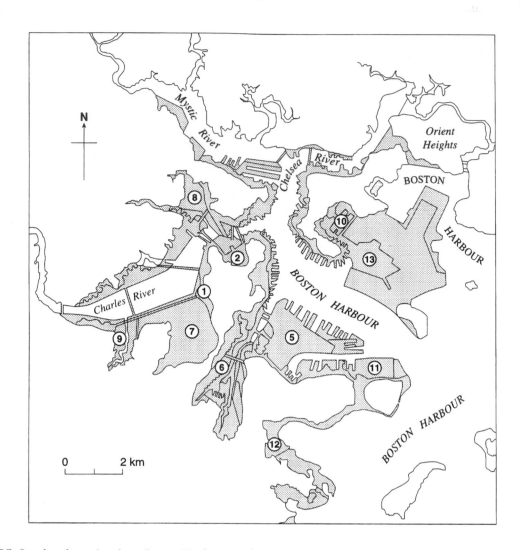

Figure 9.15 Land reclamation from Boston Harbour. Redrawn after Johnson and Lewis (1995).

Kong, which has one of the highest population densities in the world. Several million people live on a coastal strip in an area of about 30 km², hemmed in on one side by the sea and on the other by steep unstable mountains. In Hong Kong, more land reclamation, which will be at the expense of rich near-coastal fishing grounds and ecologically important wetlands, is planned for the near future. Other examples of major land reclamation schemes are for the expansion of Tokyo into Tokyo Bay, and for the new Osaka airport between the Japanese islands of Honshu and Shikoku.

Figure 9.15 shows the land reclamation in Boston from 1803 up to the present. Originally, the central region of Boston was hilly, but most of the hills were removed and the debris used as landfill to reclaim ponds and wetlands for construction. Such expansion of an urban area not only leads to degradation of the original ecosystems by the removal of vege-

tation and soil, but it also increases the cover of the land surface by impermeable materials, such as asphalt and concrete, thereby altering the hydrological characteristics of the region.

On oceanic islands during the last century, considerable areas within atolls have been reclaimed because these atolls provide the only available land for thousands of kilometres. Atolls have commonly been used for airbases and other military functions such as missile-launching sites – including test sites for nuclear explosions (see Chapter 6). Where atolls are inhabited, there has been considerable disruption to biota and the introduction of new species, frequently resulting in the extinction of endemic creatures.

Human modification of surface drainage varies in scale, from small canalisation or channelisation schemes to large-scale river regulation and redirection. These actions have consequences for the natural environment. Channelisation schemes can be both

Plate 41 *Deforestation in the Amazon.*
Courtesy of Michael Harvey/Panos Pictures.

Plate 42 *Rainforest in the Amazon cleared for cattle ranching.*
Courtesy of Rob Cousins/ Panos Pictures.

Plate 43 *Curua Forestry Station, central Amazonia, established to undertake experimental studies on methods for efficient farming and reforestation: (A) selective logging; (B) clearance of forest.*
Courtesy of Dr M. Eden.

Plate 44 *Ships
stranded in the dried-up
Aral Sea, former Soviet
Union, Central Asia.*
Courtesy of Fred Mayer/
Magnum.

advantageous (e.g. as preventive measures to reduce flood risks in threatened areas, providing irrigation waters, and in hydroelectric power schemes), or they can be detrimental (e.g. by destroying river habitats, increasing the magnitude and frequency of floods, and changing patterns of sedimentation) (Brookes 1985) (Plate 9.8). Figure 9.16 shows the possible changes associated with a channelled stream.

Since the latter half of the nineteenth century, large-scale water development schemes have become much more prevalent worldwide. In the USA, the Tennessee Valley Authority developed one of the most impressive schemes for water control, with the construction since 1933 of thirty-two large and fourteen smaller dams. They aimed to control flooding, provide water resources for domestic supply and hydroelectricity for power, and to aid river navigation (Petts 1994). The Hoover Dam on the Colorado River, with its reservoir Lake Mead, was constructed in 1935. The Colorado River, draining an area of about 8 per cent of the USA, was progressively developed from the early nineteenth century and now has nineteen dams (Figure 9.17A). The effects of the river management within the Colorado River basin have attracted much attention from environmentalists because it drains some of the most beautiful landscapes, including the Grand Canyon: also, the river life was among the most diverse in North America. Figure 9.17B and C show the reduction in discharge and sediment load since river management began. As a result of damming and water storage, less than 1 per cent of the initial river flow reaches the Colorado delta (Petts 1994). The environmental consequences include the disruption and degradation of ecological habitats along the Colorado and at its mouth. Before the development of the Colorado, the river waters flowed rapidly and fluctuations in temperature and sediment load created a variety of distinct habitats: now the river banks have stabilised there are standing pools of clear warm water with high salinity. These changes have caused the mortality of large numbers of fish and their replacement by introduced species that are more aggressive. Algal blooms are now common and pose an additional problem. New exotic vegetation has established itself along some of the stabilised flood plains to replace the endemic species.

On 26 March 1996, the US Congress allowed reservoir waters to flood the Colorado River in a controlled environmental experiment to try to help improve the aquatic and near-shore environments along the lower river. The flood lasted seven days and caused a drop in reservoir levels of about 15 cm. Initial results suggest that the experiment was successful, as the flooding produced new river bar and pool habitats. This was the first time that the US Congress had approved such an experiment with the aim of improving the environment.

The construction of river dams may have other detrimental effects; for example, locally increasing ground-water levels and increasing slope stability, increased salinisation leading to land degradation, enhanced local earthquakes, and providing new habitats for pests like mosquitoes to flourish and act as hosts for a variety of diseases (see Figure 9.18). Amongst the best-known examples of a major dam-related catastrophe occurred in 1963 associated with the Vaiont Dam in Italy, when 2,600 people were killed. This disaster happened when 0.24 km^3 of rock slipped rapidly into the reservoir to create a sharp rise in water level, which over-topped the dam. It is believed that the landslide was a result of reduced slope stability brought about through higher ground-water levels, which were a consequence of reservoir construction (Kiersch 1965).

The construction of dams can increase the risk of local earthquakes. Figure 9.19 shows the relationship between reservoir levels and earthquake frequency in the USA. Also, there are many seismically active regions of the world where large dams have been or are currently being constructed. The Tehri Dam in the Garhwal Himalaya, northern India, for example, is being constructed in a mountain region with a considerable risk of major seismic activity. Seismologists believe that an earthquake with a magnitude >7 is likely to occur in the near future, but despite this and other environmental arguments, the perceived economic imperative has meant that

Plate 9.8 *Flood channel in southern Spain. In this semi-arid environment, rainfall is rare and when it occurs it is intense, producing catastrophic flash floods. Channelisation schemes like this help to direct the flow when they occur.*

Natural Channel

Suitable water temperatures:
  adequate shading; good cover for fish
  life; minimal variation in temperatures;
  abundant leaf material input

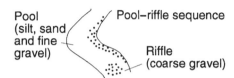

Pool (silt, sand and fine gravel)   Pool–riffle sequence   Riffle (coarse gravel)

Sorted gravels provide diversified habitats
for many stream organisms

Artificial Channel

Increased water temperatures:
  no shading; no cover for fish life;
  rapid daily and seasonal fluctuations
  in temperatures; reduced material
  input

Mostly riffle

Unsorted gravels:
  reduction in habitats; few organisms

POOL ENVIRONMENT

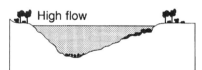

High flow

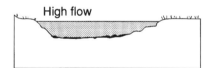

High flow

Diversity of water velocities:
  high in pools, lower in riffles. Resting
  areas abundant beneath undercut banks
  or behind large rocks, etc.

May have stream velocity higher than
  some aquatic life can withstand. Few
  or no resting places.

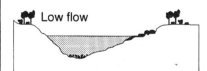

Low flow

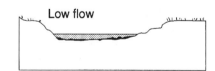

Low flow

Sufficient water depth to support fish and
other aquatic life during dry season

Insufficient depth of flow during dry
seasons to support diversity of fish
and aquatic life. Few if any pools
(all riffle)

Figure 9.16 *Comparison of natural channel morphology and hydrology with that of a channelled stream. Redrawn after Keller (1976).*

dam construction has gone ahead, creating an additional threat to the lives of millions of people who live on the Indo-Gangetic Plain.

The largest dam scheme in the world, the Three Gorges Scheme, is currently being undertaken along the course of the Yangtze River, China. After dam construction, the valley will be flooded to create a reservoir by submerging large areas of the Qutang, Wu and Xilang gorges over a total distance of about 600 km. The scheme will destroy one of China's most beautiful landscapes, and more than one million people will have to be forcibly relocated to make way for the reservoir. Part of this reservoir scheme, besides generating hydroelectric power for Shanghai, is aimed at protecting some ten million people downstream from flooding, and to open the gorges for commercial shipping. There is much controversy over the scheme, with many people arguing that the reservoir levels will be too high to contain floods, and that the scheme itself will lead to increased flooding because the silt-free waters downstream from the dam will actually increase erosion and cause greater amounts of land degradation (Pearce 1995e). The dam is being constructed in a seismically active area where landslides are frequent, and both these processes will increase the risk of catastrophic dam failure. There are also ecological considerations – the region has several endangered species, including the white flag dolphin, the Chinese alligator, the finless porpoise and the Chinese sturgeon, whose habitats are threatened by the construction scheme. Even at this late stage, an appreciation of the likely consequences may help to mitigate some of the worst potential effects of environmental degradation.

Coastal construction may also upset the balance of erosion and deposition of sediment along coasts (Table 9.3). Sediment in coastal regions is moved continuously by the activity of tidal currents and wave action. Along many coasts, the net sediment transport is roughly parallel to the coastline in a process known as **long-shore sediment drift**. A barrier such as a sea-wall, harbour or groynes may inhibit the rate of long-shore drift. The main problem with this type of coastal management is that the artificial retention of coastal and near-shore sediments in one place inevitably leads to a relative starvation of sediment elsewhere, which may have a knock-on effect along a coastline in the direction of long-shore drift. The original sediment stockpile, that is prior to coastal management using sea defences, may have acted as a natural coastal defence, protecting the coastline from erosion and associated coastal processes such as landslides (Bird 1985, Carter 1988). The southern coastline of Britain has several areas of coastal

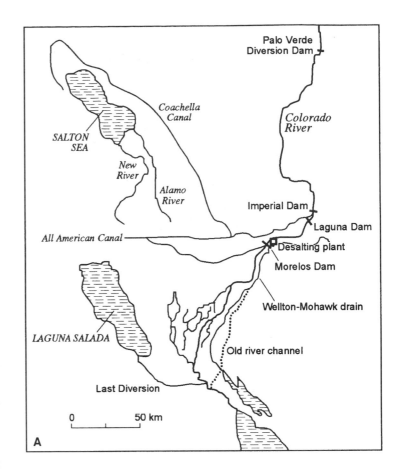

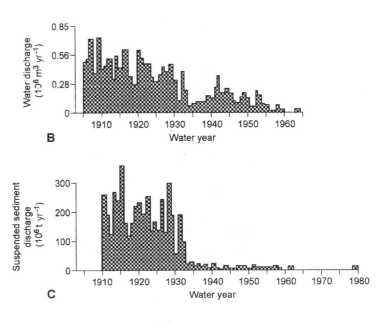

Figure 9.17 *(A) Dam constructions along the lower part of the Colorado River; (B) Discharge variations; (C) Sediment yield variations throughout the twentieth century resulting from the damming. Redrawn after Graf (1987) in Petts (1994).*

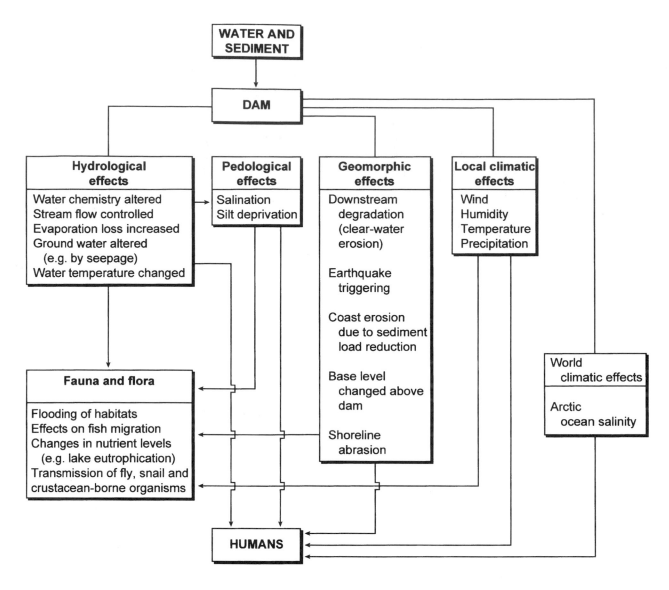

Figure 9.18 *The possible effects of dam construction on human life and various components of the environment. Redrawn after Goudie (1993b).*

retreat that have been attributed to artificial coastal constructions.

A coastal construction, such as a pier, may act as a sediment trap, with the local sediment accumulation then inhibiting sediment drift (Komar 1983). Offshore and near-shore sand bars formed in this manner may hinder shipping and can even lead to the infilling of harbours. This same problem can be turned to advantage where land reclamation is required, for example in the Miama Beach Project, Florida, which was amongst the largest of its kind that has ever been undertaken, costing, between 1976 and 1982, about US$67 million (Carter 1988).

Today there is a growing awareness that any

modifications made to one part of a coastal region can have profound effects on adjacent regions, as sediment movement is a continuous cycle of deposition and erosion. Altering sediment fluxes, or sediment budgets, may adversely affect ecological niches. There remains a need for much more research funding to encourage scientists to study sediment movement in coastal and near-shore environments, but in a manner that is integrated with ecological and other environmental considerations. An improved understanding of the importance of storms versus fair-weather processes in controlling sediment fluxes, and an appreciation of the effects of changing sea level in altering sediment transport and deposi-

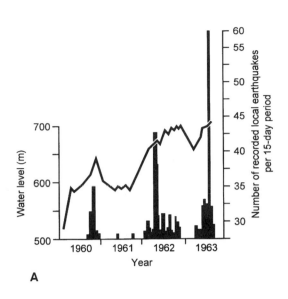

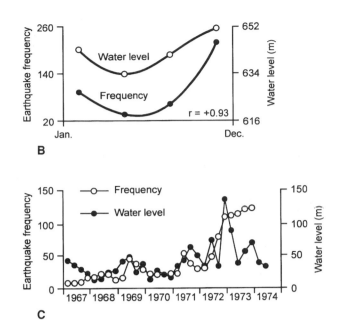

Figure 9.19 *The relationship between reservoir levels and earthquake frequencies for (A) the Vaiont Dam, Italy; (B) the Koyna Dam, India; (C) the Nurek Dam, Tajikistan. Redrawn after Goudie (1993b).*

tion patterns, will greatly help in the development and adoption of prudent coastal management programmes.

Environmental risk management and environmental impact assessment

The management of environmental risk is a complex issue, involving scientific analysis of data, technical assessments, the perception of the probability of particular events occurring and the potential seriousness of any outcomes induced by both natural processes and human activities – all placed within a socio-economic, cultural and political context. There are many definitions of risk, with little apparent consensus.

Environmental risk management attempts to prevent, control or mitigate the consequences of particular phenomena as a result of natural or human-induced events. An important concept in risk management is *safety*, i.e. determination of the tolerable levels of risk – something that is by its very nature controversial since different individuals or groups may have dissimilar perceptions of acceptable risk.

The main method of assessing risk is to undertake an **environmental impact assessment** (**EIA**). EIAs were first introduced in the US National Environmental Policy Act in 1969 to help evaluate the environmental acceptability of projects under consideration. The technique involves an amalgamation of studies based on predetermined approaches.

A useful definition of EIA is 'the evaluation of the effects likely to arise from a major project (or other action) significantly affecting the natural and man-made environment' (Wood 1995). In 1988, the UK Department of the Environment stated that EIA:

is essentially a technique for drawing together, in a systematic way, expert qualitative assessment of a project's environmental effects, and presenting the results in a way which enables the importance of the predicted effects, and the scope for modifying or mitigating them, to be properly evaluated by the relevant decision-making body before a decision is given. Environmental assessment techniques can help both developers and public authorities with environmental responsibilities to identify likely effects at an early stage, and thus to improve the quality of both project planning and decision-making.

EIAs are now undertaken in many countries, but unfortunately many prove inadequate because of the lack of true integration of the data and attitudes of the principal interested parties and/or because EIAs have been undertaken in-house by teams which have a specific vested interest in a particular outcome. Tables 9.4 and 9.5 list the criteria and basic principles for designing and evaluating EIAs. The utilisation of EIAs to improve the quality of decisions about

*Table **9.3*** *Mechanisms of human-induced erosion in coastal zones.*

Human-induced erosion zones	Effects
Beach mining for placer deposits (heavy minerals such as zircon, rutile, ilmenite and monazite)	Loss of sand from frontal dunes and beach ridges
Construction of groynes, breakwaters, jetties and other structures	Downdrift erosion
Construction of offshore breakwaters	Reduction in littoral drift
Construction of retaining walls to maintain river entrances	Interruption of littoral drift resulting in downdrift erosion
Construction of sea-walls, revetments, etc.	Wave reflection and accelerated sediment movement
Deforestation	Removal of sand by wind: sand drift
Fires	Migrating dunes and sand drift after destruction of vegetation
Grazing of sheep and cattle	Initiation of blow-outs and transgressive dunes: sand drift
Off-road recreational vehicles (dune buggies, trail bikes, etc.)	Triggering mechanisms for sand drift attendant upon removal of vegetation cover
Reclamation schemes	Changes in coastal configuration and interruption of natural processes, often causing new patterns in sediment transport
Increased recreation needs	Accelerated deterioration, and destruction, of vegetation on dunal areas, promoting erosion by wind and wave action

Source: Goudie 1993b.

proposed projects has resulted in the use of **strategic environmental assessments (SEAs)**, because of the belief that EIAs are undertaken too late in the planning process to ensure that all the impacts and alternatives relevant to sustainable development are considered (Wood 1995). The potential benefits of SEAs are listed in Table 9.6, while the key considerations in choosing an appropriate SEA technique are shown in Table 9.7.

In the UK, one of the driving forces behind risk assessment and management is the **ALARP** principle, which aims to reduce risks *as low as reasonably practicable*. Another concept is the **BPEO**, the *best*

practicable *environmental option*, or the **BATNEEC**, the *best available technique not entailing excessive costs* (Gerrard 1994). All these concepts and applications have inherent problems together with benefits. Intrinsic problems arise because of the uncertainties associated with deriving probabilistic estimates of the occurrence of particular events and due to the differing perceptions of what is an acceptable risk. EIAs and SEAs rely heavily on the integrity of those involved, an inevitably over-optimistic starting point but one, nevertheless, that should not be discouraged. For environmental risk management to succeed, there must be a comprehensive assessment of all the possible variables by independent experts, who have no clear vested interests, and assessment must be acceptable to those people who potentially (or actually) may be affected by any realisation of the risk.

Environmental auditing

The first environmental audits were adopted in the USA during the 1970s and during the past few years have become common in the USA and throughout Europe. Environmental audits provide a methodology for evaluating, at regular intervals, the environmental performance of companies, and to check compliance with environmental regulations and codes of practice. Environmental audits aim to ensure that legislation is adhered to and that fines and litigation are, wherever possible, prevented. Audits also help to alleviate concerns by the general public, create environmental awareness within and outside a company or organisation, and if necessary improve a company's or organisation's image.

Auditing programmes should have a code or standard against which to carry out the audit. In the UK, this is BS7750, the standard on environmental management systems of the British Standards Institution (BSI). This standard was launched in 1992 and amended and published in 1994. BS7750 provides standards that cover every aspect of environmental management. Similar codes have been introduced in Ireland, Spain and Canada, and they will form the basis for an international standard on environmental management systems (EMS). In 1993, the European Commission published the European Eco-Management and Audit Scheme, which provides a framework for companies and organisations to assess their own environmental impacts on a voluntary basis and to allow them to commit themselves to a policy to reduce any adverse environmental effects. Member states can, if they so wish, adopt a compulsory system for certain industries where it is deemed beneficial.

Perhaps it is important to sound a cautionary note amongst all the apparent environmental awareness and claims by companies and organisations that they wish to behave in a responsible manner. This may be true for many, but wherever the primary consideration is profit and satisfying shareholders, then a company or organisation may choose to subscribe to good practice, since it is perceived to be in its best interests to do so, but whilst doing little to reduce its environmental pollution and/or damage. Some would argue that legislation and compulsory, rather than voluntary, compliance is the best way forward. Companies tend to prevail upon governments to adopt voluntary procedures; environmentalists and environmental pressure groups tend to favour legislation and legally enforceable penalties. The issues surrounding EIAs, SEAs and environmental audits bring to the fore the debate over the proverbial carrot or stick approach.

Mineral extraction

Open-cast and underground mining activities provide humankind with the wealth that has sustained the growth of civilisation and the quality of life that is enjoyed by many. It is hard to find many items in constant daily use, let alone the luxury goods, including works of art, that have not come from the ground. Mineral extraction is an integral part of civilisation. Nothing is gained without some cost to individuals and the environment. The debate will always be focused on an acceptable level of pollution, inconvenience and cost in return for a product from mining. Different individuals and communities will have different perceptions of the degree of acceptable social and economic cost, especially if they live near the mines.

There are many problems associated with mineral extraction, including chemical pollution and disfigurement of the landscape (Plate 9.9; Table 9.8), as well as disturbance to the natural rates of Earth surface processes. Mineral extraction results in the construction of tips, increased sediment loads in rivers, soil erosion, and the pollution of water sources and adjacent land, which often leads to vegetation and soil degradation. Even as early as the Neolithic period of prehistory, humans were modifying the Earth's surface in search of its mineral wealth. Many excavations in the chalk hills of southern England represent Neolithic pits dug to extract good-quality flint for tool-making (Clark 1963). With time the

Table 9.4 *EIA system evaluation criteria.*

1 Is the EIA system based on clear and specific legal provisions?

2 Must the relevant environmental impacts of all significant actions be assessed?

3 Must evidence of the consideration, by the proponent, of the environmental impacts of reasonable alternative actions be demonstrated in the EIA process?

4 Must screening of actions for environmental significance take place?

5 Must scoping of the environmental impacts of actions take place and specific guidelines be produced?

6 Must EIA reports meet prescribed content requirements and do checks to prevent the release of inadequate EIA reports exist?

7 Must EIA reports be publicly reviewed and the proponent respond to the points raised?

8 Must the findings of the EIA report and the review be a central determinant of the decision on the action?

9 Must monitoring of action impacts be undertaken and is it linked to the earlier stages of the EIA process?

10 Must the mitigation of action impacts be considered at the various stages of the EIA process?

11 Must consultation and participation take place prior to, and following, EIA report publication?

12 Must the EIA system be monitored and, if necessary, be amended to incorporate feedback from experience?

13 Are the financial costs and time requirements of the EIA system acceptable to those involved and are they believed to be outweighed by discernible environmental benefits?

14 Does the EIA system apply to significant programmes, plans and policies, as well as to projects?

Source: Wood 1995.

extraction of metallic ores became important for the fashioning of metal tools and jewellery. There are many examples of Roman mines still visible in the landscape throughout Europe and North Africa (Cleere 1976, Rackhams 1986). Fossil fuels have also been extracted from early times. Excavations for peat as a source of fuel were widespread throughout Europe, and it is believed that the lakes and waterways of the Norfolk Broads in eastern England owe their origin to the removal of more than 25 million m^3 of peat prior to the fourteenth century (Lambert *et al.* 1970).

The coming of the Industrial Revolution saw the growth in the extraction of iron ore and coal. As a consequence, large open-cast and shaft mines were constructed, some to great depths. In some regions,

Table 9.5   *Eight basic principles for evaluating EIA processes.*

1 An effective environmental assessment process must encourage an integrated approach to the broad range of environmental considerations and be dedicated to achieving and maintaining local, national and global sustainability.

2 Assessment requirements must apply clearly and automatically to planning and decision-making on all undertakings that may have environmentally significant effects and implications for sustainability within or outside the legislating jurisdiction.

3 Environmental assessment decision-making must be aimed at identifying best options, rather than merely acceptable proposals. It must therefore require critical examination of purposes and comparative evaluation of alternatives.

4 Assessment requirements must be established in law and must be specific, mandatory and enforceable.

5 Assessment work and decision-making must be open, participative and fair.

6 Terms and conditions of approvals must be enforceable, and approvals must be followed by monitoring of effects and enforcement of compliance in implementation.

7 The environmental assessment process must be designed to facilitate efficient implementation.

8 The process must include provisions for linking assessment work into a larger regime, including the setting of overall biophysical and socio-economic objectives and the management and regulation of existing as well as proposed new activities.

Source: Gibson 1993.

for example in South Wales, so much coal was removed from underground that subsidence of large stretches of valleys occurred, causing structural damage to buildings and, locally, landslides (Bell 1988).

Disfigurement of the landscape is particularly well illustrated in large strip and open-cast mines, one of the largest being Bingham Canyon Copper Mine in Utah, which covers an area of over 7 km² and extends to a depth of more than 700 m. Goudie (1993b) estimated that the annual amounts of movement of soil and rock resulting from mineral extraction may be as much as 3,000 billion tonnes, whereas the amount of sediment carried by natural processes into the sea is far less, 24 billion tonnes each year. Other examples include the china clay workings in Cornwall, southwest England, and the quarrying for limestone on Portland Bill, southern England (Plate 9.10), where the excavation is seen as an eyesore. Others would argue that these pits give the region a very important source of employment and income, and provide a product that most people regard as useful and important.

In Scotland, the potential reserves of hard rock suitable for aggregates for road construction and concrete are enormous – principally granite, quartzite, gneiss, limestone and sandstone. Demand for these raw materials is increasing so rapidly, particularly in the densely populated areas of Europe where there are land shortages (e.g. the Netherlands and northern Germany), that they are being considered as a major export. Scotland could develop super-quarries to meet the demand, but there are many environmental objections to the construction of super-quarries, including the pollution caused by the discharge of ballast, polluted water from the quarrying operations, and damage to unspoilt countryside with valuable ecological niches. Some environmentalists also argue that super-quarries would be unnecessary if demolition rubble were to be recycled. Super-quarries, however, provide an opportunity for local employment and economic growth.

Table 9.6 *Potential benefits of strategic environmental assessment.*

Encourages the consideration of environmental objectives during policy, plan and programme-making activities within non-environmental organisations.

Facilitates consultations between authorities on, and enhances public involvement in, evaluation of environmental aspects of policy, plan and programme formulation.

May render some project EIAs redundant if impacts have been assessed adequately.

May leave examination of certain impacts to project EIAs.

Allows formulation of standard or generic mitigation measures for later projects.

Encourages consideration of alternatives often ignored or not feasible in project EIAs.

Can help determine appropriate sites for projects subsequently subject to EIA.

Allows more effective analysis of cumulative effects of both large and small projects.

Encourages and facilitates the consideration of synergistic effects.

Allows more effective consideration of ancillary or secondary effects and activities.

Facilitates consideration of long-range and delayed impacts.

Allows analysis of the impacts of policies that may not be implemented through projects.

Source: Wood and Djeddour 1992, p.7.

Table 9.7 *Key considerations in choosing SEA techniques.*

1 Will this technique or approach help to achieve the objectives of this step of the process? What is the best technique at this stage for:
  - identifying linkages?
  - estimating and forecasting effects and consequences?
  - assessing significance?
2 Does the magnitude and potential significance of the impacts warrant the level of effort required by the technique?
  - cost?
  - timing?
  - involvement of key personnel?
  - involvement of peers, outside experts and public stake-holders?
3 Is it possible and practical to utilise techniques under consideration?
  - are peers, experts and stakeholders available and willing to participate?
  - do adequate and reliable data exist?
4 Are there any other factors that may influence selection of approaches and techniques?
  - strictures of confidentiality?
  - skill levels and capacity to design and implement given techniques?
  - personal preferences of parties involved?

Source: Federal Environmental Assessment Review Office 1994.

Much of the rock waste produced by mineral extraction is piled up to form tips or spoil heaps, or is used to infill pits and produce new landforms. Some of the waste, however, may be carried into rivers to choke their courses and, ultimately, alter river drainage patterns. Unfortunately, mineral waste often contains poisonous substances such as arsenic and cadmium, which can pollute rivers and groundwater sources, and poison fish, plants and drinking water (Blunden 1991). This is particularly devastating environmentally in regions where lead is mined, such as in central Wales, and the very big copper mines, such as Ox Tedi in Papua New Guinea, where vast numbers of fish and plant life have died due to the pollutants.

Mining may be associated with extracted waste materials or overburden that is toxic. An example of this problem arose in February 1992, when water flooded the last remaining tin mine, Wheal Jane, in Cornwall, England, and flowed out of the mine workings into nearby streams and then onto the coast. These waters were rich in toxic metals, especially iron, arsenic and cadmium, which appeared as an unsightly brown, silty sludge. In March 1992, after much local public concern and media attention, the mining company was forced to pump the mine waters into a tailings-dammed pond to filter out the pollutants.

Table 9.8 *Environmental impacts of mineral extraction*

Activity	Potential impact
Excavation and ore removal	• Destruction of plant and animal habitat, human settlements and other surface features (surface mining) • Land subsidence (underground mining) Increased erosion; silting of lakes and streams • Waste generation (overburden) • Acid draining (if ore or overburden contains sulphur compounds) and metal contamination of lakes, streams and ground water
Ore concentration	• Waste generation (tailings) • Organic chemical contamination (tailings often contain residues of chemicals used in concentrators) • Acid drainage (if ore contains sulphur compounds) and metal contamination of lakes, streams and ground water
Smelting/refining	• Air pollution (substances emitted can include sulphur dioxide, arsenic, lead, cadmium and other toxic substances) • Waste generation (slag) • Impacts of producing energy (most of the energy used in extracting minerals goes into smelting and refining)

Source: Goudie 1993; Young 1992.

Plate 9.9 *Aerial view of large quarrying operation in the eastern USA.*

Mining of radioactive metals such as uranium also causes concern, as was the case during the 1950s and 1960s at Elliot Lake, Ontario, Canada. Radium is a waste product of uranium mining. The radium was deposited along with rock waste as tailings. When it rained, radium was flushed out through the mine tailings into streams or the ground-water system. Radium can cause bone cancer if ingested in drinking water or consumed, for example by eating fish. Chemical treatment was applied to the mine tailings in an attempt to remove the hazard, and the mine was then backfilled with the tailings. This method, however, has not been entirely successful and considerable debate remains over the most effective way in which to deal with radioactive mine waste.

Underground mining can also cause considerable

environmental problems. For example, mining may result in subsidence, which may be related to the extraction of rock material and water. This has been a major problem in the USA, where the Bureau of Mines estimates that over 32,000 km² has been affected, and this may well rise by 10,000 km² by the end of the century. Particularly affected areas are the Illinois basin, western Pennsylvania and northern West Virginia, where coal mining activities have caused subsidence. Below the Cheshire plain in central England, the extraction of salt has resulted in severe local subsidence and damage to buildings (Bell 1992). Subsidence can be fatal for those actually engaged in the mining operation. Mining history worldwide is littered with disasters, to the extent that they occupy an important part of prose, poetry and folk music.

Dust is yet another unpleasant and potentially harmful by-product associated with mining activities. Ill health, the destruction of vegetation, and pollution of the atmosphere and water supplies are not uncommon. This is particularly well illustrated in Sudbury, Ontario, where metal smelters release vast quantities of aerosols and gases into the atmosphere together with particulate mineral matter. Few detailed and widely publicised studies of the effects of this latter type of pollution are documented, yet it is the everyday by-product of mineral extraction.

Despite common predictions of mineral resource depletion from the 1950s to the mid-1980s, world reserves have actually increased (Hodges 1995; Table 9.9). Clearly, mineral extraction will continue and there is no reason to believe that this activity will not increase at some time in the future. Mineral extraction is an important activity that is inextricably linked to industrialisation and economic development. Although recycling of materials could reduce

Plate 9.10 *Quarrying for stone on Portland Bill, southern England.*

Table 9.9 *Worldwide annual consumption of selected metals in 1991 and reserve base in 1993. Consumption includes primary and secondary (scrap) metal, except iron, which includes only crude ore.*

Metal	Annual consumption (10³ tonnes)	Reserve base (10³ tonnes of contained metal)
Aluminium	17,194	28,000,000 *
Copper	10,714	590,000
Iron	959,609	230,000,000
Lead	5,342	130,000
Nickel	882	110,000
Tin	218	10,000
Zinc	6,993	330,000

*Bauxite (crude ore)
Source: Hodges 1995.

the overall amount of mining activity worldwide, mining will remain an important economic activity. The disfigurement of the landscape by mining activities (e.g. Plate 9.11) does not have to be long term since mines, particularly open-cast mines, can be landscaped to hide most of the activities from general view, the pits and quarries infilled, and spoil heaps either removed for infill or landscaped. Chemical pollution resulting from mining activities can be controlled through the setting of appropriate industry-wide standards and, where appropriate or deemed necessary, the introduction of legislation.

Hazardous/toxic substances and waste

In the USA, the major federal statute on solid waste is the Resources, Conservation, and Recovery Act (RCRA), passed in 1976. This Act defines 'solid' waste as having any physical form, e.g. discarded materials from industrial, commercial, mining and agricultural operations, refuse, garbage, and sludge from treatment processes and other pollution controls (OTA-BP-0–82 1992a). Under the RCRA, solid waste is defined in Section 1004(27) as:

> any garbage, refuse, sludge from a waste treatment plant, water supply treatment plant, or air pollution control facility and other discarded material, including solid, liquid, semisolid, or contained gaseous material resulting from industrial, commercial, mining, and agricultural operations, and from community activities, but does not include solid or dissolved materials in irrigation return flows or industrial discharges which are point sources subject to permits under section 1342 of Title 33, or source, special nuclear, or

byproduct material as defined by the Atomic Energy Act of 1954, as amended (68 Stat. 923).

In solid waste management, US federal efforts have focused on the disposal of so-called 'hazardous' substances, although they represent a small part of the total solid wastes. Hazardous waste is defined in the RCRA, Section 1004(5) as:

> a solid waste, or combination of solid wastes which because of its quantity, concentration, or physical, chemical, or infectious characteristics may [a] cause, or significantly contribute to an increase in mortality or an increase in serious irreversible, or incapacitating reversible illness; or [b] pose a substantial present or potential hazard to human health or the environment when improperly treated, stored, transported, or disposed of, or otherwise managed.

The Basel Convention on the Control of Trans-boundary Movement of Hazardous Wastes and their Disposal was adopted in 1989 and came into force on 5 May 1992. The convention represents a response to the problems caused by the annual worldwide production of 400 Mt of wastes that are hazardous to people or the environment because they are toxic, poisonous, explosive, corrosive, inflammable, eco-toxic and/or infectious. The Basel Convention strictly regulates the trans-boundary movement of hazardous wastes, and the signatory parties are obliged to ensure that such wastes are managed and disposed of in an environmentally acceptable manner. The main principles of the Basel Convention are (source: UNEP World Wide Web Site):

Plate 9.11 *The remains of one of the largest slate quarries in the world at Blaenau Ffestiniog in North Wales.*
Courtesy of R. Robinson-Owen.

- Trans-boundary movements of hazardous wastes should be reduced to a minimum consistent with their environmentally sound management.
- Hazardous wastes should be treated and disposed of as close as possible to their source of generation.
- Hazardous waste generation should be reduced and minimised at source.

To achieve these principles the convention's secretariat, under the direction of the UNEP, aims to control the trans-boundary movement of hazardous wastes, monitor and prevent illegal traffic, provide assistance for environmentally sound management of hazardous wastes, promote co-operation between the involved parties, and develop technical guidelines for the management of hazardous wastes.

Contaminated land

Most industrialised countries have large areas of land that have become contaminated by various substances, although very high local levels of trace metals in toxic doses do occur naturally. Issues surrounding the environmental damage and health risks associated with such land have really surfaced only in the last two decades. Redevelopment of land for housing has opened up many of these issues, e.g. in London, UK, the large housing development of Thamesmead to the east of the city was constructed on a former munitions complex of the 1,000-acre Woolwich Arsenal, closed in the 1950s.

All countries that have initiated national action plans for cleaning up contaminated land have begun with lists or registers of potentially contaminated sites. For example, in the USA, the Environmental Protection Agency has a hazard-ranking system, Canada has a national classification system, and Denmark also ranks sites potentially affected by oil pollution. Such registers enable a systematic protocol for action priorities and resource allocation, including making funds available for any clean-up.

Dealing with contaminated land, both in terms of liability and financing clean-ups, and ensuring that those responsible actually pay any costs incurred, remains a problem for governments around the world. In the USA and the EU, there are major new initiatives incorporating the 'polluter pays' principle (see Chapter 10). Many countries already have limited systems of strict liability, joint and several liability, in statutory and common law, but there is a need for more far-reaching, international legislation in order to address issues of equity, economic efficiency and liability across national borders.

In the USA, legislation was enacted by Congress in 1980 in the Comprehensive Response, Compensation and Liability Act (CERCLA), under which a federal trust fund was established (Superfund) to permit a swift response to any immediate and future health hazards caused by contaminated land. This act was partly in response to the Love Canal disaster, in which a leaking toxic waste dump led to the evacuation of hundreds of families and the contaminated area being declared a federal disaster area (Hoffman 1995). The original land owners had sold the land with the condition that they had no future liability. In 1986, the CERCLA was amended by the Superfunds Amendments and Reauthorisation Act (SARA), which included the authorisation of collecting further revenues. A 1989 report from the US auditor-general's office estimated that 135,000–425,000 sites may come within the US Environmental Protection Agency's (EPA) criteria for the treatment of contaminated land. Originally, the scheme was assumed to involve only 1,000–2,000 sites.

Superfund is a federal instrument, and many states have their own systems running in partnership with this. It is essentially a mechanism for raising money with the aim of spreading liability costs across a broad spectrum of producers, consumers and tax payers, and much of which is provided by the US Congress, i.e. under the SARA. Congress voted an extra US$8.6 billion to the fund for the period 1986–1991. Additional money included a US$2.75 billion petroleum tax, a US$2.5 billion corporate environment tax from companies with an annual turnover greater than US$2 million, US$1.5 billion from general revenues, and US$1.4 billion from a tax on chemical feed stocks. Also, US$300 million is budgeted from interest on the money in the fund, together with a further US$300 million recovered from liable parties in various clean-ups. The EPA pursues potentially responsible parties (PRPs), of which there are an estimated 14,000 notified on only 250 sites. PRPs, naturally, then pursue others and so the spiral of litigation continues. It is hard to obtain reliable figures for the enormous transaction costs of Superfund, partly since those involved are reluctant to provide information.

Under Superfund, there are four categories of PRPs who are both jointly and severally liable for any clean-up costs: the owner or operator; the prior land owner; those responsible for the disposal of hazardous substances; and those responsible for transporting the hazardous substances. Furthermore, those responsible for any contamination and pollution remain responsible at any time after the event, that is there is no statute of limitation. Such retrospectivity raises many issues, such as the extent to which it is fair to seek liability for actions that at the time of their implementation were legal but have subsequently become unlawful. Lenders' liability has proved to be amongst the most contentious issues associated with Superfund, mainly because there is no discrimination between lenders with a purely legal rather than operational relationship, even to the extent that the EPA has proposed a means of mitigating its effects. Joint and several liability for pollution has also caused considerable problems because of the difficulties in apportioning blame. So far, although Superfund has used its financial resources only for on-site costs, the EPA contains a clause permitting liability to extend to claiming 'natural resources' damage off-site.

On balance, Superfund represents a commitment by Americans to recognise and define the burden of liability for environmental damage, and to monitor pollution and clean-up. In operation, however, Superfund is associated with substantial transaction costs, estimated at 40–60 per cent, and a plethora of involved parties seeking to re-apportion liability and costs to the extent that the legal machinations are in danger of grinding Superfund to a halt.

In the EU, the US experience of the Superfund has helped to mould opinion and guide the legislators in making proposals on civil liability as a result of environmental damage caused by waste disposal ('waste' being defined as anything that is unwanted by the producer). Like its US counterpart, the proposed EU legislation focuses on the polluter pays principle, although there is discussion over 'green taxation', where potential polluters pay a pollution tax to the government, which could be used to defray the costs of any clean-up. A potential stumbling block for the EU is the avoidance of any legislation that could permit polluters to operate with relative impunity in one member state rather than another, simply because of any differences in the strictness of law enforcement within the EU. On 1 September 1989, an EU draft directive was submitted to the Commission on Civil Liability for Damage caused by Waste, which was an attempt to instigate the polluter pays principle throughout the Community. Two years later, on 28 June 1992, major amendments were made to the original draft directive, and at the time of going to press the EU has still not adopted the draft directive. The proposed directive, whilst not actually defining waste, specifically excludes oil and nuclear waste. If it is adopted, albeit with modifications, the producers and handlers of waste would be liable for damage caused to individuals, property and the environment. In contrast to the USA, it remains unclear whether banks and other financial institutions that provide support may be liable. Another significant feature of the draft directive is that common interest groups, such as Greenpeace and Friends of the Earth, can initiate lawsuits, or become involved in existing lawsuits.

Furthermore, under the EU Freedom of Information Directive, environmental and other interest groups now have more access to sensitive environmental data. It is frequently claimed that one of the major differences between the US Superfund and the EU draft directive is that the latter does not have a retroactive effect (i.e., cover 'old' pollution), but the issue of retroactive liability remains contentious, with the legal profession casting doubt on its validity.

Within the EU, legislation dealing with contaminated land varies considerably between member states. In the UK, for example, the Environmental Protection Act 1990 (EPA) provides for Waste Regulatory Authorities (WRAs) to inspect and monitor landfill sites that are no longer licensed, and to take any remedial action if the site is likely to pose a risk to human health and/or cause environmental damage. Any costs incurred can be sought from the land owner, who may not have been the operator, but essentially the EU draft directive focuses liability primarily on the producers and handlers of waste. The liability is strict in that producers of waste have a civil liability irrespective of the fault on their part, and where the producer cannot be identified then the holder of the waste becomes the deemed producer with the liability being 'joint and several'. Disposers of waste have a legitimate defence if they were deceived by the producer of waste as to the actual nature of the waste, a third party contributed to negligence through an act or omission, or *force majeure*. An important feature of any liability is that it does not impose an inequitable burden on a party to any contamination for the full clean-up costs where their contribution has been minor. For example, a producer may manufacture a relatively innocuous waste which, when mixed with a substantially more toxic substance, gives rise to serious injury or damage. In these circumstances, liability should be limited proportionately to the actual contribution to the harm. In this context, the UK House of Lords Select Committee on the European Communities (1990) crystallises the issue:

> In general, it is the role of the criminal law to enforce these norms which society requires to be respected – for instance the appropriate standards for emissions to air or water from industrial enterprises – and that of the civil law to give redress

to those who may have suffered as a result of the environmental damage – for instance when a farmer's fields have been covered in oil from a crashed tanker. If the criminal law is thus primarily a means of promoting the precautionary approach, the making good of damage or compensation for it is largely left to the civil law.

A facet of the UK Environmental Protection Act 1990 was the provision for the establishment, under Section 143 of the Act, of 'public registers of land', whereby each waste regulation authority would maintain a register recording land that is being, or has been, put to contaminative use, but excluding, for example, substances such as radon or naturally occurring arsenic. Information could be excluded from a register if it were to be regarded as commercially sensitive by the authority maintaining the register, or on appeal, by the Secretary of State, and if the information is regarded as being contrary to the interests of national security. Under the Act, each waste regulation and waste collection authority would ensure open inspection of the register by the public, at no cost. The Section 143 registers, contrary to popular belief, were not conceived to identify *actually* contaminated land, but rather to provide a database of information on sites where land use has had the *potential* to leave contamination. Commercial confidentiality is an important and controversial issue in the management of waste and environmental injury or damage, and private companies are frequently reluctant to disclose information that can be seen as potentially harming their interests.

Probably because the government was strongly influenced by various commercial pressure groups, on 24 March 1993, the Secretary of State for the Environment withdrew the proposals for statutory registers of contaminative uses of land under the 1990 Act on the grounds that the proposals would have led to the inclusion of sites not actually contaminated, while omitting others that are actually contaminated, that once registered a site could never be removed even if any contamination had been satisfactorily dealt with, and because many uncertainties remained about the liability and action that would be taken to clean up sites. The UK government plans, instead, to instigate a new wide-ranging review of the problems in this area. Prior to this withdrawal, it appears that under the Department of the Environment's revised proposals published in July 1992, the area of contaminated land would have been reduced to 10–15 per cent of that originally envisaged, estimated at between 50,000 and 10,000 ha, affecting 50,000–100,000 sites, although only a small proportion of these pose an immediate threat to public health or the environment (Fowler 1993). UK scientists at the government's Warren Spring Laboratory have completed a review of appropriate technology to clean up contaminated land, yet other civil servants are still debating a national policy for cleaning up contaminated land.

Any legislation that imposes a civil liability on polluters to pay in full the costs of any clean-up will meet strong opposition from the myriad groups with a vested interest in profit with minimum responsibility. Land registers are also unpopular with entrepreneurs, since they permit both rapid and easy access by the media, environmental monitoring groups and other interested parties to detailed information on land 'quality'. Inevitably, the introduction of land registers would make certain sites harder to sell and develop, but this should be of secondary concern to conserving the natural environment. The bottom line is that the business world faces 'Hobson's choice' of either cleaning up its act by becoming environmentally responsible, or facing financial ruin by the regulators. Indeed, within the EU, the Commission believes that the time is fast approaching for compulsory environmental insurance, partly influenced by continental European insurers. In France, Assurpol is an insurance pool of seventy-five companies, that offers global insurance coverage (except in the USA and Canada), mainly for French companies, with members having a responsibility to reinsure the risks ceded to the pool in proportion to their share of the total capacity of the pool. Assurpol, underwriting up to FFr330 million per claim (with retrocession or re-reinsurance to the Italian pool), covers liability for both accidental and non-accidental (gradual) environmental damage. The Italian counterpart, Inquinianento, is an insurance pool with a capacity of up to £20 million per claim. In the Netherlands, there is also a pool called MAS, but this deals with only relatively small claims of up to Dfl16 million per claim, and in 1992 Denmark also initiated a pool.

A relatively new issue concerns the adoption of compulsory environmental audits. In Europe, the EU is working on a proposal for a voluntary environmental audit. In the long term, environmental audits may become compulsory, something that environmental pressure groups are actively campaigning for. It is important to appreciate that unless there is international comparability in environmental legislation and any 'green tax', the net result of over-zealous, market-leading legislation is potentially to seriously disadvantage those producers who are in the countries with the strictest legislation. Such arguments must be

BOX 9.1 CONTAMINANTS

There are various definitions of contaminants and contaminated land, but contamination is not synonymous with pollution, although clearly it is a necessary condition that can lead to pollution. The types of activity that could be considered as contaminative uses of land are listed below:

Chemical industries – the production, refining and bulk storage of organic and/or inorganic chemicals, e.g. fertilisers, pesticides, paints, dyestuffs, inks, soaps, detergents, pharmaceutical products, cosmetics, toiletries, pyrotechnic materials, fireworks, recovered chemicals, etc.

Energy industries – the use of natural and/or synthetic substances in the production of energy, e.g. fossil fuels or nuclear fuel in power stations.

Extractive industries – extraction, handling and storage of substances from mines, e.g. metal ores, tailings, coal, petroleum products, etc.

Metals – processing of metals (recovery, refining, production, finishing treatments, etc.) by any means, and their use to manufacture products in processes from heavy engineering to metal working, even in small businesses in the market place. Scrap metal handling and processing are included.

Non-metals – production and/or refining of non-metals, e.g. production of cement, bricks and associated products, lime, gypsum, asbestos, fibres, ceramics, glass, vitreous enamels, etc.

Rubber industries – manufacture of synthetic and/or natural rubber products, e.g. vehicle tyres, etc.

Engineering and manufacturing – manufacture of motor vehicles, aircraft, aerospace equipment, ships, railway/tramway equipment, electrical/electronic products, mechanical engineering/industrial plant, etc.

Infrastructure – maintenance, repairing, dismantling of industrial equipment, e.g. railways/tramways and rolling stock, roads, fuel stations and road haulage vehicles, aerospace facilities and equipment, including aircraft, docks and marine vessels, etc.

Waste disposal – treating, storage and/or disposal of sewage, radioactive materials, landfill, scrap, other effluents or substances, including the cleaning of tanks and drums, etc.

Foods – manufacture/processing of animal by-products (excluding slaughterhouse butchering), animal feedstuffs, pet foods, etc.

Agriculture – burial of diseased livestock.

Textile industries – manufacture of various textiles and leathers, e.g. production of carpets or other floor coverings, including linoleum, fabrics and leather products.

Timber and timber products – treatment, including coating, of timber and timber products, e.g. wood preservatives, etc.

Paper, pulp and printing industries – paper and pulp manufacture, including processes associated with printing works, etc.

Miscellaneous – various processes such as building and industrial plant demolition, some High Street processes such as dry-cleaning, the running of laboratories, for research and/or educational purposes, etc.

---

balanced by the need for urgent government action on a whole range of environmental issues.

Finally, whilst the polluter pays principle is undeniably a worthwhile societal attitude to environmental damage, it is a salutary lesson to appreciate that in actuality obviously liable companies can procrastinate for so long over admitting any liability that before they legally accept any responsibility for an incident individuals who have been affected may have died. For example, it took until October 1991 for the Supreme Court in India to rule that the US chemical giant Union Carbide was no longer immune from criminal prosecution for the Bhopal incident in 1984, in which a poisonous cloud of methyl isocyanate killed about 3,000 people and injured another 200,000. The Bhopal victims waited seven years for justice.

Urbanisation

Since ancient times, cities have played a central part in the economic, political and cultural development of societies. Cities serve as the commercial and administrative focus for nations, and generally provide the main places for both the production and consumption of goods and services.

Currently, 45 per cent of the world's population lives in urban areas (37 per cent in **Less Developed Countries** or **LDCs**, and 73 per cent in the **More Developed Countries** or **MDCs**; ASCEND 1992). In other words, contrary to popular belief, most of the metropolitan areas where much of the world's population dwells are situated in the developed countries. Currently, about 20 per cent of the world's population inhabits metropolitan areas, and 33 per cent live in cities with populations greater than 100,000 (Angotti 1993). By the year 2000, it is estimated that 51 per cent of the world population will inhabit urban areas, rising to 65 per cent by 2025 (United Nations Population Fund 1991). In 1960, seven out of the world's ten largest urban agglomerations were in North America, Europe and Japan, with New York, London and Tokyo at the top, whereas now seven out of the top ten are in LDCs, Mexico City being the largest with a population of more than 20 million (ASCEND 1992). At 1994 rates of growth, thirteen of the world's twenty-one megacities (those with more than 10 million

BOX 9.2 CONTAMINATED SOIL CLEAN-UP TECHNOLOGIES

Soil clean-up involves the treatment to remove, stabilise or destroy contaminants using physical, solidification, chemical, biological and thermal methods. Essentially, contaminants can be disposed of in landfills or by containment. Low disposal costs have made landfill an attractive option, using either containment or attenuation of the waste, but these costs are expected to rise steeply as landfill sites become scarce, and controls on environmental pollution become more stringent. These changes are leading to the development of alternative, innovative, technologies, particularly as costs are expected to rise considerably.

Containment involves isolating contaminants from the environment to minimise any liquid (see Table 5.11) or gaseous interchange, whereas attenuation attempts to minimise the movement (e.g. by adsorption) and/or reduce the toxicity (e.g. by degradation). Currently, landfill is cost-effective and commonly practised, but design/construction problems, and long-term uncertainties about the persistence of toxic substances, can present problems for landfill. The appropriateness of various innovative technologies for the disposal of contaminants will depend on many factors, including the type of toxic substances, their concentration and distribution, the soil type and hydrological cycle in relation to the contaminants, and a risk assessment from natural hazards (e.g. earthquakes).

Physical processes do not destroy contaminants and, therefore, are often seen as the first stage of their multistage destruction and/or stabilisation. *Ex situ* processes (those away from the site of contamination) include particle separation techniques, exploiting differences in physical properties such as density, weight, grain size, magnetic susceptibility and surface chemical properties, or physical extraction (e.g. washing, steam stripping). *In situ* processes include soil washing (using aqueous solutions, acids and/or surfactants), soil vapour extraction techniques (using forced air, induced air or steam), and electroremediation (in which a direct current passes through an array of electrodes placed in the soil to induce the contaminant to flow in the pore water to the electrodes, followed by removal of the toxic solution to a water treatment plant).

Solidification processes involve commercially viable and well-proven technologies, although some techniques are at a research and development stage. Solidification technologies are grouped according to the way in which the contaminant is bound, being divided into organic and inorganic techniques: (1) organic, as thermosetting, thermoplastic micro-encapsulation or macro-encapsulation; (2) inorganic, as cement-based, lime-based, vitrified, liquid silicate or pozzolan-based (a pozzolan material acts like cement, and contains silicates or alumino-silicates that react with lime and water to form stable insoluble compounds). Organic binding systems are expensive, and at the bench-scale trial stage, whereas inorganic treatments are seen as having the best potential. Generally, it is extremely difficult to fix organic contaminants using inorganic binding processes, although cements have been routinely used to treat radioactive and other toxic wastes. Lime-based systems have also been widely used to stabilise or solidify toxic waste. Silicate-binding processes involve using powdered aluminium silicates and alkaline solutions of alkali metal polysilicates, which condense to form a high compressive-strength solid in a reaction that gives out heat energy (exothermic reaction). Vitrification, either *in situ* or *ex situ*, involves the melting of contaminated soil to produce a glass-like material, which is inert.

Chemical processes are applied to either destroy or convert the contaminants to potentially less harmful substances. Techniques include oxidation, reduction, extraction, neutralisation, hydrolysis, mobilisation, electrochemical processes, polymerisation and chemical dechlorination. For most of these processes, the contaminated soil needs to be treated as a slurry or at least with the contaminants in an aqueous solution (e.g. ground water). As with physical processes, chemical techniques can be applied either *in situ* or *ex situ*.

Biological processes (or 'biotreatment') generally use microbial biodegradation to make any contaminants benign or to remove them. Biodegradation involves the breakdown of harmful substances, which can then be left at the treatment site or removed. Biotreatments tend to be time-consuming and frequently unsuccessful in removing all of the contaminants. Also, biotreatment technologies require more research before the full implications of their use can be fully evaluated.

Thermal processes involve either incineration of the contaminants within the soil, or a two-stage process of volatilisation and pyrolysis followed by the removal or destruction of the toxic substances from a gaseous phase (by condensation or further combustion). Thermal processes are most appropriate for soils contaminated with organic substances, although some of these chemicals are difficult to incinerate, and the process is often incomplete. Some inorganic chemicals (e.g. mercury and cyanides) can be incinerated, but they may leave toxic residues in the ash, thereby posing further disposal problems. Sandy, silty, loamy and peaty soils respond to thermal processes better than clay soils, due to handling problems with the latter.

---

inhabitants) will be in the Asian–Pacific region by the year 2000. Table 9.10 shows some of the trends in urbanisation (including settlement and labour force data) for various regions between 1960 and 1990, from which the trend towards increasing global urbanisation is evident.

Urbanisation is associated with poor air quality, particularly in the megacities, because of vehicle hydrocarbon emissions, which include airborne particulates and lead (see Chapters 7 and 10).

The future of cities in the context of sustainable development (see Chapter 10) is explored in the book *The Living City: Towards a Sustainable Future*, edited by Cadman and Payne (1990), and contains a useful set of essays on these topics. In this book, Robertson (1990) examines four alternative scenarios for the future of cities:

1 'Decline and Disaster', envisaging a future in which severe breakdown follows from the failure of cities to adapt either to changes in economic function or, in the case of the Developing World, to uncontrolled expansion of the urban population;

2 'Business-as-Usual' offers a 'top-down' approach, in which cities rely upon the benefits of conventional economic development to trickle down to their more disadvantaged citizens;

3 'Hyper-Expansion', in which new technology replaces conventional employment, 'releasing' an increasing proportion of the workforce to extended periods of leisure; and

4 the 'Sane, Humane and Ecological' alternative, where changes in work and a continued trend towards decentralisation lead to the development of greater self-reliance at both a personal and an urban level.

In the fourth scenario – if it were to materialise – then greater emphasis would be placed on enabling urban dwellers to take control of their own future development, with community-led ('bottom-up') urban revival. Also, such a scenario provides more emphasis on the resourceful, self-reliant city, in which there is a minimum waste of resources, by energy conservation and recycling.

Urbanisation can cause many problems associated with changes in the hydrology of an area, for example the construction of new canals and the canalisation of rivers. These, together with buildings, paths and roads, collectively produce an impermeable surface over which water will flow. The use of storm sewers increases the rate and amount of water entering rivers. All these result in an increase in the magnitude and frequency of flooding events. The extraction of ground water for domestic use in large settlements may also cause major alterations to the land surface, such as salinisation, and other effects include ground subsidence. This is well illustrated in Venice, where the pumping of ground water for industrial purposes has caused the gradual subsidence of buildings and increased flooding during the winter.

Frequent small-amplitude floods, known as *acque alte* (high waters), have been a feature of Venice since the fourteenth century, when large-scale construction and diversion of waters began. Since the 1950s, however, floods have become more frequent and on 4 November 1966, a tidal surge of exceptional height and duration caused almost total flooding of Venice, with considerable damage to both buildings and many works of art (Ghetti and Batisse 1983). Pirazzoli (1973) showed that there had been a mean sea level rise of 6 mm per year in Venice between 1950 and 1970. This rise was primarily the result of the withdrawal of ground water for use in the city, and the dredging and embanking of the bay, which enhanced the tidal amplitude to make Venice more susceptible to flooding. Venice could be offered more protection through various combinations of: (1) raising the foundations of city buildings; (2) reducing the amplitude of the *acque alte* in schemes to dissipate the flood waters; and (3) erecting barriers against storm surges. Any effective scheme must not, however,

Table 9.10 *Urban and rural populations, settlements and labour.*

Region	Urban population as % of total 1960	Urban population as % of total 1990	Average annual population change 1960–90 (%) Urban	Average annual population change 1960–90 (%) Rural	Cities with at least 1 million % of total pop. 1960	Cities with at least 1 million % of total pop. 1990	No. of cities 1990	Total labour force 1990 (000s)	Women as % of labour force 1990	% of 1980 labour force in Agriculture	% of 1980 labour force in Industry	% of 1980 labour force in Services
World	34.2	45.2	2.8	1.3	12.2	14.8	276	2,363,547	36.1	51	21	28
Africa	18.3	33.9	4.9	2.1	5.7	9.2	24	242,784	34.4	69	12	19
North and Central America	63.2	71.4	2.0	0.7	28.7	31.8	44	189,258	37.4	12	29	58
South America	51.7	75.1	3.6	0.1	23.4	32.8	29	104,465	26.4	29	26	45
Asia	21.5	34.4	3.7	1.5	8.3	11.3	115	1,436,522	35.3	66	15	19
Europe	61.1	73.4	1.2	(0.7)	16.5	17.0	36	231,702	38.6	14	39	47
Former USSR	48.8	65.8	2.0	(0.4)	12.4	15.3	24	146,634	48.0	20	39	41
Oceania	66.3	70.6	2.0	1.3	31.8	32.2	4	12,181	37.0	20	28	52

Source: World Resources Institute 1992.

reduce the circulation of waters through the lagoon, otherwise pollution levels would increase, the ecology of parts of the lagoon would suffer, and the movement of marine lagoon and canal traffic could be severely disrupted. Attempts have been made to raise buildings by injecting concrete beneath their foundations, but this is extremely expensive and of limited success because of the small heights by which many of the old buildings can be raised without the risk of serious structural damage. In 1976, the authorities opted for a scheme to construct floodgates at the mouth of the three main channels leading to the sea (Ghetti and Batisse 1983).

Bangkok has similar problems to Venice, with rates of subsidence estimated at 3–4 cm yr^{-1}, with a maximum rate of subsidence of 10 cm yr^{-1}. As in the case of Venice, these problems are related to ground-water withdrawal from sandy aquifers, but also due to the compaction of clays because of the overlying weight of major building structures (Rau and Nutalaya 1982). Rau and Nutalaya estimate that Bangkok may subside below sea level within the next twenty years (Plate 9.12).

In his book *Metropolis 2000*, Angotti (1993) advances the view that there are essentially three categories of metropolis, which reflect general historical tendencies and planning models – although in reality every urban system contains elements of all three: (1) the US metropolis; (2) the Soviet metropolis; and (3) the dependent metropolis.

In the US metropolis, land use is segregated with the fragmentation of social groups and political institutions. The centre is densely developed and the suburbs are an urban sprawl. The population is highly mobile and dependent upon the motor vehicle, and it consumes relatively large amounts of non-renewable resources. Planning is determined

mainly by the interplay between the automobile and petroleum monopolies, together with local real estate interests. It represents the quintessential twentieth-century capitalist urban development.

In the Soviet metropolis (perhaps more aptly renamed the Russian metropolis), the archetypal twentieth-century socialist metropolis, the urban population is more integrated *vis-à-vis* its social and political structure, there is limited social mobility and little consumer choice. There is an administrative/residential centre, and relatively high-density suburbs, with transport by mass transit systems (e.g. metro and bus systems in St Petersburg and Moscow). Planning follows a highly centralised administrative/commercial structure.

The dependent metropolis reflects the particular history and dynamics of urbanisation and planning in the developing countries of Africa, Asia and Latin America (Plate 9.13). Although there are important differences between many of these metropolises, they all share a strong dependency upon the developed, capitalist world and typically are associated with considerable social deprivation and poverty. Planning strategies show a wide range of patterns, including large amounts of *ad hoc* development. Angotti (1993) argues that metropolitan planning has failed to adapt to present needs because it was rooted in the pre-metropolitan era of the industrial city and is based upon:

> simplistic notions of master planning and master building that ignore the complex division of labour and functions characteristic of the metropolis. It is strongly influenced by Utopian thinking and philosophical idealism.

The remedy advocated by Angotti is for there to be a much greater emphasis upon neighbourhood

Plate 9.12 Traditional houses in Bangkok, constructed along one of its many waterways. Rain water is collected in pots (middle left) for domestic use. With the rapid urbanisation of Bangkok, water is now extracted from the ground, which has led to large-scale subsidence, thereby threatening both traditional dwellings such as this and the modern city.

planning, which has the function of integrating the family or household with the metropolis, residence and workplace.

On 3 June 1996 in Istanbul, the United Nations opened its last major international conference of the twentieth century on the problems associated with urbanisation, particularly in Africa, Asia and Latin America – its 'city summit', more formally known as 'Habitat 2'. The conference took place in order to address the urgent need for a new environmental and social strategy to cope with the worldwide growth of megacities. In colloquial terms, the problems of 'the murky soup of poverty, homelessness, pollution, and deprivation afflicting the world's ever-expanding cities' has become known as the 'brown agenda'. According to UN figures, 600 million people are now officially homeless or living in life-threatening urban conditions, more than a billion people lack sanitation, and an additional 250 million people have no easy access to safe water. Between 30 and 60 per cent of housing in most developing country cities is

illegal, and more than 75 per cent of the homes in cities such as the Bangladesh capital, Dhaka, and Kenya's capital, Nairobi, were constructed without official permission. Dirty water causes 80 per cent of the diseases in developing countries. The UN report claims that one of the main underlying causes of the deterioration in world cities during the past decade is the economic adjustment programmes imposed by the International Monetary Fund (IMF), since they have tended to increase unemployment, poverty and homelessness. Such structural adjustment programmes were devised in the 1980s as a response by the IMF to the international debt crisis.

Developing countries with large international debts were forced to privatise and deregulate industries, reduce public spending and reduce or excise health and education subsidies. Where countries have implemented the economic structures recommended by the IMF, the knock-on effect has been increased levels of unemployment, poverty and homelessness which has led to large numbers of people flocking

Plate 9.13 *Dense urban life in Kowloon, Hong Kong. In some areas of Hong Kong, population is so dense that it is common for families of six or more to live in flats that are less than 55 m³ in area. Sleeping and working schedules have to be organised to allow people the convenience of space.*

to already overcrowded cities in a generally fruitless search for increased life chances. The net result is that many of the large cities have grown too rapidly, at a rate that is unable to absorb the numbers of people satisfactorily. The UN report links the worldwide growth in urban crime rates and the overall deterioration in the quality of city life to the processes of urbanisation. Where there is an increasing polarisation between the rich and poor – a situation that is blatant in many of the world's large cities – where inequality and social exclusion become overtly manifest, the dysfunctional aspects of urban life lead ineluctably towards increased violence. Unless societies can come to terms with the major social issues directly linked to rapid urbanisation, particularly in developing countries, then the outcome must be a very bleak one for future generations.

Perhaps a way forward in this 'brown agenda', as recommended by the UN city summit, is to devolve urban planning to citizens' groups and local authorities – to those who are directly affected by urbanisation – rather than leave it in the hands of governments and international financial organisations. These are easy recommendations to make but they are extremely difficult to implement. There are those who criticised the Istanbul conference as a talking shop without any international money being put forward to address any of the issues arising out of the conference. Whatever the criticisms levelled at this conference, however, it did provide an opportunity to raise awareness of the issues surrounding urbanisation.

Conclusions

Clearly, humankind has radically altered the Earth's surface, with accelerated impact in recent times. Daily (1995) estimates that recent human activities have led to 43 per cent of the Earth's vegetated surface now having a diminished capacity to supply benefits to humanity. The disruptive changes in land productivity have had a deleterious impact on the various biogeochemical cycles, which regulate, for example, the greenhouse gas fluxes and determine the total energy balance to the Earth's surface and atmosphere. Land degradation threatens biodiversity. It also acts as a limiting factor on economic output, a condition that is particularly affecting many developing nations. The Earth's surface and environments have an enormous capacity to recover from land degradation, but human activities need to be organised in such a manner that this potential for recovery is optimised (*cf.* Daily 1995).

There is a need to understand the Earth's natural systems and their interactive nature, together with the consequences of human activities, in order to be able to predict cause and effect more accurately. This is essential if there is to be prudent management of the Earth's resources in sympathy with the natural environment. Urbanisation tends to be an *ad hoc* process of colonisation of the natural environment. A means of reducing the actual and potential environmental damage caused by urbanisation is to plan urban centres that provide a real sense of community, with good health care and educational provision, adequate recreational open spaces, and fast and reliable mass transit systems; centres that are clean and are not dominated by private and commercial vehicle traffic, are developed on a human scale, maintain coexisting structures in scale, minimise the discrepancies between housing for the rich and the poor, and provide a rich cultural background.

Chapter 9: Key points

1 Rapid population growth has resulted in increased demands upon the Earth's resources, which has led to accelerated environmental degradation, precipitating potentially serious global climate change.

2 The human impact on land has been enormous. As land use changes, natural vegetation is cleared for agricultural use, settlements and urbanisation increase, reservoirs are created, minerals are extracted, and more land is developed for recreational purposes. Acute concern is now widely expressed over the deforestation of boreal and tropical forests, the degradation of grasslands and wetlands, and desertification. Such destruction of natural ecosystems leads to a reduction in biodiversity and impoverishment of soils. In attempts to counter the deleterious effects of land misuse in some areas, exotic plants and animals are being introduced, and indigenous fauna and flora are carefully monitored and encouraged.

3 Human impact on soils has caused considerable damage, commonly because of poor agricultural practices, excessive water extraction, poor irrigation methods (e.g. leading to salinisation), defoliation (particularly resulting in laterisation), and compaction by heavy vehicles and animals. The cumulative effects of these can be disastrous for countries whose economies are heavily dependent on agriculture. The amelioration of these poor practices and improved soil quality require an understanding of the chemistry of

soils and nutrient supply cycles. Wetlands contribute almost a quarter of the world's primary productivity and are essentially the interface between terrestrial and aquatic environments. Only recently has their value been recognised and attempts to reduce their destruction been implemented.

4 Human impact on the oceans and seas results from pollution by dumping and accidents, over-fishing, mineral extraction (e.g. phosphates) and the removal of rare and important marine life such as corals. The seas are an available resource that requires more careful research in order to avoid irreversible damage to their ecosystems, which could have a knock-on effect to the atmosphere and, ultimately, terrestrial life.

5 Environmental risk management involves evaluation of the hazards and impacts on the environment and can be partially achieved by a multi-disciplinary approach involving environmental impact assessments, environmental audits and legislation.

6 The exploitation of the Earth's resources inevitably produces waste, some of which may be hazardous/toxic (contaminants). Until the past few decades, much of this waste has been disposed of without any real concern for the damage to ecosystems, and frequently under the auspices of 'not in my back yard'. Today, as environmental issues are becoming more focused, there is much greater awareness of contaminants and contaminated land. Clean-up technologies are more readily available and preventive measures are being instigated in many countries. Many nations and international organisations are adopting the 'polluter pays' principle. Responsibility for cleaning up contaminated land has led to the introduction of legislation in countries such as the USA and throughout Europe. New and forthcoming legislation aims to identify the polluters and arrange for appropriate levels of compensation to injured parties, but because such laws are in their infancy, there are many teething problems, exemplified by the US Superfund.

Chapter 9: Further reading

Angotti, T. 1993. *Metropolis 2000: Planning, Poverty and Politics.* London: Routledge, 276 pp.
This very readable book on urbanisation offers an analysis of metropolitan development and planning in all parts of the world, and under different economic and environmental conditions. The first four chapters are devoted to an examination of metropolitan development in the United States, the former Soviet Union and the 'dependent metropolises' of the developing world. The last three chapters consider the problems of urban planning theory and practice in the metropolis and its communities. Throughout, Angotti advances the principle of 'integrated diversity' and emphasises linked neighbourhood planning with a broader vision of a planned metropolis.

Cooke, R.U. and Doornkamp, J.C. 1990. *Geomorphology in Environmental Management* (second edition). Oxford: Oxford University Press.
A comprehensive text that highlights the importance of geomorphology in environmental management and risk assessment. Suitable for college and university undergraduate students at all levels, and provides a useful reference source for teachers and researchers. Topics include mass movement, catchment studies, erosion and weathering problems, neotectonics, aeolian environments, and glacial systems.

Eden, M.J. 1989. *Land Management in Amazonia.* London: Belhaven, 269 pp.
This book considers the tropical rainforest as a global resource and its vital importance in sustaining life on Earth. The competing needs of conservation and development are assessed in Amazonia in terms of climate, geomorphology, hydrology, soils, ecology and diverse histories, and the current impact of human intervention. Case studies are presented where, for example, there are attempts to adapt resource-use systems of native peoples to encourage the more effective and less harmful exploitation of the rainforests. Conservation issues are addressed, including the role of national parks and interpretative land management. A good supplementary book for students concerned with environmental risk assessment, particularly relating to the rainforests.

Eden, M. J. and Parry, J.T. (eds) 1996. *Land Degradation in the Tropics.* London: Cassell Academic.
This is a wide-ranging, coherent and scholarly account of land degradation in the tropics. It emphasises the integration of information and theory from both the environmental and social sciences, as well as projecting the application of scholarly analysis in actual policy formulation, planning and management. A wide variety of case studies are presented under the following headings: degradation of tropical forests; degradation in the drier tropics; degradation in tropical wetlands; and urban and industrial degradation in the tropics.

Ellis, S. and Mellor, A. 1995. *Soils and Environment.* London: Routledge, 256 pp.
This book examines the ways in which soils are both influenced by, and themselves influence, the environment. It describes the analysis of soil properties, soil processes and classification. It discusses soil–human interactions and examines the relation to land systems, environmental problems and management, soil surveys, and land evaluation.

Gilpin, A. 1995. *Environmental Impact Assessment (EIA): Cutting Edge for the Twenty-first Century.* Cambridge: Cambridge University Press, 182 pp.

This is an up-to-date and well-written book on the nature of EIAs. It explains the best procedures for assessing projects and the different methods of assessing decision-making, particularly to help minimise the areas of dispute in public and private sectors for both investors and the community. Examples are taken from Europe, the Nordic countries, North America, Asia and the Pacific, providing a good international review of the use and effectiveness of EIAs. The numerous boxes, case studies, figures and tables help to clarify the discussions. It is a good text for advanced undergraduates studying environmental science, geography, planning, law and engineering, and practitioners.

Goudie, A. 1993. The *Human Impact on the Natural Environment* (fourth edition). Oxford: Blackwell Scientific, 432 pp.
A very useful undergraduate textbook that addresses the ways that human activity has changed, and is changing, the Earth's surface. The book is well illustrated and includes a comprehensive bibliography. Topics dealt with include desertification, deforestation, plant and animal invasions, marine pollution, climatic change, and environmental uncertainty.

Gradwohl, J. and Greenberg, R. 1988. *Saving Tropical Forests*. London: Earthscan, 208 pp.
A text that provides case studies from throughout the world to show how the destruction of the tropical forests might be slowed or even stopped, and how sustainable management could be achieved. A thought-provoking book for students, teachers, and policy-makers.

Grainger, A. 1990. *The Threatening Desert: Controlling Desertification*. London: Earthscan, 269 pp.
An interesting book that describes the distribution and processes of desertification, and the successes and failures that have accompanied the various attempts to combat desertification as set out by the Plan of Action resulting from the 1977 Nairobi United Nations Conference on Desertification. The book argues for a new International Plan of Action to control the increasing threat to the natural environment posed by desertification.

Huggett, R.J. 1995. *Geoecology: An Evolutionary Approach*. London: Routledge, 320 pp.
This book presents 'geoecosystems' as dynamic entities that constantly respond to external and internal effects. It combines an evolutionary and an ecological perspective, showing how animals, plants and soils interact with the terrestrial 'spheres'. The book will have wide appeal, from courses that focus more on ecology to those that emphasise the nature of the physical environment.

Ives, J.D. and Messerli, B. 1989. *The Himalayan Dilemma*. London: Routledge, 295 pp.
This book addresses the complex dynamics and environmental systems in the Himalayas and considers the problems of reconciling development and conservation. It includes a

look at the interaction between human activities and the natural environment and is a useful book for advanced undergraduate courses, teachers and policy-makers.

Morgan, R.P.C. 1986. *Soil Erosion and Conservation*. London: Longman, 298 pp.
A useful textbook aimed at undergraduate and postgraduate students who are studying soil erosion and conservation as part of any Earth science or environmental science course. The book provides an introduction to the subject, including the magnitude, frequency, rates and mechanics of wind and water erosion, erosion hazard assessment, methods of measurement, modelling and monitoring, and strategies for erosion control and conservation practices.

Parnwell, M. and Bryant, R. (eds) 1996. *Environmental Change in South-East Asia*. London: Routledge.
This book is a compilation of works by scholars, journalists, consultants and NGO activists that explore the interaction of people, politics and ecology. It explores the nature of the environmental degradation that has resulted from the rapid economic growth in Southeast Asia and the dilemmas facing policy-makers as they seek to promote sustainable development. Particular emphasis is placed on the centrality of politics to environmental change. It highlights the fatal flaws in presenting exclusively economic and ecological approaches, and the authors stress that neither the quest for sustainable development nor the process of environmental change can be understood without reference to political processes.

Poore, D. 1989. *No Timber Without Trees: Sustainability in the Tropical Forest*. London: Earthscan.
Based on a study for the International Tropical Timber Organisation, this book reviews the extent to which natural forests are being sustainably managed for timber production and how these practices could be improved. The book places timber production in the wider context of tropical rainforest conservation. Examples are drawn from Queensland, Africa, South America, the Caribbean and Asia. The book makes interesting and easy reading for students of environmental science.

Revkin, A. 1990. *The Burning Season: The Murder of Chico Mendes and the Fight for the Amazon Rain Forest*. London: Collins, 317 pp.
An inspiring, but also sad, book that emphasises the beauty of the tropical rainforests and the need for conservation in Amazonia. The book provides an ecological, historical and industrial outline of life in the forest, focusing on the life and environmental work of Chico Mendes, a rubber planter, who strove for sustainable development in the forest in which he lived. Mendes' success in reducing the exploitation of the forest by cattle ranchers cost him his life: he was murdered in 1988.

Smith, L.G. 1993. *Impact Assessment and Sustainable Resource Management*. Harlow: Longman Scientific & Technical.

This book provides an integrated approach to environmental planning, balancing academic and practical considerations. Various aspects of environmental planning include decision-making; dispute resolution; environmental law; public policy; administration; the nature of planning; and impact assessment and methodology.

Thomas, D.S.G. and Middleton, N.J. 1994. *Desertification: Exploding the Myth*. Chichester: John Wiley & Sons.
A useful text that explores the origin of the 'desertification myth', and how it has spawned multi-million dollar initiatives and came to be regarded as a leading environmental issue. The book examines the political and institutional factors that created the myth, sustaining it and protecting it against scientific criticism.

Tivy, J. 1993. *Biogeography: A Study of Plants in the Ecosphere* (third edition). Harlow: Longman.
This classic text on biogeography is essential for those undergraduates studying geography, biology, environmental studies, conservation and ecology. It explores the variations in forms and functioning of the biosphere at both the regional and global scales. It highlights the interaction between the organic and inorganic components of the ecosphere. Emphasis is placed on the importance of the plant biosphere as the primary biological product that forms the vital food link between organisms. It also emphasises the role of humans as the dominant ecological factor.

Welford, R. 1995. *Environmental Strategy and Sustainable Development*. London: Routledge, 217 pp.
An interesting debate over environmental strategy in business, providing a radical business agenda for the future. It discusses important strategies such as environmental management systems and environmental audits.

Williams, M.A.J. and Balling, R.C. 1995. *Interactions of Desertification and Climate*. London: Edward Arnold.
This book, commissioned by the UNEP and the WMO, examines current knowledge of the interactions of desertification and climate in drylands. It concludes by providing a series of useful recommendations for future dryland management.

Wood, C. 1995. *Environmental Impact Assessment: A Comparative Review*. Harlow: Longman, 337 pp.
This text provides an international review of environmental impact assessment (EIA), outlining the history, nature, proceedings, methods and criteria, and future development of EIAs. It also provides comprehensive coverage of strategic environmental assessments. It is an excellent and useful text for advanced undergraduates studying environmental science, geography, planning, law and engineering, and practitioners. Tables and boxes provide useful summaries, and diagrams help to clarify the processes involved in EIAs.

I am not yet born; O hear me.
Let not the bloodsucking bat or the rat or the stoat or the
         club-footed ghoul come near me.

I am not yet born, console me
I fear that the human race may with tall walls wall me,
         with strong drugs dope me, with wise lies lure me,
                  on black racks rack me, in blood-baths roll me.

I am not yet born; provide me
With water to dandle me, grass to grow for me, trees to talk
         to me, sky to sing to me, birds and a white light
                  in the back of my mind to guide me.

I am not yet born; forgive me
For the sins that in me the world shall commit, my words
         when they speak me, my thoughts when they think me,
                  my treason engendered by traitors beyond me,
                           my life when they murder by means of my
                                    hands, my death when they live me.

I am not yet born; rehearse me
In the parts I must play and the cues I must take when
         old men lecture me, bureaucrats hector me, mountains
                  frown at me, lovers laugh at me, the white
                           waves call me to folly and the desert calls
                                    me to doom and the beggar refuses
                                             my gift and my children curse me.

I am not yet born; O hear me,
Let not the man who is beast or who thinks he is God come near me.

I am not yet born; O fill me
With strength against those who would freeze my
         humanity, would dragoon me into a lethal automaton,
                  would make me a cog in a machine, a thing with
                           one face, a thing, and against all those who
                                    would dissipate my entirety, would
                                             blow me like thistledown hither
                                                      and thither or hither and thither
                                                               like water held in the
                                                                        hands would spill me.

Let them not make me a stone and let them not spill me.
Otherwise kill me.

Louis Macneice, 'Prayer before Birth'

Plate 45 *Strip mining
for diamonds in South
Africa. Beach sediments
which are 200 m thick
are removed to expose
gravel which contains
diamonds. The activities
are very extensive,
creating large-scale
disruption to coastal
ecosystems.*
*Courtesy of Fred Mayer/
Magnum.*

Plate 46
Chuquicamata open-pit
copper mine in Chile.
This is the largest
excavation in the world.
Courtesy of Comstock.

Plate 47 Strip mining
for coal in Fairfield,
Texas. The vast scale of
mining activity can be
visualised by comparing
the size of the crane with
the pick-up truck in the
lower left-hand corner of
the plate.
Courtesy of Comstock.

Plate 48 Intensive
mining of emeralds in
Brazil. The miners face
considerable danger from
slope-failure and flash
flooding.
Courtesy of Magnum.

The natural ecosphere under threat from human activities

The greatest challenge that confronts society and governments today, and the generations to come, is the sustainable development and intelligent management of this (some would argue over-populated) planet. In essence, societies must be able to supply sufficient food, energy, raw materials and any required manufactured products to their citizens and other nations without compromising the world's resources for future generations, and without leaving a barren wasteland of environmental degradation. Putting such lofty ideals into practice is no easy task.

Before discussing ways in which humans can manage the planet intelligently, and plan for, and implement, sustainable development, there is a need to consider the social, economic, cultural and political aspects of global environmental issues. Prominent amongst the issues that humans must come to terms with is the size of the world's population, and the levels at which sustainable development for the whole world is achievable and acceptable.

World population

The population of the world doubled from around 2.5 billion in the middle of this century to about 5 billion during 1987. By the late 1990s, world population will be about 5,292,200,000, an increase of 75 per cent since 1960, and it is projected by the United Nations Population Division to increase by 60 per cent from the 1990 figure to 8,488,600,000 by 2025, and to reach 11.3 billion by 2100 (Figures 10.1 and 10.2).

Since the eighteenth century, world population has increased eightfold and average life expectancy has doubled. Two thousand years ago, the population was 200 million, taking 1,500 years to double. In contrast, the most recent doubling has taken only twenty-seven years. The joint US National Academy of Sciences and Royal Society of London document published in February 1992, using data from the UN Population Fund's 1991 report, which noted an acceleration of population growth since 1984, and assuming a sustained decrease in fertility towards the replacement level of 2.1 offspring per woman per lifetime, stated that world population may reach 10.5 billion in 2050, with around 90 per cent of this figure concentrated in the developing countries. The United Nations Population Division predicts that 95 per cent of this expansion will take place in the less developed countries, such as India, China, Bangladesh, Pakistan, the Philippines, Indonesia, Vietnam, Iran, Mexico, Brazil, Egypt, Kenya, Tanzania, Zaire, Nigeria and Ethiopia. Although the rate of increase in world population is slowing, absolute numbers continue to increase and to exert social, environmental and economic pressure on available global resources. One expression of over-population is violence and war.

Population in the developing countries is growing most rapidly, while the developed countries have more or less stabilised (Figure 10.3). China and India have the fastest-growing populations, which are projected to rise from 1.2 billion to 1.5 billion and 935 million to 1.4 billion, respectively, by 2025 (Livernash 1995). As a consequence, these countries will experience considerable problems in supplying adequate food, fresh water and energy to their populations. Furthermore, the increased populations will exert severe stresses on the environment. In China, a considerable proportion of the population will be elderly, an inheritance from the single child per family policy. This will reduce the percentage of the population that will be in the workforce and necessitate an expansion of care and health facilities for the elderly.

World fertility and mortality data are summarised in Figure 10.4. In some developed countries, such as Germany, there has been a decrease in the birth rate over the past decade and the population is declining steadily.

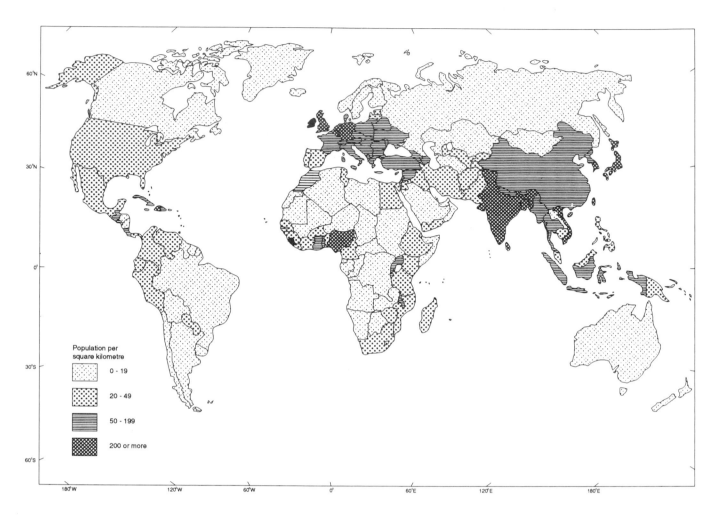

Figure 10.1 *Population density, calculated by dividing a country's population by its total surface area (land and inland water areas). Redrawn after the International World Development Report (1992).*

It is instructive to compare the estimated population growth and energy consumption for developed and less developed countries, as shown in Table 10.1. Clearly, if population growth is considered in terms of changing patterns of energy consumption, the more developed countries are growing at a faster rate than the less developed countries. Statistics similar to these could be calculated for the growth of pollution and waste disposal. Increased population, therefore, is not just a problem for less developed countries but obviously also affects developed countries.

Poor people in the developing countries have expectations that are different from those in the developed world. In developing countries, poverty forces individuals and communities to pay more attention to their immediate needs, whereas planning for future generations and global considerations

frequently seem less realistic. In developing countries, where there is greater economic marginalisation, many live on and even below subsistence levels, with starvation as an all too present threat (Plates 10.1 and 10.2). In many developing countries, life expectancy is often much lower than in the developed countries. According to United Nations estimates, in Africa and parts of Asia, child mortality rates still remain very high: in Afghanistan, 30 per cent of children die before the age of five; nineteen countries throughout Africa will suffer a 20 per cent death rate of their children before reaching five, and globally, more than 14 million children under five die each year (*World Resources 1990–91*). For those who survive, a non-existent or very basic educational provision means that the poor tend to be less educated and less articulate in making demands for better life chances. Also, the poor cannot afford a

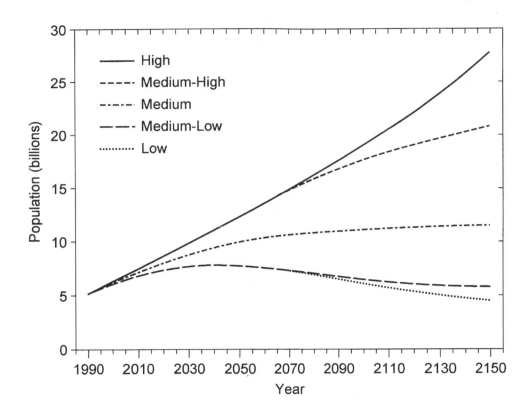

Figure 10.2 *Population projections up to 2150, based on various assumptions about the timing and rates of fertility. The low projection assumes fertility will stabilise at 1.7 children, the medium projection assumes stabilisation at 2.06 children, and the high projection assumes stabilisation at 2.5 children. Redrawn after World Resources (1994–95).*

high standard of medical provision, both preventive and curative, and therefore along with lower dietary conditions this tends to result in much higher mortality rates. All these factors, which lead to the poor being disadvantaged at all stages in their lives, is not conducive to good and prudent management, nor does it favour economic and political stability. The transition to a more capitalist, market-led economy in China and the former Soviet Union has caused major political unrest in these very large populations. Tiananmen Square in 1989, the unrest and bloodshed in Moscow in 1993, and the wars in what was Yugoslavia, epitomise the violence that erupts even in developed nations.

Throughout much of the world, the change to mechanised corporate cultivation techniques,

Figure 10.3 *Age distribution of populations of the less developed and the developed countries in 1990 compared with that projected for 2025. In the less developed countries, the population will continue to grow rapidly, with an expanding labour force. The percentage of old people will also increase with respect to the young, requiring greater care. Redrawn after Keyfitz (1989).*

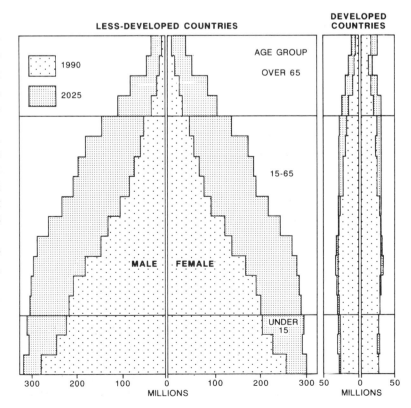

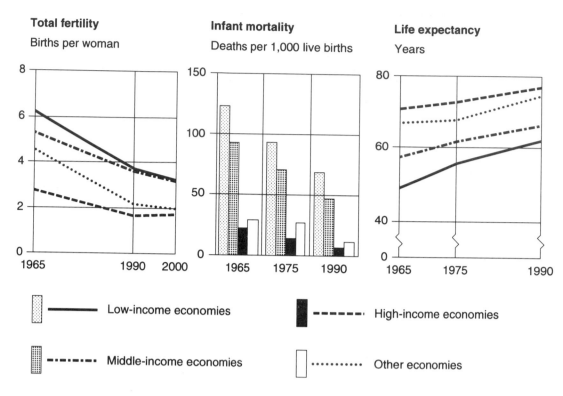

Figure 10.4 *World fertility and mortality. After the International World Bank (1992).*

replacing small family-run farms, has meant that large-scale intensive production methods are now commonplace. As a consequence of the intensive cultivation of the most fertile areas of land, using a panoply of fertilisers and insecticides, the more traditional family-run farms generally cannot compete and therefore go out of business. The result is more unemployment and larger areas of former agricultural land becoming degraded into desert.

The agricultural communities wither, and many of those who are displaced from the countryside into the cities suffer alienation and anomie. Rather than being producers, many more become dependent on others who produce. Urbanisation is a phenomenon of the twentieth century. The epitome of the urban sprawl is Mexico City, which is the largest city in the world, with a population of about 19.4 million but predicted to reach 24.4 million by the turn of

Table 10.1 *Per capita energy consumption and projected growth in per capita energy consumption at present rates (early 1990s) for developing and developed countries.*

	Less developed	More developed*
Total energy consumption (Petajoules)	76,396	261,641
Total population in 1990 (millions)	4,087.0	1,205.2
Per capita energy consumption (Petajoules)	1.8×10^{-5}	2.1×10^{-4}
Net population growth per 1,000, 1995–2000	19.2	4.0
Growth of energy consumption per person at present energy consumption rates	0.00035	0.00087

* including centrally planned countries
Source: World Resources Institute 1990–91.

Table 10.2 *Prevalence of chronic under-nutrition in developing regions.*

	1969–71		1979–81		1988–90	
Region	Millions of under-nourished	Proportion of total popu-lation %	Millions of under-nourished	Proportion of total popu-lation %	Millions of under-nourished	Proportion of total popu-lation%
Africa	101	35	128	33	168	33
Asia	751	40	645	28	528	19
Latin America	54	19	47	13	59	13
Middle East	35	22	24	12	31	12
Total developing regions	941	36	844	26	786	20

Note: Seventy-two countries with a population of less than 1 million, representing 0.6 per cent of the developing world's population, were excluded from the table totals.
Source: Food and Agriculture Organisation of the United Nations (FAO). *The State of Food and Agriculture, 1992* (FAO, Rome, 1992), p. 22.

the twenty-first century. The extreme crowding in Mexico City means that millions of people exist in squalid conditions of dire poverty, without adequate supplies of clean water and with poor sanitation. Crime and disease abound. Despite the overcrowding and bad living conditions for many, people still flock to urban centres such as Mexico City because there is at least the possibility of a job, better standards of health care and educational opportunities for the successful. For many, however, these opportunities remain unrealised. This scenario is repeated the world over in many lesser developed countries, especially, but not uniquely, where populations are large and concentrated (Plates 10.3 and 10.4).

Some cities, for example Hong Kong, have attracted large numbers of political refugees from politically unstable areas and regions under conflict. In Hong Kong, the influx of Chinese from the mainland and refugees from Vietnam ('boat people') has contributed to many urban problems in the former British colony (which reverted to Chinese rule on 1 July 1997), associated with the rapid growth of population in the 1960s and 1970s. There is growing concern today that environmental refugees will be a continuing problem in the future as sensitive areas undergo environmental changes related to population pressures and natural disasters.

An alternative perspective about urbanisation taking people away from the land is that it removes some pressure from the land as a resource, at least in terms of the growth of villages into towns. Urbanisation, however, tends to concentrate atmospheric pollution, with exhaust from motor vehicles, domestic and industrial processes, and other emissions. Urban centres tend to generate ghettos and within such poverty traps, crime becomes more prevalent.

Urbanisation favours easier access to medical care for greater numbers of people, and makes for greater efficiency by concentrating health and welfare resources: effectively such a concentration of resources supports larger populations. On the debit side, city environments, because of the greater degree of medical care and access to other social provision, can allow a population to reach levels that cause serious environmental damage and pollution.

Until the twentieth century, with the enormous strides in medical science that allowed people to live longer, for example by surviving illnesses that would previously have proved fatal, population levels did not pose the present threat to the environment. Large families were common because of the high rates of infant mortality, the shorter life span of individuals, and the desire for economic and family security in old age through a large immediate and extended family. Death, birth and longevity conspired to maintain population levels at a fairly steady state, albeit with modest increases. Seventeenth-century Europe, for example, had sufficient available land to support a larger population than actually existed at that time.

What will limit population growth? Will it be a lack of food? Can society feed the world population in say one hundred years, and what will the natural environment look like with the additional stresses as more land is utilised to produce food? Already, the world has large numbers of impoverished and starving people because of famine and war (Table 10.2), yet there are many countries with a food surplus. Food mountains and the periodic destruction of food surpluses occur in order to maintain prices and profits.

The World Commission on Environmental Development (WCED) estimates that a five- to ten-

Plate 10.1 *Scavenging on the municipal rubbish dumps is part of everyday life for the families of Communidad Veuda De Alas, El Salvador.*
Courtesy of Rhodri Jones/Oxfam.

Plate 10.3 *Large urban areas flourish in both the developed countries (as in New York (left)) and in the developing countries (e.g. makeshift dwellings in Caracas, Venezuela (right)).*
Plate on right courtesy of R. Potter.

fold increase in the world's economic activity is needed during the next fifty years to support the projected population explosion. Increased economic activity can only mean increased exploitation of raw materials, land, energy and agriculture. Cohen (1995) showed that the simple models that have been applied to non-human populations to assess **carrying capacity** cannot easily be applied to humans, because the Earth's capacity to support humans is determined not only by natural constraints, but also by human choices concerning economics, environment, culture, values, politics and demography. It is difficult to predict future resource demand based on the extrapolation of present and projected population growth, because of complex changes in the environment, such as socio-economic and cultural changes (Cuthill *et al.* 1993).

It seems that food alone may not be the rate-limiting step on the growth of the world's population. Figure 10.5 shows a bar chart with the increase in total production of cereals (dark shading) and world population (lighter legend). This diagram suggests that the world's current food production is actually greater than the rate at which world population is increasing. If food production continues at present rates, it is estimated that there should be sufficient food for the projected world population stabilised at 10 billion in 100 years. World population, however, may reach the 10 billion mark within the next fifty years (see Figure 10.2). With

figures such as these it seems reasonable to conclude that human activities are likely to exert environmental stresses to the point where recovery is not fast enough to support these large population levels. Some major natural resources may become so depleted that a lack of suitable substitutes could precipitate other problems. The exhaustion of fossil fuels such as oil, for example, will necessitate their substitution by viable alternatives. There are many scientists who do not see the exhaustion of fossil fuels as part of the ineluctable decline into a major global energy crisis from which there may be no escape other than by extreme conservation measures, but merely as a challenge for the future with 'necessity as the mother of invention'. At this time, nearing the end of the twentieth century, it seems that there are no obvious solutions and an abundance of complacency at all social and intellectual levels in society.

Economic security at a familial level correlates with lower birth rates: in general the least developed nations have the highest birth rates. Large families do indeed provide a cushioning from the vagaries of poverty for individuals at the level of the nuclear and extended family but not necessarily at the national level. Even where families in less developed countries are unable to support themselves adequately, a large family often provides emotional support and a reason for enduring poverty – the hope that children may achieve their parents' aspirations. Where a nation is

Plate 10.2 *Self-sufficiency practised at Communidad Santa Martha, El Salvador. Programmes such as this, which are self-initiated rather than imposed by external aid bodies, provide amongst the most sustainable means of development and, importantly, can empower communities to independence.*
Courtesy of Rhodri Jones/Oxfam.

Plate 10.4 *The daily commuter crush on trains in Tokyo. Urbanisation and overcrowding, causing environmental stress and associated problems, are not restricted to the poor, developing countries, but affect even the most affluent nations.*

committed to a lower birth rate, it has instigated family planning programmes. Unless individuals are economically, socially and culturally receptive to family planning, then the success of such programmes may be very limited, as has proved to be the case in Pakistan, Kenya and Nepal. In contrast, they have proved much more successful in China, Indonesia, Thailand and South Korea. One could argue that the principal reason for this success owes more to the militaristic politics of these countries and the severe penalties inflicted on those who sire a large number of offspring, rather than an internalised commitment to the policies of the respective governments.

There are those who claim that the actual population levels are not the real problem, but rather it is the use of inefficient and bad farming practices that create unnecessary food shortages. A contrary perspective is that inefficient and bad practices exist and need rectifying but should not be cited as an excuse for inaction in controlling population levels.

There are religious groups, such as the Holy See, which argue that it is morally wrong to control the growth of populations artificially through the use of contraceptives. Churches such as the Roman Catholic Church officially regard artificial means of contraception as morally wrong: their teachings are widely followed in many of the poorer Latin American countries. Governments and other policy-makers have to take this panoply of moral and ethical arguments into account when they endeavour to implement realistic action that appears to be appropriate and acceptable within a particular socio-economic milieu. There are no easy answers here. Perhaps the most important aspect in considering the control of population is to educate individuals so that they become more aware of how their child-bearing and rearing may affect their country's socio-economic fabric and environment, as well as the larger-scale global environment – the latter being of little interest to humans in dire poverty. Whatever the educational programmes, it is a self-evident truth that unless individuals perceive a personal benefit, including to their family, then it is unlikely that they will change their behavioural patterns. Self-interest is the strongest of all motives for action.

The growth of the world's population has been likened to the growth of bacteria in a laboratory culture (Clark 1989). Bacteria in cultures grow rapidly from distinct nuclei, expanding outwards and encroaching on other colonies. Where space is severely limited, as in a petri dish, the bacteria eventually die through overcrowding and the ensuing competition for inadequate food supplies. Likewise, human populations increase from centres, settlements expand in size from villages to large metropolises, creeping along communication links, continuously expanding and growing as pressures increase for living space. Will the global human population eventually die for the same reasons that cause the death of the bacteria in the laboratory petri dish? Will the expansion of the human population result in increased and larger-scale conflicts as resources and food supplies diminish? Will pollution also reduce life expectancy by as much as 50 per cent in the twenty-first century (Meadows *et al.* 1972), and limit economic growth?

The view that population growth will be self-limiting is not new. As early as 1798 the English economist, the Reverend Thomas Malthus, in his *Essay on the Principle of Population*, argued that population growth would limit itself because of the finite food supplies. He argued that if population growth exceeded a critical level, then the scarcity of food would cause famine and war, thereby reducing

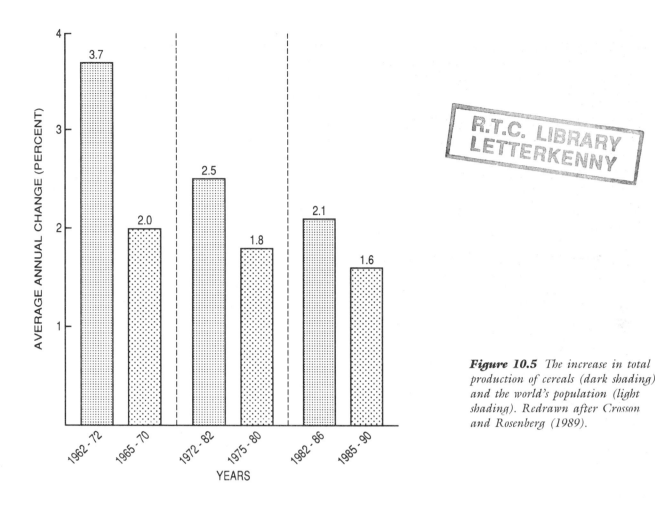

Figure 10.5 *The increase in total production of cereals (dark shading) and the world's population (light shading). Redrawn after Crosson and Rosenberg (1989).*

the population to some sort of equilibrium level. Other catastrophes, such as plagues and disease, would further limit population size.

Modern adherents to **Malthusian views**, the so-called neo-Malthusians, believe that accelerating population growth will lead to an increase in pollution and environmental degradation and effectively act to limit population growth (e.g. Ottaway 1990). Contemporary neo-Malthusian organisations, which include Population Concern, the United Nations Population Fund (UNFPA), the World Commission for Environment and Development, and Worldwatch Institute, believe that the main cause of environmental degradation is the rapid growth in world population. Contrasting views are expressed by the anti-Malthusians, who believe that it is the distribution and organisation of people, not the absolute population numbers, that cause the problems. The anti-Malthusians also argue that all too often there is an inappropriate use of technology, over-consumption and inequalities in wealth and life chances, which ultimately induce environmental degradation rather than simply over-population.

Neo-Malthusian views were revived in the late 1960s and early 1970s with the publication of Paul Ehrlich's book *The Population Bomb* and the report to the Club of Rome, *Limits to Growth* (Meadows *et al.* 1972). The latter book was based on a computer model, which predicted that world population would deplete the present world resources within the next 100 years. Ehrlich's book suggested that fast-growing population meant that the Earth's capacity to feed humans had almost been reached and a minor change in agricultural productivity, such as the failure of a monsoon in Asia, could plunge the world into political chaos, greater disorganisation, food shortages, famines and wars – a doom-and-gloom scenario of cataclysmic proportions.

A contrary perspective was put forward in 1981 by J. Simon in his book *The Ultimate Resource*. In this book, Simon suggested that population growth does not in itself present an intractable problem for the future of the Earth, but rather that an increase in population would lead directly to higher living standards and economic development. The improved lifestyle and life chances would come to fruition, so the argument runs, because humans would be forced into ever more innovation and technological advances to meet the additional need to sustain the increased population. These arguments represent a very optimistic view of human survival and embody the belief that humans will always be able to adapt to new and changing global problems.

Geographers such as Moore Lappé and Rachel Schurman, in their book *Taking Population Seriously* (1989), are also opposed to neo-Malthusian views and arguments on the basis that they are over-simplistic and deterministic. Instead, they considered what they perceived to be the causes of rapid population growth, and concluded that population problems are predominantly the result of imbalances between individuals' reproductive choices, or rather the lack of them. They emphasised the fact that many women generally have subservient roles, where they are under-valued and under-rewarded. Such socio-economic, political and cultural inferiority often leaves women with the belief that reproduction is their optimal role. Furthermore, Lappé and Schurman suggest that one of the main remedies to reduce rapid population growth is to improve socio-economic, political and cultural conditions for women, for example through improved education and direct income. They believe that such changes in the way in which society (essentially but not exclusively men) regards and treats women can encourage many more women to have smaller families.

In October 1993, many of the world's premier scientific academies met for the first time in New Delhi to consider issues such as population growth and sustainable development. Prior to this summit, the US National Academy of Sciences, the Royal Society of London, the Swedish Academy of Sciences and the Indian National Science Academy drafted a joint document that called for zero population growth within the lifetime of the present generation that are still children. It suggested means to achieve this target of zero growth and urged governments to integrate policies on population and sustainable development. The fifteen-page document, which has become known as the New Delhi Statement, was endorsed by fifty-six institutes, but many countries refused to become signatories. The New Delhi Statement has been criticised because it failed to acknowledge the central role of women in all issues concerning population and development (Jayaraman 1993b).

In September 1994, delegates from over 150 nations attended the UN-sponsored International Conference on Population and Development (ICPD) in Cairo. The greatest difference between this confer-

ence and the earlier ones in Mexico City (1984) and Bucharest (1974) was the focus on the role played by women in population growth. A 113-page document, the World Programme of Action (WPOA), outlined a new population policy aimed at stabilising global population at about 7.27 billion by 2015. Amongst the contentious options to achieve sustainable population growth were promoting modern contraceptives; promoting economic development; improving infant and child survival; improving the status of women; and educating men. The urgency for economic and political rights and education for women to help improve the future of humankind was particularly emphasised.

One of the more striking aspects of the ICPD was the presence of non-governmental organisations (NGOs), including population control activists, feminist and health/medical practitioners, academics and researchers, environmentalists, and religious groups. The most visible participants at the conference were women's health groups, who circulated a twenty-one-point document to ensure that women's perspectives and experiences were included in the ICPD (Chen *et al.* 1995). In her keynote address at the ICPD, Brundtland (1994) called for future family planning to be considered under the heading of 'reproductive health care', a strategy that included a whole range of issues surrounding reproduction that had previously been neglected by family planning programmes, i.e. the control of sexually transmissible diseases (such as HIV/AIDS), the processes of pregnancy, a reduction in mortalities in childbirth, and programmes to legalise abortion. This agenda was endorsed by the WPOA. Brundtland also emphasised that women's education is the single most important route to higher economic productivity, lower infant mortality and lower birth rates.

The WPOA, which has been tabled for approval by the UN General Assembly, balances the importance of internationally recognised human rights and the sovereignty of nations as represented by national laws, as well as development priorities that recognise different religious and ethical values, and cultural backgrounds (Sen 1995).

The United Nations Fourth Women's Conference in Beijing and the accompanying conference in Huairou for NGOs in September 1995 again drew much attention to the plight of women, focusing on issues such as reproductive freedom, economic discrimination, and even more sensitive issues such as female infanticide, rape and bride burning. Unfortunately, many of the major issues at this conference were overshadowed by the media coverage of the poor conference facilities, the denial of an

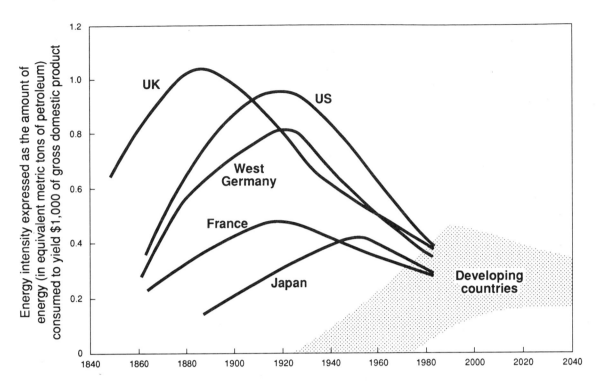

Figure 10.6 *Energy intensity versus time in industrialised and developing countries. In industrialised countries, the energy intensity ratio (ratio of energy consumption to gross domestic product) rose, then fell. Because of improvements in materials science and energy efficiency, the maxima reached by countries during industrialisation have progressively decreased with time. Developing nations can avoid repeating the history of the industrialised world by using greater energy efficiency. It is unrealistic, however, to expect developing countries to reach an energy-efficient development path very quickly, given the capital constraints and industrial weakness these countries confront. Redrawn after ASCEND (1992).*

estimated 10,000 Chinese visas, and difficulties with the participation by women from many NGOs. The denial of visas was seen as symptomatic of the denial of freedom and women's rights, an issue dealt with by Hillary Clinton in her address to the conference:

> Freedom means the right of people to assemble, organise and debate openly. It means respecting the views of those who may disagree with the view of their government.

Ideally, individuals and family groups should be able to make free choices about birth control with as much information as is possible. In reality, poor educational opportunities, and political, economic and cultural reasons conspire to deny many people both the information needed to make independent decisions and the political freedom to act upon those decisions.

The rise of the consumer society

The rate of increase in the consumption of natural resources should be a cause of concern to everyone.

Since the beginning of this century, energy consumption has increased by eighty-fold, manufacturing by one-hundred fold, and over 9 million km^2 of land (much of this previously forest) has been converted to agricultural land. Patterns of energy use are very different in the developed and developing countries (Figure 10.6). Water resources and water quality also show significant differences between the developed and developing countries (Figure 10.7).

Withdrawal of water for human activities has increased to the extent that annually more than 100 km^2 of additional land is required to meet this demand. Also, the amount of suspended sediment load in rivers has risen by 300 per cent since the eighteenth century, and over the same time industry has created more atmospheric pollution by doubling the amount of CH_4 emissions and producing an extra 25 per cent CO_2. Toxic metal pollution has shown large increases since the eighteenth century, particularly metals such as lead, cadmium and zinc (eighteen-fold), and a twofold increase in arsenic, mercury and nickel.

Gross domestic product (GDP), expressed as either an absolute figure or per capita, shows a very large

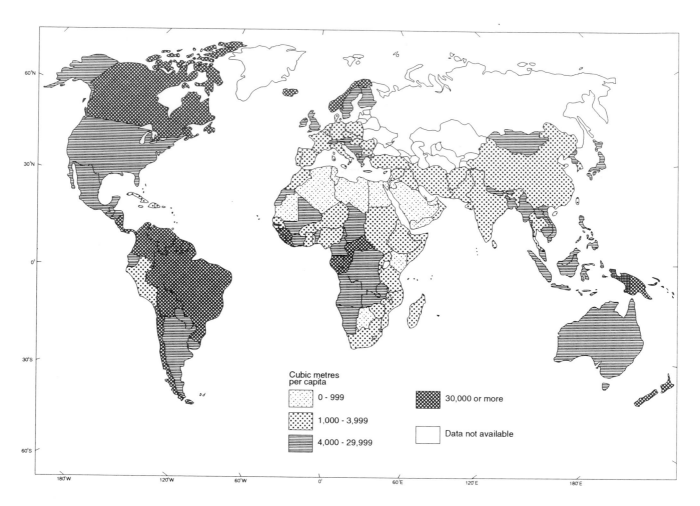

Figure 10.7 *World map showing annual renewable water resources. The average amount of water available per person per year is calculated by dividing a country's annual internal renewable water resources by its population. Redrawn after the International World Bank (1992).*

difference between the rich and poor nations (Figure 10.8). Without sufficient economic power it is extremely difficult for the developing world to improve the overall life chances and economic prosperity for individuals and a country as a whole, let alone deal adequately with atmospheric and water pollution. The disparities between rich and poor countries can be more clearly seen when comparing the gross national product (GNP) with the size of the country, as shown in Figure 10.9. Virtually any environmental indicator at different national income levels reveals the gross imbalance between rich and poor, developed and developing countries (see Figure 10.10).

Prior to the 1970s, natural resources were seen both as plentiful and essentially limitless. This misguided view changed with the publication in 1972 of the report, *Limits to Growth*, presented to the Club of Rome. The report argued that existing patterns of global resource demand and consumption would lead to a collapse of the world's socioeconomic and political systems within the next century. The report was influential in that it certainly altered many people's perceptions about sustainable growth and focused attention on the scarcity and finite nature of many natural resources. In turn, the changed economic mood led to the stockpiling of commodities throughout the 1970s and early 1980s. As a result of this siege mentality, commodity prices rose sharply, in part fuelled by the large increase in oil prices in the early 1970s. The 1975 Second Club of Rome Report, *Mankind at the Turning Point*, was much less gloomy over these issues but nevertheless warned against economic complacency.

Today, improved methods of assessing the amount of various resources, particularly non-renewable

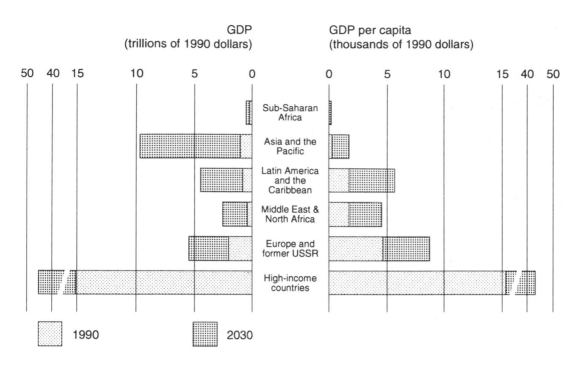

Figure 10.8 *GDP and GDP per capita in developing regions and high-income countries, 1990–2030. Data for 2030 are projections using World Bank data. Redrawn after the International World Bank (1992).*

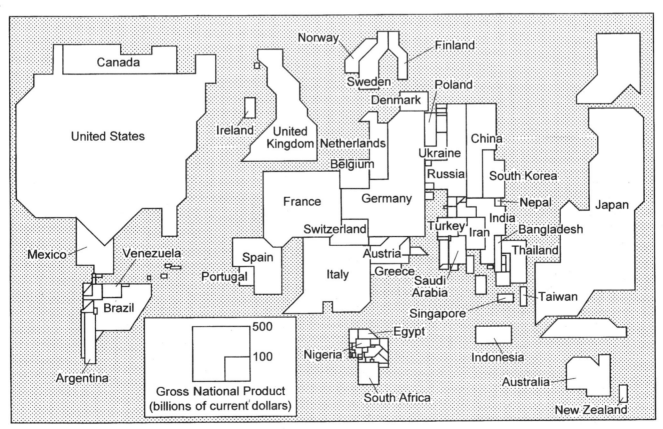

Figure 10.9 *The size of countries redrawn to correspond to their gross national products. Redrawn after World Resources Institute (1994).*

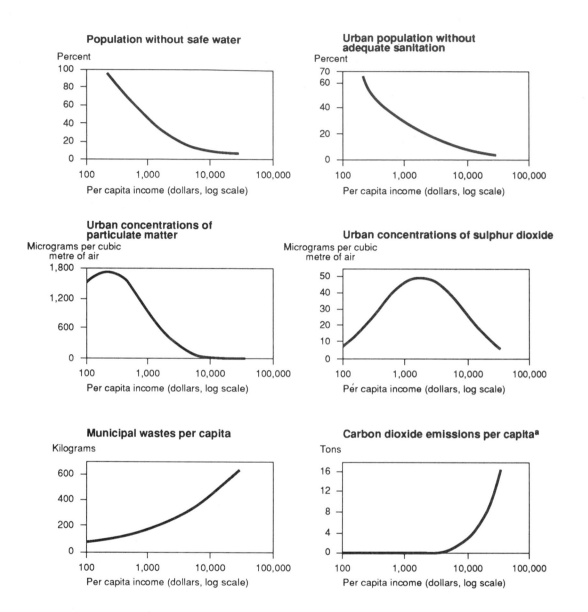

Figure 10.10 *Environmental indicators at different country income levels. Note: estimates are based on cross-country regression analysis of data from the late 1980s. a = emission levels from fossil fuels. Redrawn after the International World Bank (1992).*

energy, together with a prolonged economic recession in the West, especially in the late 1980s and up to the present day, has resulted in a subtle change of public attitudes, which now accepts that conventional fossil fuel energy resources, although finite, are projected to last longer than was originally envisaged in the 1970s. Ironically, today there is a glut of many minerals on the world market, and commodity prices have correspondingly collapsed or declined substantially since the early 1980s. On a global scale, it seems that the low demand for and low prices of many exports is part of a major world economic

recession, causing a decline in government revenues, widespread unemployment and little prospect of economic growth. The developing countries such as those in Africa and South America, have been particularly badly hit and, therefore, are finding it hard to reduce their national debts. The current world surplus in many commodity markets has caused a shift in international debate from the finite nature of many resources to an emphasis on the distribution of these resources, and the concept of sustainable development. Against this backdrop of a major economic recession, many of the Southeast Asian economies are relatively buoyant, with much better employment statistics in manufacturing and construction, together with the allied financial markets.

Many human activities waste vast amounts of energy, emit large amounts of pollutants into the atmosphere and hydrosphere as by-products from manufacturing and agriculture, and produce mountains of refuse from over-consumption. Typically, products are designed with a very limited life span and with an in-built 'throw-away' mentality that makes repair very difficult and expensive over simple replacement. Products are generally over-packaged because the manufacturers maintain that the consumer demands such presentation. It is estimated that the average person in a developed country produces 2–3 kg of refuse each day. In developing countries, this figure is less than 1 kg, much of which is recycled or utilised in other ways. Ironically, dissemination of knowledge is greatest in the developed countries concerning the consequences of human actions for the environment, and yet these countries give very small amounts of their GNP to help developing countries use available technology, capability and capital to control and manage environmental damage and pollution better (best available technology, BATNEEC). All too often, in the final analysis profit motives outweigh environmental concerns, although arguments for inaction and/or insufficient help are crafted in such a manner as to provide ostensibly valid excuses. Unfortunately, as pressure grows on developing countries to mimic the developed countries' path of progress, that is to catch up both as consumers and producers, so the consequences for the natural environment increase. Under such environmental stresses the consequences may become irreversible and catastrophic.

On a more optimistic note, however, some forms of environmental degradation are being reversed or slowing. Over the past few decades, for example, the rate of increase in human-induced extinctions of vertebrates, especially marine mammals, has declined as a result of increased pressure by organisations such as the Worldwide Fund for Nature (WWF), formerly the World Wildlife Fund. The WWF estimates that humans are still causing the extinction of approximately 1 per cent of the world's species annually; those most threatened include some species of whales and dolphins.

In terms of atmospheric pollution and environmental degradation, the release of sulphur, lead and radioactive fallout have all declined due to governmental and inter-governmental legislation, and the instigation of monitoring and control of the release of pollutants from factories. The use of lead-free petrol has been encouraged through favourable pricing and information. CFCs have been effectively banned in many industrialised countries. Acceptable levels of toxic metals and other chemicals are subject to constant review and, in many cases, downward revision.

Agro-economics

For many developed countries, the share of agriculture is frequently expressed as a proportion of GDP and is generally less than 10 per cent (Figure 10.11) – a figure that says nothing about absolute values of production. Despite modern intensive agricultural practice, with commercial crop monocultures grown in very large fields, and the intensive use of fertilisers and insecticides, one-quarter of the world's population (1.25 billion people) go hungry during at least one season of the year. At the same time, food mountains grow in the European Union, aimed at fixing food prices in wealthy countries. More than one-third of the world's population lives below the UN definition of the poverty level. Civil wars are another major global cause of environmental degradation and wasted opportunities to use land for appropriate agricultural purposes. Urbanisation also causes the increased migration of people from rural to urban areas, attracting peasants from their farms to often already overcrowded cities.

The world's population has been divided into two classes: the affluent peoples of developed countries, which dominate the Northern Hemisphere, and the poor of the developing countries, which are mainly in the Southern Hemisphere. This North–South divide is getting wider – while the North gets richer, the South grows poorer. Agro-economic policies tend to feed the nations of the Northern Hemisphere at the expense of many of those living in the Southern Hemisphere.

Data from the United Nations Environment Program (UNEP) suggest that approximately 60 per cent of the estimated 3.3 billion hectares (1 hectare = 2.47 acres) of agricultural land not in the humid regions is affected to some degree by **desertification**. The validity of these figures is open to question. The definition of desertification and the criteria used to assess it are not rigid but depend on arbitrary criteria. Nevertheless, such a large area affected by desertification as defined by UN criteria suggests that it is one of the major problems facing humans today (Crosson and Rosenberg 1989). The destruction of once fertile land through poor management can lead to accelerated erosion, waterlogging and **salinisation**

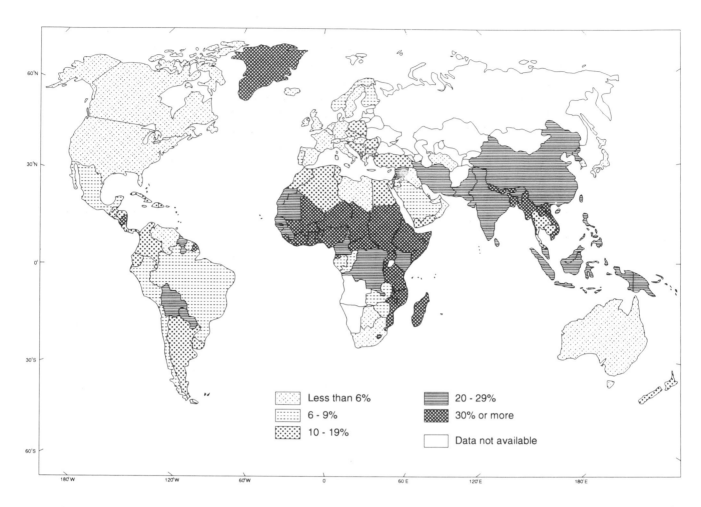

Figure 10.11 *World map showing share of agriculture in GDP, calculated by taking the value of an economy's agricultural sector and dividing it by gross domestic product. The shares say nothing about absolute values of production. For economies with high levels of subsistence farming, the share of agriculture in GDP is difficult to measure because of problems in assigning subsistence farming its appropriate value. Redrawn after the International World Bank (1992).*

of irrigated land, together with degradation of farmlands or rangelands in the arid, semi-arid and sub-humid regions. Desertification also tends to lead to a reduction in genetic diversity of both plants and animals (see Chapter 9).

Economic blocs, trade agreements and protectionism work to favour select national groups against the rest of the world. The benefits for some represent exclusion and enforced impoverishment for others. The continued growth of economic blocs with protectionist policies and associated trade barriers has created a climate of distrust between the USA and the European Union, something that has been the subject of considerable debate and diplomacy, for example in the November 1992 Uruguay Round of GATT discussions. There is also dissension within individual economic blocks. In Europe, amongst

the member states of the EU, there is a growing consensus that the EU's Common Agricultural Policy (CAP) is in need of change. In 1992, the CAP budget was 36 billion ECU (or US$46.3 billion) out of a total EU budget of 62.5 billion ECU: 6.8 per cent of the European population is engaged in farming, but they receive 55 per cent of the entire EU budget. It is because of this excessively high financial burden, amongst other subsidiary reasons, that the CAP has come in for so much criticism.

It is the responsibility of the industrialised countries to help their developing counterparts to achieve economic security. The richest 15 per cent of the world's population consumes half the global energy and uses more than a third of its fertilisers. The vast majority of people in the developing countries use little or no fertiliser, rely heavily on human and cattle power, and also use relatively little per capita energy for lighting, cooking and heating. One-quarter of the world's population consumes 80 per cent of its goods and owns three-quarters of its wealth.

On a more optimistic note, a new US-based centre intends to help developing countries to take advantage of new advances in agricultural biotechnology, with support from foundations and private companies. In April 1992, this centre, the International Service for the Acquisition of Agri-biotech Applications (ISAAA), announced that its base for the next five years is to be in the USA at Cornell University, Ithaca, in New York state. Amongst the aims of the ISAAA is the intent to help farmers to increase crop yields but, at the same time, reduce their dependence on pesticides. Figure 10.5 shows the recent percentage increase in global production of cereal crops, measured against the growing world population (also expressed as a percentage). Improved crop yields come about through the greater use of fertilisers, pesticides, the introduction of new strains of high-yield crop varieties, and better land management.

Many previous attempts to transfer technology to developing countries have failed through a lack of financial support, technical skills and infrastructure in those countries. The ISAAA hopes to remedy this by tapping the extensive knowledge held in many private companies and transferring it to developing countries where there is a politically acceptable climate and without disrupting traditional agricultural practices. The ISAAA, which has already raised millions of US dollars, will concentrate on ten target countries in Latin America, the Middle East/Africa and Asia. It has already instigated collaborative projects on plant biotechnology in Mexico, Taiwan and Costa Rica, with many other negotiations under way.

It is ironic that some countries produce enough food to feed their impoverished populations, yet lack the infrastructure and/or political means to distribute it properly. India, for example, currently produces enough food to feed its 910 million people and is now one of the top world food producers, yet 300 million Indians are still malnourished (Thompson 1995). India has amassed a two-year stockpile of grain, but as much as 10 per cent of this is wasted, spoilt, spilt or stolen: the Indian government is beginning to sell this grain abroad, but its export is hindered by inadequate port handling facilities and a lack of a well-organised infrastructure in the countryside to transfer the grain to suitable markets both at home and abroad. Conflicting government policies encourage over-production while discouraging distribution to the poor. The over-production,

however, is beginning to exhaust soil and water resources in some areas. In an attempt to rectify this situation, India is beginning to implement schemes such as Food-for-Work, Fair Price Shops and free midday meals for schoolchildren.

In developed countries, industrialisation and mechanisation are causing many modern agricultural systems to become more energy intensive. Subsidies are frequently paid to farmers in order to stop them growing certain crops. The EU provides grants and subsidies to farmers, with the consequence that overproduction of some commodity crops is encouraged, with the exploitation of marginally productive agricultural land. Such policies place additional pressure on the land, encourage the excessive use of fertilisers and pesticides and, together with unnecessary irrigation, provide a greater potential for pollution and other environmental damage. Clearly, the CAP is not the most sensible use of finance and other resources.

The Organisation for Economic Co-operation and Development (OECD) subsidises farmers in Western countries with US$300 billion each year – far in excess of the money available for soil and water improvement. The World Commission on Environment and Development (WCED) suggested that considerably more money should be used to improve the quality of the soil, water and vegetation on farmland, which would be more beneficial to farmers in the long term, instead of providing subsidies for food production. Subsidies could then offset the cost of remedial measures required to reverse any effects of land degradation, or go towards developing new land where appropriate. Unfortunately, many governments and organisations in the industrialised countries act in concert against the developing countries over agro-economic measures, which tends to impede self-sufficiency within the developing countries.

Governments should be encouraged to increase the productivity and efficiency of their farming without deleterious effects on the environment. Encouraging multiple crops and inter-cropping of nitrogen-fixing plants are measures that help to reduce the need for fertilisers and so decrease demands on irrigation. Improved irrigation techniques and a reduction in the use of pesticides, with substitutes of biologically engineered pest-resistant strains of crops, can all contribute to the prudent management of agricultural land.

Global food surpluses are a facet of the international commodity markets and provide a stark contrast to the dire poverty, malnutrition and starvation in many poor countries. These global inequalities need to be addressed much more effectively than they are currently.

Climate change and world food supply

GCMs have been used to assess the potential impact of climate change on world and regional food supply. For example, assuming a doubling of atmospheric CO_2 and a consequent temperature increase of 4.0–5.2°C, towards the upper range of the IPCC (1992) projected warming, Rosenzweig and Parry (1994) used the results from three GCMs (Table 10.3, Figures 10.12 and 10.13) to suggest that such changed global conditions would lead to only a small decrease in global crop production. Under these conditions, they also predict that the developing countries are likely to suffer most (*ibid.*).

Poverty

International Monetary Fund (IMF) figures show that for the fifteen most indebted countries, 1988 actually witnessed a reverse flow of funds from the poorer to the richer countries, from South to North of US$24.5 billion, bringing the net outflow since 1982 to US$164 billion. At the same time, between 1981 and 1987, developing country debt levels rose from US$748 billion to US$1,195 billion (figures quoted from IMF World Economic Outlook, October 1987 and April 1988, given in Huhne 1989, and Ekins 1992).

The WCED has emphasised that poverty is one of the major causes of the accelerated depletion of the Earth's resources, and the degradation of its forests, soils, species, fisheries, water and atmosphere. It estimates that an annual average national growth of 3.2–4.7 per cent is necessary to keep pace with the growing global population. Given the disparities between population growth rates throughout the world, an average growth in national income of 5 per cent would be needed in the developing countries of Asia, 5.5 per cent in Latin America and 6 per cent in Africa. These countries experienced growth of this magnitude during the 1960s and 1970s, but during the 1980s rates dropped to well below this level. The decline in GNP is blamed on population growth, deteriorating trade (often the result of protectionist policies in the industrialised countries), reduced resources and onerous long-term national debts.

Long-term national debt and its non-repayment is probably the biggest destabilising factor in reducing the growth in national income. The accumulated debt of developing countries is approximately US$1 trillion, which attracts an annual interest of US$60 billion. In 1984, the flow of money from industrial to developing countries was reversed, with more than US$43 billion exchanging hands each year. In 1988, the seventeen most debt-ridden countries paid out US$31.1 billion more than they received in aid. Without such large debts, many developing countries would find it much easier to eradicate their poverty.

The WCED believes that increased global living standards would bring about a significant reduction in population growth, something that has been demonstrably true for industrialised countries during this century. For the immediate future, the Commission proposes improved education for all and the greater empowerment of women, particularly in order to help improve the dissemination of knowledge concerning birth control and family planning, which of themselves should contribute to a reduction in population growth and therefore poverty.

The prevalence of infectious diseases is most likely to be associated with poverty. Infectious diseases not only cause immediate personal discomfort, illness and possibly death, but they affect a community and country, for example leading to decreased productivity and lower economic growth. The prevention of public health problems remains of paramount concern to the welfare of developing countries. *Comprehensive* primary health care is a broad strategy that tries to prevent public health problems. The Alma Ata Conference in 1978 promoted this health-care programme, focusing on preventive, curative and rehabilitative services. The strategy includes providing adequate clean water and food supplies; safe sanitation; immunisation against major diseases; maternal, child care and family planning advice; treatment for minor injuries and common ailments. It promotes community participation in deciding on and supporting preventive medicine programmes through health plans, training of primary health-care workers and parental education in nutrition and preventive medicine. *Selective* primary health care has been promoted by UNICEF since 1983, for example

Table 10.3 *GCM doubled CO_2 climate change scenarios.*

GCM	Year	Resolution (lat. × long.)	CO_2 (ppmbv)	Change in average global temperature/precipitation (°C)	(%)
GISS	1982	7.83° × 10°	630	4.2	11
GFDL	1988	4.4° × 7.5°	600	4.0	8
UKMO	1986	5.0° × 7.5°	640	5.2	15

GISS = Goddard Institute for Space Studies
GFDL = Geophysical Fluid Dynamics Laboratory
UKMO = United Kingdom Meteorological Office
Source: Rosenzweig and Parry (1994).

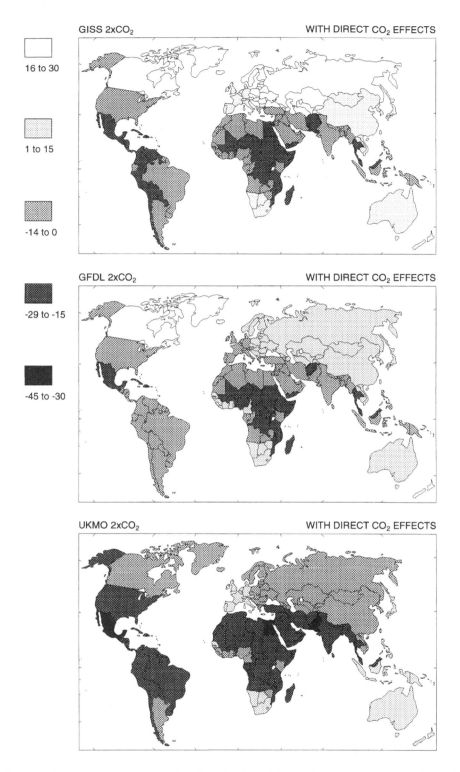

Figure 10.12 *Estimated change in average national grain yield (wheat, rice, coarse grains and protein feeds) for GISS, GFDL and UKMO climate change scenarios, showing the direct physiological effect on grain yield of the current (330 ppmbv) CO_2 concentration. Results shown are averages for countries and groups of countries in the basic linked system (BLS) world food trade model; regional variations within countries are not reflected. After Rosenzweig and Parry (1994).*

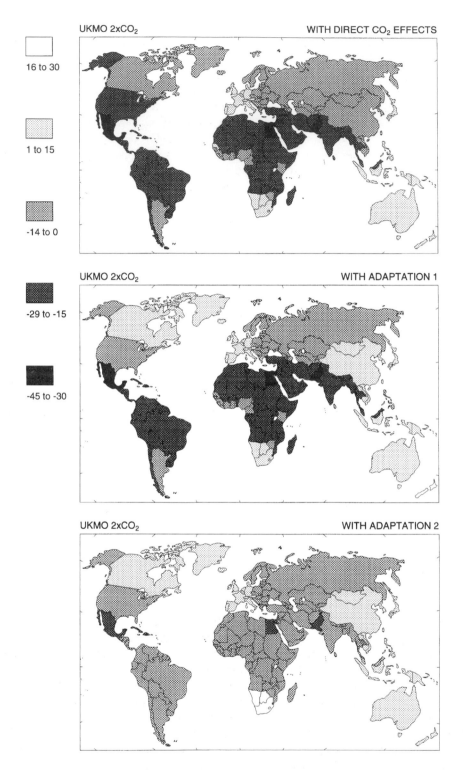

Figure 10.13 *Estimated change in average national grain yield (wheat, rice, coarse grains and protein feeds with direct 555 ppmbv CO$_2$ effects) under two levels of adaptation for the GISS, GFDL and UKMO doubled CO$_2$ climate change scenarios. Adaptation level 1 signifies minor changes to existing agricultural systems; adaptation level 2 signifies major changes. Results shown are averages for countries and groups of countries in the basic linked system (BLS) world food trade model; regional variations within countries are not reflected. After Rosenzweig and Parry (1994).*

in focusing on the monitoring of child growth, the use of oral rehydration salts for diarrhoea, the encouragement of breast feeding for infants, and immunisation. The costs of such health-care strategies are surprisingly small in relation to the possible benefits to both the individual and the wider community.

Individuals in the developed countries should feel an acute sense of shame because of the present levels of poverty in many developing countries throughout the world. Poverty is degrading to the victims and a measure of the selfishness and ignorance of those who are more fortunate. It reflects human ignorance because a truly sympathetic attitude towards poverty by others would inevitably lead to more concerted efforts to eliminate poverty. The notion of global sustainable development is considered later in this chapter, something that does not necessarily embody the elimination of poverty and famine but that often is linked to such issues.

Refugees

One of the more avoidable global problems is the forced migration of large numbers of people from their homelands because of political conflicts, war or environmental degradation. Figure 10.14 shows the world refugee situation in 1992. Over one million Rwandans must be added to this: in 1994, they fled civil war and the genocide that resulted in more than one million deaths. The concentrations of refugees from such migrations is often considerable. At Benaco, northwest Tanzania, a camp of 300,000 refugees from Burundi absorbed 410,000 Rwandans. The large displaced populations have to be fed and given shelter. In refugee camps, overcrowding and poor sanitation contribute to the spread of infectious diseases.

Together with the personal stress and problems caused by being a refugee, large numbers of refugees can cause considerable environmental damage. In 1994, for example, 850,000 Rwandan refugees and Hutu soldiers camped around the town of Goma in eastern Zaire, and partly or completely deforested about 300 km^2 of the Virunga National Park in the search for food and firewood (Pearce 1994a). During the fourteen years of Soviet occupation of Afghanistan, more than 100,000 acres of forest were cleared and 3.2 million refugees fled to Pakistan (Weinbaun 1994).

The numbers of refugees in Africa have increased considerably over the last few decades, mainly as a consequence of wars: in 1969 there were about one million, whereas today there are well over five million, mostly women, children, disabled and the elderly (Bakwesegha 1994). In Africa, refugees frequently seek sanctuary with the Organisation of African Unity (OAU), or the United Nations High Commission for Refugees (UNHCR). Convoys carrying relief supplies are often either attacked or the relief supplies disappear to supply the numerous black markets before they can reach their proper destinations.

Refugees often seek sanctuary in countries where resources are already inadequate to maintain the indigenous population. One country's problems are merely exported to the next. Even when the problems have been resolved in their native countries, it is often difficult or hazardous for the refugees to return home, especially when they represent a victimised minority tribe. Vast areas of many war-stricken regions still have unexploded bombs and land mines left behind from earlier conflicts. Afghanistan still had between nine and ten million land mines left behind after the Soviets withdrew in 1988 (Ruel 1993). The clearance of land mines is a major undertaking and one of the more hazardous problems facing countries like Afghanistan, Angola, Iraq, Cambodia, Mozambique and Somalia.

Mass migration can also occur within a single country, in which case the uprooted population is described as 'displaced'. These people endure all the same problems that refugees experience, and often additional ones such as having to cope with the trauma of civil war. The predicament in Croatia and Bosnia since the early 1990s illustrates these severe problems. The ethnic cleansing (genocide) of Muslim populations by Bosnian Serbs forced millions of people to flee from their homes in the often fruitless search for 'safe havens', which were eventually established by the UN. Many lived in siege situations, for example the Muslims in Sarajevo, who were under siege from the Bosnian Serbs. Frequent mortar attacks killed civilians in the markets and on the streets of Sarajevo, and UN aid was unable to reach the city for fear of being attacked by the Serbs. UN negotiations have done little to reduce the conflict and lessen the problems. On 30 August 1995, a week after a Sarajevo market place was shelled, killing thirty-seven and leaving eighty people wounded, NATO air strikes destroyed Serbian command posts and ammunition depots. This was the first extensive retaliation by Western peacekeeping troops against an aggressor. The problems in Bosnia are a long way from being resolved.

The problems of refugees and displaced people are not often considered a global environmental issue.

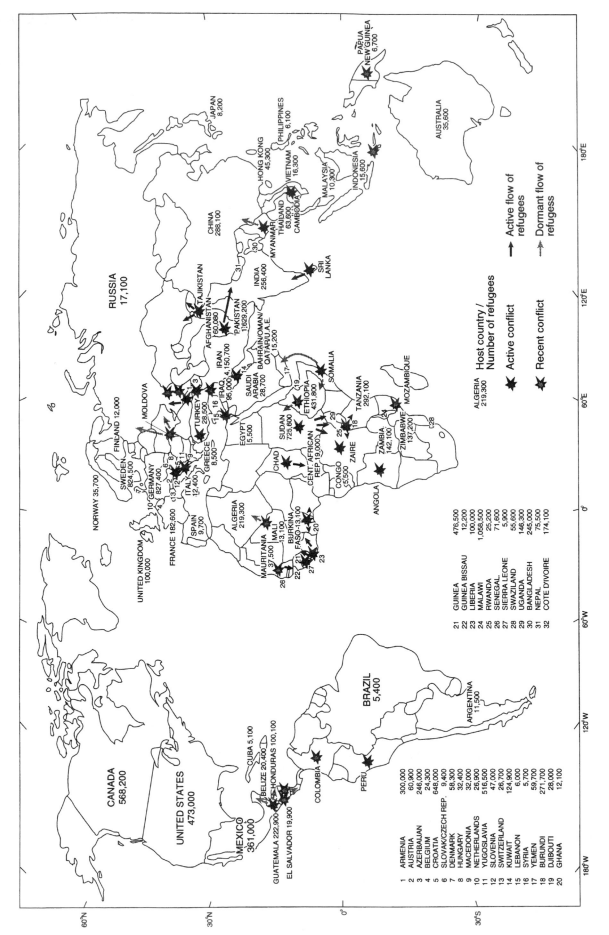

Figure 10.14 The world international refugee situation on 31 December 1992. Redrawn after UN High Commission for Refugees (1993).

Wherever large numbers of people are forced to move from their homes, there is likely to be environmental damage to both the land that these people are moving from and the land that they are being settled on. Unlike many global environmental issues, the problems of refugees and displaced people have the potential for rapid solutions, and the environmental damage is generally easily reversed in the right political climate.

Energy conservation

With increased industrialisation and higher living standards, the demand for energy will increase. This demand can only be met through increased energy production and savings through improved energy efficiency. The option favoured by many environmentalists, particularly pressure groups such as Greenpeace, is *energy conservation*. In many cases, energy efficiency and energy conservation are used as interchangeable terms, although strictly this is incorrect: efficiency involves maximising the energy output for a given input, whereas conservation may simply mean using only the least necessary amount of energy (even with inefficient sources of power and/or machinery). Increased energy production will have drastic environmental effects, probably leading to enhanced global warming as more CO_2 and other greenhouse gases are added to the atmosphere. The WCED favours energy conservation measures such as the recycling of aluminium, steel, paper and glass, and advanced technologies to reduce the consumption of energy.

The disparity between the energy used per capita in the developed and developing world is a source of some concern. The developing countries tend to have relatively modest energy demand and consumption, usually only sufficient for subsistence, whereas in developed countries energy consumption often appears profligate, with a plethora of domestic appliances such as dish-washers and air-conditioning units.

Poor countries need cheap and reliable energy. In rural areas, electrification improves the lifestyle of those in poverty. Additionally, it can reduce environmental stress, because forests suffer less rapid degradation as a result of the gathering of fuelwood, and cattle dung, which is often burned as a fuel, can be used as a fertiliser for crops. A key consideration in promoting a change in patterns of energy consumption in the developing countries is to encourage energy efficiency. Box 10.1 summarises the recommendations of the Stockholm Initiative on Energy, Environment and Sustainable Development (SEED), November 1991.

Energy-saving technologies should be encouraged. Many of the newer industrialised countries, such as Taiwan, South Korea and Brazil, already have implemented some of the new technologies, which provide them with large cost savings. Governments should also be encouraged to promote conservation policies. The USA, for example, reduced its domestic production by 23 per cent between 1973 and 1985 due to increased efficiency, and thereby demonstrated the cost-effectiveness of conservation measures, and the ease with which such a policy could be introduced. It has been estimated that in the UK energy use could be reduced by 20 per cent by using greater energy-efficiency measures.

Not only should there be government encouragement for industrial/commercial energy conservation, but domestic energy savings must be encouraged. More energy-efficient buildings should be designed and constructed. Sensible town planning can reduce the distance people have to travel between home and the workplace. Private vehicle use should be discouraged by offering acceptable alternative mass transit systems, not simply through punitive measures in inner cities and town centres. This issue is controversial because private cars account for a considerable amount of energy consumption and pollution, but many vehicle users maintain that public transport costs are too high, mass transit systems too crowded, and the routes and timetables not conducive to the switch away from private to public transport. The pedestrianisation of inner city areas needs to be undertaken sensitively, because the complete exclusion of private vehicles from shopping areas in towns and cities may discourage people from using the facilities, and encourage the development of large, out-of-town shopping complexes. Such shopping complexes tend to destroy the heart and character of towns, reduce commercial competition and employment, and be difficult to reach, particularly by the elderly and disabled.

A study by the Stockholm Environment Institute (SEI), commissioned by Greenpeace and published in February 1993, using conventional assumptions, concluded that it is technically and economically feasible to reduce current global oil use by 50 per cent within forty years, and the use of oil and other fossil fuels could be phased out entirely over the twenty-first century. In more detail, the study developed a Fossil-Free Energy Scenario (FFES) and concluded that global oil consumption could fall from 120 exajoules today to 59 exajoules in 2030; related global CO_2 emissions from oil would fall by

BOX 10.1 RECOMMENDATIONS OF THE STOCKHOLM INITIATIVE ON ENERGY, ENVIRONMENT AND SUSTAINABLE DEVELOPMENT (SEED), NOVEMBER 1991

Developing countries

1 Developing country governments should develop and implement programmes for improving power sector efficiency, both in supply and demand. These programmes should focus on greatly improved performance compatible with an integrated energy strategy and environmental sustainability.
2 Developing country governments should support efficient alternatives to capacity expansion for utilities through better utilisation of existing capabilities, and the development of independent private power facilities. Tariff reforms that make the sector creditworthy should be an integral part of such measures. However, the political, economic and social conditions in individual countries underscore the need for a country-specific approach in addressing these issues.
3 Developing country governments should strengthen financial mechanisms, institutions, and associated policies and regulations to provide innovative lending in supply- and demand-side power sector efficiency, including direct lending for private sector initiatives. Sector financing entities, including development financing institutions with portfolios in industrial modernisation, agriculture, the environment and housing, are targets for such institutional reforms.

Bilateral and multilateral institutions

1 Bilateral and multilateral institutions should dramatically alter their investment priorities to support end-use efficiency, sustainable and reliable operations and maintenance programmes, and private sector initiatives, in addition to traditional investments in supply.
2 Bilateral and multilateral institutions should provide financial and technical support to improve the legal and regulatory framework as well as the management and institutional performance of power utilities.
3 Bilateral and multilateral institutions should expand their financing to cover joint ventures in environmentally sound electric power-related technology co-operation.
4 Bilateral and multilateral institutions should provide insurance for private sector power projects to enable capital mobilisation from commercial and other markets.
5 Bilateral and multilateral institutions should commission a study investigating the lack of progress of private sector involvement in developing country power sectors.
6 Bilateral and multilateral institutions should create a fund in specific countries to support the availability and delivery of critical spare parts to ensure high system availability.

Institutional linkages

1 Bilateral and multilateral institutions should, together with developing countries, perform long-term power and environmental sector appraisals to formulate policy reform packages and investment priorities for public and private entities.
2 Existing bilateral and multilateral networks in energy and environment should be strengthened and expanded to link with developing country financing institutions and recognised centres of excellence.
3 The SEED recommendations should be widely disseminated to relevant agencies, developing country governments, and the private sector; they should also be presented to the refocused World Bank/UNDP Energy Sector Management Assistance Program (ESMAP) and its consultative group of donors and developing country representatives.

---

50 per cent by 2030, and 75 per cent by 2075. The SEI recommendations included:

1 the introduction of tough new fuel efficiency standards for all vehicles;
2 government support for public transport, and the discouragement of private vehicle use within urban areas;
3 a doubling of energy research and development budgets within ten years, the bulk of which should be used for energy efficiency and renewable energy;
4 the removal of massive subsidies on oil, such as oil exploration tax breaks;
5 the introduction of pollution taxes for oil and other fossil fuels to reflect the true costs of major oil spills and pollution damage, the effect of which would be to more than double current oil prices;
6 the establishment of a UN agency for technologies for renewable and energy efficiency (TREEs) to promote the development of these technologies through training, financial support and information;
7 multilateral development bank (MDB) lending for energy projects to be reoriented towards energy and smaller-scale renewable projects: less than 1 per cent of the World Bank power sector lending is currently used for energy-efficiency projects.

The introduction of alternative and renewable energy sources, instead of conventional fossil fuels and fuelwood, would help to reduce pollution and the amounts of CO_2, SO_x, NO_x and other harmful gases being released into the atmosphere. Such technologies include mini-hydroturbines, solar power, wind and tide power, and bio-gas. Environmentally friendly policies, alongside shifts in accepted practices or conventions, could be eased into place and stimulated with government subsidies, the cost of which can be calculated against that of having to take remedial action to restore the environment. The WCED report also suggested that many of the smaller technological schemes actually may be more appropriate to an area than conventional, harmful, practices. For example, mini-hydroelectric schemes supplying energy in a mountainous area may be more appropriate in terms of management, maintenance and cost than a large nuclear or oil-fired power station in the same region.

One of the most important considerations for sustainable development is the merger of environmental and economic decision-making bodies. The differing goals of many environmental and economic agencies commonly appear too great for the easy resolution of conflicting interests. The WCED believes that both should work together in a more synergistic manner so that potential mutual benefits become manifest. The environmental agencies have been hindered by mandates that are too narrow, small budgets, and little or no political muscle. In contradistinction, most established economic agencies have become very powerful over the years – something that is very rarely true of environmental agencies, which are often cast in the role of watchdog with power to criticise but no legal instruments to enforce environmental legislation. Penalties for damaging the environment are commonly wholly inadequate and/or derisory. The environmental agencies must have more influence and power in order to help police and enforce compliance in good environmental practice.

Working together, marketplace incentives could be developed that are both economically viable and beneficial to the environment. The introduction of energy-carbon taxes would encourage more responsible energy consumption. As well as encouraging more responsible energy consumption, there is a need to implement both 'carrots' and 'sticks' to encourage and enforce a greater respect for the environment.

Energy conservation must run hand-in-hand with the polluter pays principle (see next section). Incentives should be offered to companies that develop and manufacture energy-efficient, environmentally friendly, industrial and domestic goods and devices, for example through tax rebates and/or reduced taxation.

The polluter pays principle

Pollution could be considerably reduced through a taxation system offering incentives to organisations with a good track record of being environmentally responsible. Ideally, such measures ultimately lead to optimum, and possibly sustainable, development. The introduction of the 'Polluter Pays Principle' (PPP) is a good example of the type of legislation that is necessary. This was introduced by member countries of the OECD in 1972 and it endeavours to make industries more responsible for conserving or protecting the environment, with costs offset by the consumers. Unfortunately, many governments are still too slow to apply this principle, mainly because it is seen as too onerous to implement as it can be very difficult to make an absolute assessment of the environmental degradation produced by a specific pollution incident, let alone the long-term pollution. Atmospheric pollution, for example, may have global as well as regional effects. Who pays, and in what proportion?

The issues surrounding market-based economic instruments (MBIs) remain controversial, e.g. a pollution charge or tax, versus direct legislation to control pollution. The main advantage of market-based economic instruments is that they can help industries to allocate resources efficiently through the most economic methods, but those polluters with the largest abatement costs may decide to pay a 'pollution tax' as the cheapest option, i.e. they may see an MBI system as a 'licence to pollute'. If pollution remained at environmentally unacceptable levels under any MBI scheme, then the pollution taxes (which have to be empirical in any event) would require readjustment in order to achieve prescribed environmental standards. Ideally, any pollution tax should be set at levels at least equal to the environmental damage done, something that is no easy matter, simply because of the subjective nature of placing an actual value on various parts of the natural environment and any environmental damage.

An alternative approach to paying for cleaning up the environment and/or preventing pollution through a direct PPP is for governments to levy some form of pollution tax, which is assessed by an independent panel of scientific/technical experts and lawyers. The revenue that is raised from such a scheme could then be devoted solely to environmental pollution control and clean-up. The tax could be a blanket tax on industry with concessions for companies and organisations that meet certain (specified) targets, and additional penalty taxes could be levied against polluters for pollution incidents. Any pollution tax would have to be effective whilst not proving so punitive that many industries could not survive.

In the USA, by the mid-1970s, the issue of waste and waste disposal was regarded as one of the major socio-economic problems that needed to be confronted. As a result of political pressure, the Comprehensive Environmental Response, Compensation and Liability Act (CERCLA) 1980 was enacted, otherwise known as 'Superfund'. This Act provides for the creation of a trust fund, mainly financed from special industrial taxes, from which the money raised can be used for clean-up operations in both derelict and uncontrolled sites. If it is possible to identify the owner or operator of a site at the time of waste disposal, even extending to the waste carriers, then the federal authorities look to these parties to pay for the clean-up. In some instances, the banks have been liable as 'deemed polluters'. Ideal as the US example of the PPP may seem, there are many who regard it as a failure because the parties involved in environmental litigation have sought reimbursement from their insurers, which in turn has led to very high litigation costs between both insurers and reinsurers, to the extent that only about 12 cents in every dollar spent is actually used in the physical clean-up. In the USA, the current average cost of a Superfund clean-up is estimated as US$30 million (US RAND's Institute for Civil Justice survey). Other countries, including the EU, are looking critically at the issues surrounding the US Superfund, prior to introducing their own legislation (see Chapter 9 and POST 1992b).

Atmospheric pollution control

Worldwide, urban air pollution has decreased overall from the 1970s and early 1980s to the mid- and late 1980s in cities in the high- and middle-income countries, in contrast to the converse pattern for cities in low-income countries (Figure 10.15). Figure 10.16 shows air pollution in developing countries for selected pollutants. Atmospheric pollution represents one of the current major global problems. Even though several international conventions have taken place in recent years on global atmospheric change and ways of improving air quality, progress on these issues has been slow. In 1984, for example, nineteen countries signed an agreement to reduce SO_2 emissions by 1993, but the four biggest culprits, the USA, the UK, Poland and Spain, did not sign the agreement.

Today, the global implications of pollution are beginning to be appreciated by both scientists and the wider community, along with a recognition that countries should be responsible for more than just their own back yards, because of the trans-boundary effects of pollution. In 1988, the first world conference on the atmosphere was sponsored by the World Meteorological Organisation and UNEP, with thirty-seven countries participating in the theme of 'The Changing Atmosphere'. They called upon governments to develop global and national plans for the protection of the environment and to initiate development of an International Convention for the Protection of the Atmosphere. They also requested the introduction of taxes on fossil fuel consumption to provide money for the newly established 'World Atmosphere Fund', which could provide money for developing countries to offset the consequences of any future global warming and associated sea level rise. The 1988 conference delegates requested a reduction of CO_2 emissions and other atmospheric greenhouse gases, and adopted as one of their goals the role of promoting greater public awareness, research and the use of technologies to reduce atmospheric pollution. At the 1988 meeting, there was a push to strengthen the Montreal Protocol, which calls for a reduction in CFCs by 50 per cent by the year 2000, and to include a complete ban on CFCs after that date. The latter recommendation was pursued and addressed at an international forum in Helsinki in May 1989. At the Helsinki meeting, it was agreed by the eighty-six countries that attended the forum that CFCs should be phased out completely by the year 2000. Norway pledged 0.1 per cent of its GNP to help developing countries follow suit and offset the likely expenditure incurred while converting to the use of alternative chemicals.

The Conference of Parties (COP), which had ratified the United Nations Climate Change Convention at Rio de Janeiro in 1992, met in Berlin from 29 March to 7 April 1995. The outcome of this conference was the Berlin Mandate, which was approved, whereby the signatories agreed to return the anthropogenic emissions of greenhouse gases to 1990 levels by the year 2000. These levels of reduction, however, remain far from the recommended response strategies suggested by the IPCC and are far lower than the 20 per cent reduction requested by the Alliance of Small Island States (AOSIS). The Berlin Mandate stated that the present commitments under the Rio Climate Convention are inadequate and agreed to the establishment of a procedure that will set voluntary targets for reduction after the year 2000. A working group was established to design the protocol for the reduction of greenhouse gas emission, to be approved in 1997. The IPCC will

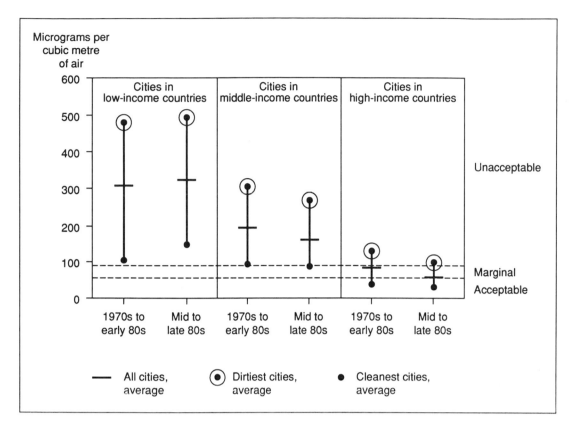

Figure 10.15 *Urban air pollution levels and trends: concentrations of suspended particulate matter across country income groups. Data are for twenty urban sites in low-income countries, fifteen urban sites in middle-income countries, and thirty-two urban sites in high-income countries. 'Cleanest cities' and 'dirtiest cities' are the first and last quartiles of sites when ranked by air quality. Periods of time series differ somewhat by site. World Health Organisation guidelines for air quality are used as the criteria for acceptability. Redrawn after the International World Bank (1992).*

act as the advisory body, and will function to look at the scientific aspects of climate change, the likely impact of such changes, and the economic implications, including an outline of future scenarios of energy use linked to social patterns. Two subsidiary bodies were also set up, the Subsidiary Body for Scientific and Technical Advice (SBTA), to convey scientific information to the COP, and the Subsidiary Body for Implementation (SBI), which will advise the COP on the national implementation of the convention (Abbott 1995a,b).

Traffic congestion in urban areas, particularly city centres, is a major contemporary problem. Figure 10.17 graphically shows the growth of traffic in Britain from 1952 to 1993 and predicted future trends. Traffic in city centres can grind to a complete standstill, termed 'gridlock'. In one of the most publicised incidents of gridlock, on 29 October 1986, a traffic accident on the San Diego freeway in Los Angeles caused an eight-hour traffic jam that involved tens of thousands of motorists. Road pricing, together with improved public transport, is needed to tackle the problem. Road pricing, which began in Singapore in 1975, is simply the imposition of a specific government tax on vehicle use and parking within urban areas. People wishing to use private vehicles must purchase a licence on a pay-and-display basis, and calculated at a daily rate. Whatever the technology and future developments in producing cleaner vehicles, people should be encouraged to use vehicles for private use less often in urban areas. Such a change in attitudes can come about only through the introduction of cheap and improved public mass transit systems in tandem with new technologies for cleaner vehicles.

Unlike New York, San Francisco, Tokyo, Paris or London, Los Angeles never had an underground railway, and the existing public transport system is inadequate. Not surprisingly, therefore, motor vehicles provide the principal means of transport, with

(A)   Index (1990=100)

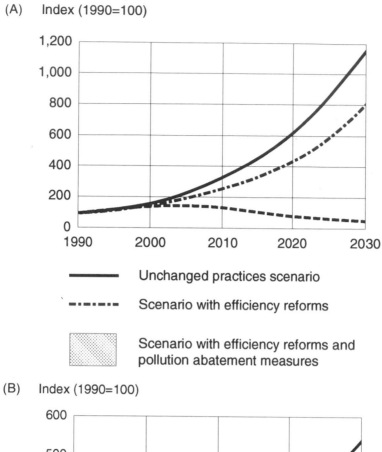

————— Unchanged practices scenario

▬ ▪ ▬ ▪ ▬ Scenario with efficiency reforms

Scenario with efficiency reforms and
pollution abatement measures

(B)   Index (1990=100)

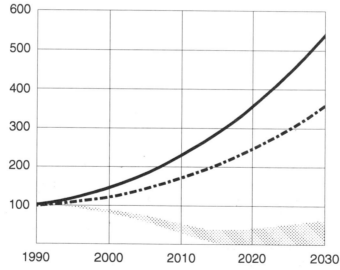

Figure 10.16 *Selected air pollutants in developing countries for three scenarios, 1990–2030. (A) Emissions of particulates from electric power generation; (B) Lead emissions from motor vehicles in urban areas in developing countries. The calculations are based on the following data and assumptions. Growth rates for per capita income and population are as given by the International World Bank (1992). Per capita income elasticity of demand for vehicle fuels equals 1.2, and fuel price and congestion price elasticities equal –0.5 and –0.6, respectively. The average life of vehicles is 15 years. Gasoline and diesel fuels each account for about half the total consumption. Efficiency reforms include congestion charges (based on data from the Singapore Area Licensing Scheme) and higher fuel taxes (assumed to rise over a 25-year period to levels now found in Europe). Pollution abatement measures include emission controls and the gradual introduction of cleaner fuels over a 25-year period. Under this scenario, lead emissions gradually drop to the bottom of the shaded band; emission levels of particulate matter, hydrocarbons, and sulphur oxides fall within the band, and nitrogen oxides are at the top. Redrawn after the International World Bank (1992).*

all of the associated atmospheric pollution and the infamous smogs. The 8.8 million people who live in Los Angeles County own (at 1993 figures) about six million motor vehicles: by 2010, the population of Los Angeles is predicted to exceed 10.2 million. In order to combat the problems of motor vehicle congestion and pollution, Los Angeles has embarked upon a programme to construct a metro system linked to a more integrated surface public transport system. The first part of this thirty-year programme came into effect on 30 January 1993, when a 7 km (Red Line) underground route with five stations was opened to connect downtown Los Angeles with Hollywood. Completion of this 46 km Red Line is projected for the year 2000. The Los Angeles metro scheme is an example of the ways in which large urban centres are attempting to combat traffic congestion and the associated poor air quality, but such projects can become viable only if people are prepared to accept a change in lifestyle. In California, tough new legislation has been introduced to reduce motor vehicle pollution (see Box 10.2).

Atmospheric pollution caused by road transport, through both individual and corporate activities, is now recognised as a major problem in many urban centres. Table 10.4 shows the contribution of road transport to air pollution in selected cities, and Table 10.5 shows the impact of motor vehicles on the environment. Figure 10.18 illustrates the types of pollutant emitted by petrol-powered vehicles.

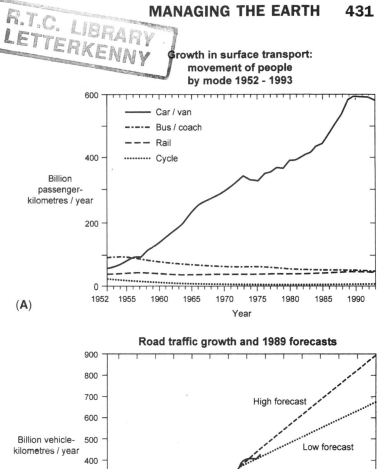

Figure 10.17 (A) Growth in surface transport: movement of people by mode 1952–1993 in Great Britain; and (B) Road traffic growth and 1989 forecasts for Great Britain. Redrawn after Royal Commission on Environmental Pollution (1994).

Sustainable development

The idea of **sustainable development** was first presented by the International Union for the Conservation of Nature (IUCN 1980) in an international forum of the World Conservation Strategy. In essence, the concept of sustainable development invokes present development of available resources without compromising the ability of future generations to meet their needs. Many people would argue that this is an abstract ideal that is impossible to achieve. How can this generation understand the needs of future generations, even before they have been born, let alone future needs before they have been formulated?

The practical application of the concept of sustainable development should involve a greater environmental awareness, by both governments and individuals. The IUCN argued that three priorities should be incorporated into all development programmes:

● the maintenance of ecological processes;
● the sustainable use of resources; and
● the maintenance of genetic diversity.

Emphasis was placed on conserving the present natural environment: because of this aspect of the report, it has been heavily criticised by some as being anti-development. Critics also argued that the report concentrated too much on attacking the symptoms of environmental degradation rather than analysing

BOX 10.2 REDUCING MOTOR VEHICLE POLLUTION

Many of the world's major cities are suffering the ever-increasing problems of serious traffic congestion and its attendant atmospheric pollution. For the motor car, California's emission laws are amongst the toughest in the world. They ensure that cars in the state emit about one-tenth of the main pollutants – carbon monoxide, nitrogen oxides and hydrocarbons – compared with the 1960s. It is these stringent standards that have led to catalytic converters and electronic fuel injection systems being commonplace in California. Necessity has proved to be the mother of invention. Catalytic converters remove pollutants from the exhaust gases before they leave the car, and electronic fuel injection systems make combustion more efficient and thus reduce the amount of pollutants. Despite these changes, California's Air Resources Board, which monitors atmospheric pollution, found that on more than 200 days in 1990 Los Angeles failed to meet the air quality guidelines. Although cars are now manufactured with cleaner emissions standards, the growth in population and increased car ownership have actually led to a deterioration in air quality in the state. The current emphasis is on encour-aging the use of cleaner petrol, known as 'reformulated gasoline' because it evaporates more slowly. In California, from November 1992, petrol companies were forced to add oxidising agents to petrol in order to reduce carbon monoxide emissions by converting it to CO_2 before emission. In 1993, this was followed by tougher controls on diesel fuel, and in 1996, vehicles were encouraged to use 'phase 2 reformulated gasoline', which will produce less benzene in the emissions.

In California, four new categories of low emission car are being demanded of the motor industry. By 1994, transitional low-emission vehicles (TLEVs), which emit half the reactive organic gases of a conventional car, were on the market, and by 1997 low-emission vehicles (LEVs), emitting one-quarter, and ultra-low-emission vehicles (ULEVs), producing one-eighth of the reactive organic gases, must also be available. In 1998, zero-emission vehicles (ZEVs) must be available, a category currently met only by electric vehicles. Electric cars, whilst much cleaner than conventional cars, are more limited in range and slower, and recharging a battery is much slower than filling up at a petrol station. In order to tackle these potentially less appealing features associated with electric cars, companies such as Chloride in the UK are developing high energy-density batteries, for example a sodium–sulphur battery.

the causes, and it was also criticised as having an anti-poor bias. The allegations of an anti-poor bias arose because in the report poverty was cited as one of the major inhibiting factors in achieving sustainable development. These criticisms led to the reformation of the IUCN in 1984 and the creation of the World Commission on Environment and Development (WCED), which became known as the Brundtland Commission. The reports of the Brundtland Commission aimed at maximising growth without jeopardising people and future resources. Emphasis was placed on economic quality being just as important as quantity.

The aims of the WCED, as stated at its inception in 1983 as an independent commission by the United Nations, are to formulate recommendations for sustainable global development (Plate 10.5). The WCED included twenty-three commissioners from twenty-two countries, chaired by Gro Harlem Brundtland of Norway. The WCED global inquiry into the state of the world involved a large amount of data analysis, the commissioning of reports by specialists, debates with world leaders and panels of experts, and public inquiries. In October 1987, the commission published its report, entitled *Our Common Future*, which emphasised that the basic needs of all people, whatever their race or creed, must be met to secure our common survival; that the poor should be given economic priority, not just through altruistic reasoning, but because they have the potential to help world economic growth by providing additional markets which can ultimately improve the world economy.

The WCED report presented strategies for sustainable development, but it identified social, institutional and political factors as the major obstacles that would hinder sustained growth. The report emphasised that an integration of both economic and ecological systems is paramount to the success of sustainable development. The commission further suggested that the ministries of finance and of the environment should not be separated in governments, but combined to share the responsibilities for development programmes. The report argued that the exploitation and depletion of natural resources should be at a rate that is not greater than the rate at which they can be replenished naturally, thereby permitting the environment to recover so that future generations will have raw materials. It was stressed that industry must play its part in replacing resources and restoring the environment. The WCED report asserts that poverty, resource depletion and environmental stress arise from basic economic disparities and because of too much sectional interest vested in political power. Sustainable development at a global level can be countenanced only if there are major changes in the management of the natural environment and the profligate exploitation of resources by human activities, something

Table 10.4 *Contribution of road transport to air pollution in selected cities.*

Region	Year	Total pollutants from all sources (10^3 tonnes)	Per cent attributable to road transport					
			CO	HC	NO_x	SO_x	Particulates	Total
Mexico City	1987	5,027	99	89	64	2	9	80
São Paulo	1981	3,150	96	83	89[a]	26	24	86
	1987	2,110	94	76	89[a]	59	22	86
Ankara	1980	690	77	73	44	3	2	57
Manila	1987	500	93	82	73	12	60	71
Kuala Lumpur	1987	435	97	95	46	1	46	79
Seoul	1983	–	15	40	60	7	35	35
Hong Kong	1987	219	–	–	75	–	44	–
Athens	1976	394	97	81	51	6	18	59
Gothenburg[b]	1980	124	96	89	70	2	50	78
London	1978	1,200	97	94	65	5	46	86
Los Angeles[b]	1976	4,698[c]	99	61	71	12	–	88
	1982	3,391[c]	99	50	64	21	–	87
Munich	1974/5	213	82	96	69	12	56	73
Osaka	1982	141	100	17	60	43	24	59
Phoenix	1986	1,240[d]	87	64	77	91	1	28

[a] Includes evaporation losses from storage and refuelling.
[b] Per cent shares apply to all transport. Motor vehicles account for 75–95% of the transport share.
[c] Excludes particulate matter.
[d] Includes 490,000 tonnes of dust from unpaved roads.
– = data not available.
Source: World Resources 1992–93. A report by the World Resources Institute in collaboration with the United Nations Environment Programme and the United Nations Development Programme (1992).

that would involve a radical new global psychology – a fresh way of thinking about economics, and personal and collective responsibility for the natural environment.

To achieve a new consensus, the WCED report proposed that changes in the present political, economic and technological systems be implemented through, for example, effective citizen participation in decision-making. Economic systems should be allowed to generate surplus commodities and provide freely exchangeable technological know-how in order to encourage self-reliance and sustainable development. The report envisaged the establishment of international organisations to endeavour to solve

Plate 10.5 *United Nations General Assembly (left) and conference room (right), UN headquarters, New York. World peace, improved life chances for all and sustainable development are dependent upon international co-operation and the implementation of internationally binding treaties and agreements forged through organisations such as the United Nations.*

Table 10.5 *The impact of motor vehicles on the environment.*

Pollutant Vehicle impact on emission	
Carbon dioxide (CO_2)	19 pounds into the atmosphere per gallon[a] 300 pounds per 15-gallon fill-up 14% of the world's CO_2 emissions from fossil fuel burning from motor vehicles
Tropospheric ozone (O_3)	Although ozone in the lower atmosphere does not emanate directly from motor vehicles, they are the major source of the ozone precursors; hydrocarbons and nitrogen oxides
Carbon monoxide (CO)	Concentrations in the lower atmosphere increase by 0.8–1.4% per year[b] 66% of OECD country emissions (78 million tonnes) from motor vehicles in 1980[c] 67% of US emissions from transportation in 1988[d]
Nitrogen oxides (NO_x)	47% of OECD country emissions (36 million tonnes) from motor vehicles in 1980[c]
Hydrocarbon compounds (HC)	39% of OECD country emissions (13 million tonnes) from motor vehicles in 1980[c]
Chlorofluorocarbons (CFCs)	54.1 tonnes consumed by US mobile air conditioners annually[e] 35.6 tonnes consumed in the US annually through leakage, service venting or accidents[e]
Diesel particulates (Tiny carbon particles hazardous to respiratory tract, visibility, and as a possible carcinogen)[f]	No overall measurements. Diesel engines emit 30–70 times more particulates than petrol-fuelled engines
Lead	90% of airborne lead from petrol vehicles.
Lead scavengers (Additives to remove lead; some (notably ethylene dibromide) may be carcinogenic)[f]	Significant amount emitted.
Aldehydes (incl. formaldehyde)	Exhaust emissions correlate with hydrocarbon (HC) emissions. Diesel engines produce a higher percentage
Benzene (identified as carcinogen)	Present in both exhaust and evaporative emissions; 70% of the total benzene emissions in the US come from vehicles
Non-diesel organics	Smaller amount per vehicle, but more mutagenic overall than diesel particles
Asbestos	Used in brake linings, clutch facings and automatic transmissions. About 22% of the total asbestos used in the US in 1984 was used in motor vehicles[f]
Metals	US EPA has identified mobile sources as significant contributors to nationwide metals inventories, including 1.4% of beryllium and 8.0% of nickel. Arsenic, manganese, cadmium and chromium may also be mobile source pollutants. High-risk hexavalent chromium does not appear to be prevalent in mobile source emissions.

Notes
[a] This figure refers to direct exhaust emission only. Transportation, refining and distribution account for perhaps 15 to 20% of total emissions.
[b] Khalil, M.A.K. and R.A. Rasmussen, 'Carbon Monoxide in the Earth's Atmosphere: Indications of a Global Increase', *Nature*, 332 (245), March 1988.
[c] Organisation for Economic Co-operation and Development, *OECD Environmental Data*, Paris, 1987.
[d] US Environmental Protection Agency, *National Air Quality and Emissions Trends Report 1988*, ref. 76, p. 56.
[e] US Environmental Protection Agency, *Regulatory Impact Analysis: Protection of Stratospheric Ozone*, Washington DC, December 1987.
[f] Carhart, B. and M. Walsh, *Potential Contributions to Ambient Concentrations of Air Toxics by Mobile Sources, Part 1*, paper presented at the 80th Annual Meeting of Air Pollution Control Association, New York, 24 June 1987.

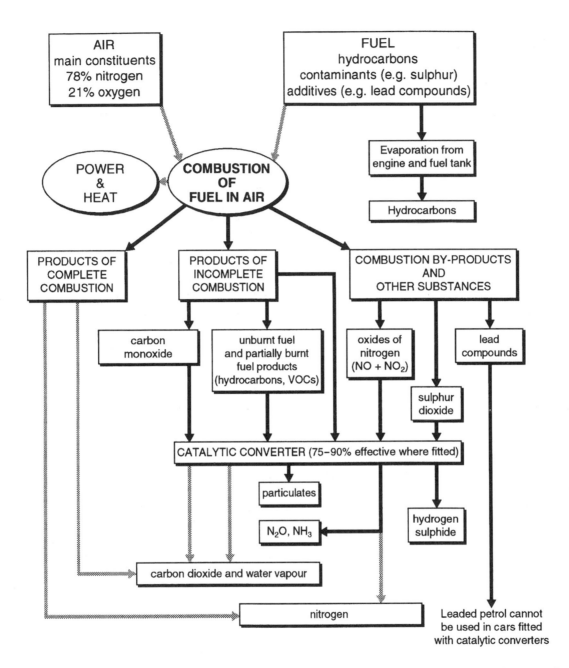

Figure 10.18 *Pollutants emitted by petrol-powered vehicles. Redrawn after Royal Commission on Environmental Pollution (1994).*

environmental problems in a truly global forum.

Our Common Future presents an ambitious and broad agenda for global sustainable development. On the debit side, the WCED report failed to identify clearly the specific barriers that inhibit the perceived action. The idealistic goals are well defined, but the mechanisms by which these could be achieved remain ambiguous. Perhaps the most significant aspect that arose from the Brundtland Report was the realisation by governments and international agencies that sustainable economic growth is untenable without

addressing the environmental consequences of any action.

In an attempt to co-ordinate the global economy, forty-four nations, dominated by the USA and the UK, met in July 1944 at Bretton Woods, New Hampshire, USA. A number of agreements were formulated, which came into effect in the following years. These agreements included the Marshall Plan for the reconstruction of Europe after the Second World War, which formed the basis for a framework for economic stability based on large-scale international income transfers. Four institutions were established: the International Bank for Reconstruction and Development (IBRD), the International Monetary Fund (IMF), the United Nations (UN) and the General Agreement on Tariffs and Trade (GATT). The IBRD, which later became known as the World Bank, was established to provide long-term international finance for investment. The IMF was to provide a source of short-term finance to help compensate for any balance of payments deficits and exchange rate fluctuations. The United Nations, formally established on 24 October 1945 with fifty-one founder countries, was to be the forum through which international decisions were to be made as well as providing the means to maintain international political and military stability. GATT was established to regulate international trade and stabilise world commodity prices. In 1961, the Organisation for Economic Co-operation and Development (OECD) was established, drawing members from twenty-five developed countries, including the original Group of Seven (the USA, Germany, the UK, Canada, France, Italy and Japan). These post-Second World War institutions, the Bretton Woods institutions, were designed to provide sustainable development, although the term was not then coined, principally for the developed world within the framework of a benign capitalist philosophy towards the developing countries.

One of the most popular definitions of sustainable development was submitted to the United Nations General Assembly in the WCED report (1987, p.43), *Our Common Future*, as 'development which meets the needs of the present without compromising the ability of future generations to meet their own needs'.

Sustainable development, for many, carries the connotation of economic growth without deleterious effects on the environment and human beings. 'Development without destruction' is another concept of sustainable development, as espoused in 1988 by the UK Department of the Environment in a report entitled *Our Common Future: a perspective by the UK on the Report of the World Commission on Environment and Development*.

The actual meaning of sustainable development is very subjective and therefore highly emotive. Seers (1977) argues that if poverty, inequality and unemployment are reduced without a loss of self-reliance, then development can be regarded as taking place. Of course, this concept is a quagmire of ambiguous meanings. One nation's perception of development is different to another's.

Most nations in Europe, and North America, would subscribe to a view of development as focused upon industrialisation, urbanisation and democratisation within a capitalist economy. Many nations in the developing countries might take a contrary view. The definition is not purely one of semantics. For international dialogue to take place at a level that can foster good will and peaceful coexistence between nations, it is imperative that each understands the other's needs. These issues are beyond the scope of this book, but they are tackled in some depth in Adam's (1990) book, *Green Development: Environment and Sustainability in the Third World*. In this book, Adams argues that sustainable development appears to be acceptable to many governments, precisely because it does not demand a radical change of policy direction. If sustainable development can only be achieved by radical political change then, because of the innately conservative nature of most political and economic systems, it seems more likely that only minimal change is likely.

Many environmental pressure groups argue that sustainable growth is impossible without deleterious effects upon the environment and on many people, particularly in the developing countries. The different attitudes of most governments and environmental pressure groups reflect fundamentally different sets of values. In many ways, the future survival of ecosystems depends upon the resolution of such diverse perceptions and goals. Development should not be restricted solely to the relatively affluent countries but seen as a global concern and global target. The United Nations represents the obvious international forum to encourage these aims and turn an ideal into some semblance of a reality.

There is a widely held perception of a North–South divide – as portrayed by Willy Brandt in reports published in 1980 and 1983, which laid the foundations for much of the Brundtland Report (1987). The South is frequently depicted as in a state of economic, political and social crisis, and subject to extreme environmental risk. The South is also portrayed as subject to problems of debt, periodic hyper-inflation, growing poverty, repeated failed crop production and falling commodity prices. In contradistinction, the North is commonly depicted as the antithesis of the South, with problems that are much less acute and that are surmountable over much shorter time spans. Although the North–South image contains some truths, these views are not strictly accurate. The Southern Hemisphere contains developed and relatively successful economies, for example those of Australia, New Zealand, and South Africa, but today the view of a North–South divide should be seen as a euphemism for a clear demarcation between those with much greater access to the means of living and the disadvantaged. It is still a truism, however, that the predominantly wealthy and developed nations remain concentrated in the Northern Hemisphere, and the less affluent developing countries in the Southern Hemisphere.

The present North–South dichotomy owes its origin to many factors, including the early imperialist exploitation and colonisation of many countries in the Southern Hemisphere in order to fuel the development and subsequent industrialisation of Europe. In the early 1980s, the developed Western world moved into a major economic recession, and with it, after 1981, the amount of foreign aid and commercial loans made available to the developing countries decreased substantially. The indebtedness of the developing countries correspondingly rocketed. Against this economic situation, the early 1980s witnessed a major failure of many cereal crops. The crop failures and food shortages in the Sahel are one example of this. Desertification and a shortage of fuelwood, together with the logging of tropical rainforests, were the attendant environmental degradation.

The North–South divide can be tackled only with a concerted effort led by the developed nations to implement much greater economic co-operation between the developed and developing nations, the transfer of scientific and technological expertise, and the rescheduling and writing-off of many debts that act like the proverbial millstone around the necks of many developing countries. This greater co-operation may come about because of entirely different factors that are very much the international currency of our time – global environmental issues.

Any proposed solutions to global environmental issues must be set within an international political framework, not only the ideological context. Many environmentalists and environmental pressure groups have failed to do this and, therefore, their messages have sounded like a bell in a vacuum. Indeed, the Brundtland Report stressed the concept of both basic needs and environmental limits set not by the environment itself but, rather, through technology and social organisation. As pointed out by Adams in his book *Green Development*, the Brundtland Report represented 'a subtle but extremely important transformation of the ecologically-based concept of sustainable development, by leading beyond concepts of physical sustainability to the socio-economic context of development' (Adams 1987, p.59). In the same book, Adams goes on to argue that 'green development' is commonly perceived as not so much about environmental management *per se*, but more often about who has the power to decide how it is managed. At its core, green development is

seen as fundamentally about self-determination for the poor.

An integral part of any global attempt at sustainable development must include foreign aid to developing countries. In 1988, the World Watch Institute estimated that the annual expenditure necessary to subsidise sustainable development and help to reduce poverty in the developing countries would be approximately US$45 billion, increasing to US$150 billion by the year 2000. Unfortunately, the present expenditure of the UNEP's environment fund is less than US$100 million, far below the necessary level of support.

In 1988, the World Bank (which has a policy of lending funds raised by floating bonds in the richer countries), and the IMF met in Berlin to discuss large aid programmes to developing countries that promote environmentally sound practices. The new Environmental Department of the World Bank, for example, will conserve land in lieu of other natural land destroyed by development projects such as a major reservoir or irrigation scheme. One such policy, known as 'Wildlands', was established to help to conserve endangered species and habitats. So far, more than forty projects in twenty-six countries aim to protect 60,000 km^2 of land, the largest of which covers 19,000 km^2 of rainforest in Rondoma Province of Amazonian Brazil.

The World Bank has also established the 'Consultant Group of International Agricultural Research' (CGIAR). This organisation is particularly concerned with increasing food production through improved agricultural techniques and the introduction of new hybrid crop strains with greater yields. The World Bank has also been working with the World Resources Institute to facilitate financial arrangements for the protection and sustainable development of habitats. Sources of revenue for such programmes could include environmental taxes levied on international trade.

Another way of looking at the problem of poverty is to consider the world's military expenditure, which collectively totals US$1 trillion annually, that is US$2.7 billion per day. Developing countries have increased their military budgets fivefold in the past twenty years. Developing countries spend large percentages of their GNP on the military; for example, Nicaragua currently spends 12.4 per cent of its GNP, Mongolia 10.4 per cent, Somalia 9.6 per cent and Ethiopia 9.3 per cent. When compared with the expenditure on health provision in the same countries, the respective figures are 4.6 per cent of GNP in Nicaragua, and less than 0.8 per cent each in Mongolia, Somalia and Ethiopia. The WCED recommended a reduction in expenditure on military resources so that the money freed could become available for more peaceful and constructive use. Achieving such a reduction in arms is not as easy. In a worst-case scenario, as environmental degradation and climatic change heighten global tensions, and as people are forced to abandon their degraded land and encroach on other already-populated lands, armed conflict might become more commonplace. The governments and military regimes in many developing countries probably will continue to regard a large investment in the military as important both to their survival in power and to ensure security across their national frontiers.

In a major study by the London Environmental Economic Centre, three major themes were identified as lying at the heart of sustainable development:

1   a consideration of an appropriate value of the environment. Evaluating the value of the environment and assessing the seriousness of resource degradation in the absence of a framework consensus is no easy matter, and no present framework exists for assessing this;
2   the need to set realistic short- and medium-term targets and mechanisms by which these can be met – this relies upon accurate predictions of the needs of future generations, especially concerning resource value changes; and
3   a need to help the most disadvantaged in society, thereby reducing the gap between rich and poor.

An Agenda of Science for Environment and Development into the twenty-first Century

The International Conference on an Agenda for Environment and Development into the twenty-first Century (ASCEND 21) was convened by the International Council of Scientific Unions (ICSU) in Vienna during the last week of November 1991. ICSU is an international organisation, primarily of scientific unions and scientific committees principally concerned with the natural sciences, and ASCEND 21 was produced, at least in part, as a preparation for the June 1992 Rio UNCED, which has become well known as the 'Earth Summit'. A summary of the influential ASCEND recommendations is given here because they provide a good example of the ways in which the professional scientific/technological community is responding to global environmental issues.

Members of the international scientific community participating in ASCEND came to a consensus on the major problems that affect the environment and hinder sustainable development. The specific areas identified as of the highest scientific priority through which the scientific community could begin to attempt to find solutions were listed as population and per capita resource consumption; depletion of agricultural/land resources; inequality and poverty; climate change; loss of biological diversity; industrialisation and waste; water scarcity; and energy consumption (ICSU 1992). ASCEND recommended the following:

- intensified research into natural and anthropogenic forces and their interrelationships, including the carrying capacity of the Earth and ways to slow population growth and reduce over-consumption;
- strengthened support for international global environmental research and observation of the total Earth system;
- research and studies at the local and regional scale on the hydrological cycle; impacts of climate change; coastal zones; loss of biodiversity; vulnerability of fragile ecosystems; and impacts of changing land use, waste, and human attitudes and behaviour;
- research into transition to a more efficient energy supply and use of materials and natural resources;
- special efforts in education and in building up of scientific institutions as well as involvement of a wide segment of the population in environment and development problem-solving;
- regular appraisals of the most urgent problems of environment and development and communication with policy-makers, the media and the public;
- establishment of a forum to link scientists and development agencies along with a strengthened partnership with organisations charged with addressing problems of environment and development;
- a wide review of environmental ethics.

These recommendations pose a challenge, not only for the international scientific community but for the whole of humankind. More specifically, if these aspirations are to be realised, then it is of paramount importance that they be adopted by those with the political and economic power to support the scientists and technologists in their endeavours to find appropriate solutions. With potential solutions, the scientific/technological community must then convince the rest of society that their solutions are morally and ethically acceptable, something that involves them in clearly defining the problem/s and translating any complex solutions into intelligible and easily understood arguments.

Earth Summit, Rio de Janeiro

On 3 June 1992, more than a hundred world leaders and 30,000 other participants held an extraordinary meeting in Rio de Janeiro for the beginning of the United Nations Conference on Environment and Development, popularly known as the Earth Summit. The summit was the brainchild of Canadian millionaire Maurice Strong, who had been a member of the Brundtland Commission and secretary-general of the first UN environmental conference in Stockholm in 1972. Amongst these leaders were US President George Bush, Chancellor Helmut Kohl of Germany, Prime Minister John Major from Britain and Prime Minister Kiichi Miyazawa of Japan. Even the Dalai Lama attended, joining the delegation of clerics, artists and green-minded parliamentarians. The central focus of the agenda to discuss the future of the Earth was treaties on biodiversity, climate change and the so-called Agenda 21 to address the problem of the twenty-first century. Amidst much media hype and world attention, delegates from the rich and poor nations met, having come to Rio with differing expectations and perceptions of the major problems facing humankind and the ways of tackling the issues. History may say that too much was expected of this summit at the time but that it marked the beginning of a continuing dialogue between the rich and poor nations over the management of the planet.

Twenty years earlier, in June 1972, the first Earth summit took place in Stockholm as the United Nations Conference on the Human Environment. After two weeks of intense negotiations, a declaration of principles and an action plan emerged. At that time, a number of present key concerns had not surfaced. For example, the hole in the ozone layer over Antarctica, let alone the hole above the Arctic or Europe, had not developed. But much of the destruction of the rainforests and threats to biodiversity were well under way. In 1972, about one-third of the Earth's tropical rainforests had been destroyed and around 0.5 per cent of the remainder was being lost each year, equivalent to 100,000 km² annually. In 1972, the world's population was 3.84 billion (72 per cent living in developing countries), whereas in 1992 it was 5.47 billion (77 per cent living in developing countries). As an example of the increasing threats to biodiversity, 1972 saw just under two million African elephants, but this had been reduced to about 600,000 in 1992, mainly because of ivory poaching. In 1972, about three million people were refugees fleeing war, a number that had risen to an estimated fifteen million by 1992. In 1992, global military spending on arms and armed

forces was projected to be just under US$800 billion (at 1988 prices), compared with US$600 billion (also at 1988 prices) (figures from UN agencies, World Resources Institute and the Worldwide Fund for Nature, reported in the *Independent*, 3 June 1992). Since 1972, some new issues have come to the fore, whilst others remain just as poignant twenty years on. The 1972 summit took place in the shadow of the Cold War, with the planet divided into rival East and West blocs, and obsessed with the nuclear arms race. In 1992, the world political stage had altered dramatically. This legacy formed part of the road to Rio.

The rationale behind the 1992 Earth Summit was that with the relaxation of Cold War tensions, combined with increased awareness of the growing ecological crisis, the conference offered a rare opportunity to persuade nations to look beyond their national interests and come to some kind of agreement over the management of the planet. Of course, countries went to Rio with very different perceptions and goals. The rich, developed countries, or North, had become accustomed to a lifestyle and share of the world's resources that they were not willing to sacrifice. The poor, developing countries, or South, for their part, were consuming irreplaceable global resources at a rate that was causing concern about the ecology of the planet and threatening sustainable development. So, at face value it seemed fairly clear that the nations of the world must abandon self-destructive practices in favour of sustainable development. This is where the simplistic arguments broke down. What is sustainable development? Who will compensate the developing countries for not destroying the tropical rainforests or for the over-exploitation of other natural, non-renewable resources? What is a fair and just level of aid to developing countries? Will the affluent North sacrifice some of its lifestyle in order to help the developing nations? And so the strands of argument ran on. There were no easy answers, and it is no wonder that many environmentalists and some nations expected too much from Rio and were disappointed by the lack of international agreement.

As the Earth Summit came to a close on 14 June, many people around the world were asking if it had all been worthwhile? The answer has to be that Rio was a qualified success. The Earth Summit ended with more than a hundred world leaders, led by Brazilian President Collor de Mello, adopting a charter for sustainable development, together with a new United Nations body to supervise its implementation. When the summit ended, 152 countries signed the Biodiversity Convention and 150 the

Climate Change Convention. The richest nation on Earth, the United States under President George Bush, refused to sign the Biodiversity Convention, claiming in a presidential election year that it would cost the American economy precious jobs and financial resources it did not have available. It was a week in which George Bush claimed that 'America's record on environmental protection is second to none, so I did not come here to apologise.' Despite this major setback for most of the delegate nations and independent environmental groups, Rio did mark an important opportunity for the developing and developed countries to express ideas together.

At Rio, plans were drawn up for a declaration of principles for the pursuit of sustainable development, known as 'Agenda 21' (for the twenty-first century), which included plans for a Desertification Treaty, a Forestry Convention and the establishment of a United Nations Sustainable Development Commission (UNCSD) to oversee its implementation. The UNCSD will also receive and provide guidance on the content and consistency of national sustainable development strategies (SDSs). The Climate Change Convention, although rather dilute in substance, committed signatories from the developed countries to setting their own targets for greenhouse gas emissions within six months, whereas developing countries were given up to three years. The world's first agreement on forests was instigated, with a set of principles on forest management and conservation, but the agreement was neither legally binding nor technically a part of Agenda 21. At the Earth Summit, the poorer nations had hoped for commitments from the richer nations for considerable additional financial aid, but received only a lukewarm response. Many developed nations still fall a long way short of the UN target aid figure of 0.7 per cent of their GNP. Rio must be viewed as the start of an ongoing international dialogue between the developed and developing nations about global environmental issues. The arguments will continue over just how successful Rio was, and it will be many years before its true historical significance can be evaluated.

Changing attitudes

If change is to come about in the ways discussed here, and societies are going to aim towards sustainable development, then the attitudes of individuals and governments will have to change. Communication and understanding between all concerned and involved groups will need to be improved. This is particularly so between scientists and policy-makers, who so often appear to speak different languages. The scientists need to deliver their arguments on a level and with language easily understood by policy-makers and the general public. Jargon should be abandoned in favour of simplicity and clarity.

When examining problems and processes, scientists often present several different scenarios, the results and conclusions being presented in a form that is too complex for policy-makers to appreciate readily and, therefore, to make decisions upon. Scientists also endeavour to be objective and non-committal. These may seem to be laudable aims, but they all too frequently mask the real subjectivity of many aspects of science, and serve only to send out ambiguous messages. But there is a real underlying problem, because most environmental issues are associated with varying degrees of uncertainty, therefore policy-makers and scientists must develop ways of working with these uncertainties.

How is change to come about? Ruckelshaus (1989) suggests that change is a three-phase process. First, world leaders need to transmit environmental values to both the public and the private sectors. Second, motivation is needed to initiate and drive those changes, and finally, and most difficult of all, institutions are needed that can translate the agreed policies into action.

The first of these phases is well under way at present. During the past year, leaders of the USA, the former USSR, the UK, France and Brazil have all made environmental statements. During the Economic Summit in Paris in July 1989, the leaders of the group of seven (G7) major industrial countries discussed environmental issues together for the first time. They called for worldwide policies to be developed to pursue the goal of sustainable development. The meeting was nicknamed the 'First Green Summit', and a seven-page document with nineteen clauses was produced, the 'Paris Communique'. Interestingly, the discussions may have been stimulated by the growing interest of politicians in environmental issues, with a real awakening of latent environmental sympathy rather than vote-catching. Hopefully, the latter interpretation is too cynical and unfair on professional politicians. This sea-change in the public consciousness was demonstrated by the large number of parliamentary seats won by environmental parties during the European election earlier that year, together with the realisation by politicians that if they want to remain in power they need to continue to attract voters to their party. There is, however, a possible cloud on the horizon to darken this optimism, because at the last G7 summit, very little was said about environmental issues, and in the last UK election such issues appeared to have been put on the proverbial back-burner. Interest and concern in environmental issues at government level may be on the wane.

Incentives to protect the natural environment, which should be introduced by governments, must include pricing policies to cover the cost of environmental degradation. Some of the methods by which governments can motivate organisations to conserve energy, restrict pollution and recycle materials have already been discussed. Other incentives might include offsetting the emission of carbon dioxide in a factory by planting trees. Citicorp, for example, a major money-lending institution, has a policy of writing off debts in exchange for areas of land in South America, which will be designated as conservation areas and made into national parks.

The development of institutions that can motivate and enforce policies has only recently begun. Unfortunately, organisations such as the UNEP, the Human Dimensions of Global Change Program (HDGCP), the WWF, the International Geosphere–Biosphere Program (IGBP), and many others, are given very small budgets and have far too little political and economic power. This situation contrasts sharply with such institutions as NATO, the World Bank and the multinational corporations. There is clearly a need for an international environmental organisation comparable to these in size, budget and power, with the aim of sustainable development of planet Earth. This organisation should integrate all the environmental facets we have discussed throughout this book, and have a truly global, not just North–South, perspective.

International environmental law

International law requires an enormous amount of trust and collective determination. It is basically a set of principles, obligations and rules binding the behaviour of the parties involved. These obligations are normally created by treaties, conventions or protocols, which are sometimes referred to as 'hard law'. These usually have enforceable targets and regulations. Other agreements are referred to as custom and framework conventions and are often called 'soft law'. These may be ambiguous and have flexible interpretations. These conventions lead to the creation of principles or declarations for international consensus. Soft laws are particularly valuable because they allow international environmental agreements to

evolve as scientific knowledge increases, and as social learning expands. They permit greater flexibility compared with hard laws, and can prove to be the optimum course of action where diplomatic manoeuvres are required in order to allow a nation to save face. Soft laws permit considerable freedom of action, so that countries can choose their own means to reach particular targets. Appendix 4 lists the main multilateral conventions concerning the environment in the form of Agenda 21 of the Rio Earth Summit, 1992. Box 10.3 provides an example of an important international convention on biodiversity.

It is important to understand the mechanism of international co-operation in the development of treaties. Table 10.6 summarises the assessment by Choucri *et al.* (1994) of the processes involved in formulating a treaty. They examined the multilateral treaty commitments by several large countries between 1920 and 1990, examining in particular whether the treaties:

1 constrained the domestic activities and sovereignty of the signatory parties;
2 addressed the problems of **global commons** – that is global resources generally deemed to be a common heritage, as opposed to territorial problems; and
3 instituted regimes requiring active collaboration.

Their results suggest that the main industrialised and developing countries are rather uniform in their attitude to treaties. They noted that the former Soviet Union has signed more treaties than the USA and Japan, and half of the treaties embody constraints on sovereignty. China, which traditionally has had a poor performance in world politics, in 1994 had signed twenty-two treaties, seven of which have only recently been ratified.

There is much consensus that environmental laws should have a strong philosophical and ethnic basis, a clear concern for the well-being of both individuals and society at large, and where possible transcend purely self-interest. The common heritage of mankind has a strong ethical basis: its doctrine obliges present generations to act as trustees of the natural and human heritage to enhance the biological and spiritual life of future generations. In the 1950s, the principle of the common heritage of mankind was incorporated into international law and has five principal elements (O'Riordan 1994a):

1 non-ownership of shared common global resources;
2 shared management of common global resources;
3 shared benefits of common global resources;

BOX 10.3 CONVENTION ON BIOLOGICAL DIVERSITY

The Convention on Biological Diversity was signed by more than 150 governments at the Rio 'Earth Summit' in 1992. It came into force on 29 December 1993 and has become the centrepiece of international efforts to conserve the Earth's biological diversity, ensure the sustainable development of its constituent parts, and promote the fair and equitable distribution of the benefits arising out of the utilisation of genetic resources.

The convention focuses on the anthropogenic influences leading to a decrease in global biological diversity, emphasising the intrinsic value of biodiversity, and of the ecological, genetic, social, economic, scientific, educational, cultural, recreational and aesthetic values of biodiversity. It is also concerned with the importance of biodiversity for evolution and for maintaining life-sustaining systems in the biosphere. The convention affirms that the conservation of biodiversity is a common concern for all humankind.

It notes the lack of information and knowledge regarding biodiversity and of the urgent need for more research to provide a basic understanding upon which to plan and implement appropriate measures to conserve global biodiversity. The document emphasises that the lack of sufficient scientific certainty should not in itself provide an excuse for postponing measures to mitigate human activities that may reduce biodiversity.

The convention acknowledges the sovereign rights of states over their own biological resources and also that states are responsible for using these resources in a sustainable manner. It also emphasises the importance of, and the need to promote, international, regional and global co-operation among states and inter-governmental organisations, as well as the non-governmental sector for the conservation of biodiversity and the sustainable use of its components. Additionally, the convention is particularly concerned with both *in situ* (in the natural habitat) and *ex situ* (outside the natural habitat) conservation of ecosystems and natural habitats, and the maintenance and recovery of viable populations of endangered species.

The convention recognises the close and traditional dependence of many indigenous and neighbouring communities embodying traditional lifestyles that use biological resources. It stresses the desirability of sharing equitable benefits from the use of traditional methods relevant to the conservation of biodiversity and its sustainability. It also recognises the role of women in conservation and sustainability and the need for full participation of women at all levels of policy-making and implementation for biodiversity conservation.

The convention acknowledges the provisions of new and additional financial resources and appropriate access to relevant technologies as likely ingredients in the world's ability to address the loss of biodiversity. Furthermore, it recognises that special provision is required to meet the needs of developing countries and that the sustainable development of biodiversity is of critical importance for meeting the food, health and other needs of the world's growing population. The convention encourages and enhances friendly relations among signatory states as a means of complementing existing international arrangements for the conservation of biodiversity and its sustainable development.

Ultimately, the convention emphasises the conservation and sustainable use of biodiversity for the benefit of all present and future generations.

Source: UNEP World Wide Web Site 1996.

---

4 use of common global resources for peaceful purposes; and
5 conservation for humanity and future generations.

In order to have effective international laws on the environment, all these treaties and agreements require the development of an acute sense of global citizenship.

A manifesto for living

There are ways of living more sensitively with the natural environment. The following represent an idealistic and provocative shopping list of the kinds of issues that we feel should at least form the basis of international discussion in an attempt to provide a global framework for sustainable development, with or without sustained growth in any single country. The aspirations of our manifesto for living are:

Feeding the world and eliminating national poverty. The first priority is the elimination of poverty, which leads to major famines. Through self-help, including freely given expertise and advice, and emergency aid where appropriate, compassion and concern for fellow human beings should ensure that starvation and malnutrition do not exist alongside extravagant affluence in the developed nations. Much more international debt incurred by the poorest developing countries could be written off. This action would have the immediate effect of providing a realistic opportunity for at least some of the poorest nations to shake off the intolerably heavy burden of debt, which impedes both the will and the means to recovery.

Control of population growth. World population is too large, and it is growing too rapidly, to permit sustainable development. The main cause of stress on the environment is over-population. There are quite simply too many people wanting more than is available, at least for those seeing a 'North' lifestyle as desirable. Even allowing for the inequitable distribution of the world's resources, the human species

Table 10.6 *Participation in multilateral environmental treaties. Numbers in brackets are percentages for each variable. See text for definition of treaties.*

Advanced countries	USA	Japan	Germany	Former USSR
Sovereignty constraint				
Domestic	18 (42)	16 (43)	33 (62)	24 (50)
Foreign	25 (58)	21 (57)	20 (38)	24 (50)
Problem-type				
Commons	28 (65)	22 (59)	32 (60)	30 (63)
Territorial	15 (45)	15 (41)	21 (40)	18 (38)
Regime-type				
Common aversions	16 (37)	16 (43)	22 (42)	19 (40)
Common interests	27 (63)	21 (57)	31 (58)	29 (60)
Total treaties in force	43	37	53	48

Developing countries (making up 44 per cent of the global population)	China	Brazil	India	Indonesia
Sovereignty constraint				
Domestic	10 (45)	14 (58)	17 (65)	9 (69)
Foreign	12 (55)	10 (42)	9 (35)	4 (31)
Problem-type				
Commons	12 (55)	11 (46)	11 (42)	6 (46)
Territorial	10 (45)	13 (54)	15 (58)	7 (54)
Regime-type				
Common aversions	11 (50)	12 (50)	14 (54)	6 (46)
Common interests	11 (50)	12 (50)	12 (46)	7 (54)
Total treaties in force	22	24	26	13

Source: Choucri *et al.* 1994.

cannot grow at the predicted rates without creating even more environmental problems.

Over-population is the most difficult global problem to tackle. It is hard to advise a couple in a very poor part of the world not to produce as many children, especially where infant mortality rates are high and they may rely on their children to provide family support later in life. The practical aspects of birth control, particularly through artificial methods, are often actively discouraged by some of the world's major religions. Perhaps the optimum way of encouraging slower population growth is by increasing the overall standard of living, or quality of life, for many people in the poorer nations. It is only through increased personal and national security that populations seem to stabilise at sensible levels.

Improving basic medical care. Longevity and good health correlate extremely closely with wealth. The richer nations enjoy a standard of health provision which is far above that available to all but a few in the poorer nations. This imbalance of the most fundamental human provision, after food, should be rectified through greater international direct aid to countries where health care is limited. Richer nations

should provide a larger proportion of their GNP as grants to train more medical personnel from the poorer nations, and supplement such bursaries with the provision of basic medical supplies to those countries where easily treatable diseases are still prevalent.

Expanding educational provision at all levels. Where possible, education should be provided free of charge at the point of demand, at all age levels and, wherever possible, in order to raise the public awareness of the role of the individual in society and his/her relationship with the natural environment. By increasing the overall educational standards in society, the potential for a more concerted effort by more individuals to take an active part in the solving of common environmental problems can be increased. Without a rudimentary education, people are more likely to act out of fear and ignorance and cannot be expected to take an active interest in national, let alone international, issues. Education is the greatest legacy that can be given to future generations of children and society in general (Plate 10.6).

Energy conservation. Natural energy resources are finite – that is they are limited. Undue waste should

Plate 10.6 *Education programme at Communidad Santa Martha, El Salvador.*
Courtesy of Rhodri Jones/Oxfam.

be avoided, and an attempt made to reduce excessive energy consumption. Renewable energy resources with minimal environmental pollution should be used where possible, and further developed to be used in preference to the non-renewable, finite resources such as fossil fuels. Technologies for energy efficiency and energy conservation should be encouraged through favourable tax regimes and other fiscal incentives. The introduction of an energy-carbon tax is an attractive means of discouraging the profligate waste of energy from conventional fossil fuels. Governments should discourage the wasteful 'disposable-commodity ethos' that is now so prevalent, and which is very wasteful of energy resources.

An estimated 50 per cent of the world's atmospheric pollution and 20 per cent of the greenhouse effect result from the motor car. In order to conserve energy and reduce the global impact of this pollution, there is a very urgent need to develop and market cleaner and more energy-efficient motor vehicles. But most important of all is the need to encourage the increased use of public mass transit systems through both highly subsidised public transport and heavy financial burdens on the use of private transport to and from large cities and other centralised places of work. Far too few large urban centres appear to be tackling the problem. Governments should have an energy policy that includes support for a diverse range of energy resources and that is sensitive to the need to maintain acceptable levels of employment.

Resource sharing. People waste energy and materials, as well as contributing unnecessary pollution to the ecosphere through the selfish use of resources. Sharing transport and using other resources more efficiently not only provide a general improvement in the quality of people's lives, but they also free so many more potential resources for others who may be less fortunate. To encourage resource sharing, countries need to introduce government and industrial incentives.

Plate 10.7 Ethiopian refugees on the Tihama Plain, North Yemen, making use of metal waste from old cars to make simple farming tools to sell in the market for food. This type of alternative technology illustrates the ingenuity of refugees and people in developing countries, their fight for survival and the potential for recycling.

Recycling resources and materials. Enormous quantities of materials are wasted. The present throw-away, disposable culture creates unnecessary pollution, and squanders energy resources and precious natural materials. This is particularly the case in the developed countries whereas in developing countries, recycling is a necessity of life (Plate 10.7).

Frosch (1995), however, suggests that there will be a new industrial revolution in the twenty-first century, an ecological one, in which manufacturers will begin to concentrate on recycling mechanisms to reduce waste and increase profits. Clean technologies should be developed to reduce recycling waste, and environmental laws enacted to enforce clean and efficient recycling. With more recycling should come the opportunity to distribute commodities more widely and, therefore, increase the life chances and quality of life for many more people and nations. As with resource sharing, government- and industry-funded incentives are necessary to stimulate more recycling of resources and materials.

International co-operation on global issues. An international, rather than solely national, perspective on global issues, especially environmental issues, must be encouraged. A full appreciation of the common fate of humanity, which is intimately and inextricably bound up in collective actions as families, neighbourhoods, villages, towns, large urban centres, counties, cantons, countries or groups of nations, must be ensured, because this leads to a more equitable distribution of resources and wealth, equality of opportunity and a harmonious coexistence with the natural environment. Such lofty ideals may not appear achievable, but they are nevertheless goals worth striving for. Through increased internationalism, individual societies and nations are more likely to understand the aspirations of others and to provide for the easy interchange of ideas and resources.

There is a real need for more effective international policing of pollution incidents, particularly where they occur in developing countries and involve multinational companies. The United Nations and World Bank are obvious instruments that could be used to bring pressure to bear on offenders and non-compliant multinational enterprises.

On an optimistic note, in October 1991 the Western developed nations and developing countries agreed to examine the ways in which GATT affects the environment. Countries might refuse to import certain goods if they are considered to have been produced in a manner harmful to the environment. At present, these sorts of import restriction are unlawful under the terms of GATT. A new international environmental code, backed up by the world's leading financial institutions and groups, could give real muscle to global concerns over the environment.

Reducing military expenditure. The futile arms race aimed at ever-refined means of destroying fellow humans is sapping valuable intellectual, financial and material resources that could be devoted to other ends, such as feeding the world, health care, education and improving the overall quality of life. Even those military strategists and politicians most committed to the arguments of keeping the peace through arms cannot fail to see the sheer waste that the arms race has created. But military expenditure will be reduced only where a nation has confidence in its future survival and a feeling of security against attack from others (Plate 10.8). Greater global co-operation across a broad front of economic, political and environmental issues has to come before any country will seriously countenance reducing its military expenditure, including nuclear weapons, and thereby free money for any pressing social and environmental needs.

Non-nuclear future. This issue is probably the most contentious item of this agenda, and it is easier to write such a slogan than it is to adopt and effect. The use of nuclear power and the manufacture of nuclear weapons, however, bequeaths an unacceptable legacy to and burden on future generations. Its polluting capacity is now well demonstrated and the concept of atoms for peace has been shown to be a chimera as the frantic arms race seems to gather pace from day to day, with the proliferation of ICBMs, SLBMs, ALBMs, MIRVs, ERWs and the Star Wars technology (see Chapter 6).

Nuclear weapons are simply too terrifying to use again. Robert McNamara, a US Defence Secretary, put the issue well when he stated that 'You cannot make a credible deterrent out of an incredible action.' The nuclear arms race has to be halted, with a reversal in the build-up of nuclear arms arsenals. Environmentally more friendly energy programmes need to be developed so that future generations are not the custodians of the radioactive waste, including contaminated processing plants, that is left behind.

The scaling down of nuclear arms arsenals, halting the nuclear arms race and the decommissioning of nuclear power stations cannot be achieved overnight. There is a need for multilateral arms reductions, with acceptable and effective verification procedures, to remove nuclear weapons. This can come about only in a climate of international trust, goodwill and co-operation. Alternative energy technology may have to be improved before existing nuclear power stations are decommissioned. Finally, many countries without easily obtainable alternative energy resources may well argue for a nuclear future. It will be incumbent on those nations who prefer a non-nuclear stance to set out realistic alternatives and to offer economic incentives to potential nuclear nations, for example other fuels and energy at competitive prices, or the technology to develop other viable energy resources. The risks of reactor meltdowns are extremely low, but we would point out that the potential long-term consequences of a major nuclear accident, albeit very unlikely, pose an unacceptable level of danger for the environment, including to humans.

Another major problem for nuclear power concerns what to do with radioactive waste. How can society ensure its safe custody and care into the future? Besides the accidents that occur in countries that, at the time of the incident, have political stability which makes any clean-up operation easier, as nuclear power plants proliferate, so too does the likelihood of such plants becoming part of politically unstable parts of the world, simply because it is often impossible to predict where these areas may develop. The demise, and probable break-up, of the Soviet Union means that safety standards in the nuclear industry may decline, thereby increasing the risk of serious nuclear accidents; Tomsk-7 is, perhaps, just such an example. And political instability within and between nations with a nuclear weapons capability, nations that previously may have been stable, will inevitably make the world a less safe place. Humans have, time and again, demonstrated an almost fatalistic inability to avoid wars, with all the human misery that accompanies them, and with such a record we really should not trust either ourselves or others with radioactive materials for ostensibly peaceful or military purposes.

Ethical investments. Those with the financial capability and power to invest money should take moral responsibility for the ways in which their money may grow. Investors should avoid providing financial loans to companies and organisations involved in polluting and destroying the natural environment, or for that matter involved in the abuse of human rights. Depending upon one's own ethics and morals, the list of acceptable investments will vary greatly, but at least individuals should recognise that it is not sufficient to make money in ignorance of the means by which this is done. Making ethical investments is not always easy, not least because initial investment portfolios may change, through buy-outs, mergers, reinvestment, etc., to include 'undesirable' activities.

Perhaps the best-known recent international example of the issues raised by the concept of ethical investments concerns the Shell Oil Company's involvement in the extraction of oil from Nigeria, where, in 1995, Ken Saro Wiwa, an internationally acclaimed poet, civil rights campaigner and environmental activist who constantly spoke out against the exploitation of his homeland, Ogoniland (in the Niger delta area), by the Nigerian government, was hanged along with fellow compatriots by the military regime after a show trial in which he was

---

Plate 10.8 *While large amounts of badly needed money are invested in arms and ammunition, inevitably less is available for social programmes to improve life in the poorer, developing nations.*
Courtesy of Rhodri Jones/Oxfam.

accused of inciting civil unrest – whilst he was already in prison! Many world leaders used diplomatic means to try and avert the hangings and condemned the show trial, and subsequently expressed concern at the Nigerian regime's murder of Saro Wiwa and the others. Also, many environmentalists accused the Shell Oil Company of effectively supporting the military regime and, by implication, the killing of Saro Wiwa and the others, because it neither publicly criticised the Nigerian leaders for their action nor implemented any economic sanctions after the action by the military regime – to the extreme of pulling out of its involvement in the Nigerian economy. Naturally, the issues surrounding this case are complex and, arguably, the Shell Oil Company felt it was not within its remit to become embroiled in internal politics. No judgement is intended in this specific example, but rather this case highlights a situation where many individuals and organisations worldwide felt that a major multinational company could have used its potential influence to greater effect but for various reasons chose not to do so.

Practising efficient and environmentally sound farming. Farming practices should be efficient but not to the detriment of the environment. Fertilisers should not pollute water resources or harm other aspects of the natural environment, for example by nitrates being leached in dangerous quantities into local water resources. The economics of farming need rationalisation, for example in the EU, where the common practice of 'set-aside' is pursued, in which, in order to maintain a relatively high market price, farmers are paid not to grow specified crops. On a global scale this seems irrational because of the intensive use of fertilisers to increase crop yields elsewhere, and as parts of the world suffer drought and food shortages alongside food mountains. Economic blocs are an inevitable consequence of any capitalist cash nexus, but organisations such as the United Nations should work to reduce the agro-economic divide between the North and South, developed and developing countries. Plucknett and Winkelmann (1995) describe possible future developments in agricultural technology that are needed and will help to increase yields while at the same time protecting the environment. These include genetic engineering, increased use of natural insecticides, conserving biodiversity to sustain genetic variety to assure robustness within agricultural systems, and the development of novel diagnostic tools to help identify viruses and other diseases on the spot.

When there is so much starvation and undernourishment throughout large parts of the world, precious agricultural land should not be wasted in rearing and grazing excessive numbers of livestock, which then require additional land for grain crops to feed the cattle, just so that people in the developed world can overeat. By consuming less meat, diets may be healthier, and the land freed by this change will allow the production of more grain crops to feed a larger proportion of the world's population.

Leaving designated natural wildernesses undeveloped and unexploited. The ecology of many as yet unspoilt and unplundered wildernesses, the Arctic and Antarctica, large tracts of the tropical rainforests and deserts, is in a precarious and fragile balance. These regions of the world often play a vital part in regulating global climate, through both positive and negative feedback mechanisms. It is through their preservation and maintenance that individuals can ensure the continued survival of life on Earth as it now exists. International treaties must be negotiated to protect these last remaining bastions of much of the planet's rare species of fauna and flora.

One recent, encouraging news story about leaving some natural wildernesses unexploited and unpolluted is over Antarctica. After two years of battles and campaigns, the anti-mining lobby seems to have won a stay of execution. In 1990, Australia, France, Belgium and Italy proposed that Antarctica be designated a World Park, a global conservation area free from exploitation. The USA and Britain opposed this suggestion, but in May 1991 Japan and Germany, both countries that the USA and Britain counted on to support their case, changed sides and undermined the pro-mining lobby. The result of this volte-face by Japan and Germany paved the way for the United States and Britain to follow suit in July, leading to the ratification of a new Antarctic Treaty in the autumn of 1991, on the thirtieth anniversary of the existing Antarctic Treaty. Indeed, on 3 July 1991, the USA was the last nation to sign the protocol after unsuccessfully holding out for an exclusion clause if a minority group wished to mine as a joint venture.

The comprehensive new treaty, ratified in Bonn on 11 October 1991, prohibits the mining and exploitation of Antarctica for the next fifty years. After this fifty-year moratorium, any nation wishing to exploit any mineral wealth in Antarctica will require the agreement of at least 75 per cent of the signatories to the treaty. The new treaty also includes safeguards to stop Antarctica being spoiled through tourism and waste disposal. There are proposals to keep tourism offshore, and to discourage the construction of hotels and encampments, which would bring their own pollution, sewage and waste problems to Antarctica. The treaty, initialed by twenty-three of the twenty-six member nations, comes into

Plate 49 *Urbanisation on a grand scale, Manhattan Island, New York. Central Park in the middle ground provides an important amenity within this concrete jungle.*

Plate 50 *Venice, which is threatened by subsidence caused by the withdrawal of ground water.*

Plate 51 *Rwandan refugees queuing for water in Goma, Zaire.* Courtesy of Betty Press/Panos Pictures.

Table 10.7 *A summary of the basic provisions of the Antarctic Treaty.*

Article I
Antarctica shall be used for peaceful purposes only. All military measures, including weapons-testing, are prohibited. Military personnel and equipment may be used, however, for scientific purposes.

Article II
Freedom of scientific investigation and co-operation shall continue.

Article III
Scientific programme plans, personnel, observations and results shall be freely exchanged.

Article IV
The Treaty does not recognise, dispute or establish territorial claims. No new claims shall be asserted while the Treaty is in force.

Article V
Nuclear explosions and disposal of radioactive wastes are prohibited.

Article VI
All land and ice shelves below 60°S are included, but high seas are covered under international law.

Article VII
Treaty-state observers have free access – including aerial observation – to any area and may inspect all stations, installations and equipment. Advance notice of all activities and of the introduction of military personnel must be given.

Article VIII
Observers under Article VII and scientific personnel under Article III are under the jurisdiction of their own states.

Article IX
Treaty states shall meet periodically to exchange information and take measures to further Treaty objectives, including the preservation and conservation of living resources. These consultative meetings shall be open to contracting parties that conduct substantial scientific research in the area.

Article X
Treaty states shall discourage activities by any country in Antarctica that are contrary to the Treaty.

Article XI
Disputes are to be settled peacefully by the parties concerned or, ultimately, by the International Court of Justice.

Article XII
After the expiration of thirty years from the date the Treaty enters into force, any member state may request a conference to review the operation of the Treaty.

Article XIII
The Treaty is subject to ratification by signatory states and is open for accession by any state that is a member of the UN or is invited by all the member states.

Article XIV
The United States is the repository of the Treaty and is responsible for providing certified copies to signatories and acceding states.

force only after it has been formally accepted by the respective governments.

The main articles of the treaty are listed in Table 10.7. Of particular note are Article I: Antarctica shall be used for peaceful purposes only. All military measures, including weapons testing, are prohibited; Article II, Freedom of scientific investigation and co-operation shall continue; Article IV, The treaty does not recognise, dispute or establish territorial claims; Article V, Nuclear explosions and disposal of radio-active wastes are prohibited; Article VII, Treaty-state observers have free access to any area and may inspect all stations, installations and equipment. In 1993, anticipating a higher profile for research in Antarctica, the US$240-million-a-year Antarctic Program, which is funded through the National Science Foundation (NSF), was upgraded from a division within the directorate to a programme within the office of the NSF director. The US Antarctica Program strives for a balance between scientific exploration and environmental protection. In the USA, concerns over the sensitive regulation and monitoring of Antarctica, and allied research, have led to an ongoing debate as to who should control the US interests on this continent, and whether this is best done through the NSF or another body such as the National Oceanic and Atmospheric Administration, or the Environmental Protection Agency.

Sadly, Greenpeace has closed its independent monitoring base on Antarctica because of the economic costs, about US$1 million per year. Without an organisation such as Greenpeace to monitor activity on Antarctica, there is a danger that aspects of the treaty could be violated without world opinion being alerted to any potential dangers. Plate 10.9A illustrates one such violation on an Antarctic base in the South Shetland Islands, while Plate 10.9B shows that the Arctic wildernesses are also being polluted by human waste. The problem remains truly global in extent.

Plate 10.9 *Large quantities of human rubbish are still being dumped from Antarctica (as shown by this photograph (left) taken in December 1992 of an Antarctic base in the South Shetland Islands), and (right) the Arctic (abandoned vehicles in the tundra landscape of northernmost Russia, photographed in August 1993).*
Plate on left courtesy of Gary Nichols.

Endpiece

In his satirical book on global problems, *All the Trouble in the World: the Lighter Side of Famine, Pestilence, Destruction and Death*, P.J. O'Rourke (1994, p.2) remarks that 'Things are better now than things have been since men began keeping track of things. Things are better than they were only a few years ago. Things are better, in fact, than they were at 9:30 this morning, thanks to Tylenol and two Bloody Marys.' Throughout his book, he examines the problems of poverty and environmental degradation and dissects the pessimistic views of many politicians, environmentalists and others. He emphasises that today is a moment of history that should be regarded as one of great hope, particularly because the Cold War is over and the world is truly beginning to communicate on a global level, addressing crucial problems in the international arena. The left-hand column of Table 10.8 illustrates some recent achievements, yet much is still to be achieved, as can be inferred from the right-hand column. Even though living conditions are better than ever for some of us, this is not the case for the majority of the people who live on our planet. There is the potential, however, to improve the living conditions of all, to make things better than ever before, if only the correct mechanisms can be implemented before the problems become irreversible.

To optimise the international effort on environmental issues, there is a need to rationalise resources by integrating the efforts of the many small organisations and pressure groups that currently exist, each with its own overheads and expensive experts, often duplicating work done elsewhere or discovering what other experts in other organisations have expensively discovered already. An international organisation with participants of ministerial rank is desperately needed to steer towards sustainable development. This is achievable if societies and individuals act quickly, efficiently and intelligently. Environmental groups should endeavour to collaborate and, perhaps, pool some of their hard-won power and influence, together with expertise and other resources. This is not an easy thing to do where organisations have established a power base that they may guard jealously.

But amongst all this suggested international collaboration at the highest and most expert levels, independent environmental pressure groups must continue to operate free from any bureaucratic structures, not least because this offers them a chance to suggest and lobby for radical solutions. Such radicalism is necessary since, even if seen as extreme, it provides a climate of debate in which nations are more likely to find sensible solutions to local and global environmental issues. Ministerial participation, as a prerequisite to all international environmental commissions and other organisations, will always have the in-built propensity for getting bogged down in side issues and then looking for solutions that tend to preserve the status quo. We would advocate greater international co-operation on global environmental issues, and at the same time strongly support the role of independent pressure groups, not least as watchdog bodies.

Finally, it is very hard for individuals and societies to move away from an anthropocentric, or exclusively human-centred, to a more biocentric, or life-

Table 10.8 *Balance sheet of human development.*

Human progress	Human deprivation
Life expectancy	
● Average life expectancy in the South increased by a third during 1960–87 and is now 80% of the North's average.	● Average life expectancy in the South is still 12 years less than in the North.
Education	
● The South now has more than five times as many students in primary education as the North: 480 million compared with 105 million. ● The South has 1.4 billion literate people, compared with nearly one billion in the North. ● Literacy rates in the South increased from 43% in 1970 to 60% in 1985.	● There are still about 100 million children of primary school age in the South not attending school. ● Nearly 900 million adults in the South are illiterate. ● Literacy rates are still only 41% in South Asia and 48% in sub-Saharan Africa.
Income	
● Average per capita income in developing countries increased by nearly 3% a year between 1965 and 1980.	● More than a billion people still live in absolute poverty. ● Per capita income in the 1980s declined by 2.4% a year in sub-Saharan Africa and 0.7% a year in Latin America.
Health	
● More than 60% of the population of the developing countries has access to health services today. ● More than 2 billion people now have access to safe, potable water.	● 1.5 billion people are still deprived of primary health care. ● 1.75 billion people still have no access to a safe source of water.
Children's health	
● Child (under five) mortality rates were halved between 1960 and 1988. ● The coverage of child immunisation increased sharply during the 1980s from 30 to 70%, saving an estimated 1.5 million lives annually.	● 1.4 million children still die each year before reaching their fifth birthday. ● Nearly 3 million children die each year from immunisable diseases.
Food and nutrition	
● The per capita average calorie supply increased by 20% between 1965 and 1985. ● Average calories supplied improved from 90% of total requirements in 1965 to 107% in 1985.	● A sixth of people in the South still go hungry every day. ● 150 million children under five (one in every three) suffer from serious malnutrition.
Sanitation	
● 1.3 billion people have access to adequate sanitary facilities.	● Nearly 3 billion people still live without adequate sanitation.
Women	
● School enrolment rates for girls have increased more than twice as fast as those for boys.	● The female literacy rate in developing countries is still only two-thirds that of males. ● The South's maternal mortality rate is 122 times that of the North's.

Source: Hewitt 1992 and UNDP 1990.

centred, perspective of the Earth. Indeed, such an approach may be impossible, simply because of our humanity, but it is certain that without a drastic change in present societal values, norms and mores, the present generation may well be counted amongst the last of humans to inhabit the Earth.

There are finite resources, yet an apparently infinite number of ways in which to squander them.

The history of life on Earth, with its record of past climate change, together with the environmental impact that humans have made over a very short time span, shows that there are lessons to be learned. People can learn only if minds are open, if individuals and governments are willing to discover the fragility and sensitivity of the natural environment before it is too late.

Chapter 10: Key points

1 The four principal components of the ecosphere under threat are the climatic system, the nutrient cycles, the hydrological cycle and biodiversity.

2 Population growth is a cause of major concern because of the stress that it imposes on the environment, although some argue that current world resources are capable of adequately sustaining an even larger global population.

3 Enormous differences in wealth, life chances, health, education and social provision exist between the developed and less developed nations. In various parts of the world, such differences in access to the means of life have initiated wars and political instability, thereby contributing to environmental stress and degradation. The number of refugees is currently increasing and constitutes not only a regional but a major global problem. Urbanisation and population pressures have concentrated pollution, poor housing, disease and poverty into large megalopolises. In many cases, these social and environmental issues can be tackled only by international co-operation, and the defrayment of much of the so-called 'Third World debt'.

4 Agro-economic problems include over-intensive land use (e.g. associated with industrial monoculture), inappropriate land use, the clearing of important natural vegetation, salinisation, laterisation, and pollution by fertilisers and pesticides such as nitrates. Methods to mitigate these effects include farming practices that concentrate upon efficient but not over-intensive crop cultivation, less emphasis on the use of environmentally harmful fertilisers and pesticides, improving soil productivity, reducing soil erosion, and stopping salinisation and desertification.

5 The rise of consumer society has led to an increased requirement for energy and natural resources. Without careful resource allocation and planning, there is a real danger that many resources may become severely depleted, something that could act as a limit to growth. As alternatives to conventional fossil fuels, renewable energy resources (e.g. solar, wind, wave, tide and biomass energy) should be encouraged, together with more research into technologies such as hydrogen energy. Without a concerted global effort to develop substantial energy supplies from renewables, there may be no alternative but to place greater reliance on nuclear energy, with its associated problems of radioactive waste disposal and the risk of major and long-term environmental pollution.

6 The concept of 'sustainable development' was introduced in 1980 by the 'World Conservation Strategy', and the arguments developed by the International Union for the Conservation of Nature. In 1984, these international groups were absorbed into the World Commission on Environment and Development, which produced the 1987 report *Our Common Future*, a document that ostensibly provided strategies for sustainable development. These strategies included reducing world poverty, improving agricultural practices, energy conservation, reducing anthropogenic greenhouse gas emissions, recycling waste, improving technologies, and reducing the disparities between rich and poor nations. The underlying arguments and strategies for sustainable development remain controversial.

7 Atmospheric pollution has become both a regional and a global issue. International agreements and conventions on atmospheric pollution control resulted in the 1984 agreement to reduce sulphur emission by 1993; the 1987 Montreal Protocol to reduce CFCs by 50 per cent by the year 2000 followed by a total ban on CFCs; the 1988 First World Conference on 'The Changing Atmosphere'; the 1989 Helsinki agreement on a total ban on CFCs by eighty-six countries by the year 2000; and the June 1992 United Nations 'Earth Summit' in Rio de Janeiro, where agreements and conventions were presented to preserve global biodiversity, and mitigate any possible global climate change precipitated by human activities: it was at Rio that Agenda 21 was signed by many nations. The Conference of Parties (COP), which had ratified the United Nations Climate Change Convention at Rio in 1992, signed the Berlin Mandate in April 1995 to return greenhouse emission to 1990 levels by the year 2000, and establish a working group directed by the IPCC to investigate strategies for reduced emissions after the year 2000.

8 In this chapter, we present a manifesto for the management of the Earth, aimed at maximising the chances of achieving global sustainable development, reducing global pollution, eliminating poverty, and increasing the life chances of individuals wherever they are born. This manifesto includes feeding the world and eliminating poverty; controlling population growth; improving basic medical care; expanding educational provisions at all levels; energy conservation; resource sharing; recycling materials and waste; international co-operation on global issues; reducing military expenditure; a non-nuclear future; efficient and environmentally sound farming practices; and preserving natural wildernesses.

Chapter 10: Further reading

Adams, W.M. 1990. *Green Development: Environment and Sustainability in the Third World*. London: Routledge, 255 pp.
A book on the problems of development and its environmental impact in the developing world. This book addresses the problems of striving for sustainable development and provides an important read for students and teachers of development and environmental studies.

Brandt, W. 1980. *North–South: A Programme for Survival*. The report of the Independent Commission on International Development Issues under the Chairmanship of Willy Brandt. London: Pan Books, 304 pp.

Brandt, W. 1983. *Common Crisis North–South: Co-operation for World Recovery*. The Brandt Commission 1983. London: Pan Books, 174 pp.
These books are important historical documents, written as a result of investigations by a group of international statesmen and leaders into the problems of inequality in the world and the failure of economic systems to tackle the issues. A spectrum of bold recommendations and reforms were proposed in order to avoid the perceived imminent world economic crisis. The authors described different elements of the global crisis in trade, energy and food supply, but concentrated on the overriding problem of how to provide the finance to help, and ways of compensating for the decline in financial liquidity to reverse the decline in trade and to raise the overall world economy. The WCED followed these reports with its publication of *Our Common Future* (see below). These books should be read by all students, teachers and policy-makers concerned with global environmental issues.

Ekins, P. 1992. *A New World Order: Grassroots Movements For Global Change*. London: Routledge, 248 pp.
A thought-provoking book on the problems associated with and resulting from war, insecurity and militarisation, poverty, the denial of human rights, and environmental destruction. Attention is given to possible solutions to these problems at a grassroots level.

Johnson, R.J., Taylor, P.J. and Watts, M. 1995. *Geographies of Global Change*. Oxford: Blackwell.
A useful book exploring geoeconomic, geopolitical, geosocial, geocultural and geoenvironmental change. In particular, it considers the collapse of socialism, the reconfiguration of North Atlantic capitalism, the hyper-mobility of capital, the rise of ferocious nationalism, global environmental change, the power of international media, and the social movements associated with population growth and international migration. This textbook provides a useful economic, political, social, cultural and ecological view of change at every geographical scale from the global to the local.

May, P. (ed.) 1996. *Environmental Management and Governance*. London: Routledge.
This book examines aspects and problems of environmental management. It considers the role of governments, at both local and national level, and the strengths and weaknesses of co-operative versus coercive environmental management. It does this through a focus on the management of natural hazards. It presents new and innovative environmental management and planning programmes, with particular focus on North America and Australia.

Mikesell, R.F. 1995. *Economic Development and the Environment*. London: Cassell.
This book examines how the environment and sustainability can be integrated with development programmes and strategics. It outlines the conceptual and theoretical issues involved in sustainable development and provides case studies to compare the successfulness of various types of development projects.

Moore Lappé, F. and Schurman, R. 1989. *Taking Population Seriously*. London: Earthscan.
This book provides a useful analysis of the reasons for population growth. The authors discuss the need to understand the underlying social and economic causes of population growth in order to implement effective population control.

Nebel, B.J. and Wright, R.T. 1993. *Environmental Science: The Way the World Works* (fourth edition). Englewood Cliffs, New Jersey: Prentice Hall, 630 pp.
A well-written and illustrated textbook, containing review questions and other exercise sections at the end of each chapter. It is divided into four parts, dealing with: (1) What ecosystems are and how they work; (2) Finding a balance between population, soil, water, and agriculture; (3) Pollution; and (4) Resources: Biota, Refuse, Energy, and Land. There is a useful bibliography and glossary at the back of this useful text for college students and teachers in environmental sciences.

Omara-Ojungu, P.H. 1992. *Resource Management in Developing Countries*. Harlow: Longman Scientific & Technical.
This text examines the problems of resource management in developing countries. It outlines the basic ecological, economic, technological and ethnological aspects of resource management. Emphasis is placed on the role of poverty as the critical problem facing resource management and development. It provides examples from Africa, Southeast Asia and Latin America.

Princen, T. and Finger, M. 1996. *Environmental NGOs in World Politics*. London: Routledge.
This book examines the importance of NGOs in world environmental politics. Four case studies, including the ivory trade ban and Great Lakes water negotiations, detail how NGOs challenge the traditional structures of world politics.

Redclift, M. 1987. *Sustainable Development: Exploring the Contradictions*. London: Methuen, 221 pp.
This book argues that the development recommendations of the 1987 WCED report need to be redirected to give greater

emphasis to local (indigenous) knowledge and experience if effective political action is to be taken to minimise any environmental damage. A book that is easily read, and that contains many interesting examples and recommendations.

Sarre, P. (ed.) 1991. *Environment, Population and Development*. London: Hodder & Stoughton, 304 pp.
A British Open University text that examines environmental issues with reference to population growth and economic and technological development. It is well illustrated and provides a good introduction for students concerned with environmental issues. Topics dealt with include: population dynamics; agriculture, productivity and sustainability; urbanisation; and behaviour and social problems.

United Nations 1994. *World Programme of Action: International Conference on Population and Development.* New York: United Nations Population Fund.
This is an important document resulting from the International Conference on Population and Development in Cairo in 1994. It comprises 16 chapters, the first two containing the preamble and principles, setting the overall tone of the document. Chapter 2 discusses general development, while Chapters 4 to 8 deal with the core issues involving the empowerment of women, families, under-served groups, and reproductive and sexual health and rights. Chapters 9 and 10 discuss migration, while Chapters 11 and 12 deal with education, and technology and research and development, respectively. Chapters 13 to 16 address national and international action, finances, and relations with non-governmental organisations. It is essential reading for all concerned with development issues.

World Commission on Environment and Development 1987. *Our Common Future*. Oxford: Oxford University Press, 400 pp.
The report of the WCED examines critical environmental and developmental problems, and contains many very useful tables and figures. Emphasis is placed on economic and ecological factors that may lead to sustainable development. This is an essential reference source for all college and university students, teachers and policy-makers concerned with environmental and development issues.

Prefix	SI symbol	Multiplication factor
exa	E	10^{18} (1,000,000,000,000,000,000)
peta	P	10^{15} (1,000,000,000,000,000)
tera	T	10^{12} (1,000,000,000,000)
giga	G	10^{9} (1,000,000,000)
mega	M	10^{6} (1,000,000)
kilo	k	10^{3} (1,000)
hecto	h	10^{2} (100)
deca	da	10

APPENDIX 2
Periodic table of chemical elements

The A and B subgroup designations, applicable to elements in rows 4, 5, 6, and 7, are those recommended by the International Union of Pure and Applied Chemistry. It should be noted that some authors and organisations use the opposite convention in distinguishing these subgroups.

† The names and symbols of elements 104–106 are those recommended by IUPAC as systematic alternatives to those suggested by the purported discoverers. Berkeley (USA) researchers have proposed Rutherfordium, Ff, for element 104 and Hahnium, Ha, for element 105. Dubna (USSR) researchers, who also claim the discovery of those elements, have proposed different names (and symbols).
*Estimated Values

Notes:
(1) Outline – synthetically prepared.
(2) Based upon carbon-12. ( ) indicates most stable or best known isotope.
(3) Entries marked with asterisks refer to the gaseous state at 273 K and 1 atm and are given in units of g/l.

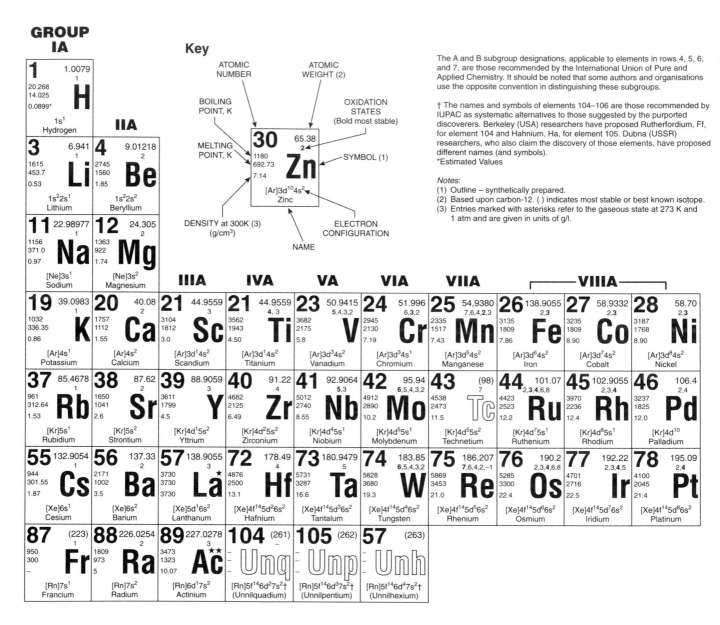

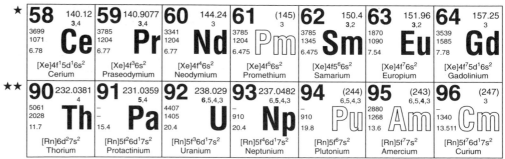

After Sargent-Welch Scientific Company, Skokie, Illinois

VII

2	4.00260
4.215	
0.95 (at 26 atm)	
0.1787*	**He**
$1s^1$	
Helium	

IIIB	IVB	VB	VIB	VIIB	

5 10.81 3	6 12.011 ±4,2	7 14.0067 ±3,5,4,2	8 15.9994 −2	9 18.998403 −1	10 20.179
4275	4470*	77.35	90.18	84.95	27.096
2300	4100*	63.14	50.35	53.48	24.553
2.34 **B**	2.62 **C**	1.251* **N**	1.429* **O**	1.696 **F**	0.901* **Ne**
$1s^1 2s^2 p$	$1s^2 2s^2 p^2$	$1s^2 2s^2 p^3$	$1s^2 2s^2 p^4$	$1s^2 2s^2 p^5$	$1s^2 2s^2 p^6$
Boron	Nitrogen	Nitrogen	Oxygen	Fluorine	Sodium

13 26.98151 3	14 28.0855 4	15 30.97376 ±3,5,4	16 32.06 ±2,4,6	17 35.453 ±1,3,5,7	18 39.948
2793	3540	550	717.75	239.1	87.30
2332*	1685	317.30	388.36	172.16	83.81
2.70 **Al**	2.33 **Si**	1.82 **P**	2.07 **S**	3.17* **Cl**	1.784* **Ar**
$[Ne]3s^1 p$	$[Ne]3s^2 p^2$	$[Ne]3s^2 p^3$	$[Ne]3s^2 p^4$	$[Ne]3s^2 p^5$	$[Ne]3s^2 p^6$
Aluminium	Silicon	Phosphorus	Sulfur	Chlorine	Argon

29 63.546 2,1	30 53.38 2	31 69.72 3	32 72.59 4	33 74.9216 ±3,5	34 78.96 −2,4,6	35 79.904 ±1,5	36 83.80
2836	1186	24??	3107	876 (subl)	4423	332.25	119.80
1357.6	692.73	302.90	1210.4	1081 (26 atm)	2523	265.90	115.78
8.96 **Cu**	7.14 **Zn**	5.81 **Ga**	5.32 **Ge**	11.5 **As**	12.2 **Se**	3.12 **Br**	3.74* **Kr**
$[Ar]3d^{10}4s^1$	$[Ar]3d^{10}4s^2$	$[Ar]3d^{10}4s^2 p^1$	$[Ar]3d^{10}4s^2 p^2$	$[Ar]3d^{10}4s^2 p^3$	$[Ar]3d^{10}4s^2 p^4$	$[Ar]3d^{10}4s^2 p^5$	$[Ar]3d^{10}4s^2 p^6$
Copper	Zinc	Gallium	Geranium	Arsenic	Selenium	Bromine	Krypton

47 107.868 1	48 112.41 2	49 114.82 3	50 118.69 4,2	51 121.75 ±3,5	52 127.60 −2,4,6	53 126.9045 −1,5,7	54 131.30
2436	104?	234?	2875	1800	1261	458.4	165.03
1234	594.18	429.76	505.06	904	722.65	386.7	161.36
10.5 **Ag**	8.65 **Cd**	7.31 **In**	7.30 **Sn**	6.68 **Sb**	6.24 **Te**	4.92 **I**	5.89* **Xe**
$[Kr]4d^{10}5s^1$	$[Kr]4d^{10}5s^2$	$[Kr]4d^{10}5s^2 p$	$[Kr]4d^{10}5s^2 p^2$	$[Kr]4d^{10}5s^2 p^3$	$[Kr]4d^{10}5s^2 p^4$	$[Kr]4d^{10}5s^2 p^5$	$[Kr]4d^{10}5s^2 p^6$
Silver	Cadmium	Indium	Tin	Antimony	Tellurium	Iodine	Xenon

79 227.0278 3,1	80 200.59 2,1	81 204.37 3,1	82 207.2 4,2	83 208.9804 3,5	84 (209) 4,2	85 (210) ±1,3,5,7	86 (222)
3130	630	1746	–	1837	1235	610	211
1337.58	234.28	????	–	544.52	527	575	202
19.3 **Au**	13.53 **Hg**	11.85 **Ti**	**Pb**	3730 **Bi**	9.8 **Po**	**At**	9.91* **Rn**
$[Xe]4f^{14}5d^{10}6s^1$	$[Xe]4f^{14}5d^{10}6s^2$	$[Xe]4f^{14}5d^{10}6s^2 p^1$	$[Xe]4f^{14}5d^{10}6s^2 p^2$	$[Xe]4f^{14}5d^{10}6s^2 p^3$	$[Xe]4f^{14}5d^{10}6s^2 p^4$	$[Xe]4f^{14}5d^{10}6s^2 p^5$	$[Xe]4f^{14}5d^{10}6s^2 p^6$
Gold	Mercury	Thallium	Lead	Bismuth	Polonium	Astatine	Radon

65 158.9254 3,4	66 152.50 1	67 161.9304 3	68 167.26	69 168.9342 3,2	70 173.04 3,2	71 174.967 3
3496	2835	2968	3136	2220	1467	3668
1630	1682	1743	1795	1818	1097	1936
8.27 **Tb**	8.54 **Dy**	8.80 **Ho**	9.05 **Er**	9.33 **Tm**	6.98 **Yb**	9.84 **Lu**
$[Xe]4f^9 6s^2$	$[Xe]4f^{10}6s^2$	$[Xe]4f^{11}6s^2$	$[Xe]4f^{12}6s^2$	$[Xe]4f^{13}6s^2$	$[Xe]4f^{14}6s^2$	$[Xe]4f^{14}5d^1 6s^2$
Terbium	Dysprosium	Holmium	Erbium	Thulium	Ytterbium	Lutetium

97 (247) 4,3	98 (251)	99 (252)	100 (257)	101 (258)	102 (259)	103 (260)
	900					
Bk	**Cf**	**Es**	**Fm**	**Md**	**No**	**Lr**
$[Rn]5f^9 7s^2$	$[Rn]5f^{10}7s^2$	$[Rn]5f^{11}7s^2$	$[Rn]5f^{12}7s^2$	$[Rn]5f^{13}7s^2$	$[Rn]5f^{14}7s^2$	$[Rn]5f^{14}6d^1 7s^2$
Berkelium	Californium	Einsteinium	Fermium	Mendelevium	Nobelium	Lawrencium

Naturally occurring radioactive isotopes are designated by a mass number although some are also manufactured. Letter m indicates an isomer of another isotope of the same mass number. Half-lives follow in parentheses, where s, min, h, d and y stand respectively for seconds, minutes, hours, days and years. The table includes mainly the longer-lived radioactive isotopes; many others have been prepared. Isotopes known to be radioactive but with half-lives exceeding 10^{12} y have not been included. Symbols describing the principal mode (or modes) of decay are as follows (these processes are generally accompanied by gamma radiation).

α alpha particle emission
β^- beta particle (electron) emission
β^+ positron emission

EC orbital electron capture
IT isomeric transition from upper to lower isomeric state
SF spontaneous fission

Element	A	Half-life / mode
Free proton	1	(15.3 min) β^-
$_1$H	3	(12.26 y) β^-
$_4$Be	7	(53.3 d) EC
	10	(1.6x10^4 y) β^-
$_6$C	11	(20.40 min) β^+
	14	(5730 y) β^-
$_9$F	18	(109.8 min) β^+
$_{11}$Na	22	(2.602 y) β^+, EC
	24	(15.02 h) β^-
$_{12}$Mg	28	(20.9 h) β^-
$_{13}$Al	26	(7.2x10^2 y) β^+, EC
$_{15}$P	32	(14.26 d) β^-
$_{16}$S	35	(87.2 d) β^-
$_{17}$Cl	36	(3.01x10^3 y) β^-
	38	(37.2 min) β^-
$_{18}$Ar	37	(35.02 d) EC
	39	(265 y) β^-
$_{19}$K	40	(1.28x10^4 y) EC
	42	(12.36 h) β^-
$_{20}$Ca	45	(165 d) β^-
$_{21}$Sc	46	(83.80 d) β^-
$_{24}$Cr	51	(27.70 d) EC
$_{25}$Mn	53	(2x10^4 y) EC
	54	(313.0 d) EC
	56	(2.578 h) β^-
$_{26}$Fe	59	(44.6 d) β^-
$_{27}$Co	56	(78.8 d) β^+, EC
	57	(270 d) βC
	58	(71.3 d) β^+, EC
	60	(5.272 y) β^-
$_{28}$Ni	57	(36.0 h) β^+, EC
	59	(8x10^4 y) EC
	63	(92 y) β^-
$_{29}$Cu	64	(12.70 h) β^-, β^+, EC
	67	(61.88 h) β^-
$_{30}$Zn	65	(244.1 d) β^+, EC
$_{31}$Ga	67	(78.2 h) EC
	72	(14.10 h) β^-
$_{32}$Ge	68	(275 d) EC
$_{33}$As	73	(80.3 d) EC
	74	(17.9 d) β^-, β^+, EC
$_{34}$Se	75	(118.5 d) β^-
	79	(6.5x10^4 y) β^-
$_{35}$Br	82	(35.34 h) β^-
$_{36}$Kr	81	(2.1x10^3 y) EC
	85	(10.72 y) β^-
$_{37}$Rb	86	(18.7 d) β^-
	87	(4.8x10^{11} y) β^-
$_{38}$Sr	90	(28.8 y) β^-
$_{39}$Y	88	(106.6 d) β^+, EC
$_{40}$Zr	93	(1.5x10^4 y) β^-
	95	(64.0 d) β^-
$_{41}$Nb	94	(2.0x10^4 y) β^-
	95	(35.15 d) β^-
$_{42}$Mo	99	(66.02 h) β^-
$_{43}$Tc	97	(2.6x10^4 y) EC
	98	(4.2x10^4 y) β^-
	99	(2.13x10^5 y) β^-
$_{44}$Ru	106	(367 d) β^-
$_{45}$Rh	101	(3.3 y) EC
$_{46}$Pd	103	(17.0 d) EC
	107	(7x10^4 y) β^-
$_{47}$Ag	108	(127 y) EC
	110	(252 d) β^-
	111	(7.45 d) β^-
$_{48}$Cd	109	(453 d) EC
$_{49}$In	114	(49.51 d) IT
$_{50}$Sn	121	(76 y) β^-
$_{51}$Sb	124	(60.20 d) β^-
	125	(2.7 y) β^-
$_{52}$Te	121	m (154 d) IT
	123	m (119.7 d) IT
	127	m (109 d) IT
$_{53}$I	129	(1.6x10^3 y) β^-
	131	(8.040 d) β^-
$_{54}$Xe	133	(5.25 d) β^-
	135	(9.10 h) β^-
$_{55}$Cs	134	(2.06 y) β^-
	135	(2.9x10^4 y) β^-
	137	(30.17 y) β^-
$_{56}$Ba	140	(12.8 d) β^-
$_{57}$La	137	(6x10^4 y) EC
	140	(40.3 h) β^-
$_{58}$Ce	144	(284 d) β^-
$_{59}$Pr	142	(19.1 h) β^-
$_{60}$Nd	147	(11.1 d) β^-
$_{61}$Pm	145	(18 y) EC
	147	(2.62 y) β^-
$_{62}$Sm	146	(7x10^3 y) α
	151	(93 y) β^-
$_{63}$Eu	152	(13 y) β^+, EC, β^-
	154	(8.5 y) β^-
$_{64}$Gd	150	(2.1x10^6 y) α
$_{65}$Tb	158	(1.2x10^3 y) EC, β^-
	160	(72.3 d) β^-
$_{67}$Ho	166	m (1.2x10^3 y) β^-
$_{69}$Tm	170	(128.6 d) β^-
	171	(1.92 y) β^-
$_{70}$Yb	169	(32.0 d) EC
	175	(4.19 d) β^-
$_{71}$Lu	176	(3.7x10^{10} y) β^-
$_{73}$Ta	182	(115.0 d) β^-
$_{74}$W	181	(140 d) EC
	185	(75.1 d) β^-
	188	(69 d) β^-
$_{75}$Re	187	(5x10^{10} y) β^-
$_{76}$Os	194	(6.0 y) β^-
$_{77}$Ir	192	(74.2 d) β^-, β^+, EC
$_{79}$Au	195	(183 d) EC
	196	(6.18 d) β^+, EC, β^-
	198	(2.696 d) β^-
	199	(3.15 d) β^-
$_{80}$Hg	203	(46.8 d) β^-
$_{81}$Tl	204	(3.77 y) β^-
$_{82}$Pb	202	(3x10^5 y) EC
	205	(3x10^7 y) EC
	210	(22.3 y) β^-, α
$_{83}$Bi	207	(38 y) EC
	208	(3.7x10^5 y) EC
	210	(5.01 d) β^-, α
	210	m (3x10^4 y) α
$_{84}$Po	208	(2.90 y) α
	209	(102 y) α
	210	(138.38 d) α
$_{85}$At	209	(5.4 h) EC, α
	210	(8.1 h) EC
	211	(7.21 h) EC, α
$_{86}$Rn	222	(3.824 d) EC, α
$_{87}$Fr	212	(19.3 min) EC, α
	222	(15 min) β^-
	223	(21.8 min) β^-
$_{88}$Ra	226	(1.60x10^3 y) α
$_{89}$Ac	227	(21.77 y) β^-
$_{90}$Th	228	(1.913 y) α
	230	(7.7x10^4 y) α
	232	(1.40x10^{10} y) α
$_{91}$Pa	231	(3.28x10^4 y) α
$_{92}$U	233	(1.59x10^3 y) α
	234	(2.44x10^5 y) α
	235	(7.04x10^8 y) α
	236	(2.34x10^7 y) α
	238	(4.47x10^9 y) α
$_{93}$Np	236	(1.1x10^5 y) EC, β^-
	237	(2.14x10^6 y) α
	239	(2.346 d) β^-
$_{94}$Pu	238	(87.75 y) α
	239	(2.41x10^4 y) α
	240	(6.54x10^3 y) α
	242	(3.8x10^5 y) α
	244	(8.3x10^7 y) α
$_{95}$Am	241	(432 y) α
	243	(7.37x10^3 y) α
$_{96}$Cm	242	(163.2 d) α
	244	(18.12 y) α
	247	(1.55x10^7 y) α
	248	(3.5x10^5 y) α, SF
$_{97}$Bk	247	(1.4x10^3 y) α
$_{98}$Cf	249	(351 y) α
	251	(900 y) α
$_{99}$Es	252	(472 d) α
	253	(20.47 d) α
	254	(276 d) α
$_{100}$Fm	255	(20.1 h) α
	257	(100.5 d) α
$_{101}$Md	258	(55 d) α
$_{102}$No	259	(58 min) α
$_{103}$Lw	260	(3.0 min) α
104	261	(65 s) α
105	262	(40 s) α
106	263	(0.9 s) α

Country	GDP/$	GDP per capita	Population 1996	change	GDP growth 1995	1996	Inflation 1995	1996
Western Europe								
Austria	$248.0bn	$30,980	8.0m	0%	2.5%	2.3%	2.5%	2.7%
Belgium	$267.2bn	$26,200	10.2m	0%	2.2%	2%	1.6%	2.2%
Denmark	$178.3bn	$34,160	5.2m	0.2%	2.8%	2.6%	2.1%	2.5%
Finland	$129.4bn	$25,330	5.1m	0%	5%	4.2%	2%	2.5%
France	$1.6trn	$27,000	58.4m	0.5%	2.7%	2.6%	1.9%	2.3%
Germany	$2.5trn	$31,170	81.3m	0%	2.6%	2.8%	2.2%	2.1%
Greece	$119.0bn	$11,500	10.4m	0%	1.7%	1.9%	9.3%	7.7%
Ireland	$64.3bn	$17,860	3.6m	0.3%	6.1%	5.5%	2.9%	3.5%
Italy	$1.2trn	$20,670	56.9m	0%	2.8%	2.7%	5.5%	6.0%
Netherlands	$431.0bn	$27,600	15.6m	0.6%	2.5%	2.7%	3%	3.2%
Norway	$605.1bn	$34,800	4.38m	0.7%	3.7%	3.2%	2.5%	2.8%
Portugal	$88.7bn	$8,960	9.9m	0%	2.7%	3%	4.8%	4.6%
Spain	$547.5bn	$13,930	39.3m	0.3%	3.1%	2.8%	4.8%	4.2%
Sweden	$225.1bn	$25,340	8.9m	0.5%	3.1%	2.5%	2.8%	2.7%
Switzerland	$296.4bn	$41,750	7.1m	0%	1.1%	1.7%	2%	2.1%
Turkey	$183.8bn	$2,890	63.7m	2.1%	4.5%	5.3%	85.9%	63.3%
UK	$1.2trn	$20,490	58.7m	0.4%	2.7%	3%	3.5%	3.0%
North America								
Canada	$611.9bn	$20,160	30.2m	1.3%	2.1%	2.5%	2.4%	2.5%
USA	$7.6trn	$28,440	266.7m	0.9%	3.0%	2.5%	2.9%	2.8%
Eastern Europe								
Bulgaria	$12.2bn	$1,460	8.4m	−0.4%	2%	3%	68%	40%
Czech Republic	$51.0bn	$4,920	10.4m	1.0%	3.8%	4.2%	9.5%	8%
Slovakia	$18.0bn	$3,390	5.3m	0%	5.6%	4.6%	10.9%	10%
Hungary	$47.8bn	$4,690	10.2m	−0.3%	1.5%	3%	28%	20%
Poland	$132.3bn	$3,420	38.7m	0.3%	5.9%	4.8%	26.5%	20%
Romania	$35.5bn	$1,565	22.7m	0%	4%	4%	32%	25%
Russia	$438.2bn	$2,960	148.2m	0%	−2%	3%	200%	80%
Ukraine	$46.0bn	$900	51.6m	0.3%	−8%	−2%	380%	100%
Asia and Pacific								
Australia	$368.8bn	$20,200	18.3m	1.1%	3.1%	2.6%	4%	3.5%
China	$636.7bn	$520	1.23bn	1.7%	9.8%	8.6%	18%	12%
Hong Kong	$165.9bn	$27,040	6.1m	−0.3%	4.5%	3.5%	9%	7.5%
India	$320.9bn	$335	956.6m	2%	4.9%	4.9%	10%	10.5%
Indonesia	$207.5bn	$1,040	199.9m	1.7%	7.1%	7.3%	9.2%	8.5%
Japan	$5.1trn	$40,500	125.7m	−0.3%	0.5%	1.4%	0%	0.1%
Kazakhstan	$16.0bn	$930	17.2m	0.9%	−10%	−3%	170%	70%
Malaysia	$97.3bn	$4,261	20.1m	2%	8.8%	8.2%	4.4%	3.9%
New Zealand	$62.5bn	$17,230	3.6m	1.1%	2%	2.5%	3.6%	1%
Pakistan	$61.9bn	$464	133.2m	2.7%	4.3%	5.3%	13.2%	11%
Philippines	$74.7bn	$1,070	70.1m	2.2%	5.4%	5.6%	8.5%	7.5%
Singapore	$92.3bn	$30,301	3.1m	2%	7.9%	7.5%	2.2%	2%

Country	GDP/$	GDP per capita	Population		GDP growth		Inflation	
			1996	change	1995	1996	1995	1996
South Korea	$523.9bn	$11,580	45.2m	0.9%	9.4%	7.7%	4.9%	4.8%
Taiwan	$310.2bn	$14,470	21.4m	0.9%	6.5%	6.5%	4.0%	3.9%
Thailand	$190.3bn	$3,110	61.1m	1.3%	8.7%	8.4%	5.4%	5%
Vietnam	$24.7bn	$325	75.7m	2.3%	9.1%	9.7%	17.5%	19%
Latin America								
Argentina	$296.3bn	$8,550	34.7m	1.2%	−1.6%	0.6%	3.4%	1.6%
Brazil	$883.3bn	$5,230	164.2m	1.6%	5.5%	3.1%	78%	31%
Chile	$64.2bn	$4,480	14.3m	1.4%	7%	6.6%	8.2%	7.7%
Colombia	$90.7bn	$2,540	35.7m	1.7%	4.8%	3.8%	21%	18.2%
Mexico	$321.2bn	$3,370	93.7m	1.8%	−4.1%	1.4%	34.8%	27%
Venezuela	$98.3bn	$4,490	21.9m	1.9%	−0.9%	1%	62%	79.5%
Africa								
Nigeria	$20.4bn	$210	98.1m	2.1%	2.6%	3.3%	50%	32%
South Africa	$143.1bn	$3,370	42.5m	2.5%	3%	4%	10.2%	12%
Zimbabwe	$7.1bn	$605	11.8m	2.8%	1.5%	3.5%	22%	18%
Middle East								
Algeria	$49.9bn	$1,730	28.8m	2.5%	3.6%	4.5%	35%	33%
Egypt	$54.9bn	$910	60.1m	2%	2.4%	3.3%	10%	9%
Iran	$59.8bn	$880	67.6m	3.2%	−2%	2%	50%	35%
Iraq	$32.0bn	$1,450	22m	2.3%	0%	0%	250%	250%
Israel	$94.3bn	$16,320	5.8m	2.8%	5.5%	5%	10%	8.5%
Jordan	$7.4bn	$1,690	4.4m	3.1%	5.7%	5.5%	3%	3.9%
Lebanon	$10.3bn	$2,710	3.8m	–	7.5%	8.5%	12%	14%
Saudi Arabia	$133.8bn	$6,930	19.3m	3.8%	1.5%	1.8%	2.5%	3.6%

Adapted from *The World in 1996*, *The Economist*, London, 146pp.

Appendix 4 reproduces the Rio Declaration on Environment and Development, associated with Agenda 21 – a programme of action for sustainable development worldwide – adopted by more than 178 governments at the United Nations Conference on Environment and Development or Earth Summit, held in Rio de Janeiro, Brazil, 3–14 June 1992. While the agreements, which were negotiated over two and a half years leading up to the Earth Summit and finalised in Rio, lack the force of international law, the adoption of the text carries with it a strong moral obligation to ensure their full implementation.

Agenda 21 represents a comprehensive blueprint for global action into the twenty-first century by governments, the United Nations organisations, development agencies, non-governmental organisations and independent sector groups, in all spheres of human activity which impacts on the environment. The United Nations Commission on Sustainable Development, set up under the aegis of the UN General Assembly in response to a request of the Rio conference and comprising government representatives, will examine the progress made in implementing Agenda 21 worldwide.

The central tenet of Agenda 21 is that humanity stands at a defining moment in history, where we face the perpetuation of disparities between and within nations, increasing poverty, ill health, and illiteracy, and the continuing deterioration of the ecosystems on which we depend for our well-being and survival. Basic human needs and improved living standards for all can be met, at least to a far greater extent than at present, by integrating environmental and developmental concerns. A global partnership is the only means of achieving such aspirations, and Agenda 21 provides a framework within which this can be attempted.

The Agenda 21 programme is set out in terms of the basis for actions, objectives, activities, and means of implementation. It is therefore a dynamic programme. Here, only the principles of the Rio Declaration on Environment and Development are printed in order to provide the reader with the spirit and aims of Agenda 21.

Rio Declaration on Environment and Development, Agenda 21

Having met at Rio de Janeiro from 3 to 14 June 1992,

Reaffirming the Declaration of the United Nations Conference on the Human Environment, adopted at Stockholm on 16 June 1972, and seeking to build upon it,

With the goal of establishing a new and equitable global partnership through the creation of new levels of co-operation among States, key sectors of societies and people,

Working towards international agreements which respect the interests of all and protect the integrity of the global environmental and developmental system,

Recognising the integral and interdependent nature of the Earth, our home,

Proclaims that:

Principle 1

Human beings are at the centre of concerns for sustainable development. They are entitled to a healthy and productive life in harmony with nature.

Principle 2

States have, in accordance with the Charter of the United Nations and the principles of international law, the sovereign right to exploit their own resources pursuant to their own environmental and developmental policies, and the responsibility to ensure that activities within their jurisdiction or control do not cause damage to the environment of other States or of areas beyond the limits of national jurisdiction.

Principle 3

The right to development must be fulfilled so as to equitably meet developmental and environmental needs of present and future generations.

Principle 4

In order to achieve sustainable development, environmental protection shall constitute an integral part of the development process and cannot be considered in isolation from it.

Principle 5

All States and all people shall co-operate in the essential task of eradicating poverty as an indispensable requirement for sustainable development, in order to decrease the disparities in standards of living and better meet the needs of the majority of the people of the world.

Principle 6

The special situation and needs of developing countries, particularly the least developed and those most environmentally vulnerable, shall be given special priority. International actions in the field of environment and development should also address the interests and needs of all countries.

Principle 7

States shall co-operate in a spirit of global partnership to conserve, protect and restore the health and integrity of the Earth's ecosystem. In view of the different contributions to global environmental degradation, States have common but differentiated responsibilities. The developed countries acknowledge the responsibility that they bear in the international pursuit of sustainable development in view of the pressures their societies place on the global environment and of the technologies and financial resources they command.

Principle 8

To achieve sustainable development and a higher quality of life for all people, States should reduce and eliminate unsustainable patterns of production and consumption and promote appropriate demographic policies.

Principle 9

States should co-operate to strengthen endogenous capacity-building for sustainable development by improving scientific understanding through exchanges of scientific and technological knowledge, and by enhancing the development, adaptation, diffusion and transfer of technologies, including new and innovative technologies.

Principle 10

Environmental issues are best handled with the participation of all concerned citizens, at the relevant level. At the national level, each individual shall have appropriate access to information concerning the environment that is held by public authorities, including information on hazardous materials and activities in their communities, and the opportunity to participate in decision-making processes. States shall facilitate and encourage public awareness and participation by making information widely available. Effective access to judicial and administrative proceedings, including redress and remedy, shall be provided.

Principle 11

States shall enact effective environmental legislation. Environmental standards, management objectives and priorities should reflect the environmental and developmental context to which they apply. Standards applied by some countries may be inappropriate and of unwarranted economic and social cost to other countries, in particular developing countries.

Principle 12

States should co-operate to promote a supportive and open international economic system that would lead to economic growth and sustainable development in all countries, to better address the problems of environmental degradation. Trade policy measures for environmental purposes should not constitute a means of arbitrary or unjustifiable discrimination or a disguised restriction on international trade. Unilateral actions to deal with environmental challenges outside the jurisdiction of the importing country should be avoided. Environmental measures addressing transboundary or global environmental problems should, as far as possible, be based on an international consensus.

Principle 13

States shall develop national law regarding liability and compensation for the victims of pollution and other environmental damage. States shall also co-operate in an expeditious and more determined manner to develop further international law regarding liability and compensation for adverse effects of environmental damage caused by activities within their jurisdiction or control to areas beyond their jurisdiction.

Principle 14

States should effectively co-operate to discourage or prevent the relocation and transfer to other States of any activities and substances that cause severe environmental degradation or are found to be harmful to human health.

Principle 15

In order to protect the environment, the precautionary approach shall be widely applied by States according to their capabilities. Where there are threats of serious or irreversible damage, lack of full scientific certainty shall not be used as a reason for postponing cost-effective measures to prevent environmental degradation.

Principle 16

National authorities should endeavour to promote the internalisation of environmental costs and the use of economic instruments, taking into account the approach that the polluter should, in principle, bear the cost of pollution, with due regard to the public interest and without distorting international trade and investment.

Principle 17

Environmental impact assessment, as a national instrument, shall be undertaken for proposed activities that are likely to have a significant adverse impact on the environment and are subject to a decision of a competent national authority.

Principle 18

States shall immediately notify other States of any natural disasters or other emergencies that are likely to produce sudden harmful effects on the environment of those States. Every effort shall be made by the international community to help States so afflicted.

Principle 19

States shall provide prior and timely notification and relevant information to potentially affected States on activities that may have a significant adverse transboundary environmental effect and shall consult with those States at an early stage and in good faith.

Principle 20

Women have a vital role in environmental management and development. Their full participation is therefore essential to achieve sustainable development.

Principle 21

The creativity, ideals and courage of the youth of the world should be mobilised to forge a global partnership in order to achieve sustainable development and ensure a better future for all.

Principle 22

Indigenous people and their communities and other local communities have a vital role in environmental management and development because of their knowledge and traditional practices. States should recognise and duly support their identity, culture and interests and enable their effective participation in the achievement of sustainable development.

Principle 23

The environment and natural resources of people under oppression, domination and occupation shall be protected.

Principle 24

Warfare is inherently destructive of sustainable development. States shall therefore respect international law providing protection for the environment in times of armed conflict and co-operate in its further development, as necessary.

Principle 25

Peace, development and environmental protection are interdependent and indivisible.

Principle 26

States shall resolve all their environmental disputes peacefully and by appropriate means in accordance with the Charter of the United Nations.

Principle 27

States and people shall co-operate in good faith and in a spirit of partnership in the fulfilment of the principles embodied in this Declaration and in the further development of international law in the field of sustainable development.

Source: *Earth Summit Agenda 21: The United Nations Programme of Action from Rio.* United Nations Publication E.93.1.11 (April 1993).

Bibliography

Abbott, A. 1995a. Climate change panel to remain main source of advice. *Nature*, 374, 584–585.

Abbott, A. 1995b. Meeting agrees on need for new targets for greenhouse gas emissions. *Nature*, 374, 584–585.

Adam, R.D. 1994. AIDS debate continues. *Nature*, 367, 212.

Adams, W.M. 1990. *Green Development: Environment and Sustainability in the Third World*. Routledge, 255 pp.

Addiscott, T. 1988. Farmers, fertilisers and the nitrate flood. *New Scientist*, 120, 1633, 50–54.

AFEAS (Alternative Fluorocarbons Environmental Acceptability Study) and PAFT (Programme for Alternative Fluorocarbon Toxicity Testing) Member Companies 1992. Information pack.

Allard, P., Carbonnelle, J., Dajlevic, D., Le Bronec, J., Morel, P., Robe, M.C., Maurenas, J.M., Faivre-Pierret, R., Martin, D., Sabrouk, J.C. and Zeltwoag, P. 1991. Eruptive and diffuse emissions of CO_2 from Mount Etna. *Nature*, 351, 387–391.

Allen, J.C., Schaffer, W.M. and Rosko, D. 1993. Chaos reduces species extinction by amplifying local population noise. *Nature*, 364, 229–232.

Allen, T. and Thomas, A. (eds) 1992. *Poverty and Development in the 1990s*. Oxford: Oxford University Press, 421 pp.

Alley, R.B., Meese, D.A., Shuman, C.A., Gow, A.J., Taylor, K.C., Grootes, P.M., White, J.W.C., Ram, M., Waddington, E.D., Mayewski, P. and Zielinski, G.A. 1993. Abrupt increase in Greenland snow accumulation at the end of the Younger Dryas event. *Nature*, 362, 527–529.

Alvarez, L.W., Alvarez, A., Asaro, F. and Michel, H.V. 1980. Extraterrestrial cause for the Cretaceous–Tertiary extinction. *Science*, 208, 1095–1108.

Alvarez, W. and Asaro, F. 1990. An extraterrestrial impact. *Scientific American*, October, 44–60.

Alvarez, W. and Muller, R.A. 1984. Evidence from crater ages for periodic impacts on the Earth. *Nature*, 308, 718–720.

American Geophysical Union 1992a. *Volcanism and Climatic Change*, UGU Special Report.

American Geophysical Union 1992b. Impacts of Mid-Atlantic sewage sludge probed. *EOS*, 73, 27–28.

American Institute of Professional Geologists, 1993. Landslides and Avalanches. In: *The Citizen's Guide to Geological Hazards*, The American Institute of Professional Geologists, Arvada, Colorado.

An Agenda of Science for Environment and Development into the 21st Century: Based on a Conference held in Vienna, Austria in November 1991. Edited by Dooge, J.C.I., Goodman, G.T., la Rivière, J.W.M., Marton-Léfevre, O'Riordan, T. and Praderie, F. 1992. Cambridge: Cambridge University Press, 331 pp.

Anderson, D.M. 1994. Red Tides. *Scientific American*, August, 52–58.

Anderson, I. 1993. Diggers at Dinosaur Cove. *New Scientist*, 13 February, 28–32.

Anderson, I. 1995. Fallout in the South Pacific. *New Scientist*, 2 September, 12–13.

Anderson, R. 1993. AIDS: trends, predictions, controversy. *Nature*, 363, 393–394.

Anderson, T.F. and Arthur, M.A. 1983. Stable isotopes of oxygen and carbon and their application to sedimentologic and environmental problems. In: *Stable Isotopes in Sedimentary Geology*, SEPM Short Course Notes 10, 1–151. Tulsa, Oklahoma: Society of Economic Paleontologists and Mineralogists.

Anderson, V. 1991. *Alternative Economic Indicators*. London: Routledge, 106 pp.

Angotti, T. 1993. *Metropolis 2000: Planning, Poverty and Politics*. London: Routledge, 276 pp.

Anthes, R.A. 1982. Tropical cyclones: their evolution, structure and effects. *American Meteorological Society Meteorological Monograph*, 19, 41.

Archibald, J.D. 1993. Were dinosaurs born losers? *New Scientist*, 13 February, 24–27.

Arizipe, L., Costanza, R. and Lutz, W. 1992. Population and natural resource use. In: Dooge, J. *et al.* (eds), *An Agenda of Science for Environment and Development into the 21st Century*. Cambridge: Cambridge University Press, 61–78.

Armishaw, R., Bardos, R.P., Dunn, R.M., Hill, J.W., Pearl, M., Rampling, T. and Wood, P.A. 1992. *Review of Innovative Contaminated Soil Clean-Up Processes*. Warren Spring Laboratory Report, UK, 231 pp.

Armstrong, B.R. 1984. Avalanche accident victims in the USA. *Ekistics*, 51, 309, 543–546.

Armstrong, S. 1988. Marooned in a mountain of manure. *New Scientist*, 26 November, 51–55.

Arnold, D. 1989. *Famine: social crisis and historical change*. Oxford: Blackwell.

Austin, J., Butchart, N. and Shine, K.P. 1992. Possibility of an Arctic ozone hole in a doubled-CO_2 climate. *Nature*, 360, 221–225.

Ayers, G.P., Penkett, S.A., Gillett, R.W., Bandy, B., Galbally, I.E., Meyer, C.P., Elsworth, C.M., Bentley, S.T. and Forgan, B.W. 1992. Evidence for photochemical control of ozone concentrations in unpolluted marine air. *Nature*, 360, 446–449.

Badr, O. and Probert, S.D. 1992. Sources of atmospheric nitrous oxide. *Applied Energy*, 42, 129–176.

Badr, O., Probert, S.D. and O'Callaghan, P.W. 1992a. Methane: a greenhouse gas in the Earth's atmosphere. *Applied Energy*, 41, 95–113.

Badr, O., Probert, S.D. and O'Callaghan, P.W. 1992b. Sinks for atmospheric methane. *Applied Energy*, 41, 137–147.

Bailey, E. 1962. *Charles Lyell*. British Men of Science. London: Nelson.

Baillie, M.G.L. and Munro, M.A.R. 1988. Irish tree-rings, Santorini and volcanic dust veils. *Nature*, 332, 344–346.

Bakan, S., Chlond, A., Cubasch, U., Feichter, J., Graf, H., Grassl, H., Hasselman, K., Kirchner, I., Latif, M., Roekner, E., Sausen, R., Schlese, U., Schriever, D., Schult, I., Schumann, U., Seilmann, F. and Welke, W. 1991. Climate response to smoke from the burning oil wells in Kuwait. *Nature*, 351, 367–371.

Baker, V. 1995. Learning from the past. *Nature*, 361, 402–440.

Bakwesegha, C.J. 1994. Forced migration in Africa and the OAU Convention. In: Adelman, H. and Sorenson, J. (eds), *African Refugees: Development Aid and Repatriation*. Boulder, Colo.: Westview Press, 3–19.

Ball, T.K., Cameron, D.G., Colman, T.B. and Roberts, P.D. 1991. Behaviour of radon in the geological environment: a review. *Quarterly Journal of Engineering Geology*, 24, 169–182.

Banks, H.O. and Richter, R.C. 1953. Sea-water intrusion into groundwater basins bordering the Californian coast and inland bays. *Transactions of the American Geophysical Union*, 34, 573–582.

Bari, M.A. and Schofield, N.J. 1992. Lowering of a shallow, saline water table by extensive eucalypt reforestation. *Journal of Hydrology*, 133, 273–291.

Barkham, J.P. 1993. For peat's sake: conservation or exploitation? *Biodiversity and Conservation*, 2, 556–566.

Barnola, J.M., Raynaud, D., Korotkevich, Y.S. and Lorius, C. 1987. Vostok ice core provides 160,000-year record of atmospheric CO_2. *Nature*, 329, 408–414.

Barrett, P.J., Adams, C.J., McIntosh, W.C., Swisher, C.C. and Wilson, G.S. 1992. Geochronological evidence supporting Antarctic deglaciation three million years ago. *Nature*, 359, 816–818.

Barry, R.G. and Chorley, R.J. 1992. *Atmosphere, Weather and Climate* (6th edition). London: Routledge, 392 pp.

Bassett, M.G. 1985. Towards a 'common language' in stratigraphy. *Episodes*, 8, 87–92.

Batey, T. 1988. *Soil Husbandry*. Aberdeen: Soil and Landuse Consultants Ltd.

Batifol, F., Boutron, C. and de Angelis, M. 1989. Changes in copper, zinc and cadmium concentration in Antarctic ice during the past 40,000 years. *Nature*, 337, 544–546.

Battarbee, R.W. 1984. Diatom analysis and the acidification of lakes. *Philosophical Transactions of the Royal Society, London*, B305, 451–477.

Battarbee, R.W. 1986. Diatom analysis. In: Berglund, B.E. (ed.), *Handbook of Holocene Palaeoecology and Palaeohydrology*, 527–570. Chichester: John Wiley & Sons Ltd.

Battarbee, R.W. 1992. Holocene lake sediments, surface water acidification and air pollution. In: Gray, J.M. (ed.), *Applications of Quaternary Research*, 101–110. *Quaternary Proceedings*, 2, Cambridge.

Battarbee, R.W. and Allott, T.E.H. 1993. Lake acidification: evidence, effects, management and reversibility. In: de Bernardi, R., Pagnotta, R. and Pugnetti, A. (eds), *Strategies for Lake Ecosystems Beyond 2000*, 319–340. Memoirs of the First Italian Idrabiol., 52.

Becker, B., Kromer, B. and Trimborn, P. 1991. A stable-isotope tree-ring timescale of the Late Glacial/Holocene boundary. *Nature*, 353, 647–649.

Bedding, J. 1989. Money down the drains. *New Scientist*, 122, 1663, 37–41.

Beder, S. 1990. Sun, surf and sewage. *New Scientist*, 14 July, 40–45.

Begley, S. 1995. Lessons of Kobe. *Newsweek*, 125, 5, 16–27.

Bekki S., Law, K.S. and Pyle, J.A. 1994. Effect of ozone depletion on atmospheric CH_4 and CO concentrations. *Nature*, 371, 595–597.

Bekki, S., Toumi, R. and Pyle, J.A. 1993. Role of sulphur photochemistry in tropical ozone changes after the eruption of Mount Pinatubo. *Nature*, 362, 331–333.

Bell, F.G. 1988. The history and techniques of coal mining and the associated effects and influence on construction. *Bulletin of the Association of Engineering Geology*, 24, 137–144.

Bell, F.G. 1992. A history of salt mining in mid-Cheshire, England, and its influence on planning. *Bulletin of the Association of Engineering Geology*, 29, 371–386.

Bell, M. and Walker, M.J.C. 1992. *Late Quaternary Environmental Change: Physical and Human Perspectives*. New York: Longman Scientific, 237 pp.

Belsky, A.J. 1990. Tree/grass ratios in East African savannas: a comparison of existing models. *Journal of Biogeography*, 17, 483–489.

Bender, M., Sowers, T., Dickson, M.-L., Orchardo, J., Grootes, P., Mayewski, P.A. and Meese, D.A. 1994. Climate correlations between Greenland and Antarctica during the past 100,000 years. *Nature*, 372, 663–666.

Bentley, S.P. and Smalley, I.J. 1984. Landslips in sensitive clays. In: Brunsden, D. and Prior, D.B. (eds), *Slope Instability*. Chichester: Wiley, 457–490.

Berkhout, F. 1991. *Radioactive Waste: Politics and Technology*. London: Routledge, 256 pp.

Bernard, A., Demaiffe, D., Mattelli, N. and Punongbayan, R.S. 1991. Anhydrite-bearing pumice from Mount Pinatubo: further evidence for the existence of sulphur-rich silicic magmas. *Nature*, 354, 139–140.

Billett, D.S.M., Lampitt, R.S., Rice, A.L. and Mantoura,

R.F.C. 1983. Seasonal sedimentation of phytoplankton to the deep-sea benthos. *Nature*, 302, 520–522.

Binns, T. 1990. Is desertification a myth? *Geography*, 75(2), 106–13.

Bird, E.F.C. 1985. *Coastline Changes: A Global Review.* Chichester: John Wiley & Sons Ltd.

Birks, H.J.B., Line, J.M., Juggins, S., Stenson, A.C. and ter Braak, C.J.F. 1990. Diatoms and pH reconstruction. *Philosophical Transactions of the Royal Society of London*, B327, 263–278.

Bland, P., Gilmor, I. and Kelley, S. 1996. Mass extinctions and asteroids, *Geoscientist*, 6, 5–6.

Blum, E. 1993. Making biodiversity conservation profitable: a case study of the Merck/INBio agreement. *Environment*, 35(4), 16–20 and 38–45.

Blunden, J. 1991. The environmental impact of mining and mineral processing. In: Blunden, J. and Reddish, A. (eds), *Energy, Resources and Environment*, Sevenoaks, Kent: Hodder & Stoughton, 79–132.

Blunden, J. and Reddish, A. (eds) 1991. *Energy, Resources and Environment.* The Open University, Hodder & Stoughton, 339 pp.

Bluth, G.J.S., Schnetzler, C.C., Krueger, A.J. and Walter, L.S. 1993. The contribution of explosive volcanism to global atmospheric sulphur dioxide concentrations. *Nature*, 366, 327–329.

Bogert, C. 1995. Bailing out California. *Newsweek*, 125, 4, 29–31.

Bond, G., Broecker, W., Labeyrie, L., Jouzel, J., Johnsen, S., McManus, J. and Bonani, G. 1993. Correlations between climate records from North Atlantic sediments and Greenland ice. *Nature*, 365, 143–147.

Bond, G., Heinrich, H., Broecker, W., Labeyrie, L., McManus, J., Andrews, J., Huon, S., Jantschik, R., Clasen, S., Simet, C., Tedesco, K., Klas, M., Bonani, G. and Ivy, S. 1992. Evidence for massive discharges of icebergs into the North Atlantic Ocean during the last glacial period. *Nature*, 360, 245–249.

Bond, G.C. and Lotti, R. 1995. Iceberg discharges into the North Atlantic on millennial time scales during the Last Glaciation. *Science*, 267, 1005–1010.

Bonney, T.G. 1895. *Charles Lyell and Modern Geology.* The Century Science Series. London: Cassell & Co.

Bormann, F.H., Likens, G.E., Fisher, D.W. and Pierce, R.S. 1968. Nutrient loss accelerated by clearcutting of a forest ecosystem. *Science*, 159, 882–884.

Botkin, D. and Keller, E. 1995. *Environmental Science: Earth as a Living Planet.* Chichester: John Wiley & Sons.

Boulton, G.S. 1993. The spectrum of Quaternary global change – understanding the past and anticipating the future. *Geoscientist*, 2, 10–12.

Boulton, G.S., Smith, G.D., Jones, A.S. and Newsome, J. 1985. Glacial geology and glaciology of the last mid-latitude ice sheets. *Journal of the Geological Society, London*, 142, 447–474.

Bowcott, O. 1996. Disaster alert after oil spill. *The Guardian*, 17 February, 1 and 3.

Bowen, D.Q., Rose, J., McCabe, A.M. and Sutherland, D.G. 1986. Correlation of Quaternary glaciations in England, Ireland, Scotland and Wales. *Quaternary Science Reviews*, 5, 299–340.

Bowler, S. 1993. Where the power lies. *New Scientist*, 23 January, 32–36.

Boyle, E.A. 1988. Cadmium: chemical tracer of deepwater palaeoceanography. *Paleoceanography*, 3, 471–489.

Boyle, G. 1996. *Renewable Energy: Power for a Sustainable Future.* Oxford: The Open University/Oxford University Press, 479 pp.

Boyle, S. 1989. More work for less energy. *New Scientist*, 5 August, 37–40.

Boyle, S. and Ardill, J. 1989. *The Greenhouse Effect.* London: Hodder & Stoughton, 298 pp.

Bradley, R.S. 1985. *Quaternary Paleoclimatology – Methods of Paleoclimatic Reconstruction.* London: Unwin Hyman, 472 pp.

Bradley, R.S. and Jones, P.D. 1992. Records of explosive volcanic eruptions over the last 500 years. In: Bradley, R.S. and Jones, P.D. (eds), *Climate Since AD 1500.* London: Routledge, 606–622.

Bradshaw, M. and Weaver, R. 1993. *Physical Geography: An Introduction to Earth Environments.* London: Mosby.

Brady, N.C. 1990. *The Nature and Properties of Soils.* London: Maxwell Macmillan.

Bragg, J.R., Prince, R.C., Harner, E.J. and Atlas, R.M. 1994. Effectiveness of bioremediation for the *Exxon Valdez* oil spill. *Nature*, 368, 413–418.

Brain, C.K. and Sillen, A. 1988. Evidence from the Swartkrans cave for the earliest use of fire. *Nature*, 336, 464–466.

Brandt, W. 1980. *North–South: A Programme for Survival.* Pan Books, London.

Brandt, W. 1983. *Common Crisis North–South: Cooperation for World Recovery.* London: Pan Books.

Branford, S. and Glock, O. 1985. *The Last Frontier: Fighting over Land in the Amazon.* London: Zed Books.

Brasseur, G. 1992. Ozone depletion, volcanic aerosols implicated. *Nature*, 359, 275–76.

Breuer, G. 1991. A strategy for the sea floor. *New Scientist*, 12 October, 34–37.

Brey, T., Klages, M., Dahm, C., Gorny, M., Gutt, J., Hain, S., Stiller, M., Arntz, W.E. Wägele, J.-W. and Zimmermann, A. 1994. Antarctic benthic diversity. *Nature*, 368, 297.

Bridger, C.A. and Helfand, L.A. 1968. Mortality from heat during July 1966 in Illinois. *International Journal of Biometeorology*, 12, 51–70.

British Petroleum 1993. *BP Statistical Review of World Energy, June 1993.* London: The British Petroleum Company plc, 37 pp.

Broecker, W.S. 1987. *How to Build a Habitable Planet.* New York: Eldigio Press.

Broecker, W.S. 1990. Salinity history of the Northern Atlantic during the last deglaciation. *Paleoceanography*, 5 (4), 459–67.

Broecker, W.S. 1994. Massive iceberg discharges as triggers for global climate change. *Nature*, 372, 421–424.

Broecker, W.S. 1995. Chaotic climate. *Scientific American*, 273(5), 62–68.

Broecker, W.S. and Denton, G.H. 1990. What drives glacial cycles? *Scientific American*, 262, 42–50.

Brookes, A. 1985. River channelization: traditional engineering methods, physical consequences and alternative practices. *Progress in Physical Geography*, 9, 44–73.

Brown, L.R. and Wolf, E.C. 1984. *Soil Erosion: Quiet Crisis in the World Economy*. Worldwatch Institute, Washington, DC, 49 pp.

Brown, P. 1992. AIDS: the challenge of the future. *New Scientist*, 18 April, *Inside Science* No. 54, 4 pp.

Brown, W. 1991. Europe's lost ozone. *New Scientist*, 131, 1779.

Browning, K.A., Allaqm, R.J., Ballard, S.P., Barnes, R.T.H., Bennetts, D.A., Maryor, R.H., Mason, P.J., McKenna, D., Mitchell, J.F.B., Senior, C.A., Slingo, A. and Smith, F.B. 1991. Environmental effects from burning oil wells in Kuwait. *Nature*, 351, 363–367.

Brundtland, G.H. 1987. *Our Common Future*. Report for the World Commission on Environment and Development. Oxford University Press, Oxford.

Brundtland, G.H. 1994. The solution to a global crisis. *Environment*, 36(10), 16–20.

Bryson, R.A. 1968. All other factors being constant: a reconciliation of several theories of climatic change. *Weatherwise*, 21, 56–61.

BSVRP (British Seismic Verification Research Project) 1989. *Quarterly Journal of the Royal Astronomical Society*, 30, 311–324.

Buffin, D. 1992. Calls to phase-out methyl bromide: a major ozone depleter, *Pesticides News*, 18, 5–11.

Burke, W.H., Denison, R.E., Hetherington, E.A., Koepnick, R.B., Nelson, H.F. and Otto, J.B. 1982. Variation of seawater $^{87}Sr/^{86}Sr$ throughout Phanerozoic time. *Geology*, 10, 516–519.

Burman, A. 1991. Saving Brazil's savannas. *New Scientist*, 2 March, 30–34.

Burton, I., Kates, W.R. and Gilbert, F.W. 1978. *The Environment as Hazard*. Oxford University Press, 240 pp.

Busalacchi, A.J. and O'Brien, J.J. 1981. Interannual variability of the equatorial Pacific in the 1960s. *Journal of Geophysical Research*, 86C, 1091.

Bush, M.B. and Flenley, J.R. 1987. The age of British chalk grasslands. *Nature*, 329, 434–436.

Butler, D.R. 1987. Snow-avalanche hazards, Southern Glacier National Park, Montana: the nature of local knowledge and individual responses. *Disasters*, 11, 214–220.

Butler, J.H., Elkins, J.W., Hall, B.D., Cummings, S.O. and Montzka, S.A. 1992. A decrease in the growth rates of atmospheric halon concentrations. *Nature*, 359, 403–405.

Byrant, E.A. 1991. *Natural Hazards*. Cambridge: Cambridge University Press, 294 pp.

Cadman, D. and Payne, G. (eds) 1990. *The Living City: Towards a Sustainable Future*. London: Routledge, 246 pp.

Caldeira, K. and Kasting, J.F. 1992. Susceptibility of the early Earth to irreversible glaciation caused by carbon dioxide clouds. *Nature*, 359, 226–228.

Calvert, S.E., Nielsen, B. and Fontugne, M.R. 1992. Evidence from nitrogen isotope ratios for enhanced productivity during formation of eastern Mediterranean sapropels. *Nature*, 359, 223–225.

Carlisle, D.B. 1995. *Dinosaurs, Diamonds, and Things from Outer Space. The Great Extinction*. Stanford: Stanford University Press.

Carlisle, D.B. and Braman, D.R. 1991. Nanometre-size diamonds in the Cretaceous/Tertiary boundary clay of Alberta. *Nature*, 352, 708–709.

Carson, R. 1962. *Silent Spring*, London: Paladin.

Carter, R.W.G. 1988. *Coastal Environments*. London: Academic Press.

Cassedy, E.S. and Grossman, P.Z. 1990. *Introduction to Energy: Resources, Technology, and Society*. Cambridge University Press, 338 pp.

Catt, J.A. 1991. Soils as indicators of Quaternary climate change in mid-latitude regions. *Geoderma*, 51, 167–187.

Catt, J.A. 1993. Quaternary geology and soil profiles. *Geoscientist*, 2, 21–24.

Caufield, C. 1986. *In the Rainforest*. London: Picador, 304 pp.

Chadwick, M.J. and Kuylenstierna, J.C.I. 1990. *The Relative Sensitivity of Ecosystems in Europe to Acidic Deposition: A Preliminary Assessment of the Sensitivity of Aquatic and Terrestrial Ecosystems*. Stockholm Environment Institute, Stockholm.

Chahine, M.T. 1992. The hydrological cycle and its influence on climate. *Nature*, 359, 373–380.

Chapman, C.R. and Morrison, D. 1994. Impacts on the Earth by asteroids and comets: assessing the hazard. *Nature*, 367, 33–40.

Chappell, J. and Shackleton, N.J. 1986. Oxygen isotopes and sea level. *Nature*, 324, 137–140.

Charles, D. 1993. In search of a better burn. *New Scientist*, 23 January, 20–25.

Charles, D.F., Battarbee, R.W., Renberg, I., Van Dam, H. and Smol, J.P. 1989. Paleoecological analysis of lake acidification trends in North America and Europe using diatoms and chrysophytes. In: Nortan, S.A., Lindberg, S.E. and Page, A.L. (eds), *Acid Precipitation, vol. 4. Soils, Aquatic Processes, and Lake Acidification*, 208–276. New York: Springer-Verlag.

Charlson, R.J., Langner, J., Rodhe, H., Leovy, C.B. and Warren, S.G. 1991. Peturbation of the Northern Hemisphere radiative balance by backscattering from anthropogenic sulphate aerosols. *Tellus*, 43A-B, 152–163.

Chen, L.C., Fitzgerald, W.M. and Bates, L. 1995. Women, politics and global management. *Environment*, 37, 1, 4–9 and 31–33.

Chesner, C.A., Rose, W.I., Deino, A., Drake, R. and Westgate, J. 1991. Eruptive history of Earth's largest Quaternary caldera (Toba, Indonesia) clarified. *Geology*, 19, 200–203.

Chester, D. 1993. *Volcanoes and Society*. London: Edward Arnold, 351 pp.

Choucri, N., Sundgren, J. and Haas, P.M. 1994. More global treaties, *Nature*, 367, 405.

Chouet, B.A. 1996. Long-period volcano seismicity: its source and use in eruption forecasting. *Nature*, 380, 309–316.

Chow, V.T. 1964. *Handbook of Applied Hydrology*. New York: McGraw-Hill.

Clapman, W.B., Jr 1973. *Natural Ecosystems*. London: Collier-Macmillan.

Clark, R.B. 1989. *Marine Pollution*. Oxford: Clarendon Press.

Clark, R.H. and Southwood, T.R.E. 1989. Risks from ionizing radiation. *Nature*, 338, 197–198.

Clark, R.R. 1963. *Grimes Graves*. London: Her Majesty's Stationery Office.

Clark, W.C. 1989. Managing Planet Earth. *Scientific American*, 261, 47–54.

Clarke, A. 1989. How green is the wind? *New Scientist*, 27 May, 62–65.

Cleere, H. 1976. Some operating parameters for Roman ironworks. *Bulletin of the Institute of Archaeology*, London, 13, 233–246.

Coffin, M.F. and Eldholm, O. 1993. Large igneous provinces. *Scientific American*, 269(4), 26–33.

Coghlan, A. 1991. Coke with pores cleans emissions from dirtiest cole. *New Scientist*, 8 June, 130, 27.

Coghlan, A. 1994. Shells tell tales on toxic sins of emission. *New Scientist*, 17 December, p.21.

Cohen, J.E. 1995. Population growth and the Earth's human carrying capacity. *Science*, 21 July, 269, 341–346.

Colhoun, E.A., Mabin, M.C.G., Adamson, D.A. and Kirk, R.M. 1992. Antarctic ice volume and contribution to sea-level fall at 20,000 yr BP from raised beaches. *Nature*, 358, 316–319.

Colodner, D.C., Boyle, E.A., Edmond, J.M. and Thomson, J. 1992. Post-depositional mobility of platinum, iridium and rhenium in marine sediments. *Nature*, 358, 402–404.

Commission of the European Communities 1992. *Briefing Paper on ALTENER Programme*.

Committee on Monitoring and Assessment of Trends in Acid Deposition 1986. *Acid Deposition: Long-Term Trends*. Washington, DC: National Academy Press, 506 pp.

CONCAWE motor vehicle emission regulations and fuel specifications – 1992 update. November 1992. Report no. 2/92 prepared for the CONCAWE Automotive Emissions Management Group by its special Task Force, AE/STF-3. Brussels.

Cook, E., Bird, T., Peterson, M., Barbetti, M., Buckley, B., D'Arrigo, R., Francey, R. and Tans, P. 1991. Climatic change in Tasmania inferred from a 1,089-year tree-ring chronology of Huon Pine. *Science*, 253, 1266–1268.

Cook, J. 1989. *Dirty Water*. Union Papersachs, London, 147 pp.

Cooke, R.U. and Doornkamp, J.C. 1990. *Geomorphology in Environmental Management* (2nd edition). Oxford: Oxford University Press.

Coope, G.R. 1986. Coleoptera analysis. In: Berglund, B.E. (ed.), *Handbook of Holocene Palaeoecology and Palaeohydrology*, 703–713. Chichester: Wiley.

Cosby *et al.* 1985. is in Battarbee papers.

Costa, J.E. and Baker, V.R. 1981. *Surficial Geology, Building with the Earth*. New York: Wiley.

Cox, P.A. 1989. *The Elements: Their Origin, Abundance and Distribution*. Oxford: Oxford Scientific Publications, 207 pp.

Cragg, C. 1993. Demanding plans for power cuts. *New Scientist*, 27 March, 13–14.

Craig, H. 1965. The measurement of oxygen isotopes in oceanographic studies and palaeotemperatures, 3–24, *Consiglio Nazionale della Richerche Laboratoriodi Geologia Nucleare*, Pisa.

Cross, M. 1993. A very dirty business. *New Scientist*, 23 January, 28–31.

Crosson, P.R. and Rosenberg, N.J. 1989. Strategies for Agriculture. *Scientific American*, 261, 128–135.

Crystal, D. (ed.) 1993. *The Cambridge Factfinder*. Cambridge: Cambridge University Press, 841 pp.

Culshaw, F. and Butler, C. 1993. *A Review of the Potential of Biodiesel as a Transport Fuel*. Energy Technology Support Unit, ETSU-R-71. London: HMSO.

Cuthill, I.C., Swaddle, J.P. and Witter, M.S. 1993. World population forecasts. *Nature*, 363, 215–216.

Daffern, T. 1983. *Avalanche Safety*. London: Diadem Books, 172 pp.

Daffern, T. 1989. *Avalanche Safety for Skiers and Climbers*. Calgary: Rocky Mountain Books.

Dahlgren, R.A. 1994. Soil acidification and nitrogen saturation from weathering of ammonium-bearing rock. *Nature*, 368, 838–840.

Daily, G.C. 1995. Restoring value to the World's degraded lands. *Science*, 269, 350–354.

Dando, W.A. 1980. *The Geography of Famine*. Washington, DC: Winston.

Dansgaard, W., Clausen, H.B., Gundestrup, N., Hammer, C.U., Johnsen, S.J., Kristinsdottir, P.M. and Reeh, N. 1982. A new Greenland deep ice core. *Science*, 218, 1273–1277.

Dansgaard, W., Johnsen, S.J., Clausen, H.B., Dahl-Jensen, D., Gundestrup, N.S., Hammer, C.U., Hvidberg, C.S., Steffesnsen, J.P., Sveinbjornsdottir, A.E., Jouzel, J. and Bond, G. 1993. Evidence for general instability of past climate from a 250-kyr ice-core record. *Nature*, 364, 218–220.

Dansgaard, W.S. 1984. Selected climates from the past and their relevance to possible future climate. In: Flohn, H. and Fantechi, R. (eds), *The Climate of Europe: Past, Present and Future*. Dordrecht: D. Reidel. 208–213.

Dansgaard, W.S. and Tauber, H. 1969. Glacier oxygen 18 content and Pleistocene ocean temperatures. *Science*, 166, 499–502.

Darby, H.C. 1956. The clearing of the woodland in Europe. In: Thomas, W.L. (ed.), *Man's Role in Changing the Face of the Earth*. Chicago: University of Chicago Press, 183–216.

Davidson, D.A. 1991. Soil erosion in the Mediterranean basin. *Geography*, 76, 71–73.

Davies-Jones, R. 1995. Tornadoes. *Scientific American*, August, 48–57.

Davis, G.R. 1990. Energy for Planet Earth. *Scientific American*, 263, 21–27.

Davis, M.B. 1976. Erosion rates and land use history in southern Michigan. *Environmental Conservation*, 3, 139–148.

Davis, R. and Castaneda, C.J. 1995. Panhandle 'like a war zone': worst storm of season pummels Florida. *USA Today*, October 6, p.3A.

Davis, S.J.M. 1987. *The Archaeology of Animals*. London: Batsford.

Davison, W., George, D.G. and Edwards, N.J.A. 1995. Controlled reversal of lake acidification by treatment with phosphate fertilizer. *Nature*, 377, 504–507.

Dawkins, R. 1986. *The Blind Watchmaker*. Longman Scientific & Technical, 332 pp.

Dawson, A.G. 1992. *Ice Age Earth: Late Quaternary Geology and Climate*. London: Routledge, 293 pp.

Dearing, J.A. 1991. Erosion and land use. In: Berglund, B.E. (ed.), *The Cultural Landscape During 6000 Years in Southern Sweden*. *Ecological Bulletins*, 41, 283–292.

Dearing, J.A. 1994. Reconstructing the history of soil erosion. In: Roberts, I. (ed.), *The Changing Global Environment*. Oxford: Blackwell Scientific, 242–261.

Decker, R. and Decker, B. 1989. *Volcanoes*. New York: W.H. Freeman & Co., 285 pp.

Delcourt, P.A. and Delcourt, H.R. 1981. Vegetation maps for eastern North America: 40,000 yr BP to the present. In: Romans, R.C. (ed.), *Geobotany II*. New York: Plenum Press, 123–165.

Denton, G.H. and Hughes, T.J. 1981. *The Last Great Ice Sheets*. New York: John Wiley & Sons, 484 pp.

Denton, G.H. *et al.* 1989. Late Weichelsian and Early Holocene. *Quaternary Research*, 31, 151–182.

Department of the Environment (UK) 1988. *Our Common Future: A Perspective by the UK on the Report of the World Commission on Environment and Development*. London: HMSO.

Department of the Environment (UK) 1992a. *Climate Change: Our National Programme for CO$_2$ Emissions*. London: HMSO.

Department of the Environment (UK) 1992b. *This Common Inheritance: Britain's Environmental Strategy*. London: HMSO.

Derbyshire, E. 1989. Mud or dust; erosion of the Chinese loess. *Geography Review*, 3, 31–35.

Derbyshire, E., Wang, J., Jin, Z., Billard, A., Egels, Y., Kasser, M., Jones, D.K.C., Muxart, T. and Owen, L. 1991. Landslides in the Gansu Loess of China. *Catena Supplement*, 20, 119–145.

Des Morais, D.J., Strauss, H., Summons, R.E. and Hayes, J.M. 1992. Carbon isotope evidence for the stepwise oxidation of the Proterozoic environment. *Nature*, 359, 605–609.

Deuser, W.G. and Ross, E.H. 1980. Seasonal change in the flux of organic carbon to the deep Sargasso Sea. *Nature*, 283, 364–365.

Dilke, F.W.W. and Gough, D.O. 1972. The solar spoon. *Nature*, 240, 262–294.

Dixon, P. 1994. Still more about AIDS. *Nature*, 367, 311.

DoE 1990. *The Householder's Guide to Radon* (2nd edition). London: UK Department of the Environment.

Dolan, J.F., Sieh, K., Rockwell, T.K., Yeats, R.S., Shaw, J., Suppe, J., Huftile, G.J. and Gath, E.M. 1995. Prospects for larger or more frequent earthquakes in the Los Angeles Metropolitan region. *Science*, 267, 199–205.

Domenico, P.A. and Schwartz, F.W. 1990. *Physical and Chemical Hydrogeology*. Chichester: John Wiley & Sons, 824 pp.

Dostrovsky, I. 1991. Chemical fuels from the Sun. *Scientific American*, 265(6), 50–66.

Douglas, I. 1989. Land degradation, soil conservation and the sediment load of the Yellow River, China:
review and assessment. *Land Degradation and Rehabilitation*, 1, 141–151.

Dowdeswell, J.A., Hamilton, G.S. and Hagen, J.O. 1991. The duration of the active phase on surge-type glaciers: contrasts between Svalbard and other regions. *Journal of Glaciology*, 37, 388–400.

Driscoll, T., Schaefer, D.A., Molot, L.A. and Dillon, P.J. 1989. *NORD Miljørapport*, 92, 1–45. Nordic Council of Ministers, Copenhagen 1989.

Dung, V.V., Giao, P.M., Chinh, N.N., Tuoc, D., Arctander, P. and Mackinnon, J. 1993. A new species of living bovid from Vietnam. *Nature*, 363, 443–444.

Durka, W., Schulze, E.-D., Gebauer, G. and Voerkellus, S. 1994. Effects of forest decline on uptake and leaching of deposited nitrate determined from ^{15}N and ^{18}O measurements. *Nature*, 372, 765–767.

Dutton, C.E. 1889. The Charleston earthquake of August 31, 1886. *U.S. Geological Survey Ninth Annual Report*, 1887–1888, 203–528.

Dutton, E.G. and Christy, J.R. 1992. Solar radiative forcing at selected locations and evidence for global lower tropospheric cooling following the eruptions of El Chichon and Pinatubo. *Geophysical Research Letters*, 19, 2313–2316.

Duvall, T.L., Jr, D'Silva, S., Jefferies, S.M., Harvey, J.W. and Schou, J. 1996. Downflows under sunspots detected by helioseismic tomography. *Nature*, 379, 235–237.

Eckel, E.B. 1970. The Alaskan earthquake, March 27, 1964: Lessons and conclusions. *U.S. Geological Survey Professional Paper*, 546.

Eckholm, E. 1975. The deterioration of mountain environments. *Science*, 189, 764–770.

Economist 1988. The vanishing jungle. 15 October, 25–28.

Eden, M.J. 1989. *Land Management in Amazonia*. London: Belhaven.

Ehrlich, P.R. 1968. *The Population Bomb*. New York. Ballentine Books.

Ehrlich, P. and Ehrlich, A. 1992. The value of biodiversity. *Ambio*, 21(3), 219–226.

Einsele, G., Ricken, W. and Seilacher, A. (eds) 1991. *Cycles and Events in Stratigraphy*. Berlin: Springer-Verlag, 955 pp.

Ekins, P. 1992. *A New World Order: Grassroots Movements for Global Change*. London: Routledge, 248 pp.

El-Sabh, M.J. and Murty, T.S. (eds) 1988. *Natural and Man-Made Hazards*. Dordrecht: D. Riedel.

England, P. 1992. Deformation of the continental crust. In: Brown, G.C., Hawkesworth, C.J. and Wilson, R.C.L. (eds), *Understanding the Earth: a New Synthesis*. Cambridge: Cambridge University Press, 275–300.

Eni 1992. Biofuels in the EEC. The Commission Proposal Effects on Eventual Development in Europe. Experience of United States and Brazil. 118 pp.

Environmental Protection Act 1990. London: HMSO.

Environmental Protection Agency 1995. *Human health benefits from sulfate reductions under Title IV of the 1990 Clean Air Act*. US Environmental Protection Agency, November 10, Final Report.

EOS, Transactions, American Geophysical Union 1992.

Impact of Mid-Atlantic Sewage Sludge Probed (editorial), 73, 27–28.

EPA 1996. Environmental Protection Agency's Worldwide Web Site: http://www.epa.gov/docs/acidrain.

Erwin, D.H. 1994. The Permo-Triassic extinction. *Nature*, 367, 231–236.

Evans, D.M. 1966. Man-made earthquakes in Denver. *Geotimes*, 10, 11–18.

Evernden, N. 1985. *The Natural Alien: Humankind and Environment* (2nd edition). Toronto: University of Toronto Press.

Fahey, D.W., Kawa, S.R., Woodbridge, E.L., Tin, P., Wilson, J.C., Jonsson, H.H., Dye, J.E., Baumgardner, D., Borrmann, S., Toohey, D.W., Avallone, L.M., Proffitt, M.H., Margitan, J., Loewenstein, M., Podolske, J.R., Salawitch, R.J., Wofsy, S.C., Ko, M.K.W., Anderson, D.E., Shoeberl, M.R. and Chan, K.R. 1993. In situ measurements constraining the role of sulphate aerosols in mid-latitude ozone depletion. *Nature*, 363, 509–514.

Fahey, D.W. *et al.* 1995. Emission measurements of the Concorde supersonic aircraft in the lower stratosphere. *Science*, 270, 70–73.

Fairbanks, R.G. 1989. A 17,000 year glacioeustatic sea level record: influence of glacial melting rates on the Younger Dryas event and deep-ocean circulation. *Nature*, 343, 637–642.

Falkowski, P.G. and Wilson, C. 1992. Phytoplankton productivity in the North Pacific Ocean since 1900 and implications for absorption of anthropogenic CO_2. *Nature*, 358, 741–743.

Fan, S.-M. and Jacob, D.J. 1992. Surface ozone depletion in Arctic spring sustained by bromine reactions on aerosols. *Nature*, 359, 522–524.

Fanning, K.A. 1989. Influence of atmospheric pollution on nutrient limitation in the ocean. *Nature*, 339, 460–462.

Farman, J. 1987. What hope for the ozone layer now? *New Scientist*, 116, 50–54.

Farman, J.C., Gardiner, B.G. and Shanklin, J.D. 1985. Large losses of total ozone in Antarctica reveal seasonal ClO_x/NO_x interaction. *Nature*, 315, 207–210.

Farrell, J.W., Pedersen, T.F., Calvert, S.E. and Nielsen, B. 1995. Glacial-interglacial changes in nutrient utilization in the equatorial Pacific Ocean. *Nature*, 377, 514–517.

Fenton, E.W. 1937. The influence of sheep on the vegetation of hill grazings in Scotland. *Journal of Ecology*, 25, 424–430.

Ferry, G. 1989. Alzheimer's and aluminium – the guesswork goes on. *New Scientist*, 121, p.1652.

Fields, B.N. 1994. AIDS: time to turn to basic science. *Nature*, 369, 95–96.

Fischer, A.G. 1982. In: Berggren, W.A. and van Couvering, J.A. (eds), *Catastrophes and Earth History*. Princeton: Princeton University Press, 129–150.

Fischer, A.G. and Arthur, M.A. 1977. Secular variations in the pelagic realm. In: Cook, H.E. and Enos, P. (eds), *Deep-water Carbonate Environments*, 19–50. Special Publication of the Society of Economic Paleontologists and Mineralogists, 25. Tulsa, Oklahoma: Society of Economic Paleontologists and Mineralogists.

Fisher, R.V. and Schmincke, H.-U. 1984. *Pyroclastic Rocks*. Springer-Verlag, Berlin, 472 pp.

Flegal, A.R., Maring, H. and Niemeyer, S. 1993. Anthropogenic lead in Antarctic sea water. *Nature*, 365, 242–244.

Flood, M. 1991. *Energy Without End*. Friends of the Earth. 76 pp.

Flowers, E.C., McCormick, R.A. and Kurfis, K.R. 1969. Atmospheric turbidity over the United States, 1961–1966. *Journal of Applied Meteorology*, 8, 955–962.

Foucault, A. and Stanley, D.J. 1989. Late Quaternary palaeoclimatic oscillations in East Africa recorded by heavy minerals in the Nile Delta. *Nature*, 339, 44–46.

Fowler, D. 1993. A land laid waste. *Surveyor*, 19–21. 25 February issue.

Freeth, S. 1992. The deadly cloud hanging over Cameroon. *New Scientist*, 15 August, 23–27.

Friday, L. and Laskey, R. (eds) 1989. *The Fragile Environment: the Darwin College Lectures*. Cambridge University Press, 198 pp.

Friends of the Earth 1992. *Energy for a Future: Friends of the Earth's Evidence to the Government's Review of Energy Policy*. London: Friends of the Earth, 39 pp.

Fritts, H.C. 1976. *Tree Rings and Climate*. London, Academic Press.

Fronval, T., Jansen, E., Bloemendal, J. and Johnsen, S. 1995. Oceanic evidence for coherent fluctuations in Fennoscandian and Laurentide ice sheets on millennium timescales. *Nature*, 374, 443–446.

Frosch, R.A. 1995. The industrial ecology of the 21st century. *Scientific American*, September, 178–181.

Frosch, R.A. and Gallopoulos, N.E. 1989. Strategies for manufacturing. *Scientific American*, 261, 144–152.

Frutiger, H. 1980. Swiss avalanche hazard maps. *Journal of Glaciology*, 26(94), 518–519.

Fujita, T.T. 1973. Tornadoes around the world. *Weatherwise*, 26, 56–62 and 79–83.

Fulkerson, W., Judkins, R.R. and Sanghvi, M.K. 1990. Energy from fossil fuels. *Scientific American*, 263, 83–89.

Fuller, M.L. 1912. *The New Madrid Earthquake*. United States Geological Survey, Bulletin 494, 120 pp. Reprinted in 1988 by the Central United States Earthquake Consortium, Memphis, Tennessee.

Furth, H.P. 1995. Fusion. *Scientific American*, September, 174–177.

Galloway, J.N. 1990. The intercontinental transport of sulfur and nitrogen. In: Knap, A.H. (ed.), *The Long Range Atmospheric Transport of Natural and Contaminant Substances*. Netherlands: Kluwer Academic Publishing, pp. 87–104.

Galloway, J.N. and Rodhe, H. 1991. Regional atmospheric budgets of S and N fluxes: how well can they be quantified? In: Last, F.T. and Watling, R. (eds), *Acidic Deposition: its Nature and Impacts*. Edinburgh: The Royal Society of Edinburgh, 61–80.

Ganeshram, R.S., Pedersen, T.F., Calvert, S.E. and Murray, J.W. 1995. Large changes in oceanic nutrient inventories from glacial to interglacial periods. *Nature*, 376, 755–758.

Gao, G. 1993. The temperatures and oxygen-isotope composition of early Devonian oceans. *Nature,* 361, 712–714.

Gasse, F., Tehet, R., Durand, A., Gibert, E. and Fontes, J.-C. 1990. The arid–humid transition in the Sahara and Sahel during the last deglaciation. *Nature,* 346, 141–146.

Gates, D.M. 1993. *Climate Change and its Biological Consequences.* Sunderland, Massachusetts: Sinauer Associates, Inc., 280 pp.

Gavaghan, H. 1989. The problems in policing short-range missiles. *New Scientist,* 3 June, 35.

Genin, A., Lazar, B. and Brenner, S. 1995. Vertical mixing and coral death in the Red Sea following the eruption of Mount Pinatubo. *Nature,* 377, 507–510.

Genthon, C., Barnola, J.M., Raynaud, D., Lorius, C., Jouzel, J., Barkov, N.I., Korotkevich, Y.S. and Kotlyakov, V.M. 1987. Vostok ice core: climatic response to CO_2 and orbital forcing changes over the last climatic cycle. *Nature,* 329, 414–419.

Gentry, A.H. and Lopez-Parodi, J. 1980. Deforestation and increased flooding of the upper Amazon. *Science,* 210, 1354–1356.

Gerlach, T. 1991. Etna's greenhouse pump. *Nature,* 351, 352–353.

Gerrard, S. 1994. Environmental risk management. In: O'Riordan, T. (ed.) *Environmental Science for Environmental Management,* 296–316. Harlow: Longman Scientific & Technical.

Ghadiri, H. and Payne, D. 1980. A study of soil splash using cine photography. In: Boodt, M. de & Gabriels, D. (eds), *Assessment of Erosion.* Chichester: Wiley, 185–192.

Ghetti, A. and Batisse, M. 1983. The overall protection of Venice and its lagoon. *Nature and Resources,* 19, 7–19.

Gibbons, J.H., Blair, P.D. and Gwin, H.L. 1989. Strategies for energy use. *Scientific American,* 261, 136–143.

Gibson, R.B. 1993. Environmental assessment design: lessons from the Canadian experience. *The Environmental Professional,* 15, 12–24.

Gillham, C.A., Leech, P.K. and Eggleston, H.S. 1992. *UK Emissions of Air Pollutants 1970–1990.* National Atmospheric Emissions Inventory. Warren Spring Laboratory, UK Department of Environment Report LR 887 (AP).

Gilliland, R.L. 1989. Solar evolution. *Palaeogeography, Palaeoclimatology, Palaeoecology (Global and Planetary Change Section),* 75, 35–55.

Glacken, C.J. 1967. *Traces on the Rhodian Shore: Nature and Culture in Western Thought from Ancient Times to the End of the Eighteenth Century.* Berkeley: University of California Press, 763 pp.

Gleason, D.F. and Wellington, G.M. 1993. Ultraviolet radiation and coral bleaching. *Nature,* 365, 836–838.

Gleick, J. 1987. *Chaos.* Heinemann, 352 pp.

Gleick, P.H. (ed.) 1993. *Water in Crisis.* Oxford: Oxford University Press, 25–39.

Gleick, P.H. 1994. Water, war and peace in the Middle East. *Environment,* 36(3), 6–15 and 35–42.

Glenn, C.R. and Kelts, K. 1991. Sedimentary rhythms in lake deposits. In: Einsele, G., Ricken, W. and Seilacher, A. (eds), *Cycles and Events in Stratigraphy,* Berlin: Springer Verlag, 188–221.

Golnaraghi, M. and Kaul, R. 1995. Responding to ENSO. *Environment,* 37(1), 16–20 and 38–44.

Goodman, D. and Redclift, M. 1991. *Refashioning Nature: Food, Ecology and Culture.* Routledge, 279 pp.

Gordon, A. and Suzuki, D. 1991. *It's a Matter of Survival.* London: Harper Collins, 278 pp.

Gordon, D., Smart, P.L., Ford, D.C., Andrews, J.N., Atkinson, T.C., Rowe, P.J. and Christopher, N.S.J. 1989. Dating the Late Pleistocene Interglacial and Interstadial Periods in the UK. *Quaternary Research,* 31, 14–26.

Goslar, T., Arnold, M., Bard, E., Kuc, T., Pazdur, M.F., Raiska-Jasiewiczowa, M., Rozanski, K., Tisnerat, N., Walanus, A., Eicik, B. and Wieckowski, K. 1995. High concentration of atmospheric ^{14}C during the Younger Dryas cold episode. *Nature,* 377, 414–417.

Goss, E.G. 1991. The hurricane dilemma in the United States. *Episodes,* 14, 36–45.

Goudie, A.S. 1986. *The Human Impact on the Environment.* Oxford: Basil Blackwell.

Goudie, A.S. 1992. *Environmental Change* (3rd edition). Oxford: Clarendon Press, 329 pp.

Goudie, A. 1993a. *The Nature of the Environment.* (3rd edition). Oxford: Blackwell, 397 pp.

Goudie, A. 1993b. *The Human Impact.* Oxford: Blackwell.

Gough, D.O. 1981. *Solar Physics,* 74, 21–34.

Gould, S.J. 1991. *Wonderful Life: the Burgess Shale and the Nature of History.* London: Penguin Books, 347 pp.

Gourlay, K.A. 1988. *Poisoners of the Seas.* London: Zed Book Ltd, 256 pp.

Government of Hong Kong 1992. Final Report of the Commission of Inquiry into Rainstorm Disasters 1992. Hong Kong: Hong Kong Government Printer.

Gradwohl, J. and Greenberg, R. 1988. *Saving Tropical Forests.* London: Earthscan Publications.

Graedel, T.E. and Crutzen, P.J. 1993. *Atmospheric Change: An Earth System Perspective.* New York: W.H. Freeman & Co., 446 pp.

Graham, J.B., Dudley, R., Aguilar, N.M. and Gans, C. 1995. Implications of the late Palaeozoic oxygen pulse for physiology and evolution. *Nature,* 375, 117–120.

Grainger, A. 1990. *The Threatening Desert: Controlling Desertification.* London: Earthscan Publications.

Green, G.M. and Sussman, R.W. 1990. Deforestation history of the eastern rain forests of Madagascar from satellite images. *Science,* 248, 212–215.

Greenland Ice-core Project (GRIP) Members 1993. Climate instability during the last interglacial period recorded in the GRIP ice core. *Nature,* 364, 203–207.

Greenpeace 1990. *The Greenpeace Report, Global Warming.* Oxford: Oxford University Press.

Gregory, K. and Rowlands, H. 1990. Have global hazards increased? *Geographical Review,* 4, 35–38.

Gribbin, J. 1988. *The Hole in the Sky.* Reading: Corgi Books, 160 pp.

Plate 52 *United Nations convoy on a highway from Belgrade to Sarajevo.*
Courtesy of Chris Stowers/ Panos Pictures.

Plate 53 *Berlin Climate Conference in progress on 29 March 1995.*
Courtesy of Greenpeace/ Langrock/Zenit.

Plate 54 *A bomb disposal expert clears mines in Cambodia. Land mines are one of the greatest hazards facing many countries when refugees wish to return and resettle in lands which were former war zones.*
Courtesy of Nic Dunlop/Panos Pictures.

Gribbin, J. 1989. The end of the ice ages? *New Scientist*, 17 June, 48–52.

Gribbin, J. 1991. Climate now. *New Scientist*, Inside Science Number 44, 16 March, 1–4.

Griffin, B. 1994. AIDS debate continues. *Nature*, 367, 212.

Grigg, D. 1995. *An Introduction to Agricultural Geography* (2nd edition). London: Routledge, 217 pp.

Gross, E.G. 1991. The hurricane dilemma in the United States. *Episodes*, 14 36–45.

Grove, J.M. 1979. The glacial history of the Holocene. *Progress in Physical Geography*, 3, 1–54.

Grove, J.M. 1987. Glacier fluctuations and hazards. *Geographical Journal*, 153, 3, 351–369.

Grove, J.M. 1988. *The Little Ice Age*. London: Methuen, 489 pp.

Grun, R. and Stringer, C.B. 1991. Electron spin resonance dating and the evolution of modern humans. *Archaeometry*, 33, 153–199.

Guiot, J., Pons, A., de Beaulieu, J.L. and Reille, M. 1989. A 140,000-year continental climate reconstruction from two European pollen records. *Nature*, 338, 309–314.

Gundersen, P. 1992. *NORD Miljørapport*, 41, 55–110. Nordic Council of Ministers, Copenhagen 1992.

Haas, J.E., Kates, R.W. and Bowden, M.J. (eds) 1977. *Reconstruction Following Disasters*. Cambridge, Mass: MIT Press.

Hall, D.O., Rosillo-Calle, F. and de Groot, P. 1992. Biomass energy. *Energy Policy*, 20, 62–73.

Hamer, M. 1992. Down came the drought. *New Scientist*, 2 May, 22–23.

Hamer, M. 1994. Dying from too much dust. *New Scientist*, 12 March.

Hammer, C.U., Clausen, H.B. and Dansgaard, W. 1980. Greenland ice sheet evidence of postglacial volcanism and its climatic impact. *Nature*, 288, 230–235.

Hammer, C.U., Clausen, H.B. and Dansgaard, W. 1981. Past volcanism and climate revealed by Greenland ice cores. *Journal of Volcanology and Geothermal Research*, 11, 3–11.

Hammer, C.U., Clausen, H.B., Friedrich, W.L. and Tauber, H. 1987. The Minoan eruption of Santorini in Greece dated 1645 BC? *Nature*, 328, 517–519.

Hanes, T.L. 1971. Succession after fire in the chaparral of southern California. *Ecological Monographs*, 41, 27–52.

Hansen, J.E. and Lacis, A.A. 1990. Sun and dust versus greenhouse gases: an assessment of their relative roles in global climate change. *Nature*, 346, 713–719.

Harlan, J.R. 1986. Plant domestication: diffuse origins and diffusion. In: Barigozzi, C. (ed.), *The Origins and Domestication of Cultivated Plants*. Amsterdam: Elsevier, 21–34.

Harland, W.B. *et al.* 1989. *A Geological Time Scale*. Cambridge: Cambridge University Press.

Harriman, R. and Morrison, B.R.S. 1982. The ecology of streams draining forested and non-forested catchments in an area of Scotland subject to acid precipitation. *Hydrobiologia*, 88, 251–263.

Harriman, R., Morrison, B.R.S., Caines, L.A., Collen, P. and Watt, A.W. 1987. Long term changes in fish populations of acid steams and lochs in Galloway, south-west Scotland. *Water, Air and Soil Pollution*, 32, 89–112.

Harrison, L. 1992. Europe gets clean away. *Windpower Monthly*, September, 18–19.

Harte, J., Holdren, C., Schneider, R. and Shirley, C. 1991. *Toxics A to Z: A Guide to Everyday Pollution Hazards*. Berkeley: University of California Press.

Hassard, J. 1992. Arms and the ban. *New Scientist*, 28 November, 136, 38–41.

Hawking, S.W. 1988. *A Brief History of Time: From the Big Bang to Black Holes*. London: Bantam Press, 198 pp.

Hebbeln, D., Dokken, T., Andersen, E.S., Hald, M. and Elverhøl, A. 1994. Moisture supply for northern ice-sheet growth during the Last Glacial Maximum. *Nature*, 370, 357–360.

Heinrich, H. 1988. *Quaternary Research*, 29, 143–152.

Heliker, C. 1991. *Volcanic and Seismic Hazards on the Island of Hawaii*, 48 pp. US Department of the Interior/US Geological Survey, Denver. US Government Printing Office.

Heller, F., and Lui, T.S. 1984. Magnetism of the Chinese loess deposits. *Journal of Geophysical Research*, 77, 125–141.

Henderson-Sellers, A. and Robinson, P.J. 1986. *Contemporary Climatology*. London: Longman Scientific & Technical, 439 pp.

Herndl, G.J., Müller-Niklas, G. and Frick, J. 1993. Major role of ultraviolet-B in controlling bacterioplankton growth in the surface layer of the ocean. *Nature*, 361, 717–719.

Hewitt, T. 1992. Developing countries – 1945 to 1990. In: Allen, T. and Thomas, A. (eds), *Poverty and Development in the 1990s*. Oxford: Oxford University Press, 221–237.

Hewitt, T. and Smyth, I. 1992. Is the world overpopulated? In: Allen, T. and Thomas, A. (eds), *Poverty and Development in the 1990s*. Oxford: Oxford University Press, 78–96.

Hey, R.D. 1990. Environmental river engineering. *Journal of Water and Environmental Management*, 4(4), 335–340.

Hibler, C.P. and Hancock, C.M. 1990. Waterborne giardiasis. In: McFeters, G.A. (ed.), *Drinking Water Microbiology*. New York: Springer-Verlag.

Hilderbrand, A.R., Penfield, G.T., Kring, D.A., Pilkington, M., Camargo, A.Z., Jacobsen, S.B. and Boynton, W.V. 1991. Chicxulub Crater: a possible Cretaceous/Tertiary boundary impact crater on the Yucatan Peninsula, Mexico. *Geology*, 19, 867–871.

Hinrichsen, D. 1990. *Our Common Seas: Coasts in Crisis*. London: Earthscan Publications.

Hiscock, K. 1994. Groundwater pollution and protection. In: O'Riordan, T. (eds), *Environmental Science for Environmental Management*, 246–262. Harlow: Longman Scientific.

Hodges, C.A. 1995. Mineral resources, environmental issues and land use. *Science*, 2 June, 268, 1305–1312.

Hoffert, M.I. 1992. Climatic sensitivity, climatic feedbacks and policy implications. In: Mintzer,

I.M. (ed.), *Confronting Climate Change: Risks, Implications and Responses.* Cambridge: Cambridge University Press, 33–54.

Hoffman, A. 1985. Patterns of family extinction depend on definition and geological timescale. *Nature,* 315, 659–662.

Hoffman, A.J. 1995. An easy rebirth at Love Canal. *Environment,* 37(2), 4–9 and 25–31.

Hoffman, A. and Ghiold, J. 1985. Randomness in the pattern of 'mass extinctions' and 'waves of originations'. *Geological Magazine,* 122(1), 1–4.

Hofmann, D.J., Deshler, T.L., Aimedieu, P., Matthews, W.A., Johnston, P.V., Kondo, Y., Sheldon, W.R., Byrne, G.J. and Benbrook, J.R. 1989. Stratospheric clouds and ozone depletion in the Arctic during January 1989. *Nature,* 340, 117–121.

Hofmann, D.J., Oltmans, S.J., Harris, J.M., Solomon, S., Deshler, T. and Johnson, B.J. 1992. Observation and possible causes of new ozone depletion in Antarctica in 1991. *Nature,* 359, 283–287.

Holdren, J.P. and Pachauri, R.K. 1992. Energy. In: *An Agenda of Science for Environment and Development into the 21st Century.* Based on a Conference held in Vienna, Austria in November 1991. Cambridge University Press. 103–118.

Holdsworth, G. 1986. Evidence for a link between atmospheric thermonuclear detonations and nitric acid. *Nature,* 324, 551–554.

Hollin, J.T. 1969. Ice-sheet surges and the geological record. *Canadian Journal of Earth Sciences,* 5, 903–910.

Holt, S.J. 1993. Harvesting of whales. *Nature,* 361, 391.

Homewood, B. 1993. Will Brazil's cars go on the wagon? *New Scientist,* 9 January, 22–24.

Hornung, M., Stevens, P.A. and Reynolds, B. 1986. The impact of pasture improvement on the soil solution chemistry of some stagnopodzols in mid-Wales. *Soil Use and Management,* 2, 18–26.

Hough, S.E. 1995. Earthquakes in the Los Angeles metropolitan region: a possible fractal distribution of rupture size. *Science,* 267, 211–213.

Houghton, J.T., Meira Filho, L.G., Lee, H., Callander, B.A., Haites, E., Harris, N. and Maskell, K. (eds) (Intergovernmental Panel on Climate Change) 1995. *Climate Change 1994. Radiative Forcing of Climate Change and An Evaluation of the IPCC IS92 Emission Scenario.* Cambridge: Cambridge University Press, 339 pp.

House of Commons Energy Committee Fourth Report 1990. *The Cost of Nuclear Power* (2 volumes). London: HMSO.

House of Commons Energy Committee Fourth Report 1992. *Renewable Energy* (3 volumes). London: HMSO.

House of Commons Environment Committee First Report 1992. *The Government's Proposals for an Environment Agency.* London: HMSO.

House of Commons Trade and Industry Committee First Report 1993. *British Energy Policy and the Market for Coal.* London: HMSO.

House of Lords Select Committee on the European Communities 25th Report 1990. *Paying for Pollution: Civil Liability for Damage Caused by Waste.* London: HMSO.

Housner, G.W. 1987. *Confronting Natural Disasters: An International Decade for Natural Hazard Reduction.* Washington DC: National Academy Press.

Hovan, S.A., Rea, D.K., Pisias, N.G. and Shackleton, N.J. 1989. A direct link between the China loess and marine records: aeolian flux to the north Pacific. *Nature,* 340, 296–298.

Howe, C.W. 1991. An evaluation of US air and water policies. *Environment,* 33(7), 10–15 and 32–36.

Hsu, K.J. 1983. *The Mediterranean Was a Desert.* New Jersey: Princeton University Press, 197 pp.

Huggett, R.J. 1995. *Geoecology: an Evolutionary Approach.* London: Routledge, 320 pp.

Hughen, K.A., Overpeck, J.T., Peterson, L.C. and Trumbore, S. 1996. Rapid climate changes in the tropical Atlantic region during the last deglaciation. *Nature,* 380, 51–54.

Hughes, T.P. 1994. Catastrophes, phase shifts, and large-scale degradation of a Caribbean coral reef. *Science,* 265, 1547–1551.

Huhne, C. 1989. Some lessons of the debt crisis: never again? In: O'Brien, R. and Datta, T. (eds), *International Economics and Financial Markets: the AMEX Bank Review Prize Essays 1988.* Oxford University Press, Oxford.

Hulme, M. 1989. Is environmental degradation causing drought in the Sahel? An assessment from the recent empirical research. *Geography,* 74, 38–46.

Hulme, M. and Kelly, P.M. 1993. Exploring the links between desertification and climate change. *Environment,* 35(6), 4–11 and 39–46.

ICSU 1992. International Council of Scientific Unions: *An Agenda of Science for Environment and Development into the 21st Century.* Cambridge: Cambridge University Press, 331 pp.

Igarashi, G., Saeki, S., Takahata, N., Sumikawa, K., Tasaka, S., Sasaki, Y., Takahashi, M. and Sano, Y. 1995. Ground-water radon anomaly before the Kobe earthquake in Japan. *Science,* 269, 60–61.

Imbrie, J. and Imbrie, K.P. 1979. *Ice Ages: Solving the Mystery.* Harvard: Harvard University Press, 224 pp.

Imbrie, J., van Donk, J. and Kipp, N.G. 1973. Paleoclimatic investigation of a late Pleistocene Caribbean deep-sea core: comparison of isotopic and faunal methods. *Quaternary Research,* 3, 10–38.

Imeson, A.C. 1971. Heather burning and soil erosion on the North Yorkshire Moors. *Journal of Applied Ecology,* 8, 537–542.

Intergovernmental Panel on Climatic Change (IPCC) 1990. *Climate Change.* Cambridge: Cambridge University Press.

Intergovernmental Panel on Climatic Change (IPCC) 1992. *Climate Change 1992: The Supplementary Report to the IPCC Scientific Assessment.* Houghton, J.T., Callander, B.H. and Varney, S.K. (eds). Cambridge: Cambridge University Press.

Ives, J.D. and Messerli B. 1989. *The Himalayan Dilemma.* London: Routledge, 295 pp.

Iwata, G.K. 1991. Interactions between aquaculture and the environment. *Critical Reviews in Environmental Control,* 21, 177–216.

Jackson, L.E. 1987. Debris flow hazard in the Canadian Rocky Mountains. *Geological Survey of Canada*, Paper, 86–11.

Jacobs, D.K. and Sahagian, D.L. 1993. Climate-induced fluctuations in sea level during non-glacial times. *Nature*, 361, 710–712.

Jacobs, G.A., Hurlburt, H.E., Kindle, J.C., Metzger, E.J., Mitchell, J.L., Teague, W.J. and Wallcraft, A.J. 1994. Decade-scale trans-Pacific propagation and warming effects of an El Niño anomaly. *Nature*, 370, 360–363.

Jacobs, S.S. 1992. Is the Antarctic ice sheet growing? *Nature*, 360, 29–33.

Janz, B. 1968. Southern Alberta's paralysing snowstorms in April 1967. *Weatherwise*, 21(2), 70–75 and 94.

Jayaraman, K.S. 1993a. India seeks to learn the lessons of the Maharashtra earthquake. *Nature*, 365, 593.

Jayaraman, K.S. 1993b. Science academies call for global goal of zero population growth. *Nature*, 366, 3.

Jelgersma, S. 1966. Sea level changes in the last 10,000 years. *International Symposium of World Climate from 8000–0 BC*, 54–69. Royal Meteorological Society.

Jennings, J.S. 1996. *The Millennium and Beyond – Some Issues Which Will Shape Our Future*. Royal Dutch/Shell Group of companies, Shell Briefing Service. UK: Billington Press, 12 pp.

Jenny, H. 1941. *Factors in Soil Formation*. New York: McGraw-Hill.

Jenny, H. 1980. *The Soil Resource, Origin and Behaviour*. New York: Springer-Verlag.

Johansson, T.D. *et al.* (eds) 1993. *Renewable Energy for Fuels and Electricity*. Washington DC and Calif.: Island Press.

Johnsen, S.J., Clausen, H.B., Dansgaard, W., Fuhrer, K., Gunerstrup, N., Hammer, C.U., Iversen, P., Jouzel, J., Stauffer, B. and Steffensen, J.P. 1992. Irregular glacial interstadials recorded in a new Greenland ice core. *Nature*, 359, 311–313.

Johnson, C., Henshaw, J. and McInnes, G. 1992. Impact of aircraft and surface emissions of nitrogen oxides on tropospheric ozone and global warming. *Nature*, 355, 69–71.

Johnson, D.L. and Lewis, L.A. 1995. *Land Degradation: Creation and Destruction*. Oxford: Blackwell, 335 pp.

Johnson, D.W., Kilsby, C.G., McKenna, D.S., Saunders, R.W., Jenkins, G.J., Smith, F.B. and Foot, J.S. 1991. Airborne observations of the physical and chemical characteristics of the Kuwait oil smoke plume. *Nature*, 353, 617–621.

Jones, V.J., Stevenson, A.C. and Battarbee, R.W. 1989. The acidification of lakes in Galloway, south-west Scotland: a diatom and pollen study of the post-glacial history of the Round Loch of Glenhead. *Journal of Ecology*, 77, 1–23.

Journal of the Geological Society, London, 1986. *Geo-chemical Aspects of Acid Rain*. Thematic set of papers, 619–720.

Journal of the Geological Society, London, 1991, *Monitoring Active Volcanoes*. Thematic set of papers. 148, 561–593.

Jouzel, J., Lorius, C., Petit, J.R., Genthon, C., Barkov, N.I., Kotlyakov, V.M. and Petrov, V.M. 1987. Vostok ice core: a continuous isotope temperature record over the last climatic cycle (160,000 years). *Nature*, 329, 403–407.

Joyce, C. 1991. Ozone hole linked to sea temperatures. *New Scientist*, 131, 22.

Kanamori, H. and Kikuchi, M. 1993. The 1992 Nicaragua earthquake: a slow tsunami earthquake associated with subducted sediments. *Nature*, 361, 714–716.

Karl, T.R., Knight, R.W. and Plummer, N. 1995. Trends in high-frequency climate variability in the twentieth century. *Nature*, 377, 217–220.

Kasting, J.F. 1989. Long-term stability of the Earth's climate. *Palaeogeography, Palaeoclimatology, Palaeoecology*, 75, 83–95.

Keeling, C.D., Baconstow, R.B., Carter, A.F., Piper, S.R., Whorf, T.P., Heimann, M., Mook, W.G. and Roeloffgen, H. 1989. A three-dimensional model of atmospheric CO_2 transport based on observed winds: I. Analysis of observational data. In: Peterson, D.H. (ed.), *Aspects of Climate Variability in the Pacific and the Western Americas*, 165–236. Washington DC: AGU Geophysical Monograph, 55.

Keeling, R.F. and Shertz, S. 1992. Seasonal and interannual variations in atmospheric oxygen and implications for the global carbon cycle. *Nature*, 358, 723–727.

Keigwin, L.D., Curry, W.B., Lehman, S.J. and Johnsen, S. 1994. The role of the deep ocean in North Atlantic climate change between 70 and 130 kyr ago. *Nature*, 371, 323–326.

Keller, E.A. 1976. Channelisation: environmental, geomorphic and engineering aspects. In: Coates, D.R. (ed.), *Geomorphology and Engineering*. Stroudsburg: Dowden, Hutchinson & Ross, 115–140.

Kelly, P.M., Munro, M.A.R., Hughes, M.K. and Goodess, C.M. 1989. Climate and signature years in west European oaks. *Nature*, 340, 57–60.

Kemp, A.E. and Baldauf, J.G. 1993. Vast Neogene laminated diatom mat deposits from the eastern equatorial Pacific Ocean. *Nature*, 362, 141–144.

Kemp, D.D. 1990. *Global Environmental Issues: a Climatological Approach*. Routledge, 220 pp.

Kemp, D.D. 1994. *Global Environmental Issues: a Climatological Approach*. (2nd edition). London: Routledge.

Kennett, J.P. 1982. *Marine Geology*. Englewood Cliffs, New Jersey: Prentice Hall, 813 pp.

Kennett, J.P. and Ingram, B.L. 1995. A 20,000-year record of ocean circulation and climate change from the Santa Barbara basin. *Nature*, 377, 510–514.

Kerr, R.A. 1994. Climate modeling's fudge factor comes under fire. *Science*, 265, 1528.

Kerr, R.A. 1995a. At quadrennial geophysics fest, earth scientists think globally, *Science*, 269, 477–478.

Kerr, R.A. 1995b. Bigger jolts are on the way for southern California. *Science*, 267, 176–177.

Kerr, R.A. 1995c. Earth's surface may move itself. *Science*, 269, 1214–1215.

Kerr, R.A. 1995d. Faraway tsunami hints at a really big northwest quake. *Science*, 267, 962.

Keyfitz, N. 1989. The growing human population. *Scientific American*, 261, 118–126.

Kiehl, J.T. and Briegleb, B.P. 1993. The relative role of sulphate aerosols and greenhouse gases in climate forcing. *Science*, 260, 311–314.

Kiersch, G.A. 1965. The Vaiont Reservoir disaster. *Mineral Information Service*, 18, 129–138.

King, C.-Y., Koizumi, N. and Kitagawa, Y. 1995. Hydrogeological anomalies and the 1995 Kobe earthquake. *Science*, 269, 38–39.

Kissinger, H.A. 1982. *Years of Upheaval*. London: Weidenfeld & Nicolson and Michael Joseph, 1,283 pp.

Kitchell, J.A. and Pena, D. 1984. Periodicity of extinctions in the geologic past: deterministic versus stochastic explanations. *Science*, 226, 689–692.

Knapp, B.J. 1979. *Soil Processes*. London: George Allen & Unwin.

Knox, J.C. 1993. Large increases in flood magnitude in response to modest changes in climate. *Nature*, 361, 430–432.

Komar, P.D. 1983. Coastal erosion in response to the construction of jetties and breakwaters. In: Komar, P.D. (ed.), *Handbook of Coastal Processes and Erosion*. Boca Raton, Florida, CRC Press, 191–204.

Kotilainen, A.T. and Shackleton, N.J. 1995. Rapid climate variability in the North Pacific Ocean during the past 95,000 years. *Nature*, 377, 323–326.

Krogh, T.E., Kamo, S.L. and Bohor, B.F. 1993. Fingerprinting the K/T impact site and determining the time of impact by U-Pb dating of single shocked zircons from distal ejecta. *Earth and Planetary Science Letters*, 119, 425–429.

Kudrass, H.R., Erienkeuser, H., Vollbrecht, R. and Weiss, W. 1991. Global nature of the Younger Dryas cooling event inferred from oxygen isotope data from Sulu Sea cores. *Nature*, 349, 406–409.

Kukla, G. 1987. Loess stratigraphy in central China. *Quaternary Science Review*, 6, 191–219.

Kukla, G., Heller, F., Lui, X.M., Xu, T.C., Lui, T.S. and An, Z.S. 1988. Pleistocene climates in China dated by magnetic susceptibility. *Geology*, 16, 811–814.

Kukla, G., An, Z.S., Melice, J.L., Gavin, J. and Xiao, J.L. 1990. Magnetic susceptibility record of Chinese loess. *Transactions of the Royal Society of Edinburgh: Earth Sciences*, 81, 263–288.

Kumar, C., Patel, N. and Bloembergen, N. 1987. Strategic defense and directed-energy weapons. *Scientific American*, 257(3), 31–37.

Kumar, N., Anderson, R.F., Mortlock, R.A., Froelich, P.N., Kubik, P., Dittrich-Hannen, B. and Suter, M. 1995. Increased biological productivity and export production in the glacial Southern Ocean. *Nature*, 378, 675–680.

Lacerda, L.D. and Salomons, W. 1991. *Mercury in the Amazon*. Dutch Ministry of Housing, Physical Planning and Environmental Report, Institute of Soil Fertility, Haren.

Lachenbruch, A.H. and Marshall, B.V. 1986. Changing climate: geothermal evidence from permafrost in the Alaskan Arctic. *Science*, 234, 689–696.

LaMarche, V.C. and Hirschboech, K.K. 1984. Frost rings in trees as records of major volcanic eruptions. *Nature*, 307, 121–126.

Lamb, H.F., Gasse, F., Benkaddour, A., El Hamouti, N., van der Kaars, S., Perkins, W.T., Pearce, N.J. and Roberts, C.N. 1995. Relation between century-scale Holocene arid intervals in tropical and temperate zones. *Nature*, 373, 134–137.

Lamb, H.H. 1972. *Climate: Present, Past and Future*, vol. 1. London: Methuen.

Lambert, J.H., Jennings, J.N., Smith, C.T., Green, C. and Hutchinson, J.N. 1970. The making of the Broads: a reconsideration of their origin in the light of new evidence. *Royal Geographical Society Research Series*, 3.

Landsberg, J.P., McDonald, B. and Watt, F. 1992. Absence of aluminium in neuritic plague cores in Alzheimer's disease. *Nature*, 360, 65–68.

Langer, J., Rodhe, H., Crutzen, P.J. and Zimmermann, P. 1992. Anthropogenic influence on the distribution of tropospheric sulphate aerosol. *Nature*, 359, 712–716.

Larsen, E., Gulliksen, S., Lauritzen, S.-E., Lie, R., Lovlie, R. and Mangerud, J. 1987. Cave stratigraphy in western Norway: multiple Weichselian glaciations and interstadial vertebrate fauna. *Boreas*, 16, 267–292.

Last, F.T. 1991. Critique. In: Last, F.T. and Watling, R. (eds), *Acidic Deposition: its Nature and Impacts*. Edinburgh: The Royal Society of Edinburgh, 273–324.

Last, F.T. and Watling, R. (eds) 1991. *Acidic Deposition: its Nature and Impacts*. Edinburgh: The Royal Society of Edinburgh, 343 pp.

Lawson, D.E. 1986. Response of permafrost terrain to disturbance: a synthesis of observations from northern Alaska, U.S.A. *Arctic and Alpine Research*, 18(1), 1–17.

Lawson *et al.* 1908. The California earthquake of April 18, 1906. Report of State Earthquake Investigation Commission: Carnegie University Institution of Washington Publication 87, 1, Atlas.

Le Guenno, B. 1995. Emerging viruses. *Scientific American*, October, 56–64.

Le Roy LaDurie, E. 1975. Histoire de la France Rurale / sous la Direction de Georges Duby et Armand Wallon. 309.44 HIS.

Lean, G. 1996. Dangerous fuel may go into Milford Haven. *Independent on Sunday*, 25 February, 11.

Lee, J. and Manning, L. 1995. Environmental lung disease. *New Scientist*, Inside Science 84, 16 September.

Legeckis, R. 1977. Longwaves in the eastern equatorial Pacific Ocean: a view from a geostationary satellite. *Science*, 197, 1179.

Leggett, J., Pepper, W.J. and Swart, R.J. 1992. Emissions scenarios for IPCC: an update. In: Houghton, J.T., Callander, B.A. and Varney, S.K. (eds), *Climate Change 1992: The Supplementary Report to the IPCC Scientific Assessment*. Cambridge: Cambridge University Press, 69–95.

Legrand, M., Feniet-Saigne, C., Saltzman, E.S., Germain, C., Barkov, N.I. and Petrov, V.N. 1991. Ice-core record of oceanic emissions of dimethyl-sulphide during the last climate cycle. *Nature*, 350, 144–146.

Lehman, S.J. and Kelgwin, L.D. 1992. Sudden changes in North Atlantic circulation during the last deglaciation. *Nature*, 356, 757–762.

Lents, J.M. and Kelly, W.J. 1993. Clearing the air in Los Angeles. *Scientific American*, 269(4), 18–25.

Leopold, L.B. 1968. *Hydrology for Urban Land Planning – A Guidebook on the Hydrological Effects of Urban Land Use*. US Geological Survey Circular, 620.

Lester, J.N. 1990. *Sewage and Sewage Sludge Treatment* (Chapter 3), 33–62.

Leuenberger, M. and Siegenthaler, U. 1992. Ice-age atmospheric concentration of nitrous oxide from an Antarctic ice core. *Nature*, 360, 449–451.

Leuenberger, M., Siegenthaler, U. and Langway, C.C. 1992. Carbon isotope composition of atmospheric CO_2 during the last ice age from an Antarctic ice core. *Nature*, 357, 488–490.

Linell, K.A. 1973. Long-term effects of vegetation cover on permafrost stability in an area of discontinuous permafrost. In: *North American Contributions, Permafrost Second International Conference, Yakutsk, USSR, 1973*. Washington, DC: National Academy of Sciences, 688–693.

Linsley, B.K. 1996. Oxygen-isotope record of sea level and climate variations in the Sulu Sea over the past 150,000 years. *Nature*, 380, 234–237.

Liu, H.-S. 1992. Frequency variations of the Earth's obliquity and the 100-kyr ice-age cycles. *Nature*, 358, 397–399.

Liu, T., Tang, W.Q. and Fu, L.-L. 1995. Recent warming event in the Pacific may be an El Niño. *EOS, Transactions of the American Geophysical Union*, 76, 429 and 437.

Liuxiuming, X.M., Shaw, J., Lui, T.H., Heller, F. and Yuan, B.Y. 1992. Magnetic mineralogy of Chinese loess and its significance. *Geophysical Journal International*, 108, 301–308.

Livernash, R. 1995. The future of populous economies: China and India shape their destinies. *Environment*, 37(6), 7–11 and 25–34.

Lorius, C., Barkov, N.I., Jouzel, J., Korotkevich, Y.S., Kotlyakov, V.M. and Raynaud, D. 1988. Antarctic ice core: CO_2 and change over the last climatic cycle. *EOS*, 68, 681–684.

Lorius, C., Jouzel, J., Ritz, C., Merlivat, L., Barkov, N.I., Korotkevich, Y.S. and Kotlyakov, V.M. 1985. A 150,000-year climatic record from Antarctic ice. *Nature*, 310, 591–596.

Lovelock, J.E. 1988. *The Ages of Gaia: a Biography of our Living Earth*. Oxford: Oxford University Press, 252 pp.

Lowe, J.J. and Walker, M.J.C. 1984. *Reconstructing Quaternary Environments*. Longman, London, 389 pp.

Lowenthal, D. 1965. *Man and Nature* by George Perkins Marsh, first published in 1864. Cambridge, Mass.: Harvard University Press.

Mabogunje, A.L. 1995. The environmental challenges in Sub-Saharan Africa. *Environment*, 37(4), 4–9 and 31–35.

MacAyeal, D.R. 1992. Irregular oscillations of the West Antarctic ice sheet. *Nature*, 359, 29–32.

Macfarlane, M.J. 1976. *Laterites and landscape*. London: Academic Press.

Macilwain, C. 1993. Conservationists fear defeat on revised flood control policies. *Nature*, 365, 478.

Macilwain, C. 1994. Nature, not levees, blamed for flood. *Nature*, 369, 348.

MacKenzie, D. 1991. Energy answers for North and South. *New Scientist*, 16 February, 48–51.

MacKenzie, D. 1995. The cod that disappeared. *New Scientist*, 16 September, 24–29.

MacNeill, J. 1989. Strategies for Sustainable Economic Development. *Scientific American*, 261, 154–165.

Maddox, J. 1989. The biggest greenhouse still intact. *Nature*, 338, 111.

Maddox, J. 1993. Where the AIDS virus hides away. *Nature*, 362, 287.

Maitlis, P. and Rourke, J. 1993. Rich seams for chemicals. *New Scientist*, 23 January, 37–41.

Malthus, T. 1798. *An Essay on the Principle of Population and a Summary View of the Principle of Population*, A. Field (ed.), 1970. Harmondsworth: Pelican.

Manabe, S. and Stouffer, R.J. 1993. Century-scale effects of increased atmospheric CO_2 on the ocean–atmosphere system. *Nature*, 364, 215–218.

Manabe, S. and Stouffer, R.J. 1995. Simulation of abrupt climate change induced by freshwater input to the North Atlantic Ocean. *Nature*, 378, 165–167.

Mannion, A.M. 1991. *Global Environmental Change*. Harlow: Longman Scientific & Technical.

Marrow, J.E., Coombs, J. and Lees, E.W. 1987. *An Assessment of Bio-Ethanol as Transport Fuel in the UK*. ETSU-R44 (volume 1). London: HMSO.

Marrow, J.E. and Coombs, J. 1990. *An Assessment of Bio-Ethanol as Transport Fuel in the UK*. ETSU-R55 (volume 2). London: HMSO.

Marsh, W.M. and Grossa, J.M. 1996. *Environmental Geography: Science, Land Use, and Earth Systems*. Chichester: John Wiley & Sons Ltd, 426 pp.

Martin, J. 1990. Glacial–Interglacial CO_2 change: the iron hypothesis. *Paleoceanography*, 5, 1–13.

Martin, J.H. *et al.* (44 authors) 1994. Testing the iron hypothesis in ecosystems of the equatorial Pacific Ocean. *Nature*, 371, 123–129.

Martin, J.H., Fitzwater, S.E. and Gordon, R.M. 1990. *Global Biogeochemical Cycles*, 4, 5–12.

Martin, J.H., Gordon, R.M. and Fitzwater, S.E. 1991. *Limnology & Oceanography*, 36, 1793–1802.

Martin, P.S. 1984. Prehistoric overkill; a glacial model. In: Martin, P.S. and Klein, R.G. (eds), *Quaternary Extinctions: A Prehistoric Revolution*. Tucson, Arizona: University of Arizona Press.

Martinson, D.G. *et al.* 1987. *Quaternary Research*, 27, 1–29.

Martyn, C.N., Barker, D.J.P., Osmond, C., Harris, E.C., Edwardson, J.A. and Lacey, R.F. 1989. Geographical relation between Alzheimer's disease and aluminium in drinking water. *Lancet* (14 January), 1, 8624, 59–62.

Mather, J.R. and Yoshioka, G.A. 1968. The role of climate in the distribution of vegetation. *Annals of the Association of American Geographers*, 58, 29–41.

Mathews, W.H. and Clague, J.J. 1993. The record of jokulhlaups from Summit Lake, northwestern British Columbia. *Canadian Journal of Earth Sciences*, 30, 499 508.

Mayewski, P.A. and Jeschke, P.A. 1979. Himalayan and

Trans-Himalayan glacier fluctuations since AD 1812. *Arctic and Alpine Research*, 11(3), 267–287.

Maurits la Rivire, J.W. 1989. Threats to the world's water, *Scientific American*, 261, 80–94.

McCormack, J. 1989. *Acid Earth: the Global Threat of Acid Pollution*. London: Earthscan Publications, 225 pp.

McCormick, M.P. and Velga, R.E. 1992. *Journal of Geophysical Research Letters*, 9, 155–158.

McCormick, M.P., Thomason, L.W. and Trepte, C.R. 1995. Atmospheric effects of the Mt. Pinatubo eruption. *Nature*, 373, 399–404.

McGuire, W.J. (ed.) 1991. Thematic set of papers: Monitoring active volcanoes. *Journal of the Geological Society, London*, 148, 561–593.

McIlveen, R. 1992. *Fundamentals of Weather and Climate*. London: Chapman & Hall, 497 pp.

McLaughlin, S.B. and Downing, D.J. 1995. Interactive effects of ambient ozone and climate measured on growth of mature forest trees. *Nature*, 374, 252–254.

McManus, J.F., Bond, G.C., Broecker, W.S., Johnsen, S., Labeyrie, L. and Higgins, S. 1994. High-resolution climate records from the North Atlantic during the last interglacial. *Nature*, 371, 326–329.

Meade, R.B. 1991. Reservoirs and earthquakes. *Engineering Geology*, 30, 245–262.

Meade, R.H. and Trimble, S.W. 1974. Changes in sediment loads in rivers of the Atlantic drainage of the United States since 1900. *Publications of the International Association of Hydrological Science*, 113, 99–104.

Meadows, D.L., Randers, J. and Behrens, W.W., III 1972. *The Limits to Growth: A Report to the Club of Rome's Project on the Predicament of Mankind.* New York: Potomac Associates.

Meese, D. *et al.* 1994. *Preliminary Depth-Age Scale of the GISP2 Ice Core*. Special Report 94–1. Cold Regions Research and Engineering Laboratory, Hanover, New Hampshire.

Meko, D.M. 1992. Dendroclimatic evidence from the Great Plains of the United States. In: Bradley, R.S. and Jones, P.D. (eds), *Climate since AD 1500*. London: Routledge.

Meyers, S.D. and O'Brien, J.J. 1995. Pacific Ocean influences atmospheric carbon dioxide. *EOS, Transactions of the American Geophysical Union*, 76, 533 and 537.

Micklin, P.P. 1988. Desiccation of the Aral Sea. *Science*, 241, 1170–1176.

Micklin, P.P. 1992. The Aral crisis: introduction to the special issue. *Post-Soviet Geography*, 33, 5, 269–282.

Middleton, N. 1995. *The Global Casino: an Introduction to Environmental Issues*. London: Edward Arnold, 332 pp.

Midgley, C. and Nuttall, N. 1996. Sea Empress oil disaster 'as bad as Torrey Canyon'. *The Times*, 27 February, 3.

Miller, E.K., Blum, J.D. and Friedland, A.J. 1993. Determination of soil exchangeable-cation loss and weathering rates using Sr isotopes. *Nature*, 362, 438–441.

Miller, G.H. and de Vernal, A. 1992. Will greenhouse

warming lead to Northern Hemisphere ice-sheet growth? *Nature*, 355, 244–246.

Miller, J.M. and Ball, T.K. 1969. *Second Progress Report on the Measurement of Radon in Soil Air as a Prospecting Technique*. Institute of Geological Sciences, Metalliferous Minerals and Applied Geochemistry Unit Report 28. Keyworth: British Geological Survey.

Miller, J.M. and Ostle, D. 1973. Radon measurements in uranium prospecting. In: *Uranium Exploration Methods* (Conference Volume). Vienna: International Atomic Energy Agency, 229–239.

Milliman, J.D. *et al.* 1987. Man's influence on the erosion and transport of sediment by Asian rivers: the Yellow River example. *Journal of Geology*, 95, 751–62.

Milne, R. 1989. North Sea algae threaten British coasts. *New Scientist*, 122, 1663, 37–41.

Milne, R. 1991. Renewable energy – Britain's untrapped resource. *New Scientist*, 12 October, 132, 20–21.

Mintzer, I.M. (ed.) 1992. *Confronting Climate Change: Risks, Implications and Responses*. Cambridge: Cambridge University Press, 382 pp.

Mitchell, J.M., Jr 1976. An overview of climatic variability and its causal mechanisms. *Quaternary Research*, 6, 481–493.

Mitchell, R.B. 1995. Lessons from international oil pollution. *Environment*, 37(4), 10–15 and 36–41.

Mohnen, V.A. 1988. The challenge of acid rain. *Scientific American*, August issue, 14–22.

Montgomery, H., Pessagno, E., Soegaard, K., Smith, C., Munoz, I. and Pessagno, J. 1992. Misconceptions concerning the Cretaceous/Tertiary boundary at the Brazos River, Falls County, Texas. *Earth and Planetary Science Letters*, 109, 593–600.

Mook, W. and Woillard, G. 1982. Carbon-14 dates at Grand Pile. Correlation of land and sea chronologies. *Science*, 215, 159–161.

Moore Lappé, F. and Schurman, R. 1989. *Taking Population Seriously*. London: Earthscan.

Morgan, G.M. 1973. General description of the hail problem in the Po valley of northern Italy. *Journal of Applied Meteorology*, 12(2), 338–353.

Morgan, H. and Simms, D.L. 1988. *Setting trigger concentrations for contaminated land*. In: Vol. 1 Contaminated Soil '88, Second International TNO Conference on Contaminated Soil, 11–15 April 1988, Hamburg. Dordrecht: Kluwer Academic.

Morgan, R.P.C. 1986. *Soil Erosion and Conservation*. New York: Longman Scientific & Technical, 298 pp.

Morgan, V.I., Goodwin, I.D., Etheridge, D.M. and Wookey, C.W. 1991. Evidence from Antarctic ice cores for recent increases in snow accumulation. *Nature*, 354, 58–60.

Mortimer, N. 1989. *Evidence to the House of Commons Select Committee on Energy Bearing on the Energy Policy Implications of the Greenhouse Effect and Proof of Evidence*, Friends of the Earth 9, to the Hinckley Point C public inquiry. London: Friends of the Earth.

Muir, H. 1995. Striking back at lightning. *New Scientist*, 7 October, 26–30.

Mungall, C. and McLaren, D.J. (eds) 1990. *Planet Under Stress: The Challenge of Global Change*. Oxford: Oxford University Press, 344 pp.

Muniz, I.P. 1991. Freshwater acidification: its effects on species and communities of freshwater microbes, plants and animals. In: Last, F.T. and Watling, R. (eds), *Acidic Deposition: its Nature and Impacts*, Edinburgh: The Royal Society of Edinburgh, 227–254.

Murozumi, M., Chow, T.J. and Patterson, C.C. 1969. *Geochimica Cosmochimica Acta*, 33, 1247–1294.

Murphy, A.P. 1991. Chemical removal of nitrate from water. *Nature*, 350, 223–225.

Myers, M.F. and White, G.F. 1993. The challenge of the Mississippi floods. *Environment*, 38(10), 5–9.

Myers, N. 1986. Environmental repercussions of deforestation in the Himalayas. *Journal of World Forestry Resource Management*, 2, 63–72.

Myers, N. 1988a. Tropical forests: much more than stocks of wood. *Journal of Tropical Ecology*, 4, 209–221.

Myers, N. 1988b. Threatened biotas: 'hot-spots' in tropical forests. *The Environmentalist*, 8, 187–208.

Myers, N. 1990. The biodiversity challenge: expanded hotspot analysis. *The Environmentalist*, 10(4), 1–14.

Myers, N. 1995a. Environmental unknowns. *Science*, 21 July, 269, 358–360.

Myers, N. 1995b. The world's forests: need for a policy appraisal. *Science*, 268, 823–824.

Mysak, L.A. and Lin, C.A. 1990. The tempering seas. In: Mungall, C. and McLaren, D.J. (eds), *Planet Under Stress*. Oxford: Oxford University Press, 134–148.

Naeem, S., Thompson, L.J., Lawler, S.P., Lawton, J.H. and Woodfin, R.M. 1994. Declining biodiversity can alter the performance of ecosystems. *Nature*, 368, 734–737.

NAS (National Academy of Sciences) 1986. *Acid Deposition: Long-Term Trends*. Committee on Monitoring and Assessment of Trends in Acid Deposition.

Nash, L. 1993. Water quality and health. In: Gleick, P.H. (ed.), *Water in Crisis*. Oxford: Oxford University Press, 25–39.

Nebel, B.J. and Wright, R.T. 1993. *Environmental Science: The Way the World Works* (4th edition). Englewood Cliffs, NJ: Prentice-Hall.

Nelson, B.K., MacLeod, G.K. and Ward, P.D. 1991. Rapid change in strontium isotopic composition of sea water before the Cretaceous/Tertiary boundary. *Nature*, 351, 644–647.

Nelson, R. 1988. Dryland management: the 'desertification' problem. *Environment Department Working Paper*, No. 8. Washington: World Bank.

New Scientist, 1989. Winds of change. Climate in a spin. *New Scientist*, 25 March, 121, 27.

Newhouse, J. 1989. *The Nuclear Age: from Hiroshima to Star Wars*. London: Michael Joseph, 486 pp.

Nilsson, S. and Pitt, D. 1991. *Mountain World in Danger: Climate Change in the Forests and Mountains of Europe*. London: Earthscan Publications.

Ninkovich, D., Shackleton, N.J., Abdel-Monem, A., Obradovich, J.D. and Izett, G. 1978. K-Ar age of the late Pleistocene eruption of Toba, north Sumatra. *Nature*, 276, 574–577.

Nisbet, E.G. and Fowler, C.M.R. 1995. Is metal disposal toxic to deep oceans? *Nature*, 375, 715.

Noma, E. and Glass, A.L. 1987. Mass extinction pattern: result of chance. *Geological Magazine*, 124(4), 319–322.

Northup, R.R., Yu, Z., Dahlgren, R.A. and Vogt, K.A. 1995. Polyphenol control of nitrogen release from pine litter. *Nature*, 377, 227–229.

Nowak, M.A. and McMichael, A.J. 1995. How HIV defeats the immune system. *Scientific American*, August, 58–65.

Nriagu, J.O. 1993. Legacy of mercury pollution. *Nature*, 363, 589–590.

Nuclear Power and the Greenhouse Effect 1990. United Kingdom Atomic Energy Authority booklet, 30 pp.

Obradovich, J., Snee, L.W. and Izett, G.A. 1989. Is there more than one glassy layer in the late Eocene? *Geological Society of America Abstracts with Programs*, 21, 134.

OECD 1987. *Pricing of Water Services*. Paris: Organisation for Economic Co-operation and Development.

Oechel, W.C., Hastings, S.J., Vourlitis, G., Jenkins, M., Riechers, G. and Grulke, N. 1993. Recent changes of Arctic tundra ecosystems from a net carbon dioxide sink to a source. *Nature*, 361, 520–523.

Oechsli, F.W. and Buechly, R.W. 1970. Excess mortality associated with three Los Angeles September hot spells. *Environmental Research*, 3, 277–284.

Oeschger, H. and Mintzer, I.M. 1992. Lessons from the ice cores: rapid climate changes during the last 160,000 years. In: Mintzer, I.M. (ed.), *Confronting Climate Change: Risks, Implications and Responses*. Cambridge: Cambridge University Press, 55–64.

Office of Technology Assessment 1984. Protecting the nation's groundwater from contamination. *Office of Technology Assessment, OTA-0-233*: Washington, DC, US Congress, 244 pp.

Officer, C. 1993. Victims of volcanoes. *New Scientist*, 20 February, 34–38.

OFWAT 1990. *Paying for Water – A Time for Decisions*. Birmingham: Office of Water Services.

O'Hara, S.L., Street-Perrott, F.A. and Burt, T.P. 1993. Accelerated soil erosion around a Mexican highland lake caused by prehispanic agriculture. *Nature*, 362, 48–51.

Oldfield, F. and Clark, R.L. 1990. Lake sediment-based studies of soil erosion. In: Broadman, J., Foster, I.D.L. and Dearing, J.A. (eds), *Soil Erosion on Agricultural Land*, New York: John Wiley and Sons Ltd, 201–208.

Oliver, D. 1991. The house that came in from the cold. *New Scientist*, 9 March, 45–49.

Oltmans, S.J. and Levy, H. 1992. Seasonal cycle of surface ozone over the western North Atlantic. *Nature*, 358, 392–394.

O'Neill, B. 1989. Falling out over water. *New Scientist*, 21 October, 60–64.

Opdyke, N.D., Glass, B., Hays, J.D. and Foster, J. 1966. Paleomagnetic study of Antarctic deep-sea cores. *Science*, 154, 349–357.

Open University 1991. *Case Studies in Oceanography and Marine Affairs*. Oxford: Pergamon Press.

O'Riordan, T. (ed.) 1994a. *Environmental Science for Environmental Management.* Harlow: Longman Scientific, 369 pp.

O'Riordan, T. 1994b. The global environment debate. In: O'Riordan, T. (ed.) *Environmental Science for Environmental Management.* Harlow: Longman Scientific, 16–29.

O'Rourke, P.J. 1994. *All the Trouble in the World: the Lighter Side of Famine, Pestilence, Destruction and Death.* London: Picador, 340 pp.

Ottaway, R. 1990. *Less People, Less Pollution: An Answer to Environmental Decline Caused by the World's Population Explosion.* London: Bow Group.

Owen, J.A. and Ward, M.N. 1989. Forecasting Sahel rainfall. *Weather*, 44, 57–64.

Owen, L.A. 1995. Shaping the Himalayas. *Geographical Magazine*, 112(2), 23–25.

Owen, L.A. 1996. High roads high risks. *Geographical Magazine*, 118(1), 12–15.

Owen, L.A., Sharma, M. and Bigwood, R. 1995. Mass movement hazard in the Garhwal Himalaya: the effects of the 20 October 1991 Garhwal earthquake and the July–August 1992 monsoon season. In: McGregor, D.F.M. and Thompson, D.A. (eds), *Geomorphology and Land Management in a Changing Environment.* Chichester: John Wiley & Sons, 69–88.

Page, R.A., Boore, D.M., Bucknam, R.C. and Thatcher, W.R. 1992. *Goals, Opportunities, and Priorities for the USGS Earthquake Hazards Reduction Program.* US Department of the Interior/US Geological Survey, Denver, Circular 1079. Washington: US Government Printing Office, 60pp.

Paillard, D. and Lebeyrie, L. 1994. Role of the thermohaline circulation in the abrupt warming after Heinrich events. *Nature*, 372, 162–164.

Pain, S. 1989a. Globe set green agenda rolling in the Hague. *New Scientist*, 18 March, 33.

Pain, S. 1989b. Greenhouse warming at nuclear inquiry. *New Scientist*, 121, 1656, 33.

Pakiser, L.C. 1991. *Earthquakes.* US Department of the Interior/US Geological Survey, Denver. Washington: US Government Printing Office, 20 pp.

Pakiser, L.C., Eaton, J.P., Healy, J.H. and Raleigh, C.B. 1969. Earthquake prediction and control. *Science*, 166, 1467–1474.

Panel on Seismic Hazard Analysis 1988. *Probabilistic Seismic Hazard Analysis.* Washington DC: National Academic Press.

Park, C. 1992. *Tropical Rainforests.* London: Routledge, 188 pp.

Parliamentary Office of Science and Technology (POST) 1992a. *Clean Coal Technology.* UK Parliamentary Office of Science and Technology, Briefing Note 38.

Parliamentary Office of Science and Technology (POST) 1992b. *The Polluter Pays Principle and Cost Recovery Charging.* UK Parliamentary Office of Science and Technology, 35 pp.

Parliamentary Office of Science and Technology (POST) 1993a. *Dealing with Drought: Environmental and Technical Aspects of Water Shortages*, 82 pp.

Parliamentary Office of Science and Technology (POST) 1993b. *Biofuels for Transport.* UK Parliamentary Office of Science and Technology, Briefing Note 41.

Parry, M. 1990. *Climate Change and World Agriculture.* London: Earthscan Publications, 157 pp.

Patel, T. 1995. France claims Mururoa rocks are blast-proof. *New Scientist*, 30 September, 7.

Patterson, C. and Smith, A.B. 1987. Is the periodicity of extinctions a taxonomic artefact? *Nature*, 330, 248–251.

Patterson, C. and Smith, A.B. 1989. Periodicity in extinction: the role of systematics. *Ecology*, 70(4), 802–811.

Patterson, W. 1989. Energy issues another challenge. *New Scientist*, 28 January, 45–50.

Pearce, D., Markandyn, A. and Barber, E. 1989. *Blueprint for a Green Economy.* London: Earthscan.

Pearce, F. 1987. *Acid Rain.* Penguin Books, 162 pp.

Pearce, F. 1989a. Felled trees deal double blow to global warming. *New Scientist*, 123, 1682, 25.

Pearce, F. 1989b. Methane: the hidden greenhouse gas. *New Scientist*, 122, 1663, 37–41.

Pearce, F. 1990. Whatever happened to acid rain? *New Scientist*, 15 September 1990, 57–60.

Pearce, F. 1992. Mirage of the shifting sands. *New Scientist*, 12 December, 38–42.

Pearce, F. 1993a. When the tide comes in. *New Scientist*, 2 January, 22–27.

Pearce, F. 1993b. How Britain hides its acid soil. *New Scientist*, 27 February, 29–33.

Pearce, F. 1993c. A long dry season. *New Scientist*, 17 July, 15–16.

Pearce, F. 1994a. Soldiers lay waste to Africa's oldest park. *New Scientist*, 3 December, 4.

Pearce, F. 1994b. Water in the war zone. *New Scientist*, 17 December, 13–14.

Pearce, F. 1994c. Dam truths on the Danube. *New Scientist*, 17 September, 27–31.

Pearce, F. 1995a. Dead in the water. *New Scientist*, 4 February, 26–31.

Pearce, F. 1995b. Raising the Dead Sea. *New Scientist*, 22 July, 33–37.

Pearce, F. 1995c. Poisoned waters. *New Scientist*, 21 October, 29–33.

Pearce, F. 1995d. Rockall mud richer than rainforest. *New Scientist*, 16 September, 8.

Pearce, F. 1995e. The biggest dam in the world. *New Scientist*, 28 January, 25–29.

Peck, A.J. 1978. Salinization of non-irrigated soils and associated streams: a review. *Australian Journal of Soil Research*, 16, 157–168.

Peltier, W.R. 1990. Our fragile inheritance. In: Mungall, C. and McLaren, D.J. (eds), *Planet Under Stress.* Oxford: Oxford University Press, 80–95.

Pendick, D. 1995. And here is the eruption forecast. *New Scientist*, 7 January, 26–29.

Peters, C.M., Gentry, A.H. and Mendelsohn, R.O. 1989. Valuation of an Amazonian rainforest. *Nature*, 339, 655–656.

Petit-Maire, N., Fontugne, M. and Rouland, C. 1991. Atmospheric methane ratio and environmental changes in the Sahara and Sahel during the last 130 k yrs. *Paleogeography, Paleoclimatology and Paleoecology*, 86, 197–204.

Peto, J. 1990. Radon and the risks of cancer. *Nature*, 345, 389–390.

Petts, G. 1994. Large-scale river regulation. In: Roberts, N. (ed.), *The Changing Global Environment*, Oxford: Blackwell Scientific, 262–284.

Peucker-Ehrenbrink *et al.* 1995. The marine $^{187}Os/^{186}Os$ record of the past 80 million years. *Earth and Planetary Science Letters*, 130, 155–167.

Pickering, K.T., Soh, W. and Taira, A. 1991. Scale of tsunami-generated sedimentary structures in deep water. *Journal of the Geological Society of London*, 148, 211–214.

Pidgeon, J.D. and Soane, B.D. 1978. Soil structure and strength relations following tillage, zero tillage and wheel traffic in Scotland. In: Emerson, W.W., Bond, R.D. and Dexter, A.R. (eds), *Modification to Soil Structure*. Chichester: John Wiley and Sons, 371–378.

Pielou, E.C. 1991. *After the Ice Age*. Chicago: The University of Chicago Press.

Pimental, D., Harvey, C., Resosudarmo, P., Sinclair, K., Kurz, D., McNair, M., Crist, S., Shpritz, L., Fitton, L., Saffouri, R. and Blair, R. 1995. Environmental and economic costs of soil erosion and conservation benefits. *Science*, 267, 1117–1123.

Pimm, S.L., Russell, G.J., Gittleman, J.L. and Brooks, T.M. 1995. The future of biodiversity. *Science*, 21 July, 269, 347–350.

Pirazzoli, P. 1973. Inondations et niveaux marins a Venise. *Mem Lab. Geomorph. École Pratique Hautes*, Et 22, Dinard.

Plafker, G. and Ericksen, G.E. 1978. Nevados Husascaran avalanches, Peru. In: Voight, B. (ed.), *Rockslides and Avalanches, 1*. Oxford: Elsevier Scientific, 277–314.

Plucknett, D.L. and Winkelmann, D.L. 1995. Technology for sustainable agriculture. *Scientific American*, September, 182–186.

Poag, C.W., Powars, D.S., Poppe, L.J., Mixon, R.B., Edwards, L.C., Folger, D.W. and Bruce, S. 1992. Deep Sea Drilling Project Site 612 bolide event: new evidence of a late Eocene impact-wave deposit and a possible impact site, U.S. east coast. *Geology*, 20, 771–774.

Polzin, K.L., Speer, K.G., Toole, J.M. and Schmitt, R.W. 1996. Intense mixing of Antarctic Bottom Water in the equatorial Atlantic Ocean. *Nature*, 380, 54–57.

Poore, D. 1989. *No Timber Without Trees: Sustainability in the Tropical Forest*. London: Earthscan Publications.

Porritt, J. and Winner, D. 1988. *The Coming of the Greens*. London: Fontana Paperbacks, 287 pp.

Porter, S.C. 1986. Pattern and forcing of northern hemisphere glacier variation during the last millennium. *Quaternary Research*, 26, 27–48.

Porter, S.C. and An, Z. 1995. Correlation between climate events in the North Atlantic and China during the last glaciation. *Nature*, 375, 305–308.

Porter, S.C. *et al.* 1976. Lyell Centenary Issue. *British Journal of History of Science*, 9(36), 91–242.

Postel, S. 1993. Water and agriculture. In: Gleick, P.H. (ed.), *Water in Crisis*. Oxford: Oxford University Press, 56–66.

Powell, F.A. 1983. Bushfire weather. *Weatherwise*, 36(3), 126.

PPP. Greenland ice melting to give 6 m rise in sea level reference. Global and Planetary Change 90, 385–394.

Prentice, I.C. 1993. Process and production. *Nature*, 363, 209.

Prentice, I.C., Cramer, W., Harrison, S.P., Leemans, R., Monserud, R.A. and Solomon, A.M. 1992. A global biome model based on plant physiology and dominance, soil properties and climate. *Journal of Biogeography*, 19, 117–134.

Preston, R. 1994. *The Hot Zone*. New York: Anchor.

Price, L.W. 1972. The periglacial environment, permafrost, and man. *Association of American Geographers*, Resource Paper No. 14.

Price, M. 1985. Introducing groundwater. London: Chapman & Hall, 195 pp.

Price, M. 1991. Water from the ground. *New Scientist*, Inside Science, 16 February, 42.

Proposal for a Council Directive on excise duties on motor fuels from agricultural sources, 92/C 73/04, COM(92) 36 final, 5 March 1992. *Official Journal of the European Communities*, No. C73/6 (24.3.92).

Prospero, J.M., Savoie, D.L., Saltzman, E.S. and Larsen, R. 1991. Impact of oceanic sources of biogenic sulphur on sulphate aerosol concentrations at Mawson, Antarctica. *Nature*, 350, 221–223.

Purseglove, J. 1991. Liberty, ecology, modernity. *New Scientist*, 28 September, 45–48.

Pyne, S.J. 1982. *Fire in America: a Cultural History of Wildfire and Rural Fire*. Princeton University Press.

Rackham, O. 1986. *The History of the Countryside*. London: Dent.

Radford, T. 1990. *The Crisis of Life on Earth: Our Legacy from the Second Millennium*. London: Thorsons Publishing Group, 224 pp.

Rahmstorf, S. 1994. Rapid climate transitions in a coupled ocean–atmosphere model. *Nature*, 372, 82–85.

Rahmstorf, S. 1995. Bifurcations of the Atlantic thermohaline circulation in response to changes in the hydrological cycle. *Nature*, 378, 145–149.

Raleigh, B. *et al.* 1977. Prediction of the Haicheng earthquake. *Transactions of the American Geophysical Union*, 58(5), 236–272.

Ramage, J. 1988. *Energy: A Guidebook*. Oxford: Oxford University Press.

Ramanathan, V. and Collins, W. 1991. Thermodynamic regulation of ocean warming by cirrus clouds deduced from observations of the 1987 El Niño. *Nature*, 351, 27–32.

Rampino, M.R. and Self, S. 1992. Volcanic winter and accelerated glaciation following the Toba super-eruption. *Nature*, 359, 50–52.

Rampino, M.R. and Stothers, R.B. 1984. Terrestrial mass extinctions, cometry impacts and the Sun's motion perpendicular to the galactic plane. *Nature*, 308, 709–712.

Rau, J.L. and Nutalaya, P. 1982. Geomorphology and land subsidence in Bangkok, Thailand. In: Craig, R.G. and Crafts, J.L. (eds), *Applied Geomorphology*. Mackays & Chatham Ltd, 181–201.

Raup, D.M. and Sepkoski, J.J., Jr 1984. Periodicity of extinctions in the geologic past. *Proceedings of the*

National Academy of Sciences, U.S.A., 81, 801–805.

Raup, D.M. and Sepkoski, J.J., Jr 1986. Periodic extinctions of families and genera. *Science*, 231, 833–836.

Ravven, W. 1991. Asteroid impact emptied Gulf of Mexico. *New Scientist*, 129, 1762, 14.

Raymo, M.E. and Ruddiman, W.F. 1992. Tectonic forcing of late Cenozoic climate. *Nature*, 359, 117–122.

Redclift, M. 1987. *Sustainable Development: Exploring the Contradictions*. London: Methuen.

Revkin, A. 1990. *The Burning Season: The Murder of Chico Mendes and the Fight for the Amazon Rain Forest*. Collins, 317 pp.

Richards, K. 1989. All gas and garbage. *New Scientist*, 3 June, 38–41.

Ridley, M. 1993. Cleaning up with cheap technology. *New Scientist*, 23 January, 26–27.

Rippon, S. 1984 *Nuclear Energy*. London: Heinemann.

Roberts, N. 1989. *The Holocene: An Environmental History*. Oxford: Basil Blackwell.

Roberts, N. (ed.) 1994. *The Changing Global Environment*. Oxford: Blackwell, 531 pp.

Robertson, J. 1990. Alternative futures of cities. In: Cadman, D. and Payne, G. (eds), *The Living City: Towards a Sustainable Future*. London: Routledge, 127–135.

Robin, G. 1977. Ice cores and climate change. *Philosophical Transactions of the Royal Society of London*, B280, 143–168.

Rodhe, H.E., Cowling, E., Galbally, I., Galloway, J. and Herrera, R. 1988. Acidification and regional air pollution in the tropics. In: Rodhe, H.E. and Herrera, R. (eds), *Acidification in Tropical Countries*. Chichester: Wiley, 3–39.

Rogers, P., Lydon, P. and Seckler, D. 1989. *Easter Waters Study: Strategies to Manage Flood and Drought in the Ganges–Brahmaputra Basin*. US Agency for International Development, Washington DC.

Rooney, J.F. 1967. The urban snow hazard in the United States: an appraisal of disruption. *Geographical Review*, 57, 538–559.

Rosenzweig, C. and Parry, M.L. 1994. Potential impact of climate change on world food supply. *Nature*, 367, 133–138.

Rosman, K.J.R., Chisholm, W., Boutron, C.F., Candelone, J.P. and Görlach, U. 1993. Isotopic evidence for the source of lead in Greenland snows since the late 1960s. *Nature*, 362, 333–335.

Rowlands, S.A., Hall, A.K., McCormick, P.G., Street, R., Hart, R.J., Ebell, G.F. and Donecker, P. 1994. Destruction of toxic materials. *Nature*, 367, 223.

Royal Commission on Environmental Pollution 1994. *Transport and the Environment*. Report 18, London: HMSO.

Ruckelshaus, W.D. 1989. Towards a Sustainable World. *Scientific American*, 261, 166–175.

Ruddiman, W.F. and Kutzbach, J.E. 1991. Plateau uplift and climatic change. *Scientific American*, 264 (3), 42–50.

Ruddiman, W.F. and McIntyre, A. 1981. The North

Atlantic during the last deglaciation. *Palaeogeography, Palaeoclimatology, Palaeoecology*, 35, 145–214.

Ruel, S. 1993. The scourge of land mines. *UN Focus*, October, 4 pp.

Russell, J.M., III, Luo, M., Cicerone, R.J. and Deaver, L.E. 1996. Satellite confirmation of the dominance of chlorofluoro-carbons in the global stratospheric chlorine budget. *Nature*, 379, 526–529.

Safina, C. 1995. The world's imperiled fish. *Scientific American*, 273(5), 46–53.

Sagan, C. and Turco, R. 1990. *A Path Where No Man Thought: Nuclear Winter and the End of the Arms Race*. London: Century Press.

Sagan, C., Thompson, W.R., Carlson, R., Gurnett, D. and Hord, C. 1993. A search for life on Earth from Galileo spacecraft. *Nature*, 365, 715–721.

Sarnthein, M. 1978. Sand deserts during the Glacial Maximum and climatic optimum. *Nature*, 273, 43–46.

Sarre, P. 1978. The diffusion of Dutch elm disease. *Area*, 10(2), 81–85.

Sarre, P. (ed.) 1991. *Environment, Population and Development*. London: Hodder & Stoughton.

Sayles, F.L., Martin, W.R. and Deuser, W.G. 1994. Response of benthic oxygen demand to particulate organic carbon supply in the deep sea near Bermuda. *Nature*, 371, 686–689.

Schaaf, M. and Thurow, J. 1995. Late Pleistocene–Holocene climatic cycles recorded in Santa Barbara Basin sediments: interpretation of colour density logs from Site 893. In: Kennett, J.P., Baldauf, J.G. and Lyle, M. (eds), *Proceedings of the Ocean Drilling Program, Scientific Results*. College Station, Texas: Ocean Drilling Program, 146, 31–44.

Schidlowski, M. 1988. A 3,800-million-year isotopic record of life from carbon in sedimentary rocks. *Nature*, 333, 313–318.

Schlesinger, M.E. and Jiang, X. 1991. Revised projection of future greenhouse warming. *Nature*, 350, 219–221.

Schlesinger, M.E. and Ramankutty, N. 1992. Implications for global warming of intercycle solar irradiance variations. *Nature*, 360, 330–333.

Schwartz, R.D. and James, P.B. 1984. Periodic mass extinctions and the Sun's oscillation about the galactic plane. *Nature*, 308, 712–713.

Schweingruber, F.H. 1989. *Tree Rings: Basics and Applications of Dendrochronology*. London: Kluwer Academic Publishing, 276 pp.

Scientific American 1990. Special Issue: *Energy for Planet Earth*, 263, 20–115.

Scoging, H. 1993. The assessment of desertification. *Geography*, 339, 78, 2, 190–193.

Scourse, J.D. 1993. Quaternary sea-level change and coastal management. *Geoscientist*, 2, 13–16.

Scurlock, J. and Hall, D. 1991. The Carbon Cycle. *New Scientist*, 2 November, Inside Science No. 51.

Seers, D. 1977. The new meaning of development. *International Development Review*, 3, 2–7.

Selby 1985. *Earth's Changing Surface*. Oxford: Clarendon Press.

Self, S., Pampino, M.A. and Bonbera, J.J. 1981. The

possible effects of large 19th and 20th century volcanic eruptions on zonal and hemispherical surface temperatures. *Journal of Volcanology and Geothermal Research*, 11, 41–60.

Semtner, A.J. 1984. The climatic response of the Arctic Ocean to Soviet river diversions. *Climatic Change*, 6, 109–130.

Sen, A. 1981. *Poverty and Famine: An Essay on Entitlement and Deprivation*. Oxford: Oxford University Press.

Sen, G. 1995. The world programme of action: a new paradigm for population policy. *Environment*, 37(1), 10–15 and 34–37.

Sepkoski, J.J., Jr 1990. Mass extinction: processes: periodicity. In: Briggs, D.E.G. and Crowther, P.R. (eds), *Palaeobiology – A Synthesis*. Oxford: Blackwell Scientific Publications.

Sepkoski, J.J., Jr and Raup, D.M. 1986. Periodicity in marine extinction events. In: Elliot, D.K. (ed.), *Dynamics of Extinction*. New York: John Wiley & Sons Ltd, 3–36.

Shackleton, N.J. 1987. Oxygen isotopes, ice volume and sea level. *Quaternary Science Reviews*, 6, 183–190.

Shackleton, N.J. 1990. Estimating atmospheric CO_2. *Nature*, 347, 427–428.

Shackleton, N.J. and Opdyke, N.D. 1973. Oxygen isotope and palaeomagnetic stratigraphy of equatorial Pacific core V28–238: oxygen isotope temperatures and ice volumes on a 10^5 and 10^6 year scale. *Quaternary Research*, 3, 39–55.

Shackleton, N.J., Backman, J., Zimmerman, H., Kent, D.V., Hall, M.A., Roberts, D.G., Schnitker, D., Baldauf, J.G., Desprairies, A., Homrighausner, R., Huddleston, P., Keen, J.B., Kaltenbach, A.J., Krumsiek, K.A.O., Morton, A.C., Westberg-Smith, J. 1984. Oxygen isotope calibration of the onset of ice-rafting and history of glaciations in the North Atlantic region. *Nature*, 307, 620–623.

Shackleton, N.J., Berger, A. and Pettier, W.R. 1990. An alternative astronomical calibration of the lower Pleistocene timescale based on ODP Site 677. *Transactions of the Royal Society of Edinburgh, Earth Sciences*, 81, 251–261.

Shackleton, N.J., Hall, M.A., Line, J. and Shuxi, C. 1983. Carbon isotope data in core V19–30 confirm reduced carbon dioxide contentration in the ice age atmosphere. *Nature*, 306, 319–322.

Sharp, A.D., Trudgill, S.T., Cooke, R.U., Price, C.A., Crabtree, R.W., Pickles, A.M. and Smith, D.I. 1982. Weathering of the balustrade on St Paul's Cathedral, London. *Earth Surface Processes and Landforms*, 7, 387–389.

Sharpton, V.L., Dalrymple, G.B., Marin, L.E., Ryder, G., Schuraytz, B.C. and Urrutia-Fucugauchi, J. 1992. New links between the Chicxulub impact structure and the Cretaceous/Tertiary boundary. *Nature*, 359, 819–821.

Sheehan, M.J. 1988. *Arms Control: Theory and Practice*. Oxford: Blackwell Scientific Publications, 188 pp.

Shen, G.T., Boyle, E.A. and Lea, D.W. 1987. Cadmium in corals as a tracer of historic upwelling and industrial fallout. *Nature*, 328, 794–796.

Shenton, J. and Gildemeister, V. 1994. Still more about AIDS. *Nature*, 367, 311.

Sheppard, C. and Price, A. 1991. Will marine life survive the Gulf War? *New Scientist*, 9 March, 36–40.

Shoemaker, E.M., Wolfe, R.F. and Shoemaker, C.S. 1990. Asteroid and comet flux in the neighbourhood of the Earth. *Geological Society of America, Special Publication*, 247, 155–170.

Siegenthaler, U. and Joos, F. 1992. Use of a simple model for studying the oceanic tracer distributions and the global carbon cycle. *Tellus*, 44B, 186–207.

Sievering, H., Boatman, J., Gorman, E., Kim, Y., Anderson, L., Ennis, G., Luria, M. and Pandis, S. 1992. Removal of sulphur from the marine boundary layer by ozone oxidation in sea-salt aerosols. *Nature*, 360, 571–573.

Simon, J. 1981. *The Ultimate Resource*. Oxford: Martin Robertson.

Simonich, S.L. and Hites, R.A. 1995. Global distribution of persistent organochlorine compounds. *Science*, 269, 1851–1854.

Simpson, S. 1990. *The Times Guide to the Environment*. Times Books, 224 pp.

Singh, O.N., Borchers, R., Fabian, P., Lal, S. and Subbaraya, B.H. 1988. Measurements of atmospheric BrO_x radicals in the tropical and mid-latitude atmosphere. *Nature*, 334, 593–595.

Sioli, H. 1992. The effects of deforestation in Amazonia. *Geographical Journal*, 151, 197–203.

Skole, D. and Tucker, C. 1993. Tropical deforestation and habitat fragmentation in the Amazon: satellite data from 1978 to 1988. *Science*, 260, 1905–1910.

Slingo, T. 1989. Wetter clouds dampen global greenhouse warming. *Nature*, 341, 104.

Small, R.D. 1991. Environmental impact of fires in Kuwait. *Nature*, 350, 11–12.

Smit, J. 1994. Blind tests and muddy waters. *Nature*, 368, 809–810.

Smith, K. 1992. *Environmental Hazards*. London: Routledge.

Smith, K. 1993. Riverine flood hazard. *Geography*, 339, 78, 2, 182–185.

Smith, K.L., Jr 1987. *Oceanography*, 32, 201–220.

Smith, P.M. and Warr, K. (eds) 1991. *Global Environmental Issues*. The Open University. London: Hodder & Stoughton, 294 pp.

Smith, R.A. 1852. *Memoirs and Proceedings of the Manchester Literary and Philosophical Society*, 2, 207–217.

Sowers, T., Bender, M., Labeyrie, L., Martinso, D., Jouzel, J., Raynald, D., Pichon, J.J. and Korotkeritch, Y.S. 1993. A 135,000-year Vostok-Specmap Common Temporal Framework. *Palaeoceanography*, 8, 737–766.

Spanier, E. and Galil, B.S. 1991. Lessepsian migration: a continuous biogeographical process. *Endeavour*, 15, 102–106.

Spencer, R.W. and Christy, J.R. 1990. Precise monitoring of global temperature trends from satellites. *Science*, 247, 1558–1562.

Spinks, P. 1990. Plug into the Sun. *New Scientist*, 22 September, 48–51.

Stauffer, B., Lochbronner, E., Oeschger, H. and Schwander, J. 1988. Methane concentration in the glacial atmosphere was only half that of pre-industrial Holocene. *Nature*, 332, 812–814.

Stauffer, P.H., Nishimura, S. and Batchelor, B.C. 1980.

Volcanic ash in Malaya from a catastrophic eruption of Toba, Sumatra, 30,000 years ago. In: Nishimura, S. (ed.), *Physical Geology of the Indonesian Island Arcs.* Kyoto, 156–164.

Sterling, C. 1976. Nepal. *Atlantic Monthly*, 238(4), 14–25.

Stockholm Environment Institute 1993. *Energy Without Oil: The Technical and Economic Feasibility of Phasing out Global Oil Use.*

Stoddart, D.R. 1968. Catastrophic human interference with coral atoll ecosystems. *Geography*, 53, 25–40.

Stoiber, R.E., Williams, S.N. and Huebert, B. 1987. *Journal of Volcanology and Geothermal Research*, 33, 1–8.

Stone, B. 1995. A world of viruses. *Newsweek*, 22 May, 18–19.

Stone, R. 1995. If the mercury soars, so may health hazards. *Science*, 267, 957–958.

Stott, P. 1991. Recent trends in the ecology and management of the world's savanna formations. *Progress in Physical Geography*, 15, 18–28.

Stott, P. 1994. Savanna landscapes and global environmental change. In: Roberts, N. (ed.), *The Changing Global Environment.* Oxford: Blackwell Scientific Publications, 287–303.

Street-Perrott, F.A. and Perrott, R.A. 1990. Abrupt climatic fluctuations in the tropics: the influence of Atlantic Ocean circulation. *Nature*, 343, 607–612.

Strum, M. and Benson, C.S. 1985. A history of jokulhlaups from Strandline Lake, Alaska, USA. *Journal of Glaciology*, 31, 272–280.

Stuart, A.J. 1993. Death of the megafauna: mass extinction in the Pleistocene. *Geoscientist*, 2, 17–20.

Sugden, D. and Hulton, N. 1994. Ice volumes and climate change. In: Roberts, I. (ed.), *The Changing Global Environment*, Oxford: Blackwell Scientific, 150–172.

Surveyor 1992. Cutting the cost of soil reclamation. 29 January issue, 1–12.

Swanell, R.P.J. and Head, I.M. 1994. Bioremediation comes of age. *Nature*, 368, 396.

Swanson, T. 1992. Economics of a biodiversity convention. *Ambio*, 21(3), 250–258.

Swinburne, N. 1993. It came from outer space. *New Scientist*, 20 February, 28–32.

Swisher C.C., III, Grajales-Nishimura, J.M., Mantanari, A., Margolis, S.V., Claeys, P., Alvarez, W., Renne, P., Cedillo-Pardo, E., Maurrasse, F. J.-M., Curtis, G.H., Smil, S. and Williams, M.O. 1992. Coeval $^{40}Ar/^{39}Ar$ ages of 65.0 million years ago from Chicxulub Crater melt rock and Cretaceous–Tertiary boundary tektites. *Science*, 257, 954–958.

Swiss Federal Laboratories for Materials Testing and Research 1992. *Untersuchung des Emissionsverhaltens eines Nutzfahrzeugmotors bei Betrieb mit Rapsolmethylester.*

Thaddeus, P. and Chanan, G.A. 1985. Cometry impacts, molecular clouds, and the motion of the Sun perpendicular to the galactic plane. *Nature*, 314, 73–75.

The British Seismic Verification Research Project (BSVRP) 1989. *Quarterly Journal of the Royal Astronomical Society*, 30, 311–324.

The Greenhouse Conspiracy, 1990. Channel 4 Television transcript, edited by Derek Jones, 28 pp.

The Greenpeace Report, Global Warming, 1990. Oxford: Oxford University Press, 554 pp.

This Common Inheritance: The Second Report, Britain's Environmental Strategy 1992. UK Government White Paper. London: HMSO, 192 pp.

Thomas, D.S.G. and Middleton, N.J. 1993. Salinization: new perspectives on a major desertification issue. *Journal of Arid Environments*, 24, 95–105.

Thompson, D. 1995. Problems of plenty. *Time*, August 21, 22–23.

Thompson, R.D. 1989. Short-term climate change: evidence, cause, environmental consequences and strategies for action. *Progress in Physical Geography*, 13, 315–347.

Thompson, R.D. 1992. The changing atmosphere and its impact on planet Earth. In: Mannion, A.M. and Bowlby, S.R. (eds), *Environmental Issues in the 1990s.* Chichester: John Wiley & Sons Ltd, 61–96.

Thorarinsson, S. 1974. *Vötnin Strí_. Saga Skei_arárhlaupa og Grímsvatnagosa.* Reykjavík: B_kaútgáfa Menningarsjó_s, 255 pp.

Tilling, R.I. and Lipman, P.W. 1993. Lessons in reducing volcano risk. *Nature*, 364, 277–280.

Tilman, D. and Downing, J.A. 1994. Biodiversity and stability in grasslands. *Nature*, 367, 363–365.

Tivy, J. 1982. *Biogeography.* Harlow: Longman.

Toon, O.B. and Tolbert, M.A. 1995. Spectroscopic evidence against nitric acid trihydrate in polar stratospheric clouds. *Nature*, 375, 218–221.

Torbett, M.V. 1989. Solar system and galactic influences on the stability of the Earth. *Palaeogeography, Palaeoclimatology, Palaeoecology (Global and Planetary Change Section)*, 75, 3–33.

Tsunogai, U. and Wakita, H. 1995. Precursory chemical changes in ground water: Kobe earthquake, Japan. *Science*, 269, 61–63.

Tucker, C.J., Dregne, H.E. and Newcomb, W.W. 1991. Expansion and contraction of the Sahara desert from 1980 to 1990. *Science*, 253, 299–301.

Tufnell, L. 1984. *Glacier Hazards.* London: Longman.

Tuomisto, H., Ruokolainen, K., Kalliola, R., Linna, A., Danjoy, W. and Rodriguez, Z. 1995. Dissecting Amazonian biodiversity. *Science*, 269, 63–66.

Turco, R.P., Toon, O.B., Park, C., Whitten, R.C., Pollack, J.B. and Noerdlinger, P. 1981. Tunguska meteor fall of 1908: effects on stratospheric ozone. *Science*, 214, 19–23.

Turco, R.P., Toon, O.B., Ackerman, T.P., Pollack, J.B. and Sagan, C. 1984. The climatic effects of nuclear war. *Scientific American*, 251(2), 33–43.

Turner, R.E. and Rabalais, N.N. 1994. Coastal eutrophication near the Mississippi river delta. *Nature*, 368, 619–621.

Tyler, C. 1990. The sense of sustainability. *Geographical Magazine*, February, 8–13.

Tyndall, J. 1861. *Philosophical Magazine*, 22, 161.

UK Department of the Environment, 1995. *Climate Change: The UK Programme. Progress Report on Carbon Dioxide Emissions.* Department of the Environment, Central Office of Information.

United Nations Development Programme 1990. *Human Development Report 1990.* Oxford: Oxford University Press.

United Nations Environmental Programme 1992. *World Atlas of Thematic Indicators of Desertification*. London: Arnold.

United Nations Food and Agriculture Organisation 1995. *The State of the World's Fisheries and Aquiculture*. United Nations Press.

United Nations High Commission for Refugees 1993. *The State of the World's Refugees: The Challenge for Protection*. London: Penguin Books.

United Nations Office of Disaster Relief 1984. *Disaster prevention and mitigation, Vol. 11 Preparedness Aspects*. New York: United Nations Office of Disaster Relief Co-ordinator.

United Nations Population Fund 1991. *The State of the World Population*.

Urey, H.C., Lowenstam, H.A., Epstein, S. and McKinney, C.R. 1951. Measurement of paleotemperatures and temperatures of the Upper Cretaceous of England, Denmark, and the south-eastern United States. *Bulletin of the Geological Society of America*, 62, 399–416.

US Congress, Office of Technology Assessment 1990. *Replacing Gasoline: Alternative Fuels for Light-Duty Vehicles*. OTA-E-364. Washington DC: US Government Printing Office, 136 pp.

US Congress, Office of Technology Assessment 1991. *U.S. Oil Import Vulnerability: The Technical Replacement Capability*. OTA-E-503. Washington DC: US Government Printing Office, 136 pp.

US Congress, Office of Technology Assessment 1992a. *Managing Industrial Solid Wastes from Manufacturing, Mining, Oil and Gas Production, and Utility Coal Combustion – Background Paper*. OTA-BP-0-82. Washington DC: US Government Printing Office, 130 pp.

U.S. Congress, Office of Technology Assessment 1992b. *Fueling Development: Energy Technologies for Developing Countries*. OTA-E-516. Washington DC: US Government Printing Office, 336 pp.

van Andel, T.H. 1994. *New Views on an Old Planet: A History of Global Change*. Cambridge: Cambridge University Press, 439 pp.

van Andel, T.H., Zangger, E. and Demitrack, A. 1990. Land use and soil erosion in prehistoric and historical Greece. *Journal of Field Archaeology*, 17, 379–396.

van Geen, A., Luoma, S.N., Fuller, C.C., Anima, R., Clifton, H.E. and Trumbore, S. 1992. Evidence from Cd/Ca ratios in foraminifera for greater upwelling off California 4,000 years ago. *Nature*, 358, 54–56.

Vandal, G.M., Fitzgerald, W.F., Boutron, C.F. and Candelone, J.-P. 1993. Variations in mercury deposition to Antarctica over the past 34,000 years. *Nature*, 362, 621–623.

Varnes, D.J. 1978. Slope movements and types and processes. In: *Landslides: Analysis and Control*, Special Report 176, 11–33, Transportation Research Board/National Academy of Sciences, Washington, DC.

Vaughan, D. 1993. Chasing the rogue icebergs. *New Scientist*, 9 January, 24–27.

Vaughan, D.G. and Doake, C.S.M. 1996. Recent atmospheric warming and the retreat of ice shelves on the Antarctic Peninsula. *Nature*, 379, 328–331.

Veiga, M.M., Meech, J.A. and Onate, N. 1994. Mercury pollution from deforestation. *Nature*, 368, 816–817.

Veum, T., Jansen, E., Arnold, M., Beyer, I. and Duplessy, J.-C. 1992. Water mass exchange between the North Atlantic and the Norwegian Sea during the past 28,000 years. *Nature*, 356, 783–785.

Vine, F.J. and Matthews, D.H. 1963. Magnetic anomalies over ocean ridges. *Nature*, 199, 947–999.

Vogel, J.S., Cornell, W., Nelson, D.E. and Southon, J.R. 1990. Vesuvius/Avellino, one possible source of seventeenth century BC climatic disturbances. *Nature*, 344, 534–537.

Vogelmann, A.M., Ackerman, T.P. and Turco, D.P. 1992. Enhancement in biologically effective ultraviolet radiation following volcanic eruptions. *Nature*, 359, 47–49.

Voight, B. 1990. The 1985 Nevado del Ruiz volcano catastrophe: anatomy and reprospection. *Journal of Volcanology and Geothermal Research*, 42, 151–188.

Volz, A. and Kley, D. 1988. Evaluation of Montsouris series of ozone measurements made in the nineteenth century. *Nature*, 332, 240–242.

Von Gunten, H.R. and Lienert, C. 1993. Decreased metal concentrations in ground water caused by controls of phosphate emissions. *Nature*, 364, 220–222.

Wadleigh, M.A. and Velzer, J. 1992. $^{18}O/^{16}O$ and $^{13}C/^{12}C$ in lower Palaeozoic articulate brachiopods. Implications for isotopic composition of seawater. *Geochimica Cosmochimica Acta*, 56, 431–443.

Waldbott, G.L. 1978. *Health Effects of Environmental Pollutants*. Saint Louis: C.V. Mosby.

Walker, A.S. 1992. *Deserts: Geology and Resources*. US Department of the Interior/US Geological Survey, Denver. Washington: US Government Printing Office, 60 pp.

Walker, H.J., Coleman, J.M., Roberts, H.H. and Tye, R.S. 1987. Wetland loss in Louisiana. *Geografiska Annaler*, 69A(1), 189–200.

Ward, R. 1978. *Floods: A Geographical Perspective*. London: Macmillan.

Wardley-Smith, J. (ed.) 1976. *The Control of Oil Pollution on the Sea and Inland Waters: The Effect of Oil Spills on the Marine Environment and Methods of Dealing with Them*. London: Graham & Trotman, 254 pp.

Warren, A. and Maizels, J.K. 1976. *Ecological Change and Desertification*. London: University College Press.

Water Services Association 1992. *Waterfacts*. Richmond, UK: Print Management Systems Ltd, 65 pp.

Waters, J.W., Froidevaux, L., Read, W.G., Manney, G.L., Elson, L.S., Flower, D.A., Jarnot, R.F. and Harwood, R.S. 1993. Stratospheric ClO and ozone from the Microwave Limb Sounder on the Upper Atmosphere Research Satellite. *Nature*, 362, 597–602.

Watkins, N.D., Sparks, R.S.J., Sigurdsson, H., Huang, T.C., Federman, A., Carey, S. and Ninkovich, D. 1978. Volume and extent of the Minoan tephra from Santorini Volcano: new evidence from deep-sea cores. *Nature*, 271, 122–126.

Watson, A. 1991. Gaia. *New Scientist*, 6 July, Inside Science No. 48, 4 pp.

Watson, A.J. and Lovelock, J.E. 1983. Biological homeostasis of the global environment: the parable of the 'daisy' world. *Tellus*, 35B, 282–289.

Webster, J.K. 1986. *The Complete Australian Bushfire Book*. Melbourne: Nelson.

Weinbaum, M.G. 1994. *Pakistan and Afghanistan: Resistance and Reconstruction*. Boulder, Colo.: Westview Press, 190 pp.

Welford, R. 1995. *Environmental Strategy and Sustainable Development*. London: Routledge, 217 pp.

Wells, N.A. and Andriamihaja, B. 1993. The initiation and growth of gullies in Madagascar: are humans to blame? *Geomorphology*, 8, 1–46.

Welsh Office 1969. *A Selection of Technical Reports Submitted to the Aberfan Tribunal*. London: HMSO.

Wertine, T.A. 1973. Pyrotechnology: man's first industrial uses of fire. *American Scientist*, 61, 670–682.

Wesson, R.L., Helley, E.J., Lajoie, K.R. and Wenthworth, C.M. 1975. Faults and future earthquakes. In: Borcherdt, R.D. (ed.), *Studies for Seismic Zonation of the San Francisco Bay Region*. US Geological Survey Professional Paper, 941-A, A5-A30.

Wesson, R.L. and Wallace, R.E. 1985. Predicting the next great earthquake in California. *Scientific American*, 252, 2, 35–43.

Westrich, H.R. and Gerlach, T.M. 1992. Magmatic gas source for stratospheric SO_2 cloud from the June 15, 1991 eruption of Mount Pinatubo. *Geology*, 20, 867–870.

Which? magazine report, August 1991. River pollution, 458–461.

White, G.F. (ed.) 1974. *Natural Hazards*. Oxford: Oxford University Press.

White, I.D., Mottershead, D.N. and Harrison, S.J. 1984. *Environmental Systems*. London: Allen & Unwin, 495 pp.

Whitmore, D.P. and Jackson, A.A. 1984. Are periodic mass extinctions driven by a distant solar companion? *Nature*, 308, 713–715.

Whittow, J. 1980. *Disasters: The Anatomy of Environmental Hazards*. London: Penguin Books, 411 pp.

Wigley, T.M.L. 1981. Climate and paleoclimate: what we can learn about solar luminosity variations. *Solar Physics*, 74, 435–471.

Wigley, T.M.L. 1989. Possible climate change due to SO_2-derived cloud condensation nuclei. *Nature*, 339, 365–367.

Wigley, T.M.L. 1993. Balancing the global carbon budget. Implications for projections of future carbon dioxide concentration changes. *Tellus*, 45B, 409–425.

Wigley, T.M.L. and Raper, S.C.B. 1992. Implications for climate and sea level of revised IPCC emissions scenarios. *Nature*, 357, 293–300.

Wigley, T.M.L., Richels, R. and Edmonds, J.A. 1996. Economic and environmental choices in the stabilization of atmospheric CO_2 concentrations. *Nature*, 379, 240–243.

Wignall, P. 1992. The day the world nearly died. *New Scientist*, 25 January, 51–55.

Wilkinson, W.B. and Brassington, F.C. 1991. Rising groundwater levels – an international problem. In: Downing, R.A. and Wilkinson, W.B. (eds), *Applied Groundwater Hydrology – A British Perspective*. Oxford: Clarendon Press, 35–53.

Williams, M. 1989. *Americans and their Forests*. Cambridge: Cambridge University Press.

Williams, M. 1990. Understanding wetlands. In: Williams, M. (ed.), *Wetlands: A Threatened Landscape*. Institute of British Geographers Special Publication 25. Oxford: Blackwell Scientific Publications, 1–41.

Williams, M.J., Dunkerley, D.L., de Deckker, P., Kershaw, A.P. and Stokes, T. 1993. *Quaternary Environments*. London: Edward Arnold, 329 pp.

Williams, P. and Woessner, P.N. 1996. The real threat of nuclear smuggling. *Scientific American*, 274(1), 26–30.

Williams, R.S. and Moore, J.G. 1983. *Man against Volcano: The Eruption of Heimaey, Vestmannaeyjar, Iceland*. US Geological Survey.

Williams, S.J., Dodd, K. and Gohn, K.K. 1991. *Coasts in Crisis*. US Department of the Interior/US Geological Survey Circular 1075. US Government Printing Office, 32 pp.

Williams, W.D. and Aladin, N.V. 1991. The Aral Sea: recent limnological changes and their conservation significance. *Aquatic Conservation*, 1, 3–23.

Williamson, P. and Gribbin, J. 1991. How plankton change the climate. *New Scientist*, 16 March, 48–52.

Wilson, A.T. 1964. Origin of ice ages: an ice shelf theory for Pleistocene glaciation. *Nature*, 201, 147–149.

Wilson, A.T. 1969. The climatic effects of large-scale surges of ice sheets. *Canadian Journal of Earth Sciences*, 6, 911.

Wilson, E.O. 1989. Threats to biodiversity. *Scientific American*, 261(3), 60–66.

Wolfe, J.A. 1991. Palaeobotanical evidence for a June 'impact winter' at the Cretaceous/Tertiary boundary. *Nature*, 352, 420–423.

Wolfenden, J. and Mansfield, T.A. 1991. Physiological disturbances in plants caused by air pollutants. In: Last, F.T. and Watling, R. (eds), *Acidic Deposition: Its Nature and Impacts*. Edinburgh: The Royal Society of Edinburgh, 117–138.

Wolman, M.G. 1971. Evaluating alternative techniques of floodplain mapping. *Water Resources Research*, 7, 1383–1392.

Wolman, M.G. and Miller, J.P. 1960. Magnitude and frequency of forces in geomorphic processes. *Journal of Geology*, 68, 54–74.

Wolman, M.G. and Schick, A.P. 1967. Effects of construction on fluvial sediments, urban and suburban areas of Maryland. *Water Resources Research*, 3(2), 451–464.

Wood, B. 1992. Origin and evolution of the genus *Homo*. *Nature*, 355, 783–790.

Wood, C. 1995. *Environmental Impact Assessment: A Comparative Review*. Harlow: Longman, 337 pp.

Wood, C. and Djeddour, M. 1992. Strategic environmental assessment: EA of policies, plans and programmes. *Impact Assessment Bulletin*, 10, 3–22.

Woods, A.W. and Kenneth, W. 1991. Dimensions and dynamics of co-ignimbrite eruption columns. *Nature*, 350, 225–227.

Woolbach, W.S., Lewis, R.S. and Anders, E. 1991. Cretaceous extinctions: evidence for wildfires and search for meteoric material. *Science,* 230, 167–170.

World Bank 1979. *Nepal: Development Performance and Prospects. A World Bank Country Study, South Asia Regional Office.* Washington, DC: World Bank, 123 pp.

World Commission on Environment and Development 1987. *Our Common Future.* Oxford: Oxford University Press, 400 pp.

World Development Report 1992: *Development and the Environment* 1992. Published for the World Bank, Oxford University Press, 308 pp.

World Energy Council 1992. *Energy for Tomorrow's World – The Realities, the Real Options and the Agenda for Achievement.* Draft Summary Global Report.

World Resources Institute 1990. A Report by The World Resources Institute in collaboration with The United Nations Environment Programme and The United Nations Development Programme 1990. *World Resources 1990–91.* Oxford: Oxford University Press, 383 pp.

World Resources Institute 1994. *World Resources 1994–1995.* Oxford: Oxford University Press, 400 pp.

Wright, R.F. and Hauhs, M. 1991. Reversibility of acidification: soils and surface waters. In: Last, F.T. and Watling, R. (eds), *Acidic Deposition: Its Nature and Impacts.* Edinburgh: The Royal Society of Edinburgh, 169–191.

Wuethrich, B. 1995. El Niño goes critical. *New Scientist*, 4 February, 33–35.

Yaalon, D.H. and Yaron, B. 1966. Framework for man-made soil changes: an outline of metapedogenesis. *Soil Science*, 102(4), 272–277.

Yearley, S. 1991. *The Green Case: A Sociology of Environmental Issues, Arguments and Politics.* Routledge, 197 pp.

Yergin, D. 1991. *The Prize: The Quest for Oil, Money, and Power.* London: Simon & Schuster, 885 pp.

Young, G.J., Dooge, J.C.I. and Rodda, J.C. 1994. *Global Water Resource Issues.* Cambridge: Cambridge University Press.

Young, J.E. 1992. *Mining the Earth.* Worldwatch Paper 109, 1–53.

Zahnie, K. and Grinspoon, D. 1990. Comet dust as a source of amino acids at the Cretaceous/Tertiary boundary. *Nature*, 348, 157–160.

Zimmerman, P. 1989. A new resource for arms control. *New Scientist*, 23 September, 38–43.

Zwally, H.J., Brenner, A.C., Major, J.A., Bindschadler, R.A. and Marsh, J.G. 1989. Growth of the Greenland ice sheet: measurement. *Science*, 246, 1587–1589.

Glossary

Note: not all the words in this glossary appear in the text, but they are included here in order to make as complete a listing of useful terms as is possible.

abiotic – non-living, relating to factors and/or things that are independent/separate from living things.

accretionary prism – thick wedge-shaped body of sediments formed by both tectonic and sedimentary processes, and associated with the off-scraping of material above a subducting oceanic plate.

acetogenic bacteria – bacteria that convert sugars into fatty acids.

acid – chemical substance that releases hydrogen ions when dissolved in water, or an aqueous solution containing an excess of hydrogen ions.

acid rain – rain and snow with a pH of less than 5.6. Acid rain strictly refers to the wet deposition of acids and acid-forming substances.

acid susceptibility – capacity of a water body to become acidified.

acid-neutralising capacity (ANC) – ability of a water body to reduce (or neutralise) the acidity of incoming acid water.

acidic deposition – falling of acids and acid-forming substances from the atmosphere onto the surface of the Earth. Acid rain is a type of acidic deposition.

acidification – increase in acidity (lower pH) in a water body.

actinides – fourteen chemical elements in the final period of the periodic table of elements (thorium, protactinium, uranium, neptunium, plutonium, americium, curium, berkelium, californium, einsteinium, fermium, mendelevium, nobelium, lawrencium).

activated sludge – sludge containing living organisms feeding on the solids to encourage its breakdown, and which are recycled during secondary treatment of sewage.

active coke – pellets of treated coal (coke) approximately 5 mm in diameter that are used to catalyse reactions to help remove pollutants in the generation of power in coal-fired power stations.

acute radiation syndrome (ARS) – symptoms resulting from intensive irradiation of the body such as nausea, vomiting, abdominal pain, fever, dehydration, loss of hair, infection, haemorrhage, damage to bone marrow and cancers, notably leukaemia and breast cancer.

acute toxicity – effect of a single high-level exposure to specified chemical/s, e.g. due to accidental release.

adaption (evolutionary/ecological) – changes in the function and/or structure of a system to produce greater life chances through survivability and reproduction.

aeolian – pertaining to wind processes, landforms or sediments.

aerosol – solid or liquid particles suspended or dispersed in a gas.

AIDS – Acquired Immune Deficiency Syndrome.

albedo – measure of the reflectivity of a body or surface, often used to describe the ability of the Earth's cloud cover to reflect incoming solar radiation.

algal bloom – proliferation of algae in water bodies as a result of changes in water chemistry and temperature.

alpha particle – particle emitted from the nucleus of an atom during radioactive decay; an alpha particle has an atomic mass of 4 and is equivalent to a helium nucleus.

Alzheimer's disease – illness that leads to senile dementia, and for which the causes remain poorly understood.

americium-241 – radioactive isotope of americium.

amphiboles – group of dark-coloured rock-forming minerals comprising iron-magnesium silicates.

anaerobic – without free oxygen.

anaerobic digestion – breakdown of organic matter by organisms in an oxygen-free environment.

aniline point – the minimum temperature for a complete mixing of aniline and materials such as gasoline, used in some specifications to indicate the aromatic content of oils and to calculate the approximate heat of combustion. Aniline is a compound in the $C_6H_5NH_2$ group, e.g. benzanilide $C_6H_5NHCOC_6H_5$.

anion – atom that by virtue of the imbalance in electrical forces has a net negative charge.

anoxic – oxygen-deficient environments.

anthracite – dark hard coal composed of 92–98 per cent carbon.

anthropogenic – human-influenced processes or forms.

anticyclone – an area or system of atmospheric high pressure.

aquifer – body of rock at depth, which is capable of storing ground water.

asteroids (asteroid belt) – small, possibly fragmented, planetesimals that orbit around the Sun between the orbits of Mars and Jupiter.

asthenosphere – layer within the Earth's upper mantle that extends from depths of 5–50 km to *c.* 300 km and is characterised by a lower mechanical strength and lower resistance to deformation than the region above the crust.

atmosphere – gaseous layer surrounding the Earth and bound to it by gravitational attraction.

atmospheric fluidised-bed combustion – new technique used in coal-fired power stations to reduce noxious gaseous emissions, involving passing the gases through a bed of coal and limestone which becomes supported (fluidised) by the upward flow of the gas in a fluidised-bed combustion furnace.

aurora australis – illumination of the sky in the Southern Hemisphere, sometimes in brilliant colours, resulting from high-speed solar particles entering the ionosphere and releasing electrons from air molecules. The re-establishment of air molecules leads to the emission of light.

aurora borealis – illumination of the sky in the Northern Hemisphere, sometimes in brilliant colours, resulting from high-speed solar particles entering the ionosphere and releasing electrons from air molecules. The re-establishment of air molecules leads to the emission of light.

autotroph – organism that can synthesise the organic substances it requires entirely from inorganic nutrients, by obtaining energy from light and/or various inorganic substances. The main autotrophs are green plants.

background radiation – radioactivity from non-human sources.

badlands – intensely dissected landscape produced by natural or human-influenced erosion.

base (alkali) – any chemical substance which releases hydroxyl ions (OH^-) when dissolved in water, or an aqueous solution containing an excess of hydroxyl ions.

BATNEEC – 'best available techniques not entailing excessive cost.'

becquerel (Bq) – unit by which radioactivity is measured. 1 Bq = 1 atomic disintegration per second.

bedload – the sediment load carried along very near to the bottom of a flowing current (e.g. river bed or sea floor) rather than in suspension, and which tends to be the coarser and heavier grain-size fraction.

benthic – pertaining to bottom-dwelling organisms.

benzene (C_6H_6) – aromatic hydrocarbon widely used in industry.

beta activity – release of beta particles (the emission of an electron from the nucleus) during the radioactive decay of an element.

Big Bang – explosion that marked the creation of the Universe, which probably occurred between 15,000 million and 20,000 million years ago.

biochemical oxygen demand (BOD) – amount of oxygen used ('demanded') in chemical/biological processes during the digestion or oxidation of wastes. The potential environmental impacts of wastes are frequently expressed in terms of their BOD value.

biodegradable – refers to a compound that can be decomposed and/or disintegrated by biological processes. Antonym is 'non-biodegradable'.

biodiversity – pertaining to the number, variety and variability of living organisms and their habitats.

bio-fuel – substance produced by organic activity that can be used as an energy source, either in a pure form or refined and blended with conventional fossil fuels, such as petroleum.

bio-gas – gas mixture arising from anaerobic digestion of organic matter, and comprising about two-thirds methane, one-third carbon dioxide, and minor amounts of other gases. The methane content in bio-gas makes it useful as a fuel/energy resource.

biogenic – pertaining to organic origin.

biomass – mass of biological matter present per plant or animal, per community, or per unit area. Total dry organic matter or stored energy content of living organisms in a specified area.

biome – ecosystem linked through similar climatic conditions and vegetation, e.g. tropical rainforest, desert, high-latitude tundra.

biomineralisation – formation of minerals by living organisms.

biosphere – layer at the interface of the Earth's crust, ocean and atmosphere where life is found, i.e. the total ecosystem of the Earth.

biosynthesis – chemical reactions promoted by an organism in order to make new chemicals.

biota – general term for any specific or all living organisms and their associated ecosystems.

biotic province – a biogeographical division in which faunas are distinct from other regions.

bioturbation – burrowing activity of organisms, which helps to mix up the sediment and soil layers.

bitumen – black to dark brown solid or semi-solid thermoplastic material possessing waterproofing and adhesive properties, obtained from processing crude oil. It is a complex combination of higher molecular weight organic compounds containing a relatively high percentage of hydrocarbons with carbon numbers greater than C_{25} and a high carbon to oxygen ratio – there are also trace amounts of metals such as nickel, iron and vanadium.

bituminous coal – coal containing between 78 and 90 per cent carbon, and which has a moderately high calorific value.

bivalve – mollusc with a shell composed of two distinct parts (valves), which are generally similar. The two halves of the shell, which house the animal, are joined by a flexible muscle or ligament.

black smoke – visible smoke comprising particulates, and in an engine it is formed from pyrolysis and incomplete combustion of fuel, typically under high loads.

Black smokers – submarine springs that form chimneys of sulphides, typically of zinc, iron, copper and molybdenum, and oxides of manganese, which are present along the mid-ocean ridges at depths of about 2.5 km.

Boreal forest – Northern coniferous forest (including spruce, fir and hemlock) approximately bounded in the north by the July 10°C average isotherm and transitional to the tundra. It is synonymous with *taiga*.

BP – before the present day.

British thermal unit (BTU) – amount of heat required to raise the temperature of 1 pound weight (1 lb) of water through 1° Fahrenheit (1°F).

bromeliad – any member of the pineapple family,

typically with fleshy, spiny-leafed epiphytes (plants not rooted in the soil, but growing above ground level, usually on other plants), and notable for their ability to hold water in the cup-shaped centre of the leaf rosette. Bromeliads are particularly common in the canopy of tropical rainforests.

buffer – chemical that can maintain the pH of a solution by reacting with the excess acid or alkali (base). Limestone is a natural buffer that helps stabilise the pH of ground water and soil close to neutral.

buffering chemical reaction – chemical reaction that reduces the likelihood of the solution changing its pH.

butterfly effect – highly variable knock-on effect (positive feedback) or output produced by a system due to subtle change in the initial inputs.

caatinga – form of dry thorny woodland found in northeast Brazil.

caesium-137 – radioactive isotope of caesium.

calorie – amount of heat required to raise 1 gram of water through 1° Celsius (1°C). When used in connection with food, the units are typically kilocalories, which is the amount of heat required to raise 1 litre of water through 1°C.

calving – process that involves the fracturing and break-up of ice sheets or ice caps where they enter the sea or a lake, and which leads to the formation of icebergs.

carbon cycle – transfer of carbon, one of six basic elements (CHONPS = carbon, hydrogen, oxygen, nitrogen, phosphorus and sulphur) for life, through the biosphere, hydrosphere, atmosphere and lithosphere. For conversion purposes, 1 million tonnes carbon (MtC) = 3.67 million tonnes of CO_2.

carbon dioxide (CO_2) – naturally occurring gas molecule comprising two atoms of oxygen and one atom of carbon. CO_2 is produced during respiration by humans, and is used during photosynthesis by plants.

carbon monoxide (CO) – gas molecule with one oxygen atom and one carbon atom. CO is highly toxic to humans and many other organisms.

carbon tax – tax levied on fossil fuels in proportion to the amount of CO_2 produced during combustion.

carbonate compensation depth (CCD) – depth in the oceans where sea water is under-saturated with respect to $CaCO_3$; this leads to the dissolution of carbonate sediment as it falls through the water column or sits on the sea floor. CCD varies between oceans.

carbonic acid (H_2CO_3) – acidic rainwater or snow formed by the water combining with atmospheric CO_2.

carcinogenic – having the potential to cause cancer in animals, but typically used to imply an ability to cause cancer in humans.

carrying capacity – the maximum population that can be maintained by a habitat or ecosystem without degrading the ability of that habitat or ecosystem to maintain that population in the future.

catalyst – substance that may initiate or speed up a chemical reaction without itself being changed during the reaction. In biological reactions, enzymes act as catalysts.

catalytic converter – device fitted to motor vehicle exhaust systems to reduce the emission of pollutants.

catastrophe theory – the hypothesis that important and widespread changes in the physical environment are brought about by major, relatively brief, and sudden events.

cation – atom that has lost one or more electrons from its orbiting shells and thus has a net positive charge.

cation exchange – substitution of one cation for another of a different element in a mineral structure.

caveat emptor principle – Latin: 'let the buyer beware'; the principle in which, for example, responsibility for any environmental harm caused by contaminated land passes from the vendor to the purchaser.

cellulose – complex carbohydrate that forms the basic structural component of plant cell walls. Cellulose comprises glucose molecules and cannot be digested by humans.

centre pivot irrigation – irrigation system utilising a rotating spray arm up to several hundred metres long and supported by wheels pivoting about a central well from which water is pumped. These systems are commonly used to irrigate desert regions artificially.

cerrado – form of savannah vegetation comprising predominantly grasses and small trees, found in Brazil.

cetane number – indicator of how easily a fuel ignites under compression, i.e. the ignition quality. Higher number means easier ignition.

channelisation – straightening of natural stream channels by the construction of artificial banks.

chaos – unpredictable or random processes and their consequences.

chaos theory – theory explaining phenomena as being a consequence of inherent randomness in a system. There are mathematical models to simulate chaotic systems.

chemical weathering – see *weathering*.

chitinous exoskeleton – external part of an animal, usually an insect or a crustacean, comprising a hard organic substance called chitin.

chlorinated hydrocarbon – organic compound containing chlorine.

chlorofluorocarbons (CFCs) – organic compounds comprising carbon, chlorine and fluorine atoms, commonly used in industrial processes and manufacturing, that are very stable in the troposphere and degraded in the stratosphere by solar radiation, which releases the chlorine, which may contribute to ozone depletion.

chlorophyll – green pigment in plant tissues, which is essential for photosynthesis.

chondritic meteorite – iron-rich meteorite.

chronic toxicity – assessment conducted over the lifetime of test animals to evaluate any late-in-life toxicity caused by exposure to specified chemical/s.

circumpolar vortex – meandering belts of winds associated with the sub-polar front.

clay minerals – group of finely crystalline layered silicate minerals.

claystone – fine-grained rock comprising grains less than 0.0039 mm in diameter.

cloud seeding – artificial addition of condensation nuclei such as silver iodide crystals into the atmosphere to help produce rain.

CO_2 emissions – CO_2 emissions are commonly measured according to the carbon content, in millions of tonnes of carbon (MtC), where 1 tonne of carbon is equivalent to 3.67 (or 44/12) tonnes of carbon dioxide.

coastal set-up – term used to describe the meteorological conditions, often associated with storms, along the coast.

coccolithophorida – microscopic algae that secrete calcareous shells comprising round platelets known as coccoliths. These coccoliths reached their acme in the Cretaceous, when their deposition led to the formation of extensive and thick chalk deposits.

coesite – form (polymorph) of quartz formed under high pressure.

cogeneration – simultaneous production of both electrical energy and heat, e.g. for domestic/commercial, industrial or other purposes.

coleoptera – order of insects comprising the beetles.

collagen – group of proteins that have great tensile strength and are present in tendons, ligaments, connective tissues of the skin, dentin and cartilages.

combustion-dust loading – increased quantities of dust in the atmosphere produced as a by-product of burning fossil fuels.

comet – small object in Space, with a nucleus less than a few kilometres in diameter, composed of dry ice, frozen methane, frozen ammonia and water ice, with small dust particles embedded within it.

confining layer – stratum of rock that inhibits the movement of ground water.

continental drift – lateral movement of continental plates around the Earth, a theory superseded by the plate tectonic theory, which ascribes this plate movement to sea-floor spreading.

continental plate – rigid outer layer of the Earth, the lithosphere, which averages about 40 km in thickness and may form relatively stable and long-lived parts of the crust. It constitutes the continents and has an average silica (SiO_2) content of 65 per cent.

convergent plate boundaries – boundaries between discrete plates, where two plates are colliding; they include the collision between two oceanic plates, two continental plates, and an oceanic and a continental plate.

core – the central zone of the Earth, composed of nickel and iron.

Coriolis force – an apparent force acting on moving objects, notably winds and ocean currents, which results from the Earth's rotation. The force is proportional to the speed and latitude of the moving feature, varying from zero at the equator to a maximum at the poles.

Cretaceous – period of geological time from about 146 Ma to 65 Ma.

critical load – maximum load of a pollutant that the environment can sustain before damage occurs.

crust – the outermost solid layer of the Earth, averaging about 5 km in thickness beneath the oceans and approximately 60 km beneath mountain ranges.

cryptosporidia – intestinal parasite that causes diarrhoea and vomiting in humans.

cybernetics – study of regulating and self-regulating mechanisms in nature and technology.

cyclone/depression – an enclosed area of low pressure within the atmosphere.

deforestation – conversion of forest land to other uses, e.g. pasture, cropland, etc.

deglaciation – period of time when glaciers start to retreat at the end of a glacial or stadial.

demand side management – planning, implementation and monitoring of utility activities to encourage customers/users to modify their pattern and total consumption of energy use/electricity.

dendritic cells – cells with branching or tree-like forms.

dendrochronology – time correlation based on the width of annual growth rings of trees.

dendroecology – science of the width of annual growth rings of trees in order to interpret specific ecological events that resulted in changes in the tree's ability to photosynthesise and fix carbon.

desalination plant – plant for purifying sea water and converting it into high-quality drinking water, using distillation and other techniques.

desertification – spread of desert-like conditions in arid or semi-arid lands due to climatic change or human influence.

desulfomaculum – organic compound that aids bacterial degradation of oil.

desulfovibrio – organic compound that aids bacterial degradation of oil.

deuterium – isotope of hydrogen.

Devensian – last glacial stage in Britain.

diatoms – microscopic unicellular algae that secrete siliceous walls.

dichloro-diphenyl-trichloro-ethane (DDT) – poisonous organic compound that is used as a pesticide.

dilatancy–diffusion model – theory that is used to explain phenomena and events that occur before and during an earthquake.

diluvial theory – theory that attributed landforms, sediments and fossils to the Noahian deluge.

dimethyl sulphide – biologically produced organic chemical that may act as condensation nuclei.

divergent oceanic spreading ridges (mid-ocean ridges) – see *spreading centres*.

DNA – Deoxyribonucleic acid, an organic compound comprising two polymer strands wrapped round each other in a double helix; DNA carries the genetic information of living organisms.

dolomite – rock/mineral comprising carbonate of calcium and magnesium = $CaMg(CO_3)_2$.

dose–response relationship – relationship between the level of a pollutant and the environmental impact.

downstream flood – river flood that becomes progressively larger down-valley.

drought – condition of dryness because of a lack of precipitation.

dry deposition – the direct transfer of gases and particles to surfaces, whether of leaves, soil or building materials.

dust bowl – region where dust is actively being deflated

from the ground and transported, especially North American prairies.

dust pneumonia – see *pneumoconiosis*.

Dutch elm disease – widespread fungoid killer of elm trees, first described in the Netherlands and introduced into Europe from Asia during World War II.

dynamic equilibrium – situation that is fluctuating about some apparent average state, where the average state itself is also changing through time.

Earth-surface processes – processes acting on the Earth's surface which includes river, lake, sea, slope, biological and atmospheric processes.

ecological niche – occupation of space by a community in a particular environmental setting.

ecology – study of living organisms' habits, modes of life and relations to their surroundings.

ecosphere – all-encompassing realm that includes the atmosphere, hydrosphere, lithosphere and biosphere.

ecosystem – community with interacting organisms of different species, and their relationships with the associated chemical and physical systems.

El Niño event – appearance of unusually warm water off the coasts of Peru, Ecuador and Chile, which causes major shifts in the general circulation of the atmosphere.

electrical resistivity – ability of a substance to resist the flow of an electrical current through it.

electromagnetic spectrum – range of electrical and magnetic radiation of varying wavelengths that travels at the same velocity as light.

endemic – pertaining to all factors in a given area.

endemic species – a species that is native to a particular region.

endemism – a situation in which a species or taxonomic group is restricted to a particular geographical region, due to factors such as isolation or climate.

endothermic reaction – chemical reaction that involves the absorption of heat, which may cool down the surrounding environment.

entropy – degree of disorder in any physical or chemical system; the greater the entropy, the greater the inherent disorder in a system.

EPA – (US) Environmental Protection Agency, which is responsible for managing federal efforts to control air and water quality, reduce radiation and pesticide hazards, regulate the disposal of hazardous waste, and undertake/sponsor environmental research. (UK) Environmental Protection Act.

epidemiology – study of the pattern of diseases and/or other harmful effects produced by toxic substances in various groups of people, with the purpose of understanding the reasons for certain individuals/groups of people being more susceptible than others to ill health. An important aspect is the use of statistical techniques in such investigations.

EU – European Union. Previously referred to as the EC, or European Community, EEC or European Economic Community.

eukaryotic organism – organism that requires O_2 to biosynthesise.

eustasy – global change in sea level.

eutrophic – pertaining to lakes, ponds or rivers that abound in plant nutrients and are therefore highly productive.

eutrophication – addition of nutrients to water bodies, which increases their productivity.

evaporite – water-soluble mineral that has been deposited by precipitation from saline water as a result of evaporation, e.g. halite (NaCl), gypsum ($CaSO_4.2H_2O$) and anhydrite ($CaSO_4$).

evapotranspiration – diffusion of water vapour into the atmosphere from vegetated surfaces.

exajoule – 10^{18} joules (see *joule*).

FAO – United Nations Food and Agricultural Organisation.

famine – acute starvation associated with a sharp increase in death rates.

fault (geological) – crack or fissure in rock produced by Earth (tectonic) movement along which displacement has occurred.

fission (nuclear) – splitting of larger atoms into two or more lighter elements, with the release of energy.

flash flooding – flood event commonly associated with ephemeral streams in arid and semi-arid environments. The flood is characterised by a nearly instantaneous rise in discharge, which progresses downstream as waves, e.g. as bores (solitary waves).

fluidised-bed combustion furnaces – see *atmospheric fluidised-bed combustion*.

fluvial – pertaining to river processes, landforms or sediments.

fold – geological strata or alignments of minerals that have been deformed by compressional forces into bends.

food chain – transfer of food from one type of organism to another in sequence.

food web – transfer of food from one type of organism to another within a complex community of organisms.

foraminifera – single-celled (protozoan) micro-organisms that secrete calcareous skeletons and drift in the seas and oceans.

fossil fuels – energy sources in the form of buried organic matter that have generally undergone chemical and physical changes. Common fossil fuels include petroleum, natural gas (mainly methane), coal and peat.

fractal geometry – type of scale-invariant geometry with a non-integral number of dimensions.

Fujita intensity scale – scale that describes the damage associated with a tornado in relation to the velocity of the rotating spiral of wind.

fumarole – small vent associated with volcanic centres through which liquids erupt.

fungicide – chemical that kills fungi.

G7 nations – economic grouping of the world's seven richest nations: Canada, France, Germany, Italy, Japan, the UK, and the USA. Currently, discussions are taking place to make Russia a full member.

Gaia Hypothesis – hypothesis developed by the scientist James Lovelock, which suggests the Earth is a self-regulating system, like a living organism, able to maintain its climate, atmosphere, soil and ocean composition in a stable balance favourable for life.

galaxy – cluster of stars held together by stong mutual gravitational attraction. The Milky Way is a galaxy containing our Solar System; it is seen from the Earth as a band of stars.

gasohol – blended alcohol and conventional petroleum products, e.g. a blend of 90 per cent gasoline + 10 per cent bio-ethanol.

GATT – General Agreement on Tariffs and Trade. GATT is an international mechanism to control and regulate economic growth through legislative/fiscal policy. The last round, the Uruguay Round, of negotiations lasted from 1988 to 1993.

GDP – Gross Domestic Product.

general circulation models (GCMs) – simulation of atmospheric circulation involving a system of equations used to describe atmospheric and ocean-water motion, the heat exchange and fluxes within this system, and the consequences. GCMs usually require the solution of these equations on a high-speed computer or super-computer.

geochronology – measurement of time intervals or dating of events on a geological time scale.

geodesy – science of the shape and size of the Earth by survey and mathematical means.

geoid – shape of the Earth at mean sea level.

geomorphologist – scientist who studies landforms.

geosphere – the Earth.

geostrophic wind – a wind whose strength and direction is a balance between the pressure gradient force and the Coriolis force. In the upper troposphere and lower stratosphere, it tends to parallel the isobars and has a velocity that often exceeds several hundred kilometres per hour.

geothermal energy – utilisation of the Earth's internal heat energy to generate energy.

geothermal gradient – rate of change of temperature with depth, generally used in relation to the Earth's crust.

giardia – parasite that lives in the human gut lining and causes dysentery.

glacial – cold stage during an ice age when ice sheets, glaciers, permafrost and sea ice are more extensive.

glacio-isostatic rebound – uplift of regions in which the lithosphere was previously depressed by the weight of former glaciers.

glacio-marine – pertaining to marine environment, landform or sediments that are/were influenced by glacial processes.

global commons – a concept that endorses the view that world resources and environment belong to all humankind, and should be managed and protected on an international scale.

global warming potential (GWP) – effect that a given amount of a trace gas can have on forcing climate compared with the effect of the same amount of CO_2.

globalisation – the forging of linkages and interconnections between states and societies that make up the modern world system, in which events, decisions and activities in one part of the world can have consequences for individuals and communities in other parts of the world.

glycerine – hydrocarbon belonging to the alcohol family used in the manufacture of a number of commercial products including cosmetics, soaps and nitroglycerin.

GNP – Gross National Product.

Gondwana – large continent that existed in the Southern Hemisphere, and which split up about 300 Ma (the Late Palaeozoic) to form Africa, Australia, Antarctica, South America and India.

greenhouse effect – effect analogous to that which is supposed to operate in a greenhouse, whereby the Earth's surface is maintained at a much higher temperature than the approximate balance conditions with the solar radiation reaching the Earth's surface.

greenhouse gas – any gas which absorbs infrared radiation in the atmosphere. These gases include carbon dioxide, methane, water vapour, nitrous oxide, ozone and chlorofluorocarbons.

gypsum – mineral (hydrated calcium sulphate, $CaSO_4.2H_2O$) precipitated from a saline solution.

Hadley cell – global wind circulation comprising air that rises over low latitudes due to convection, and cools and descends over the subtropical anticyclone belt, resulting in the trade winds, which blow towards the equator.

haemorrhagic fever – illness produced by a virus, which causes a fever, followed by a period in which the patient deteriorates and superficial bleeding develops, where blood seeps from vessels under the skin and bruises appear; other cardiovascular, digestive, renal and neurological complications may follow.

half-life – time required for 50 per cent of the atoms of a radioactive isotope to decay to a different element/group of elements, with the associated emission of various sub-atomic particles and the release of energy.

halite – mineral (NaCl) precipitated from a saline solution.

Halocarbons – organic compounds that contain chlorine and bromine, and which are important stratospheric ozone-depleters; examples of these gases include CFC-11 (CCl_3F) and HCFC-22 ($CHClF_2$).

halons – organic compounds containing bromine and fluorine. These chemicals play a part in stratospheric ozone depletion.

hard water – water in which certain minerals, particularly calcium carbonate, are dissolved and which tend to precipitate out as a 'scum', e.g. to fur up kettles, water pipes, and other domestic/industrial appliances/machinery.

heat engine – name given to the mechanism by which tropical cyclones develop and are maintained.

heat-island effect – relative warmth of a city compared with the surrounding countryside, controlled by urban activity.

heavy metals – metallic elements with high atomic masses, such as mercury, lead, arsenic, tin, cadmium, cobalt, selenium, copper and manganese.

hectare – 2.47 acres.

Heinrich layers – layers of ice-rafted sediment which are present in cores collected from the North Atlantic.

hepatitis B – disease that causes inflammation of the liver; hepatitis B is transmitted by infection, including sexually.

herbicide – chemical that kills or inhibits the growth of a plant.

herbivore – plant-eating animal.

heterogeneous reaction – reaction of chemicals in different states, for example as between gas and liquid, gas and solid, or solid and liquid.

HIV – human immuno-deficiency virus.

hl – hectolitre = 100 litres.

Holocene – period of time after the last major glaciation (*c.* 11,000 years BP) to the present.

hominid – creature of the family Hominidae (primates) of which only one species exists today – *Homo Sapiens*.

hominoids – upstanding bipedal human-like apes, which are generally considered to be the ancestors of modern humans.

hot spot – region of relatively high heat flux on the surface of the Earth, caused by anomalously hot magma rising towards the surface from the mantle as a plume, causing volcanic/igneous activity at the Earth's surface, uplift, and possibly the splitting apart of the oceanic or continental crust.

hot-spots – areas that have exceptional concentrations of species with high levels of endemism.

hurricane – name given to tropical cyclone that originates in the Caribbean or mid-Atlantic Ocean.

hydroelectric power (hydro-power) – generation of electricity by utilisation of water power.

hydrological cycle – continuous movement of water (vapour, liquid and solid) on, in and above the Earth's surface.

hydrological flowpath – direction of flow of water beneath the ground.

hydrology – study of the movement of water (vapour, liquid and solid) on, in and above the Earth's surface.

hydroperoxyl radical (HO_2^{3-}) – a negatively charged molecule comprising one atom of hydrogen and two atoms of oxygen.

hydrosphere – the Earth's water layer, which includes the liquid, solid and gaseous phases.

hydrothermal systems – hot fluids, usually water, that are rich in dissolved gases, nutrients and metals. These originate from within the lithosphere and emerge at the Earth's surface, commonly on the sea floor, where there are zones of particularly high heat flow, notably associated with volcanic areas.

hydroxyl radical (OH^-) – negatively charged ion comprising one atom of hydrogen and one atom of oxygen.

hypersaline – extremely saline (of water).

Ice Age – period in Earth's history when ice sheets were extensive and sea-ice and permafrost were widespread in mid and high latitudes.

ice shelf – floating sheet of ice attached to the coast, which is nourished by snow falling onto its surface and/or by land-based glaciers.

icehouse effect – conditions that lead to global cooling, the opposite effect to greenhouse warming.

ignimbrite – welded or non-welded pyroclastic rock (formed by explosive fragmentation of magma and/or previously solid rock during volcanic eruptions), comprising mainly pumice and ash.

impact matrix – multidimensional array used to show the effects of policy actions on the environment.

impact winter – global cooling resulting from the reduction of solar radiation induced by increased atmospheric dust as a result of ejected debris and global fires, e.g. as might be caused by a meteorite impact on the Earth.

infrared radiation – form of electromagnetic radiation with a longer wavelength than visible light.

interglacial – warm stage between glacials in an ice age, when glaciers retreated, sea ice and permafrost were of limited extent, and tundra was replaced by forest.

interstadial – warm period during a glacial stage.

intertropical convergence zone (ITCZ) – zone of nearly continuous atmospheric low pressure with light and variable winds, high humidity and intermittent heavy rain showers, which is present near the equator.

inversion layer – level in the atmosphere that prevents the vertical mixing of air. Such a layer typically shows increasing temperature upwards, the reverse to the normal upward cooling.

iridium anomaly – high concentration of the element iridium in sediments.

irrigation – supply of water, usually via channels, to agricultural land.

island arc – a curved line of islands produced by subduction-related volcanism.

isostasy – condition of hydrostatic equilibrium between sections of the lithosphere with respect to the underlying asthenospheric mantle.

isotopes – different forms of an element with identical chemical properties by virtue of having the same atomic number, i.e. they have the same number of protons in the nucleus but a different number of neutrons.

joint – fissure in rock along which little or no relative face-parallel displacement has occurred, and which is formed by the relaxation of crustal stresses.

jokulhlaup – catastrophic flood formed by the drainage of a sub-glacial or ice-dammed lake.

joule (J) – unit of energy expended in order to do work, equivalent to a newton metre (Nm). 1 newton = force required to accelerate 1 kg through 1 m per second per second ($m\ s^{-2}$). 1 watt of energy (W) is equivalent to 1 joule per second.

Jurassic – period of geological time from 190 Ma to 135 Ma.

Kelvin waves – internal gravity waves with a long wave length that progresses along the thermocline. They are similar to surface wind waves in that the main force is gravity. Kelvin waves cross the Pacific Ocean, for example, within a few months.

kerogen – hydrocarbon occurring in crude oil and formed by the breakdown of organic matter.

kinetic energy – energy possessed by matter by virtue of its motion, e.g. heat, wave energy.

lacustrine – pertaining to lake environments, processes or landforms.

lahar – landslide comprising volcanic material, usually generated during a volcanic eruption.

lapse rate – rate of decrease of temperature per unit height in the atmosphere, usually about 6.5°C km^{-1}.

Last Glacial Maximum – period of time

(*c.* 25,000–16,000 years BP) coinciding with the maximum extent of glaciers in the last major glaciation.

laterisation – enrichment of sesquioxides of aluminium and/or iron in a soil, which leads to the formation of laterites.

Laurentide ice sheet – ice sheet that formed during the last glaciation (known in North America as the 'Wisconsin'), which covered vast areas of North America.

lava flow – molten rock, which originates from the mantle and which flows on and across the Earth's surface.

Law of Gravitation (Newton's) – states that a gravitational force exists between two bodies which is proportional to the product of their masses and inversely proportional to the square of their distance apart.

LDC – 'less developed country', equates with 'developing country' or 'Third World'.

leguminous plants – plants that have bulbous growths (legumes) on their roots that contain symbiotic bacteria which fix nitrogen to form nitrates.

leptospirosis (sewerman's disease) – disease contracted from contact with rat's urine.

life-cycle cost (operating costs) – the cost of goods and/or services over an entire life cycle.

light year – distance light travels in one year.

lignite – organic-rich deposit containing about 70 per cent carbon.

limestone – rock mainly comprising the mineral calcium carbonate ($CaCO_3$), formed either directly by precipitation from solution or from the accumulation of detrital organic or inorganic $CaCO_3$.

lipid – substance that has similar properties to fat.

liquefied petroleum gas (LPG) – comprises mainly propane, butane and isobutane, but may contain unsaturated C_3 and C_4 hydrocarbons. Its main use is as a fuel, but it also has widespread use as an aerosol propellant, and as a chemical feedstock.

lithosphere – Earth's crust and upper portion of the mantle, which together constitute a layer of relative mechanical strength compared with the more easily deformable asthenosphere below.

Little Ice Age – cold period during the seventeenth century, when glaciers advanced throughout the world.

load – in an engine, the resistance that is overcome by the torque delivered (to which it is numerically equivalent).

loess – silt size (63–2 μm) sediment deposited by wind.

long-shore sediment drift (littoral drift) – transport of sediment by wave action (sub-)parallel to a coastline.

low-velocity zone – the zone in the upper mantle within which seismic waves are slowed and partially absorbed.

lymphocyte – white blood cells, which are of fundamental importance in the immune system.

macrophages – stationary cells within living tissue, e.g. in the lymph nodes, spleen, bone marrow, and alveoli in the lungs, which engulf and destroy bacteria, playing an important role in the immune system.

magma – molten rock material found beneath the Earth's crust, and from which volcanic rocks are formed.

magmatic – processes that involve the formation, movement, emplacement or crystallisation of magma.

malaria – intermittent and remittent fever transmitted by the bite of a mosquito, which conveys the parasite that transmits the disease.

Malthusian views – ideas expressed by Reverend Thomas Malthus and others, who believe that world population can be kept in check by disease, wars and natural disasters.

manganese nodules – spherical precipitation comprising manganese, iron and lower concentrations of nickel, copper, cobalt and molybdenum present on the ocean floor, associated with areas of high heat flow and black smokers.

mantle – internal layer of the Earth extending from 5–50 km below the surface to a depth of 2,900 km. The uppermost part is essentially rigid (as a very high-viscosity fluid, the lower part of the lithosphere), and the lower part of the asthenosphere is partially molten.

mass movement – movement of material downslope under the influence of gravity.

MDC – 'more developed country', equates with 'developed country' or 'industrialised country'.

meltdown – process in a nuclear reactor where the uncontrolled accumulation of heat increases the temperature of the core until it becomes (goes) critical and literally melts down. If the meltdown is not totally confined within the reactor containment building, core meltdown can release extremely high and dangerous levels of radioactivity, thousands of times greater than was generated by the nuclear fission bomb that destroyed Hiroshima.

Messinian salinity crisis – period of time (Messinian) approximately 5 Ma when the waters in the Mediterranean almost dried up, which increased the salinity of the waters and led to the precipitation of large salt deposits.

metamorphism – processes by which the composition, structure and texture of rocks are altered by the action of heat and pressure.

meteorite – extraterrestrial material that may fall to Earth if it travels across the Earth's orbit. Each year, thousands of small meteorites enter the Earth's atmosphere, but burn up before reaching the ground. Meteorites comprise relatively primitive matter in the Solar System and therefore provide scientists with an opportunity to study the early history of the Solar System.

meteorology – science of the atmosphere concerned with the study of its dynamics and composition.

methaemoglobinaemia – disease responsible for 'blue baby' birth.

methanogenic bacteria – bacteria that convert organic acids into biological gases, notably methane (CH_4).

methyl chloroform – organic compound used as a solvent and cleaning fluid, which may act as a greenhouse gas.

methyl isocyanate – poisonous organic cyanide compound.

microfossils – extremely small to microscopic remains of past organisms. They include foraminifera, diatoms, Coleoptera and pollen.

microglial cells – small cells of the central nervous system.

mid-ocean ridge – submarine, only locally subaerial, linear mountain chain with a central rift valley marking the boundaries between two oceanic plates that are moving apart, and along which basaltic rocks are being formed by the creation of new sea floor.

Milankovitch cycles – natural fluctuations in the Earth's orbital parameters and named after the person who first clearly elucidated these, i.e. changes in the Earth's precession, obliquity and eccentricity in orbit around the Sun, and which lead to cyclically varying solar flux, and therefore induce global climate change.

millisievert (mSv) – a measure of the radiation dose received by an individual, i.e. the amount of energy given up by radiation in a particular mass of body tissue through which it passes. The exact relationship between 1 Bq and 1 mSv is complex and depends, amongst other factors, on the type of radiation (alpha, beta or gamma) and the sensitivity of different types of tissue.

Mohorovicic discontinuity (Moho) – the seismic discontinuity that marks the boundary between the Earth's crust and mantle.

monoculture – cultivation of a single-species crop.

mutually assured destruction (MAD) – the certain annihilation of the participants in a nuclear war. MAD assumes that a pre-emptive strike by one side will still result in complete destruction.

NASA – US National Aeronautics and Space Administration.

natural bitumen – similar in physical properties to bitumen (see *bitumen*), it is naturally occurring and has a different composition to synthetic bitumen.

negative feedback – return of a fraction of an output from a system to the input of the system that stabilises the subsequent output of the system.

nematodes – a class of worms that vary in size from about 1 mm to 5 cm.

Neolithic (New Stone Age) – ancient cultural stage or level of human development characterised by stone tools shaped by polishing or grinding.

neoplastic disease – medical condition involving the formation of a neoplasm or tumour.

nephos – name given by the people of Athens to a mixture of atmospheric toxins that formed in Athens on 1 October 1991.

NIMBY – acronym for 'not in my back yard', used to epitomise the widely held public attitude that undesirable facilities (nuclear power plants, waste disposal utilities, chemical plants, etc.) should be sited away from their homes and/or workplaces and, rather, near somebody else.

nitrate – chemical compound containing nitrogen, an essential nutrient for life.

nitrous oxide (N_2O) – nitrogen oxide commonly derived from the use of fertilisers, or the combustion of fossil fuels or biomass. Nitrous oxide causes concern because tropospheric N_2O is a greenhouse gas, whereas stratospheric N_2O contributes to ozone depletion.

NOAA – National Oceanic and Atmosphere Administration.

North Atlantic Deep Water (NADW) – cold, dense deep-ocean current that travels southwards in the North Atlantic.

nuclear fission – splitting of a heavy atomic nucleus, e.g. uranium or plutonium, into two fragments of roughly equal mass and releasing large amounts of energy.

nuclear fusion – nuclear reactions between light elements to form heavier ones, whilst releasing large amounts of energy.

nuclear reprocessing plant – industrial site where radioactive substances, commonly in the form of spent nuclear reactor rods, are refined into either more concentrated material and/or made less harmful.

nuclear winter – severe deterioration of climate that might take place as a result of multiple nuclear explosions, which may generate great fires and wind; large quantities of smoke and dust may be ejected into the upper atmosphere and remain there for a period of months to years, causing prolonged darkness and reduced incoming solar radiation, resulting in extreme cooling of the Earth's surface, possibly to −15 to −25°C.

nucleus – centre of an atom containing neutrons and protons.

occult deposition – the deposition by impaction of cloud/fog droplets commonly containing appreciably larger concentrations of major ions than the biggest drops of rain (at the same sites).

OPEC – Organisation of Petroleum Exporting Countries.

orbital parameters – pertaining to the Earth's rotation around the Sun, and including its precession (19,000–23,000 years), obliquity (40,000 years) and eccentricity (100,000 and 400,000 years).

Organisation for Economic Co-operation and Development (OECD) – includes most of the world's industrialised market economies, i.e. Australia, Austria, Belgium, Canada, Denmark, Finland, France, Germany, Greece, Iceland, Ireland, Italy, Japan, Luxembourg, the Netherlands, New Zealand, Norway, Portugal, Spain, Sweden, Switzerland, Turkey, the United Kingdom, and the USA.

organochloride – organic compound containing chloride.

organotin – chemical compound containing tin, used primarily as a fungicide.

orogeny – tectonic mechanism that creates continental mountain chains.

oxygenic photosynthesis – biological activity in which O_2 is released to the atmosphere from the splitting up of water molecules.

ozone (O_3) – form of oxygen, each molecule of which comprises three atoms of oxygen.

ozone layer – zone in the atmosphere between 15 and 45 km above the Earth's surface that contains ozone, which reaches concentrations of 1 in 10^5 parts at 35 km altitude. This layer is also referred to as the ozonesphere.

ozonesphere – see *ozone layer*.

palaeoceanography – the study of past oceans, their configurations, chemistry and dynamics.

palaeoclimate – past climate.

palaeoclimatology – the study of past climates, their rates of change and dynamics.

palaeoenvironmental – pertaining to past environments.

palaeolatitude – former latitudinal position of a region that has experienced plate movement because of sea-floor spreading.

palaeomagnetic anomaly – a deviation from the average strength of the Earth's magnetic field.

palaeontology – the study of past creatures (fossils), which includes actual remains, remains replaced by mineral matter and the trails, tracks and burrows of past creatures.

palaeosol – ancient, fossil or relict soil or soil horizon.

palaeotemperature – past temperature.

palynologist – scientist who studies pollen.

Pangaea – large continent (super-continent) that existed in the Northern Hemisphere, and which split up during the Late Palaeozoic to form Asia, Europe and North America.

pathogen – disease-producing organism.

pedology – study of soils.

permafrost – ground that persists below 0°C for two or more years.

permeability – capacity of a rock or soil to transmit fluid.

Permian – period of geological time from 280 Ma to 225 Ma.

petajoule – unit of energy. 1 million tonnes of oil equivalent = 41.87 petajoules.

pH scale – logarithmic scale that provides a measure of the acidity (inversely related to the concentration of hydrogen ions) in a solution. pH7 is neutral, while decreasing values indicate increased acidity, and values greater than 7 represent increasing alkalinity (basicity).

pheromones – hormones that are produced by organisms to attract the opposite sex.

phosphate – chemical compound containing phosphorus, an essential nutrient for life.

phosphate nodule – concretionary growth, commonly formed early in the burial history of sediments, containing exclusively or mainly phosphate (see *phosphate*).

photic zone – surface layer of a lake, sea or ocean above the maximum depth to which light penetrates.

photochemical – chemical substance that is sensitive to light, and may involve a change in its chemical composition or energy state when bombarded with solar energy.

photochemical smog – poor air quality caused by sunlight catalysing chemical reactions mainly with nitrogen compounds and hydrocarbons, commonly to produce a reddish-yellow-brown haze. Photochemical smogs commonly form on very warm, sunny days in large urban areas subjected to large amounts of exhaust emissions from motor vehicles.

photon – a quantum of light energy.

photosynthesis – biological process in which plants convert CO_2 and H_2O to carbohydrates and release O_2.

photosynthetic strategy – means by which organisms synthesise chlorophyll using sunlight.

photo-voltaic cell – device capable of changing solar energy directly into electricity.

planetary accretion – the formation of planets by the joining of small solid bodies, which originated from interstellar gas and dust, during the early history of the Solar System.

plankton – drifting or floating organic life, chiefly microscopic, found at various depths in seas, lakes and rivers.

plate tectonics – theory that explains the nature of the Earth's surface in terms of continental and oceanic plates (currently, there are eight major and several minor lithospheric plates, which move relative to each other because of convection cells in the Earth's mantle, causing the creation of new oceanic crust at spreading centres, and its destruction at subduction zones).

Pleistocene Period – interval of geological time that spanned from 1.64 Ma to 10,000 years BP and experienced widespread glaciation.

plume (mantle plume) – see *hot spot*.

plutonium – radioactive element used as a fuel in certain nuclear reactors and as an ingredient in nuclear weapons.

pneumoconiosis (dust pneumonia) – lung disease caused by inhaling fine dust, common in coal miners.

poliomyelitis – inflammation of the grey matter in the spinal cord which may lead to paralysis.

pollutant – a substance that causes pollution, typically because it reaches concentrations or levels that pose an environmental problem.

polychlorinated biphenyls (PCBs) – organic compounds containing chlorine and a phenyl group, which are used in the manufacture of paints, plastics, adhesives, hydraulic fluids and electrical components, and which are toxic to humans and other organisms.

polycyclic aromatic hydrocarbons – organic compounds with carbon ring-chains, and which contribute to atmospheric pollution.

population density – number of individuals per unit area.

population explosion – term used to describe an exponential increase in population where conditions favour a very large birth and survival rate.

porosity – ratio of void volume to bulk volume of rock or soil.

positive feedback – return of a fraction of an output from a system to the input of the system, which strengthens the subsequent output of the system.

positive point sources – sharp points that have a net positive electrical charge.

pozzolan material – substance acting like cement, and containing silicates or aluminosilicates, which react with lime and water to form stable insoluble compounds; used in disposal of toxic waste.

ppb – parts per billion. Unit of measurement typically used to define very low concentrations of chemical elements and compounds.

ppm – parts per million. Unit of measurement typically used to define low concentrations of chemical elements and compounds.

precipitation – all forms of moisture that condense in the atmosphere and are deposited on the Earth's surface.

precursors – events or phenomena that occur a short time before an earthquake or volcanic eruption.

primary energy – includes fossil fuels (coal, crude oil, gas), and biomass in a raw state prior to processing into a form suitable for energy consumption.

Proterozoic – period of geological time from 2,500 Ma to 550 Ma.

punctuated evolution – non-gradual, abrupt development of organisms leading to the generation of new species.

pyrmnestophyte algae – see *coccolithophorida*.

pyrotoxins – toxic chemicals produced by fires.

pyroxenes – group of rock-forming silicate minerals that are rich in iron and magnesium.

Quaternary Period – geological time from 2.5 Ma or 1.64 Ma BP to present, which was characterised by widespread glaciations.

radiation – transmission of electromagnetic energy from a body to its surroundings.

radiative forcing – a measure of the ability of greenhouse gases to perturb the heat balance in a simplified model of the Earth–atmosphere system.

radioactivity – process of emitting subatomic particles and energy.

radiocarbon dating – technique used to determine the age of a material in years by measuring the decay of the ^{14}C isotope present within that material.

radioisotope – isotope of a chemical element that is naturally unstable, and tends to become more stable by the emission of radioactive particles, e.g. alpha, beta or gamma radiation.

radiometric – prefix to age, or the technique used to date a substance in years by determining the relative proportions of radioactive isotopes and their decay products within that substance.

radionucleide – atomic nucleus of element that is capable of breaking down into a new isotope by radioactive decay.

radium – radioactive chemical element (source of alpha particles) with a half-life of 1,620 years and present in naturally occurring uranium ores as a result of the radioactive decay of the uranium.

radon – naturally occurring radioactive gas that may build up in houses and other buildings in sufficient concentrations to pose a serious health risk.

rain gauge – instrument used to measure the quantity of rain falling over a period of time at a particular location.

raised beach – emerged shoreline represented by stranded beach deposits, marine shells, and wave-cut platforms backed by former sea cliffs.

rape (oilseed rape) – cereal that produces an oilseed that can be used to make rape methyl ester, which can be used to power diesel engines.

rape methyl ester (RME) – ester produced from oilseed rape, and which is used, for example, as motor transport fuel, raw or blended.

recurrence interval – time period between successive earthquake or flooding events in a particular area.

redox – reactions involving the loss or gain of electrons.

retention ponds – artificial ponds that are constructed to collect flood waters to allow the controlled release of water into the main stream.

retrovirus – virus that has its genetic material in the form of ribonucleic acid.

reverse transcriptase – an enzyme found in retroviruses that catalyses synthesis of DNA.

risk – probability of an adverse outcome, or (sometimes) the likelihood attached to different outcomes.

RNA (ribonucleic acid) – molecule involved in the synthesis of protein.

röentgen – unit of invisible electromagnetic radiation.

salinisation – concentration of salts in the upper layers of a soil due to the drawing of water upwards by the evaporation of near-surface waters.

sapropel – mud or ooze comprising decomposing organic material, usually present in aqueous environments.

saturated zone – a subsurface zone in which all the rock openings are filled with water.

savannah – grassland region in the tropics or subtropics.

schistosomiasis (bilharzia) – chronic disease found in residents of the tropics, produced by the presence of a flatworm parasite in the blood and bladder.

scleractinian corals (hexacoralla) – group of corals (multicellular organisms that secrete a calcareous skeleton), which first evolved in the Middle Triassic and still exist today; they form an important component of coral reefs.

sea-floor spreading – movement of oceanic crust by plate tectonic processes away from ocean ridge spreading centres.

sedimentary ironstone – rock with a high concentration of iron oxides, iron sulphides and/or iron silicates, which were deposited by precipitation and deposition of detrital sediments.

seismic – pertaining to earthquake activity.

seismic gap – zone or layer within a tectonically active region that has not experienced a large earthquake during historical times or the recent past.

seismic wave – a package of elastic strain energy that travels away from a seismic source, e.g. an earthquake.

seismologist – scientist who studies the dynamics of earthquakes, using them to aid the elucidation of the Earth's interior.

seismometer – instrument used for detecting earthquakes.

self-purification – natural process in which waste is degraded by microbes in water.

sensitive clays – fine material (<2 mm) that easily deforms and fails under stress.

sensitivity analysis – analysis of the effects on an appraisal of varying key assumptions and variables.

seroarchaeology – the science of the history and development of AIDS.

sewage – liquid waste products from a community.

sewerage – network of pipes and associated appliances for the collection and transport of sewage.

siderophile elements – elements (e.g. Ni, Cr, Co, Ag) that are soluble in molten iron (Fe).

silicate minerals – chemical substances that contain silicon and oxygen atomic structures: these include the amphiboles, pyroxenes, feldspars, micas and clay minerals.

smart materials – synthetic materials comprising polymers that are able to change their physical properties (e.g. density, shape, tensile strength, colour, etc.) in response to an external stimulus such as the passage of an electric current, temperature, pressure, stress, strain rate, etc. Smart materials are very much at a research and development stage.

soft water – water that lacks or has very low concentrations of dissolved calcium and/or magnesium salts, commonly carbonates.

solar energy – electromagnetic energy produced by the Sun.

solar flux – flow of radiation from the Sun to the Earth.

South-East Trade Winds – strong winds that blow from a subtropical high-pressure region around 30° south of the equator.

species – basic unit of classification of living things, which may be defined by morphology, the ability for members of the species to interbreed and by ecological requirements.

speleothem – deposit, usually of calcite, or less commonly silica, gypsum or ice, formed by the precipitation from water that has seeped through the ground and which, upon contact with the air, generally in a cave, results in the formation of features such as stalactites, stalagmites and flowstone.

spheroids – approximately spherical, sand-size material believed to result from the crystallisation at high temperatures of material melted by a meteorite impact and rapidly ejected into the air or water. Spheroids typically consist of the mineral feldspar.

spike – term used in Earth sciences to denote a datum time against which other events may be measured and/or other geological units correlated, e.g. 'golden spikes' refer to internationally recognised and correlatable events such as major extinctions and/or radiations of species.

sporopollenin – material comprising the waxy outer layer of a pollen grain.

spreading centre (oceanic) – linear, generally submerged, ridge along a plate margin, which represents a zone where the crust is forming and moving away from the ridge.

SSSI – 'site of special scientific interest', area designated by the UK government's advisers as requiring conservation.

stadial – cold period during an interglacial stage.

stishovite – variety (polymorph) of quartz (SiO_2) formed under high pressures.

storm sewer – channel or pipe that drains rain water into a river, lake or the sea.

strain – deformation of a material resulting from an applied force.

strain gauge – instrument used to measure ground deformation.

stratigraphy – the study of the order and arrangement of geological strata.

stratosphere – layer of the atmosphere above the troposphere, ranging in altitude from 8–15 km at the lower boundary to about 50 km at the top, accounting for about 10 per cent of the mass of the atmosphere. Between 15 and 35 km altitude, the air contains relatively high concentrations of naturally occurring ozone, up to 10 ppm, referred to as the 'ozone layer'. Temperature increases with altitude in the stratosphere (from about –50 to 0°C), and vertical mixing is relatively slow. The stratosphere is essentially unaffected by weather.

stratospheric ozone (high-level ozone) – triatomic molecules of oxygen, which are present in the atmosphere at heights of between 15 and 45 km.

stream gauge – instrument used to measure the amount of water passing a particular location.

stress – force exerted per unit area.

stromatolites – organic-sedimentary structures produced by sediment trapping, binding and/or precipitation as a result of the growth and metabolic activity of micro-organisms, principally algae, which live in the seas, marshes or lakes. They still form today, but they reached their acme during the Proterozoic, when they formed mounds several hundred metres across and tens of metres high.

strontium-90 – radioactive isotope of strontium.

subduction – the product of re-absorption of oceanic lithosphere at destructive plate margins.

subtropical anticyclones – high-pressure atmospheric systems in the subtropics, which develop divergent winds.

sunspots – areas of above-average temperature seen on the Sun's surface, and which develop and decay periodically.

Superfund – popular name for the US Comprehensive Environmental Response, Compensation, and Liability Act (CERCLA) of 1980, which provides a mechanism and funding to help clean up potentially dangerous hazardous waste sites.

superinsulator – substance that conducts very little heat.

surficial glacial deposits – veneer of sediments of glacial origin present on the land surface.

surging glacier – glacier that flows at a velocity which is an order of magnitude higher than normal.

suspended solids – solid pollutants suspended as particles in air or water.

sustainable development – development of available resources without compromising the ability of future generations to meet their needs.

symbiosis – mutually beneficial relationship between organisms.

symbiotic bacteria – bacteria that live in leguminous plants and obtain nutrients from the plants and fix nitrogen to form nitrates, which in turn can be used by the plant.

syngas – mix of H_2 and CO used to synthesise liquid fuels such as pure hydrogen, methanol and gasoline.

system – set of interrelated parts, which includes components (elements) and states, and the relationships between the elements and states.

systematics – methodological study of classification, such as the classification of the diversity of biological systems.

systematist – scientist who studies systematics.

tectonic processes – deformation processes (strain) that are a consequence of continental and oceanic plate movements; they include vulcanicity, earthquakes, mountain building, subsidence, and crustal extension and rotation.

tectonics – study of the structures of the Earth's lithosphere and the processes involving stress and strain that form them.

tektites – spherules of glass generated by a meteorite impact.

temperature inversion – an atmospheric condition in which the environmental lapse rate is reversed and temperature increases vertically through a given layer of the atmosphere.

tephra – general term used for all pyroclastic deposits produced by explosive volcanism.

terawatt (TW) – unit of energy output or consumption. 1TW = 1 million kW.

Tertiary – period of geological time that spans from the Cretaceous (65 Ma) to the Quaternary Period (1.64 Ma BP).

thermocline – zone of water within a lake or the sea that marks a sharp temperature change, typically from an upper warm surface layer to cooler water below.

thermohaline circulation – movement of ocean water induced by the interaction of waters with different temperatures and salinity.

thermokarst – degradation of permafrost resulting in waterlogging, lake formation and subsidence.

thermophobic – cold-seeking.

threshold – physical and/or chemical state across which there is a sudden change of conditions brought about by the increase of an input to a critical level.

till – sediment deposited by glacial ice.

tillite – sediment, deposited by glacial ice, that has become lithified.

tornado – violent, rotating storm with winds of up to 100 m s^{-1} circulating round a funnel cloud some 100 m in diameter.

traction load – material moved by a river along a river bed.

trade winds – tropical easterly winds that blow from the subtropical high-pressure areas in latitudes 30–40° north and south.

transform plate boundary – the boundary between two lithospheric plates that are sliding past each other.

transpiration – loss of water vapour from the cells of plants.

treated sludge – sewage sludge that has undergone biological, chemical or heat treatment, long-term storage, or any other appropriate process, so as to significantly reduce its fermentability and the health hazards resulting from its use.

Triassic – period of geological time from 225 Ma to 190 Ma.

tributyltin (TBT) – type of organotin used as an anti-fouling paint on boats.

tritium – an isotope of hydrogen (^{3}H), comprising two neutrons and one proton in the nucleus of the atom. Tritium is typically made artificially, but it does occur naturally in very small amounts.

tropical cyclone – intense cyclonic vortex that forms in tropical oceans and has winds that reach 33 m s^{-1}.

troposphere – lower portion of the atmosphere, which extends to heights of 8–15 km above the Earth's surface, accounting for about 90 per cent of the atmospheric mass, and in which most of the weather processes take place, causing it to be well mixed. Temperature decreases with altitude.

tropospheric ozone – triatomic oxygen that forms and is present near ground level to heights of about 12–15 km in the atmosphere.

tsunami – sea wave caused by submarine earthquake or landslide.

tundra – biologically defined region that is treeless and marshy and usually has permanently frozen subsoil; typical of northernmost Eurasia, the Arctic and parts of the Antarctic.

typhoid – infectious fever with severe intestinal irritation.

typhoon – tropical cyclone that originates in the west Pacific Ocean.

ultraviolet radiation – electromagnetic radiation of short wavelength.

unsaturated zone – a subsurface zone in which rock openings are filled partly with air and partly with water.

upstream flood – flood whose effects become progressively dissipated down-valley.

upwelling – movement of deep water towards the sea surface, often rich in nutrients and allowing increased productivity in surface waters, e.g. of phytoplankton.

uranium/thorium dating – geochemical technique that compares the ratio of uranium and its daughter element, thorium, to provide an absolute age of a rock or fossil.

urbanisation – establishment and/or growth of cities and towns.

ventilation – term used in oceanography to refer to the circulation of oxygen-rich waters in the world's oceans.

verification – checking by third parties of the accuracy or truthfulness of experiments or assertions by others, and commonly used in connection with nuclear arms and other international treaties.

volcanic – pertaining to processes, materials or landforms that are the direct consequence of the activities associated with volcanoes.

volcanic ash – fragmented rock or crystals that are ejected by volcanic eruptions.

volcanic basement – crustal rocks that are volcanic in origin.

volcano – vent or opening through which magma, ash or volatiles erupt onto the Earth's surface and into the atmosphere.

volcanism – eruption of magma, ash or volatiles onto or above the Earth's surface.

Wadati–Benioff zone – a distinct earthquake zone that begins at an ocean trench and slopes landward and downwards into the Earth at an angle of between 30 and 60°, associated with a subduction zone.

Wallace realm – a zoogeographical division with a distinct fauna, demarcated by Alfred Russel Wallace.

water cycle – the complete loop system involving water movement from evaporation, precipitation and run-off through the surface and/or ground.

water table – surface above an unconfined body of ground water that has totally filled fissures and pores in bedrock or soil.

watershed – ridge of ground from which surface waters flow in different directions, eventually to collect in different streams; watersheds delimit the area of catchment for particular rivers.

watt (W) – unit of power. For electrical energy, units expressed as We; for thermal energy, Wt. 1,000 watts = 1 kW. 1 unit of electricity contains 1 kWh of energy.

wavelength – distance between two successive crests or troughs of wave form.

weathering – decomposition (chemical weathering) and/or disintegration (physical/mechanical weathering) of rock *in situ* by chemical and/or physical processes.

Weichselian – last glacial stage in mainland Europe.

westerlies – mid-latitude winds, which blow from the southwest.

wet deposition – the incorporation of particles and gases into rain and snow, which deposit by gravity – 'acid rain' in the strict scientific sense.

wetland – ecosystem constantly containing surface water, and which is commonly regularly flooded, e.g. marshes and swamps.

white smoke – an aerosol of totally or partially unburnt fuel emitted from an engine; normally emitted only when the engine and ambient air are cold: in worn engines, lubricating oil may provide a source.

Wisconsin – last glacial stage in North America.

xeroscraping – ecosystem containing drought-resistant plants.

Younger Dryas Stadial – cold period *c.* 11,000 years BP during the early part of the present interglacial (Holocene).

zooplankton – small aquatic animals, most of which are microscopic, comprising mature and larval stages of many animal groups, including the Protozoa, Crustacea and Mollusca.

Index

Note: References to boxes are in **bold**; those to plates, tables and figures are in *italics*.